Praise for *The Terror Timeline* from the Jersey Girls

"Paul Thompson's exhaustive and richly detailed research has now provided the world with a veritable treasure trove of 9/11 information. If you want to know everything about 9/11, you must read this book. If you want to better understand 9/11, you must read this book. Our intelligence agencies should be recruiting people like Paul Thompson, because he's brilliant."

—KRISTEN BREITWEISER, *co-chair, September 11th Advocates; member, Family Steering Committee; and wife of Ronald M. Breitweiser, WTC Tower Two*

"Having lost loved ones on 9/11, we had the passion and drive to research and follow through with our quest for answers, but until we stumbled across Paul Thompson's exquisitely detailed and well-sourced timeline on the Internet, much of the information that we found out about the events surrounding 9/11 were out of context or unverifiable. Paul's timeline gave us a much-needed measure of clarity when our lives were filled with ambiguities."

—PATTY CASAZZA, *co-chair, September 11th Advocates; member, Family Steering Committee; and wife of John Casazza, WTC Tower One*

"Paul Thompson's timeline is an invaluable tool and a must read for anyone who really wants to understand the events of 9/11. One can only hope that the 9/11 Independent Commission's report, with its unprecedented access to millions of documents, is as detailed and thorough as this citizen's account is. If the many questions raised by this book are ever answered, it would go a long way toward making us all safer."

—MINDY KLEINBERG, *co-chair, September 11th Advocates; member, Family Steering Committee; and wife of Alan Kleinberg, WTC Tower One*

"For all of us who lost our loved ones on 9/11, and for anyone who seeks the truth about what happened on that awful September day, Paul Thompson's timeline is where all research into the subject should begin. I am eternally grateful for Paul's meticulous work and have referred to Thompson's timeline since the beginning of my quest for answers into my husband Ken's death. Thank you, Paul."

—LORIE VAN AUKEN, *co-founder, September 11th Advocates; member, Family Steering Committee; and wife of Kenneth Van Auken, WTC Tower One*

YEAR BY YEAR, DAY BY DAY, MINUTE BY MINUTE: A COMPREHENSIVE

THE TERROR

CHRONICLE OF THE ROAD TO 9/11—AND AMERICA'S RESPONSE

TIMELINE

Paul Thompson

and the Center for Cooperative Research

HARPER

NEW YORK • LONDON • TORONTO • SYDNEY

TO MY PARENTS

who instilled in me a desire for truth and justice.

TO THE VICTIMS OF 9/11, EVERYWHERE IN THE WORLD

May we learn the full truth and act wisely upon that knowledge,
so that others may be spared similar suffering in years to come.

Contents

Foreword by Peter Lance

In the fall of 2002, as I was researching the intelligence failures that dated back twelve years prior to the 9/11 attacks, I had no idea that halfway around the world, in New Zealand, a young American Stanford graduate was preparing his own history of the same events. My book, *1000 Years for Revenge,* ultimately focused on the negligence in the New York office of the FBI. It began with the words of the Roman poet Juvenal: "Quis custodiet ipsos custodies?": *Who is guarding the guardians themselves?* Back then it was a difficult question to answer, since so much of the factual database on al-Qaeda and what our government knew about it on the road to 9/11 was classified. The challenge was to piece together a mosaic of the truth from the thousands of pages of trial records, news stories, books, and other open-source references I could get my hands on.

A year later, as I began researching my next book—on the failure of the 9/11 Commission to get at the full truth behind the 9/11 attacks—I found the answer to Juvenal's question: The man guarding the guardians is Paul Thompson. By assembling a meticulous timeline tracing back the history of Islamic terrorism to 1979, he and his coworkers at the California-based Center for Cooperative Research provide an objective database against which the nation's spy agencies can be judged.

With almost no budget and through simple access to the Internet, Thompson and his colleagues have done what the 9/11 Commission itself failed to do, fulfilling the intent of Congress to "make a full and complete accounting of the circumstances surrounding the attacks." Further, he's supported his findings with mainstream media sources in what amounts to an almost week-by-week history of the events that preceded the attacks.

When the dots on that terrible day are lined up, as they are in this book, the reader is confronted with a jaw-dropping illustration of government negligence at its worst.

It's impossible to judge the quality of the 9/11 Commission's work fully when it comes to evidence that has been classified by the government. Since vast amounts of data remains "in the black," we're forced to rely on the commission's staff to determine who knew what and when. And, as evidenced in my new book, *Cover Up*, their record is vastly incomplete.

What we *can* rely on is Thompson's vast history—sourced to a massive collection of articles that, when printed out in their entirety, fill more than one hundred two-inch-thick three-ring binders. Thompson's sourcing is drawn entirely from the mainstream media, including reports from the *New York Times, Washington Post, Time,* and *Newsweek,* the major broadcast and cable networks, and hundreds of other sources.

By examining those stories, many of which describe events in real time as they happened, we are presented with a stunning collection of facts, all of it unencumbered by government spin. This is crucial, because the Bush administration's record on even so well-documented a moment of time as the day of 9/11 itself has changed in the thirty-one months between that Tuesday morning and June 17, 2004, when the 9/11 Commission completed its last public hearing. Going back and examining the events as they were chronicled by print and electronic media sources during that period offers us an independent method of judging the performance of not just this presidency, but the two that came before it.

Years ago, *Washington Post* publisher Philip Graham called daily journalism "the first rough draft of history." Paul Thompson's timeline realizes this definition at the highest possible level. It allows anyone with access to the Internet or a bookstore to "audit the auditors"—in effect, to measure the work of our intelligence agencies against the best historical record there is: the day-to-day description of events by reporters worldwide.

Just a single citation offers a sense of how effective the timeline can be as an independent yardstick in measuring the negligence of our intelligence and defense agencies. The 9/11 Commission's Staff Statement No. 17, which deals with the defense failures on the day of 9/11, notes that the first signal to FAA flight controllers that American Airlines Flight 11 may have been hijacked came at 8:13 A.M.

Here's the relevant passage from Paul's timeline, which references both the 9/11 Commission report and several other news stories:

8:13 A.M.: Flight 11 Hijacked, but Pilot Makes No Distress Call

The last routine communication takes place between ground control and the pilots of Flight 11 around this time. Flight controller Pete Zalewski is handling the flight. The pilot responds when told to turn right, but immediately afterward fails to respond to a command to climb. Zalewski repeatedly tries to reach the pilot, even using the emergency frequency, but gets no response. [BOSTON GLOBE, 11/23/01; NEW YORK TIMES, 10/16/01; MSNBC, 9/11/02 (B); 9/11 COMMISSION REPORT, 6/17/04] Flight 11 is apparently hijacked around this time.

The 9/11 Commission Staff Statement says that "In addition to making notifications within the FAA, Boston Center took the initiative, at 8:34, to contact the military through the FAA's Cape Cod facility. They also tried to obtain assistance from a former alert site in Atlantic City, unaware it had been phased out."

In effect, the commission says, it took 21 minutes for the FAA to contact the North American Aerospace Command (NORAD) via Otis Air National Guard base on Cape Cod—and that, although attempts were made to get help from a former NORAD early warning site in New Jersey, that site had been "phased out."

There it is: The only reference in the report of the last official body to evaluate the FAA's and NORAD's response to the Atlantic City facility. But if you consult Paul Thompson's timeline you'll find this:

8:34 A.M.: Atlantic City Fighters Not Reached; Not Redeployed Until Much Later

Around this time, Boston flight control attempts to contact an Atlantic City, New Jersey, air base, to send fighters after Flight 11. For decades, the air base had two fighters on 24-hour alert status, but this changed in 1998 due to budget cutbacks. The flight controllers do not realize this, and apparently try in vain to reach someone. Two F-16s from this base are practicing bombing runs over an empty stretch of the Pine Barrens near Atlantic City. Only eight minutes away from New York City, they are not alerted to the emerging crisis. Shortly after the second WTC crash at 9:03 A.M., the two F-16s are ordered to land and are refitted with air-to-air missiles, then sent aloft. However, the pilots re-launch more than an hour after the second crash. They are apparently sent to Washington, but do not reach there until almost 11:00 A.M. After 9/11, one newspaper questions why NORAD "left what seems to be a yawning gap in the midsection of its air defenses on the East Coast—a gap with New York City at the center." [*BERGEN RECORD*, 12/5/03; 9/11 COMMISSION REPORT, 6/17/04] Had these two fighters been notified at 8:37 A.M. or before, they could have reached New York City before Flight 11.

In other words, the Atlantic City facility was still active. In fact, two F-16s from the 177th Fighter Wing in Pomona, New Jersey, were not only in the air, they were *eight minutes* from Lower Manhattan—and would have been able, if properly notified, to get to the World Trade Center's North Tower before AA Flight 11 slammed into it.

Even more revealing, the online version of Thompson's timeline allows you to click right through to reporter Mike Kelly's December 5, 2003, *Bergen Record* story, where you'll see that both Governor Tom Kean, the 9/11 Commission chairman, and John Farmer, one of his lead investigators, are quoted—establishing that they were aware of the F-16s' presence.

"We want to know why jets at Pomona were decommissioned," Farmer is quoted as saying in the *Record* story, which went on to ask why fighters at bases closer to New York than Cape Cod weren't scrambled. "That's a big question," Kean said in the story.

Yet, despite the fact that these two top 9/11 Commission officials were aware of the issue, the staff left out this crucial piece of information from its final statement. Worse, the commission statement created the impression that the Atlantic City facility was inactive, when in fact it had two planes in the air at the time of the attacks.

This is only one of hundreds of such important and provocative leads highlighted by Thompson's research.

"For months and months now we have relied on Paul Thompson's timelines for the real truth behind the attack," says Lorie van Auken, whose husband, Ken, died on the 107th floor of the North Tower after it was struck by AA Flight 11. "Paul and the people at the Center for Cooperative Research have performed an invaluable service by documenting the public body of knowledge leading up to and beyond 9/11."

Now, thanks to the ReganBooks division of HarperCollins, Thompson's remarkable database is available in printed form. From the first entry, marking the invasion of Soviet troops into Afghanistan on December 26, 1979, through the latest Afghanistan entries chronicling more war, chaos, and misguided U.S. intervention in that unfortunate country, Paul's timeline offers a chilling reminder of the old French adage *Plus c'est changes, plus c'est la meme chose:* the more things change, the more they stay the same.

In the end, though, information is power. And the work of Paul Thompson and the Center for Cooperative Research provides interested readers with the ultimate owner's manual to their government and its errant conduct of the war on terrorism.

PETER LANCE
New York City, August 2004

Introduction

Never could I have imagined I would end up writing a book like this. Probably like most Americans on September 10, 2001, I knew theoretically that terrorism was a threat, but I had no special expertise on the subject. On September 11, 2001, along with millions of Americans, I was shocked and horrified as I watched the events unfold. And, like most Americans, as the weeks and months passed after September 11, I paid some attention to stories about 9/11, but did not give them special attention.

I first became deeply interested in 9/11 in May 2002, when a series of stories appeared suggesting that the Bush administration had received more warnings before 9/11 than they had admitted. First, it was revealed that one month before 9/11, President Bush had received a memo entitled "Bin Laden Determined to Strike in U.S." Next, FBI agent Coleen Rowley divulged that her Minnesota FBI unit's attempt to obtain a search warrant a few weeks before 9/11 for the information on the computer of suspicious flight school student Zacarias Moussaoui was rebuffed by FBI headquarters. Finally, when FBI agent Ken Williams's July 2001 memo—which had suggested that terrorists might be training in U.S. flight schools—surfaced, I began to wonder just how much the Bush administration *had* known beforehand, and just how fully and accurately the media had reported this story.

I turned to the Internet for more information. Not surprisingly, given the innumerable unanswered questions and conflicting reports, dozens of 9/11 conspiracy-theory websites had sprouted on the Internet. While the quality of these sites varied greatly, the best of these sites were not conspiracy-minded at all. They simply highlighted unanswered questions and demanded answers. I noticed that most of the facts that inspired people to post these sites came from reputable mainstream sources such as the *New York Times*, CBS News, the *Washington Post*, *Newsweek*, and the *New Yorker*. As might be expected, given the absence of information on certain issues, some people tended to theorize about the "missing" facts or about why certain things had occurred. What interested me, however, wasn't the theories so much as the facts being quoted. There was much that I'd never heard of.

I kept digging, and came across Michael Ruppert's "From the Wilderness" website, containing a small but excellent timeline that was well-sourced from mainstream media news reports. The more I dug into this timeline and its source articles, the more surprised I was, and the more I wanted to know. Each question I tried to answer seemed to give rise to three new questions that needed answers as well. Soon I began adding to Ruppert's timeline myself, to try to make sense of all the confusing information. I already had a full-time job, but found myself working long hours in my free time, in what had become an obsession—an attempt to find out the truth of what really happened, both before and on 9/11.

Before long, I had posted my own timeline on the Internet. As this site began to attract other readers, they began sending me additional information, and people started volunteering their help. And so the work grew—and grew. By mid-2004, the timeline had more than 1,400 entries. This book is based on that research.

As I delved into the thousands of articles written about 9/11, I was struck by the many articles that were simultaneously good and bad. They were good in that they reported the latest news developments. But too often it seemed that the stories were missing important context. It was as if many new stories existed in a vacuum. In one notable example, I read numerous articles reporting on the Bush administration's insistence that prior to 9/11 they had no idea that al-Qaeda would ever use airplanes as flying weapons. Yet these articles generally failed to remind the reader of the many July 2001 reports that authorities at the G-8 summit in Genoa, Italy, had feared that al-Qaeda would attack the summit with airplanes guided by suicide pilots—and that Bush himself had spent the night on board an aircraft carrier so as to not be targeted in such an attack. How could Bush deny that he was aware of the possibility that terrorists would use airplanes as flying weapons after that summit?

Had the earlier events simply been forgotten? Did the reporters assume that the reader would remember those earlier events? Were the reporters (or their editors) too afraid to publicly contradict the official pronouncements? Whatever the reason, it seemed to me that the reader could not be well informed unless new stories were put into proper historical context.

I also discovered that new or significant information was often mentioned only in passing, deep within a given article, with little or no follow-up. For example, a CBS *60 Minutes* story titled "The Plot" focused mostly on information that had already been covered elsewhere. But there was a scintillating new detail buried halfway through the long story: "Just days after [lead hijacker Mohamed] Atta returned to the U.S. from Spain, Egyptian intelligence in Cairo says it received a report from one of its operatives in Afghanistan that 20 al-Qaeda members had slipped into the U.S. and four of them had received flight training on Cessnas. To the Egyptians, pilots of small planes didn't sound terribly alarming, but they passed on the message to the CIA anyway, fully expecting Washington to request information. The request never came." [CBS NEWS, 10/9/02]

I found this highly significant: Egypt had provided the broad outlines of the 9/11 plot to the CIA prior to the attacks, but the CIA had not been interested. I waited for follow-up stories the next day, scouring the Associated Press, Reuters, and other news outlets, but found none. The Internet was buzzing with the news, but it seemed forgotten—or deemed irrelevant somehow—by the mainstream media. I have since seen this phenomenon occur countless times in the mainstream media coverage of 9/11: important "dots" are mentioned somewhere, but the dots are not connected, and important dots remain obscure. So, as I expanded my timeline, I tried to connect the dots. Where two accounts conflicted, I compiled both versions. Where new reports supplemented information regarding an event that had been reported on previously, I added the new information to my summary of that event.

This search for the facts, rather than theories or government spin, and for the whole truth rather than unrelated "dots," drove me first to create the 9/11 timeline website; now it has given rise to this book. Certain 9/11 myths—such as the myth that the attacks were unpreventable, or that airplanes could be used as flying bombs was unthinkable prior to 9/11—will continue to persist as long as most Americans remain unaware of the facts, as long we are not reminded of the past.

Such myths are dangerous, and continue to have profound negative consequences for the United States. Early 2004 polls indicated that a majority of Americans still believed that former Iraqi leader Saddam Hussein had a hand in the 9/11 attacks, and a majority continued to believe that at least some of the hijackers were Iraqi. Would a majority of the public have supported the war in Iraq in 2003 had the public been better informed about 9/11?

Yet finding the truth about 9/11 is no easy task. A significant amount of misinformation and even disinformation is reported on a weekly basis. For example, the *London Times,* considered by many to be one of the most reputable newspapers in the world, reported in late 2001 that hijacker Mohamed Atta met with an Iraqi spy in the Czech Republic, that the spy gave Atta a flask of anthrax, and that the spy met him on at least three other occasions. According to the report, Atta met with Iraqi agents in Germany, Spain, and Italy. Supposedly Iraq was training hundreds of al-Qaeda operatives and was providing flight simulator training. [LONDON TIMES, 10/27/01] We now know that this story is inaccurate in virtually every single detail. Even the single alleged meeting between Atta and the spy, so often touted by senior Bush administration officials to garner support for the Iraqi War, has now been wholly discredited.

In the course of developing the 9/11 timeline, I have tried to weed out the most obviously erroneous accounts as new or better evidence becomes available. I have also relied almost exclusively on mainstream U.S. and foreign news sources, and government documents. Of the thousands of facts presented in this book, however, it is inevitable that some will be inaccurate. To assist the reader, this book uses a rather unorthodox citation method (as shown in the CBS News and *London Times* examples above). It is my hope that providing the referenced source immediately after the information will help the reader to judge the veracity of the material, and to locate the original source article upon which a particular entry is based. (Please note that source articles referenced in this book that have been posted on the Internet are directly accessible from links in the online version of this timeline, available at www.cooperativeresearch.org.)

I have attempted here to compile the most comprehensive collection of 9/11-related facts that are publicly available as of mid-2004. Although this work is based on thousands of source documents, it is inevitably incomplete. I am sure I have missed some "dots." Other information, important in its own right, seemed tangential to the themes of this book, and so was not included. Still other information calls for closer examination than can fairly be provided here. For example, the effect of 9/11 on civil liberties could itself easily fill a book this size. Finally, this work remains incomplete because the public record remains incomplete. While substantial information relating to the 9/11 Commission is included here, their work was only being completed as this book went to press, and details of their final report could not be addressed or placed in a full and proper context in this forum. Additionally, the Bush administration continues to withhold substantial relevant information relating to the events surrounding 9/11. More facts are continually uncovered, however, and we expect that important new revelations will be included in future editions of this book.

There is no doubt that 9/11 is one of the pivotal events of our time. It already has changed the course of history. But what exactly *was* 9/11? What happened? Why did it happen? How could it have happened? Why wasn't it prevented? Who in our government failed in their constitutional duty to "protect and defend" us from these attacks? And how can we ensure that such failures will not occur again?

We must find the answers to these questions. The United States will remain vulnerable—we are not safe and we will not be safe—until we can answer these questions, until we call to account those who have failed us, and until we repair institutional dysfunctions that contributed to this failure.

I believe the public has not been told the truth about 9/11. The quest for the truth is not yet complete, and the work must continue. It is my hope that this book provides a benchmark of what we know today, and what we have yet to learn.

STYLISTIC NOTES

There are a few stylistic issues to keep in mind while reading. This book is written from an American point of view. For example, references to a Secretary of State or Defense Secretary refer to positions in the U.S. government unless stated otherwise.

Subject Matter Categorization

This book is organized by subject-matter category. Quite often, however, a particular fact or event may affect more than one issue. For example, some counterterrorism efforts before 9/11 bear directly on the issue of what the U.S. knew, prior to 9/11, about the terrorist threat. Likewise, Bush administration claims regarding what was known prior to 9/11 bear equally on pre-9/11 warning signs and post-9/11 investigations. Generally, events have been included in the chapter most closely related or most affected by the event, and are not repeated in different categories, although events deemed particularly relevant are referenced in each related chapter. In order to obtain the full picture of the interplay of events, however, one must review all of the chapters, because in the end, they are all related. Note also that references are generally only made in chapter introductions if the material is not referred to in any of the chapter's entries.

Background Information and Abbreviations

Due to space considerations, background information regarding people, places, or facts is generally not provided, and a certain level of knowledge is assumed. For example, the country of Yemen might be mentioned without giving its exact geographic location. Additionally, certain frequently mentioned organizations are referred to using their familiar acronyms, and frequently mentioned individuals are referred to simply by job title and last name. It may be helpful to refer to the following reference list while reading this book:

Organizations

CIA	U.S. Central Intelligence Agency
DEA	U.S. Drug Enforcement Administration
FAA	U.S. Federal Aviation Administration
FDA	U.S. Food and Drug Administration
FBI	U.S. Federal Bureau of Investigation
FEMA	U.S. Federal Emergency Management Agency
ISI	Inter-Services Intelligence, Pakistan's intelligence agency
Mossad	Israel's intelligence agency
NORAD	U.S. North American Aerospace Defense Command
NSA	U.S. National Security Agency
SEC	U.S. Security and Exchange Commission
Taliban	The rulers of Afghanistan, 1996–2001
WTC	World Trade Center

Important individuals

Ashcroft	John Ashcroft, U.S. Attorney General, Bush administration
bin Laden	Osama bin Laden (unless otherwise noted), leader of al-Qaeda terrorist organization
Bush	George W. Bush, U.S. President, January 2001–present (President George H.W. Bush, U.S. President January 1989–January 1993, is always referenced using his full name)
Cheney	Richard "Dick" Cheney, U.S. Vice President, January 2001–present
Clarke	Richard Clarke, National Coordinator for Counterterrorism, 1998–October 2001. Usually referred to as counterterrorism "tsar" Richard Clarke
Clinton	Bill Clinton, U.S. President, January 1993–January 2001

Mahmood	Lieutenant General Mahmood Ahmad, Director of the ISI
Mueller	Robert Mueller, Director of the FBI, July 2001–present
Musharraf	General Pervez Musharraf, President of Pakistan, 1999–present
Powell	Colin Powell, U.S. Secretary of State, January 2001–present
Rice	Condoleezza Rice, U.S. National Security Adviser, January 2001–present
Rumsfeld	Donald Rumsfeld, U.S. Secretary of Defense, January 2001–present
Tenet	George Tenet, Director of the CIA, 1997–2004

Variations in spelling of names

Many names in this book have multiple commonly used spellings. Arabic names in particular often are translated into English in different ways, and this can cause confusion. For example, Osama bin Laden is the preferred spelling in this book, but his name can be spelled Usama bin Laden, Osama bin Ladin, Ussama bin Ladin, and so forth. In another example, the terrorist Ramzi bin al-Shibh is frequently referred to as Ramzi Binalshibh or Ramzi bin al Shibh. There are a great many names in this book, and it is confusing enough to keep track of them all without also having to keep track of all the different spellings. Therefore, this book uses one spelling exclusively, even when quoting from original sources that use alternative spellings.

Tenses

Each entry has been written in the present tense. Where appropriate, quotes have been modified (and so noted) to the present tense.

PART I
BEFORE 9/11

1.

Warning Signs

U.S. officials, especially those in the Bush administration, have repeatedly insisted that they had no evidence that Osama bin Laden was planning an attack inside the U.S. For example, President Bush insists, "Had I known there was going to be an attack on America I would have moved mountains to stop the attack." [NEW YORK TIMES, 4/18/04] Officials claim that hints of an attack—to the extent they existed—always pointed overseas, and also claim that no one thought that terrorists would use airplanes as flying weapons against U.S. symbols. For example, National Security Adviser Condoleezza Rice said in mid-2002 that, "Even in retrospect," the Bush administration had no hints of such a form of attack. [WHITE HOUSE, 5/16/02]

What are the facts as we know them today? What did U.S. government officials know, and when did they know it? Understanding these warning signs—and the U.S. government's response to the signs—is paramount to understanding how 9/11 could have happened and, equally important, whether it can happen again. This chapter chronicles, based on the current public record, what people in our government *did* know and when they knew it. (Even more warning signs are detailed elsewhere in the book, as in chapter 3 on counterterrorism efforts before 9/11, chapter 7 on hijackers Nawaf Alhazmi and Khalid Almihdhar, and chapter 10 on Zacarias Moussaoui.)

The seeds of 9/11 were planted long before the Bush administration came to power. In its quest to win the Cold War, the U.S. spent billions of dollars funding the mujahedeen resistance in Afghanistan, seeking, literally, to spend the Soviet Union to death. Focused on victory, the U.S. directly and indirectly transformed the ill-armed and overwhelmed Afghan freedom fighters into a well-armed, increasingly effective international "army." That many in this "army" were anti-Western, fundamentalist Islamists seemed irrelevant at the time, as they were fighting a common enemy.

The U.S. eventually won the Cold War—but at what price? In Afghanistan, the communist government was eventually replaced with the tyrannical Taliban. Many from the mujahedeen resistance reconstituted into al-Qaeda and similar organizations. Their guerilla-style jihad, so effective against the Soviets, morphed into terrorist jihad against the U.S. and other Western countries.

By the time Bill Clinton became president in 1993, bin Laden had already sponsored a number of successful attacks on American targets. During the Clinton administration, he sponsored more. After the simultaneous bombing of two U.S. embassies in Africa in 1998, and the U.S.'s failed retaliation attempt against him, Osama bin Laden became a household name around the world.

Each successive attack bin Laden perpetrated was a warning sign—that his primary target was the U.S., that his primary weapon was terrorism, and that he had the resources necessary to wage his war. Each successive attack foiled was yet another warning sign—of what he planned to do, how he planned to do it, and what he likely would try to do again.

One might fairly argue that hindsight is always 20/20, that Monday morning quarterbacking is simply unfair. Thus, in reviewing these warning signs, one should consider at least the following: How reliable would the warning have sounded at the time? From whom did the information originate—from an anonymous stranger, a friendly foreign government, or our own governmental surveillance? How, if at all, was the new information matched up with other facts known at the time? Who received the information, and what did they do with it; how far up the chain-of-command did the information travel, and was that information shared with other relevant parties or agencies?

Unfortunately, some of these questions cannot be answered at this time. All too often, the information that is available only raises more questions than it answers. The Bush administration's continued secrecy in the name of "national security" only exacerbates this problem, leading many to cry cover-up, others to charge criminal negligence, and still others to consider conspiracy.

The purpose of this book is not to speculate or to theorize, but to compile what is known today. And what is known is quite troubling. The warning signs were innumerable, the U.S. government's response often inexplicable.

December 26, 1979: Soviet Forces, Lured in by the CIA, Invade Afghanistan

The Soviet Union invades Afghanistan. They will withdraw in 1989 after a brutal ten-year war. It has been commonly believed that the invasion was unprovoked. However, in a 1998 interview, Zbigniew Brzezinski, President Jimmy Carter's National Security Adviser, will reveal that the CIA began destabilizing the pro-Soviet Afghan government six months earlier in a deliberate attempt to get the Soviets to invade and have their own Vietnam-type costly war: "What is most important to the history of the world? The Taliban or the collapse of the Soviet empire? Some stirred-up Moslems or the liberation of Central Europe and the end of the Cold War?" [LE NOUVEL OBSERVATEUR, 1/98; MIRROR, 1/29/02] The U.S. and Saudi Arabia give a huge amount of money (estimates range up to $40 billion total for the war) to support the mujahedeen guerrilla fighters opposing the Russians. Most of the money is managed by the ISI, Pakistan's intelligence agency. [NATION, 2/15/99]

Early 1980: Osama bin Laden, with Saudi Backing, Supports Afghan Rebels

Osama bin Laden begins providing financial, organizational, and engineering aid for the mujahedeen in Afghanistan, with the advice and support of the Saudi royal family. [NEW YORKER, 11/5/01] Some, including Richard Clarke, counterterrorism "tsar" during the Clinton and George W. Bush administrations, believe he was handpicked for the job by Prince Turki al-Faisal, head of Saudi Arabia's secret service. [SUNDAY TIMES, 8/25/02; NEW YORKER, 11/5/01] The Pakistani ISI want a Saudi prince as a public demonstration of the commitment of the Saudi royal family and as a way to ensure royal funds for the anti-Soviet forces. The agency fails to get royalty, but bin Laden, with his family's influential ties, is good enough for the ISI. [MIAMI HERALD, 9/24/01] (Clarke will argue later that the Saudis and other Muslim governments used the Afghan war in an attempt to get rid of their own misfits and troublemakers.) This multinational force later coalesces into al-Qaeda. [AGAINST ALL ENEMIES, BY RICHARD CLARKE, 3/04, P. 52]

**Osama
bin Laden**

1984–1994: U.S. Supports Militant Textbooks for Afghanistan

The U.S., through USAID and the University of Nebraska, spends millions of dollars developing and printing textbooks for Afghan schoolchildren. The textbooks are filled with violent images and militant

Islamic teachings, part of covert attempts to spur resistance to the Soviet occupation. For instance, children are taught to count with illustrations showing tanks, missiles, and land mines. Lacking any alternative, millions of these textbooks are used long after 1994; the Taliban are still using them in 2001. In 2002, the U.S. will start producing less violent versions of the same books, which President Bush says will have "respect for human dignity, instead of indoctrinating students with fanaticism and bigotry." (He will fail to mention who created those earlier books. [*WASHINGTON POST*, 3/23/02; CANADIAN BROADCASTING CORP., 5/6/02]

March 1985: U.S. Escalates War in Afghanistan

The CIA, British MI6 (Britain's intelligence agency), and the Pakistani ISI agree to launch guerrilla attacks from Afghanistan into then Soviet-controlled Tajikistan and Uzbekistan, attacking military installations, factories, and storage depots within Soviet territory, and do so until the end of the war. The CIA also begins supporting the ISI in recruiting radical Muslims from around the world to come to Pakistan and fight with the Afghan mujahedeen. The CIA gives subversive literature and Korans to the ISI, who carry them into the Soviet Union. Eventually, around 35,000 Muslim radicals from 43 Islamic countries will fight with the Afghan mujahedeen. Tens of thousands more will study in the hundreds of new madrassas funded by the ISI and CIA in Pakistan. Their main logistical base is in the Pakistani city of Peshawar. [*WASHINGTON POST*, 7/19/92; *PITTSBURGH POST-GAZETTE*, 9/23/01; *HONOLULU STAR-BULLETIN*, 9/23/01; *THE HINDU*, 9/27/01] In the late 1980s, Pakistani President Benazir Bhutto, feeling the mujahedeen network has grown too strong, tells President George H. W. Bush, "You are creating a Frankenstein." However, the warning goes unheeded. [*NEWSWEEK*, 9/24/01] By 1993, President Bhutto tells Egyptian President Hasni Mubarak that Peshawar is under de facto control of the mujahedeen, and unsuccessfully asks for military help in reasserting Pakistani control over the city. Thousands of mujahedeen fighters return to their home countries after the war is over and engage in multiple acts of terrorism. One Western diplomat notes these thousands would never have been trained or united without U.S. help, and says, "The consequences for all of us are astronomical." [*ATLANTIC MONTHLY*, 5/96]

1985–1989: Precursor to al-Qaeda Puts Down U.S. Roots

Sheikh Abdullah Azzam, bin Laden's mentor, makes repeated trips to the U.S. and other countries, building up his organization, Makhtab al-Khidimat (MAK), also known as the Services Office. Branches of the MAK open in over 30 U.S. cities, as Muslim-Americans donate millions of dollars to

support the Afghan war against the Soviet Union. Azzam is assassinated in a car bomb attack in late 1989. Some U.S. intelligence officials believe bin Laden ordered the killing. Bin Laden soon takes over the MAK, which morphs into al-Qaeda. His followers take over MAK's offices in the U.S., and they become financial conduits for al-Qaeda terrorist operations. [*1000 YEARS FOR REVENGE*, BY PETER LANCE, 9/03, PP. 40–41]

**Sheikh
Abdullah Azzam**

1986: Bin Laden Works with CIA, at Least Indirectly

The CIA, ISI, and bin Laden build the Khost tunnel complex in Afghanistan. This will be a major target of bombing and fighting when the U.S. attacks the Taliban in 2001. [PITTSBURGH POST-GAZETTE, 9/23/01; HONOLULU STAR-BULLETIN, 9/23/01; THE HINDU, 9/27/01] It will be reported in June 2001 that "bin Laden worked closely with Saudi, Pakistani, and U.S. intelligence services to recruit mujahedeen from many Muslim countries," but this information has not been reported much since 9/11. [UPI, 6/14/01] A CIA spokesperson will later claim, "For the record, you should know that the CIA never employed, paid, or maintained any relationship whatsoever with bin Laden." [ANANOVA, 10/31/01]

September 1986: CIA Provides Afghan Rebels Stinger Missiles

Worried that the Soviets are winning the war in Afghanistan, the U.S. decides to train and arm the mujahedeen with Stinger missiles. The Soviets are forced to stop using the attack helicopters that were being used to devastating effect. Some claim the Stingers turn the tide of the war and lead directly to Soviet withdrawal. Now the mujahedeen are better trained and armed than ever before. [GHOST WARS, BY STEVE COLL, 2/04, PP. 11, 149–51; AGAINST ALL ENEMIES, BY RICHARD CLARKE, 3/04, PP. 48–50]

September 1987–March 1989: Head U.S. Consular Official
Claims He's Told to Issue Visas to Unqualified Applicants

Michael Springmann, head U.S. consular official in Jeddah, Saudi Arabia, later claims that during this period he is "repeatedly told to issue visas to unqualified applicants." He turns them down, but is repeatedly overruled by superiors. Springmann loudly complains to numerous government offices, but no action is taken. He is fired and his files on these applicants are destroyed. He later learns that recruits from many countries fighting for bin Laden against Russia in Afghanistan were funneled through the Jeddah office to get visas to come to the U.S., where the recruits would travel to train for the Afghan war. According to Springmann, the Jeddah consulate was run by the CIA and staffed almost entirely by intelligence agents. This visa system may have continued at least through 9/11, and 15 of the 19 9/11 hijackers received their visas through Jeddah, possibly as part of this program. [BBC, 11/6/01; ASSOCIATED PRESS, 7/17/02 (B); FOX NEWS, 7/18/02]

August 11, 1988: Bin Laden Forms al-Qaeda

Bin Laden conducts a meeting to discuss "the establishment of a new military group," according to notes that are found later. Over time, this group becomes known as al-Qaeda, roughly meaning "the base" or "the foundation." [ASSOCIATED PRESS, 2/19/03 (B)] It will take U.S. intelligence years even to realize a group named al-Qaeda exists.

February 15, 1989: Soviet Forces Withdraw from Afghanistan

Soviet forces withdraw from Afghanistan, but Afghan communists retain control of Kabul, the capital, until April 1992. [WASHINGTON POST, 7/19/92] Richard Clarke, a counterterrorism official during the Reagan and

George H. W. Bush administrations and the counterterrorism "tsar" by 9/11, later claims that the huge amount of U.S. aid provided to Afghanistan drops off drastically as soon as the Soviets withdraw, abandoning the country to civil war and chaos. The new powers in Afghanistan are tribal chiefs, the Pakistani ISI, and the Arab war veterans coalescing into al-Qaeda. [AGAINST ALL ENEMIES, BY RICHARD CLARKE, 3/04, PP. 52–53]

July 1990: Blind Sheikh on Terrorist Watch List Enters U.S.

Despite being on a U.S. terrorist watch list for three years, radical Muslim leader Sheikh Omar Abdul-Rahman enters the U.S. on a "much-disputed" tourist visa issued by an undercover CIA agent. [VILLAGE VOICE, 3/30/93; ATLANTIC MONTHLY, 5/96; 1000 YEARS FOR REVENGE, BY PETER LANCE, 9/03, P. 42] Abdul-Rahman was heavily involved with the CIA and Pakistani ISI efforts to defeat the Soviets in Afghanistan, and became famous traveling all over the world for five years recruiting new mujahedeen. However, he never hid his prime goals to overthrow the governments of the U.S. and Egypt. [ATLANTIC MONTHLY, 5/96]

Sheikh Omar Abdul-Rahman

He is "infamous throughout the Arab world for his alleged role in the assassination of Egyptian president Anwar el-Sadat." Abdul-Rahman immediately begins setting up a terrorist network in the U.S. [VILLAGE VOICE, 3/30/93] He is believed to have befriended bin Laden while in Afghanistan, and bin Laden secretly pays Abdul-Rahman's U.S. living expenses. [ATLANTIC MONTHLY, 5/96; ABC NEWS, 8/16/02] Abdul-Rahman's ties to the assassination of Rabbi Meir Kahane in 1990 are later ignored. As one FBI agent will say in 1993, he is "hands-off. It was no accident that the sheikh obtained a visa and that he is still in the country. He's here under the banner of national security, the State Department, the NSA, and the CIA." According to a very high-ranking Egyptian official, Abdul-Rahman continues to assist the CIA in recruiting new mujahedeen after moving to the U.S.: "We begged America not to coddle the sheikh." Egyptian intelligence warns the U.S. that Abdul-Rahman is planning new terrorist attacks, and on November 12, 1992, terrorists connected to him machine-gun a busload of Western tourists in Egypt. Still, he will continue to live freely in New York City. [VILLAGE VOICE, 3/30/93] He will finally be arrested in 1993 and convicted of assisting in the 1993 WTC bombing. [ATLANTIC MONTHLY, 5/96]

November 5, 1990: First bin Laden–Related Terror Attack on U.S.

Egyptian-American El Sayyid Nosair assassinates controversial right-wing Zionist leader Rabbi Meir Kahane. Kahane's organization, the Jewish Defense League, was linked to dozens of bombings and is ranked by the FBI as the most lethal domestic terrorist group in the U.S. at the time. Nosair is captured after a police shoot-out. An FBI informant says he saw Nosair meeting with Muslim leader Sheikh Omar Abdul-Rahman a few days before the attack, and evidence indicating a wider plot with additional targets is found. [VILLAGE VOICE, 3/30/93] Files found in Nosair's possession give details of a terrorist cell, mention al-Qaeda, and discuss the destruction of tall U.S. buildings. Incredibly, this vital information is not translated until years later. [ABC NEWS, 8/16/02] Instead, within 12 hours of the assassination, New York police declare the assassination the work of a "lone gunman" and they stick with that story. In Nosair's subsequent trial, prosecutors will choose not to introduce his incriminating possessions or his confession as evidence, and an apparent "open-and-shut case" will end with his acquittal. However, he will be sentenced to 22 years on other lesser charges. [VILLAGE VOICE, 3/30/93] Bin Laden contributes to Nosair's defense fund. Many of those involved in Kahane's assassination will plan the 1993 WTC bombing. As one FBI agent puts it, "The fact is that in 1990, myself and my detectives, we had in our office in handcuffs, the people who blew up the World Trade Center in '93. We were told to release them." [ABC NEWS, 8/16/02]

1992–1996: Bin Laden Attacks U.S. Interests Using Sudanese Base

With a personal fortune of around $250 million (estimates range from $50 to $800 million [MIAMI HERALD, 9/24/01]), bin Laden begins plotting terrorist attacks against the U.S. from his new base in Sudan. The first attack kills two tourists in Yemen at the end of 1992. [NEW YORKER, 1/24/00] The CIA learns of his involvement in that attack in 1993, and learns that same year that he is channeling money to Egyptian extremists. U.S. intelligence also learns that by January 1994 he is financing at least three terrorist training camps in North Sudan. [NEW YORK TIMES, 8/14/96; PBS FRONTLINE 9/01; 9/11 CONGRESSIONAL INQUIRY 7/24/03 (B)]

September 1, 1992: U.S. Misses Opportunity to Stop
First WTC Bombing and Discover al-Qaeda

Terrorists Ahmad Ajaj and Ramzi Yousef enter the U.S. together. Ajaj is arrested at Kennedy Airport in New York City. Yousef is not arrested, and later, he masterminds the 1993 bombing of the WTC. "The U.S. government was pretty sure Ajaj was a terrorist from the moment he stepped foot on U.S. soil," because his "suitcases were stuffed with fake passports, fake IDs and a cheat sheet on how to lie to U.S. immigration inspectors," plus "two handwritten notebooks filled with bomb recipes, six bomb-making manuals, four how-to videotapes concerning weaponry, and an advanced guide to surveillance training." However, Ajaj is charged only with passport fraud, and serves a six-month sentence. From prison, Ajaj frequently calls Yousef and others in the 1993 WTC bombing plot, but no one translates the calls until long after the bombing. [LOS ANGELES TIMES, 10/14/01] Ajaj is released from prison three days after the WTC bombing, but is later rearrested and sentenced to more than 100 years in prison. [LOS ANGELES TIMES, 10/14/01] One of the manuals seized from Ajaj is horribly mistranslated for the trial. For instance, the title page is said to say "The Basic Rule," published in 1982, when in fact the title says "Al-Qaeda" (which means "the base" in English), published in 1989. Investigators later complain that a proper translation could have shown an early connection between al-Qaeda and the WTC bombing. [NEW YORK TIMES, 1/14/01]

December 1992: First Realization That bin Laden Is Behind an Attack on a U.S. Target

A bomb explodes in a hotel in Aden, Yemen, killing two tourists. U.S. soldiers had just left the hotel for Somalia. [MIAMI HERALD, 9/24/01] U.S. intelligence is still unaware of bin Laden's funding of the Rabbi Meir Kahane assassination in 1990. However, it will conclude in April 1993 that "[Bin Laden] almost certainly played a role" in this attack. [9/11 CONGRESSIONAL INQUIRY, 7/24/03 (B)]

1993–1998: Al-Qaeda Double Agent Is Arrested and
Then Released; Tells Secrets About al-Qaeda

Canadian police arrest Ali Mohamed, a high-ranking al-Qaeda figure. However, they release him when the FBI says he is a U.S. agent. [GLOBE AND MAIL, 11/22/01] Mohamed, a former U.S. Army sergeant, will con-

tinue to work for al-Qaeda for a number of years. He trains bin Laden's personal bodyguards and trains a terrorist cell in Kenya that later blows up the U.S. embassy there. Meanwhile, between 1993 and 1997 he tells secrets to the FBI about al-Qaeda's operations. He is re-arrested in late 1998 and subsequently convicted for his role in the 1998 U.S. embassy bombing in Kenya. [CNN, 10/30/98; INDEPENDENT, 11/1/98] "For five years he was moving back and forth between the U.S. and Afghanistan. It's impossible the CIA thought he was going there as a tourist. If the CIA hadn't caught on to him, it should be

Ali Mohamed

dissolved and its budget used for something worthwhile," according to a former Egyptian intelligence officer. [WALL STREET JOURNAL, 11/26/01]

1993 (B): Early Evidence of Osama's Interest in Airplanes and Training Pilots

Bin Laden buys a jet from the U.S. military in Arizona. The U.S. military approves the transaction. The aircraft is later used to transport missiles from Pakistan that kill American Special Forces in Somalia. He also has some of his followers begin training as pilots in U.S. flight schools. These initial flight trainings will be abandoned when their details are later revealed in a court case about Operation Bojinka in January 1995. [SUNDAY HERALD, 9/16/01]

1993 (C): Expert Panel Predicts Terrorists Will Use
Planes as Weapons on Symbolic U.S. Targets

An expert panel commissioned by the Pentagon postulates that an airplane could be used as a missile to bomb national landmarks. However, the panel does not publish this idea in its "Terror 2000" Report. [WASHINGTON POST, 10/2/01] One of the authors of the report later says, "We were told by the Department of Defense not to put it in . . . and I said, 'It's unclassified, everything is available.' In addition, they said, "We don't want it released, because you can't handle a crisis before it becomes a crisis. And no one is going to believe you.'" [ABC NEWS, 2/20/02] However, in 1994, one of the panel's experts will write in Futurist magazine, "Targets such as the World Trade Center not only provide the requisite casualties but, because of their symbolic nature, provide more bang for the buck. In order to maximize their odds for success, terrorist groups will likely consider mounting multiple, simultaneous operations with the aim of overtaxing a government's ability to respond, as well as demonstrating their professionalism and reach." [WASHINGTON POST, 10/2/01]

January 20, 1993: Bill Clinton Inaugurated

Bill Clinton replaces George H. W. Bush as U.S. President. Clinton remains U.S. president until 2001.

February 26, 1993: WTC Is Bombed but Does Not Collapse, as Hoped

An attempt to topple the WTC fails, but six people are killed in the misfired blast. Analysts later determine that had the terrorists not made a minor error in the placement of the bomb, both towers could have fallen and up to 50,000 people could have been killed. Ramzi Yousef, who has close ties to bin Laden, organizes the attempt. [CONGRESSIONAL HEARINGS, 2/24/98] The *New York Times* later reports on Emad Salem, an undercover agent who will be the key government witness in the trial against Yousef. Salem testifies that the FBI knew about the attack beforehand and told him they would thwart it by substituting a harmless powder for the explosives. However, an FBI supervisor called off this plan, and the bombing was not stopped. [NEW YORK TIMES, 10/28/93] Other suspects were ineptly investigated before the bombing as early as 1990. Several of the bombers were trained by the CIA to fight in the Afghan war, and the CIA later concludes, in inter-

Damage caused by the World Trade Center explosion in 1993

nal documents, that it was "partly culpable" for this bombing. [INDEPENDENT, 11/1/98] U.S. officials later state that the overall mastermind of the 9/11 attacks, Khalid Shaikh Mohammed, is a close relative, probably an uncle, of Yousef. [INDEPENDENT, 6/6/02; LOS ANGELES TIMES, 9/1/02] One of the attackers even left a message found by investigators stating, "Next time, it will be very precise." [ASSOCIATED PRESS, 9/30/01]

June 24, 1993: New York Landmark Bombing Plot Is Foiled

Eight people are arrested, foiling a plot to bomb several New York City landmarks. The targets were the United Nations building, 26 Federal Plaza, and the Lincoln and Holland tunnels. The plotters are connected to Ramzi Yousef and Sheikh Omar Abdul-Rahman. If the bombing, planned for later in the year, had been successful, thousands would have died. [9/11 CONGRESSIONAL INQUIRY, 7/24/03 (B)]

October 3–4, 1993: Al-Qaeda and Pakistani Leader Support Somalia Attack on U.S. Soldiers

Eighteen U.S. soldiers are attacked and killed in Mogadishu, Somalia, in a spontaneous gun battle following an unsuccessful attempt by U.S. Army Rangers to snatch a local warlord. (This event later becomes the subject of the movie *Black Hawk Down*.) A 1998 U.S. indictment will charge bin Laden and his followers with training the attackers. [PBS FRONTLINE, 10/3/02 (C)] The link between bin Laden and the Somali killers of U.S. soldiers appears to be Pakistani terrorist Maulana Masood Azhar. [LOS ANGELES TIMES, 2/25/02] Azhar is associated with Pakistan's ISI. He will be imprisoned briefly after 9/11 and then released.

1994–September 2001: Evidence of Terrorist Connections in Arizona Obtained by Local Agents Repeatedly Ignored by FBI Headquarters

By 1990, Arizona became one of the main centers in the U.S. for radical Muslims, and it remains so throughout this period. However, terrorism remains a low priority for the Phoenix, Arizona, FBI office.

Around 1990, hijacker Hani Hanjour moves to Arizona for the first time and he spends much of the next decade in the state. The FBI apparently remains oblivious of Hanjour, though one FBI informant claims that by 1998 they "knew everything about the guy." [NEW YORK TIMES, 6/19/02] In 1994, the Phoenix FBI office uncovers startling evidence connecting Arizona to radical Muslim terrorists. The office videotapes two men trying to recruit a Phoenix FBI informant to be a suicide bomber. One of the men is linked to terrorist leader Sheikh Omar Abdul-Rahman. [LOS ANGELES TIMES, 5/26/02; NEW YORK TIMES, 6/19/02] In 1998, the office's international terrorism squad investigates a possible Middle Eastern extremist taking flight lessons at a Phoenix airport. FBI agent Ken Williams initiates an investigation into the possibility of terrorists learning to fly aircraft, but he has no easy way to query a central FBI database about similar cases. Because of this and other FBI communication problems, he remains unaware of most U.S. intelligence reports about the potential use of airplanes as weapons, as well as other, specific FBI warnings issued in 1998 and 1999 concerning terrorists training at U.S. flight schools. [9/11 CONGRESSIONAL INQUIRY, 7/24/03]

1994: Al-Qaeda-Connected Group Tries to Fly Airplane into Eiffel Tower; Similar Attempts Elsewhere

Three separate incidents this year involve people hijacking airplanes and/or crashing them into buildings. A disgruntled Federal Express worker tries to crash a DC-10 into a company building in Memphis but the crew overpowers him. A lone pilot crashes a small plane onto the White House grounds, just missing the President's bedroom. An Air France flight is hijacked by a terrorist group linked to al-Qaeda, with the aim of crashing it into the Eiffel Tower, but French Special Forces storm the plane before it takes off. [NEW YORK TIMES, 10/3/01] *Time* magazine details the Eiffel Tower suicide plan in a cover story. [TIME, 1/9/05; 1000 YEARS FOR REVENGE, BY PETER LANCE, 9/03, P. 258]

December 12, 1994: Operation Bojinka Trial Run Fails, but Kills One

Terrorist Ramzi Yousef attempts a trial run of Operation Bojinka, planting a small bomb on a Philippine Airlines flight to Tokyo, and disembarking on a stopover before the bomb is detonated. The bomb explodes, killing one man and injuring several others. It would have successfully caused the plane to crash if not for the heroic efforts of the pilot. [LOS ANGELES TIMES, 9/1/02; 9/11 CONGRESSIONAL INQUIRY, 9/18/02]

January 6, 1995: Pope Assassination and Bojinka Plot to Bomb a Dozen Airplanes Is Foiled

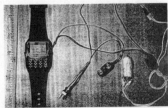

One of Ramzi Yousef's timers seized by Philippine police in January 1995

Responding to an apartment fire, Philippine investigators uncover an al-Qaeda plot to assassinate the Pope that is scheduled to take place when he visits the Philippines one week later. While investigating that scheme, they also uncover Operation Bojinka, planned by the same people: 1993 WTC bomber Ramzi Yousef and 9/11 mastermind Khalid Shaikh Mohammed. [INDEPENDENT, 6/6/02; LOS ANGELES TIMES,

6/24/02; LOS ANGELES TIMES, 9/1/02] The first phase of the plan is to explode 11 or 12 passenger planes over the Pacific Ocean. [AGENCE FRANCE-PRESSE, 12/8/01] Had this plot been successful, up to 4,000 people would have been killed in planes flying to Los Angeles, San Francisco, Honolulu, and New York. [INSIGHT, 5/27/02] All the bombs would be planted at about the same time, but some would be timed to go off weeks or even months later. Presumably worldwide air travel could be interrupted for months. [1000 YEARS FOR REVENGE, BY PETER LANCE, 9/03, PP. 260–61] This phase of Operation Bojinka was scheduled to go forward just two weeks later on January 21. [THE CELL, BY JOHN MILLER, MICHAEL STONE, AND CHRIS MITCHELL, 8/02, P. 124, INSIGHT, 5/27/02]

January 20, 1995: First Hints of Bojinka "Second Wave" Revealed

Philippine and U.S. investigators learn that Ramzi Yousef, Khalid Shaikh Mohammed, and their fellow plotters were actually planning three different terror attacks when they were foiled in early January. In addition to the planned assassination of the Pope, and the first phase of Operation Bojinka previously discovered, the terrorists also planned to crash about a dozen passenger planes into prominent U.S. buildings. It is often mistakenly believed that there is one Bojinka plan to blow up some planes and crash others into buildings, but in fact these different forms of attack are to take place in two separate phases. [1000 YEARS FOR REVENGE, BY PETER LANCE, 9/03, P. 259] Philippine investigator Colonel Rodolfo Mendoza learns about this second phase through the examination of recently captured Bojinka plotter Abdul Hakim Murad.

Abdul Hakim Murad

On January 20, Mendoza writes a memo about Murad's latest confession, saying, "With regards to their plan to dive-crash a commercial aircraft at the CIA headquarters, subject alleged that the idea of doing same came out during his casual conversation with [Yousef] and there is no specific plan yet for its execution. What the subject [has] in his mind is that he will board any American commercial aircraft pretending to be an ordinary passenger. Then he will hijack said aircraft, control its cockpit, and dive it at the CIA headquarters. He will use no bomb or explosives. It is simply a suicidal mission that he is very much willing to execute." [1000 YEARS FOR REVENGE, BY PETER LANCE, 9/03, PP. 277–78; INSIGHT, 5/27/02]

February 1995: Bojinka "Second Wave" Fully Revealed to Philippines Investigators; Information Given to U.S.

As Colonel Mendoza, the Philippines investigator, continues to interrogate Operation Bojinka plotter Abdul Hakim Murad, details of a post-Bojinka "second wave" emerge. Author Peter Lance calls this phase "a virtual blueprint of the 9/11 attacks." Murad reveals a plan to hijack commercial airliners at some point after the effect of Bojinka dies down. Murad himself had been training in the U.S. for this plot. He names the buildings that would be targeted for attack: CIA headquarters, the Pentagon, an unidentified nuclear power plant, the Transamerica Tower in San Francisco, the Sears

Col. Rodolfo Mendoza

Tower, and the World Trade Center. Murad continues to reveal more information about this plot until he is handed over to the FBI in April. [1000 YEARS FOR REVENGE, BY PETER LANCE, 9/03, PP. 278–80] He identifies approximately ten other men who met him at the flight schools or were getting similar training. They came from Sudan, United Arab Emirates, Saudi Arabia, and Pakistan. Apparently none of these pilots match the names of any of the 9/11 hijackers. However, he also gives information pointing to the terrorist Hambali through a front company named Konsonjaya. Hambali later hosts an important al-Qaeda meeting attended by two of the 9/11 hijackers. [ASSOCIATED PRESS, 3/5/02 (B)] Colonel Mendoza even makes a flow chart connecting many key players together, including bin Laden, bin Laden's brother-in-law Mohammed Jamal Khalifa, Ramzi Yousef, and 9/11 mastermind Khalid Shaikh Mohammed (named as Salem Ali a.k.a. Mohmad). Philippine authorities later claim that they provide all of this information to U.S. authorities, but the U.S. fails to follow up on any of it. [1000 YEARS FOR REVENGE, BY PETER LANCE, 9/03, PP. 303–4] Khalifa is in U.S. custody and released even after the Philippine authorities provide this information about him.

February 7, 1995: Yousef Is Arrested and Talks, but Hides Operation Bojinka Second Wave and bin Laden Ties

Ramzi Yousef is arrested in Pakistan. At the time, Khalid Shaikh Mohammed is staying in the same building, and brazenly gives an interview to *Time* magazine as "Khalid Sheikh," describing Yousef's capture. [1000 YEARS FOR REVENGE, BY PETER LANCE, 9/03, PP. 328] Yousef had recruited Istaique Parker to implement a limited version of Operation Bojinka. Parker was to place bombs on board two flights bound from

Ramzi Yousef

Bangkok to the U.S., but got cold feet and instead turned in Yousef. [1000 YEARS FOR REVENGE, BY PETER LANCE, 9/03, PP. 284-85] The next day, as Yousef is flying over New York City on his way to a prison cell, an FBI agent says to him, "You see the Trade Centers down there, they're still standing, aren't they?" Yousef responds, "They wouldn't be if I had enough money and enough explosives." [MSNBC, 9/23/01; THE CELL, BY JOHN MILLER, MICHAEL STONE, AND CHRIS MITCHELL, 8/02, P. 135] Yousef also soon admits to ties with Wali Khan Shah, who fought with bin Laden in Afghanistan, and Mohammed Jamal Khalifa, one of bin Laden's brothers-in-law, who is being held by the U.S. at the time. Despite Yousef's confession, Khalifa is released a few months later. Although Yousef talks freely, he makes no direct mention of bin Laden, or the planned second wave of Operation Bojinka that closely parallels the later 9/11 plot. [1000 YEARS FOR REVENGE, BY PETER LANCE, 9/03, PP. 297–98]

Spring 1995: More Evidence That the WTC Remains a Terrorist Target

In the wake of uncovering the Operation Bojinka plot, Philippine authorities find a letter on a computer disc written by the terrorists who had planned the failed 1993 WTC bombing. This letter apparently was never sent, but its contents are revealed in 1998 congressional testimony. [CONGRESSIONAL HEARINGS, 2/24/98] The Manila police chief also reports discovering a statement from bin Laden around this time that, although they failed to blow up the WTC in 1993, "on the second attempt they would be successful." [AGENCE FRANCE-PRESSE, 9/13/01]

April 3, 1995: *Time* Magazine, Senator
Highlight Author's Flying Bomb Idea

Time magazine's cover story reports on the potential for terrorists to kill thousands in highly destructive acts. Senator Sam Nunn (D) outlines a scenario in which terrorists destroy the U.S. Capitol Building by crashing a radio-controlled airplane into it. "It's not far-fetched," he says. His idea was taken from Tom Clancy's book *Debt of Honour* published in August 1994. [TIME, 4/3/95] High-ranking al-Qaeda leaders will claim later that Flight 93's target was the Capitol Building. [GUARDIAN, 9/9/02]

May 11, 1995: FBI Memo Fails to Mention
Operation Bojinka "Second Wave"

FBI agents, having held Operation Bojinka plotter Abdul Hakim Murad for about a month, write a memo containing what they have learned from interrogating him. The memo contains many interesting revelations, including that Ramzi Yousef, a mastermind of the 1993 World Trade Center bombing, "wanted to return to the United States in the future to bomb the World Trade Center a second time." However, this memo does not contain a word about the second wave of Operation Bojinka—to fly about 12 hijacked airplanes into prominent U.S. buildings—even though Murad had recently fully confessed this plot to Philippines investigators, who claim they turned over tapes, transcripts, and reports with Murad's confessions of the plot to the U.S. when they handed over Murad. It has not been explained why this plot is not mentioned in the FBI's summary of Murad's interrogation. [1000 YEARS FOR REVENGE, BY PETER LANCE, 9/03, PP. 280–82] After 9/11, a Philippine investigator will refer to this third plot when he says of the 9/11 attacks, "It's Bojinka. We told the Americans everything about Bojinka. Why didn't they pay attention?" [WASHINGTON POST, 9/23/01] In an interview after 9/11, Khalid Shaikh Mohammed will claim that the 9/11 attacks were a refinement and resurrection of this plot. [AUSTRALIAN, 9/9/02]

July 1995: U.S. Intelligence Report Concludes
Terrorists Intent on Attacking Inside U.S.

A U.S. National Intelligence Estimate concludes that the most likely terrorist threat will come from emerging "transient" terrorist groupings that are more fluid and multinational than older organizations and state-sponsored surrogates. This "new terrorist phenomenon" is made up of loose affiliations of Islamist extremists violently angry at the U.S. Lacking strong organization, they get weapons, money, and support from an assortment of governments, factions, and individual benefactors. [9/11 COMMISSION REPORT, 3/24/04] The estimate warns that terrorists are intent on striking specific targets inside the U.S., especially landmark buildings in Washington and New York. In 1997, the intelligence estimate is updated with bin Laden mentioned on the first page as an emerging threat and points out he might be interested in attacks inside the U.S. However, this new estimate is only two sentences long and lacks any strategic analysis on how to address the threat. [ASSOCIATED PRESS, 4/16/04]

November 13, 1995: Al-Qaeda Bombing in Saudi Arabia,
U.S. Realizes bin Laden Is More Than Financier

Two truck bombs kill five Americans and two Indians in the U.S.-operated Saudi National Guard training center in Riyadh, Saudi Arabia. Al-Qaeda is blamed for the attacks. [ASSOCIATED PRESS, 8/19/02] The attack changes U.S. investigators' views of bin Laden from terrorist financier to terrorist leader. [THE CELL. BY JOHN MILLER, MICHAEL STONE, AND CHRIS MITCHELL, 8/02, P. 150] The Vinnell Corporation, thought by some experts to be a CIA front, owns the facility attacked. [LONDON TIMES, 5/14/03]

1996: FBI Fumbles Flight School Investigation

Finding a business card for a U.S. flight school in the possession of Operation Bojinka plotter Abdul Hakim Murad, the FBI investigates the U.S. flight schools Murad attended. [WASHINGTON POST, 9/23/01] He had trained at about six flight schools off and on, starting in 1990. Apparently, the FBI closes the investigation when they fail to find any other potential suspects. [INSIGHT, 5/27/02] However, Murad had already confessed to Philippine authorities the names of about ten other al-Qaeda operatives learning to fly in the U.S., and they had asserted that they provided this information to the U.S. The U.S. will fail to follow up on it before 9/11. [ASSOCIATED PRESS, 3/5/02 (B)]

January 1996: Muslim Extremists Plan
Suicide Attack on White House

U.S. intelligence obtains information concerning a suicide attack on the White House planned by individuals connected with Sheikh Omar Abdul-Rahman and a key al-Qaeda operative. The plan is to fly from Afghanistan to the U.S. and crash into the White House. [9/11 CONGRESSIONAL INQUIRY 9/18/02]

April 1996–March 1997: Yousef Communicates with Terrorists from
Within Maximum Security Prison Using Telephone Provided by FBI

Ramzi Yousef, mastermind along with Khalid Shaikh Mohammed of the 1993 WTC bombing and the Operation Bojinka plots, is in a maximum-security prison, sentenced to hundreds of years of prison time for his terror plots. However, he can communicate with Gregory Scarpa Jr., a mob figure in the cell next to him. The FBI sets up a sting operation with Scarpa's cooperation to learn more of what and whom Yousef knows. Scarpa is given a telephone, and he allows Yousef to use it. However, Yousef uses the sting operation for his own ends, communicating with terrorists on the outside in code language without giving away their identities. He attempts to find passports to get co-conspirators into the U.S., and there is some discussion about imminent attacks on U.S. passenger jets. Realizing the scheme has backfired, the FBI terminates the telephone sting in late 1996, but Yousef manages to keep communicating with the outside world for several more months. [1000 YEARS FOR REVENGE. BY PETER LANCE, 9/03, PP. 280–82; NEW YORK DAILY NEWS, 9/24/00; NEW YORK DAILY NEWS, 1/21/02]

June 25, 1996: Khobar Towers Are Bombed;
Culprit Is Unclear

Explosions destroy the Khobar Towers in Dhahran, Saudi Arabia, killing 19 American soldiers and wounding 500. [CNN, 6/26/96] Saudi officials later interrogate the suspects, declare them guilty, and execute them—without letting the FBI talk to them. [PBS *FRONTLINE*, 2001, *IRISH TIMES*, 11/19/01] Saudis blame Hezbollah, the Iranian-influenced group, but U.S. investigators still believe bin Laden was somehow involved. [*SEATTLE TIMES*, 10/29/01] Bin Laden admits instigating the attacks in a 1998 interview. [*MIAMI HERALD*, 9/24/01] Ironically, the bin Laden family is later awarded the contract to rebuild the installation. [*NEW YORKER*, 11/5/01] In 1997, Canada catches one of the Khobar Tower attackers and extradites him to the U.S. However, in 1999, he is shipped back to Saudi Arabia before he can reveal what he knows about al-Qaeda

Destruction at the Khobar Towers, Dhahran, Saudi Arabia

and the Saudis. One anonymous insider calls it, "President Clinton's parting kiss to the Saudis." [*BEST DEMOCRACY MONEY CAN BUY*, BY GREG PALAST, 2/03, P. 102] In June 2001, a U.S. grand jury will indict 13 Saudis for the bombing. According to the indictment, Iran and Hezbollah were also involved in the attack. [9/11 CONGRESSIONAL INQUIRY, 7/24/03 (B)]

July 6–August 11, 1996: "Atlanta Rules" Established to Protect
Against Terrorists Using Planes as Flying Weapons

U.S. officials identify crop dusters and suicide flights as potential terrorist weapons that could threaten the Olympic Games in Atlanta, Georgia. They take steps to prevent any air attacks. They ban planes from getting too close to Olympic events. During the games, they deploy Black Hawk helicopters and U.S. Customs Service jets to intercept suspicious aircraft over the Olympic venues. Agents monitor crop-duster flights within hundreds of miles of downtown Atlanta. They place armed fighter jets on standby at local air bases. Flights to Atlanta get special passenger screening. Law enforcement agents also fan out to regional airports throughout northern Georgia "to make sure nobody hijacked a small aircraft and tried to attack one of the venues," says Woody Johnson, the FBI agent in charge. Counterterrorism "tsar" Richard Clarke uses this same security blanket approach to other major events, referring to the approach as "Atlanta Rules." [*CHICAGO TRIBUNE*, 11/18/01; *WALL STREET JOURNAL* 4/1/04; *AGAINST ALL ENEMIES*, BY RICHARD CLARKE, 3/04, PP. 108–09]

August 1996: Bin Laden Calls for Attack
on Western Targets in Arabia

Bin Laden issues a public *fatwa*, or religious decree, authorizing attacks on Western military targets in the Arabian Peninsula. This eliminates any doubts that bin Laden is merely a financier of terrorist attacks, rather than an active terrorist. [9/11 CONGRESSIONAL INQUIRY, 9/18/02]

September 5, 1996: Yousef Trial Ignores Bojinka "9/11 Blueprint" Plot

Terrorist Ramzi Yousef and two other defendants, Abdul Hakim Murad and Wali Khan Amin Shah, are convicted of crimes relating to Operation Bojinka. [CNN, 9/5/96] In the nearly 6,000-page transcript of the three-month Bojinka trial, there is not a single mention of the "second wave" of Bojinka that closely paralleled the 9/11 plot. Interrogations by Philippine investigator Colonel Rodolfo Mendoza had exposed the details of this plot quite clearly. However, not only does the FBI not call Mendoza to testify, but his name is not even mentioned in the trial, not even by his assistant, who does testify. "The FBI seemed to be going out of its way to avoid even a hint of the plot that was ultimately carried out on 9/11," notes author Peter Lance. [*1000 YEARS FOR REVENGE*, BY PETER LANCE, 9/03, PP. 350–51]

October 1996: Iranian Hijacking Plot Uncovered

U.S. intelligence learns of an Iranian plot to hijack a Japanese plane over Israel and crash it into Tel Aviv. While the plot was never carried out, it is one more example of intelligence agencies being aware that planes could be used as suicide weapons. [9/11 CONGRESSIONAL INQUIRY, 9/18/02]

November 24, 1996: Passenger Plane Suicide Attack Narrowly Averted

A hijacked airliner crashes into the shallow waters off the coast of a resort in the Comoros Islands

Several Ethiopians take over a passenger airliner and let it run out of fuel. Hijackers fight with the pilot as the hijackers try to steer the plane into a resort on a Comoros Islands beach in the Indian Ocean, but seconds before reaching the resort the pilot is able to crash the plane into shallow waters instead, 500 yards short of the resort. One hundred and twenty-three of the 175 passengers and crew die. [*NEW YORK TIMES*, 11/25/96; *AUSTRALIAN*, 11/26/96; *HOUSTON CHRONICLE*, 11/26/96]

1997: Possible Unmanned Aerial Attacks Raise Concerns at FBI, CIA

FBI and CIA are concerned that an unnamed terrorist group, which has apparently purchased an unmanned aerial vehicle (UAV), will use it for terrorist attacks. At the time, the agencies believed that the only reason to use this UAV would be for either reconnaissance or attack. The primary concern is that terrorists will use the UAV to attack outside the United States, for example, by flying a UAV into a U.S. Embassy or a visiting U.S. delegation. [9/11 CONGRESSIONAL INQUIRY, 9/18/02]

January 20, 1997: Clinton Re-Inaugurated; "Atlanta Rules" Applied at This and Other Events

Bill Clinton is re-inaugurated as President. An extensive set of security measures to prevent airplanes as weapons crashing into the inauguration is used. These measures, first used in the 1996 Atlanta Olympics and thus referred to as the "Atlanta Rules," include the closing of nearby airspace, the use of intercept helicopters, the basing of armed fighters nearby, and more. This plan will later be used for the 1999 North Atlantic Treaty Organization's 50th anniversary celebration in Washington, the 2000 Republican convention in Philadelphia, the 2000 Democratic convention in New York, and the George W. Bush inauguration in 2001. [AGAINST ALL ENEMIES, BY RICHARD CLARKE, 3/04, PP. 110–11, WALL STREET JOURNAL, 4/1/04] At some point near the end of the Clinton administration, the Secret Service and Customs Service agree to create a permanent air defense unit to protect Washington. However, these agencies are part of the Treasury Department, and the leadership there refuses to fund the idea. The permanent unit is not created until after 9/11. [WALL STREET JOURNAL 4/1/04]

February 12, 1997: Vice President Gore's Aviation Security Report Released

The White House Commission on Aviation Safety and Security, led by Vice President Al Gore, issues its final report, which highlights the risk of terrorist attacks in the U.S. The report references Operation Bojinka, the failed plot to bomb twelve American airliners out of the sky over the Pacific Ocean, and calls for increased aviation security. The commission reports that [it] "believes that terrorist attacks on civil aviation are directed at the United States, and that there should be an ongoing federal commitment to reducing the threats that they pose." [GORE COMMISSION, 2/12/97] However, the report has little practical effect: "Federal bureaucracy and airline lobbying [slow] and [weaken] a set of safety improvements recommended by a presidential commission—including one that a top airline industry official now says might have prevented the September 11 terror attacks." [LOS ANGELES TIMES, 10/6/01]

1998: Hijacking Proposed to Obtain Release of Blind Sheikh

A son of Sheikh Omar Abdul-Rahman, the al-Qaeda leader convicted in 1995 of conspiring to blow up tunnels and other New York City landmarks, is heard to say that the best way to free his father from a U.S. prison might be to hijack an American plane and exchange the hostages. This is mentioned in President Bush's August 2001 briefing titled "Bin Laden Determined to Strike in U.S." [WASHINGTON POST, 5/18/02 (B)]

1998 (B): Indonesia Gives U.S. Warning of 9/11 Attack?

Hendropriyono, the Indonesian chief of intelligence, will later claim that, "[w]e had intelligence predicting the September 11 attacks three years before it happened but nobody believed us." He says Indonesian intelligence agents identify bin Laden as the leader of the group plotting the attack and that the U.S. disregards the warning, but otherwise offers no additional details. The Associated Press notes, "Indonesia's intelligence services are not renowned for their accuracy." [ASSOCIATED PRESS, 7/9/03 (C)]

February 22, 1998: Bin Laden Expands Fatwa Against U.S., Allies

Bin Laden issues a *fatwa*, declaring it the religious duty of all Muslims "to kill the Americans and their allies—civilians and military . . . in any country in which it is possible." [PBS FRONTLINE, 2001; SUNDAY HERALD, 9/16/01; COMPLETE TEXT OF THE FATWA FROM AL-QUDS AL-ARABI, 2/23/98] This is an expansion of an earlier *fatwa* issued in August 1996, which had called for attacks in the Arabian Peninsula only.

May 15, 1998: Oklahoma FBI Memo Warns of Potential
Terrorist-Related Flight Training; No Investigation Ensues

An FBI pilot sends his supervisor in the Oklahoma City FBI office a memo warning that he has observed "large numbers of Middle Eastern males receiving flight training at Oklahoma airports in recent months." The memo, titled "Weapons of Mass Destruction," further states this "may be related to planned terrorist activity" and "light planes would be an ideal means of spreading chemicals or biological agents." The memo does not call for an investigation, and none occurs. [9/11 CONGRESSIONAL INQUIRY, 7/24/03 (B), NEWSOK, 5/29/02] The memo is "sent to the bureau's Weapons of Mass Destruction unit and forgotten." [NEW YORK DAILY NEWS 9/25/02] In 1999, it is learned that an al-Qaeda agent has studied flight training in Norman, Oklahoma. Hijackers Mohamed Atta and Marwan Alshehhi will briefly visit the same school in 2000; Zacarias Moussaoui trains at the school in 2001.

After May 15, 1998: FBI Again Ignores Warnings About
Terrorists Planning to Obtain U.S. Pilot Training

The FBI receives reports that a terrorist organization might be planning to bring students to the U.S. for flight training, at some point in 1998 after the May 15 memo warns about possible terrorists training at U.S. flight schools. [NEW YORK DAILY NEWS, 9/25/02] The FBI is aware that people connected to this unnamed organization have performed surveillance and security tests at airports in the U.S. and made comments suggesting an intention to target civil aviation. Apparently, this warning is not shared with other FBI offices or the FAA, and a connection with the Oklahoma warning is not made; a similar warning will follow in 1999. [9/11 CONGRESSIONAL INQUIRY, 7/24/03 (B)]

May 26, 1998: Bin Laden Promises to Bring Jihad to U.S.

Bin Laden discusses "bringing the war home to America," in a press conference from Afghanistan. [9/11 CONGRESSIONAL INQUIRY, 9/18/02] He indicates the results of his jihad will be "visible" within weeks. [9/11 CONGRESSIONAL INQUIRY, 7/24/03 (B)] Two U.S. embassies will be bombed in August.

May 28, 1998: Bin Laden Wants to Use Missiles Against U.S. Aircraft

Bin Laden indicates he may attack a U.S. military passenger aircraft using antiaircraft missiles, in an interview with ABC News reporter John Miller. In the subsequent media coverage, Miller repeatedly refers to bin Laden as "the world's most dangerous terrorist," and "the most dangerous man in the world." [ABC NEWS, 5/28/98; ABC NEWS, 6/12/98; ESQUIRE, 2/99; 9/11 CONGRESSIONAL INQUIRY, 7/24/03 (B)]

June 1998: U.S. Learns That bin Laden Is
Considering Attacks Against Washington, New York

U.S. intelligence obtains information from several sources that bin Laden is considering attacks in the U.S., including Washington and New York. This information is given to senior U.S. officials in July

1998. [9/11 CONGRESSIONAL INQUIRY, 9/18/02] Information mentions an attack in Washington, probably against public places. U.S. intelligence assumes that bin Laden places a high priority on conducting attacks in the U.S. More information about a planned al-Qaeda attack on a Washington government facility will be uncovered in the spring of 1999. [9/11 CONGRESSIONAL INQUIRY, 7/24/03; 9/11 CONGRESSIONAL INQUIRY, 7/24/03 (B)]

August 1998: CIA Warns That Terrorists
Plan to Fly Bomb-Laden Plane into WTC

In an intelligence report, the CIA asserts that Arab terrorists plan to fly bomb-laden aircraft from a foreign country into the WTC. The FBI and the FAA do not take the threat seriously because of the state of aviation in that country. (The country has not been disclosed.) Later, other intelligence information connects this group to al-Qaeda. [NEW YORK TIMES, 9/18/02; 9/11 CONGRESSIONAL INQUIRY, 9/18/02] An FBI spokesperson says the report "was not ignored, it was thoroughly investigated by numerous agencies" and found to be unrelated to al-Qaeda. [WASHINGTON POST, 9/19/02 (B)] However, it is later determined that the group in fact did have ties to al-Qaeda. [NEW YORK TIMES, 9/18/02; 9/11 CONGRESSIONAL INQUIRY, 7/24/03]

August 4, 1998: Threat Precedes Embassy Bombings

The Egyptian Islamic Jihad, a terror group that has joined forces with al-Qaeda, issues a statement threatening to retaliate against the U.S. for its involvement rounding up three of its members helping Muslim forces fight in Albania. The group announces, "We wish to inform the Americans . . . of preparations for a response which we hope they read with care, because we shall write it with the help of God in the language they understand." U.S. embassy bombings follow three days later. [CNN, 1/01]

August 7, 1998: Terrorists Bomb U.S.
Embassies in Kenya and Tanzania

Two U.S. embassies in Africa are bombed almost simultaneously. The attack in Nairobi, Kenya, kills 213 people, including 12 U.S. nationals, and injures more than 4,500. The attack in Dar es Salaam, Tanzania, kills 11 and injures 85. The attack is blamed on al-Qaeda. [PBS FRONTLINE, 2001] The attack shows al-Qaeda has a capability for simultaneous attacks. A third attack against the U.S. embassy in Uganda fails. [ASSOCIATED PRESS, 9/25/98]

A victim of the Nairobi, Kenya, U.S. embassy bombing is carried away on a stretcher

Late August 1998: Al-Qaeda Planning
U.S. Attack, but Not Yet Ready

The FBI learns that al-Qaeda is planning an attack on the U.S., but "things are not ready yet. We don't have everything prepared," according to a captured member of the al-Qaeda cell that bombed the U.S. embassy in Kenya. [USA TODAY, 8/29/02]

September 1998: Bin Laden's Next Operations
May Involve Crashing Airplane into U.S. Airport

U.S. intelligence uncovers information that bin Laden's next operation could possibly involve crashing an aircraft loaded with explosives into a U.S. airport. This information is provided to senior U.S. officials. [9/11 CONGRESSIONAL INQUIRY, 9/18/02; *WASHINGTON POST*, 9/19/02 (B)]

October–November 1998: Al-Qaeda
U.S.-based Recruiting Efforts Uncovered

U.S. intelligence learns al-Qaeda is trying to establish a terrorist cell within the U.S. There are indications they might be trying to recruit U.S. citizens. In the next month, there is information that a terror cell in the United Arab Emirates is attempting to recruit a group of five to seven young men from the U.S. to travel to the Middle East for training. This is part of a plan to strike a U.S. domestic target. [9/11 CONGRESSIONAL INQUIRY, 9/18/02; 9/11 CONGRESSIONAL INQUIRY, 7/24/03 (B)]

October 8, 1998: FAA Warns of al-Qaeda
Threat to U.S. Civil Aviation

The FAA issues the first of three 1998 warnings to U.S. airports and airlines urging a "high degree of vigilance" against threats to U.S. civil aviation from al-Qaeda. It specifically warns against a possible terrorist hijacking "at a metropolitan airport in the Eastern United States." The information is based on statements made by bin Laden and other Islamic leaders and intelligence information obtained after the U.S. cruise missile attacks in August. All three warnings come in late 1998, well before 9/11. [*BOSTON GLOBE*, 5/26/02]

Autumn 1998: Rumors of bin Laden Plot Involving
Aircraft in New York and Washington Surface Again

U.S. intelligence hears of a bin Laden plot involving aircraft in the New York and Washington areas. [9/11 CONGRESSIONAL INQUIRY, 9/18/02; *NEW YORK TIMES*, 9/18/02] In December it will learn that al-Qaeda plans to hijack U.S. aircraft are proceeding well and that two individuals have successfully evaded checkpoints in a dry run at a New York airport. [9/11 CONGRESSIONAL INQUIRY, 7/24/03 (B)]

November 1998: Turkish Extremists' Plan to Crash Airplane
into Famous Tomb Uncovered

U.S. intelligence learns that a Turkish extremist group named Kaplancilar had planned a suicide attack. The conspirators, who were arrested, planned to crash an airplane packed with explosives into a famous tomb during a government ceremony. The Turkish press said the group had cooperated with bin Laden and the FBI includes this incident in a bin Laden database. [9/11 CONGRESSIONAL INQUIRY, 9/18/02; 9/11 CONGRESSIONAL INQUIRY, 7/24/03]

December 1, 1998: Bin Laden Actively Planning Attacks Inside U.S.

According to a U.S. intelligence assessment, "[bin Laden] is actively planning against U.S. targets and already may have positioned operatives for at least one operation. . . . Multiple reports indicate [he] is keenly interested in striking the U.S. on its own soil . . . al-Qaeda is recruiting operatives for attacks in the U.S. but has not yet identified potential targets." Later in the month, a classified document prepared by the CIA and signed by President Clinton states: "The intelligence community has strong indications that bin Laden intends to conduct or sponsor attacks inside the U.S." [9/11 CONGRESSIONAL INQUIRY, 9/18/02; *WASHINGTON POST*, 9/19/02 (B); 9/11 CONGRESSIONAL INQUIRY, 7/24/03; 9/11 CONGRESSIONAL INQUIRY, 7/24/03 (B)]

December 21, 1998: Bin Laden May Be Planning Attacks on New York and Washington

In a *Time* magazine cover story entitled "The Hunt for Osama," it is reported that intelligence sources "have evidence that bin Laden may be planning his boldest move yet—a strike on Washington or possibly New York City in an eye-for-an-eye retaliation. 'We've hit his headquarters, now he hits ours,' says a State Department aide." [TIME, 12/21/98]

1999: British Intelligence Warns al-Qaeda Plans to Use Aircraft, Possibly as Flying Bombs

MI6, the British intelligence agency, gives a secret report to liaison staff at the U.S. embassy in London. The reports states that al-Qaeda has plans to use "commercial aircraft" in "unconventional ways," "possibly as flying bombs." [SUNDAY TIMES, 6/9/02]

1999: FBI Learns of Terrorists' Plans to Send Students to U.S. for Aviation Training; Investigation Opportunity Bungled

The FBI receives reports that a terrorist organization is planning to send students to the U.S. for aviation training. The organization's name remains classified, but apparently it is a different organization than one mentioned in a very similar warning the year before. The purpose of this training is unknown, but the organization viewed the plan as "particularly important" and it approved open-ended funding for it. The Counterterrorism Section at FBI headquarters issues a notice instructing 24 field offices to pay close attention to Islamic students from the target country engaged in aviation training. Ken Williams's squad at the Phoenix FBI office receives this notice, although Williams does not recall reading it. Williams will later write a memo on this very topic in July 2001. The 9/11 Congressional Inquiry later concludes, "There is no indication that field offices conducted any investigation after receiving the communication." [9/11 CONGRESSIONAL INQUIRY, 7/24/03 (B)] However, an analyst at FBI headquarters conducts a study and determines that each year there are about 600 Middle Eastern students attending the slightly over 1,000 U.S. flight schools. [NEW YORK TIMES, 5/4/02; 9/11 CONGRESSIONAL INQUIRY, 7/24/03 (B)] In November 2000, a notice will be issued to the field offices, stating that it has uncovered no indication that the terrorist group is recruiting students. Apparently, Williams will not see this notice either. [9/11 CONGRESSIONAL INQUIRY, 7/24/03 (B)]

February 1999: Pilot Suicide Squad Rumored in Iraq

U.S. Intelligence obtains information that Iraq has formed a suicide pilot unit that it plans to use against British and U.S. forces in the Persian Gulf. The CIA comments that this report is highly unlikely and is probably disinformation. [9/11 CONGRESSIONAL INQUIRY, 7/24/03]

March 1999: Plot to Use Hang Glide Bomb Tested, Thwarted

U.S. intelligence learns of plans by an al-Qaeda member who is also a U.S. citizen to fly a hang glider into the Egyptian Presidential Palace and then detonate the explosives he is carrying. The individual, who received hang glider training in the U.S., brings a hang glider back to Afghanistan, but various problems arise during the testing of the glider. This unnamed person is subsequently arrested and is in custody abroad. [9/11 CONGRESSIONAL INQUIRY, 9/18/02]

Spring 1999: U.S. Uncovers bin Laden Plans to Attack Washington

U.S. intelligence learns of a planned bin Laden attack on a U.S. government facility in Washington (the specific facility targeted has not been identified). [9/11 CONGRESSIONAL INQUIRY, 9/18/02; *NEW YORK TIMES*, 9/18/02]

June 1999: CIA Reports That bin Laden Plans Attack in U.S.

In testimony before the Senate Select Committee on Intelligence and in a briefing to House Permanent Select Committee on Intelligence staffers one month later, the chief of the CIA's Counter Terrorism Center describes reports that bin Laden and his associates are planning attacks in the U.S. [9/11 CONGRESSIONAL INQUIRY 9/18/02]

July 14, 1999: FBI Hears Pakistani ISI Agent Promise to Attack WTC

See chapter 12.

September 1999: U.S. Report Predicts Spectacular Attack on Washington

A report prepared for U.S. intelligence entitled the "Sociology and Psychology of Terrorism" is completed. It states, "Al-Qaeda's expected retaliation for the U.S. cruise missile attack . . . could take several forms of terrorist attack in the nation's capital. Al-Qaeda could detonate a Chechen-type building-buster bomb at a federal building. Suicide bomber(s) belonging to al-Qaeda's Martyrdom Battalion could crash-land an aircraft packed with high explosives (C-4 and Semtex) into the Pentagon, the headquarters of the Central Intelligence Agency (CIA), or the White House. Whatever form an attack may take, bin Laden will most likely retaliate in a spectacular way." The report is by the National Intelligence Council, which advises the President and U.S. intelligence on emerging threats. [ASSOCIATED PRESS, 4/18/02] The Bush administration later claims to have never heard of this report until May 2002, despite the fact that it had been publicly posted on the Internet since 1999, and "widely shared within the government" according to the *New York Times*. [CNN, 5/18/02; *NEW YORK TIMES*, 5/18/02]

September 1999 (B): Bin Laden to Attack in U.S.—
Possibly in California and New York City

U.S. intelligence obtains information that bin Laden and others are planning a terrorist act in the U.S., possibly against specific landmarks in California and New York City. The reliability of the source is unknown. [9/11 CONGRESSIONAL INQUIRY, 9/18/02]

September 1999 (C): FBI Investigates Flight School Attendee Connected to bin Laden

Agents from Oklahoma City FBI office visit the Airman Flight School in Norman, Oklahoma to investigate Ihab Ali, who has already been identified as bin Laden's former personal pilot. Ali attended the school in 1993 and is later named as an unindicted coconspirator in the 1998 U.S. Embassy bombing in Kenya. [CNN, 10/16/01; BOSTON GLOBE, 9/18/01; 9/11 CONGRESSIONAL INQUIRY 10/17/02] When Ali was arrested in May 1999, he was working as a taxi driver in Orlando, Florida. Investigators discover recent ties between him and high-ranking al-Qaeda leaders, and suspect he was a "sleeper" agent. [ST. PETERSBURG TIMES, 10/28/01] However, the FBI agent visiting the school is not given most background details about him. [9/11 CONGRESSIONAL INQUIRY, 7/24/03 (B)] It is not known if these investigators are aware of a terrorist flight school warning given by the Oklahoma City FBI office in 1998. Hijackers Mohamed Atta and Marwan Alshehhi later visit the Airman school in July 2000 but ultimately decide to train in Florida instead. [BOSTON GLOBE, 9/18/01] Al-Qaeda agent Zacarias Moussaoui takes flight lessons at Airman in February 2001. One of the FBI agents sent to visit the school at this time visits it again in August 2001 asking about Moussaoui, but he fails to make a connection between the two visits.

September 15, 1999: Bipartisan Commission Concludes
Terrorist Attack Will Occur on U.S. Soil, Killing Many

The first phase of the U.S. Commission on National Security/21st Century, co-chaired by former Senators Gary Hart (D) and Warren Rudman (R), is issued. It concludes: "America will be attacked by terrorists using weapons of mass destruction and Americans will lose their lives on American soil, possibly in large numbers." [USCNS REPORT, 9/15/99]

October 5, 1999: Bin Laden Might Be Planning Major Attack in U.S.

The highly respected Jane's Terrorism and Security Monitor reports that U.S. intelligence is worried that bin Laden is planning a major terrorist attack on U.S. soil. They are said to be particularly concerned about some kind of attack on New York, and they have recommended stepped-up security at the New York Stock Exchange and the Federal Reserve. [NEWSMAX, 10/5/99]

2000: Attempted Flight Simulator Purchase Hints at Pilot Training

At some point during this year, an FBI internal memo states that a Middle Eastern nation has been trying to purchase a flight simulator in violation of U.S. restrictions. The FBI refuses to disclose the date or details of this memo. [LOS ANGELES TIMES, 5/30/02]

October 31, 1999: Suicide Pilot Crashes Commercial Airliner into Ocean

EgyptAir Flight 990 crashes into the ocean off the coast of Massachusetts, killing all 217 people on board. It is immediately suspected that one of the pilots purposely crashed the plane, and this is the eventual conclusion of a National Transportation Safety Board investigation. Thirty-three Egyptian military officers were aboard the plane, leading to suspicions of a terrorist motive, but no connections between the suicide pilot and terrorism can be found. [ASSOCIATED PRESS, 11/2/99; ASSOCIATED PRESS, 1/21/00; *ATLANTIC MONTHLY*, 11/01; *AVIATION WEEK AND SPACE TECHNOLOGY*, 3/25/02]

2000–September 11, 2001: NORAD Stages Exercises with Similarities to 9/11 Attack

In a roughly two year period before the 9/11 attacks, NORAD conducts regional war game exercises simulating hijacked airliners used as weapons to crash into targets and cause mass casualties. One of the imagined targets is the World Trade Center. In another exercise, jets perform a mock shoot-down over the Atlantic Ocean of a jet laden with chemical poisons headed toward the U.S. A third exercise scheduled for April 2001 has the Pentagon as the target, but apparently, that drill is not run after officials say it is too unrealistic. NORAD confirms, "[n]umerous types of civilian and military aircraft were used as mock hijacked aircraft" in these drills. [*USA TODAY*, 4/18/04] At some undetermined point before 9/11, a regional exercise simulates the crash of a foreign airplane flying into the U.S. and crashing into a famous U.S. building. The building is not known, but it is said not to be either the WTC or the Pentagon. This exercise "involved some flying of military aircraft as well as a command post exercise in which communications procedures were practiced in an office environment." NORAD later states that prior to 9/11, it "normally conducted four major exercises a year, most of which included a hijack scenario." [CNN, 4/19/04]

January 3, 2000: Al-Qaeda Attack on USS *The Sullivans* Fails; Remains Undiscovered

An al-Qaeda attack on USS *The Sullivans* in Yemen's Aden harbor fails when their boat filled with explosives sinks. The attack remains undiscovered, and a duplication of the attack by the same people later successfully hits the USS *Cole* in October 2000. [PBS *FRONTLINE*, 10/3/02 (C)]

March 2000: U.S. Intelligence Learns bin Laden May Target Statue of Liberty, Skyscrapers, Other Sites

U.S. intelligence obtains information about the types of targets that bin Laden's network might strike. The Statue of Liberty is specially mentioned, as are skyscrapers, ports, airports, and nuclear power plants. [9/11 CONGRESSIONAL INQUIRY, 9/18/02]

March 2000 (B): U.S. Intelligence Learns al-Qaeda May Attack West Coast

U.S. Intelligence Community obtains information suggesting al-Qaeda is planning attacks in specific West Coast areas, possibly involving the assassination of several public officials. [9/11 CONGRESSIONAL INQUIRY,

7/24/03 (B)] While these attacks do not materialize, this is the same month the CIA learns that two known al-Qaeda terrorists have just flown to Los Angeles (see chapter 7).

April 2000: Would-Be Hijacker Tells FBI About Plot to Fly Plane into U.S. Building

Niaz Khan, a British citizen originally from Pakistan, is recruited into an al-Qaeda plot. He is flown to Lahore, Pakistan, and then trains in a compound there for a week with others on how to hijack passenger airplanes. He trains on a mock cockpit of a 767 aircraft (an airplane type used on 9/11). He is taught hijacking techniques, including how to smuggle guns and other weapons through airport security and how to get into a cockpit. He is then flown to the U.S. and told to meet with a contact. He says, "They said I would live there for a while and meet some other people and we would hijack a plane from JFK and fly it into a building." [LONDON TIMES, 5/9/04] He has "no doubt" this is the 9/11 plot. However, Khan slips away and gambles away the money given to him by al-Qaeda. Afraid he would be killed for betraying al-Qaeda, he turns himself in to the FBI. For three weeks, FBI counterterrorism agents in Newark, New Jersey interview him. [MSNBC, 6/3/04; OBSERVER, 6/6/04] One FBI agent recalls, "We were incredulous. Flying a plane into a building sounded crazy but we polygraphed him and he passed." [LONDON TIMES, 5/9/04] A former FBI official says the FBI agents believed Khan and aggressively tried to follow every lead in the case, but word came from FBI headquarters saying, "Return him to London and forget about it." He is returned to Britain and handed over to British authorities. However, the British only interview him for about two hours, and then release him. He is surprised that authorities never ask for his help in identifying where he was trained in Pakistan, even after 9/11. [MSNBC, 6/3/04] His case is mentioned in the 2002 9/11 Congressional Inquiry report, but the plot is apparently mistakenly described as an attempt to hijack a plane and fly it to Afghanistan. [9/11 CONGRESSIONAL INQUIRY, 9/18/02]

April 17, 2000: Arizona FBI Agent Initiates Investigation into Flight School Students, but Faces Delays

Arizona FBI agent Ken Williams gets a tip that makes him suspicious that some flight students might be terrorists. [NEW YORK TIMES, 6/19/02] It appears that flight school student Zacaria Soubra is seen at a shooting range with a known jihad veteran. [LOS ANGELES TIMES, 10/28/01 (C)] On this day, Williams starts a formal investigation into Soubra. [ARIZONA REPUBLIC, 7/24/03] Soubra is the main focus of Williams's later memo. But Williams's work is greatly slowed because of internal politics and personal disputes. When he finally returns to this case in December 2000, he and all the other agents on the international-terrorism squad are diverted to work on a high-profile arson case. Says James Hauswirth, another Arizona agent, "[Williams] fought it. Why take your best terrorism investigator and put him on an arson case? He didn't have a choice." The arson case is finally solved in June 2001 and Williams once again returns to the issue of terrorist flight school students. His memo comes out one month later instead of some time in 2000. Hauswirth writes a letter to FBI Director Mueller in late 2001, complaining, "[Terrorism] has always been the lowest priority in the division; it still is the lowest priority in the division." Others concur that the Arizona FBI placed a low priority on terrorism cases. [LOS ANGELES TIMES 5/26/02; NEW YORK TIMES, 6/19/02]

June 2000: Multiple Web Domains Related to 2001 and/or WTC Terrorist Attack Are Registered

Around this time, a number of very suspicious web domains are registered, including the following: attackamerica.com, attackonamerica.com, attackontwintowers.com, august11horror.com, august11terror.com, horrorinamerica.com, horrorinnewyork.com, nycterroriststrike.com, pearlharborinmanhattan.com, terrorattack2001.com, towerofhorror.com, tradetowerstrike.com, worldtradecenter929.com, worldtradecenterbombs.com, worldtradetowerattack.com, worldtradetowerstrike.com, and wterroristattack2001.com. A counterterrorism expert says, "It's unbelievable that [the registration company] would register these domain names" and "if they did make a comment to the FBI, it's unbelievable that the FBI didn't react to it." Several of the names mention 2001 and, apparently, there were no other websites mentioning other years. Registering a site requires a credit card, so presumably, this story could provide leads, but it is unclear what leads the FBI gets from this, if any. No sites are active on 9/11. [CYBERCAST NEWS SERVICE, 9/19/01] All of the domain name registrations had expired around June 2001. [CYBERCAST NEWS SERVICE, 9/20/01] This story is later called an "urban legend," but the debunkers are later themselves criticized. [INSIGHT, 3/11/02]

August 12, 2000: Italian Intelligence Wiretap of al-Qaeda Terrorist Cell Reveals Massive Aircraft-based Strike

Italian intelligence successfully wiretap the al-Qaeda terrorist cell in Milan, Italy from late 1999 until summer 2001. [BOSTON GLOBE, 8/4/02] In a wiretapped conversation from this day, suspected Yemeni terrorist Abdulsalam Ali Abdulrahman tells wanted Egyptian terrorist Es Sayed about a massive strike against the enemies of Islam involving aircraft and the sky, a blow that "will be written about in all the newspapers of the world. This will be one of those strikes that will never be forgotten. . . . This is a terrifying thing. This is a thing that will spread from south to north, from east to west: The person who came up with this program is a madman from a madhouse, a madman but a genius." In another conversation, Abdulrahman tells Es Sayed: "I'm studying airplanes. I hope, God willing, that I can bring you a window or a piece of an airplane the next time we see each other." The comment is followed by laughter. Beginning in October 2000, FBI experts will help Italian police analyze the intercepts and warnings. Neither Italy nor the FBI understands their meaning until after 9/11, but apparently, the Italians understand enough to give the U.S. an attack warning in March 2001. [LOS ANGELES TIMES, 5/29/02; GUARDIAN, 5/30/02; WASHINGTON POST, 5/31/02] The Milan cell "is believed to have created a cottage industry in supplying false passports and other bogus documents." [BOSTON GLOBE, 8/4/02]

September 2000: Al-Qaeda Agent Testifies of Pilot Training

L'Houssaine Kherchtou arrives in the U.S. to testify against other al-Qaeda agents. He reveals that from 1992 to 1995 he trained in Nairobi, Kenya, to be a pilot for al-Qaeda. His training stopped when he left al-Qaeda in 1995. [STATE DEPARTMENT, 2/22/01]

October 24–26, 2000: Pentagon Practices Emergency Drill
for Possible Hijacked Airliner Crash into Building

Pentagon officials carry out a "detailed" emergency drill based upon the crashing of a hijacked airliner into the Pentagon. [MDW NEWS SERVICE, 11/3/00; MIRROR, 5/24/02] The Pentagon is such an obvious target that, "For years, staff at the Pentagon joked that they worked at 'Ground Zero,' the spot at which an incoming nuclear missile aimed at America's defenses would explode. There is even a snack bar of that name in the central court- yard of the five-sided building, America's most obvious military bullseye." [DAILY TELEGRAPH 9/16/01] After 9/11, a Pentagon spokesperson will claim: "The Pentagon was simply not aware that this aircraft was coming our way, and I doubt prior to Tuesday's event, anyone would have expected anything like that here." [NEWSDAY, 9/23/01]

A plane crash is simulated inside the cardboard court- yard of a model Pentagon

December 2000: National Intelligence Estimate Report
Downplays Threat to Domestic Aviation

A classified section of the yearly National Intelligence Estimate report given to Congress downplays any threat to domestic aviation. It says that FBI investigations confirm domestic and international terrorist groups are operating within the U.S. but they are focusing primarily on fundraising, recruiting new members, and disseminating propaganda. While international terrorists have conducted attacks on U.S. soil, these acts represent anomalies from their traditional targeting which focuses on U.S. interests over- seas. [9/11 CONGRESSIONAL INQUIRY, 7/24/03]

January 24, 2001: Italians Hear of "Brothers" Going to U.S.
for "Very, Very Secret" Plan, Other Clues

On this day, Italian intelligence hears an interesting wiretapped conversation eerily similar to the one from August 12, 2000. This one occurs between terrorists Es Sayed and Ben Soltane Adel, two mem- bers of al-Qaeda's Milan cell. Adel asks, in reference to fake documents, "Will these work for the broth- ers who are going to the United States?" Sayed responds angrily, saying "Don't ever say those words again, not even joking! . . . If it's necessary . . . whatever place we may be, come up and talk in my ear, because these are very important things. You must know . . . that this plan is very, very secret, as if you were protecting the security of the state." This is only one of many clues found from the Italian wiretaps and passed on to U.S. intelligence in March 2001. However, they apparently are not properly understood until after 9/11. Adel is later arrested and convicted of belonging to a terrorist cell, and Es Sayed flees to Afghanistan in July 2001. [GUARDIAN, 5/30/02]

February–July 2001: Trial Presents FBI with Information About Pilot Training Scheme

A trial is held in New York City for four defendants charged with involvement in the 1998 U.S. African embassy bombings. All are ultimately convicted. Testimony reveals that two bin Laden operatives had received pilot training in Texas and Oklahoma and another had been asked to take lessons. One bin Laden aide becomes a government witness and gives the FBI detailed information about a pilot training scheme. This new information does not lead to any new FBI investigations into the matter. [WASHINGTON POST, 9/23/01]

February 7, 2001: Tenet Warns Congress About bin Laden

CIA Director Tenet warns Congress in open testimony that "[t]he threat from terrorism is real, it is immediate, and it is evolving." He says bin Laden and his global network remains "the most immediate and serious threat" to U.S. interests. "Since 1998 bin Laden has declared that all U.S. citizens are legitimate targets," he says, adding that bin Laden "is capable of planning multiple attacks with little or no warning." [ASSOCIATED PRESS, 2/7/01; SUNDAY HERALD, 9/23/01]

March 2001: Italians Advise U.S. About al-Qaeda Wiretaps

The Italian government gives the U.S. information about possible attacks based on apartment wiretaps in the Italian city of Milan. [FOX NEWS, 5/17/02] Presumably, the information includes a discussion between two al-Qaeda agents talking about a "very, very secret" plan to forge documents "for the brothers who are going to the United States." The warning may also have mentioned a wiretap the previous August involving one of the same people that discussed a massive strike against the enemies of Islam involving aircraft. Two months later, wiretaps of the same Milan cell also reveal a plot to attack a summit of world leaders.

March 2001 (B): Al-Qaeda to Attack Inside the U.S. in April

An intelligence source claims that a group of al-Qaeda operatives is planning to conduct an unspecified attack inside the U.S. in April. One of the operatives allegedly resides in the U.S. There are also reports of planned attacks in California and New York State for the same month, though whether this is reference to the same plot is unclear. [9/11 CONGRESSIONAL INQUIRY, 7/24/03]

March 4, 2001: Television Show Eerily Envisions 9/11 Attacks

Contradicting the later claim that no one could have envisioned the 9/11 attacks, a short-lived Fox television program called *The Lone Gunmen* airs a pilot episode in which terrorists try to fly an airplane into the WTC. The heroes save the day and the airplane narrowly misses the building. There are no terrorists on board the aircraft; they use remote control technology to steer the plane. Ratings are good for the show, yet the eerie coincidence is barely mentioned after 9/11. Says one media columnist, "This seems to be collective amnesia of the highest order." [TV GUIDE, 6/21/02] In the show, the heroes also determine, "The terrorist group responsible was actually a faction of our own government. These malefactors were seeking to stimulate arms manufacturing in the lean years following the end of the Cold War by bringing down a plane in New York City and fomenting fears of terrorism." [MYERS REPORT, 6/20/02]

April 2001: Military Considers Exercise
Simulating Flying Airplane into Pentagon

NORAD is planning to conduct a training exercise named Positive Force. Some Special Operations personnel trained to think like terrorists unsuccessfully propose adding a scenario simulating "an event having a terrorist group hijack a commercial airliner and fly it into the Pentagon." Military higher-ups and White House officials reject the exercise as either "too unrealistic" or too disconnected to the original intent of the exercise. The proposal comes shortly before the exercise, which takes place this month.

[BOSTON HERALD, 4/14/04; GUARDIAN, 4/15/04; WASHINGTON POST, 4/14/04 (G); NEW YORK TIMES, 4/14/04]

April 2001 (B): Speculation That Commercial Pilots
Could Be Potential Terrorists

A source with terrorist connections speculates to U.S. intelligence that "bin Laden would be interested in commercial pilots as potential terrorists." The source warns that the U.S. should not focus only on embassy bombings, because terrorists are seeking "spectacular and traumatic" attacks along the lines of the WTC bombing in 1993. Because the source was offering personal speculation and not hard information, the information is not disseminated widely. [9/11 CONGRESSIONAL INQUIRY, 9/18/02; NEW YORK TIMES, 9/18/02]

April 2001 (C): FBI Translators Point to Explicit Warning from Afghanistan

FBI translators Sibel Edmonds and Behrooz Sarshar will later claim to know of an important warning given to the FBI at this time. In their accounts, a reliable informant on the FBI's payroll for at least ten years tells two FBI agents that sources in Afghanistan have heard of an al-Qaeda plot to attack the U.S. and Europe in a suicide mission involving airplanes. Al-Qaeda agents, already in place inside the U.S., are being trained as pilots. By some accounts, the names of prominent U.S. cities are mentioned. It is unclear if this warning reaches FBI headquarters or beyond. The two translators later privately testify to the 9/11 Commission. [WORLDNETDAILY, 4/6/04; VILLAGE VOICE, 4/14/04; SALON, 3/26/04] Sarshar's notes of the interview indicate that the informant claimed his information came from Iran, Afghanistan, and Hamburg, Germany (the location of the primary 9/11 al-Qaeda cell). However, anonymous FBI officials claim the warning was very vague and doubtful. [CHICAGO TRIBUNE, 7/21/04] In reference to this warning and apparently others, Edmonds says, "President Bush said they had no specific information about September 11, and that's accurate. However, there was specific information about use of airplanes, that an attack was on the way two or three months beforehand, and that several people were already in the country by May of 2001. They should've alerted the people to the threat we were facing." [SALON, 3/26/04] She adds, "There was general information about the time-frame, about methods to be used but not specifically about how they would be used and about people being in place and who was ordering these sorts of terror attacks. There were other cities that were mentioned. Major cities with skyscrapers." [INDEPENDENT, 4/2/04]

April–May 2001: Bush, Cheney Receive Numerous al-Qaeda Warnings

President George W. Bush, Vice President Dick Cheney, and national security aides are given briefing papers headlined, "Bin Laden Planning Multiple Operations," "Bin Laden Public Profile May Presage Attack," and "Bin Laden Network's Plans Advancing." The exact contents of these briefings remain classified, but apparently, none specifically mentions a domestic U.S. attack. [NEW YORK TIMES, 4/18/04]

April 18, 2001: FAA Warns Airlines About Middle Eastern Hijackers

The FAA sends a warning to U.S. airlines that Middle Eastern terrorists could try to hijack or blow up a U.S. plane and that carriers should "demonstrate a high degree of alertness." The warning stems from the April 6, 2001, conviction of Ahmed Ressam over a failed plot to blow up Los Angeles International Airport during the millennium celebrations. This warning expires on July 31, 2001. [ASSOCIATED PRESS, 5/18/02] This is one of a number of general warnings issues to airlines, but it is more specific than usual. [CNN, 3/02; CNN, 5/17/02]

April 24, 2001: U.S. Military Planned for Terrorist Action Against Americans in 1960s

James Bamford's book, *Body of Secrets,* reveals a secret U.S. government plan named Operation Northwoods. All details of the plan come from declassified military documents. [ASSOCIATED PRESS, 4/24/01; BALTIMORE SUN, 4/24/01; ABC NEWS, 5/1/01; WASHINGTON POST, 4/26/01] The heads of the U.S. military, all five Joint Chiefs of Staff, proposed in a 1962 memo to commit terrorist acts against Americans and blame Cuba to create a pretext for invasion. Says one document, "We could develop a Communist Cuban terror campaign in the Miami area, in other Florida cities and even in Washington. . . . We could blow up a U.S. ship in Guantanamo Bay and blame Cuba. Casualty lists in U.S. newspapers would cause a helpful wave of indignation." In March 1962, Lyman L. Lemnitzer, Chairman of the Joint Chiefs of Staff, presented the Operation Northwoods plan to President John Kennedy and Defense Secretary Robert McNamara. The plan was rejected. Lemnitzer then sought to destroy all evidence of the plan. [BALTIMORE SUN, 4/24/01; ABC NEWS, 5/1/01] Lemnitzer was replaced a few months later, but the Joint Chiefs continued to plan "pretext" operations at least through 1963. [ABC NEWS, 5/1/01] One suggestion in the plan was to create a remote-controlled drone duplicate of a real civilian aircraft. The real aircraft would be loaded with "selected passengers, all boarded under carefully prepared aliases," and then take off with the drone duplicate simultaneously taking off near by. The aircraft with passengers would secretly land at a U.S. military base while the drone continues along the other plane's flight path. The drone would then be destroyed over Cuba in a way that places the blame on Cuban fighter aircraft. [HARPER'S, 7/1/01] Bamford says, "Here we are, 40 years afterward, and it's only now coming out. You just wonder what is going to be exposed 40 years from now." [INSIGHT, 7/30/01] Some 9/11 skeptics later claim that the 9/11 attacks could have been orchestrated by elements of the U.S. government, and see Northwoods as an example of how top U.S. officials could hatch such a plot. [OAKLAND TRIBUNE, 3/27/04]

May 2001: Report Warns of al-Qaeda Infiltration from Canada

U.S. intelligence obtains information that al-Qaeda is planning to infiltrate the U.S. from Canada and carry out a terrorist operation using high explosives. The report does not say exactly where, when, or how an attack might occur. Two months later, the information is shared with the FBI, the INS, the U.S. Customs Service, and the State Department, and it will be shared with President Bush in August. [9/11 CONGRESSIONAL INQUIRY 9/18/02; WASHINGTON POST 9/19/02 (B)]

May 2001 (B): Bin Laden Associates Head West, Prepare for Martyrdom

The Defense Department gains and shares information indicating that seven people associated with bin Laden have departed from various locations for Canada, Britain, and the U.S. The next month, the CIA learns that key operatives in al-Qaeda are disappearing while others are preparing for martyrdom. [9/11 CONGRESSIONAL INQUIRY, 9/18/02; WASHINGTON POST, 9/19/02 (B)]

May 2001 (C): Iranian Tells of Plot to Attack WTC

An Iranian in custody in New York City tells local police of a plot to attack the World Trade Center. No more details are known. [FOX NEWS 5/17/02]

May 2001 (D): Medics Train for Airplane Hitting Pentagon

U.S. Medicine magazine later reports: "Though the Department of Defense had no capability in place to protect the Pentagon from an ersatz guided missile in the form of a hijacked 757 airliner, DoD [Department of Defense] medical personnel trained for exactly that scenario in May." The Tri-Service DiLorenzo Health Care Clinic and the Air Force Flight Medicine Clinic train inside the Pentagon this month "to fine-tune their emergency preparedness." [U.S. MEDICINE, 10/01]

May–July 2001: NSA Picks Up Word of Imminent Terrorist Attacks

Over a two-month period, the NSA reports "at least 33 communications indicating a possible, imminent terrorist attack." None of these reports provide any specific information on where, when, or how an attack might occur. These reports are widely disseminated to other intelligence agencies. [9/11 CONGRESSIONAL INQUIRY, 9/18/02; MSNBC, 9/18/02] National Security Adviser Rice later reads what she calls "chatter that was picked up in [2001's] spring and summer. 'Unbelievable news coming in weeks,' said one. 'Big event—there will be a very, very, very, very big uproar.' 'There will be attacks in the near future.'" [WASHINGTON POST, 4/8/04 (C)] The NSA Director later claims that all of the warnings were red herrings. [NSA DIRECTOR CONGRESSIONAL TESTIMONY, 10/17/02]

May–July 2001 (B): 9/11 Attacks Originally Planned for Early Date

In 2001, bin Laden apparently pressures Khalid Shaikh Mohammed for an attack date earlier then 9/11. According to information obtained from the 9/11 Commission (apparently based on a prison interrogation of Mohammed), bin Laden first requests an attack date of May 12, 2001, the seven-month anniversary of the *Cole* bombing. Then, when bin Laden learns from the media that Israel's Prime Minister Ariel Sharon would be visiting the White House in June or July 2001, he attempts once more to accelerate the operation to coincide with his visit. [9/11 COMMISSION, 6/16/04 (B)] The surge of warnings around this time possibly could be related to these original preparations.

May 29, 2001: U.S. Citizens Overseas Cautioned

The State Department issues an overseas caution connected to the conviction of defendants in the bombing of the U.S. embassies in Kenya and Tanzania. That warning says "Americans citizens abroad may be the target of a terrorist threat from extremist groups" with links to bin Laden. The warning will be reissued on June 22. [CNN, 6/23/01]

May 30, 2001: Yemenis Are Caught Taking Suspicious New York Photos

Two Yemeni men are detained after guards see them taking photos at 26 Federal Plaza in New York City. They are questioned by INS agents and let go. A few days later, their confiscated film is developed, showing photos of security checkpoints, police posts, and surveillance cameras of federal buildings, including the FBI's counterterrorism office. The two men are later interviewed by the FBI and determined not to be terrorists. However, they had taken the pictures on behalf of a third person living in Indiana. By the time the FBI looks for him, he has fled the country and his documentation is found to be based on a false alias. In 2004, it is reportedly still unknown whether the third man is a terrorist. The famous briefing given to President George W. Bush on August 6, 200, will mention the incident, warning that the FBI is investigating "suspicious activity in this country consistent with the preparations for hijackings or other types of attacks, including recent surveillance of federal buildings in New York." When Bush's August 6 briefing is released in 2004, a White House fact sheet fails to mention the still missing third man. [NEW YORK POST, 7/1/01; NEW YORK POST, 9/16/01; WASHINGTON POST, 5/16/04]

May 31, 2001: "Tightly Organized System" of al-Qaeda Cells Found in U.S.

The *Wall Street Journal* summarizes tens of thousands of pages of evidence disclosed in a recently concluded trial of al-Qaeda terrorists. They are called "a riveting view onto the shadowy world of al-Qaeda." The documents reveal numerous connections between al-Qaeda and specific front companies and charities. They even detail a "tightly organized system of cells in an array of American cities, including Brooklyn, N.Y.; Orlando, Fla.; Dallas, Tex.; Santa Clara, Calif.; Columbia, Mo., and Herndon, Va." The 9/11 hijackers had ties to many of these same cities and charities. [WALL STREET JOURNAL, 5/31/01]

June 2001: Germans Warn of Plan to Use Aircraft
as Missiles on U.S. and Israeli Symbols

German intelligence warns the CIA, Britain's intelligence agency, and Israel's Mossad that Middle East-ern terrorists are planning to hijack commercial aircraft to use as weapons to attack "American and Israeli symbols, which stand out." A later article quotes unnamed German intelligence sources who state the information was coming from Echelon surveillance technology, and that British intelligence had access to the same warnings. However, there were other informational sources, including specific information and hints given to, but not reported by, Western and Near Eastern news media six months before 9/11. [FRANKFURTER ALLGEMEINE ZEITUNG, 9/11/01; WASHINGTON POST, 9/14/01; FOX NEWS, 5/17/02]

June 2001 (B): U.S. Intelligence Warns
of Spectacular Attacks by al-Qaeda Associates

U.S. intelligence issues a terrorist threat advisory, warning U.S. government agencies that there is a high probability of an imminent terrorist attack against U.S. interests: "Sunni extremists associated with al-Qaeda are most likely to attempt spectacular attacks resulting in numerous casualties." The advisory mentions the Arabian Peninsula, Israel, and Italy as possible targets for an attack. Afterwards, intelligence information provided to senior U.S. leaders continues to indicate that al-Qaeda expects near-term attacks to have dramatic consequences on governments or cause major casualties. [9/11 CONGRESSIONAL INQUIRY, 9/18/02]

June 2001 (C): CIA Fears Terrorists Will Strike on Fourth of July.

The CIA provides senior U.S. policy makers with a classified warning of a potential attack against U.S. interests that is thought to be tied to Fourth of July celebrations in the U.S. [SUNDAY HERALD, 9/23/01]

June–July 2001: Terrorist Threat Reports Surge,
Frustration with White House Grows

Terrorist threat reports, already high in the preceding months, surge even higher. President Bush, Vice President Cheney, and national security aides are given briefing papers with headlines such as "Bin Laden Threats Are Real" and "Bin Laden Planning High Profile Attacks." The exact contents of these briefings remain classified, but according to the 9/11 Commission they consistently predict upcoming attacks that will occur "on a catastrophic level, indicating that they would cause the world to be in tur-moil, consisting of possible multiple—but not necessarily simultaneous—attacks." CIA Director Tenet later recalls that by late July the warnings coming in could not get any worse. He feels that President Bush and other officials grasp the urgency of what they are being told. [9/11 COMMISSION REPORT, 4/13/04 (B)] But Deputy CIA Director John McLaughlin later states that he feels a great tension, peaking these months, between the Bush administration's apparent misunderstanding of terrorism issues and his sense of great urgency. McLaughlin and others are frustrated when inexperienced Bush officials question the validity of certain intelligence findings. Two unnamed, veteran Counter Terrorism Center officers deeply involved in bin Laden issues are so worried about an impending disaster that they consider resigning and going public with their concerns. [9/11 COMMISSION, 3/24/04 (C)] Dale Watson, head of counterterrorism at the FBI, wishes he had "500 analysts looking at Osama bin Laden threat information instead of two." [9/11 COMMISSION REPORT, 4/13/04 (B)]

June 1–2, 2001: Military Conducts and Plans Exercises Similar to the 9/11 Attacks

Amalgam Virgo 01, a multi-agency planning exercise sponsored by NORAD (the North American Aerospace Defense Command, in charge of defending U.S. airspace), involves the hypothetical scenario

Bin Laden is pictured on the cover of the first Amalgam Virgo exercise

of a cruise missile launched by "a rogue [government] or somebody" from a barge off the East Coast. Bin Laden is pictured on the cover of the proposal for the exercise. [AMERICAN FORCES PRESS SERVICE, 6/4/02] The exercise takes place at Tyndall Air Force Base in Florida. [GLOBAL SECURITY, 4/14/02] Amalgam Virgo 02 is already being planned before 9/11. The plan involves two simultaneous commercial aircraft hijackings. One, a Delta 757, with actual Delta pilots and actors posing as passengers, flies from Salt Lake City to Honolulu. It is "hijacked" by FBI agents posing as terrorists. The other is a DC-9 hijacked by Canadian police near Vancouver. Fighters are to respond and possibly "mock" shoot down the aircraft. [CNN, 6/4/02 (B), AFPS, 6/4/02] 9/11 Commissioner Richard Ben-Veniste notes that this planned exercise shows that despite frequent comments to the contrary, the military had considered simultaneous hijackings before 9/11. [9/11 COMMISSION, 5/23/03]

June 4, 2001: Illegal Afghans Overheard Discussing New York City Hijacking Attack

At some point in 2000, three men claiming to be Afghans but using Pakistani passports entered the Cayman Islands, possibly illegally. [MIAMI HERALD, 9/20/01] In late 2000, Cayman and British investigators began a yearlong probe of these men, which lasts until 9/11. [LOS ANGELES TIMES, 9/20/01] They are overheard discussing hijacking attacks in New York City during this period. On this day, they are taken into custody, questioned, and released some time later. This information is forwarded to U.S. intelligence. [FOX NEWS, 5/17/02] In late August, a letter to a Cayman radio station will allege these same men are agents of bin Laden "organizing a major terrorist act against the U.S. via an airline or airlines." [MIAMI HERALD, 9/20/01; LOS ANGELES TIMES, 9/20/01]

June 13, 2001: Bin Laden Wants to Assassinate
Bush with an Explosives-Filled Airplane

Egyptian President Hosni Mubarak later claims that Egyptian intelligence discovers a "communiqué from bin Laden saying he wanted to assassinate President Bush and other G8 heads of state during their summit in Italy" on this day. The communiqué specifically mentions this would be done via "an airplane stuffed with explosives." The U.S. and Italy are sent urgent warnings of this. [NEW YORK TIMES, 9/26/01]

Mubarak claims that Egyptian intelligence officials informed American intelligence officers between March and May 2001 that an Egyptian agent had penetrated al-Qaeda. Presumably, this explains how Egypt is able to give the U.S. these warnings. [NEW YORK TIMES, 6/4/02]

June 20, 2001: *Time* Magazine Mentions al-Qaeda Using Planes as Weapons

Time magazine reports: "For sheer diabolical genius (of the Hollywood variety), nothing came close to the reports that European security services are preparing to counter a bin Laden attempt to assassinate President Bush at next month's G8 summit in Genoa, Italy. According to German intelligence sources, the plot involved bin Laden paying German neo-Nazis to fly remote-controlled model aircraft packed with Semtex into the conference hall and blow the leaders of the industrialized world to smithereens. (Paging Jerry Bruckheimer . . .)." [TIME, 6/20/01] This report follows warnings given by Egypt the week before. In addition, there are more warnings before the summit in July. James Hatfield, author of an unflattering book on Bush called *Fortunate Son*, repeats the claim in print a few days later, writing: "German intelligence services have stated that bin Laden is covertly financing neo-Nazi skinhead groups throughout Europe to launch another terrorist attack at a high-profile American target." [ONLINE JOURNAL, 7/3/01] Two weeks later, Hatfield apparently commits suicide. However, there is widespread speculation that his death was payback for his revelation of Bush's cocaine use in the 1970s. [SALON, 7/20/01]

June 21, 2001: Senior al-Qaeda Officials Say "Important Surprises" Coming Soon

A reporter for the Middle East Broadcasting Company interviews bin Laden. Keeping a promise made to Taliban leader Mullah Omar, bin Laden does not say anything substantive, but Ayman al-Zawahiri and other top al-Qaeda leaders promise that "[the] coming weeks will hold important surprises that will target American and Israeli interests in the world." [ASSOCIATED PRESS, 6/24/01; ASSOCIATED PRESS, 6/25/01] The reporter says, "There is a major state of mobilization among the Osama bin Laden forces. It seems that there is a race of who will strike first. Will it be the United States or Osama bin Laden?" [REUTERS, 6/23/01] After 9/11, the reporter concludes, "I am 100 percent sure of this, and it was absolutely clear they had brought me there to hear this message." [A PRETEXT FOR WAR, BY JAMES BAMFORD, 6/04, P. 236] The reporter is also shown a several-months-old videotape with bin Laden declaring to his followers, "It's time to penetrate America and Israel and hit him them where it hurts most." [CNN, 6/21/01] Author James Bamford theorizes that the original 9/11 plot involved a simultaneous attack on Israel and that shoe bomber Richard Reid may have originally wanted to target an Israeli aircraft around this time. For instance, Reid flies to Tel Aviv, Israel on July 12, 2001, to test if airline security would check his shoes for bombs. [A PRETEXT FOR WAR, BY JAMES BAMFORD, 6/04, PP. 236–39]

June 26, 2001: State Department Issues Worldwide Caution; Military on Alert Overseas

The State Department issues a worldwide caution warning American citizens of possible attacks. [CNN, 3/02] Also around this time, U.S. military forces in the Persian Gulf are placed on heightened alert and naval ships there are sent out to sea, and other defensive steps are taken overseas. This is in response to a recent warning the week before where an al-Qaeda video was shown, saying, "It's time to penetrate America and Israel and hit them where it hurts most." However, as author James Bamford later notes, "No precautions were ever taken within the United States, only overseas." [A PRETEXT FOR WAR, BY JAMES BAMFORD, 6/04, P. 241]

June 28, 2001: Tenet Warns of Imminent al-Qaeda Attack

CIA Director Tenet writes an intelligence summary for National Security Adviser Rice: "It is highly likely that a significant al-Qaeda attack is in the near future, within several weeks." A highly classified analysis at this time adds, "Most of the al-Qaeda network is anticipating an attack. Al-Qaeda's overt publicity has also raised expectations among its rank and file, and its donors." [WASHINGTON POST, 5/17/02] Apparently, the same analysis also adds, "Based on a review of all source reporting over the last five months, we believe that [bin Laden] will launch a significant terrorist attack against U.S. and/or Israeli interests in the coming weeks. The attack will be spectacular and designed to inflict mass casualties against U.S. facilities or interests. Attack preparations have been made. Attack will occur with little or no warning." [9/11 CONGRESSIONAL INQUIRY, 7/24/03 (B)] This warning is shared with "senior Bush administration officials" in early July. [9/11 CONGRESSIONAL INQUIRY, 9/18/02] Apparently, these warnings are largely based on a warning given by al-Qaeda leaders to a reporter a few days earlier. Counterterrorism "tsar" Richard Clarke also later asserts that Tenet tells him around this time, "It's my sixth sense, but I feel it coming. This is going to be the big one." [AGAINST ALL ENEMIES, BY RICHARD CLARKE, 3/04, P. 235]

Summer 2001: Threat Alerts Increase to Record High

Congressman Porter Goss (R), Chairman of the House Intelligence Committee, later says on the intelligence monitoring of terrorist groups, "the chatter level [goes] way off the charts" around this time and stays high until 9/11. Given Goss's history as a CIA operative, presumably he is kept "in the know" to some extent. [LOS ANGELES TIMES, 5/18/02] A later Congressional report states: "Some individuals within the intelligence community have suggested that the increase in threat reporting was unprecedented, at least in terms of their own experience." [9/11 CONGRESSIONAL INQUIRY, 9/18/02] Two counterterrorism officials later describe the alerts of this summer as "the most urgent in decades." [9/11 CONGRESSIONAL INQUIRY, 9/18/02]

**CIA Director
George Tenet**

Summer 2001 (B): Tenet Believes "Something Is Happening"

Deputy Secretary of State Richard Armitage later claims that at this time, CIA Director "Tenet [is] around town literally pounding on desks saying, something is happening, this is an unprecedented level of threat information. He didn't know where it was going to happen, but he knew that it was coming." [9/11 CONGRESSIONAL INQUIRY, 7/24/03]

Summer 2001 (C): Bin Laden Speech Mentions
20 Martyrs in Upcoming Attack; Other Hints of Attack Spread Widely

Word begins to spread within al-Qaeda that an attack against the U.S. is imminent, according to later prison interrogations of Khalid Shaikh Mohammed. Many within al-Qaeda are aware that Mohammed has been preparing operatives to go to the U.S. Additionally, bin Laden makes several remarks hinting at an upcoming attack, spawning rumors throughout Muslim extremist circles worldwide. For instance, in a recorded speech at the al Faruq training camp in Afghanistan, bin Laden specifically urges trainees to pray for the success of an upcoming attack involving 20 martyrs. [9/11 COMMISSION REPORT, 6/16/04 (B)] There are other indications that knowledge of the attacks spreads in Afghanistan. The *Daily Telegraph* later reports that "the idea of an attack on a skyscraper [is] discussed among [bin Laden's] supporters in Kabul." At some unspecified point before 9/11, a neighbor in Kabul sees diagrams showing a skyscraper attack in a house known as a "nerve center" for al-Qaeda activity. [DAILY TELEGRAPH, 11/16/01] U.S. soldiers will later find forged visas, altered passports, listings of Florida flight schools and registration papers for a flight simulator in al-Qaeda houses in Afghanistan. [NEW YORK TIMES, 12/6/01] A bin Laden bodyguard later claims that in May 2001 he hears bin Laden tell people in Afghanistan that the U.S. would be hit with a terrorist attack, and thousands would die. [GUARDIAN, 11/28/01]

July 2001: CIA Learns Impending Attack Widely Known in Afghanistan

The CIA hears an individual who had recently been in Afghanistan say, "Everyone is talking about an impending attack." [9/11 CONGRESSIONAL INQUIRY, 9/18/02; WASHINGTON POST, 9/19/02]

July 2001 (B): India Warns U.S. of Possible Terror Attacks

India gives the U.S. general intelligence on possible terror attacks; details are not known. U.S. government officials later confirm that Indian intelligence had information "that two Islamist radicals with ties to Osama bin Laden were discussing an attack on the White House," but apparently, this particular report is not given to the U.S. until two days after 9/11. [FOX NEWS, 5/17/02]

July 1, 2001: Senators Warn of Terrorist Incident Within Three Months

Senators Dianne Feinstein (D) and Richard Shelby (R), both members of the 9/11 Congressional Inquiry, appear on CNN's "Late Edition with Wolf Blitzer," and warn of potential attacks by bin Laden. Feinstein says, "One of the things that has begun to concern me very much as to whether we really have our house in order, intelligence staff have told me that there is a major probability of a terrorist incident within the next three months." [CNN, 3/02]

July 10, 2001: FBI Agent Sends Memo Warning That
"Inordinate Number" of Muslim Extremists Are Learning to Fly in Arizona

Phoenix Arizona, FBI agent Ken Williams sends a memorandum warning about suspicious activities involving a group of Middle Eastern men taking flight training lessons in Arizona. The memo is titled:

"Zakaria Mustapha Soubra; IT-OTHER (Islamic Army of the Caucasus)," because it focuses on Zakaria Soubra, a Lebanese flight student in Prescott, Arizona, and his connection with a terror group in Chechnya that has ties to al-Qaeda. It is subtitled: "Osama bin Laden and Al-Muhjiroun supporters attending civil aviation universities/colleges in Arizona." [FORTUNE, 5/22/02; ARIZONA REPUBLIC, 7/24/03] The memo is based on an investigation Williams had initiated in 2000, but had trouble pursuing because of the low priority the Arizona FBI office gave terror investigations. In the memo, Williams does the following:

- Names nine other suspect students from Pakistan, India, Kenya, Algeria, the United Arab Emirates, and Saudi Arabia. [DIE ZEIT, 10/1/02] Hijacker Hani Hanjour, attending flight school in Arizona in early 2001, is not mentioned in the memo, but one of his acquaintances is. Another person on the list is later arrested in Pakistan in March 2002 with al-Qaeda leader Abu Zubaida. [9/11 CONGRESSIONAL INQUIRY, 7/24/03; WASHINGTON POST, 7/25/03 (C)]

- Notes that he interviewed some of these students, and heard some of them make hostile comments about the U.S. Additionally, he noticed that they were suspiciously well informed about security measures at U.S. airports. [DIE ZEIT 10/1/02]

- Notes an increasing, "inordinate number of individuals of investigative interest" taking flight lessons in Arizona. [DIE ZEIT, 10/1/02; 9/11 CONGRESSIONAL INQUIRY, 7/24/03]

- Suspects that some of the ten people he has investigated are connected to al-Qaeda. [9/11 CONGRESSIONAL INQUIRY, 7/24/03] He discovered that one of them was communicating through an intermediary with Abu Zubaida. [SAN JOSE MERCURY NEWS, 5/23/02] Potentially this is the same member of the list mentioned above who is later captured with Abu Zubaida.

- Discusses connections between several of the students and a radical group called Al-Muhajiroun. [SAN JOSE MERCURY NEWS, 5/23/02] This group supported bin Laden, and issued a *fatwa*, or call to arms, that included airports on a list of acceptable terror targets. [ASSOCIATED PRESS, 5/22/02] Soubra, the main focus of the memo, is a member of Al-Muhajiroun and an outspoken radical, but he is later cleared of any ties to terrorism. [LOS ANGELES TIMES, 10/28/01 (C)]

- Warns of a possible "effort by Osama bin Laden to send students to the U.S. to attend civil aviation universities and colleges" [FORTUNE, 5/22/02], so they can later hijack aircraft. [DIE ZEIT, 10/1/02]

- Recommends that the "FBI should accumulate a listing of civil aviation universities/colleges around the country. FBI field offices with these types of schools in their area should establish appropriate liaison. FBI [headquarters] should discuss this matter with other elements of the U.S. intelligence community and task the community for any information that supports Phoenix's suspicions." [ARIZONA REPUBLIC, 7/24/03] (The FBI has already done this, but because of poor FBI communications, Williams is not aware of the report.)

- Recommends that the FBI ask the State Department to provide visa data on flight school students from Middle Eastern countries, which will facilitate FBI tracking efforts. [NEW YORK TIMES, 5/4/02]

The memo is emailed to six people at FBI headquarters in the bin Laden and radical fundamentalist units, and to two people in the FBI New York field office. [9/11 CONGRESSIONAL INQUIRY, 7/24/03] He also shares some concerns with the CIA. [SAN JOSE MERCURY NEWS, 5/23/02] However, the memo is merely marked "routine," rather than "urgent." It is generally ignored, not shared with other FBI offices, and the recommendations are not taken. One colleague in New York replies at the time that the memo is "speculative and not very significant." [DIE ZEIT, 10/1/02; 9/11 CONGRESSIONAL INQUIRY, 7/24/03] Williams is unaware of many FBI investigations and leads that could have given weight to his memo. Authorities later claim that Williams was only pursuing a hunch, but one familiar with classified information says, "This was not a vague hunch. He was doing a case on these guys." [SAN JOSE MERCURY NEWS, 5/23/02]

Mid-July 2001: More G-8 Summit
Warnings Describe Planes as Flying Bomb

U.S. intelligence reports another spike in warnings related to the July 20–22 G-8 summit in Genoa, Italy. The reports include specific threats discovered by the head of Russia's Federal Bodyguard Service that al-Qaeda will try to kill Bush as he attends the summit. [CNN, 3/02] Two days before the summit begins, the BBC reports: "The huge force of officers and equipment which has been assembled to deal with unrest has been spurred on by a warning that supporters of Saudi dissident Osama bin Laden might attempt an air attack on some of the world leaders present." [BBC, 7/18/01] The attack is called off.

Mid-July 2001 (B): Tenet Warns Rice About "Major Attack"

CIA Director Tenet has a special meeting with National Security Adviser Rice and her aides about al-Qaeda. Says one official at the meeting, "[Tenet] briefed [Rice] that there was going to be a major attack." Another at the meeting says Tenet displays a huge wall chart showing dozens of threats. Tenet does not rule out a domestic attack but says an overseas attack is more likely. [TIME, 8/4/02]

July 16, 2001: British Spy Agencies Warn al-Qaeda
Is in "The Final Stages" of Attack in the West

British spy agencies send a report to British Prime Minister Tony Blair and other top officials warning that al-Qaeda is in "the final stages" of preparing a terrorist attack in the West. The prediction is "based on intelligence gleaned not just from [British intelligence] but also from U.S. agencies, including the CIA and the National Security Agency," which cooperate with the British. "The contents of the July 16 warning would have been passed to the Americans, Whitehall sources confirmed." The report states there is "an acute awareness" that the attack is "a very serious threat." [LONDON TIMES, 6/14/02]

July 18, 2001: FBI, FAA Issues More Warnings

The FBI issues another warning to domestic law enforcement agencies about threats stemming from the convictions in the millennium bomb plot trial. The FAA also issues a warning, telling the airlines to "use the highest level of caution." [CNN, 3/02]

July 20–22, 2001: During G-8 Summit, Italian Military
Prepare Against Attack from the Sky

The G8 summit is held in Genoa, Italy. Acting on previous warnings that al-Qaeda would attempt to kill Bush and other leaders, Italian authorities surround the summit with antiaircraft guns. They keep fighters in the air and close off local airspace to all planes. [LOS ANGELES TIMES, 9/27/01] The warnings are taken so seriously that Bush stays overnight on an aircraft carrier offshore, and other world leaders stay on a luxury ship. [CNN, 7/18/01] No attack occurs. U.S. officials at the time state that the warnings were "unsubstantiated" but after 9/11, they will claim success in preventing an attack. [LOS ANGELES TIMES 9/27/01]

July 31, 2001: FAA Issues General Hijacking Warning

The FAA issues another warning to U.S. airlines, citing no specific targets but saying, "terror groups are known to be planning and training for hijackings, and we ask you therefore to use caution." These alerts will expire by 9/11. Note that pilots and flight attendants later claim they were never told about warnings such as these. The airlines also disagree about the content of pre-9/11 warnings generally. For instance, American Airlines states these warnings were "extremely general in nature and did not identify a specific threat or recommend any specific security enhancements." The text of these warnings remains classified. [CNN, 3/02; ANANOVA, 5/17/02]

Late July 2001: Taliban Foreign Minister Tries to
Warn U.S. and UN of Huge Attack Inside the U.S.

Taliban Foreign Minister Wakil Ahmed Muttawakil learns that bin Laden is planning a "huge attack" on targets inside America. The attack is imminent, and will kill thousands. He learns this from Tahir Yildash, leader of the rebel Islamic Movement of Uzbekistan (IMU), which is allied with al-Qaeda at the time. Muttawakil sends an emissary to pass this information on to the U.S. consul general, and another U.S. official, "possibly from the intelligence services," also attends the meeting. The message is not taken very seriously; one source blames this on "warning fatigue" from too many warnings. In addition, the emissary supposedly is from the Foreign Ministry, but did not say the message came from Muttawakil himself. The emissary then takes the message to the Kabul offices of UNSMA, the political wing of the UN. They also fail to take the warning seriously. [INDEPENDENT, 9/7/02: REUTERS, 9/7/02]

Late July 2001 (B): Ex-House Judiciary Committee's Chief Investigator
Tries to Warn About Plans to Strike Buildings in New York

David Schippers, noted conservative Chicago lawyer and the House Judiciary Committee's chief investigator in the Clinton impeachment trial, claims two days after 9/11 that he had tried to warn federal authorities about plans to strike buildings in lower Manhattan. Schippers says, "I was trying to get people to listen to me because I had heard that the terrorists had set up a three-pronged attack:" an American airplane, the bombing of a federal building in the heartland and a massive attack in lower Manhattan. He tries contacting Attorney General Ashcroft, the White House, and even the House managers with whom he had worked, but nobody returns his phone calls. "People thought I was crazy. What I was doing was I was calling everybody I knew telling them that this has happened," he says. "I'm telling you the more I see of the stuff that's coming out, if the FBI had even been awake they would have seen it." He also claims to know of ignored warnings about the 1995 Oklahoma City bombing, and evidence that Middle Easterners were connected with that attack. [INDIANAPOLIS STAR, 5/18/02] Other mainstream sources have apparently shied away from Schippers' story, but he has added details in an interview on the partisan Alex Jones Show. He claims that FBI agents in Chicago and Minnesota first contact him and tell him that a terrorist attack is going to occur in lower Manhattan. A group of these agents reportedly wants to testify about what they know, but first want legal protection from government retribution. [ALEX JONES SHOW, 10/10/01]

Late July 2001 (C): Argentina Relays Warning to the U.S.

Argentina's Jewish community receives warnings of a major terrorist attack against the United States, Argentina, or France from "a foreign intelligence source." The warning is then relayed to the Argentine security authorities. It is agreed to keep the warning secret in order to avoid panic while reinforcing security at Jewish sites in the country. Says a Jewish leader, "It was a concrete warning that an attack of major proportion would take place, and it came from a reliable intelligence source. And I understand the Americans were told about it." Argentina has a large Jewish community that has been bombed in the past, and has been an area of al-Qaeda activity. [FORWARD, 5/31/02]

Late July 2001 (D): Egypt Warns CIA of 20 al-Qaeda Operatives in U.S.;
Four Training to Fly; CIA Is Not Interested

CBS later reports, in a long story on another topic: "Just days after [Mohamed] Atta return[s] to the U.S. from Spain, Egyptian intelligence in Cairo says it received a report from one of its operatives in Afghanistan that 20 al-Qaeda members had slipped into the U.S. and four of them had received flight training on Cessnas. To the Egyptians, pilots of small planes didn't sound terribly alarming, but they [pass] on the message to the CIA anyway, fully expecting Washington to request information. The request never [comes]." [CBS NEWS, 10/9/02] This appears to be just one of several accurate Egyptian warnings from their informants inside al-Qaeda.

Late Summer 2001: Jordan Warns U.S. That Aircraft
Will Be Used in Major Attack Inside the U.S.

Jordanian intelligence (the GID) makes a communications intercept deemed so important that King Abdullah's men relay it to Washington, probably through the CIA station in Amman. To make doubly sure the message gets through it is passed through an Arab intermediary to a German intelligence agent. The message states that a major attack, code named "The Big Wedding," is planned inside the U.S. and that aircraft will be used. "When it became clear that the information was embarrassing to Bush administration officials and congressmen who at first denied that there had been any such warnings before September 11, senior Jordanian officials backed away from their earlier confirmations." The *Christian Science Monitor* calls the story "confidently authenticated" even though Jordan has backed away from it.

[*INTERNATIONAL HERALD TRIBUNE*, 5/21/02; *CHRISTIAN SCIENCE MONITOR*, 5/23/02]

Late Summer 2001 (B): U.S. Intelligence Learns al-Qaeda Is
Considering Mounting Terrorist Operations in the U.S.

U.S. intelligence learns that an al-Qaeda operative is considering mounting terrorist operations in the U.S. There is no information on the timing or specific targets. [9/11 CONGRESSIONAL INQUIRY, 9/18/02]

August 2001: Moroccan Informant Warns U.S. of
Large Scale, Imminent Attack in New York

According to simultaneous reports in a French magazine and a Moroccan newspaper, a Moroccan agent named Hassan Dabou has penetrated al-Qaeda to the point of getting close to bin Laden by this time. Dabou claims he learns that bin Laden is "very disappointed" that the 1993 bombing had not toppled the WTC, and plans "large scale operations in New York in the summer or fall of 2001." Dabou is called to the U.S. to report this information directly, and in so doing blows his cover, losing his ability to gather more intelligence. The *International Herald Tribune* later calls the story "not proved beyond a doubt" but intriguing, and asks the CIA to confirm or deny the story. The CIA has refused to do so.

[AGENCE FRANCE PRESSE, 11/22/01; *INTERNATIONAL HERALD TRIBUNE*, 5/21/02; *LONDON TIMES*, 6/12/02]

August 2001 (B): Russia Warns U.S. of Suicide Pilots

Russian President Vladimir Putin warns the U.S. that suicide pilots are training for attacks on U.S. targets. [FOX NEWS, 5/17/02] The head of Russian intelligence also later states, "We had clearly warned them" on several occasions, but they "did not pay the necessary attention." [AGENCE FRANCE-PRESSE, 9/16/01] A Russian newspaper on September 12, 2001 claims, "Russian Intelligence agents know the organizers and executors of these terrorist attacks. More than that, Moscow warned Washington about preparation to these actions a couple of weeks before they happened." Interestingly, the article claims that at least two of the terrorists were Muslim radicals from Uzbekistan. [*IZVESTIA*, 9/12/01]

August 2001 (C): CIA Warns of Possible Attack Inside U.S.

The CIA issues a report warning the White House, Pentagon, and State Department that bin Laden is intent on launching a terrorist attack soon, possibly inside the U.S. [SUNDAY HERALD, 9/23/01]

August 2001 (D): U.S. Learns of Plot to Crash Airplane into U.S. Embassy in Nairobi

U.S. intelligence learns of a plot to either bomb the U.S. embassy in Nairobi, Kenya, from an airplane or crash an airplane into it. Two people who were reportedly acting on instructions from bin Laden met in October 2000 to discuss this plot. [9/11 CONGRESSIONAL INQUIRY, 9/18/02]

August 2001 (E): Persian Gulf Informant Gives Ex-CIA Agent Information About "Spectacular Terrorist Operation"

Former CIA agent Robert Baer is advising a prince in a Persian Gulf royal family, when a military associate of this prince passes information to him about a "spectacular terrorist operation" that will take place shortly. He is given a computer record of around 600 secret al-Qaeda operatives in Saudi Arabia and Yemen. The list includes ten names that will be placed on the FBI's most wanted terrorists list after 9/11. He is also given evidence that a Saudi merchant family had funded the USS *Cole* bombing on October 12, 2000, and that the Yemeni government is covering up information related to that bombing. At the military officer's request, he offers all this information to the Saudi Arabian government. However, an aide to the Saudi defense minister, Prince Sultan, refuses to look at the list or to pass the names on (Sultan is later sued for his complicity in the 9/11 plot in August 2002). Baer also passes the information on to a senior CIA official and the CIA's Counter Terrorism Center, but there is no response or action. Portions of Baer's book describing his experience are blacked out, having been censored by the CIA. [FINANCIAL TIMES, 1/12/02; SEE NO EVIL, BY BILL GERTZ, PP. 55–58]

Between August 2001 and September 11, 2001: NORAD Runs Hijacking Simulations

A NORAD exercise, planned in July 2001 and conducted some time afterward but apparently before 9/11, involves real planes from airports in Utah and Washington State that simulate a hijacking. Those planes are escorted by U.S. and Canadian aircraft to airfields in British Columbia and Alaska. *USA Today* notes that this is an exception to NORAD's claim that they focused only on external threats to the U.S. and did not consider the possibility of threats arising from within the U.S. [USA TODAY, 4/18/04]

Early August 2001: Government Informant Warns Congressmen of Plan to Attack the WTC

Randy Glass, a former con artist turned government informant, later claims that he contacts the staff of Senator Bob Graham [D] and Representative Robert Wexler [D] at this time and warns them of a plan to attack the WTC, but his warnings are ignored. [PALM BEACH POST, 10/17/02] Glass also tells the media at this time that his recently concluded informant work has "far greater ramifications than have so far been revealed," and "potentially, thousands of lives [are] at risk." [SOUTH FLORIDA SUN-SENTINEL, 8/7/01] Glass was a key informant in a sting operation involving ISI agents who were illegally trying to purchase sophisticated U.S. military weaponry in return for cash and heroin. He later claims that in July 1999, one ISI agent named Rajaa Gulum Abbas pointed to the WTC and said, "Those towers are coming down. [PALM BEACH POST, 10/17/02] Most details remain sealed, but Glass points out that his sentencing document dated June 15, 2001, lists threats against the WTC and Americans. [WPBF CHANNEL 25, 8/5/02] Florida State Senator Ron Klein, who had dealings with Glass before 9/11, later says he is surprised it took so many months for the U.S. to listen to Glass: "Shame on us." [PALM BEACH POST, 10/17/02] Klein recalls getting a warning from Glass, though he cannot recall if it mentions the WTC specifically. He says he was told U.S. intelligence agencies would look into it. [WPTV, 10/7/02] Senator Graham later acknowledges that his office had contact with Glass before 9/11, and was told about a WTC attack: "I was concerned about that and a dozen other pieces of information which emanated from the summer of 2001." However, Graham will say that he personally was unaware of Glass's information until after 9/11. [PALM BEACH POST, 10/17/02] In October 2002, Glass will testify under oath before a private session of the 9/11 Congressional Inquiry, stating, "I told [the inquiry] I have specific evidence, and I can document it." [PALM BEACH POST, 10/17/02]

Early August 2001 (B): CIA's Concern over Planned bin Laden Strikes Inside U.S. Are Heightened

The Associated Press later reports that the "CIA had developed general information a month before the attacks that heightened concerns that bin Laden and his followers were increasingly determined to strike on U.S. soil." A CIA official affirmed, "[t]here was something specific in early August that said to us that [bin Laden] was determined in striking on U.S. soil." [ASSOCIATED PRESS, 10/3/01]

Early August 2001 (C): Britain Warns U.S. Again; Specifies Multiple Airplane Hijackings

Britain gives the U.S. another warning about an al-Qaeda attack. The previous British warning on July 16, 2001, was vague as to method, but this warning specifies multiple airplane hijackings. This warning is said to reach President Bush. [SUNDAY HERALD, 5/19/02]

August 1, 2001: Actor Communicates Concerns to Stewardess That Airplane Will Be Hijacked; Warning Forwarded to the FAA

Actor James Woods, flying first class on an airplane, notices four Arabic-looking men, the only other people in the first class section. He concludes they are terrorists, acting very strangely (for instance, only talking in whispers). [BOSTON GLOBE, 11/23/01] He tells a flight attendant, "I think this plane is going to be hijacked," adding, "I know how serious it is to say this." He conveys his worries to the pilots, and they assure him that the cockpit would be locked. [NEW YORKER, 5/27/02] The flight staff later notifies the FAA about these suspicious individuals. Though the government will not discuss this event, it is highly unlikely that any action is taken regarding the flight staff's worries. [NEW YORKER, 5/27/02] Woods is not interviewed by the FBI until after 9/11. Woods says the FBI believes that all four men took part in the 9/11 attacks, and the flight he was on was a practice flight for them. [O'REILLY FACTOR, 2/14/02] Woods believes one was Khalid Almihdhar and another was Hamza Alghamdi. [NEW YORKER, 5/27/02] The FBI later reports that this may have been one of a dozen test run flights starting as early as January. Flight attendants and passengers on other flights later recall men looking like the hijackers who took pictures of the cockpit aboard flights and/or took notes. [ASSOCIATED PRESS, 5/29/02] The FBI has not been able to find any evidence of hijackers on the flight manifest for Woods' flight. [NEW YORKER, 5/27/02]

August 1, 2001 (B): FBI Reissues Warning That Overseas Law Enforcement Agencies May Be Targets

With the approaching third anniversary of the U.S. embassy bombings in Africa, the FBI reissues a warning that overseas law enforcement agencies may be targets. [CNN, 3/02]

August 6, 2001: Bush Receives Memo: "Bin Laden Determined to Strike in U.S."

See chapter 3.

August 15, 2001: CIA Counterterrorism Head: "We Are Going to Be Struck Soon"

Cofer Black, head of the CIA's Counter Terrorism Center, says in a speech to the Department of Defense's annual Convention of Counterterrorism, "We are going to be struck soon, many Americans are going to die, and it could be in the U.S." Black later complains that top leaders are unwilling to act at this time unless they are given "such things as the attack is coming within the next few days and here is what they are going to hit." [9/11 CONGRESSIONAL INQUIRY, 9/26/02 (B)]

August 15–28, 2001: Moussaoui Arrest Raises Serious Concerns of Airplane-based Terrorist Attack with Local FBI; Washington Headquarters Ignores Pleas for Search Warrant Until After 9/11

See chapter 10.

August 16, 2001: FAA Issues Warning; Airlines Say Warning Not Received

The FAA issues a warning to airlines concerning disguised weapons. According to later testimony by National Security Adviser Rice, the FAA is concerned about reports that the terrorists have made breakthroughs in disguising weapons as cell phones, key chains, and pens [CNN, 3/02; REUTERS, 5/16/02] However, the major airlines later deny receiving such notification. For instance, a Delta spokesperson states: "We were not aware of any warnings or notifications of any specific threats." [FOX NEWS, 5/16/02]

August 21, 2001: Inmate Warns of Impending Attack in New York

Walid Arkeh, a Jordanian serving time in a Florida prison, is interviewed by FBI agents after warning the government of an impending terrorist attack. He had been in a British jail from September 2000 to July 2001, and while there had befriended three inmates, Khalid al-Fawwaz, Adel Abdel Bary and Ibrahim Eidarous. U.S. prosecutors charge, "the three men ran a London storefront that served as a

Walid Arkeh

cover for al-Qaeda operations and acted as a conduit for communications between bin Laden and his network." [ORLANDO SENTINEL, 10/30/02] Al-Fawwaz was bin Laden's press agent in London, and bin Laden had called him over 200 times before al-Fawwaz was arrested in 1998. [FINANCIAL TIMES, 11/29/01 (B), SUNDAY TIMES, 3/24/02] The other two had worked in the same office as al-Fawwaz. All three had been indicted as co-conspirators with bin Laden in the August 1998 U.S. embassy bombings. Arkeh tells the FBI that he had learned from these three that "something big [is] going to happen in New York City," and that they call the 1993 attack on the WTC "unfinished business." Tampa FBI agents determine that he had associated with these al-Qaeda agents, but nonetheless they do not believe him. According to Arkeh, one agent responds to his "something big" warning by saying: "Is that all you have? That's old news." The agents fail to learn more from him. On September 9, concerned that time is running out, a fellow prisoner will try to arrange a meeting, but nothing happens before 9/11. The Tampa FBI agents have a second interview with him hours after the 9/11 attacks, but even long after 9/11 they claim he cannot be believed. On January 6, 2002, the Tampa FBI issued a statement: "The information [was] vetted to FBI New York, the Acting Special Agent in Charge of the Tampa Division and the United States Attorney for the Middle District of Florida. All agreed the information provided by this individual was vague and unsubstantiated . . . Mr. Arkeh did not provide information that had any bearing on the FBI preventing September 11." [ORLANDO SENTINEL, 1/6/02; ORLANDO SENTINEL, 10/30/02] However, a different group of FBI agents will interview him in May 2002 and find his information is credible.

August 27, 2001: Spanish Police Tape Phone Calls
Indicating Aviation-Based Plans to Attack U.S.

Spanish police tape a series of cryptic, coded phone calls from a caller in Britain using the codename "Shakur" to Barakat Yarkas (also known as Abu Dahdah), the leader of a Spanish al-Qaeda cell presumably visited by Mohamed Atta in July. A Spanish judge claims that a call by Shakur on this day shows foreknowledge of the 9/11 attacks. Shakur says that he is "giving classes" and that "in our classes, we have entered the field of aviation, and we have even cut the bird's throat." Another possible translation is, "We are even going to cut the eagle's throat," which would be a clearer metaphor for the U.S. [OBSERVER, 11/25/01; GUARDIAN, 2/14/02] In a Spanish indictment, the unknown Shakur is described as "a presumed member of the September 11 suicide commandos." [INTERNATIONAL HERALD TRIBUNE, 11/21/01] The Spanish terrorist cell led by Yarkas is allegedly a hub of financing, recruitment, and support services for al-Qaeda in Europe. Yarkas's phone number is later also found in the address book of Said Bahaji, and he had ties with Mohammed

Haydar Zammar and Mamoun Darkazanli. All three are associates of Atta in Hamburg. [LOS ANGELES TIMES, 11/23/01] Yarkas also "reportedly met with bin Laden twice and was in close contact with" top deputy Muhammad Atef. [WASHINGTON POST, 11/19/01] On November 11, 2001, Yarkas and ten other Spaniards will be arrested and charged with al-Qaeda terrorist activity. [INTERNATIONAL HERALD TRIBUNE, 11/21/01]

August 29, 2001: Cayman Islands Letter Warns of "Major Terrorist Act Against U.S. via an Airline or Airlines"

Three men from either Pakistan or Afghanistan living in the Cayman Islands are briefly arrested in June 2001 for discussing hijacking attacks in New York City. On this day, a Cayman Islands radio station receives an unsigned letter claiming these same three men are agents of bin Laden. The anonymous author warns that they "are organizing a major terrorist act against the U.S. via an airline or airlines." The letter is forwarded to a Cayman government official but no action is taken until after 9/11. When the Cayman government notifies the U.S. is unknown. Many criminals and/or businesses use the Cayman Islands as a safe, no tax, no-questions-asked haven to keep their money. The author of the letter meets with the FBI shortly after 9/11, and claims his information was a "premonition of sorts." The three men are later arrested. What has happened to them since their arrest is unclear. [MIAMI HERALD, 9/20/01; LOS ANGELES TIMES, 9/20/01; MSNBC, 9/23/01]

August 30–September 4, 2001: Egypt Warns al-Qaeda Is in Advanced Stages of Planning Significant Attack on U.S.

According to Egyptian President Hasni Mubarak, Egyptian intelligence warns American officials that bin Laden's network is in the advanced stages of executing a significant operation against an American target, probably within the U.S. [ASSOCIATED PRESS, 12/7/01; NEW YORK TIMES, 6/4/02] He says he learned this information from an agent working inside al-Qaeda. U.S. officials deny receiving any such warning from Egypt. [ABC NEWS, 6/4/02]

Late August 2001: Bin Laden Boasts in Interview of "Very, Very Big" Strike Against U.S.

In an interview with the London-based newspaper *al-Quds al-Arabi*, bin Laden boasts that he is planning a "very, very big" and "unprecedented" strike against the U.S. The interview is not publicly released until after 9/11, however, so it is unclear if U.S. intelligence is aware of this before 9/11. [INDEPENDENT, 9/17/01; ABC NEWS, 9/12/01]

Late August 2001 (B): Hussein Puts His Troops on Highest Military Alert Since Gulf War

A *Daily Telegraph* article later claims that Iraq leader Saddam Hussein puts his troops on their highest military alert since the Gulf War. A CIA official states that there was nothing obvious to warrant this move: "He was clearly expecting a massive attack and it leads you to wonder why." Hussein apparently makes a number of other moves suggesting foreknowledge, and the article strongly suggests Iraqi complicity in the 9/11 attacks. [DAILY TELEGRAPH, 9/23/01 (B)] Iraq will later be sued by 9/11 victims' relatives on the grounds that they had 9/11 foreknowledge but did not warn the U.S.

Late August 2001 (C): French Warning to U.S. Echoes Earlier Israeli Warning

French intelligence gives a general terrorist warning to the U.S.; apparently, its contents echo an Israeli warning from earlier in the month (see chapter 14). [FOX NEWS, 5/17/02]

Early September 2001: NSA Intercepts Phone Calls
from bin Laden's Chief of Operations to the U.S.

The NSA intercepts "multiple phone calls from Abu Zubaida, bin Laden's chief of operations, to the United States." The timing and information contained in these intercepted phone calls has not been disclosed. [ABC NEWS, 2/18/02]

Early September 2001 (B): Bin Laden's Intercepted
Phone Calls Discuss an Operation in the U.S. Around 9/11 Date

According to British inside sources, "shortly before September 11," bin Laden contacts an associate thought to be in Pakistan. The conversation refers to an incident that will take place in the U.S. on, or around 9/11, and discusses possible repercussions. In another conversation, bin Laden contacts an associate thought to be in Afghanistan. They discuss the scale and effect of a forthcoming operation; bin Laden praises his colleague for his part in the planning. Neither conversation specifically mentions the WTC or Pentagon, but investigators have no doubt the 9/11 attacks were being discussed. The British government has obliquely made reference to these intercepts: "There is evidence of a very specific nature relating to the guilt of bin Laden and his associates that is too sensitive to release." These intercepts haven't been made public in British Prime Minister Tony Blair's presentation of al-Qaeda's guilt because "releasing full details could compromise the source or method of the intercepts." [SUNDAY TIMES, 10/7/01]

Early September 2001 (C): Iranian Inmate in Germany
Warns of Imminent Attack on WTC

An Iranian man known as Ali S. in a German jail waiting deportation repeatedly phones U.S. law enforcement to warn of an imminent attack on the WTC in early September. He calls it "an attack that will change the world." After a month of badgering his prison guards, he is finally able to call the White House 14 times in the days before the attack. He then tries to send a fax to President Bush, but is denied permission hours before the 9/11 attacks. German police later confirm the calls. Prosecutors later say Ali had no foreknowledge and his forebodings were just a strange coincidence. They say he is mentally unstable. Similar warnings also come from a Moroccan man being held in a Brazilian jail. [DEUTSCHE PRESSE-AGENTUR, 9/13/01; OTTAWA CITIZEN, 9/17/01; ANANOVA, 9/14/01; SUNDAY HERALD, 9/16/01]

Early September 2001 (D): Bin Laden Moves Training Bases

One article later suggests that bin Laden moves his training bases in Afghanistan "in the days before the attacks." [*PHILADELPHIA INQUIRER*, 9/16/01] These bases are under close military satellite surveillance.

September 1, 2001: American Airlines Issues Internal Memo Warning of Imposters

Around this date, American Airlines sends out an internal memo warning its employees to be on the lookout for impostors after one of its crews had uniforms and ID badges stolen in Rome, Italy, in April. [*REUTERS*, 9/14/01; *BOSTON GLOBE*, 9/18/01] It is later reported that two of the hijackers on Flight 11 use these stolen ID's to board the plane. [*SUNDAY HERALD*, 9/16/01] On 9/11, a man is arrested with three Yemen passports (all using different names) and two Lufthansa crew uniforms. [*CHICAGO SUN TIMES*, 9/22/01] It is also reported that when Mohamed Atta takes a flight from Portland, Maine, to Boston on the morning of 9/11, his bags are not transferred to his hijacked flight, and remain in Boston. Later, airline uniforms are found inside. [*BOSTON GLOBE*, 9/18/01] Boston's Logan Airport had been repeatedly fined for failing to run background checks on their employees, and many other serious violations. [*CNN*, 10/12/01]

September 3, 2001: Author Is Banned from Internal U.S. Flights
Because of FAA Concern "Something About to Happen"

Author Salman Rushdie, the target of death threats from radical Muslims for years, is banned by U.S. authorities from taking internal U.S. flights. He says the FAA told his publisher the reason was that it had "intelligence of something about to happen." One newspaper states, "The FAA confirmed that it stepped up security measures concerning Mr. Rushdie but refused to give a reason." [*LONDON TIMES*, 9/27/01]

September 7, 2001: Priest Is Told of Plot to
Attack U.S. and Britain Using Hijacked Airplanes

At a wedding in Todi, Italy, Father Jean-Marie Benjamin is told of a plot to attack the U.S. and Britain using hijacked airplanes as weapons. He is not told specifics regarding time or place. He immediately passes what he knows to a judge and several politicians. He later states, "Although I am friendly with many Muslims, I wondered why they were telling me, specifically. I felt it my duty to inform the Italian government." Benjamin has been called "one of the West's most knowledgeable experts on the Muslim world." Two days after 9/11, he meets with the Italian Foreign Minister on this topic. He says he learned the attack on Britain failed at the last minute. [*ZENIT*, 9/16/01] It is not known if the Italian government warns the U.S. government of this before 9/11.

September 7, 2001 (B): State Department Issues Overseas Warning

The State Department issues a little noticed warning, alerting against an attack by al-Qaeda. However, the warning focuses on a threat to American citizens overseas, and particularly focuses on threats to U.S. military personnel in Asia. [*U.S. STATE DEPARTMENT*, 9/7/01] Such warnings are issued periodically and are usually so vague few pay attention to them. In any event, most airlines and officials claim that they did not see this warning until after 9/11. [SAN FRANCISCO CHRONICLE, 9/14/01]

Before September 9, 2001: Northern Alliance Has "Limited Knowledge" of Attack; Warns the West

Declassified Defense Intelligence Agency documents from November 2001 suggest that Northern Alliance leader General Ahmed Shah Massoud had gained "limited knowledge" "regarding the intentions of [al-Qaeda] to perform a terrorist act against the U.S. on a scale larger than the 1998 bombing of the U.S. Embassies in Kenya and Tanzania." It further points out he may have been assassinated on September 9, 2001, because he "began to warn the West." The documents are heavily censored, and specifics are lacking, but Massoud did made an oblique public warning before European Parliament earlier in the year. [AGENCE FRANCE-PRESSE, 9/14/03; PAKISTAN TRIBUNE, 9/13/03]

September 9, 2001: Congressman Foresees "Something Terrible" Will Happen in Wake of Massoud Assassination

Congressman Dana Rohrabacher (R), who has long experience in Afghanistan and even fought with the mujahedeen there, later claims he immediately sees the assassination of Northern Alliance leader Ahmed Shah Massoud as a sign that "something terrible [is] about to happen." He is only able to make an appointment to meet with top White House and National Security Council officials for 2:30 P.M. on 9/11. The events of that morning make the meeting moot. [SPEECH TO THE HOUSE OF REPRESENTATIVES, 9/17/01]

September 9, 2001 (B): Osama Tells His Stepmother That "Big News" Will Come in Two Days

It is later reported that on this day, bin Laden calls his stepmother and says, "In two days, you're going to hear big news and you're not going to hear from me for a while." U.S. officials later tell CNN that "in recent years they've been able to monitor some of bin Laden's telephone communications with his [step]mother. Bin Laden at the time was using a satellite telephone, and the signals were intercepted and sometimes recorded." [NEW YORK TIMES, 10/2/01] Stepmother Al-Khalifa bin Laden, who raised Osama bin Laden after his natural mother died, is apparently waiting in Damascus, Syria, to meet Osama there, so he calls to cancel the meeting. [SUNDAY HERALD, 10/7/01] They had met periodically in recent years. Before 9/11, to impress important visitors, NSA analysts would occasionally play audio tapes of bin Laden talking to his stepmother. The next day government officials say about the call, "I would view those reports with skepticism." [CNN, 10/2/01]

September 10, 2001: Alarm Bells Sound over Unusual Trading in U.S. Stock Options Market

According to CBS News, in the afternoon before the attack, "alarm bells were sounding over unusual trading in the U.S. stock options market." It has been documented that the CIA, the Mossad, and many other intelligence agencies monitor stock trading in real time using highly advanced programs such as Promis. Both the FBI and the Justice Department have confirmed the use of such programs for U.S. intelligence gathering through at least this summer. This would confirm that the CIA should have had additional advance warning of imminent attacks against American and United Airlines planes. [CBS NEWS, 9/19/01] There are even allegations that bin Laden was able to get a copy of Promis. [FOX NEWS, 10/16/01]

September 10, 2001 (B): NSA Intercepts: "The Match Is About to Begin" and "Tomorrow Is Zero Hour"

At least two messages in Arabic are intercepted by the NSA. One states "The match is about to begin" and the other states "Tomorrow is zero hour." Later reports translate the first message as "The match begins tomorrow." [REUTERS, 9/9/02] The messages were sent between someone in Saudi Arabia and someone in Afghanistan. The NSA claims that they weren't translated until September 12, and that even if they had been translated in time, "they gave no clues that authorities could have acted on." [ABC NEWS, 6/7/02; REUTERS, 6/19/02] These messages turn out to be only two of about 30 pre-9/11 communications from suspected al-Qaeda operatives or other militants referring to an imminent event. An anonymous official says of these messages, including the "Tomorrow is zero hour" message, "You can't dismiss any of them, but it does not tell you tomorrow is the day." [REUTERS, 9/9/02] There is a later attempt to explain the messages away by suggesting they refer to the killing of Afghani opposition leader Ahmed Shah Massoud the day before. [REUTERS, 10/17/02]

September 10, 2001 (C): U.S. Intercepts: "Watch the News" and "Tomorrow Will Be a Great Day for Us"

U.S. officials later admit American agents had infiltrated al-Qaeda cells in the U.S., though how many agents and how long they had been in al-Qaeda remains a mystery. On this day, electronic intercepts connected to these undercover agents hear messages such as, "Watch the news" and "Tomorrow will be a great day for us." When asked why these messages did not lead to boosted security or warnings the next day, officials refer to them as "needles in a haystack." What other leads may have come from this prior to this day are not revealed. [USA TODAY, 6/4/02] At least until February 2002, the official story was that the "CIA failed to penetrate al-Qaeda with a single agent." [ABC NEWS, 2/18/02]

September 10, 2001 (D): U.S. Generals Warned Not to Fly on Morning of 9/11

According to a *Newsweek* report on September 13, "[t]he state of alert had been high during the past two weeks, and a particularly urgent warning may have been received the night before the attacks, causing some top Pentagon brass to cancel a trip. Why that same information was not available to the 266 people who died aboard the four hijacked commercial aircraft may become a hot topic on the Hill." [NEWSWEEK, 9/13/01] Far from becoming a hot topic, the only additional media mention of this story is in the next issue of *Newsweek*: "a group of top Pentagon officials suddenly canceled travel plans for the next morning, apparently because of security concerns." [NEWSWEEK, 9/17/01]

September 10, 2001 (E): Intelligence Intercepts Show al-Qaeda Agents Ordered to Return to Afghanistan by This Date

In a major post-9/11 speech, British Prime Minister Tony Blair claims that "shortly before September 11, bin Laden told associates that he had a major operation against America under preparation, [and] a range of people were warned to return back to Afghanistan because of action on or around September 11." His claims come from a British document of telephone intercepts and interrogations revealing al-Qaeda orders to return to Afghanistan by September 10. [CNN, 10/4/01; TIME, 10/5/01] However, Blair may have the direction incorrect, since would-be hijacker Ramzi bin al-Shibh later claims that he is the one who passes to bin Laden the date the attacks will happen and warns others to evacuate. [AUSTRALIAN, 9/9/02]

Before September 11, 2001: "We're Ready to Go," "Big Thing Coming" Not Analyzed Until After 9/11

Though the NSA specializes in intercepting communications, the CIA and FBI intercept as well. After 9/11, CIA and FBI officials discover messages with phrases like, "There is a big thing coming," "They're going to pay the price," and "We're ready to go." Supposedly, most or all of these intercepted messages are not analyzed until after 9/11. [NEWSWEEK, 9/24/01]

Author's note: The remaining entries in this chapter chronicle key Bush administration officials' responses to reports of pre-9/11 warning signs, as described above, and the growing public questions as to whether 9/11 could have been prevented. For additional information on this subject, please see chapter 19 on Post-9/11 Investigations and chapter 20 on Other Post-9/11 Events.

September 12, 2001: U.S. Denies Any Hints of bin Laden Plot to Attack in U.S.

The government's initial response to the 9/11 attacks is that it had no evidence whatsoever that bin Laden planned an attack in the U.S. "There was a ton of stuff, but it all pointed to an attack abroad," says one official. Furthermore, in the 24 hours after the attack, investigators would have been searching through "mountains of information." However, "the vast electronic 'take' on bin Laden, said officials who requested anonymity, contained no hints of a pending terror campaign in the United States itself, no orders to subordinates, no electronic fund transfers, no reports from underlings on their surveillance of the airports in Boston, Newark, and Washington." [MIAMI HERALD, 9/12/01]

September 14, 2001: FBI Director Caught in "Whopper"

FBI Director Mueller describes reports that several of the hijackers had received flight training in the U.S. as "news, quite obviously," adding, "If we had understood that to be the case, we would have—perhaps one could have averted this." It is later discovered that contrary to Mueller's claims, the FBI had interviewed various flight school staffs about Middle Eastern terrorists on numerous occasions, from 1996 until a few weeks

FBI Director Robert Mueller

before 9/11. [*WASHINGTON POST*, 9/23/01; *BOSTON GLOBE*, 9/18/01] Three days later, he says, "There were no warning signs that I'm aware of that would indicate this type of operation in the country." [DEPARTMENT OF JUSTICE TRANSCRIPT, 9/17/01] *Slate* magazine later contrasts this with numerous other contradictory statements and articles, and awards Mueller the "Whopper of the Week." [*SLATE*, 5/17/02]

September 16, 2001: Bush Claim That Using Planes as Missiles Was Impossible to Predict Is Contradicted by Former CIA Official

President Bush says, "Never (in) anybody's thought processes . . . about how to protect America did we ever think that the evil doers would fly not one but four commercial aircraft into precious U.S. targets . . . never." [NATO, 9/16/01] A month later, Paul Pillar, the former deputy director of the CIA's counterterrorist center, says, "The idea of commandeering an aircraft and crashing it into the ground and causing high casualties, sure we've thought of it." [*LOS ANGELES TIMES*, 10/14/01]

February 6, 2002 (B): Tenet Is "Proud" of CIA's Handling of 9/11

CIA Director Tenet tells a Senate hearing that there was no 9/11 intelligence failure. When asked about the CIA record on 9/11, he says, "We are proud of that record." He also states that the 9/11 plot was "in the heads of three or four people" and thus nearly impossible to prevent. [*USA TODAY*, 2/7/02]

May 8, 2002: FBI Could Not Have Foreseen 9/11, Declares Director

FBI Director Mueller states, "[T]here was nothing the agency could have done to anticipate and prevent the [9/11] attacks." [9/11 CONGRESSIONAL INQUIRY, 9/18/02]

May 15, 2002: Bush's August 6, 2001, Warning Is Leaked to Public

The Bush administration is embarrassed when the CBS Evening News reveals that President Bush had been warned about al-Qaeda domestic attacks in August 2001 (see chapter 3). Bush had repeatedly said that he had "no warning" of any kind. Press Secretary Ari Fleischer states unequivocally that while Bush had been warned of possible hijackings, "[t]he president did not—not—receive information about the use of airplanes as missiles by suicide bombers." [*NEW YORK TIMES*, 5/15/02; *WASHINGTON POST*, 5/16/02] "Until the attack took place, I think it's fair to say that no one envisioned that as a possibility." [MSNBC, 9/18/02] Fleischer claims the August memo was titled "Bin Laden Determined to Strike the U.S.," but the real title is soon found to end with " . . . Strike in U.S." [*WASHINGTON POST*, 5/18/02 (B)] The *Guardian* will state a few days later, "the memo left little doubt that the hijacked airliners were intended for use as missiles and that intended targets were to be inside the U.S." It further states that, "now, as the columnist Joe Conason points out in the current edition of the *New York Observer*, 'conspiracy' begins to take over from 'incompetence' as a likely explanation for the failure to heed—and then inform the public about—warnings that might have averted the worst disaster in the nation's history." [*GUARDIAN*, 5/19/02]

May 16, 2002: Cheney Warns Democrats Against Criticizing Handling of Pre-9/11 Warnings

In the wake of new information on what President Bush knew, Vice President Cheney states, "[M]y Democratic friends in Congress . . . need to be very cautious not to seek political advantage by making incendiary suggestions, as were made by some today, that the White House had advance information that would have prevented the tragic attacks of 9/11." He calls such criticism "thoroughly irresponsible . . . in time of war" and states that any serious probe of 9/11 foreknowledge would be tantamount to giving "aid and comfort" to the enemy. [WASHINGTON POST, 5/17/02 (C)]

May 16, 2002 (B): Nobody Predicted 9/11–Style Attacks, Says Rice

National Security Adviser Rice states, "I don't think anybody could have predicted that these people would take an airplane and slam it into the World Trade Center, take another one and slam it into the Pentagon, that they would try to use an airplane as a missile," adding that "even in retrospect" there was "nothing" to suggest that. [WHITE HOUSE, 5/16/02] Contradicting Rice's claims, former CIA Deputy Director John Gannon acknowledges that such a scenario has long been taken seriously by U.S. intelligence: "If you ask anybody could terrorists convert a plane into a missile? [N]obody would have ruled that out." Rice also states, "The overwhelming bulk of the evidence was that this was an attack that was likely to take place overseas." [MSNBC, 5/17/02] *Slate* awards Rice the "Whopper of the Week" when the title of Bush's August 6 briefing is revealed: "Bin Laden Determined to Strike in U.S." [SLATE, 5/23/02] Rice later concedes that "somebody did imagine it" but says she did not know about such intelligence until well after this conference. [ASSOCIATED PRESS, 9/21/02]

May 16, 2002 (C): Airlines Claim No Knowledge of Possible Terrorist Attacks

In response to all of the revelations about what was known before 9/11, the major airlines hold a press conference saying they were never warned of a specific hijacking threat, and were not told to tighten security. For instance, an American Airlines spokesman states that the airline "received no specific information from the U.S. government advising the carrier of a potential terrorist hijacking in the United States in the months prior to September 11, 2001. American receives FAA security information bulletins periodically, but the bulletins were extremely general in nature and did not identify a specific threat or recommend any specific security enhancements." [MIAMI HERALD, 5/17/02] The FAA gave 15 warnings to the airlines between January and August 2001; but about one general security warning a month had been common for a long time. [CNN, 5/17/02] Even a government official called these warnings "standard fare." [MIAMI HERALD, 5/17/02]

May 21–22, 2002: Prisoner Told FBI of Imminent al-Qaeda Attacks

Walid Arkeh, a prisoner in Florida, is interviewed by a group of FBI agents in New York City. The agents seek information regarding the 1988 U.S. embassy bombings and are there to interview him about information he learned from three al-Qaeda prisoners he had befriended. During the interview, Arkeh claims that, in August 2001, he told the FBI that al-Qaeda was likely to attack the WTC and other targets soon, but he was dismissed. After 9/11, his warning still was not taken seriously by the local FBI. The New York FBI agents are stunned. One says to him: "Let me tell you something. If you know what happened in New York, we are all in deep shit. We are in deep trouble." Arkeh tells the agents that these prisoners hinted that the WTC would be attacked, and targets in Washington were mentioned as well. However, they did not tell him a date or that airplanes would be used. The New York FBI later informs him that they found his information credible. [ORLANDO SENTINEL, 10/30/02] Arkeh is later deported to Jordan despite a Responsible Cooperators Program promising visas to those who provided important terrorist information. (It is unclear whether any one ever has been given a reward through this program.) [ORLANDO SENTINEL, 11/10/02; ORLANDO SENTINEL, 1/11/03; ORLANDO SENTINEL, 3/12/03]

June 4, 2002: Bush Acknowledges Agencies Made Mistakes, Continues to Insist That 9/11 Could Not Have Been Prevented

For the first time, Bush concedes that his intelligence agencies had problems: "In terms of whether or not the FBI and the CIA were communicating properly, I think it is clear that they weren't." [LONDON TIMES, 6/5/02] However, in an address to the nation three days later, President Bush still maintains, "Based on everything I've seen, I do not believe anyone could have prevented the horror of September the 11th." [SYDNEY MORNING HERALD, 6/8/02] Days earlier, *Newsweek* reported that the FBI had prepared a detailed chart showing how agents could have uncovered the terrorist plot if the CIA had told them what it knew about the hijackers Nawaf Alhazmi and Khalid Almihdhar sooner. (FBI Director Mueller denies the existence of such a chart. [WASHINGTON POST, 6/3/02]) One FBI official says, "There's no question we could have tied all 19 hijackers together." [NEWSWEEK, 6/2/02] Attorney General Ashcroft also says it is unlikely better intelligence could have stopped the attacks. [WASHINGTON POST, 6/3/02]

June 18, 2002: FBI Director Maintains 9/11 Attacks Could Not Have Been Prevented

FBI Director Mueller testifies before the Congressional 9/11 inquiry. His testimony is made public in September 2002. [ASSOCIATED PRESS, 9/26/02] Mueller claims that with the possible exception of Zacarias Moussaoui, "[t]o this day we have found no one in the United States except the actual hijackers who knew of the plot and we have found nothing they did while in the United States that triggered a specific response about them." [CONGRESSIONAL INTELLIGENCE COMMITTEE, 9/26/02] The 9/11 Congressional Inquiry will later conclude near the end of 2002 that some hijackers had contact inside the U.S. with individuals known to the FBI, and the hijackers "were not as isolated during their time in the United States as has been previously suggested." [LOS ANGELES TIMES, 12/12/02] Mueller also claims, "There were no slip-ups. Discipline never broke down. They gave no hint to those around them what they were about." [CONGRESSIONAL INTELLIGENCE COMMITTEE, 9/26/02]

**NSA Director
Michael
Hayden**

October 17, 2002: NSA Denies Having Indications of 9/11 Planning

NSA Director Michael Hayden testifies before a Congressional inquiry that the "NSA had no [indications] that al-Qaeda was specifically targeting New York and Washington . . . or even that it was planning an attack on U.S. soil." Before 9/11, the "NSA had no knowledge . . . that any of the attackers were in the United States." Supposedly, a post-9/11 NSA review found no intercepts of calls involving any of the 19 hijackers. [REUTERS, 10/17/02; *USA TODAY*, 10/18/02; NSA DIRECTOR CONGRESSIONAL TESTIMONY, 10/17/02] (See chapter 19.)

January 22, 2003: CIA Chief Says Intelligence Was Insufficient to Prevent 9/11

CIA Deputy Director for Operations James Pavitt says he is convinced that all the intelligence the CIA had on September 11, 2001 could not have prevented the 9/11 attacks. "It was not as some have suggested, a simple matter of connecting the dots," he claims. [REUTERS, 1/23/03]

April 13, 2004: Bush Continues to Insist That 9/11 Could Not Have Been Prevented

In a press conference, President Bush states, "We knew he [Osama bin Laden] had designs on us, we knew he hated us. But there was nobody in our government, and I don't think [in] the prior government, that could envision flying airplanes into buildings on such a massive scale." [*GUARDIAN*, 4/15/04] He also says, "Had I any inkling whatsoever that the people were going to fly airplanes into buildings, we would have moved heaven and earth to save the country." [WHITE HOUSE, 4/13/04; *NEW YORK TIMES*, 4/18/04 (C)] Two days earlier, he said, "Had I known there was going to be an attack on America I would have moved mountains to stop the attack." [*NEW YORK TIMES*, 4/18/04] In July 2004, he will claim even more generally, "Had we had any inkling whatsoever that terrorists were about to attack our country, we would have moved heaven and earth to protect America." [*NEW JERSEY STAR-LEDGER*, 7/22/04]

2.

Insider Trading and Other Foreknowledge

Bush administration officials have claimed that the 9/11 plot was such a closely held secret that the U.S. government could not reasonably have been expected to uncover it in time to prevent it. For example, FBI Director Robert Mueller has contended that "to this day we have found no one in the United States except the actual hijackers who knew of the plot . . . " with the possible exception of Zacarias Moussaoui. [9/11 CONGRESSIONAL INQUIRY, 9/26/02] CIA Director George Tenet has echoed this position, insisting that the plot was "in the heads of three or four people" and even some of the hijackers may not have known what they were going to do. [USA TODAY, 2/7/02]

What are the facts? As chronicled in this chapter, in the weeks and days leading up to 9/11, children and adults alike were aware of the pending attacks and even discussed them in detail—from Seattle to Dallas and from Daytona Beach to Jersey City, in elementary schools, in taxis, in mosques, and on the Internet. Osama bin Laden apparently mentioned details of the attacks to large crowds, and rumors of the attack had spread in some fundamentalist Muslim circles throughout the world. [9/11 COMMISSION. 6/16/04 (B)]

Indeed, significant evidence exists indicating that some who had foreknowledge may have tried to profit financially from such knowledge. In the weeks and days leading up to 9/11, highly irregular stock trading transactions took place, involving stocks in United Airlines, American Airlines, and a few of the largest WTC tenants. There appear to have been even bigger trades in gold, oil, and U.S. Treasury bonds. Stock trading irregularities occurred not just in the U.S., but in European and Pacific region stock markets as well. While the FBI has concluded that there was no insider trading, German investigators found "almost irrefutable proof of insider trading." [MIAMI HERALD, 9/24/01] And in 2004, a University of Illinois professor published the results of his statistical analysis of the pre-9/11 trade activity; his finding was that it is highly unlikely that the trades could have been the result of random market fluctuations. [CHICAGO TRIBUNE, 4/25/04] The SEC and Secret Service initiated an investigation shortly after 9/11. However, as of mid-2004, neither agency has released any findings, or any information regarding the status of the investigations. Investigations in other countries are also apparently still ongoing.

Furthermore, as mentioned in chapter 1, in the afternoon before the attack, "alarm bells were sounding over unusual trading in the U.S. stock options market." [CBS NEWS, 9/19/01] So even if there was no terrorist insider trading, the perception that there was should have led to action. Whatever the investigations into the trading actions finally conclude, once again what we already know raises serious questions. How is it possible that the Bush administration, which insists that al-Qaeda and terrorism was a "top priority," failed to learn about or act upon such a remarkably poorly kept "secret"?

July 16, 2001: New York Taxi Driver Tells of E-mails
Warning Imminent al-Qaeda Attack on New York and Washington

A *Village Voice* reporter is told by a New York taxi driver, "You know, I am leaving the country and going home to Egypt sometime in late August or September. I have gotten e-mails from people I know saying that Osama bin Laden has planned big terrorist attacks for New York and Washington for that time. It will not be safe here then." He does in fact return to Egypt for that time. The FBI, which is not told about this lead until after 9/11, interrogates and then releases him. He claims that many others knew what he knew prior to 9/11. [VILLAGE VOICE, 9/25/02 (B)]

August 6, 2001: Suspicious Trading of Companies
Affected by 9/11 May Begin by This Date

Insider trading based on advanced knowledge of the 9/11 attacks may have begun on this date, if not earlier. Investigators later discover a large number of put option purchases (a speculation that the stock will go down) that expire on September 30 at the Chicago Board Options Exchange are bought on this date. If exercised, these options would have led to large profits. One analyst later says, "From what I'm hearing, it's more than coincidence." [REUTERS, 9/20/01]

Early September 2001: Seattle Security Guard Tells Friend of Impending Attack

A few days before 9/11, a Seattle security guard of Middle Eastern descent tells an East Coast friend on the phone that terrorists will soon attack the U.S. After 9/11, the friend tells the FBI, and passes a lie detector test. The security guard refuses to cooperate with the FBI or take a lie detector test. He is not arrested—apparently the FBI determines that while he may have had 9/11 foreknowledge, he was not involved in the plot. [SEATTLE POST-INTELLIGENCER, 10/12/01]

Early September 2001 (B): New York Mosque Warning:
Stay Out of Lower Manhattan on 9/11

Shortly before 9/11, people attending a New York mosque are warned to stay out of lower Manhattan on 9/11. The FBI's Joint Terrorist Task Force interviews dozens of members of the mosque, who confirm the story. The mosque leadership denies any advanced knowledge and the case apparently remains unsolved. [NEW YORK DAILY NEWS, 10/12/01]

Early September 2001 (C): Rumors in New York City's Arab-American Community About Attacks

A veteran detective involved with post-9/11 investigations later claims that rumors in New York City's Arab-American community about the 9/11 attacks are common in the days beforehand. The story "had been out on the street" and the number of leads turning up later is so "overwhelming" that it is difficult to tell who knows about the attacks from secondhand sources and who knows about it from someone who may have been a participant. After 9/11, tracking leads regarding Middle Eastern employees who did not show up for work on 9/11 are "a serious and major priority." [JOURNAL NEWS, 10/11/01]

Early September 2001 (D): NYSE Sees "Unusually Heavy Trading in Airline and Related Stocks"

The Securities and Exchange Commission (SEC) later announces that they are investigating the trading of shares of 38 companies in the days just before 9/11. The *San Francisco Chronicle* reports that the New York Stock Exchange sees "unusually heavy trading in airline and related stocks several days before the attacks." All 38 companies logically stand to be heavily affected by the attacks. They include parent companies of major airlines American, Continental, Delta, Northwest, Southwest, United, and US Airways as well as cruise lines Carnival and Royal Caribbean, aircraft maker Boeing and defense contractor Lockheed Martin. The SEC is also looking into suspicious short selling of numerous insurance company stocks, but, to date, no details of this investigation have been released. [ASSOCIATED PRESS, 10/2/01; SAN FRANCISCO CHRONICLE, 10/3/01]

Early September 2001 (E): Sharp Increase in Short Selling of American and United Airlines Stocks

There is a sharp increase in short selling of the stocks of American and United Airlines on the New York Stock Exchange prior to 9/11. A short sell is a bet that a particular stock will drop. There is an increase of 40 percent of short selling over the previous month for these two airlines, compared to an 11 percent increase for other big airlines and one percent for the exchange overall. A significant profit is to be made: United stock drops 43 percent and American drops 39 percent the first day the market reopens after the attack. Short selling of Munich Re, the world's largest reinsurer, is also later noted by German investigators. Inquiries into short selling millions of Munich Re shares are made in France days before the attacks. [REUTERS, 9/20/01; SAN FRANCISCO CHRONICLE, 9/22/01] Munich Re stock will plummet after the attacks, as they claim the attacks will cost them $2 billion. [DOW JONES BUSINESS NEWS, 9/20/01] There is also suspicious trading activity involving reinsurers Swiss Reinsurance and AXA. These trades are especially curious because the insurance sector "is one of the brightest spots in a very difficult market" at this time. [LOS

ANGELES TIMES, 9/19/01] There is also a short spike on Dutch airline KLM stock three to seven days before 9/11, reaching historically unprecedented levels. [*USA TODAY*, 9/26/01]

Early September 2001 (F): "Almost Irrefutable Proof of Insider Trading" in Germany

German central bank president Ernst Welteke later reports that a study by his bank indicates, "There are ever clearer signs that there were activities on international financial markets that must have been carried out with the necessary expert knowledge," not only in shares of heavily affected industries such as airlines and insurance companies, but also in gold and oil. [*DAILY TELEGRAPH*, 9/23/01 (C)] His researchers have found "almost irrefutable proof of insider trading." [*MIAMI HERALD*, 9/24/01] "If you look at movements in markets before and after the attack, it makes your brow furrow. But it is extremely difficult to really verify it." Nevertheless, he believes that "in one or the other case it will be possible to pinpoint the source." [*FOX NEWS*, 9/22/01] Welteke reports "a fundamentally inexplicable rise" in oil prices before the attacks [*MIAMI HERALD*, 9/24/01] and then a further rise of 13 percent the day after the attacks. Gold rises nonstop for days after the attacks. [*DAILY TELEGRAPH*, 9/23/01 (C)]

Ernst Welteke

Early September 2001 (G): Suspicion of Insider Trading in Many Other Countries

Numerous other overseas investigations into insider trading before 9/11 are later established. There are investigations in Belgium, France, Switzerland, Luxembourg, Monte Carlo, Cyprus, and other countries. There are particularly strong suspicions British markets are manipulated. Italy will later investigate suspicious share movements on the day of the attack, as well as the previous day. Japan will also look into the trading of futures contracts. [*FOX NEWS*, 9/22/01; *CNN*, 9/24/01; *BBC* 9/18/01 (B)] The British will take just two weeks to conclude that their markets were not manipulated. [*MARKETPLACE RADIO REPORT*, 10/17/01]

Early September 2001 (H): Unusually High Volume Trade of U.S. Treasury Note Purchases

After 9/11, both the SEC and the Secret Service announce probes into an unusually high volume trade of five-year U.S. Treasury note purchases around this time. These transactions include a single $5 billion trade. The *Wall Street Journal* explains: "Five-year Treasury notes are among the best investments in the event of a world crisis, especially one that hits the U.S. The notes are prized for their safety and their backing by the U.S. government, and usually rally when investors flee riskier investments, such as stocks." The value of these notes has risen sharply since the events of September 11. The article also points out that with these notes, "tracks would be hard to spot." [*WALL STREET JOURNAL*, 10/2/01]

September 6, 2001: New York Student Forecasts Destruction of WTC

Antoinette DiLorenzo, teaching English to a class of Pakistani immigrants, asks a student gazing out the window, "What are you looking at?" The student points towards the WTC, and says, "Do you see those two buildings? They won't be standing there next week." At the time, nothing is thought of it, but on September 13, the FBI will interview all the people in the classroom and confirm the event. The FBI later places the boy's family under surveillance but apparently is unable to find a connection to the 9/11 plot. An MSNBC reporter later sets out to disprove this "urban myth," but to his surprise, finds all the details of the story are confirmed. The fact that the family members are recent immigrants from Pakistan might mean the information came from Pakistan. [MSNBC, 10/12/01] Supposedly, on November 9, 2001, the same student predicts there will be a plane crash on November 12. On that day, American Airlines Flight 587 will crash on takeoff from New York, killing 260 people. Investigators will later determine that the crash is accidental. One official at the school later says many Arab-American students have come forward with their own stories about having prior knowledge before 9/11: "Kids are telling us that the attacks didn't surprise them. This was a nicely protected little secret that circulated in the community around here." [INSIGHT, 9/10/02]

September 6, 2001 (B): Bin Laden
Allegedly Informed of Exact Attack Date

According to a later interview with would-be hijacker Ramzi bin al-Shibh, a courier sent by bin al-Shibh tells bin Laden on this day when the 9/11 attacks will take place. [AUSTRALIAN, 9/9/02] However, there are doubts about this interview (see chapter 20).

September 6–10, 2001: Suspicious Trading of Put Option
Contracts on American and United Airlines Occur

Suspicious trading occurs on the stock of American and United, the two airlines hijacked in the 9/11 attacks. "Between 6 and 7 September, the Chicago Board Options Exchange [sees] purchases of 4,744 put option contracts [a speculation that the stock will go down] in UAL versus 396 call options—where a speculator bets on a price rising. Holders of the put options would [net] a profit of $5 million once the carrier's share price [dive] after September 11. On September 10, 4,516 put options in American Airlines, the other airline involved in the hijackings, [are] purchased in Chicago. This compares with a mere 748 call options in American purchased that day. Investigators cannot help but notice that no other airlines [see] such trading in their put options." One analyst later says, "I saw put-call numbers higher than I've ever seen in ten years of following the markets, particularly the options markets." [ASSOCIATED PRESS, 9/18/01; SAN FRANCISCO CHRONICLE, 9/19/01] "To the embarrassment of investigators, it has also [learned] that the firm used to buy many of the 'put' options . . . on United Airlines stock was headed until 1998 by 'Buzzy' Krongard, now executive director of the CIA." Krongard was chairman of Alex Brown Inc., which was bought by Deutsche Bank. "His last post before resigning to take his senior role in the CIA was to head Bankers Trust—Alex Brown's private client business, dealing with the accounts and investments of wealthy customers around the world." [INDEPENDENT, 10/14/01]

September 6–10, 2001 (B): Suspicious Trading on Stocks of Two Large WTC Tenants

The Chicago Board Options Exchange sees suspicious trading on Merrill Lynch and Morgan Stanley, two of the largest WTC tenants. In the first week of September, an average of 27 put option contracts in its shares are bought each day. Then the total for the three days before the attacks is 2,157. Merrill Lynch, another WTC tenant, see 12,215 put options bought between September 7–10, when the previous days had seen averages of 252 contracts a day. [INDEPENDENT, 10/14/01] Dylan Ratigan of Bloomberg Business News, speaking of the trading on Morgan Stanley and other companies, says, "This would be one of the most extraordinary coincidences in the history of mankind if it was a coincidence." [ABC NEWS, 9/20/01]

September 9, 2001: Internet Forum Message Warns of 9/11 Attack

A message is posted on Alsaha.com, a website based in Dubai, United Arab Emirates, warning of the 9/11 attack. It proclaims that in the next two days, a "big surprise" is coming from the Saudi Arabian region of Asir, the remote, mountainous province that produced most of the 19 hijackers who struck on September 11. Since 9/11, the FBI and CIA have closely monitored this website as "a kind of terrorist early-warning system" due to its popularity with Muslim fundamentalists. However, it is doubtful if they were monitoring the site before 9/11, or noticed this message. [NEWSWEEK, 5/25/03]

September 10, 2001: Dallas Fifth Grader Forecasts World War III

A fifth grader in Dallas, Texas, casually tells his teacher, "Tomorrow, World War III will begin. It will begin in the United States, and the United States will lose." The teacher reports the comment to the FBI, but does not know if they act on it at the time. The student skips the next two days of school. The event may be completely coincidental, but the newspaper that reports the story also notes that two charities located in an adjacent suburb have been under investigation based on suspected fund-raising activities for Islamic terrorist organizations. [HOUSTON CHRONICLE, 9/19/01] The FBI investigates and decides "no further investigation [is] warranted." [HOUSTON CHRONICLE, 10/1/01]

September 10, 2001 (B): New Jersey Student Warns
Teacher to Stay Away from Lower Manhattan

A sixth-grade student of Middle Eastern descent in Jersey City, New Jersey, says something that alarms his teacher at Martin Luther King Jr. Elementary School. "Essentially, he [warns] her to stay away from lower Manhattan because something bad [is] going to happen," says Sgt. Edgar Martinez, deputy director of police services for the Jersey City Police Department. [INSIGHT, 9/10/02]

September 10, 2001 (C): Suspicious Trading on
United Airlines Stock Occurs at Pacific Exchange

The trading ratio on United Airlines is 25 times greater than normal at the Pacific Exchange. Pacific Exchange officials later decline to state whether this abnormality is being investigated. [SAN FRANCISCO CHRONICLE, 9/19/01]

September 10, 2001 (D): Trader Makes Suspicious
Investments Moves; Later Accused of 9/11 Foreknowledge

Amr Elgindy orders his broker to liquidate his children's $300,000 trust account fearing a sudden crash in the market. He also tells his stockbroker that the Dow Jones average, then at 9,600, will fall to below 3,000. Elgindy is arrested in San Diego in May 2002, along with FBI agents Jeffrey Royer and Lynn Wingate, who, according to Government prosecutors, were using their FBI positions to obtain inside information on various corporations. They also questioned whether Elgindy had foreknowledge of the 9/11 attacks. [NEW YORK TIMES, 5/23/02; TIMES OF LONDON, 5/30/02; NEW YORK TIMES, 6/8/02] A report published in the *San Diego Tribune,* however, casts some doubt on the government's allegations. [SAN DIEGO UNION-TRIBUNE, 6/16/02]

September 10, 2001 (E): Florida Man
Predicts Imminent American Bloodshed

In a bar in Daytona Beach, Florida the night before the 9/11 attacks, three men make anti-American sentiments and talk of impending bloodshed. One says, "Wait 'til tomorrow. America is going to see bloodshed." These are not any of the hijackers, since they had all left Florida by this time. [MSNBC 9/23/01; ASSOCIATED PRESS, 9/14/01]

September 10, 2001 (F): Hours Before Attacks,
San Francisco Mayor Receives Warning

Eight hours prior to the attacks, San Francisco Mayor Willie Brown receives a warning from "my security people at the airport," advising him to be cautious in traveling. [SAN FRANCISCO CHRONICLE, 9/12/01] Later reports claim that this is because someone saw the State Department warning of September 7, which focused on the threat to military personnel in Asia. Brown is scheduled to fly to New York the next morning. [SAN FRANCISCO CHRON-ICLE, 9/14/01; SAN FRANCISCO CHRONICLE, 9/12/01; U.S. STATE DEPARTMENT, 9/7/01] The source of the warning, and why it was personally issued to Brown, remains unknown.

San Francisco Mayor Willie Brown

September 11, 2001: More Than $100 Million Rushed from WTC

Data recovery experts later looking at 32 hard drives salvaged from the 9/11 attacks discover a surge in credit card transactions from the WTC in the hours before and during the attacks. Unusually large sums of money are rushed through computers even as the disaster unfolds. Investigators later say, "There is a suspicion that some people had advance knowledge of the approximate time of the plane crashes in order to move out amounts exceeding $100 million. They thought that the records of their transactions could not be traced after the mainframes were destroyed." [REUTERS, 12/19/01]

September 14, 2001: Deutsche Bank Exec Resigns,
Prompting Speculations of 9/11 Connection

Mayo Shattuck III resigns, effective immediately, as head of the Alex Brown unit of Deutsche Bank. No reason is given. Some speculate later that this could have to do with the role of Deutsche Bank in the pre-9/11 purchases of put options on the stock of companies most affected by 9/11. Deutsche Bank is also one of the four banks most used by the bin Laden family. [NEW YORK TIMES, 9/15/01; WALL STREET JOURNAL, 9/27/01]

September 29, 2001: $2.5 Million
in Airline Options Go Unclaimed

$2.5 million in put options on American Airlines and United Airlines are reported unclaimed. This is likely the result of the suspension in trading on the New York Stock Exchange after the attacks which gave the SEC time to be waiting if the owners showed up to redeem their put options placed the week before the 9/11 attacks. [SAN FRANCISCO CHRONICLE, 9/29/01]

October 16, 2001: Bin Laden Cleared of Insider Trading in Britain

"The Financial Services Authority—Britain's main financial regulator—has cleared bin Laden and his henchmen of insider trading. There has been a widespread suspicion that members of the al-Qaeda organization had cashed in on the U.S. attacks, dumping airline, aerospace and insurance company shares before September 11th. The Authority says that after a thorough investigation, it has found no hard evidence of any such deals in London." [MARKETPLACE RADIO REPORT, 10/17/01] On September 24, Belgium's Financial Minister had claimed there were strong suspicions that British markets may have been used for 9/11-related insider trading in early September. [CNN, 9/24/01]

June 3, 2002: The Results of 9/11 Related
Insider Trading Inquiries Are Still Unknown

A rare follow-up article about insider trading based on 9/11 foreknowledge confirms that numerous inquiries in the U.S. and around the world are still ongoing. However, "all are treating these inquiries as if they were state secrets." The author speculates: "The silence from the investigating camps could mean any of several things: Either terrorists are responsible for the puts on the airline stocks; others besides terrorists had foreknowledge; the puts were just lucky bets by credible investors; or, there is nothing whatsoever to support the insider-trading rumors." [INSIGHT, 6/3/02] Another article notes that Deutsche Bank Alex Brown, the American investment banking arm of German giant Deutsche Bank, purchased at least some of these options. Deutsche Bank Alex Brown was once headed by 'Buzzy' Krongard, who is currently Executive Director of the Central Intelligence Agency (CIA). "This fact may not be significant. And then again, it may. After all, there has traditionally been a close link between the CIA, big banks, and the brokerage business." [BUSINESS LINE, 2/11/02]

September 10, 2003: SEC, Others Still Keep Mum About
Insider Trading Investigations

Slate reports that two years after the 9/11 attacks, neither the Chicago Board Options Exchange nor the Securities and Exchange Commission will make any comment about their investigations into insider trading before 9/11. "Neither has announced any conclusion. The SEC has not filed any complaint alleging illegal activity, nor has the Justice Department announced any investigation or prosecution. . . . So, unless the SEC decides to file a complaint—unlikely at this late stage—we may never know what they learned about terror trading." [SLATE, 9/10/03]

September 19, 2003: FBI Finds No Insider Trading

Spokesperson Paul Bresson announces that the FBI has concluded that there was no insider trading in U.S. securities markets by people with advance knowledge of the 9/11 attacks. According to Bresson, the "vast majority" of a preattack surge of trading in options that bet on a drop in the stock of AMR Corp., which owns American Airlines, and UAL Corp., which owns United Airlines, was conducted by investment hedge funds implementing bearish investment strategies or hedging a line position of common stock and was not linked to terrorists. [ST. PETERSBURG TIMES, 9/19/03; WASHINGTON POST, 9/19/03] Independent research will indicate otherwise a few months later.

April 25, 2004: Academic Paper Determines
9/11 Insider Trading Not Due to Chance

Allen Poteshman, a professor of finance at the University of Illinois, publishes a paper demonstrating that the insider trading in options on United and American airline stocks indicates someone profited from foreknowledge of 9/11. Poteshman concludes, "There is evidence of unusual option market activity in the days leading up to September 11." [CHICAGO TRIBUNE 4/25/04; POTESHMAN STUDY]

3.

Counterterrorism Before 9/11

Prior to 1993, few Americans worried about the threat of radical Muslim terrorism. Bin Laden had not been identified as a terrorist, and U.S. officials were still unaware of al-Qaeda (although it had been formed in 1988).

This changed in early 1993, six weeks into President Clinton's first term, when terrorists bombed the World Trade Center, killing six people. Thereafter, the Clinton administration substantially increased its counterterrorism efforts. As the al-Qaeda threat grew, the Clinton administration's focus sharpened. Counterterrorism efforts during this period had mixed results. In June 1993, a plot to bomb New York City landmarks was foiled, but in June 1996, al-Qaeda bombed the Khobar Towers in Saudi Arabia, killing 19 U.S. soldiers and wounding 500. Bin Laden was identified as a terrorist and al-Qaeda was belatedly discovered, while other leads were dropped or missed altogether. Al-Qaeda succeeded in bombing two U.S. Embassies in 1998 (killing 224 people), but it was prevented from carrying out its planned millennium bombings. One can argue about the effectiveness of the Clinton administration's counterterrorism efforts. However, by all accounts, President Clinton and his administration were actively focused on—if not obsessed with—the looming al-Qaeda threat by the end of 2000.

The Bush administration insists that counterterrorism remained a top priority in the first eight months of 2001 [9/11 COMMISSION. 04/08/04], but it continues to withhold any information that could support this assertion. The public record reflects that the intense focus on terrorism in place at the end of the Clinton administration dropped dramatically under the Bush administration. With few exceptions, little attention was paid to terrorism, even as the number of warnings reached unprecedented levels. For example, in 2003, the 9/11 Commission asked Bush's Transportation Secretary, Norman Mineta, "Did this higher level of [terrorist] chatter [before 9/11] . . . result in any action across the government? I take it your answer is no."" He replied, "That's correct." [ASSOCIATED PRESS, 5/23/03 (C)] A new, more aggressive al-Qaeda policy, drafted in December 2000, languished during the first eight months of 2001, and was not yet finalized in early September. Deputy Defense Secretary Wolfowitz reportedly wondered, "Who cares about a little terrorist in Afghanistan?" [NEWSWEEK, 3/22/04]

Clearly, the 9/11 attacks represent a colossal failure by U.S. intelligence. However, to what extent did this failure result from government agency dysfunction that transcended administrations? From the actions or inactions of prior administrations? From the Bush administration's seemingly utter lack of focus on the terrorist threat?

**President
Bill Clinton**

January 20, 1993: Bill Clinton Inaugurated

Bill Clinton replaces George H. W. Bush as U.S. president. He remains president until 2001.

March 1994: FBI Begins to Focus
on Radical Fundamentalism

The FBI creates the Radical Fundamentalist Unit to investigate international radical fundamentalism, including al-Qaeda. (An FBI unit focusing on bin Laden will not be created until 1999.) [9/11 CONGRESSIONAL INQUIRY, 7/24/03]

1995: U.S. Declines to Accept Sudanese Files on al-Qaeda Leaders

The government of Sudan offers the U.S. all its files on bin Laden, who had been living in Sudan since 1991 partly because there were no visa requirements to live there. Sudanese officials had been monitoring him, collecting a "vast intelligence database on Osama bin Laden and more than 200 leading members of his al-Qaeda terrorist network . . . [The U.S. was] offered thick files, with photographs and detailed biographies of many of his principal cadres, and vital information about al-Qaeda's financial interests in many parts of the globe." The U.S. apparently declines to accept the Sudanese government's offer. After 9/11, a U.S. agent who has seen the files on bin Laden's men in Khartoum says that some files were "an inch and a half thick." [GUARDIAN, 9/30/01]

January 24, 1995: Clinton Tries to Stop Terrorist Funding with Executive Order

President Clinton issues Executive Order No. 12947, making it a felony to raise or transfer funds to designated terrorist groups or their front organizations. [EXECUTIVE ORDER, 01/24/95; AGAINST ALL ENEMIES, BY RICHARD CLARKE, 3/04, P. 98]

March 1995: U.S. Ignores Information About al-Qaeda in Terrorist Manual

Belgian investigators find a CD-ROM of an al-Qaeda terrorist manual and begin translating it a few months later. Versions of the manual are circulated widely and are seized by the police all over Europe.

A former CIA official claims the CIA does not obtain a copy of the manual until the end of 1999: "The truth is, they missed for years the largest terrorist guide ever written." He blames CIA reluctance to scrutinize its support for the anti-Soviet jihad in the 1980s. The CIA, however, claims that the manual is not that important, and that in any case it had copies for years. [NEW YORK TIMES, 1/14/01; CBS NEWS, 2/20/02]

April 19, 1995: Oklahoma City Bombing; Possible Middle East Connection

The Alfred P. Murrah Federal Building in Oklahoma City is bombed. U.S. citizen Timothy McVeigh is convicted of the bombing, but some maintain there is a Middle Eastern connection. For instance, Richard Clarke, counterterrorism "tsar" during the Clinton and George W. Bush administrations, says the possibility is intriguing and he has been unable to disprove it. [AGAINST ALL ENEMIES, BY RICHARD CLARKE, 3/04, P. 127] The bombing leads to a surge in concern about terrorism. The Antiterrorism and Effective Death Penalty Act becomes law as a result of such concern. However, many anti-terrorism provisions Clinton seeks are not approved by the Republican-controlled Congress. Many politicians agree with the National Rifle Association that proposed restrictions on bomb-making would infringe on the constitutional right to bear arms. [AGAINST ALL ENEMIES, BY RICHARD CLARKE, 3/04, PP. 98–99]

June 1995: U.S. Considers Bombing bin Laden
for Sponsoring Assassination Attempt

There is a failed assassination attempt on Egyptian President Hosni Mubarak as he visits Ethiopia. The CIA concludes bin Laden authorized the operation, and they plan a retaliation attack. [9/11 CONGRESSIONAL INQUIRY, 7/24/03 (B)] Evidence suggests that the government of Sudan and Hassan al-Turabi, Sudan's leader, know where bin Laden is living and helped support the plot. The United Nations Security Council places sanctions on Sudan as a result. The U.S. examines options for attacking bin Laden and/or Turabi's facilities in the Sudanese capital. The options developed by the U.S. military are rejected for being unstealthy and a de facto war on Sudan. In the ensuing months, there are reports of Egyptian covert operations against bin Laden and an Egyptian military build-up on the Sudanese border. These factors influence bin Laden's decision to move to Afghanistan in 1996. [AGAINST ALL ENEMIES, BY RICHARD CLARKE, 3/04, PP. 140–41]

1996: British Intelligence and al-Qaeda Allegedly
Cooperate in Plot to Assassinate Libyan Leader

Anas al-Liby

Al-Muqatila, a cover for a Libyan al-Qaeda terrorist cell, tries to kill Libyan leader Colonel Mu'ammar al-Qadhafi. Al-Qadhafi survives, but several terrorists and innocent bystanders are killed. [DAWN, 10/30/02] According to David Shayler, a member of the British intelligence agency MI5, and Jean-Charles Brisard and Guillaume Dasquié, authors of the controversial book *The Forbidden Truth*, the British intelligence agency MI6 pays al-Qaeda the equivalent of $160,000 to help fund this assassination attempt.

Shayler later goes to prison for revealing this information and the British press is banned from discussing the case. [NEW YORK TIMES, 8/5/98; OBSERVER, 11/10/02] Well after the failed attempt, the British allegedly continue to support al-Muqatila—for instance, the group openly publishes a newsletter from a London office. [THE FORBIDDEN TRUTH, BY JEAN-CHARLES BRISARD, GUILLAUME DASQUIÉ, AND WAYNE MADSEN, 5/02 EDITION, PP. 97–98] Anas al-Liby, a member of the group, is given political asylum in Britain and lives there until May 2000. There is now a $25 million reward for his capture. [OBSERVER, 11/10/02; FBI MOST WANTED TERRORISTS LIST]

Early 1996: CIA Forms New Counterterrorism bin Laden Unit

The CIA's Counter Terrorism Center creates a special unit focusing specifically on bin Laden. About 10 to 15 individuals are assigned to the unit initially. This grows to about 35 to 40 by 9/11. [9/11 CONGRESSIONAL INQUIRY, 9/18/02] The unit is set up "largely because of evidence linking [bin Laden] to the 1993 bombing of the WTC." [WASHINGTON POST, 10/3/01 (C)] By early 1997, the unit is certain that bin Laden is not just a financier but an organizer of terrorist activity. It knows that al-Qaeda has a military committee planning operations against U.S. interests worldwide. Although this information is disseminated in many reports, the unit's sense of alarm about bin Laden isn't widely shared or understood within the intelligence and policy communities. Employees in the unit feel their zeal attracts ridicule from their peers. [9/11 COMMISSION, 3/24/04 (C)]

Early 1996–October 1998: U.S. Tracks bin Laden's Satellite Phone Calls

During this period, bin Laden and Muhammed Atef, his military commander, use a satellite phone provided by a friend to direct al-Qaeda's operations. Its use is discontinued two months after a U.S. missile strike against bin Laden's camps on August 20, 1998, when an unnamed senior official boasts that the U.S. can track his movements through the use of the phone. [SUNDAY TIMES, 3/24/02; SENATOR SHELBY 9/11 CONGRESSIONAL INQUIRY REPORT, 12/11/02] Records show "Britain was at the heart of the terrorist's planning for his worldwide campaign of murder and destruction:" 260 calls were made to 27 phone numbers in Britain. The other countries called were Yemen (over 200 calls), Sudan (131), Iran (106), Azerbaijan (67), Pakistan (59), Saudi Arabia (57), a ship in the Indian Ocean (13), the U.S. (6), Italy (6), Malaysia (4), and Senegal (2). "The most surprising omission is Iraq, with not a single call recorded." [SUNDAY TIMES, 3/24/02]

March 1996: U.S., Sudan Squabble over bin Laden's Fate

The U.S. pressures Sudan to do something about bin Laden, who is currently based in that country. According to some accounts, Sudan readily agrees, not wanting to be labeled a terrorist nation. Sudan's defense minister engages in secret negotiations with the CIA in Washington. Sudan offers to extradite bin Laden to anywhere he might stand trial. Some accounts claim that Sudan offers bin Laden to the U.S., but the U.S. decides not to take him because they do not have enough evidence at the time to charge him with a crime. [VILLAGE VOICE, 10/31/01; WASHINGTON POST, 10/3/01] Richard Clarke, counterterrorism "tsar" for both Clinton and George W. Bush, calls this story a "fable" invented by the Sudanese and Americans friendly to Sudan. He points out that bin Laden "was an ideological blood brother, family friend, and benefactor" to Sudanese leader Hassan al-Turabi, so any offers to hand him over may have been disingenuous. [AGAINST ALL ENEMIES, BY RICHARD CLARKE, 3/04, PP. 142–43] (CIA Director Tenet later denies that Sudan made any "direct offers to hand over bin Laden." [9/11 CONGRESSIONAL INQUIRY, 10/17/02]) The U.S. reportedly asks Saudi Arabia, Egypt, and Jordan to accept bin Laden into custody, but is refused by all three governments. [GHOST WARS, BY STEVE COLL, 2/04, PP. 323] The 9/11 Commission later claims it finds no evidence that Sudan offers bin Laden directly to the U.S., but it does find evidence that Saudi Arabia is discussed as an option. [9/11 COMMISSION REPORT, 3/23/04] U.S. officials insist that bin Laden leave the country for anywhere but Somalia. One U.S. intelligence source in the region later states: "We kidnap minor drug czars and bring them back in burlap bags. Somebody didn't want this to happen." [VILLAGE VOICE, 10/31/01; WASHINGTON POST, 10/3/01]

April 1996: U.S. and Britain Again Decline to Accept al-Qaeda Files

As in 1995, the U.S. again rejects Sudan's offer to turn over voluminous files about bin Laden and al-Qaeda. An American involved in the secret negotiations later says that the U.S. could have used Sudan's offer to keep an eye on bin Laden, but that another arm of the federal government blocks the efforts. "I've never seen a brick wall like that before. Somebody let this slip up," he says. "We could have dismantled his operations and put a cage on top. It was not a matter of arresting bin Laden but of access to information. That's the story, and that's what could have prevented September 11. I knew it would come back to haunt us." [VILLAGE VOICE, 10/31/01; WASHINGTON POST, 10/3/01] Around this time, MI6, the British intelligence agency, also rebuffs Sudan's offer to provide al-Qaeda intelligence. Sudan makes a standing offer: "If someone from MI6 comes to us and declares himself, the next day he can be in [the capital city] Khartoum." A Sudanese government source later adds, "We have been saying this for years." The offer is not taken up until after 9/11. [GUARDIAN, 9/30/01]

May 18, 1996: Sudan Expels bin Laden;
U.S. Fails to Stop His Flight to Afghanistan

After Sudan asks bin Laden to leave the country, he moves to Afghanistan. He departs along with many other al-Qaeda members, plus much money and resources. Bin Laden flies to Afghanistan in a C-130 transport plane with an entourage of about 150 men, women, and children, stopping in Doha, Qatar, to refuel, where governmental officials greet him warmly. [LOS ANGELES TIMES, 9/1/02; GHOST WARS, BY STEVE COLL, 2/04, P. 325] The U.S. knows in advance that bin Laden is going to Afghanistan, but does nothing to stop him. Elfatih Erwa, Sudan's minister of state for defense at the time, later says in an interview, "We warned [the U.S.]. In Sudan, bin Laden and his money were under our control. But we knew that if he went to Afghanistan no one could control him. The U.S. didn't care; they just didn't want him in Somalia. It's crazy." [VILLAGE VOICE, 10/31/01; WASHINGTON POST, 10/3/01]

June 1996: Informant Exposes al-Qaeda Secrets to U.S.; No Apparent Response Ensues

Jamal al-Fadl, an al-Qaeda operative from al-Qaeda's first meeting in the late 1980s until 1995, tells the U.S. everything he knows about al-Qaeda. Before al-Fadl's debriefings, U.S. intelligence had amassed thick files on bin Laden and his associates and contacts. However, they had had no idea how the many pieces fit together. "Al-Fadl was the Rosetta Stone," an official says. "After al-Fadl, everything fell into place." [THE CELL, BY JOHN MILLER, MICHAEL STONE, AND CHRIS MITCHELL, 8/02, PP. 154-65] By late 1996, based largely on al-Fadl's information, the CIA definitively confirms that bin Laden is more of a terrorist than just a terrorist financier. They also learn the term "al-Qaeda" for the first time. [9/11 CONGRESSIONAL INQUIRY, 7/24/03 (B)] Yet the U.S. will not take "bin Laden or al-Qaeda all that seriously" until after the bombing of U.S. embassies in Africa in 1998. [THE CELL, BY JOHN MILLER, MICHAEL STONE, AND CHRIS MITCHELL, 8/02, P. 213] It takes two years to turn al-Fadl's information into the first U.S. indictment of bin Laden. [NEW YORK TIMES, 09/30/01 (B); 9/11 CONGRESSIONAL INQUIRY, 7/24/03 (B); PBS FRONTLINE, 9/01] One person that al-Fadl describes in detail is Wali Khan Shah, one of the plotters of the Operation Bojinka plot (see chapter 1). U.S. intelligence learns that Shah has al-Qaeda ties. Author Peter Lance notes that U.S. intelligence should have concluded that Shah's fellow Operation Bojinka plotter, Khalid Shaikh Mohammed, also has al-Qaeda ties. However, there is no new effort to find Mohammed, and he later goes on to mastermind the 9/11 attacks. [1000 YEARS FOR REVENGE, BY PETER LANCE, 9/03, PP. 330–31]

Wali Khan Shah

July 17, 1996: TWA Flight 800 Crashes; Counterterrorism Funding Boosted in Response

TWA Flight 800 crashes off the coast of Long Island, New York, killing the 230 people on board. The cause of the crash is debated for a long time afterward, and terrorism is considered a possibility. With this accident in mind, President Clinton requests, and Congress approves, over $1 billion in counterterrorism-related funding in September 1996. [AGAINST ALL ENEMIES, BY RICHARD CLARKE, 3/04, P. 130]

1997: Al-Qaeda Still Not Recognized as Terrorist Organization by State Department

While the State Department listed bin Laden as a financier of terror in its 1996 survey of terrorism, al-Qaeda is not included on the 1997 official U.S. list of terrorist organizations subject to various sanctions. Al-Qaeda will not be officially recognized as a terrorist organization until 1998. [NEW YORK TIMES, 12/30/01]

1998: FAA Testing Reveals Frightening Airport Security Lapses; Little Done in Response Except Small Penalties

The FAA creates "Red Teams"—small, secretive teams traveling to airports and attempting to foil their security systems—in response to the 1988 bombing of a Pan Am 747 over Scotland. According to later reports, the Red Team conducts extensive testing of screening checkpoints at a large number of domestic airports in 1998. The results were frightening: "We were successful in getting major weapons—guns and bombs—through screening checkpoints with relative ease, at least 85 percent of the time in most cases. At one airport, we had a 97 percent success rate in breaching the screening checkpoint. . . . The individuals who occupied the highest seats of authority in the FAA were fully aware of this highly vulnerable state of aviation security and did nothing." [NEW YORK TIMES, 2/27/02] In 1999, the New York Port Authority and major airlines at Boston's Logan Airport will be "fined a total of $178,000 for at least 136 security violations [between 1999–2001]. In the majority of incidents, screeners hired by the airlines for checkpoints in terminals routinely [fail] to detect test items, such as pipe bombs and guns." [ASSOCIATED PRESS, 9/12/01 (C)]

April 15, 1998: Libya Issues First Arrest Warrant for bin Laden

The first Interpol (international police) arrest warrant for bin Laden is issued—by Libya. [OBSERVER, 11/10/02] According to the authors of the controversial book *The Forbidden Truth*, British and U.S. intelligence agencies play down the arrest warrant, and have the public version of the warrant stripped of important information, such as the summary of charges and the fact that Libya requested the warrant. At this point, no Western country has yet issued a warrant for bin Laden, even though he publicly called for attacks on Western targets beginning in 1996. The arrest warrant is issued for the 1994 murder of two German antiterrorism agents. Allegedly, Britain and the U.S. aren't interested in catching bin Laden at this time due to his involvement with Britain in attempts to assassinate Libyan leader Colonel Mu'ammar al-Qadhafi in 1996. [THE FORBIDDEN TRUTH, BY JEAN-CHARLES BRISARD, GUILLAUME DASQUIÉ, AND WAYNE MADSEN, 5/02 EDITION, PP. 97–98.]

May 1998: FBI Gives Counterterrorism Top Priority but No Extra Resources

The FBI issues a strategic, five-year plan that designates national and economic security, including counterterrorism, as its top priority for the first time. However, it is later determined that neither personnel nor resources are shifted accordingly. FBI counterterrorism spending remains constant from this point until 9/11. Only about six percent of the FBI's agent work force is assigned to counterterrorism on 9/11. [9/11 COMMISSION REPORT, 4/13/04; NEW YORK TIMES, 4/18/04]

May 22, 1998: Clinton Creates Counterterrorism Post, Selects Richard Clarke

President Clinton creates the new post of National Coordinator for Counterterrorism. He names Richard Clarke for the job, and Clarke soon becomes known as the counterterrorism "tsar." [WASHINGTON POST, 4/20/00] This is outlined in a new presidential directive on counterterrorism that also outlines goals of fighting terrorism and attempts to strengthen interagency coordination of counterterrorism efforts. [9/11 COMMISSION REPORT, 3/24/04 (D)]

June 8, 1998: Grand Jury Issues Sealed Indictment Against bin Laden and Others

A U.S. grand jury issues a sealed indictment, charging bin Laden and other al-Qaeda leaders with conspiracy to attack the United States. [PBS FRONTLINE, 10/3/02 (C)] The grand jury took two years to reach an indictment, largely based on information from Jamal al-Fadl, a former al-Qaeda operative. [NEW YORK TIMES, 09/30/01

(B), 9/11 CONGRESSIONAL INQUIRY, 7/24/03 (B), PBS FRONTLINE, 9/01] This secret indictment will be superseded by a public one issued in November 1998. [PBS FRONTLINE, 10/3/02 (C)]

September 1998: Memo Outlines al-Qaeda's U.S. Infrastructure

U.S. intelligence authors a memorandum detailing al-Qaeda's infrastructure in the U.S. This memo, which includes information regarding al-Qaeda's use of fronts for terrorist activities [9/11 CONGRESSIONAL INQUIRY, 9/18/02], is provided to senior U.S. officials. [9/11 CONGRESSIONAL INQUIRY, 7/24/03]

November 4, 1998: U.S. Issues Public Indictment of bin Laden, Others for Embassy Bombings

The U.S. publicly indicts bin Laden, Mohammed Atef, and others for the U.S. embassy bombings in Kenya and Tanzania. Bin Laden had been secretly indicted on different charges earlier in the year in June. Record $5 million rewards are announced for information leading to his arrest and the arrest of Mohammed Atef. [PBS FRONTLINE, 2001] Shortly thereafter, bin Laden allocates $9 million in reward money for the assassinations of four U.S. government officials in response to the reward on him. A year later, it is learned that the Secretary of State, Defense Secretary, FBI Director, and CIA Director are the targets. [9/11 CONGRESSIONAL INQUIRY, 9/18/02; MSNBC, 9/18/02; 9/11 CONGRESSIONAL INQUIRY, 7/24/03]

December 4, 1998: CIA Issues Ineffective "Declaration of War" on al-Qaeda

CIA Director Tenet issues a "declaration of war" on al-Qaeda, in a memorandum circulated in the intelligence community. This is ten months after bin Laden's *fatwa* on the U.S., which is called a "de facto declaration of war" by a senior U.S. official in 1999. Tenet says, "We must now enter a new phase in our effort against bin Laden. . . . each day we all acknowledge that retaliation is inevitable and that its scope may be far larger than we have previously experienced . . . We are at war . . . I want no resources or people spared in this efforts [sic], either inside CIA or the [larger intelligence] community." Yet a Congressional joint committee later finds that few FBI agents ever hear of the declaration. Tenet's fervor does not "reach the level in the field that is critical so [FBI agents] know what their priorities are." In addition, even as the counterterrorism budget continues to grow generally, there is no massive shift in budget or personnel until after 9/11. For example, the number of CIA personnel assigned to the Counter Terrorism Center (CTC) stays roughly constant until 9/11, then nearly doubles from approximately 400 to approximately 800 in the wake of 9/11. The number of CTC analysts focusing on al-Qaeda rises from three in 1999 to five by 9/11. [NEW YORK TIMES, 9/18/02; 9/11 CONGRESSIONAL INQUIRY, 9/18/02]

Late 1998: Al-Qaeda Leader Located in Sudan, but U.S. Does Not Try to Capture Him

Intelligence agents learn Mohammed Atef (also known as Abu Hafs)—head of Islamic Jihad and one of the top three leaders of al-Qaeda [ABC NEWS, 11/17/01]—is staying in a particular hotel room in Khartoum,

Sudan. White House officials ask that Atef be killed or captured and interrogated. International capture operations of terrorists have become routine by the mid-1990s, but in this case, both the Defense Department and the CIA are against it, although Atef does not even have bodyguards. The CIA puts the operation in the "too hard to do box," according to one former official. The CIA says it is incapable of conducting such an operation in Sudan, but in the same year, the CIA conducts another spy mission in the same city. [NEW YORK TIMES, 12/30/01; AGAINST ALL ENEMIES, BY RICHARD CLARKE, 3/04, PP. 143–46] A plan is eventually made to seize him, but by then he has left the country. [NEW YORK TIMES, 12/30/01] Atef is considered a top planner of the 9/11 attacks, and is later killed in a bombing raid in November 2001.

1999: FBI Creates bin Laden Unit

The FBI creates its own unit to focus specifically on bin Laden, three years after the CIA created such a special unit. By 9/11, 17 to 19 people are working in this unit out of over 11,000 FBI staff. [9/11 CONGRESSIONAL INQUIRY, 9/18/02]

Early 1999: Memo Calls for New Approach on bin Laden; Focuses on State-Sponsorship, Money Trail

State Department Coordinator for Counterterrorism Michael Sheehan writes a memo calling for a new approach in containing bin Laden. He urges a series of actions the U.S. could take toward Pakistan, Afghanistan, Saudi Arabia, the United Arab Emirates, and Yemen to persuade them to help isolate al-Qaeda. He calls Pakistan the key country and urges that terrorism be made the central issue with them. He advises the U.S. to work with all these countries to curb money laundering. However, a former official says Sheehan's plan lands "with a resounding thud." Pakistan continues to "feign cooperation but [does] little" about its support for the Taliban. [NEW YORK TIMES, 10/29/01]

June 7, 1999: Bin Laden Finally Makes FBI's "10 Most Wanted"

The FBI puts bin Laden on its "10 Most Wanted List." This is almost a year and a half after bin Laden's "declaration of war" against the U.S. on February 22, 1998, and about six months after the CIA's "declaration of war against al-Qaeda" in December 1998. It is also three years after an internal State Department document connected bin Laden to financing and planning numerous terrorist attacks. [PBS FRONTLINE 10/3/02 (C), 9/11 CONGRESSIONAL INQUIRY 7/24/03 (B)]

June 8, 1999: New York Emergency Command Center Opened in WTC Building 7

New York City Mayor Rudolph Giuliani opens a $13 million emergency command center on the 23rd floor of World Trade Center Building 7. [NEWSDAY, 9/12/01] The center is intended to coordinate responses to various emergencies, including natural disasters like hurricanes or floods and terrorist attacks. The 50,000 square foot center has reinforced, bulletproof and bomb-resistant walls, its own air supply and water tank, and three backup generators. This command center is to be staffed around the clock and is intended as a meeting place for city leaders in the event of an act of terrorism. [CNN, 6/7/99; LONDON TIMES 9/12/01] The center is ridiculed as "Rudy's bunker." [TIME, 12/31/01] Most controversial is the 6,000–gallon fuel tank. In 1998 and 1999, Fire Department officials warn that the fuel tank violates city fire codes and poses a hazard. According to one Fire Department memorandum, if the tank were to catch fire it could produce "disaster." Building 7 will be destroyed late in the day on 9/11; some suspect this tank helps explains why. [NEW YORK TIMES, 12/20/01]

October 8, 1999: Al-Qaeda Declared Foreign Terrorist Organization

The Secretary of State finally legally declares al-Qaeda a foreign terrorist organization that is threatening to the U.S. [9/11 CONGRESSIONAL INQUIRY, 7/24/03 (B); 9/11 COMMISSION REPORT, 3/24/04]

November 30, 1999: Jordan Thwarts al-Qaeda Plot

Jordanian officials successfully uncover an al-Qaeda plot to blow up the Radisson Hotel in Amman, Jordan, and other sites on January 1, 2000. [PBS FRONTLINE, 10/3/02 (C)] A call between al-Qaeda leader Abu Zubaida and a suspected Jordanian terrorist exposes the plot. In the call, Zubaida states, "The grooms are ready for the big wedding." [SEATTLE TIMES, 6/23/02] This call reflects an extremely poor code system, because the FBI had already determined in the wake of the 1998 U.S. embassy bombings that "wedding" was the al-Qaeda code word for bomb. [THE CELL, BY JOHN MILLER, MICHAEL STONE, AND CHRIS MITCHELL, 8/02, P. 214] Furthermore, it appears al-Qaeda fails to later change the system, because the code-name for the 9/11 attack is also "The Big Wedding." [CHICAGO TRIBUNE, 9/5/02]

Early December 1999: U.S. Takes Action to
Stop al-Qaeda Millennium Bombing Plot

The CIA learns from the Jordanian government about an al-Qaeda millennium bombing plot. Counterterrorism "tsar" Richard Clarke is told of this, and he implements a plan to neutralize the threat. [AGAINST ALL ENEMIES, BY RICHARD CLARKE, 3/04, PP. 205, 211] The plan, approved by President Clinton, focuses on harassing and disrupting al-Qaeda members throughout the world. The FBI is put on heightened alert, counterterrorism teams are dispatched overseas, a formal ultimatum is given to the Taliban to keep al-Qaeda under control, and friendly intelligence agencies are asked to help. There are Cabinet-level meetings nearly every day dealing with terrorism. [WASHINGTON POST, 4/20/00; ASSOCIATED PRESS, 6/28/02] All U.S. embassies, military bases, police departments, and other agencies are given a warning to be on the lookout for signs of an al-Qaeda millennium attack. One alert border agent responds by arresting terrorist Ahmed Ressam, which leads to the unraveling of several bombing plots. No terror attacks occur. However, Clarke claims the FBI generally remains unhelpful. For example, around this time the FBI says there are no websites in the U.S. soliciting volunteers for training in Afghanistan or money for terrorist front groups. Clarke has a private citizen check to see if this is true, and within days, he is given a long list of such websites. The FBI and Justice Department apparently fail to do anything with the information. [NEWSWEEK, 3/31/04 (B)]

December 14, 1999: Al-Qaeda Terrorist
Planning LAX Attack Is Arrested

Al-Qaeda terrorist Ahmed Ressam is arrested in Port Angeles, Washington, attempting to enter the U.S. with components of explosive devices. 130 pounds of bomb-making chemicals and detonator components are found inside his rental car. He subsequently admits he planned to bomb Los Angeles International Airport on December 31, 1999. [NEW YORK TIMES, 12/30/01] Alert border patrol agent Diana Dean stops him; she and other agents nationwide had been warned recently to look for suspicious activity. Ressam's bombing would have been part of a wave of attacks against U.S. targets over the New Year's weekend. He is later connected to al-Qaeda and convicted. [9/11 CONGRESSIONAL INQUIRY 9/18/02; PBS FRONTLINE, 10/3/02]

Diana Dean

December 14–31, 1999: FBI Thwarts
Additional Millennium Attack Plots

In the wake of the arrest of Ahmed Ressam, FBI investigators work frantically to uncover more millennium plots before they are likely to take place at the end of the year. Documents found with Ressam lead to co-conspirators in New York, then Boston and Seattle. Enough people are arrested to prevent any attacks. Counterterrorism "tsar" Richard Clarke later says, "I think a lot of the FBI leadership for the first time realized that . . . there probably were al-Qaeda people in the United States. They realized that only after they looked at the results of the investigation of the millennium bombing plot." [PBS FRONTLINE, 10/3/02] Yet Clinton's National Security Adviser Sandy Berger says, "Until the very end of our time in office, the view we received from the [FBI] was that al-Qaeda had limited capacity to operate in the U.S. and any presence here was under surveillance." No analysis is done before 9/11 to investigate just how big that presence might be. [WASHINGTON POST, 9/20/02]

December 31, 1999–January 1, 2000:
Attacks Against American Targets Avoided Through Alerts and Luck

Earlier in December, the CIA estimated that al-Qaeda would launch between five and 15 attacks against American targets around the world over the New Year's weekend. "Because the U.S. is [bin Laden]'s ultimate goal . . . we must assume that several of these targets will be in the U.S. . . ." [9/11 CONGRESSIONAL INQUIRY, 7/24/03 (B), TIME, 8/4/02] Since late 1999, there has been intelligence that targets in Washington and New York would be attacked at this time. [9/11 CONGRESSIONAL INQUIRY, 9/18/02] There in fact are a number of planned attacks, including bomb attacks on the Boston and Los Angeles airports, a hotel in Jordan, and a naval ship in Yemen. However, all of the attacks are foiled, thanks to alerts and luck. [WASHINGTON POST, 1/20/02]

March 2000: FBI Agent Apparently Destroys bin Laden E-Mail Intercepts

An FBI agent, angry over a glitch in an e-mail tracking program that has somehow mixed innocent non-targeted e-mails with those belonging to al-Qaeda, reportedly accidentally destroys all of the FBI's Denver-based intercepts of bin Laden's colleagues in a terrorist investigation. The tracking program is called Carnivore. However, the explanation sounds dubious and it is flatly contradicted in the same article: "A Justice Department official, speaking on condition of anonymity, said Tuesday night that the e-mails were not destroyed." [ASSOCIATED PRESS, 5/28/02]

March 2000 (B): Clinton Attempt to Fight Terrorism
Financing Defeated by Republican

The Clinton administration begins a push to fight terrorism financing by introducing a tough anti-money laundering bill. The bill faces tough opposition, mostly from Republicans and lobbyists who enjoy the anonymity of offshore banking, which would be affected by the legislation. Despite passing the House Banking Committee by a vote of 31 to 1 in July 2000, Senator Phil Gramm (R) refuses to let the bill come up for a vote in his Senate Banking Committee. [TIME, 10/15/01] Other efforts begun at this time to fight terrorism financing are later stymied by the new Bush administration in February 2001.

March 10, 2000: Review of Counterterrorism Efforts Show Continued Worries

National Security Adviser Sandy Berger chairs a Cabinet-level meeting to review the wave of attempted terror attacks around the millennium. There are counterterrorism reports that disruption efforts "have not put too much of a dent" into bin Laden's overseas network, and that it is feared "sleeper cells" of terrorists have taken root in the U.S. Some ideas, like expanding the number of Joint Terrorism Task Forces across the U.S., are adopted. Others, like a centralized translation unit for domestic intercepts, are not. [9/11 COMMISSION REPORT, 3/24/04 (D)] In July 2004, it is revealed that the Justice Department is investigating Berger for taking classified documents relating to this review effort out of a secure reading room in 2003. Most of the documents are returned, but a few apparently are lost. [ASSOCIATED PRESS, 7/20/04; WASHINGTON POST, 7/22/04]

April 20, 2000: Some Complain Clinton
Administration Focusing Too Much on Terrorism

The *Washington Post* writes, "With little fanfare, [President Clinton] has begun to articulate a new national security doctrine in which terrorists and other 'enemies of the nation-state' are coming to occupy the position once filled by a monolithic communist superpower." In his January 2000 State of the Union address, President Clinton predicts that terrorists and organized criminals will pose "the major security threat" to the U.S. in coming decades. However, some claim that a "preoccupation with bin Laden has caused errors in judgment." National Security Adviser Sandy Berger counters that the threat of large-scale terrorist attacks on U.S. soil is "a reality, not a perception. . . . We would be irresponsible if we did not take this seriously." Counterterrorism "tsar" Richard Clarke predicts that the U.S.'s new ene-mies "will come after our weakness, our Achilles heel, which is largely here in the United States." [WASH-INGTON POST, 4/20/00]

April 30, 2000: State Department Issues
Counterterrorism Report Focused on South Asia

The State Department issues its annual report describing the U.S. attempt to combat terrorism. For the first time, it focuses on South Asia. The *New York Times* notes, "The report reserves its harshest criticism for Afghanistan" and "is also severely critical of Pakistan." However, neither country is placed on the offi-cial list of countries sponsoring terrorism, which has remained unchanged since 1993. [NEW YORK TIMES, 4/30/00]

May 2000: CIA and FBI Again Reject Sudan's Offer to Provide al-Qaeda Files

The CIA and FBI send a joint investigative team to Sudan to investigate whether that country is a spon-sor of terrorism. It determines that it is not, but the U.S. does not take Sudan off its official list of ter-rorist states. As in 1995 and 1996, Sudan offers to hand over its voluminous files on al-Qaeda, and the offer is again turned down. [GUARDIAN, 9/30/01]

September 2000: Candidate George W. Bush
Promises Emphasis on Countering Terrorism in U.S.

George W. Bush, campaigning for president, writes in an article, "There is more to be done preparing here at home. I will put a high priority on detecting and responding to terrorism on our soil." [NATIONAL GUARD MAGAZINE, 9/00] This repeats verbatim comments made in a speech a year before at the start of the presidential campaign [CITADEL, 9/23/99], and in both cases the context is about weapons of mass destruc-tion. However, after 9/11, now President Bush will say of bin Laden: "I knew he was a menace and I knew he was a problem. I was prepared to look at a plan that would be a thoughtful plan that would bring him to justice, and would have given the order to do that. I have no hesitancy about going after him. But I didn't feel that sense of urgency." [WASHINGTON POST, 5/17/02]

September–October 2000: Predator Flights over
Afghanistan Are Initiated—Then Halted

An unmanned spy plane called the Predator begins flying over Afghanistan, showing incomparably detailed real-time video and photographs of the movements of what appears to be bin Laden and his aides. It flies successfully over Afghanistan 16 times. [9/11 COMMISSION, 3/24/04] President Clinton is impressed by a two-minute video of bin Laden crossing a street heading toward a mosque. Bin Laden is surrounded by a team of a dozen armed men creating a professional forward security perimeter as he moves. The Predator has been used since 1996, in the Balkans and Iraq. One Predator crashes on take-off and another is chased by a fighter, but it apparently identifies bin Laden on three occasions. Its use is stopped in Afghanistan after a few trials, mostly because seasonal winds are picking up. It is agreed to resume the flights in the spring, but the Predator fails to fly over Afghanistan again until after 9/11. [AGAINST ALL ENE-

Footage from a Predator drone apparently shows bin Laden surrounded by security

MIES, BY RICHARD CLARKE, 3/04, PP. 220–21; WASHINGTON POST, 12/19/01] On September 15, 2001, CIA Director Tenet apparently inaccurately tells President Bush, "The unmanned Predator surveillance aircraft that was now armed with Hellfire missiles had been operating for more than a year out of Uzbekistan to provide real-time video of Afghanistan." [WASHINGTON POST, 1/29/02]

September 15–October 1, 2000: Sydney Olympics Officials' Top Concern: Airliner-Based al-Qaeda Attack

Olympics officials later reveal, "A fully loaded, fueled airliner crashing into the opening ceremony before a worldwide television audience at the Sydney Olympics is one of the greatest security fears for the Games." During the Olympics, Australia has six planes in the sky at all times ready to intercept any wayward aircraft. In fact, "IOC officials [say] the scenario of a plane crash during the opening ceremony was uppermost in their security planning at every Olympics since terrorists struck in Munich in 1972." Bin Laden is considered the number-one threat. [SYDNEY MORNING HERALD, 9/20/01] These security measures are similar to those used in the 1996 Atlanta Olympics and other events, including Clinton's second inauguration. Similar planning is already underway before 9/11 for the 2002 Winter Olympics in Utah. [WALL STREET JOURNAL, 4/1/04]

October 12, 2000: USS *Cole* Bombed by al-Qaeda Terrorists; Investigation Thwarted

The USS *Cole* is bombed in the Aden, Yemen, harbor by al-Qaeda terrorists. Seventeen U.S. soldiers are killed. [ABC NEWS, 10/13/00] The Prime Minister of Yemen at the time later claims that hijacker "Khalid Almihdhar was one of the *Cole* perpetrators, involved in preparations. He was in Yemen at the time and stayed after the *Cole* bombing for a while, then he left." [GUARDIAN, 10/15/01] John O'Neill and his team of 200 hundred FBI investigators enter Yemen two days later, but are unable to accomplish much due to restrictions placed on them and because of tensions with U.S. Ambassador Barbara Bodine. All but about 50 investigators are forced to leave by the end of October. Even though O'Neill's boss visits and finds that Bodine is O'Neill's "only detractor," O'Neill and much of his team are forced to leave in November, and the investigation stalls without his personal relationships to top Yemeni officials. [NEW YORKER,

Damage to the USS *Cole*

1/14/02; SUNDAY TIMES, 2/3/02; THE CELL, BY JOHN MILLER, MICHAEL STONE, AND CHRIS MITCHELL, 8/02, P. 237] Increased security threats force the reduced FBI team still in Yemen to withdraw altogether in June 2001. [PBS FRONTLINE, 10/3/02 (B)] The *Sunday Times* later notes, "The failure in Yemen may have blocked off lines of investigation that could have led directly to the terrorists preparing for September 11." [SUNDAY TIMES, 2/3/02]

December 2000: Incoming Bush Administration Briefed on
Terrorism Threat; Apparently Ignores Recommendations

CIA Director Tenet and other top CIA officials brief President-elect Bush, Vice President-elect Cheney, National Security Adviser Rice, and other incoming national security officials on al-Qaeda and covert action programs in Afghanistan. Deputy Director for Operations James Pavitt recalls conveying that bin Laden is one of the gravest threats to the country. Bush asks whether killing bin Laden would end the problem. Pavitt says he answers that killing bin Laden would have an impact but not stop the threat. The CIA recommends the most important action to combat al-Qaeda is to arm the Predator drone and use it over Afghanistan. [9/11 COMMISSION. 3/24/04; REUTERS, 3/24/04 (B)] However, while the drone is soon armed, Bush never gives the order to use it in Afghanistan until after 9/11.

December 20, 2000: Clarke Plan to Neutralize al-Qaeda
Deferred Pending Administration Transition

Counterterrorism "tsar" Richard Clarke submits a plan to "roll back" al-Qaeda over a period of three to five years until it is ineffectual. [9/11 COMMISSION REPORT, 3/24/04 (D)] The main component is a dramatic increase in covert aid to the Northern Alliance in Afghanistan to first tie down the terrorists and then "eliminate the sanctuary" for bin Laden. Financial support for terrorist activities will be systematically attacked, nations fighting al-Qaeda will be given aid to defeat them, and the U.S. will plan for direct military and covert action in Afghanistan. The plan will cost several hundred million dollars. However, since there are only a few weeks left before the Bush administration takes over, it is decided to defer the decision until the new administration is in place. One senior Clinton official later says, "We would be handing [the Bush administration] a war when they took office on January 20. That wasn't going to happen." However, the plan is rejected by the Bush administration and no action is taken. According to one senior Bush administration official, the proposal amounts to "everything we've done since 9/11." [TIME, 8/4/02] Russia's President Vladimir Putin later claims he tried to egg on the previous Clinton administration—without success—to act militarily against the whole Taliban regime: "Washington's reaction at the time really amazed me. They shrugged their shoulders and said matter-of-factly: 'We can't do anything because the Taliban does not want to turn him over.'" [GUARDIAN, 9/22/01]

Early 2001: Bush Staffers Less Concerned with Terrorism

Clinton and Bush staff overlap for several months while new Bush appointees are appointed and confirmed. Clinton holdovers seem more concerned about al-Qaeda than the new Bush staffers. For instance, according to a colleague, Sandy Berger, Clinton's National Security Adviser, had become "totally preoccupied" with fears of a domestic terror attack. [NEWSWEEK, 5/27/02] Brian Sheridan, Clinton's outgoing Deputy Defense Secretary for Special Operations and Low-Intensity Conflict, is astonished when his offers during the transition to bring the new military leadership up to speed on terrorism are brushed aside. "I offered to brief anyone, any time on any topic. Never took it up." [LOS ANGELES TIMES, 3/30/04] Army Lieutenant General Donald

Kerrick, Deputy National Security Adviser and manager of Clinton's NSC (National Security Council) staff, still remains at the NSC nearly four months after Bush took office. He later notes that while Clinton's advisers met "nearly weekly" on terrorism by the end of his term, he does not detect the same kind of focus with the new Bush advisers: "That's not being derogatory. It's just a fact. I didn't detect any activity but what [Clinton holdover] [Richard] Clarke and the CSG [Counterterrorism and Security Group] were doing." [WASHINGTON POST, 1/20/02] Kerrick submits a memo to the new people at the NSC, warning, "We are going to be struck again." He says, "They never responded. It was not high on their priority list. I was never invited to one meeting. They never asked me to do anything. They were not focusing. They didn't see terrorism as the big megaissue that the Clinton administration saw it as." Kerrick adds, "They were gambling nothing would happen." [LOS ANGELES TIMES, 3/30/04] Bush's first Joint Chiefs of Staff Chairman, Henry Shelton, later says terrorism was relegated "to the back burner" until 9/11. [WASHINGTON POST, 10/2/02]

Early 2001 (B): Taliban Disinformation Project Is Cancelled

The heads of the U.S. military, the Joint Chiefs of Staff, have become frustrated by the lack of CIA disinformation operations to create dissent among the Taliban, and at the very end of the Clinton administration, they begin to develop a Taliban disinformation project of their own, which is to go into effect in 2001. When they are briefed, the Defense Department's new leaders kill the project. According to Joint Chiefs of Staff Chairman Henry Shelton, "[Defense Secretary] Rumsfeld and Deputy [Defense] Secretary Paul Wolfowitz were against the Joint Staff having the lead on this." They consider this a distraction from their core military missions. As far as Rumsfeld is concerned, "This terrorism thing was out there, but it didn't happen today, so maybe it belongs lower on the list . . . so it gets defused over a long period of time." [LOS ANGELES TIMES, 3/30/04]

January–August 2001: General Warnings to Airlines Have Little Impact

The FAA issues 15 general warnings issued to airlines in these months. The airlines have been getting an average of more than one warning a month for a long time. [CNN, 3/02; CNN, 5/17/02] As one newspaper later reports, "there were so many [warnings] that airline officials grew numb to them." [ST. PETERSBURG TIMES, 9/23/02] Bush administration officials later state that the terror information they are receiving is so vague that tighter security does not seem required. [ASSOCIATED PRESS, 5/18/02] However, it seems that even these general warnings are never passed on to airline employees. Rosemary Dillard, a supervisor for American Airlines, states, "My job was supervision over all the flight attendants who flew out of National, Baltimore, or Dulles. In the summer of 2001, we had absolutely no warnings about any threats of hijackings or terrorism, from the airline or from the FAA." [NEW YORK OBSERVER, 6/17/04] The content of these seemingly harmless warnings remain classified after 9/11. They are said to be exempted from public disclosure by a federal statute that covers "information that would be detrimental to the security of transportation if disclosed." [NEW YORK OBSERVER, 6/17/04]

January 20–September 10, 2001: Bush Briefed on al-Qaeda over 40 Times

National Security Adviser Rice later testifies to the 9/11 Commission that in the first eight months of Bush's presidency before 9/11, "the president receive[s] at these [Presidential Daily Briefings] more than 40 briefing items on al-Qaeda, and 13 of those [are] in response to questions he or his top advisers posed." [9/11 COMMISSION, 4/8/04] However, CIA Director Tenet claims that none of the warnings specifically indicates terrorists plan to fly hijacked commercial aircraft into buildings in the U.S. [NEW YORK TIMES, 4/4/04]

Early January 2001: Al-Qaeda Threat Highlighted for Powell

Counterterrorism "tsar" Richard Clarke briefs Secretary of State Powell about the al-Qaeda threat. He urges decisive and quick action against al-Qaeda. Powell meets with the CSG (Counterterrorism and Security Group) containing senior counterterrorism officials from many agencies. He sees that all members of the group agree al-Qaeda is an important threat. For instance, Deputy Defense Secretary Brian Sheridan says to Powell, "Make al-Qaeda your number one priority." [AGAINST ALL ENEMIES, BY RICHARD CLARKE, 3/04, PP. 227–30]

January 3, 2001: Clarke Briefs Rice on
al-Qaeda Threat; Keeps Job but Loses Power

Richard Clarke, counterterrorism "tsar" for the Clinton administration, briefs National Security Adviser Rice and her deputy, Steve Hadley, about al-Qaeda. [WASHINGTON POST, 1/20/02] Outgoing National Security Adviser Sandy Berger makes an unusual appearance at the start of the meeting, saying to Rice, "I'm coming to this briefing to underscore how important I think this subject is." He claims that he tells Rice during the transition between administrations, "I believe that the Bush administration will spend more time on terrorism generally, and on al-Qaeda specifically, than any other subject." Clarke presents his plan to "roll back" al-Qaeda that he had given to the outgoing Clinton administration a couple of weeks earlier. [TIME, 8/4/02] He gets the impression that Rice has never heard the term al-Qaeda before. Rice decides this day to retain Clarke and his staff, but downgrades his official position as National Coordinator for Counterterrorism. While he is still known as the counterterrorism "tsar," he has less power and now reports to deputy secretaries instead of attending Cabinet-level meetings. He no longer is able to send memos directly to the president. [AGAINST ALL ENEMIES, BY RICHARD CLARKE, 3/04; PP. 227–30, GUARDIAN, 3/25/04]

January 10, 2001–September 4, 2001:
Armed Predator Drone Is Readied, but Unused

Even before President Bush's official inauguration, Clinton holdover counterterrorism "tsar" Richard Clarke pushes National Security Adviser Rice and other incoming Bush officials to resume Predator drone flights over Afghanistan (originally carried out in September and October 2000) in an attempt to find and assassinate bin Laden. [WASHINGTON POST, 1/20/02; CBS NEWS, 6/25/03] On January 10, Rice is shown a video clip of bin Laden filmed by a Predator drone the year before. [WASHINGTON POST, 1/20/02] Clarke learns of an Air Force plan to arm the Predator. The original plan calls for three years of testing, but Clarke pushes so hard that the armed Predator is ready in three months. [NEW YORKER, 7/28/03] A Hellfire missile is successfully test fired from a Predator on February 16, 2001. [CBS NEWS, 6/25/03] In early June, a duplicate of the brick house where bin Laden is believed to be living in Kandahar, Afghanistan, is built in Nevada, and destroyed by a Predator missile. The test shows that the missile fired from

A Predator drone

miles away would have killed anyone in the building, and one participant calls this the long sought after "holy grail" that could kill bin Laden within minutes of finding him. [WASHINGTON POST, 1/20/02] Clarke repeatedly advocates using the Predator, armed or unarmed. However, bureaucratic infighting between the CIA and the Air Force over who would pay for it and take responsibility delays its use. Clarke later says, "Every time we were ready to use it, the CIA would change its mind." [NEW YORKER, 7/28/03] Rice and Deputy National Security Adviser Steve Hadley decide to delay reconnaissance flights until the armed version is ready. In July 2001, Hadley directs the military to have armed Predators ready to deploy no later than September 1. [9/11 COMMISSION REPORT, 3/24/04 (D)] The issue comes to a head in early September, but even then, a decision to use the Predator is delayed. [NEW YORKER 7/28/03]

January 21, 2001: George W. Bush Inaugurated

George W. Bush is inaugurated as president, replacing President Bill Clinton.

January 25, 2001: Clarke Presents Plan to
Roll Back al-Qaeda, but Response Is Delayed

Counterterrorism "tsar" Richard Clarke submits a proposal to National Security Adviser Rice and "urgently" asks for a Cabinet-level meeting on the al-Qaeda threat. [AGAINST ALL ENE-MIES, BY RICHARD CLARKE, 3/04, PP. 230–31] He forwards his December 2000 strategy paper and a copy of his 1998 "Delenda Plan" (see chapter 4). He lays out a proposed agenda for urgent action:

Richard Clarke

- Approve covert assistance to Ahmed Shah Massoud's Northern Alliance fighting the Taliban. [9/11 COMMISSION REPORT, 3/24/04 (D)]

- Significantly increase funding for CIA counterterrorism activity. [9/11 COMMISSION REPORT, 3/24/04 (D)]

- Respond to the USS *Cole* bombing with an attack on al-Qaeda. (The link between al-Qaeda and that bombing had been assumed for months and is confirmed in the media two days later.) According to the *Washington Post*, "Clarke argue[s] that the camps [are] can't-miss targets, and they [matter]. The facilities [amount] to conveyor belts for al-Qaeda's human capital, with raw recruits arriving and trained fighters departing either for front lines against the Northern Alliance, the Afghan rebel coalition, or against American interests somewhere else. The U.S. government had whole libraries of images filmed over Tarnak Qila and its sister camp, Garmabat Ghar, 19 miles farther west. Why watch al-Qaeda train several thousand men a year and then chase them around the world when they left?" No retaliation is taken on these camps until after 9/11. [WASHINGTON POST, 1/20/02]

- Go forward with new Predator drone reconnaissance missions in the spring and use an armed version when it is ready. [9/11 COMMISSION REPORT, 3/24/04 (D)]

- Step up the fight against terrorist fundraising. [9/11 COMMISSION REPORT, 3/24/04 (D)]

- Be aware that al-Qaeda sleeper cells in the U.S. are not just a potential threat, but are a "major threat in being." Additionally, more attacks have almost certainly been set in motion.
[PBS FRONTLINE, 10/3/02; WASHINGTON POST, 1/20/02]

Rice's response to Clarke's proposal is that the Cabinet will not address the issue until it has been "framed" at the deputy secretary level. However, this initial deputy meeting is not given high priority and it does not take place until April 2001. [AGAINST ALL ENEMIES, BY RICHARD CLARKE, 3/04, PP. 230–31] Henry Shelton, Joint Chiefs of Staff Chairman until 9/11, says, "The squeaky wheel was Dick Clarke, but he wasn't at the top of their priority list, so the lights went out for a few months. Dick did a pretty good job because he's abrasive as hell, but given the [bureaucratic] level he was at" there was no progress. [LOS ANGELES TIMES, 3/30/04; THE AGE OF SACRED TERROR, BY DANIEL BENJAMIN AND STEVEN SIMON, 10/02, PP. 335–36] Some counterterrorism officials think the new administration responds slowly simply because Clarke's proposal originally came from the Clinton administration. [TIME, 8/4/02] For instance, Thomas Maertenson, on the National Security Council in both the Clinton and Bush administrations, says, "They really believed their campaign rhetoric about the Clinton administration. So anything [that administration] did was bad, and the Bushies were not going to repeat it." [NEW YORK TIMES, 3/24/04; MINNEAPOLIS STAR-TRIBUNE, 3/25/04]

January 27, 2001 (B): Al-Qaeda's Role in USS *Cole* Bombing Triggers No Immediate Response

The *Washington Post* reports that the U.S. has confirmed the link between al-Qaeda and the October 2000 USS *Cole* bombing. [WASHINGTON POST, 1/27/01] This conclusion is stated without hedge in a February 9 briefing for Vice President Cheney. [WASHINGTON POST, 1/20/02] In the wake of that bombing, Bush stated on the campaign trail, "I hope that we can gather enough intelligence to figure out who did the act and take the necessary action. . . . There must be a consequence." [WASHINGTON POST, 1/20/02] Deputy Defense Secretary Paul Wolfowitz later complains that by the time the new administration is in place, the *Cole* bombing was "stale." Defense Secretary Rumsfeld concurs, stating that too much time had passed to respond. [9/11 COMMISSION REPORT, 3/24/04 (B)] The new Bush administration fails to resume the covert deployment of cruise missile submarines and gunships on six-hour alert near Afghanistan's borders that had begun under President Clinton. The standby force gave Clinton the option of an immediate strike against targets in Afghanistan harboring al-Qaeda's top leadership. This failure makes a possible assassination of bin Laden much more difficult. [WASHINGTON POST, 1/20/02]

January 31, 2001: Bipartisan Commission Issues Final Report on Terrorism, but Conclusions Are Ignored

The final report of the U.S. Commission on National Security/21st Century, co-chaired by former Senators Gary Hart (D), and Warren Rudman (R) is issued. The bipartisan report was put together in 1998 by then-President Bill Clinton and then-House Speaker Newt Gingrich. Hart and Rudman personally brief National Security Adviser Rice, Defense Secretary Rumsfeld, and Secretary of State Powell on their findings. The report has 50 recommendations on how to combat terrorism in the U.S., but all of them are ignored by the Bush administration. According to Senator Hart, Congress begins to take the commission's suggestions seriously in March and April, and legislation is introduced to implement some of the recommendations. Then, "Frankly, the White House shut it down . . . The president said 'Please wait, we're going to turn this over to the vice president' . . . and so Congress moved on to other things, like tax cuts and the issue of the day." The White House announces in May that it will have Vice President Cheney study the potential problem of domestic terrorism despite the fact that this commission had just studied the issue for 2 ½ years. Interestingly, both this commission and the Bush administration were already assuming a new cabinet level National Homeland Security Agency would be enacted eventually, even as the public remained unaware of the term and the concept. [SALON, 9/12/01; SALON, 4/2/04] Hart is incredulous that neither he nor any of the other members of this commission are ever asked to testify before the 9/11 Commission. [SALON, 4/6/04]

February 2001: Bush Administration Abandons Global Crackdown on Terrorist Funding

According to *Time* magazine, "The U.S. was all set to join a global crackdown on criminal and terrorist money havens [in early 2001]. Thirty industrial nations were ready to tighten the screws on offshore financial centers like Liechtenstein and Antigua, whose banks have the potential to hide and often help launder billions of dollars for drug cartels, global crime syndicates—and groups like Osama bin Laden's al-Qaeda organization. Then the Bush administration took office." [TIME, 10/15/01] After pressure from the powerful banking lobby, the Treasury Department under Paul O'Neill halts U.S. cooperation with these international efforts begun in 2000 by the Clinton administration. Clinton had created a National Terrorist Asset Tracking Center in his last budget, but under O'Neill no funding for the center is provided and the tracking of terrorist financing slows down. [FOREIGN AFFAIRS, 7/01; TIME, 10/15/01]

Early February 2001: Clarke Urges Cheney to Take Action Against al-Qaeda

Counterterrorism "tsar" Richard Clarke briefs Vice President Cheney about the al-Qaeda threat. He urges decisive and quick action against al-Qaeda. Cheney soon visits CIA headquarters for more information about al-Qaeda. However, at later high-level meetings Cheney fails to bring up al-Qaeda as a priority issue. [*AGAINST ALL ENEMIES,* BY RICHARD CLARKE, 3/04, PP. 227–30; *TIME,* 8/4/02]

February 9, 2001: Bin Laden's Financial Network Laid Bare

U.S. officials claim significant progress in defeating bin Laden's financial network, despite significant difficulties. It is claimed that, "bin Laden's financial and operational networks has been 'completely mapped' in secret documents shared by the State Department, CIA, and Treasury Department, with much of the mapping completed in detail by mid-1997." [UPI, 2/9/01] Reporter Greg Palast later notes that when the U.S. freezes the assets of terrorist organizations in late September 2001, U.S. investigators likely knew much about the finances of those organizations but took no action before 9/11. [*SANTA FE NEW MEXICAN,* 3/20/03]

February 13, 2001: NSA Breaks al-Qaeda's Secret Codes

UPI, while covering a trial of bin Laden's al-Qaeda followers, reports that the NSA has broken bin Laden's encrypted communications. U.S. officials confirm "codes were broken." Presumably, this happened some time earlier and the codes have been changed by this time. [UPI, 2/13/01]

February 26, 2001: Bush Administration "Paying No Attention" to Terrorism

Paul Bremer, who will be appointed the U.S. administrator of Iraq in 2003, says in a speech that the Bush administration is "paying no attention" to terrorism. "What they will do is stagger along until there's a major incident and then suddenly say, 'Oh my God, shouldn't we be organized to deal with this.'" Bremer speaks shortly after chairing the National Commission on Terrorism, a bipartisan body formed during the Clinton administration. [ASSOCIATED PRESS, 4/29/04]

March 7, 2001: Plan to Fight al-Qaeda Considered, but with "Little Urgency"

Deputy National Security Adviser Steve Hadley chairs an informal meeting of some counterparts from other agencies to discuss al-Qaeda. They begin a broad review of the government's approach to al-Qaeda and Afghanistan. According to the *New York Times,* the approach is "two-pronged and included a crisis warning effort to deal with immediate threats and longer-range planning by senior officials to put into place a comprehensive strategy to eradicate al-Qaeda." Counterterrorism "tsar" Richard Clarke again pushes for immediate decisions on assisting Ahmed Shah Massoud and his Northern Alliance in Afghanistan. Hadley suggests dealing with this as part of the broad review. Clarke supports a larger program, but he warns that delay risks the Alliance's defeat. Clarke also advocates using the armed Predator drone. However, despite an increasing number of alarming warnings following this meeting, there is little follow-up. "By June, a draft of a presidential directive authorizing an ambitious covert action plan is circulating through the upper echelons of the administration, but there seem[s] little urgency about putting the plan into effect." [*NEW YORK TIMES,* 4/4/04; *NEW YORK TIMES,* 3/24/04 (D); 9/11 COMMISSION REPORT, 3/24/04; 9/11 COMMISSION REPORT, 3/24/04 (D)]

March 26, 2001: CIA Benefits from Major Software Improvements

The *Washington Post* reports on major improvements of the CIA's intelligence gathering capability "in recent years." A new program called Oasis uses "automated speech recognition" technology to turn audio feeds into formatted, searchable text. It can distinguish one voice from another and differentiates "speaker 1" from "speaker 2" in transcripts. Software called Fluent performs "cross lingual" searches, translates difficult languages like Chinese and Japanese (apparently such software is much better than similar publicly available software), and even automatically assesses the contextual importance. Other new software can turn a suspect's "life story into a three-dimensional diagram of linked phone calls, bank deposits and plane trips," while still other software can efficiently and quickly process vast amounts of video, audio and written data. [WASHINGTON POST, 3/26/01] However, the government will later report that a number of messages about the 9/11 attacks, such as one stating "tomorrow is the zero hour," are not translated until after 9/11 because analysts were "too swamped." [ABC NEWS, 6/7/02]

Spring 2001: Ashcroft Doesn't Want FBI Director to Talk About Terrorism

Attorney General Ashcroft talks with FBI Director Louis Freeh before an annual meeting of special agents. Ashcroft lays out his priorities, which according to one participant is "basically violent crime and drugs." Freeh bluntly replies that those are not his priorities and he talks about counterterrorism. "Ashcroft does not want to hear about it," says one witness. [NEWSWEEK, 5/27/02]

April 4, 2001: Bugging Techniques Reach New Heights

The BBC reports on advances in electronic surveillance. The U.S.'s global surveillance program, Echelon, has become particularly effective in monitoring mobile phones, recording millions of calls simultaneously and checking them against a powerful search engine designed to pick out key words that might represent a security threat. Laser microphones can pick up conversations from up to a kilometer away by monitoring window vibrations. If a bug is attached to a computer keyboard, it is possible to monitor exactly what is being keyed in, because every key on a computer has a unique sound when depressed. [BBC, 4/4/01]

April 30, 2001: Wolfowitz in Deputy Secretary Meeting: "Who Cares About [bin Laden]?"

The Bush administration finally has its first Deputy Secretary-level meeting on terrorism. [TIME, 8/4/02] According to counterterrorism "tsar" Richard Clarke, he advocates that the Northern Alliance needs to be supported in the war against the Taliban, and the Predator drone flights need to resume over Afghanistan so bin Laden can be targeted. [AGAINST ALL ENEMIES, BY RICHARD CLARKE, 3/04, P. 231] Deputy Defense Secretary Paul Wolfowitz says the focus on al-Qaeda is wrong. He states, "I just don't understand why we are beginning by talking about this one man bin Laden," and "Who cares about a little terrorist in Afghanistan?" Wolfowitz insists the focus should be Iraqi-sponsored terrorism instead. He claims the 1993 attack on the WTC must have been done with help from Iraq, and rejects the CIA's assertion that there has been no Iraqi-sponsored terrorism against the U.S. since 1993. (A spokesperson for Wolfowitz later calls Clarke's account a "fabrication.") [AGAINST ALL ENEMIES, BY RICHARD CLARKE, 3/04, PP. 30, 231, NEWSWEEK, 3/22/04] Wolfowitz repeats these sentiments immediately after 9/11 and tries to argue that the U.S. should attack Iraq. Deputy Secretary of State Richard Armitage agrees with Clarke that al-Qaeda is an important threat. Deputy National Security Adviser Steve Hadley, chairing the meeting, brokers a compromise between Wolfowitz and the others. The group agrees to hold additional meetings focusing on al-Qaeda first (in June and July), but then later look at other terrorism, including any Iraqi terrorism. [AGAINST ALL ENEMIES, BY RICHARD CLARKE, 3/04, PP. 30, 231–32] Vice President Cheney's Chief of Staff I. Lewis "Scooter" Libby and Deputy CIA Director John McLaughlin also attend the hour-long meeting. [TIME, 8/4/02]

April 30, 2001 (B): Annual Terrorism Report Says Focusing on bin Laden Is "Mistake"

The U.S. State Department issues its annual report on terrorism. The report cites the role of the Taliban in Afghanistan, and notes the Taliban "continued to provide safe haven for international terrorists, particularly Saudi exile Osama bin Laden and his network." However, as CNN describes it, "Unlike last year's report, bin Laden's al-Qaeda organization is mentioned, but the 2000 report does not contain a photograph of bin Laden or a lengthy description of him and the group. A senior State Department official told CNN that the U.S. government made a mistake last year by focusing too tightly on bin Laden and 'personalizing terrorism . . . describing parts of the elephant and not the whole beast.'" [CNN, 4/30/01]

May 2001: Warnings About Impending Terrorist Attack Fail to Alert White House

Around this time, intercepts from Afghanistan warn that al-Qaeda could attack an American target in late June or on the July 4 holiday. However, the White House's Counterterrorism and Security Group does not meet to discuss this prospect. This group also fails to meet after intelligence analysts overhear conversations from an al-Qaeda cell in Milan suggesting that bin Laden's agents might be plotting to kill Bush at the European summit in Genoa, Italy, in late July. In fact, the group hardly meets at all. By comparison, the Counterterrorism and Security Group met two or three times a week between 1998 and 2000 under Clinton. [NEW YORK TIMES, 12/30/01]

May 2001 (B): Bush, Who Has Yet to Take Any Action Against al-Qaeda, Is Tired of "Swatting at Flies"

It is claimed that after a routine briefing by CIA Director Tenet to President Bush regarding the hunt for al-Qaeda leader Abu Zubaida, Bush complains to National Security Adviser Rice that he is tired of "swatting at flies" and wants a comprehensive plan for attacking terrorism. Counterterrorism "tsar" Richard Clarke already has such a plan, but it has been mired in bureaucratic deadlock since January. After this, progress remains slow. [TIME 8/4/02; 9/11 COMMISSION REPORT, 3/24/04 (D)]

May 8, 2001: Cheney Heads Task Force Responding to
Domestic Attacks, but No Action Is Taken Before 9/11

Bush entrusts Vice President Cheney to head the new Office of National Preparedness, a part of FEMA. This office is supposed to oversee a "national effort" to coordinate all federal programs for responding to domestic attacks. Cheney informs the press: "One of our biggest threats as a nation" may include "a terrorist organization overseas. We need to look at this whole area, oftentimes referred to as homeland defense." The focus is on state-funded terrorists using weapons of mass destruction, and neither bin Laden nor al-Qaeda is mentioned. [NEW YORK TIMES, 7/8/02] Cheney's task force is supposed to report to Congress by October 1, 2001, after a review by the National Security Council. Bush states that he "will periodically chair a meeting of the National Security Council to review these efforts." [WASHINGTON POST, 1/20/02] In July, two senators send draft counterterrorism legislation to Cheney's office, but a day before 9/11, they are told it might be another six months before he gets to it. The task force is just beginning to hire staff a few days before 9/11. Former Senator Gary Hart (D) later implies that this task force is created to prevent Congress from enacting counterterrorism legislation pro- posed by a bipartisan commission he had co-chaired in January. [SALON, 4/2/04; SALON, 4/6/04]

Vice President Richard "Dick" Cheney

May 10, 2001: Ashcroft Omits Counterterrorism from List of Goals

Attorney General Ashcroft sends a letter to department heads telling them the Justice Department's new agenda. He cites seven goals, but counterterrorism is not one of them. Yet just one day earlier, he testified before Congress and said of counterterrorism, "The Department of Justice has no higher priority." [NEW YORK TIMES, 2/28/02] Dale Watson, head of the FBI's Counterterrorism Division, recalls nearly falling out of his chair when he sees counterterrorism not mentioned as a goal. [9/11 COMMISSION REPORT, 4/13/04] In August, a strategic plan is distributed, listing the same seven goals and 36 objectives. Thirteen objectives are high- lighted, but the single objective relating to counterterrorism is not highlighted. [NEW YORK TIMES, 2/28/02]

June 2001: Clarke Asks for Different Job as
White House Fails to Share His Urgency

Counterterrorism "tsar" Richard Clarke asks for a transfer to start a new national program on cyber security. His request is granted, and he is to change jobs in early October 2001. He makes the change despite the 9/11 attacks. He claims that he tells National Security Adviser Rice and her deputy Steve Hadley, "Perhaps I have become too close to the terrorism issue. I have worked it for ten years and to me it seems like a very important issue, but maybe I'm becoming like Captain Ahab with bin Laden as the White Whale. Maybe you need someone less obsessive about it." [AGAINST ALL ENEMIES, BY RICHARD CLARKE, 3/04, PP. 25–26; WHITE HOUSE, 10/9/01] He later claims, "My view was that this administration, while it listened to me, either didn't believe me that there was an urgent problem or was unprepared to act as though there were an urgent problem. And I thought, if the administration doesn't believe its national coordinator for counterterrorism when he says there's an urgent problem, and if it's unprepared to act as though there's an urgent problem, then probably I should get another job." [NEW YORK TIMES, 3/24/04]

Early June 2001: Counterterrorism Plan Circulated, but Contingency Plans Are Not Created

Deputy National Security Adviser Steve Hadley circulates a draft presidential directive on policy toward al-Qaeda. Counterterrorism "tsar" Richard Clarke and his staff regard the new approach as essentially the same as the proposal that they developed in December 2000 and presented to the Bush administration in January 2001. The draft has the goal of eliminating al-Qaeda as a threat over a multi-year period, and calls for funding through 2006. It has a section calling for the development of contingency military plans against al-Qaeda and the Taliban. Hadley contacts Deputy Defense Secretary Paul Wolfowitz to tell him these contingency plans will be needed soon. However, no such plans are developed before 9/11. Defense Secretary Rumsfeld and others later admit that the contingency plans available immediately after 9/11 are unsatisfactory. [9/11 COMMISSION REPORT, 3/24/04 (B), 9/11 COMMISSION REPORT, 3/24/04 (D)] The draft is now discussed in three more deputy-level meetings.

Steve Hadley

June 1, 2001: New Policy: Only Defense Secretary May Approve Fighter Jet Launch

According to the *New York Observer* and government documents, the procedure for dealing with hijackings within the United States changes on this date. It requires that, with the exception of "immediate responses," requests for [military] assistance must be forwarded to the Defense Secretary [Donald Rumsfeld] for approval. Rumsfeld later claims that protection against a domestic terrorist attack is not his responsibility; it is instead "a law-enforcement issue." [NEW YORK OBSERVER, 6/17/04; CJCSI, 7/31/97; CJCSI, 6/1/01]

June 9, 2001: FBI Agent Writes Memo Claiming His Agency Is Not Trying to Catch Known Terrorists Living in the U.S.

Robert Wright, an FBI agent who spent ten years investigating terrorist funding, writes a memo that slams the FBI. He states, "Knowing what I know, I can confidently say that until the investigative responsibilities for terrorism are transferred from the FBI, I will not feel safe . . . The FBI has proven for the past decade it cannot identify and prevent acts of terrorism against the United States and its citizens at home and abroad. Even worse, there is virtually no effort on the part of the FBI's International Terrorism Unit to neutralize known and suspected international terrorists living in the United States." [CYBERCAST NEWS SERVICE, 5/30/02] He claims the "FBI was merely gathering intelligence so they would know who to arrest when a terrorist attack occurred" rather than actually trying to stop the attacks. [UPI, 5/30/02] Wright's shocking allegations are largely ignored when they first become public a year later. He is asked on CNN's Crossfire, one of the few outlets to cover the story at all, "Mr. Wright, your charges against the FBI are really more disturbing, more serious, than [Coleen] Rowley's [on August 28, 2001]. Why is it, do you think, that you have been ignored by the media, ignored by the congressional committees, and no attention has been paid to your allegations?" The *Village Voice* says that the problem is partly because he went to the FBI and asked permission to speak publicly instead of going straight to the media as Rowley did. The FBI put severe limits on what details Wright can divulge. He is now suing them. [*VILLAGE VOICE*, 6/19/02]

June 13, 2001: Counterterrorism Not Part of Bush Defense Plan

At President Bush's first meeting with NATO heads of state in Brussels, Belgium, Bush outlines his five top defense issues. Missile defense is at the top of the list. Terrorism is not mentioned at all. This is consistent with his other statements before 9/11. Almost the only time he ever publicly mentions al-Qaeda or bin Laden before 9/11 is later in the month, in a letter that renews Clinton administration sanctions on the Taliban. [CNN, 6/13/01; *WASHINGTON POST*, 4/1/04] He only speaks publicly about the dangers of terrorism once before 9/11, in May, except for several mentions in the context of promoting a missile defense shield. [*WASHINGTON POST*, 1/20/02]

June 27–July 16, 2001: Counterterrorism Plan
Delayed with More Deputies Meetings

The first Bush administration deputy-secretary-level meeting on terrorism in late April is followed by three more deputy meetings. Each meeting focuses on one issue: one meeting is about al-Qaeda, one about the Pakistani situation, and one on Indo-Pakistani relations. Counterterrorism "tsar" Richard Clarke's plan to roll back al-Qaeda, which has been discussed at these meetings, is worked on some more, and is finally approved by National Security Adviser Rice and the deputies on August 13. It now can move to the Cabinet level before finally reaching President Bush. The Cabinet-level meeting is scheduled for later in August, but too many participants are on vacation, so the meeting takes place in early September. [*WASHINGTON POST*, 1/20/02; 9/11 COMMISSION REPORT, 3/24/04; 9/11 COMMISSION REPORT, 3/24/04 (D)]

Late June 2001: FAA Disregards Recommended Antiterrorist Measures

Counterterrorism "tsar" Richard Clarke gives a direct warning to the FAA to increase security measures in light of an impending terrorist attack. The FAA refuses to take such measures. [*NEW YORKER*, 1/14/02]

Summer 2001: Military Plans Reducing
Domestic Air Defenses Still Further

During this period, apparently, there are only 14 fighter planes on active alert to defend the continental U.S. (and six more defending Canada and Alaska). [*BERGEN RECORD*, 12/5/03] However, in the months before 9/11, rather than increase the number, the Pentagon was planning to reduce the number still further. Just after 9/11, the *Los Angeles Times* will report, "While defense officials say a decision had not yet been made, a reduction in air defenses had been gaining currency in recent months among task forces assigned by [Defense Secretary] Rumsfeld to put together recommendations for a reassessment of the military." By comparison, in the Cold War atmosphere of the 1950s, the U.S. had thousands of fighters on alert throughout the U.S. [LOS ANGELES TIMES, 9/15/01 (B)] As late as 1998, there were 175 fighters on alert status. [BERGEN RECORD, 12/5/03] Also during this time, FAA officials try to dispense with "primary" radars altogether, so that if a plane were to turn its transponder off, no radar could see it. NORAD rejects the proposal. [AVIATION WEEK AND SPACE TECHNOLOGY, 6/3/02]

Summer 2001 (B): Classified al-Qaeda Surveillance Program Curtailed

According to *Newsweek*, the Justice Department curtails "a highly classified program called 'Catcher's Mitt' to monitor al-Qaeda suspects in the United States." This is apparently because a federal judge severely chastised the FBI for improperly seeking permission to wiretap terrorists. [NEWSWEEK, 3/22/04]

July 2, 2001: FBI Warns of Possible al-Qaeda Attacks; Little Action Results

The FBI issues a warning of possible al-Qaeda attacks to law enforcement agencies, stating, "[T]here are threats to be worried about overseas. While we cannot foresee attacks domestically, we cannot rule them out." It further states, "[T]he FBI has no information indicating a credible threat of terrorist attack in the United States." It asks law enforcement agencies to "exercise vigilance" and "report suspicious activities" to the FBI. Two weeks later, acting FBI Director Thomas Pickard has a conference call with all field office heads mentioning the heightened threat. However, FBI personnel later fail to recall any heightened sense of threat from summer 2001. Only those in the New York field office took any action or recall this later. [CNN, 3/02 (H); 9/11 COMMISSION REPORT, 4/13/04 (B)]

July 3, 2001: Rare Discussion Takes Place Between
National Security Advisers on Terrorism

This is one of only two dates that Bush's national security leadership discusses terrorism. (The other discussion occurs on September 4.) Apparently, the topic is only mentioned in passing and is not the focus of the meeting. This group, made up of the national security adviser, CIA director, defense secretary, secretary of state, joint chiefs of staff chairman and others, met around 100 times before 9/11 to discuss a variety of topics, but apparently rarely terrorism. The White House "aggressively defended the level of attention [to terrorism], given only scattered hints of al-Qaeda activity." This lack of discussion stands in sharp contrast to the Clinton administration and public comments by the Bush administration. [TIME, 8/4/02] Bush said in February 2001, "I will put a high priority on detecting and responding to terrorism on our soil." A few weeks earlier, Tenet told Congress, "The threat from terrorism is real, it is immediate, and it is evolving." [ASSOCIATED PRESS, 6/28/02]

July 3, 2001 (B): Tenet Makes Urgent Request

CIA Director Tenet makes an urgent special request to 20 friendly foreign intelligence services, asking for the arrests of anyone on a list of known al-Qaeda operatives. [WASHINGTON POST, 5/17/02]

July 4–14, 2001: Bin Laden Reportedly Receives Lifesaving Treatment in Dubai, Said to Meet with CIA While There

Bin Laden, America's most wanted criminal with a $5 million bounty on his head, supposedly receives lifesaving treatment for renal failure from American specialist Dr. Terry Callaway at the American hospital in Dubai, United Arab Emirates. He is possibly accompanied by Dr. Ayman al-Zawahiri (who is said to be bin Laden's personal physician, al-Qaeda's second-in-command, and leader of Egypt's Islamic Jihad), plus several bodyguards. Callaway supposedly treated bin Laden in 1996 and 1998, also in Dubai. Callaway later refuses to answer any questions on this matter. [LE FIGARO, 10/31/01; AGENCE FRANCE-PRESSE, 11/1/01; LONDON TIMES, 11/01/01] During his stay, bin Laden is visited by "several members of his family and Saudi personalities," including Prince Turki al Faisal, then head of Saudi intelligence. [GUARDIAN, 11/1/01] On July 12, bin Laden reportedly meets with CIA agent Larry Mitchell in the hospital. Mitchell apparently lives in Dubai as an Arab specialist under the cover of being a consular agent. The CIA, the Dubai hospital, and even bin Laden deny the story. The two news organizations that broke the story, *Le Figaro* and Radio France International, stand by their reporting. [LE FIGARO, 10/31/01; RADIO FRANCE INTERNATIONAL, 11/1/01] The explosive story is widely reported in Europe, but there are only two, small wire service stories on it in the U.S. [UPI, 11/1/01; REUTERS, 11/10/01] The *Guardian* claims that the story originated from French intelligence, "which is keen to reveal the ambiguous role of the CIA, and to restrain Washington from extending the war to Iraq and elsewhere." The *Guardian* adds that during his stay bin Laden is also visited by a second CIA officer. [GUARDIAN, 11/1/01] On July 15, Larry Mitchell reportedly returns to CIA headquarters to report on his meeting with bin Laden. [RADIO FRANCE INTERNATIONAL, 11/1/01] French terrorism expert Antoine Sfeir says the story of this meeting has been verified and is not surprising: It "is nothing extraordinary. Bin Laden maintained contacts with the CIA up to 1998. These contacts have not ceased since bin Laden settled in Afghanistan. Up to the last moment, CIA agents hoped that bin Laden would return to the fold of the U.S., as was the case before 1989." [LE FIGARO, 11/1/01]

July 5, 2001: Clarke Warns of "Something Really Spectacular" FAA and FBI Respond Poorly

At the request of National Security Adviser Rice and White House Chief of Staff Andrew Card, counterterrorism "tsar" Richard Clarke leads a meeting of the Counterterrorism and Security Group, attended by officials from a dozen federal agencies. They discuss intelligence regarding terrorism threats and potential attacks on U.S. installations overseas. Two attendees recall Clarke stating that "something really spectacular is going to happen here, and it's going to happen soon." One who attended the meeting later calls the evidence that "something spectacular" is being planned by al-Qaeda "very gripping." [TIME, 8/4/02; WASHINGTON POST, 5/17/02] Clarke directs every counterterrorist office to cancel vacations, defer non-vital travel, put off scheduled exercises, and place domestic rapid-response teams on much shorter alert. By early August, all of these emergency measures are no longer in effect. [CNN, 3/02; WASHINGTON POST, 5/17/02] The FAA issues general threat advisories, but neither the FAA's top administrator nor Transportation Secretary

Norman Mineta is aware of an increased threat level. [*NEW YORK TIMES*, 4/18/04] Clarke says rhetorically that he wants to know if a sparrow has fallen from a tree. A senior FBI official attends the meeting and promises a redoubling of efforts. However, just five days later, when FBI agent Ken Williams sends off his memo speculating that al-Qaeda may be training operatives as pilots in the U.S., the FBI fails to share this information with any other agency. [*WASHINGTON POST*, 5/17/02; *AGAINST ALL ENEMIES*, BY RICHARD CLARKE, 3/04, PP. 236–37]

July 6, 2001: Clarke Briefs Senior
Security Officials on al-Qaeda Threat

One day after heading a meeting on al-Qaeda with the Counterterrorism and Security Group, counterterrorism "tsar" Richard Clarke heads a similar meeting at the White House with senior security officials at the FAA, Immigration, Secret Service, Coast Guard, Customs, and other agencies. The CIA and FBI give briefings on the growing al-Qaeda threat. The CIA says al-Qaeda members "believe the upcoming attack will be 'spectacular,' qualitatively different from anything they have done to date." [9/11 COMMISSION REPORT, 3/24/04 (D)] Then Clarke says, "You've just heard that CIA thinks al-Qaeda is planning a major attack on us. So do I. You heard CIA say it would probably be in Israel or Saudi Arabia. Maybe. But maybe it will be here. Just because there is no evidence that says that it will be here, does not mean it will be overseas. They may try to hit us at home. You have to assume that is what they are going to do. Cancel summer vacations, schedule overtime, have your terrorist reaction teams on alert to move fast. Tell me, tell each other, about anything unusual." [*AGAINST ALL ENEMIES*, BY RICHARD CLARKE, 3/04, P. 236]

July 12, 2001: Ashcroft Reputedly Uninterested in Terrorism

On July 5, the CIA briefs Attorney General Ashcroft on the al-Qaeda threat, warning that a significant terrorist attack is imminent, and a strike could occur at any time. [9/11 COMMISSION REPORT, 4/13/04 (B)] On this day, acting FBI Director Tom Pickard briefs Ashcroft about the terror threat inside the U.S. Pickard later swears under oath that Ashcroft tells him, "he did not want to hear about this anymore." Ashcroft, also under oath, later categorically denies the allegation, saying, "I did never speak to him saying that I didn't want to hear about terrorism." However, Ruben Garcia, head of the Criminal Division, and another senior FBI official corroborate Pickard's account. Ashcroft's account is supported by his top aide, but another official in Ashcroft's office who could also support Ashcroft's account says he cannot remember what happened. Pickard briefs Ashcroft on terrorism four more times that summer, but he never mentions al-Qaeda to Ashcroft again before 9/11. [MSNBC, 6/22/04] Pickard later makes an appeal to Ashcroft for more counterterrorism funding; Ashcroft rejects the appeal on September 10, 2001. [9/11 COMMISSION REPORT, 4/13/04] Pickard later says, "Before September 11th, I couldn't get half an hour on terrorism with Ashcroft. He was only interested in three things: guns, drugs, and civil rights." [*THE CELL*, BY JOHN MILLER, MICHAEL STONE, AND CHRIS MITCHELL, 8/02, P. 293]

July 26, 2001: Ashcroft Stops Flying Commercial Airlines; Refuses to Explain Why

CBS News reports that Attorney General Ashcroft has stopped flying commercial airlines due to a threat assessment, but "neither the FBI nor the Justice Department . . . would identify [to CBS] what the threat was, when it was detected or who made it." [CBS NEWS, 7/26/01] "Ashcroft demonstrated an amazing lack of curiosity when asked if he knew anything about the threat. 'Frankly, I don't,' he told reporters." [SAN FRANCISCO CHRONICLE, 6/3/02] It is later reported that he stopped flying in July based on threat assessments made on May 8 and June 19. In May 2002, it is claimed the threat assessment had nothing to do with al-Qaeda, but Ashcroft walked out of his office rather than answer questions about it. [ASSOCIATED PRESS, 5/16/02] The *San Francisco Chronicle* concludes, "The FBI obviously knew something was in the wind. . . . The FBI did advise Ashcroft to stay off commercial aircraft. The rest of us just had to take our chances." [SAN FRANCISCO CHRONICLE, 6/3/02] CBS's Dan Rather later asks of this warning: "Why wasn't it shared with the public at large?" [WASHINGTON POST, 5/27/02]

Attorney General John Ashcroft

July 27, 2001: Rice Briefed on Terrorist Threats, Advised to Keep Ready

Counterterrorism "tsar" Richard Clarke reports to National Security Adviser Rice and her deputy Steve Hadley that the spike in intelligence indicating a near-term attack appears to have ceased, but he urges them to keep readiness high. Intelligence indicates that an attack has been postponed for a few months. [9/11 COMMISSION REPORT, 3/24/04 (D)] In early August, CIA Director Tenet also reports that intelligence suggests that whatever terrorist activity might have been originally planned has been delayed. [9/11 COMMISSION, 3/24/04 (C)]

August 2001: Bush Administration Rejects Plan to Capture Al-Zawahiri

The U.S. receives intelligence that bin Laden's right-hand man, Ayman al-Zawahiri, is receiving medical treatment at a clinic in San'a, Yemen. However, the Bush administration rejects a plan to capture him, as officials are not 100 percent sure the patient is al-Zawahiri. Officials later regret the missed opportunity. [ABC NEWS, 2/20/02]

Ayman al-Zawahiri

August 4, 2001: Nothing New in Bush Letter to Pakistani President

President Bush sends a letter to Pakistani President Pervez Musharraf, warning him about supporting the Taliban. However, the tone is similar to past requests dating to the Clinton administration. There had been some discussion that U.S. policy toward Pakistan should change. For instance, at the end of June, counterterrorism "tsar" Richard Clarke "urged that the United States [should] think about what it would do after the next attack, and then take that position with Pakistan now, before the attack." Deputy Secretary of State Richard Armitage later acknowledges that a new approach to Pakistan is not yet implemented by 9/11. [9/11 COMMISSION REPORT, 3/24/04]

August 4–30, 2001: Bush Nearly Sets Record
for Longest Presidential Vacation

President Bush spends most of August 2001 at his Crawford, Texas, ranch, nearly setting a record for the longest presidential vacation. While it is billed a "working vacation," news organizations report that Bush is doing "nothing much" aside from his regular daily intelligence briefings. [ABC NEWS, 8/3/01; WASHINGTON POST, 8/7/01; SALON, 8/29/01] One such unusually long briefing at the start of his trip is a warning that bin Laden is planning to attack in the U.S., but Bush spends the rest of that day fishing. By the end of his trip, Bush has spent 42 percent of his presidency at vacation spots or en route. [WASHINGTON POST, 8/7/01] At the time, a poll shows that 55 percent of Americans say Bush is taking too much time off. [USA TODAY, 8/7/01] Vice President Cheney also spends the entire month in a remote location in Wyoming. [JACKSON HOLE NEWS AND GUIDE, 8/15/01]

August 6, 2001: Bush Briefing Titled
"Bin Laden Determined to Strike in U.S."

President Bush receives a classified intelligence briefing at his Crawford, Texas ranch indicating that bin Laden might be planning to hijack commercial airliners. The memo provided to him is titled "Bin

Laden Determined to Strike in U.S." The entire memo focuses on the possibility of terrorist attacks inside the U.S. [NEWSWEEK, 5/27/02; NEW YORK TIMES, 5/15/02] Incredibly, the *New York Times* later reports that Bush "[breaks] off from work early and [spends] most of the day fishing." [NEW YORK TIMES, 5/25/02] The existence of this memo is kept secret, until it is leaked in May 2002, causing a storm of controversy. While National Security Adviser Rice claims the memo is only one and a half pages long; other accounts state it is 11 1/2 pages instead of the usual two or three. [NEWSWEEK, 5/27/02; NEW YORK TIMES, 5/15/02; DIE ZEIT, 10/1/02] She disingenuously asserts that, "It was an analytic report that talked about [bin Laden]'s methods of operation, talked about what he had done historically, in 1997, in 1998. . . . I want to reiterate, it was not a warning. There was no specific time, place, or method mentioned." [WHITE HOUSE, 5/16/02] A page and a half of the contents are released on April 10, 2004, after Rice testifies before the 9/11 Commission. [WASHINGTON POST, 4/10/04] Rice testifies that the memo is mostly historic regarding bin Laden's previous activities, and she says it contains no specific information that would have prevented an attack. The memo, as released, includes at least the following information:

President Bush at his Crawford, Texas, ranch on August 6, 2001

• Bin Laden "has wanted to conduct attacks inside the U.S." since 1997. "Members of al-Qaeda, including some U.S. citizens, [have] resided in or travelled to the U.S. for years and the group apparently maintain[s] a support structure" in the U.S.

- "A clandestine source said in 1998 that a bin Laden cell was recruiting Muslim-American youth for attacks."

- According to an Egyptian Islamic Jihad operative in 1998, "Bin Laden was planning . . . to exploit the operative's access to the U.S. to mount a terrorist strike."

- A discussion of the arrest of Ahmed Ressam (who plotted to attack the Los Angeles Airport at the turn of the millennium) and the 1998 U.S. embassy bombings.

- Uncorroborated information obtained in 1998 "that Bin Laden wanted to hijack a U.S. aircraft to gain the release" of Sheikh Omar Abdul-Rahman and other U.S.-held extremists.

- The CIA and FBI are investigating a call to the U.S. Embassy in the United Arab Emirates "saying that a group of Bin Laden supporters was in the U.S. planning attacks with explosives."

- "FBI information since [1998] indicates patterns of suspicious activity in [the U.S.] consistent with preparations for hijackings or other types of attacks, including recent surveillance of federal buildings in New York" by two Yemeni men who were detained on May 30, 2001.

- "The FBI is conducting approximately 70 full field investigations in the U.S. that it considered bin Laden-related." [WASHINGTON POST, 04/10/04 (B); 9/11 CONGRESSIONAL INQUIRY, 9/18/02; 9/11 CONGRESSIONAL INQUIRY, 7/24/03]

(The 9/11 Congressional inquiry calls it "a closely held intelligence report for senior government officials" presented in early August 2001.) [9/11 CONGRESSIONAL INQUIRY, 7/24/03]

August 6, 2001 (B): Perle's Concern About Iraq, North Korea, and Iran Before 9/11 Becomes "Axis of Evil" Afterward

Richard Perle, head of the Defense Policy Board and foreign policy adviser to Bush, is asked about new challenges now that the Cold War is over. He cites three: "We're concerned about Saddam Hussein, we're concerned about the North Koreans, about some future Iranian government that may have the weapon they're now trying so hard to acquire . . ." [AUSTRALIAN BROADCASTING CORP., 8/6/01] Note that these three nations are the same three named in Bush's famous January 2002 "axis of evil" speech. [CNN, 1/29/02]

August 6, 2001 (C): Justice Department Reaffirms "Wall" Policy

In testimony before the 9/11 Commission, Attorney General Ashcroft complains, "[T]he single greatest structural cause for September 11 was the wall that segregated criminal investigators and intelligence agents." However, on this day, Ashcroft's Assistant Attorney General, Larry Thompson, writes a memo reaffirming the policy that is later criticized as this "wall." [9/11 COMMISSION THOMPSON TESTIMONY, 12/8/03; WASHINGTON POST, 4/18/04]

August 7, 2001: Version of Bush's al-Qaeda
Briefing Is Incomplete, Poorly Distributed

One day after Bush receives a Presidential Daily Briefing entitled, "Bin Laden Determined to Strike in U.S.," a version of the same material is given to other top government officials. However, this Senior Executive Intelligence Brief or SEIB, does not contain the most important information from Bush's briefing. It does not mention that there are 70 FBI investigations into possible al-Qaeda activity, does not mention a May 2001 threat of U.S.-based explosives attacks, and does not mention FBI concerns about recent surveillance of buildings in New York City. Typically, this type of memo "goes to scores of Cabinet-agency officials from the assistant secretary level up and does not include raw intelligence or sensitive information about ongoing law enforcement matters," according to the Associated Press. Some members of Congress later express concern that policy makers were given an incomplete view of the terrorist threat. [ASSOCIATED PRESS, 4/13/04 (B)]

August 19, 2001: FBI's Best al-Qaeda Expert Under
Investigation for Trivial Issue, His Resignation Soon Follows

The *New York Times* reports that counterterrorism expert John O'Neill is under investigation for an incident involving a missing briefcase. [NEW YORK TIMES, 8/19/01] In July 2000, he misplaced a briefcase con-

John O'Neill

taining important classified information, but it was found a couple of hours later still locked and untouched. Why such a trivial issue would come up over a year later and be published in the *New York Times* seems entirely due to politics. Says the *New Yorker*, "The leak seemed to be timed to destroy O'Neill's chance of being confirmed for [a National Security Council] job," and force him into retirement. A high-ranking colleague says the leak was "somebody being pretty vicious to John." [NEW YORKER, 1/14/02] John O'Neill suspects his enemy Tom Pickard, then interim director of the FBI, orchestrated the article. [PBS FRONTLINE, 10/3/02 (B)] The *New Yorker* later speculates that with the retirement of FBI Director Freeh in June, it appears O'Neill lost his friends in high places, and the new FBI Director wanted him replaced with a Bush ally. [NEW YORKER, 1/14/02] O'Neill resigns a few days later.

August 22, 2001: O'Neill Quits FBI in Frustration; Misses Important Warnings

Counterterrorism expert John O'Neill resigns from the FBI. He says it is partly because of the recent power play against him, but also because of repeated obstruction of his investigations into al-Qaeda. [NEW YORKER, 1/14/02] In his last act, he signs papers ordering FBI investigators back to Yemen to resume the USS *Cole* investigation, now that Barbara Bodine is leaving as ambassador (they arrive a couple days before 9/11). He never hears the CIA warning about hijackers Nawaf Alhazmi and Khalid Almihdhar sent out just one day later. Because he fell out of favor a few months earlier, he also is never told about Ken Williams' flight school memo from July 10, 2001, or the arrest of Zacarias Moussaoui on August 15, 2001 [PBS FRONTLINE, 10/3/02 (D)]; nor did he attend a June meeting when the CIA revealed some of what it knew about Alhazmi and Almihdhar. [PBS *FRONTLINE*, 10/3/02] The FBI New York office eventually hears of Walid Arkeh's warning that the WTC would be attacked, but presumably not in time for O'Neill to hear it.

August 23, 2001: O'Neill Begins Job as Head of Security at the WTC

John O'Neill begins his new job as head of security at the WTC. [*NEW YORKER*, 1/14/02] A friend says to him, "Well, that will be an easy job. They're not going to bomb that place again." O'Neill replies, "Well actually they've always wanted to finish that job. I think they're going to try again." On September 10, he moves into his new office on the 34th floor of the North Tower. That night, he tells colleague Jerry Hauer, "We're due for something big. I don't like the way things are lining up in Afghanistan" (a probable reference to the assassination of Afghan leader Ahmed Shah Massoud the day before). O'Neill will be killed in the 9/11 attack. [PBS *FRONTLINE*, 10/3/02 (D)]

August 29, 2001: Bush Vows Security Is His First Responsibility

President Bush says, "We recognize it's a dangerous world. I know this nation still has enemies, and we cannot expect them to be idle. And that's why security is my first responsibility. And I will not permit any course that leaves America undefended." [9/11 CONGRESSIONAL INQUIRY, 9/18/02]

September 4, 2001: Clarke Memo: Imagine Hundreds of Dead Due to Government Inaction

Hours before the only significant Bush administration Cabinet-level meeting on terrorism before 9/11, counterterrorism "tsar" Richard Clarke writes a critical memo to National Security Adviser Rice. He criticizes the Defense Department for reluctance to use force against al-Qaeda and the CIA for impeding the deployment of unmanned Predator drones to hunt for bin Laden. According to the *Washington Post*, the memo urges "officials to imagine a day when hundreds of Americans lay dead from a terrorist attack and ask themselves what more they could have done." [*WASHINGTON POST*, 3/24/04; *WASHINGTON POST*, 3/25/04 (B); 9/11 COMMISSION REPORT, 3/24/04 (D)]

September 4, 2001 (B): Cabinet-Rank Advisers Discuss Terrorism, Approve Revised Version of Clarke's Eight Month-Old-Plan

President Bush's cabinet-rank advisers discuss terrorism for the second of only two times before 9/11. [*WASHINGTON POST*, 5/17/02] National Security Adviser Rice chairs the meeting; neither President Bush nor Vice President Cheney attends. Counterterrorism "tsar" Richard Clarke later says that in this meeting, he and CIA Director Tenet speak passionately about the al-Qaeda threat. No one disagrees that the threat is serious. Secretary of State Powell outlines a plan to put pressure on Pakistan to stop supporting al-Qaeda. Defense Secretary Rumsfeld appears to be more interested in Iraq. The only debate is over whether to fly the armed Predator drone over Afghanistan to attack al-Qaeda. [*AGAINST ALL ENEMIES*, BY RICHARD CLARKE, 3/04, PP. 237–38] Clarke's earlier plans to "roll back" al-Qaeda have been discussed and honed in many meetings and are now presented as a formal National Security Presidential Directive. The directive is "apparently" approved, though the process of turning it into official policy is still not done. [9/11 COMMISSION REPORT, 3/24/04 (D)] There is later disagreement over just how different the directive presented is from Clarke's earlier plans. For instance, some claim the directive aims not just to "roll back" al-Qaeda, but also to "eliminate" it altogether. [TIME, 8/4/02] However, Clarke notes that even though he wanted to use the word "eliminate," the approved directive merely aims to "significantly erode" al-Qaeda. The word "eliminate" is only added after 9/11. [WASHINGTON POST, 3/25/04 (B)] The *Washington Post* notes that the directive approved on this day "did not differ substantially from Clinton's policy." [WASHINGTON POST, 3/27/04] *Time* magazine later comments, "The fight against terrorism was one of the casualties of the transition, as Washington spent eight months going over and over a document whose outline had long been clear." [TIME, 8/4/02] The primary change from Clarke's draft original draft is that the approved plan calls for more direct financial and logistical support to the Northern Alliance and other anti-Taliban groups. The plan also calls for drafting plans for possible U.S. military involvement, "but those differences were largely theoretical; administration officials told the [9/11 Commission's] investigators that the plan's overall timeline was at least three years, and it did not include firm deadlines, military plans, or significant funding at the time of the September 11, 2001, attacks." [WASHINGTON POST, 3/27/04; REUTERS, 4/2/04]

September 4, 2001 (C): Debate Heats Up over Predator Drone; Decision Again Delayed

Attendees to an important Cabinet-level meeting on terrorism have a heated debate over what to do with the Predator drone. Counterterrorism "tsar" Richard Clarke has been repeatedly pushing for the use of the Predator over Afghanistan (in either armed or unarmed versions), and he again argues for its immediate use. Everyone agrees that the armed Predator capability is needed, but there are disputes over who will manage and/or pay for it. CIA Director Tenet says his agency will operate the armed Predator "over my dead body." [WASHINGTON POST, 10/2/02] Clarke recalls, "The Air Force said it wasn't their job to fly planes to collect intelligence. No one around the table seemed to have a can-do attitude. Everyone seemed to have an excuse." [NEW YORKER, 7/28/03] National Security Adviser Rice concludes that the armed Predator is not ready (even though it had been proven in tests during the summer), but she also presses Tenet to reconsider his opposition to immediately resume reconnaissance flights, suspended since September the year before. After the meeting, Tenet agrees to proceed with such flights. [9/11 COMMISSION, 3/24/04 (C), 9/11 COMMISSION REPORT, 3/24/04 (D)] The armed Predator is activated just days after 9/11, showing that it was ready to be used after all. [ASSOCIATED PRESS, 6/25/03]

September 6, 2001: WTC Security Scaled Back in Week and Day Before 9/11 Attack

Security personnel at the WTC are working extra-long shifts because of numerous phone threats. However, on this day bomb-sniffing dogs are abruptly removed. Security further drops right before 9/11. WTC guard Daria Coard says in an interview later on the day of 9/11: "Today was the first day there was not the extra security." [NEWSDAY, 9/12/01 (B)] Janitor William Rodriguez later claims that he saw hijacker Mohand Alshehri inside the WTC in June 2001. [NEW YORK DAILY NEWS, 6/15/04]

September 6, 2001 (B): Senator Hart Sees No "Sense of Urgency" from Rice on Terrorism

Former Senator Gary Hart (D), one of the two co-chairs of a comprehensive, bipartisan report on terrorism released in January 2001, meets with National Security Adviser Rice to see if the Bush administration is implementing the report's recommendations. After giving a grave warning, he recalls her response: "She didn't seem to feel a terrible sense of urgency. Her response was simply 'I'll talk to the vice president about it.' . . . Even at this late date, nothing was being done inside the White House." [SALON, 4/2/04]

September 9, 2001: Bush's First Budget Has Gaps for Counterterrorism Funding

President Bush's first budget calls for $13.6 billion on counterterrorism programs, compared with $12.8 billion in President Clinton's last budget and $2 billion ten years earlier. However, there are gaps between what military commanders say they need to combat terrorists and what they are slated to receive. The Senate Armed Services Committee tries to fill those gaps with $600 million diverted from ballistic missile defense, but on this day Defense Secretary Rumsfeld threatens to urge a veto if the Senate proceeds to shift the money. [KNIGHT RIDDER 9/27/01; TIME 8/4/02; WASHINGTON POST 1/20/02]

September 9–11, 2001: NORAD Begins Northern Vigilance Fighter Exercise

NORAD begins Operation Northern Vigilance. This military exercise deploys fighters to Alaska and Northern Canada to monitor a Russian air force exercise in the Russian arctic, apparently focused in the Barents Sea above Norway. This operation is still ongoing on 9/11. It is not clear how many fighters have been diverted from the continental U.S. on 9/11 to take part in this exercise. [NORAD, 9/9/01; CANADIAN BROADCASTING CORP., 11/27/01] On the morning of 9/11, the Russians are asked to cancel their exercise and do so. Northern Vigilance is then canceled around 9 A.M. [TORONTO STAR, 12/9/01; NATIONAL POST, 10/19/02]

September 10, 2001: Deputies Still Putting Final Touches on Three-Year Plan to Stop al-Qaeda

Another deputies meeting further considers policy toward Afghanistan and Pakistan, and makes further revisions to the National Security Presidential Directive regarding al-Qaeda. [9/11 COMMISSION REPORT, 3/24/04 (D)] By the end of the meeting, a formal, three-phase strategy is agreed upon. An envoy is to go to Afghanistan and give the Taliban another chance to expel bin Laden. If this fails, more pressure will be put on the Taliban, including more support for the Northern Alliance and other groups. If the Taliban still refuse to change, the U.S. will try to overthrow the Taliban through more direct action. The timeframe for this strategy is about three years. [9/11 COMMISSION REPORT, 3/24/04] CIA Director Tenet is formally tasked to draw up new authorities for the covert action program envisioned, and request funding to implement it. [9/11 COMMISSION, 3/24/04 (C)] The directive is then to be sent to National Security Adviser Rice for approval. President Bush is apparently aware of the directive and prepared to sign it (though he hasn't attended any of the meetings about it), but he does not sign it until October. [MSNBC, 5/16/02; *LOS ANGE-LES TIMES*, 5/18/02; *WASHINGTON POST*, 4/1/04]

September 10, 2001 (B): Rumsfeld Announces Defense Department "Cannot Track $2.3 Trillion in Transactions"

In a speech to the Department of Defense, Defense Secretary Rumsfeld announces that the Department of Defense "cannot track $2.3 trillion in transactions." CBS later calculates that 25 percent of the yearly defense budget is unaccounted for, and quotes a long-time defense budget analyst: "[Their] numbers are pie in the sky. The books are cooked routinely year after year." Coverage of this rather shocking story is nearly nonexistent given the events of the next day. [DEFENSE DEPARTMENT, 9/10/01; CBS NEWS, 1/29/02]

September 10, 2001 (C): Ashcroft Opposes Counterterrorism Funding

Attorney General Ashcroft rejects a proposed $58 million increase in financing for the bureau's counterterrorism programs. On the same day, he sends a request for budget increases to the White House. It covers 68 programs—but none of them relate to counterterrorism. He also sends a memorandum to his heads of departments, stating his seven priorities—none of them relate to counterterrorism. [*NEW YORK TIMES*, 6/1/02; *GUARDIAN*, 5/21/02] He further proposes cutting a program that gives state and local counterterrorism grants for equipment like radios and preparedness training from $109 million to $44 million. Yet Ashcroft stopped flying public airplanes in July due to an as yet undisclosed terrorist threat, and in a July speech he proclaimed, "Our No. 1 priority is the prevention of terrorist attacks." [*NEW YORK TIMES*, 2/28/02]

September 10, 2001 (D): Cheney's Domestic Terrorism Task Force Finally Beginning to Hire Staff

The domestic terrorism task force announced by President Bush and Vice President Cheney in May 2001 is just gearing up. Cheney appointed Admiral Steve Abbot to lead the task force in June, but he does not receive his White House security pass until now. Abbot has only hired two staffers and been working full time for a few days prior to 9/11. The task force was to have reported to Congress by October 1, 2001, a date they could not have met. [*CONGRESSIONAL QUARTERLY*, 4/15/04; *NEW YORK TIMES*, 12/27/01]

September 10, 2001 (E): Review of Counterterrorism Legislation May Take Six Months, Says Cheney Aide

Senator Dianne Feinstein (D), who, with Senator Jon Kyl (R), has sent a copy of draft legislation on counterterrorism and national defense to Vice President Cheney's office on July 20, is told by Cheney's top aide on this day "that it might be another six months before he would be able to review the material." [*NEWSWEEK*, 5/27/02]

September 10, 2001 (F): U.S. Has Only 32
Air Marshals—and None on Domestic Flights

The number of U.S. air marshals (specially trained, plainclothes armed federal agents deployed on airliners) has shrunk from about 2,000 during the Cold War to 32 by this date. None are deployed on domestic flights. The number is later increased to about 2,000, but it would take about 120,000 marshals at a cost of $10 billion a year to protect all daily flights to, from, or within the U.S. [LOS ANGELES TIMES, 1/14/02]

Before September 11, 2001: CIA, FBI Lack Counterterrorism Resources, and Focus

Just prior to 9/11, the CIA and FBI do not have enough staff working on al-Qaeda. Only 17 to 19 people are working in the FBI's special unit focusing on bin Laden and al-Qaeda. [9/11 CONGRESSIONAL INQUIRY, 9/18/02] The FBI has a $4.3 billion anti-terrorism budget, but of its 27,000 employees, just 153 are devoted to terrorism analysis. [SYDNEY MORNING HERALD, 6/8/02] The FBI's "analytic expertise has been 'gutted' by transfers to operational units" and only one strategic analyst is assigned full time to al-Qaeda. The FBI office in New York is very aware of the threat from bin Laden, but many branch offices remain largely unaware. [9/11 CONGRESSIONAL INQUIRY, 9/18/02] A senior FBI official later tells Congress that there are fewer FBI agents assigned to counterterrorism on this day than in August 1998, when the U.S. embassy bombings in Africa made bin Laden a household name. [NEW YORK TIMES, 9/22/02] The CIA has only about 35 to 40 people assigned to their special bin Laden unit. It has five strategic analysts working full time on al-Qaeda. [9/11 CONGRESSIONAL INQUIRY, 9/18/02] The CIA and FBI later complain that some of these figures are misleading. [NEW YORK TIMES, 9/18/02] "Individuals in both the CIA and FBI units . . . reported being seriously overwhelmed by the volume of information and workload prior to September 11, 2001." Despite numerous warnings that planes could be used as weapons, such a possibility was never studied, and a congressional report later blames lack of staff as a major reason for this. [9/11 CONGRESSIONAL INQUIRY, 9/18/02] Senator Patrick Leahy (D) also notes, "Between the Department of Justice and the FBI, they had a whole task force working on finding a couple of houses of prostitution in New Orleans. They had one on al-Qaeda." [CBS NEWS, 9/25/02]

Before September 11, 2001 (B): Key Counterterrorism Position Still Unfilled

The position of Deputy Secretary for Special Operations and Low-Intensity Conflict, the Defense Department post traditionally dealing the most with counterterrorism, still has not been filled since being vacated in January 2001 when Bush became president. Aides to Defense Secretary Rumsfeld later tell the 9/11 Commission that "the new [Defense Department] team was focused on other issues" and not counterterrorism. [NEWSWEEK, 3/24/04]

September 11, 2001: The 9/11 Attack: 3,000 Die in New York City and Washington, D.C.

The 9/11 attack: Four planes are hijacked, two crash into the WTC, one into the Pentagon, and one crashes into the Pennsylvania countryside. Nearly 3,000 people are killed. According to officials, the entire U.S. is defended by only 14 fighters (two planes each in seven military bases). [DALLAS MORNING NEWS,

9/16/01], and none of the fighters are at "bases close to two obvious terrorist targets—Washington, D.C., and New York City." A defense official says, "I don't think any of us envisioned an internal air threat by big aircraft. I don't know of anybody that ever thought through that." [NEWSDAY, 9/23/01]

September 11, 2001 (B): Planned Rice Speech
on Threats Contains No Mention of al-Qaeda

National Security Adviser Rice is scheduled to deliver a speech claiming to address "the threats and problems of today and the day after, not the world of yesterday." The speech is never given due to the 9/11 attacks earlier in the day, but the text is later leaked to the media. The *Washington Post* calls the speech "telling insight into the administration's thinking" because it promotes missile defense and contains no mention of al-Qaeda, bin Laden or Islamic extremist groups. The only mention of terrorism is in the context of the danger of rogue nations such as Iraq. In fact, there are almost no public mentions of bin Laden or al-Qaeda by Bush or other top Bush administration officials before 9/11, and the focus instead is on missile defense. [WASHINGTON POST, 4/1/04; WASHINGTON POST, 4/1/04 (D)]

4.

The Hunt for bin Laden

There are many questions about Osama bin Laden. Who is he exactly? How did he become a terrorist? Why has he become so popular in the Muslim world? These questions are very important, but largely beyond the scope of this book. As this book focuses on the failures that led to 9/11, this chapter focuses on U.S. intelligence failures surrounding bin Laden before 9/11.

Shortly after the Soviet-Afghan War began in 1979, the Saudi-born bin Laden traveled to Afghanistan to join the mujahedeen resistance. There he became well-known in the Islamic world by funding and fighting alongside the Afghans. He left Afghanistan in the late 1980s, after the resistance successfully drove the Soviets from the country. In 1996 he returned, formed a strong alliance with the fundamentalist Taliban ruling Afghanistan, and developed the global terrorist organization al-Qaeda.

Since at least as early as August 1998, when al-Qaeda blew up two U.S. embassies in Africa, the U.S. has sought—but failed—to kill bin Laden. The fact that he survived a missile strike only increased his stature in the Islamist world. Subsequent attempts, to the extent they were made, failed as well. This chapter focuses on U.S. efforts to target bin Laden prior to 9/11, and to explore why these efforts were not successful.

It is not known what impact killing bin Laden would have had upon the 9/11 plot. Anecdotal evidence from interrogations of 9/11 mastermind Khalid Shaikh Mohammed suggests that bin Laden championed the 9/11 attacks when other top al-Qaeda leaders were against it. Even in mid-2001 he gave the go-ahead over considerable internal opposition, based on Taliban leader Mullah Omar's wish not to be targeted after such an attack. [9/11 COMMISSION REPORT, 6/16/04 (B)] The validity of such information is questionable, especially since Mohammed was tortured and may have wanted to exaggerate bin Laden's role in the attack planning. But it at least suggests the possibility that the 9/11 attacks could have been stopped had bin Laden been killed as late as 2001.

Many questions remain unanswered. Did the CIA or military leadership drag its feet in the hunt for bin Laden, and if so, why? Why did the U.S. not develop closer, stronger ties to the Northern Alliance, a natural enemy of the Taliban and their close ally, al-Qaeda? Why did U.S. intelligence agents fail to penetrate al-Qaeda in Afghanistan? Was there a change in attitude or approach to al-Qaeda from the Clinton to the Bush administrations? The consequences of this failure to date have already been tragic. The potential future consequences are unimaginable.

September 1994: ISI Creates the Taliban, Assists Afghanistan Conquest

See chapter 12.

1997: CIA Re-opens Afghanistan Operations

Special CIA paramilitary teams enter Afghanistan again in 1997. [WASHINGTON POST, 11/18/01] (The CIA's anti-Soviet covert operations officially ended in January 1992. [GHOST WARS, BY STEVE COLL, 2/04, P. 233]) Around 1998 there will be a push to recruit more agents capable of operating or traveling in Afghanistan. Many locals are recruited, including some Taliban military leaders. However, apparently none is close to bin Laden. This problem is not fixed in succeeding years. [WASHINGTON POST, 2/22/04; 9/11 COMMISSION, 3/24/04 (C)]

May 26, 1997: Taliban Government Is Officially Recognized by Saudis

The Saudi government becomes the first country to extend formal recognition of the Taliban government of Afghanistan. Pakistan and the United Arab Emirates will follow suit. On 9/11, these three countries are the only countries that officially recognize the Taliban. [9/11 CONGRESSIONAL INQUIRY, 7/24/03 (B)]

1998: U.S. and Uzbekistan Conduct Joint Operations Against Taliban

Beginning in 1998, if not before, Uzbekistan and the U.S. conduct joint covert operations against Afghanistan's Taliban regime and bin Laden. [TIMES OF INDIA, 10/14/01; WASHINGTON POST, 10/14/01]

June 1998: U.S. Develops Plan to Capture bin Laden

In 1997 and early 1998, the U.S. had developed a plan to capture bin Laden in Afghanistan. A CIA-owned aircraft was stationed in a nearby country, ready to land on a remote landing strip long enough to pick him up. However, problems with having to hold bin Laden too long in Afghanistan made the operation unlikely. The plan morphs into using a team of Afghan informants to kidnap bin Laden from inside his heavily defended farm. In this month, the plan is given to CIA Director Tenet for approval, but he rejects it without showing it to President Clinton. It is thought unlikely to succeed and the Afghan allies are considered unreliable. [AGAINST ALL ENEMIES, BY RICHARD CLARKE, 3/04, PP. 220–21; WASHINGTON POST, 2/22/04] It is later speculated that the airstrip used for these purposes is occupied and used as a base of operations early in the post-9/11 Afghan war. [WASHINGTON POST, 12/19/01]

August 7, 1998: Two U.S. Embassies in Africa Bombed

See chapter 1.

Mid-August 1998: Clinton Authorizes Assassination of bin Laden

President Clinton signs a Memorandum of Notification, which authorizes the CIA to plan the capture of bin Laden using force. The CIA draws up detailed profiles of bin Laden's daily routines, where he sleeps, and his travel arrangements. The assassination never happens, supposedly because of inadequate intelligence. However, as one officer later says, "you can keep setting the bar higher and higher, so that nothing ever gets done." An officer who helped draw up the plans says, "We were ready to move" but "we were not allowed to do it because of this stubborn policy of risk avoidance . . . It is a disgrace." [PHILADELPHIA INQUIRER, 9/16/01] Additional memoranda quickly follow that authorize the assassination of up to ten other al-Qaeda leaders, and authorize the shooting down of private aircraft containing bin Laden. [WASHINGTON POST, 12/19/01] However, "These directives [lead] to nothing." [NEW YORKER, 7/28/03]

Mid-August 1998–2000: U.S. Submarines Ready to Attack bin Laden

Within days of the U.S. African embassy bombings, the U.S. permanently stations two submarines, reportedly in the Indian Ocean, ready to hit al-Qaeda with cruise missiles on short notice. Missiles are fired from these subs later in the month in a failed attempt to assassinate bin Laden. Six to ten hours' advance warning is now needed to review the decision, program the cruise missiles, and have them reach their target. However, in the rare opportunities when the possibility of attacking bin Laden occurs, CIA Director Tenet says the information is not reliable enough and the attack cannot go forward. [WASHINGTON POST, 12/19/01; NEW YORK TIMES, 12/30/01] At some point in 2000, the submarines are withdrawn, apparently because the Navy wants to use them for other purposes. Therefore, when the unmanned Predator spy plane flies over Afghanistan in late 2000 and identifies bin Laden, there is no way to capitalize on that opportunity. [AGAINST ALL ENEMIES, BY RICHARD CLARKE, 3/04, PP. 220–21] The Bush administration fails to resume the submarine patrol. Lacking any means to attack bin Laden, military plans to strike at him are no longer updated after March 2001. [9/11 COMMISSION REPORT, 3/24/04 (B)]

August 20, 1998: U.S. Fires on al-Qaeda's
Afghan Training Camps, Sudanese Facility

The U.S. fires 66 missiles at six training camps in Afghanistan and 13 missiles at a pharmaceutical factory in Khartoum, Sudan in retaliation for the U.S. embassy bombings. [WASHINGTON POST, 10/3/01 (C)] The U.S. insists the attacks are aimed at terrorists "not supported by any state," despite obvious evidence to the contrary. About 30 people are killed in the Afghanistan attacks, but no important al-Qaeda figures die. [OBSERVER, 8/23/98; NEW YORKER, 1/24/00] A soil sample is said to show that the pharmaceutical factory was producing chemical weapons, but many doubts about the sample later arise. Some U.S. officials later admit they had no direct link between bin Laden and the factory at the time of the attack. [NEW YORK TIMES, 9/21/98; NEW YORKER, 10/12/98] The U.S. later unfreezes the bank accounts of the nominal factory owner and takes other concilia-tory actions, but admits no wrongdoing. It is later learned that of the six camps targeted in Afghanistan, only four were hit, and of those, only one had connections to bin Laden. Two of the camps belonged to the ISI, and five ISI officers and some twenty trainees are killed. Clinton declares that the missiles were aimed at a "gather-ing of key terrorist leaders," but it is later revealed that the referenced meet-ing took place a month earlier, in Pakistan. [OBSERVER 8/23/98, NEW YORKER, 1/24/00] Counterterrorism "tsar" Richard Clarke claims he was promised by the Navy that it would fire their missiles from below the ocean surface. However, in

El Shifa Plant in Sudan

fact, many destroyers fired their missiles from the surface. [AGAINST ALL ENEMIES, BY RICHARD CLARKE, 3/04, PP. 188–89] He adds, "not only did they use surface ships—they brought additional ones in, because every captain wants to be able to say he fired the cruise missile." [NEW YORKER, 7/28/03] As a result, the ISI (or bin Laden sympathiz-ers within) had many hours to alert bin Laden. Clarke says he believes that "if the [ISI] wanted to capture bin Laden or tell us where he was, they could have done so with little effort. They did not cooperate with us because ISI saw al-Qaeda as helpful in pressuring India, particularly in Kashmir." [AGAINST ALL ENEMIES, BY RICHARD CLARKE, 3/04, PP. 188–89]

August 24, 1998: Bombed Training Camps Were Built by U.S. and Allies

The *New York Times* reports that the training camps recently attacked by the U.S. in Afghanistan were built by the U.S. and its allies, years before. The U.S. and Saudi Arabia gave the Afghans between $6 billion and $40 billion to fight the Soviets in the 1980s. Many of the people targeted by the missile attacks were trained and equipped by the CIA years before. [NEW YORK TIMES, 8/24/98]

August 27, 1998: "Delenda Plan" to Combat al-Qaeda Is Prepared

Following the cruise missile attack on al-Qaeda targets on August 20, immediate plans are made for fol-low up attacks to make sure bin Laden is killed. However, on this day, Defense Secretary William Cohen is advised that available targets are not promising. Some question the use of expensive missiles to hit very primitive training camps, and there is the concern that if bin Laden is not killed, his stature will only grow further. As discussions continue, counterterrorism "tsar" Richard Clarke prepares a plan he calls "Delenda," which means "to destroy" in Latin. His idea is to have regular, small strikes in Afghanistan whenever the intelligence warrants it. The plan is rejected. Counterterrorism officials in the Defense Secretary's office independently create a similar plan, but it too is rejected. [9/11 COMMISSION REPORT, 3/24/04 (B)] The Delenda Plan also calls for diplomacy against the Taliban, covert action focused in Afghanistan, and financial measures to freeze bin Laden–related funds. These aspects are not formally adopted, but they guide future efforts. [9/11 COMMISSION REPORT 3/24/04 (D)]

September 23, 1998: U.S. Administration Officials
Confirm No Direct bin Laden Link to Sudanese Factory

Senior Clinton administration officials admit they had no evidence directly linking bin Laden to the Al Shifa factory at the time of retaliatory strikes on August 20. However, intelligence officials assert that they found financial transactions between bin Laden and the Military Industrial Corporation—a company run by the Sudan's government. [*NEW YORK TIMES*, 9/23/98; PBS *FRONTLINE*, 2001]

October 1998: Military Analyst Goes Where
Spies Fail to Go, but Her Efforts Are Rejected

Julie Sirrs, a military analyst for the Defense Intelligence Agency (DIA), travels to Afghanistan. Fluent in local languages and knowledgeable about the culture, she had made a previous undercover trip there in October 1997. She is surprised that the CIA was not interested in sending in agents after the failed missile attack on bin Laden in August 1998, so she returns at this time. Traveling undercover, she meets with Northern Alliance leader Ahmed Shah Massoud. She sees a terrorist training center. Sirrs claims, "The Taliban's brutal regime was being kept in power significantly by bin Laden's money, plus the nar-

cotics trade, while [Massoud's] resistance was surviving on a shoestring. With even a little aid to the Afghan resistance, we could have pushed the Taliban out of power. But there was great reluctance by the State Department and the CIA to undertake that." She partly blames the interest of the U.S. government and the oil company Unocal to see the Taliban achieve political stability to enable a trans-Afghanistan pipeline (see chapter 5). She claims, "Massoud told me he had proof that Unocal had provided money that helped the Taliban take Kabul." She also states, "The State Department didn't want to

Julie Sirrs

have anything to do with Afghan resistance, or even, politically, to reveal that there was any viable option to the Taliban." After two weeks, she returns with a treasure trove of maps, photographs, and interviews. [*NEW YORK OBSERVER*, 3/11/04; ABC NEWS, 2/18/02] By interviewing captured al-Qaeda operatives, she learns that the official Afghanistan airline, Ariana Airlines, is being used to ferry weapons and drugs, and learns that bin Laden goes hunting with "rich Saudis and top Taliban officials." [*LOS ANGELES TIMES*, 11/18/01 (B)] When she returns from Afghanistan, her material is confiscated and she is accused of being a spy. Says one senior colleague, "She had gotten the proper clearances to go, and she came back with valuable information," but high level officials "were so intent on getting rid of her, the last thing they wanted to pay attention to was any information she had." She is cleared of wrongdoing, but her security clearance is pulled. She eventually quits the DIA in frustration. [*NEW YORK OBSERVER*, 3/11/04; ABC NEWS, 2/18/02] She claims that U.S. intelligence on bin Laden and the Taliban relied too heavily on the ISI for its information. [ABC NEWS, 2/18/02 (B)]

December 1998: U.S. Locates bin Laden; Declines to Strike

U.S. intelligence learns that bin Laden is staying at a particular location in Afghanistan, and missile strikes are readied against him. However, principal advisers to President Clinton agree not to recommend a strike because of doubts about the intelligence and worries about collateral damage. In the wake of this incident, officials attempt to find alternatives to cruise missiles, such a precision strike aircraft. However, U.S. Central Command chief General Anthony Zinni is apparently opposed to deployment of these aircraft near Afghanistan, and they are not deployed. [9/11 COMMISSION REPORT, 3/24/04 (B)]

Late 1998: Clinton Signs More Directives Authorizing
CIA to Plan bin Laden Assassination

President Clinton signs additional, more explicit directives authorizing the CIA to plan the assassination of bin Laden. The initial emphasis is on capturing bin Laden and only killing him if the capture attempt is unsuccessful. The military is unhappy about this, so Clinton continues to sign additional directives before leaving office, each one authorizing the use of lethal force more clearly than the one before. [WASHINGTON POST, 2/22/04 (B)]

Late 1998 (B): Failed Missile Attack Said to
Increase bin Laden's Stature in Muslim World

According to reports, the failed August 1998 U.S. missile attack against bin Laden on August 20, 1998 has greatly elevated bin Laden's stature in the Muslim world. A U.S. defense analyst later states, "I think that raid really helped elevate bin Laden's reputation in a big way, building him up in the Muslim world. . . . My sense is that because the attack was so limited and incompetent, we turned this guy into a folk hero." [WASHINGTON POST, 10/3/01 (C)] An Asia Times article published just prior to 9/11 suggests that because of the failed attack, "a very strong Muslim lobby emerge[s] to protect [bin Laden's] interests. This includes Saudi Crown Prince Abdullah, as well as senior Pakistani generals. Prince Abdullah has good relations with bin Laden as both are disciples of slain Doctor Abdullah Azzam." [ASIA TIMES, 8/22/01] In early 1999, Pakistani President Musharraf complains that by demonizing bin Laden, the U.S. has turned him into a cult hero. The U.S. decides to play down the importance of bin Laden. [UPI, 6/14/01]

Late 1998–2000: U.S. Administration Officials
Seek Ground-Based Plan to Kill bin Laden

National Security Adviser Sandy Berger and Secretary of State Madeleine Albright repeatedly seek consideration of a "boots on the ground" option to kill bin Laden, using the elite Delta Force. Clinton also supports the idea, telling Joint Chiefs of Staff Chairman Henry Shelton, "You know, it would scare the [expletive] out of al-Qaeda if suddenly a bunch of black ninjas rappelled out of helicopters into the middle of their camp." However, Shelton says he wants "nothing to do" with such an idea. He calls it naive, and ridicules it as "going Hollywood." He says he would need a large force, not just a small team. [WASHINGTON POST, 12/19/01] U.S. Central Command chief General Anthony Zinni is considered the chief opponent to the "boots on the ground" idea. [WASHINGTON POST, 10/2/02] Clinton orders "formal

Henry Shelton

planning for a mission to capture the al-Qaeda leadership." Reports are contradictory, but some claim Clinton was told such plans were drawn up when in fact they were not. [TIME, 8/4/02; WASHINGTON POST, 10/2/02] In any event, no such plans are implemented.

1999: Joint CIA-NSA Project Taps into al-Qaeda's Tactical Radios

A joint project team run by the CIA and NSA slips into Afghanistan and places listening devices within range of al-Qaeda's tactical radios. [WASHINGTON POST, 12/19/01]

1999 (B): CIA Purportedly Establishes Network of
Agents Throughout Afghanistan, Central Asia

CIA Director Tenet later claims that in this year, the CIA establishes a network of local agents throughout Afghanistan and other countries aimed at capturing bin Laden and his deputies. [UPI, 10/17/02] Tenet states that by 9/11, "a map would show that these collection programs and human networks were in place in such numbers to nearly cover Afghanistan. This array means that, when the military campaign to topple the Taliban and destroy al-Qaeda [begins in October 2001], we [are] able to support it with an enormous body of information and a large stable of assets." [9/11 CONGRESSIONAL INQUIRY, 10/17/02]

February 1999: Bin Laden Missile Strike Called Off
for Fear of Hitting Persian Gulf Royalty

Intelligence reports foresee the presence of bin Laden at a desert hunting camp in Afghanistan for about a week. Information on his presence appears reliable, so preparations are made to target his location with cruise missiles. However, intelligence also puts an official aircraft of the United Arab Emirates (UAE) and members of the royal family from that country in the same location. Bin Laden is hunting with the Emirati royals, as he did with leaders from the UAE and Saudi Arabia on other occasions. Policy makers are concerned that a strike might kill a prince or other senior officials, so the strike never happens. A top UAE official at the time denies that high-level officials are there, but evidence subsequently confirms their presence. [9/11 COMMISSION REPORT, 3/24/04 (B)]

May 1999: U.S. Intelligence Provides
bin Laden's Location; CIA Fails to Strike

U.S. intelligence obtains detailed reporting on where bin Laden is located for five consecutive nights. CIA Director Tenet decides against acting three times, because of concerns about collateral damage and worries about the veracity of the single source of information. Frustration mounts. One CIA official writes to a colleague in the field, "having a chance to get [bin Laden] three times in 36 hours and foregoing the chance each time has made me a bit angry. . . ." There is one more opportunity to strike bin Laden in July 1999, but after that there is apparently no intelligence good enough to justify considering a strike. [9/11 COMMISSION REPORT, 3/24/04]

October 1999: Joint U.S.-ISI Operation to Kill Osama Falters

The CIA readies an operation to capture or kill bin Laden, secretly training and equipping approximately 60 commandos from the Pakistani ISI. Pakistan supposedly agrees to this plan in return for the lifting of economic sanctions and more economic aid. The plan is ready to go by this month, but it is aborted because on October 12, General Musharraf takes control of Pakistan in a coup. Musharraf refuses to continue the operation despite the promise of substantial rewards. [WASHINGTON POST, 10/3/01 (C)] Some U.S. officials later say the CIA was tricked, that the ISI just feigned to cooperate as a stalling tactic, and never intended to get bin Laden. [NEW YORK TIMES, 10/29/01]

October 1999 (B): CIA Considers Increased Aid to Northern Alliance

Worried about intercepts showing a growing likelihood of al-Qaeda attacks around the millennium, the CIA steps up ties with Ahmed Shah Massoud, leader of the Northern Alliance fighting the Taliban. The CIA sends a team of agents to his headquarters in a remote part of northern Afghanistan, seeking his help to capture or kill bin Laden. Massoud complains that the U.S. is too focused on bin Laden, and isn't interested in the root problems of Taliban, Saudi, and Pakistani support for terrorism that is propping him up. He agrees to help nonetheless, and the CIA gives him more aid in return. However, the U.S. is officially neutral in the Afghan civil war and the agents are prohibited from giving any aid that would "fundamentally alter the Afghan battlefield." [WASHINGTON POST, 2/23/04] DIA agent Julie Sirrs, newly retired, is at Massoud's headquarters at the same time as the CIA team. She gathers valuable intelligence from captured al-Qaeda soldiers while the CIA agents stay in their guesthouse. She publishes much of what she learned on this trip and other trips in the summer of 2001. [WASHINGTON POST, 2/28/04]

**General
Ahmed Shah
Massoud**

November 14, 1999: Limited UN Sanctions on Afghanistan

United Nations sanctions against Afghanistan take effect. The sanctions freeze Taliban assets and impose an air embargo on Ariana Airlines in an effort to force the Taliban to hand over bin Laden. [BBC, 2/6/00] However, Ariana keeps its illegal trade network flying, until stricter sanctions ground it in 2001.

Spring 2000: CIA Paramilitary Teams Begin Working with Anti-Taliban Forces

Around this time, special CIA paramilitary teams begin "working with tribes and warlords in southern Afghanistan" and help "create a significant new network in the region of the Taliban's greatest strength." [WASHINGTON POST, 11/18/01]

July 2000: Potential Informant Ignored by Australian and U.S. Authorities

Jack Roche, an Australian Caucasian Muslim, tries to inform on al-Qaeda for Australia or the U.S., but is ignored. In April, Roche returned from a trip to Afghanistan, Pakistan, and Malaysia, where he took an explosives training course and met with bin Laden, Mohammed Atef, Khalid Shaikh Mohammed, and other top al-Qaeda leaders. In Pakistan, Mohammed discussed attacking U.S. jets in Australia and  gave Roche money to start an al-Qaeda cell in Australia. Roche also met Hambali in Malaysia and was given more money there. Early this month, he tries to call the U.S. embassy in Australia, but they ignore him. He then tries to contact the Australian intelligence agency several times, but they too ignore him. In September 2000, his housemate also tries to contact Australian intelligence about what he has learned from Roche but his call is ignored as well. Australian Prime Minister John Howard later acknowledges that authorities made a "very serious mistake" in ignoring Roche, though he also downplays the importance of Roche's information. Roche is later sentenced to nine years in prison for conspiring with al-Qaeda to blow up an Israeli embassy. [BBC, 6/1/04; LOS ANGELES TIMES, 6/7/04]

Jack Roche

October 12, 2002: USS *Cole* Bombed by al-Qaeda Terrorists

See chapter 3.

Late Autumn 2000: CIA Support for Massoud Weakens

Covert CIA support for Ahmed Shah Massoud, the Northern Alliance guerrilla leader fighting the Taliban, is minimal and fraying. In the wake of the USS *Cole* bombing, the CIA develops a plan where the U.S. would increase support for Massoud if he produces strong intelligence about bin Laden's whereabouts. Counterterrorism "tsar" Richard Clarke outlines this CIA proposal to National Security Adviser Sandy Berger, but Berger rejects it. Aid to Massoud continues to languish under the new Bush administration, until Clarke's proposal (slightly modified) is tentatively approved a week before 9/11. [WASHINGTON POST, 2/23/04]

November 2000: Taliban Allegedly Offers to Hand bin Laden to U.S. Officials

In 1999, Kabir Mohabbat, an Afghan-American businessman, had initiated conversations about bin Laden between the U.S. government and the Taliban. According to Mohabbat, the Taliban were ready to hand bin Laden over to a third country, or the International Court of Justice, in exchange for having the U.S.-led sanctions against Afghanistan lifted. (Elmar Brok, a German member of the European Parliament, later confirms that he helps Mohabbat make contact with the U.S. government in 1999.) The initial talks lead to a secret meeting this month between Taliban ministers and U.S. officials in a Frankfurt hotel. Taliban Foreign Minister Wakil Ahmed Muttawakil reportedly says in the meeting, "You can have him whenever the Americans are ready. Name us a country and we will extradite him." However, after this face-to-face meeting, further discussions are never held because, Brok believes, a "political decision" has been made by U.S. officials not to continue the negotiations. He does not clarify when he believes such a decision was made. [REUTERS, 6/5/04]

November 7, 2000: Plans to Target bin Laden Delayed Pending 2000 Election

In the wake of the USS *Cole* bombing, National Security Adviser Sandy Berger meets with Defense Secretary William Cohen to discuss a new approach to targeting bin Laden. Berger says, "We've been hit many times, and we'll be hit again. Yet we have no option beyond cruise missiles." He once again brings up the idea of a "boots on the ground" option—a Delta Force special operation to get bin Laden. A plan is drawn up but the order to execute it is never given. Cohen and Joint Chiefs of Staff Chairman Henry Shelton oppose the plan. By December 21, the CIA reports that it strongly suspects that al-Qaeda was behind the bombing, but fails to definitively make that conclusion. That makes such an attack politically difficult. Says a former senior Clinton aide, "If we had done anything, say, two weeks before the election, we'd be accused of helping [presidential candidate] Al Gore." [*TIME*, 8/4/02; 9/11 COMMISSION REPORT, 3/24/04 (D)]

December 2000: CIA Develops Plan to Increase Support to Massoud, Strike bin Laden

The CIA's Counter Terrorism Center develops a plan to strike at bin Laden in Afghanistan called the "Blue Sky memo." It recommends increased support to anti-Taliban groups and especially a major effort to back Ahmed Shah Massoud's Northern Alliance, to tie down al-Qaeda personnel before they leave Afghanistan. No action is taken in the last few weeks of the Clinton administration; the CIA presses the ideas unsuccessfully early in the new Bush administration. [9/11 COMMISSION REPORT, 3/24/04 (C)] The National Security Council counterterrorism staff also prepares a strategy paper, incorporating ideas from the Blue Sky memo. [9/11 COMMISSION REPORT, 3/24/04 (D)]

December 19, 2000: U.S. Seeks Taliban Overthrow; Considers Russia-U.S. Invasion of Afghanistan

The *Washington Post* reports, "The United States has quietly begun to align itself with those in the Russian government calling for military action against Afghanistan and has toyed with the idea of a new raid to wipe out Osama bin Laden. Until it backed off under local pressure, it went so far as to explore whether a Central Asian country would permit the use of its territory for such a purpose." Russia and the U.S. are discussing "what kind of government should replace the Taliban. Thus, while claiming to oppose a military solution to the Afghan problem, the United States is now talking about the overthrow of a regime that controls nearly the entire country, in the hope it can be replaced with a hypothetical government that does not exist even on paper." [*WASHINGTON POST*, 12/19/00] It appears that all pre-9/11 plans to invade Afghanistan involve attacking from the north with Russia.

February 2001: U.S. Fails to Back Plan to Overthrow Taliban

Abdul Haq, a famous Afghan leader of the mujahedeen, convinces Robert McFarlane, National Security Adviser under President Ronald Reagan, that Haq and about 50 fellow commanders could lead a force to start a revolt against the Taliban in Southern Afghanistan. However, Haq wants to do this under the authority of Zahir Shah, the popular former king of Afghanistan, whom the U.S. does not support. The CIA fails to give any support to Haq. Says one CIA official to McFarlane a few months later, "We don't yet have our marching orders concerning U.S. policy; it may be that we will end up dealing with the Taliban." Haq goes ahead with his plans without U.S. support, and is killed in October. [LOS ANGELES TIMES, 10/28/01 (B); WALL STREET JOURNAL, 11/2/01]

March 2001: U.S. and Taliban Discuss Handing over bin Laden

Taliban envoy Rahmatullah Hashimi meets with reporters, middle-ranking State Department bureaucrats, and private Afghanistan experts in Washington. He carries a gift carpet and a letter from Afghan leader Mullah Omar for President Bush. He discusses turning bin Laden over, but the U.S. wants to be handed bin Laden and the Taliban want to turn him over to some third country. A CIA official later says, "We never heard what they were trying to say. We had no common language. Ours was, 'Give up bin Laden.' They were saying, 'Do something to help us give him up.' . . . I have no doubts they wanted to get rid of him. He was a pain in the neck." Others claim the Taliban were never sincere. About 20 more meetings on giving up bin Laden take place up until 9/11, all fruitless. [WASHINGTON POST, 10/29/01] Allegedly, Hashimi also proposes that the Taliban would hold bin Laden in one location long enough for the U.S. to locate and kill him. However, this offer is refused. This report, however, comes from Laila Helms, daughter of former CIA director Richard Helms, who is doing public relations for the Taliban at the time. While it's interesting that this information came out before 9/11, one must be skeptical, since Helms' job was public relations for the Taliban. [VILLAGE VOICE, 6/6/01]

March 1, 2001: Taliban Blow Up Giant Buddha Statues, Disregard International Opinion

The Taliban begins blowing up two giant stone Buddhas of Bamiyan (ancient statues carved into an Afghan mountainside, which are considered priceless treasures). They face great international condemnation in response, but no longer seem to be courting international recognition. Apparently, even ISI efforts to dissuade them fail. [TIME, 8/4/02; TIME, 8/4/02]

March 7, 2001: Russia Submits Report on bin Laden to UN Security Council, U.S. Fails to Act

The Russian Permanent Mission at the United Nations secretly submits "an unprecedentedly detailed report" to the UN Security Council about bin Laden, his whereabouts, details of his al-Qaeda network, Afghan drug running, and Taliban connections in Pakistan. The report provides "a listing of all bin Laden's bases, his government contacts and foreign advisers," and enough information to potentially locate and kill him. The U.S. fails to act. Alex Standish, the editor of the highly respected *Jane's Intelligence Review*, concludes that the attacks of 9/11 were less of an American intelligence failure than the result of "a political decision not to act against bin Laden." [JANE'S INTELLIGENCE REVIEW, 10/5/01]

March 8, 2001: U.S. Declines to Freeze al-Qaeda's
Assets Despite Call from UN and EU

The United Nations and the European Union direct their members to freeze the assets of some al-Qaeda leaders, including Sa'd Al-Sharif, bin Laden's brother-in-law and the head of his finances, but the U.S. does not do so until after 9/11. [GUARDIAN, 10/13/01 (B)] For a time, the U.S. claims that Sa'd Al-Sharif helped fund the 9/11 attacks, but the situation is highly confused and his role is doubtful.

March 15, 2001: India, Iran, Russia, and U.S. Work in Concert to Remove Taliban

Jane's Intelligence Review reports that the U.S. is working with India, Iran, and Russia "in a concerted front against Afghanistan's Taliban regime." India is supplying the Northern Alliance with military equipment, advisers, and helicopter technicians and both India and Russia are using bases in Tajikistan and Uzbekistan for their operation. [JANE'S INTELLIGENCE REVIEW, 3/15/01]

April 6, 2001: Rebel Leader Warns Europe
and U.S. About Imminent Terrorist Attacks

Ahmed Shah Massoud, leader of the Northern Alliance fighting the Taliban in Afghanistan, has been trying to get aid from the U.S. but his people are only allowed to meet with low level U.S. officials. In an attempt to get his message across, he addresses the European Parliament: "If President Bush doesn't help us, these terrorists will damage the U.S. and Europe very soon." [DAWN, 4/7/04; TIME, 8/4/02] Massoud also meets privately with some CIA officials while in Europe. He tells them that his guerrilla war against the Taliban is faltering and unless the U.S. gives a significant amount of aid, the Taliban will conquer all of Afghanistan. No more aid is forthcoming. [WASHINGTON POST, 2/23/04]

May 2001: U.S. Gives Taliban Millions

Secretary of State Powell announces that the U.S. is granting $43 million in aid to the Taliban government, purportedly to assist hungry farmers who are starving since the destruction of their opium crop occurred in January on orders of the Taliban. [LOS ANGELES TIMES, 5/22/01] This follows $113 million given by the U.S. in 2000 for humanitarian aid. [STATE DEPARTMENT FACT SHEET, 12/11/01] A *Newsday* editorial notes that the Taliban "are a decidedly odd choice for an outright gift . . . Why are we sending these people money—so much that Washington is, in effect, the biggest donor of aid to the Taliban regime?" [NEWSDAY, 5/29/01]

May 2001 (B): U.S. Military Drafts Scenario for Afghan Operation

General William Kernan, commander in chief of the Joint Forces Command, later mentions: "The details of Operation Enduring Freedom in Afghanistan which fought the Taliban and al-Qaeda after the September 11 attacks, were largely taken from a scenario examined by Central Command in May 2001." [AGENCE FRANCE-PRESSE, 7/23/02]

May–June 2001: Muslim Convert Inadvertently Learns of 9/11 Plot

John Walker Lindh, a young Caucasian man from California who has converted to Islam, travels to Peshawar, Pakistan, in an attempt to fight for Islamic causes. He had been studying the Koran for about six months elsewhere in Pakistan, but otherwise had no particularly special training, qualifications, or

connections. Within days, he is accepted into al-Qaeda and sent to the al Faruq training camp in Afghanistan. Seven other U.S. citizens are already training there. He inadvertently learns details of the 9/11 attacks. In June, he is told by an instructor that "bin Laden had sent forth some fifty people to carry out twenty suicide terrorist operations against the United States and Israel." He learns that the 9/11 plot is to consist of five attacks, not the four that actually occur. The other fifteen operations are to take place

John Walker Lindh

later. He is asked if he wants to participate in a suicide mission, but declines. [A PRETEXT FOR WAR, BY JAMES BAMFORD, 6/04, PP. 234-36; GETTING AWAY WITH MURDER, BY RICHARD D. MAHONEY, 6/04, PP. 162, 216] Author James Bamford comments, "The decision to keep CIA employees at arm's length from [al-Qaeda] was a serious mistake. At the same moment the CIA was convinced al-Qaeda was impenetrable, a number of American citizens were secretly joining al-Qaeda in Afghanistan—and being welcomed with open arms." [A PRETEXT FOR WAR, BY JAMES BAMFORD, 6/04, P. 161]

June 2001: U.S. Still Fails to Aid Taliban Resistance

The U.S. considers substantially aiding Ahmed Shah Massoud and his Northern Alliance. As one counterterrorism official put it, "You keep [al-Qaeda terrorists] on the front lines in Afghanistan. Hopefully you're killing them in the process, and they're not leaving Afghanistan to plot terrorist operations." A former U.S. special envoy to the Afghan resistance visits Massoud this month. Massoud gives him "all the intelligence he [has] on al-Qaeda" in the hopes of getting some support in return. However, he gets nothing more than token amounts and his organization isn't even given "legitimate resistance movement" status. [TIME, 8/4/02]

Early June 2001: Taliban Leader Claims Interest in Resolving bin Laden Issue

Reclusive Taliban leader Mullah Omar says the Taliban would like to resolve the bin Laden issue, so there can be "an easing and then lifting of UN sanctions that are strangling and killing the people of [Afghanistan]." [UPI, 6/14/01]

June 26, 2001: U.S., Russia, and Regional Powers Cooperate to Oust Taliban

An Indian magazine reports more details of the cooperative efforts of the U.S., India, Russia, Tajikistan, Uzbekistan, and Iran against the Taliban regime: "India and Iran will 'facilitate' U.S. and Russian plans for 'limited military action' against the Taliban if the contemplated tough new economic sanctions don't bend Afghanistan's fundamentalist regime." Earlier in the month, Russian President Vladimir Putin told a meeting of the Confederation of Independent States that military action against the Taliban may happen, possibly with Russian involvement using bases and forces from Uzbekistan and Tajikistan as well. [INDIAREACTS, 6/26/01]

Summer 2001: FBI Neglects Chance to Infiltrate al-Qaeda Training Camp

A confidential informant tells an FBI field office agent that he has been invited to a commando-training course at a camp operated by al-Qaeda in Afghanistan. The information is passed up to FBI headquarters, which rejects the idea of infiltrating the camp. An "asset validation" of the informant, a routine but critical exercise to determine whether information from the source was reliable, is also not done. The FBI later has no comment on the story. [U.S. NEWS AND WORLD REPORT, 6/10/02]

Late Summer 2001: U.S. Contingency Plans to Attack Afghanistan

According to a later *Guardian* report, "reliable western military sources say a U.S. contingency plan exist[s] on paper by the end of the summer to attack Afghanistan from the north." [GUARDIAN, 9/26/01]

Mid-August 2001: Afghan Leader Organizes
Taliban Resistance Without U.S. Support

Abdul Haq, a famous Afghan leader of the mujahedeen, returns to Peshawar, Pakistan, from the U.S. Having failed to gain U.S. support, except for that of some private individuals such as former National Security Adviser Robert McFarlane, Haq begins organizing subversive operations in Afghanistan. [LOS ANGELES TIMES, 10/28/01 (B); WALL STREET JOURNAL, 11/2/01] He is later killed entering Afghanistan in October 2001, after his position is reportedly betrayed to the Taliban by the ISI.

August 30, 2001: Osama Reportedly Named Commander of Afghanistan Army

It is reported in Russia and Pakistan that the Taliban has named bin Laden commander of the Afghanistan army. [UPI, 8/30/01]

September 9, 2001: Northern Alliance Leader Massoud
Is Assassinated in Anticipation of 9/11 Attack

General Ahmed Shah Massoud, the leader of Afghanistan's Northern Alliance, is assassinated by two al-Qaeda agents posing as Moroccan journalists. [TIME, 8/4/02] A legendary mujahedeen commander and a brilliant tactician, Massoud had pledged to bring freedom and democracy to Afghanistan. The BBC says the next day, "General Massoud's death might well have meant the end of the [Northern] alliance" because there clearly was no figure with his skills and popularity to replace him. [BBC, 9/10/01; BBC, 9/10/01 (B)] "With Massoud out of the way, the Taliban and al-Qaeda would be rid of their most effective opponent and be in a stronger position to resist the American onslaught." [ST. PETERSBURG TIMES, 9/9/02] It appears the assassination was supposed to happen earlier: the "journalists" waited for three weeks in Northern Alliance territory to meet Massoud. Finally on September 8, an aide says they "were so worried and excitable they were begging us." They were granted an interview after threatening to leave if the interview did not happen in the next 24 hours. Meanwhile, the Taliban army (together with elements of the Pakistani army) had massed for an offensive against the Northern Alliance in the previous weeks, but the offensive began only hours after the assassination. Massoud was killed that day but Northern Alliance leaders pretend for several days that Massoud was only injured in order to keep the Northern Alliance army's morale up, and they are able to stave off total defeat. The timing of the assassination and the actions of the Taliban army suggest that the 9/11 attacks were known to the Taliban leadership. [TIME, 8/4/02] Though it is not widely reported, the Northern Alliance releases a statement the next day: "Ahmed Shah Massoud was the target of an assassination attempt organized by the Pakistani [intelligence service] ISI and Osama bin Laden." [RADIO FREE EUROPE, 9/10/01; NEWSDAY, 9/15/01; REUTERS, 10/4/01] This suggests that the ISI may also have had prior knowledge of the attack plans.

Before September 11, 2001:
Congressman Says U.S. Intelligence Not Interested in
Informant Who Could Pinpoint bin Laden's Location

Congressman Dana Rohrabacher (R), who claims to have made many secret trips into Afghanistan and to have fought with the mujahedeen, later describes to Congress a missed opportunity to capture bin Laden. He claims that "a few years" before 9/11, he is contacted by someone he knows and trusts from the 1980s Afghan war, who claims he could pinpoint bin Laden's location. Rohrabacher passes this information to the CIA, but the informant isn't contacted. After some weeks, Rohrabacher uses his influence to set up a meeting with agents in the CIA, NSA, and FBI. Yet even then, the informant is not contacted, until weeks later, and then only in a "disinterested" way. Rohrabacher concludes, "that our intelligence services knew about the location of bin Laden several times but were not permitted to attack him . . . because of decisions made by people higher up." [HOUSE OF REPRESENTATIVES SPEECH, 9/17/01]

5.

Pipeline Politics

Throughout the 1990s it was believed that Central Asia contained vast natural resource wealth, with virtually unlimited potential for oil and gas production. The challenge was geography. Bounded by the high Tibetan plateau on one side and the American-hostile nation of Iran on the other, Afghanistan seemed to provide an excellent option for a pipeline to transport oil and gas from the landlocked Caspian Sea basin, across Pakistan, to the Indian Ocean.

The U.S.-based Unocal Corporation led the way in plans to build a trans-Afghanistan gas pipeline. Afghanistan was economically devastated from many years of war, and had little to offer to Western companies except for pipelines crossing the country. In India, not far away, the notorious Enron Corporation was building a massive power plant in the town of Dabhol, with little to power it except gas that would come from the trans-Afghanistan pipeline. It was the company's single biggest investment. Given later allegations of questionable ties between President Bush and Enron executives such as Ken Lay, it is an unanswered question whether Bush's diplomatic stance with Afghanistan might have been influenced by Enron's increasingly desperate attempts to turn a profit on its Indian power plant. The company ultimately failed and went bankrupt.

Although it now appears that prior estimates may have been overestimated and plans for a trans-Afghanistan pipeline are stalled, the U.S. remains intensely interested in the various pipeline projects currently contemplated in the Caspian Sea Region, including a trans-Afghan pipeline. Notably, President Bush has appointed Zalmay Khalilzad as Washington's ambassador to Kabul. Khalilzad, an Afghan, previously worked under Deputy Defense Secretary Paul Wolfowitz; "It was Khalilzad—when he was a huge Taliban fan—who conducted the risk analysis for Unocal (Union Oil Company of California) for the infamous proposed $2 billion, 1,500 kilometer-long Turkmenistan-Afghanistan-Pakistan [TAP] gas pipeline." [ASIA TIMES, 12/23/03]

Along with Russia, China, and other countries, multiple U.S. companies are engaged in developing the oil fields of that region. And, as noted in the Asia Times, while "a dreamer would see harmony between Russia, the U.S., and China all engaged in a sensible exploitation of Central Asia's oil and gas and minerals. . . . it doesn't look like it's going to be this way. The perception in Islam . . . is that America is using the 'war on terror' to exclusively advance its own strategic oil and gas interests." [ASIA TIMES, 12/23/03]

This chapter focuses on what role, if any, U.S. interest in the region's oil reserves played in U.S. foreign policy in Afghanistan before 9/11. Did the U.S. turn a blind eye toward the dangers posed by the Taliban, focusing instead on the presumed economic benefits of a pipeline? If so, why? Did Enron—and its Dabhol power plant in India—play any role? What role did Unocal play, if any, in U.S. policy toward that region?

Some believe that answers to these questions may lie in the documents of Vice President Cheney's early 2001 Energy Task Force. Two groups sued even before 9/11 to have access to these papers, but the issue remains tied up in courts.

(For additional information relating to U.S. interest in oil, and how that may have affected decisions made by the Bush administration both before and after 9/11, see chapter 16.)

1991–1997: Oil Investment in Central Asia Follows Soviet Collapse

The Soviet Union collapses in 1991, creating several new nations in Central Asia. Major U.S. oil companies, including ExxonMobil, Texaco, Unocal, BP Amoco, Shell, and Enron, directly invest billions in these Central Asian nations, bribing heads of state to secure equity rights in the huge oil reserves in these regions. The oil companies commit to $35 billion in future direct investments in Kazakhstan. It is believed at the time that these oil fields will have an estimated $6 trillion potential value. U.S. companies own approximately 75 percent of the rights. These companies, however, face the problem of having to pay exorbitant prices to Russia for use of the Russian pipelines to get the oil out. [NEW YORKER, 7/9/01; ASIA TIMES, 1/26/02]

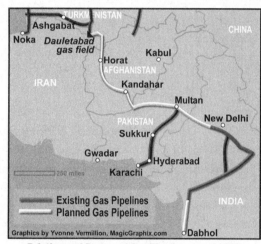

Existing and Proposed Gas Pipelines as of 1997

November 1993: Enron Power Plant Creates Demand for an Afghanistan Pipeline

The Indian government approves construction of Enron's Dabhol power plant, located near Bombay on the west coast of India. Enron has invested $3 billion, the largest single foreign investment in India's history. Enron owns 65 percent of the Dabhol liquefied natural gas power plant, intended to provide one-fifth of India's energy needs by 1997. [ASIA TIMES, 1/18/01; INDIAN EXPRESS, 2/27/00] It is the largest gas-fired power plant in the world. Earlier in the year, the World Bank concluded that the plant was "not economically viable" and refused to invest in it. [NEW YORK TIMES, 3/20/01]

1995–November 2001: U.S. Lobbies India Over Enron Power Plant

Enron's $3 billion Dabhol, India power plant runs into trouble in 1995 when the Indian government temporarily cancels an agreement. The plant is projected to get its energy from the proposed Afghan pipeline and deliver it to the Indian government. Enron leader Ken Lay travels to India with Commerce

Secretary Ron Brown the same year, and heavy lobbying by U.S. officials continue in subsequent years. By summer 2001, the National Security Council leads a "Dabhol Working Group" with officials from various Cabinet agencies to get the plant completed and functioning. U.S. pressure on India intensifies until shortly before Enron files for bankruptcy in December 2001. U.S. officials later claim their lobbying merely supported the $640 million of U.S. government investment in the plant. But critics say the plant received unusually strong support under both the Clinton and Bush administrations. [NEW YORK DAILY NEWS. 1/18/02; WASHINGTON POST, 1/19/02]

September–October 1995: Unocal Obtains Turkmenistan Pipeline Deal

Oil company Unocal signs an $8 billion deal with Turkmenistan to construct two pipelines (one for oil, one for gas), as part of a larger plan for two pipelines intended to transport oil and gas from Turkmenistan through Afghanistan and into Pakistan. Before proceeding further, however, Unocal needs to execute agreements with Pakistan and Afghanistan; Pakistan and Ahmed Shah Massoud's government in Afghanistan, however, have already signed a pipeline deal with an Argentinean company. Henry Kissinger, hired as speaker for a special dinner in New York to announce the Turkmenistan pipeline deal, expresses skepticism that it will succeed, referring to the Unocal plan as "the triumph of hope over experience." Unocal will later open an office in Kabul, weeks after the Taliban capture of the capital in late 1996 and will interact with the Taliban, seeking support for its pipeline until at least December 1997. [GHOST WARS, BY STEVE COLL, PP. 301–13, 329, 338, 364-66]

December 1995: Caspian Sea Said to Contain
Two-Thirds of World's Known Oil Reserves

The American Petroleum Institute asserts that the states bordering the Caspian Sea, north of Afghanistan, contain two-thirds of the world's known reserves, or 659 billion barrels. Such numbers spur demand for an Afghan pipeline. However, by April 1997, estimates drop to 179 billion barrels. [MIDDLE EAST JOURNAL, 9/22/00] This is still substantial, but the estimates continue to drop in future years.

May 1996: U.S. Seeks Stability in Afghanistan for Unocal Pipeline

Robin Raphel, Deputy Secretary of State for South Asia, speaks to the Russian Deputy Foreign Minister about Afghanistan. She says that the U.S. government "now hopes that peace in the region will facilitate U.S. business interests," such as the proposed Unocal gas pipeline from Turkmenistan through Afghanistan to Pakistan. [GHOST WARS, BY STEVE COLL, 2/04, P. 330]

June 24, 1996: Uzbekistan Cuts a Deal with Enron

Uzbekistan signs a deal with Enron "that could lead to joint development of the Central Asian nation's potentially rich natural gas fields." [HOUSTON CHRONICLE, 6/25/96] The $1.3 billion venture teams Enron with the state companies of Russia and Uzbekistan. [HOUSTON CHRONICLE, 6/30/96] On July 8, 1996, the U.S. government agrees to give $400 million to help Enron and an Uzbeki state company develop these natural gas fields. [OIL AND GAS JOURNAL, 7/8/96]

September 27, 1996: Victorious Taliban Supported
by Pakistan; Viewed by U.S., Unocal as "Stabilizing Force"

The Taliban conquer Kabul [ASSOCIATED PRESS, 8/19/02], establishing control over much of Afghanistan. A surge in the Taliban's military successes at this time is later attributed to an increase in direct military assistance from Pakistan's ISI. [NEW YORK TIMES, 12/8/01] The oil company Unocal is hopeful that the Taliban will stabilize Afghanistan and allow its pipeline plans to go forward. According to some reports, "preliminary agreement [on the pipeline] was reached between the [Taliban and Unocal] long before the fall of Kabul. . . . Oil industry insiders say the dream of securing a pipeline across Afghanistan is the main reason why Pakistan, a close political ally of America's, has been so supportive of the Taliban, and why America has quietly acquiesced in its conquest of Afghanistan." [DAILY TELEGRAPH, 10/11/96] The 9/11 Commission later concludes that some State Department diplomats are willing to "give the Taliban a chance" because it might be able to bring stability to Afghanistan, which would allow a Unocal oil pipeline to be built through the country. [9/11 COMMISSION REPORT, 3/24/04]

October 11, 1996: Afghan Pipeline Gets Media Attention

The *Daily Telegraph* publishes an interesting article about pipeline politics in Afghanistan. "Behind the tribal clashes that have scarred Afghanistan lies one of the great prizes of the 21st century, the fabulous energy reserves of Central Asia. . . . 'The deposits are huge,' said a diplomat from the region. 'Kazakhstan alone may have more oil than Saudi Arabia. Turkmenistan is already known to have the fifth largest gas reserves in the world.'" [DAILY TELEGRAPH, 10/11/96]

August 1997: CIA Monitors Central Asia for Oil Reserves

The CIA creates a secret task force to monitor Central Asia's politics and gauge its wealth. Covert CIA officers, some well-trained petroleum engineers, travel through southern Russia, Azerbaijan, Kazakhstan, and Turkmenistan to sniff out potential oil reserves. [TIME, 5/4/98]

October 27, 1997: Halliburton Announces Turkmenistan Project;
Unocal and Delta Oil Form Consortium

Halliburton, a company headed by CEO—and future Vice President—Dick Cheney, announces a new agreement to provide technical services and drilling for Turkmenistan. The press release mentions, "Halliburton has been providing a variety of services in Turkmenistan for the past five years." On the same day, a consortium to build a pipeline through Afghanistan is formed. It is called CentGas, and the two main partners are Unocal and Delta Oil of Saudi Arabia. [HALLIBURTON PRESS RELEASE, 10/27/97; CENTGAS PRESS RELEASE, 10/27/97]

November 1997: Enron and bin Laden Family Team Up for Project

Industry newsletter reports that Saudi Arabia has abandoned plans for open bids on a $2 billion power plant near Mecca, deciding that the government will build it instead. Interestingly, one of the bids was made by a consortium of Enron, the Saudi Binladin Group (run by Osama's family), and Italy's Ansaldo Energia. [ALEXANDER'S GAS AND OIL CONNECTIONS, 1/22/98]

December 1997: Unocal Establishes Pipeline Training Facility
Near bin Laden's Compound

Unocal pays University of Nebraska $900,000 to set up a training facility near bin Laden's Kandahar compound, to train 400 Afghani teachers, electricians, carpenters and pipe fitters in anticipation of using them for their pipeline in Afghanistan. 150 students are already attending classes. [DAILY TELEGRAPH, 12/14/97, GHOST WARS, BY STEVE COLL, 2/04, P. 364]

December 4, 1997: Taliban Representatives Visit Unocal in Texas

Representatives of the Taliban are invited guests to the Texas headquarters of Unocal to negotiate their support for the pipeline. Future President George W. Bush is Governor of Texas at the time. The Taliban appear to agree to a $2 billion pipeline deal, but will do the deal only if the U.S. officially recognizes the Taliban regime. The Taliban meet with U.S. officials. According to the *Daily Telegraph*, "the U.S. government, which in the past has branded the Taliban's policies against women and children 'despicable,' appears anxious to please the fundamentalists to clinch the lucrative pipeline contract." A BBC regional correspondent says that "the proposal to build a pipeline across Afghanistan is part of an international scramble to profit from developing the rich energy resources of the Caspian Sea." [BBC, 12/4/97, DAILY TELEGRAPH, 12/14/97]

February 12, 1998: Unocal VP Advocates Afghan Pipeline Before Congress

Unocal Vice President John J. Maresca—later to become a Special Ambassador to Afghanistan—testifies before the House of Representatives that until a single, unified, friendly government is in place in Afghanistan the trans-Afghani pipeline will not be built. He suggests that with a pipeline through Afghanistan, the Caspian basin could produce 20 percent of all the non-OPEC oil in the world by 2010. [HOUSE INTERNATIONAL RELATIONS COMMITTEE TESTIMONY, 2/12/98]

Early 1998: U.S. Official Meets with Taliban; Promote Afghan Pipeline

Bill Richardson, the U.S. Ambassador to the UN, meets Taliban officials in Kabul. (All such meetings are illegal, because the U.S. still officially recognizes the government the Taliban ousted as the legitimate rulers of Afghanistan.) U.S. officials at the time call the oil and gas pipeline project a "fabulous opportunity" and are especially motivated by the "prospect of circumventing Iran, which offers another route for the pipeline." [BOSTON GLOBE, 9/20/01]

August 7, 1998: Terrorists Bomb U.S. Embassies in Kenya and Tanzania

See chapter 1.

June 1998: Enron Shuts Down Uzbekistan Pipeline Project

Enron's agreement from 1996 to develop natural gas with Uzbekistan is not renewed. Enron closes its office there. The reason for the "failure of Enron's flagship project" is an inability to get the natural gas out of the region. Uzbekistan's production is "well below capacity" and only 10 percent of its production is being exported, all to other countries in the region. The hope was to use a pipeline through Afghanistan, but "Uzbekistan is extremely concerned at the growing strength of the Taliban and its potential impact on stability in Uzbekistan, making any future cooperation on a pipeline project which benefits the Taliban unlikely." A $12 billion pipeline through China is being considered as one solution, but that wouldn't be completed until the end of the next decade at the earliest. [ALEXANDER'S GAS AND OIL CONNECTIONS, 10/12/98]

June 23, 1998: Future VP Cheney Raves About Caspian Sea Opportunities

Future Vice President Cheney, currently working for the Halliburton energy company, states, "I can't think of a time when we've had a region emerge as suddenly to become as strategically significant as the Caspian. It's almost as if the opportunities have arisen overnight." [CATO INSTITUTE LIBRARY; CHICAGO TRIBUNE, 8/10/00]

August 9, 1998: Northern Alliance Stronghold Conquered by Taliban; Pipeline Project Now Looks Promising

The Northern Alliance capital of Afghanistan, Mazar-i-Sharif is conquered by the Taliban. Military support of Pakistan's ISI plays a large role; there is even an intercept of an ISI officer stating, "My boys and I are riding into Mazar-i-Sharif." [NEW YORK TIMES, 12/8/01] This victory gives the Taliban control of 90 percent of Afghanistan, including the entire proposed pipeline route. CentGas, the consortium behind the gas pipeline that would run through Afghanistan, is now "ready to proceed. Its main partners are the American oil firm Unocal and Delta Oil of Saudi Arabia, plus Hyundai of South Korea, two Japanese companies, a Pakistani conglomerate and the Turkmen government." However, the pipeline cannot be financed unless the government is officially recognized. "Diplomatic sources said the Taliban's offensive was well prepared and deliberately scheduled two months ahead of the next UN meeting" where members are to decide whether the Taliban should be recognized. [DAILY TELEGRAPH, 8/13/98]

October 1998: Military Analyst Travels to Afghanistan; Sees Taliban–al-Qaeda Alliance; Later Alleges Unocal Assisted Taliban

See chapter 4.

December 5, 1998: Unocal Abandons Afghan Pipeline Project

Unocal announces it is withdrawing from the CentGas pipeline consortium, and closing three of its four offices in Central Asia. Clinton refuses to extend diplomatic recognition to the Taliban, making business there legally problematic. A concern that Clinton will lose support among women voters for upholding the Taliban plays a role in the cancellation. [NEW YORK TIMES, 12/5/98]

Late 1998: Taliban Stall Pipeline Negotiations

During the investigation of the August 7, 1998, U.S. embassy bombings, FBI counterterrorism expert John O'Neill finds a memo by al-Qaeda leader Mohammed Atef on a computer. The memo shows that bin Laden's group has a keen interest in and detailed knowledge of negotiations between the Taliban and the U.S. over an oil and gas pipeline through Afghanistan. Atef's analysis suggests that the Taliban are not sincere in wanting a pipeline, but are dragging out negotiations to keep Western powers at bay. [SALON, 6/5/02]

June 1999: Enron Plans Power Plant with bin Laden Family

Enron announces an agreement to build a $140 million power plant in the Gaza Strip, between Israel and Egypt. One of the major financiers for the project is the Saudi Binladin Group, a company owned by bin Laden's family. This is the second attempted project between these two companies. Ninety percent complete, the construction will be halted because of Palestinian-Israeli violence and then Enron's bankruptcy. [WASHINGTON POST, 3/2/02]

July 4, 1999: Executive Order Issued Against Taliban

With the chances of a pipeline deal with the Taliban looking increasingly unlikely, President Clinton finally issues an executive order prohibiting commercial transactions with the Taliban. The order also freezes the Taliban's U.S. assets. Clinton blames the Taliban for harboring bin Laden. [EXECUTIVE ORDER, 7/4/99; CNN, 7/6/99]

December 20, 1999: Iran Said to Be Supporting Conflict in Afghanistan to Further Their Own Pipeline Plans

The BBC explains one reason why the Northern Alliance has been able to hold out for so long in its civil war against the Taliban in Afghanistan: "Iran has stirred up the fighting in order to make sure an international oil pipeline [goes] through its territory and not through Afghanistan." [BBC, 12/20/99]

January 21, 2001: Bush Administration Takes Over; Many Have Oil Industry Connections

George W. Bush is inaugurated as the 43rd U.S. President, replacing Clinton. The only Cabinet-level figure to remain permanently in office is CIA Director Tenet, appointed in 1997 and reputedly, a long-time friend of George H. W. Bush. FBI Director Louis Freeh stays on until June 2001. Numerous figures in

Bush's administration have been directly employed in the oil industry, including Bush, Vice President Cheney, and National Security Adviser Rice. Rice had been on Chevron's Board of Directors since 1991, and even had a Chevron oil tanker named after her. [SALON, 11/19/01] It is later revealed that Cheney is still being paid up to $1 million a year in "deferred payments" from Halliburton, the oil company he headed. [GUARDIAN, 3/12/03] Enron's ties also reach deep into the administration. [WASHINGTON POST, 1/18/02] The "Atlanta Rules" are applied and security precautions are taken to guard against the possibility of an attack on the inauguration ceremony using an airplane as a flying weapon (see chapter 1). [WALL STREET JOURNAL, 4/1/04]

May 2001: Cheney's Energy Plan Foresees Government
Helping U.S. Companies Expand Into New Markets

Vice President Cheney's national energy plan is released to the public. It calls for expanded oil and gas drilling on public land and easing regulatory barriers to building nuclear power plants. [ASSOCIATED PRESS, 12/9/02] There are several interesting points, little noticed at the time. It suggests that the U.S. cannot depend exclusively on traditional sources of supply to provide the growing amount of oil that it needs. It will also have to obtain substantial supplies from new sources, such as the Caspian states, Russia, and Africa. It also notes that the U.S. cannot rely on market forces alone to gain access to these added supplies, but will also require a significant effort on the part of government officials to overcome foreign resistance to the outward reach of American energy companies. [JAPAN TODAY, 4/30/02] The plan was largely decided through Cheney's secretive Energy Task Force. Both before and after this, Cheney and other Task Force officials meet with Enron executives (including one meeting a month and a half before Enron declares bankruptcy in December 2001). Two separate lawsuits are later filed to reveal details of how the government's energy policy was formed and whether Enron or other players may have influenced it, but as of mid-2004 the Bush administration has successfully resisted all efforts to release these documents. [ASSOCIATED PRESS, 12/9/02]

May 23, 2001: Former Unocal Employee Becomes Member of NSC

Zalmay Khalilzad is appointed Special Assistant to the President and Senior Director for Gulf, Southwest Asia and Other Regional Issues on the National Security Council. Khalilzad was an official in the Reagan and George H. W. Bush administrations. During the Clinton years, he worked for Unocal. After 9/11, he is appointed as special envoy to Afghanistan. [INDEPENDENT, 1/10/02; STATE DEPARTMENT PROFILE, 2001]

Zalmay K. Halilzad

June 2001: Enron Shuts Down Expensive
Indian Plant Afghan Pipeline Fails to Materialize

Enron's power plant in Dabhol, India, is shut down. The failure of the $3 billion plant, Enron's largest investment, contributes to Enron's bankruptcy in December. Earlier in the year, India stopped paying its bill for the energy from the plant, because energy from the plant cost three times the usual rates. [NEW

YORK TIMES, 3/20/01] Enron had hoped to feed the plant with cheap Central Asian gas, but this hope was dashed when a gas pipeline through Afghanistan was not completed. The larger part of the plant is still only 90 percent complete when construction stops around this time. [NEW YORK TIMES, 3/20/01] Enron executives meet with the Commerce Secretary about its troubled Dabhol power plant during this year [NEW YORK TIMES, 2/21/02], and Vice President Cheney lobbies the leader of India's main opposition party about the plant this month. [NEW YORK TIMES, 2/21/02]

June 27, 2001: India and Pakistan Discuss
Building Pipeline Project Through Iran

The *Wall Street Journal* reports that Pakistan and India are discussing jointly building a gas pipeline from Central Asian gas fields through Iran to circumvent the difficulties of building the pipeline through Afghanistan. Iran has been secretly supporting the Northern Alliance to keep Afghanistan divided so no pipelines could be put through it. [WALL STREET JOURNAL, 6/27/01]

July 21, 2001: U.S. Official Threatens Possible
Military Action Against Taliban by October if Pipeline Is Not Pursued

Three former American officials, Tom Simons (former U.S. Ambassador to Pakistan), Karl Inderfurth (former Deputy Secretary of State for South Asian Affairs) and Lee Coldren (former State Department expert on South Asia) meet with Pakistani and Russian intelligence officers in a Berlin hotel. [SALON, 8/16/02] This is the third of a series of back-channel conferences called "brainstorming on Afghanistan." Taliban representatives sat in on previous meetings, but boycotted this one due to worsening tensions. However, the Pakistani ISI relays information from the meeting to the Taliban. [GUARDIAN, 9/22/01] At the meeting, Coldren passes on a message from Bush officials. He later says, "I think there was some discussion of the fact that the United States was so disgusted with the Taliban that they might be considering some military action." [GUARDIAN, 9/26/01] Accounts vary, but former Pakistani Foreign Secretary Niaz Naik later says he is told by senior American officials at the meeting that military action to overthrow the Taliban in Afghanistan is planned to "take place before the snows started falling in Afghanistan, by the middle of October at the latest." The goal is to kill or capture both bin Laden and Taliban leader Mullah Omar, topple the Taliban regime and install a transitional government of moderate Afghans in its place. Uzbekistan and Russia would also participate. Naik also says, "It was doubtful that Washington would drop its plan even if bin Laden were to be surrendered immediately by the Taliban." [BBC, 9/18/01] One specific threat made at this meeting is that the Taliban can choose between "carpets of bombs"—an invasion—or "carpets of gold"— the pipeline. [THE FORBIDDEN TRUTH, BY JEAN-CHARLES BRISARD, GUILLAUME DASQUIÉ, AND WAYNE MADSEN, 7/02, P. 43] Naik contends that Tom Simons made the "carpets" statement. Simons claims, "It's possible that a mischievous American participant, after several drinks, may have thought it smart to evoke gold carpets and carpet bombs. Even Americans can't resist the temptation to be mischievous." Naik and the other American participants deny that the pipeline was an issue at the meeting. [SALON, 8/16/02]

August 2, 2001: U.S. Official Secretly Meets Taliban Ambassador in Last Attempt to Secure Pipeline Deal

Christina Rocca, Director of Asian Affairs at the State Department, secretly meets the Taliban ambassador in Islamabad, apparently in a last ditch attempt to secure a pipeline deal. Rocca was previously in charge of contacts with Islamic guerrilla groups at the CIA, and oversaw the delivery of Stinger missiles to Afghan mujahedeen in the 1980s. [IRISH TIMES, 11/19/01; SALON, 2/8/02; THE FORBIDDEN TRUTH, BY JEAN-CHARLES BRISARD, GUILLAUME DASQUIÉ, AND WAYNE MADSEN, 7/02, P. 45]

October 5, 2001: Study Reveals "Significant Oil and Gas Deposits" in Afghanistan

Contrary to popular belief, Afghanistan "has significant oil and gas deposits. During the Soviets' decade-long occupation of Afghanistan, Moscow estimated Afghanistan's proven and probable natural gas reserves at around five trillion cubic feet and production reached 275 million cubic feet per day in the mid-1970s." Nonstop war since has prevented further exploitation. [ASIA TIMES, 10/5/01] A later article suggests that the country may also have as much copper as Chile, the world's largest producer, and significant deposits of coal, emeralds, tungsten, lead, zinc, uranium ore and more. Estimates of Afghanistan's natural wealth may even be understated, because surveys were conducted decades ago, using less-advanced methods and covering limited territory. [HOUSTON CHRONICLE, 12/23/01]

October 9, 2001: Afghan Pipeline Idea Is Revived

U.S. Ambassador Wendy Chamberlin meets with the Pakistani oil minister. She is briefed on the gas pipeline project from Turkmenistan, across Afghanistan, to Pakistan, which appears to be revived "in view of recent geopolitical developments"—in other words, the 9/11 attacks. [FRONTIER POST, 10/10/01]

December 2, 2001: Enron Files for Bankruptcy

Enron files for Chapter 11 bankruptcy—the biggest bankruptcy in history up to that date. [BBC, 1/10/02] However, in 2002 Enron will reorganize as a pipeline company and will continue working on its controversial Dabhol power plant. [HOUSTON BUSINESS JOURNAL, 3/15/02]

December 8, 2001: U.S. Oil Companies to Invest $200 Billion in Kazakhstan

During a visit to Kazakhstan in Central Asia, Secretary of State Powell states that U.S. oil companies are likely to invest $200 billion in Kazakhstan alone in the next five to ten years. [NEW YORK TIMES, 12/15/01]

December 22, 2001: Karzai Assumes Power in Afghanistan

Afghani Prime Minister Hamid Karzai and his transitional government assume power in Afghanistan. It had been revealed a few weeks before that he had been a paid consultant for Unocal, as well as Deputy Foreign Minister for the Taliban. [LE MONDE, 12/13/01; CNN, 12/22/01 (B)]

Hamid Karzai

January 1, 2002: Ex-Unocal Employee Becomes U.S. Special Envoy to Afghanistan

Zalmay Khalilzad, already a Special Assistant to the President, is appointed by Bush as a special envoy to Afghanistan. [BBC, 1/1/02] In his former role as Unocal adviser, Khalilzad participated in negotiations with the Taliban to build a pipeline through Afghanistan. He also wrote op-eds in the *Washington Post* in 1997 supporting the Taliban regime, back when Unocal was hoping to work with the Taliban. [INDEPENDENT, 1/10/02]

February 9, 2002: Pakistani and Afghan Leaders Revive Afghanistan Pipeline Idea

Pakistani President Musharraf and Afghan leader Hamid Karzai announce their agreement to "cooperate in all spheres of activity" including the proposed Central Asian pipeline, which they call "in the interest of both countries." [IRISH TIMES, 2/9/02; GULF NEWS, 2/9/02]

February 14, 2002: U.S. Military Bases Line Afghan Pipeline Route

The Israeli newspaper *Ma'ariv* notes: "If one looks at the map of the big American bases created [in the Afghan war], one is struck by the fact that they are completely identical to the route of the projected oil pipeline to the Indian Ocean." *Ma'ariv* also states, "Osama bin Laden did not comprehend that his actions serve American interests . . . If I were a believer in conspiracy theory, I would think that bin Laden is an American agent. Not being one I can only wonder at the coincidence." [CHICAGO TRIBUNE, 3/18/02]

May 30, 2002: Afghan, Turkmen, and Pakistani Leaders Sign Pipeline Deal

Afghanistan's interim leader, Hamid Karzai, Turkmenistan's President Niyazov, and Pakistani President Musharraf meet in Islamabad and sign a memorandum of understanding on the trans-Afghanistan gas

pipeline project. [ALEXANDER'S GAS AND OIL CONNECTIONS, 6/8/02; DAWN, 5/31/02] Afghan leader Hamid Karzai (who formerly worked for Unocal) calls Unocal the "lead company" in building the pipeline. [BBC, 5/13/02] The *Los Angeles Times* comments, "To some here, it looked like the fix was in for Unocal when President Bush named a former Unocal consultant, Zalmay Khalilzad, as his special envoy to Afghanistan late last year." [LOS ANGELES TIMES, 5/30/02 (B)]

Turkmenistan's President Niyazov, and Pakistani President Musharraf

November 1, 2002: Caspian Oil Potential Was Wildly Overestimated

Steven Mann, Director of the State Department's Caspian Basin Energy Policy Office, points out that the Caspian Sea nations contain 50 billion barrels of proven oil reserves. [ASSOCIATED PRESS, 11/1/02] "Caspian oil represents four percent of the world reserves. It will never dominate the world markets, but it will have an important role to play," said Mann. He concludes that the Caspian Sea energy "will not be a second Persian Gulf." [ASSOCIATED PRESS, 11/1/02] This estimation comes as a great surprise. The Energy Information Agency (EIA) had predicted that the Caspian region would contain in excess of 200 billion barrels of oil. [ENERGY, 1/1/03] Two months later, the magazine *Energy* will comment, "What ever happened to all the talk of a new oil utopia in the Caspian Sea and Central Asia? Word was that Caspian-Central Asian oil reserves would dwarf the Middle East. Yet, in the year since the Afghan War began, it seems that all the rumors of Caspian riches have died out and the center of oil interest has returned once again to Saudi Arabia and Iraq." [ENERGY, 1/1/03]

December 27, 2002: Afghanistan, Pakistan, and Turkmenistan Agree on Building Pipeline

Afghanistan, Pakistan and Turkmenistan reach an agreement in principle to build the Trans-Afghanistan Pipeline, a $3.2 billion project that has been delayed for many years. Skeptics say the project would require an indefinite foreign military presence in Afghanistan. [ASSOCIATED PRESS, 12/26/02; BBC, 12/27/02; BBC, 5/30/02] As of mid-2004, construction has yet to begin.

PART II
THE TERRORISTS

6.

Al-Qaeda in Germany

According to the FBI, "much of what took place on [9/11] originated during the mid-1990s when [hijackers] Mohamed Atta, Marwan Alshehhi, and Ziad Jarrah moved to Germany, eventually settling in Hamburg, and began to associate with Islamic extremists." While the men were not radical when they arrived in Germany, they became so while there, after they met a group of radical Islamists at the al Quds mosque in Hamburg, Germany. [THE 9/11 INVESTIGATIONS, BY STEVEN STRASSER, 5/04, PP. 414-45] These hijackers are said to have been the pilots for three of the hijacked airplanes. Thus, Germany is a logical point to begin understanding the hijackers—and the development of the 9/11 plot.

This chapter does not detail the backgrounds of the hijackers, nor does it detail all of their movements and actions while in Germany. Rather, it focuses on information relating to the issue of whether the 9/11 plot could have been stopped in Germany.

By 1999, U.S. and German intelligence had been investigating an al-Qaeda cell in Hamburg, of which the three German-based hijackers were members for some time. U.S. intelligence began investigating an associate of Mohammed Haydar Zammar, one of al-Qaeda's Hamburg cell members, as early as 1993. [THE 9/11 INVESTIGATIONS, BY STEVEN STRASSER, 5/04, P. 416] Associates of the hijackers, such as Mohammed Haydar Zammar, Mounir el Motassadeq, Mamoun Darkazanli, and Said Bahaji were periodically investigated. Wiretaps and personal surveillance were used. The intelligence obtained through these investigations could have led to the 9/11 hijackers. Indeed, according to some reports, chief hijacker Mohamed Atta was under direct surveillance by the CIA during this period. While it appears that in most cases Germany shared key intelligence with the U.S., the U.S. repeatedly failed to disclose its own intelligence to Germany.

It is not yet fully known what the German and/or U.S. governments knew about the Hamburg al-Qaeda cell and its members who would become 9/11 hijackers. German prosecutor Matthias Krauss, who investigated the Hamburg al-Qaeda cell, was scheduled to testify before the 9/11 Commission in June 2004 about both pre-9/11 communication problems between German and U.S. intelligence officials and the U.S. government's cooperation with foreign governments prosecuting suspected terrorists in the post-9/11 period. However, he unexpectedly cancelled at the last minute. [ASSOCIATED PRESS, 6/15/04] Documents showing what and when German officials knew of the Hamburg cell have been mysteriously lost. [HAMBURGER ABENDBLATT, 1/29/04] U.S. officials, while appearing before the 9/11 Commission, have been less than forthcoming on this issue. Based on what is known, however, it is difficult to understand both governments' failures to take action on the intelligence they had obtained, both before and after 9/11.

While it hasn't been confirmed, as this book goes to press the British *Observer* reports that German prosecutors are preparing to drop all the most serious charges against Mounir el Motassadeq, the only man convicted for the 9/11 attacks anywhere in the world. Prosecutors fear that crucial American evidence was obtained by torturing detainees. The key to the case was testimony from U.S. prisoners Ramzi bin al-Shibh and Khalid Shaikh Mohammed. A senior German intelligence official claims that the U.S. has now given the Germans their interrogations records, but they are so lacking in details as to be virtually useless. He said, "Their contents may be information and they may be disinformation." [OBSERVER, 7/18/04]

If Motassadeq is released, it will represent yet another failure. There are said to be over a dozen 9/11 plotters still in Germany. [CHICAGO TRIBUNE, 2/25/03] Yet none of these, nor any other members of the Hamburg cell, have been convicted of any 9/11-related crimes. The U.S. failed to give the German government vital information on the Hamburg cell before 9/11, and apparently continues to withhold information that would lead to convictions today. Why?

March 1997: German Government
Investigates Hamburg al-Qaeda Cell

Investigation of al-Qaeda contacts in Hamburg by the Office for the Protection of the Constitution (BfV), Germany's domestic intelligence service, begins at least by this time (Germany refuses to disclose additional details). [NEW YORK TIMES, 1/18/03] Telephone intercepts show that a German investigation into Mohammed Haydar Zammar is taking place this month. It is later believed that Zammar, a German of Syrian origin, is a part of the Hamburg al-Qaeda cell. [LOS ANGELES TIMES, 1/14/03] He later claims he recruited Mohamed Atta and others into the cell. [WASHINGTON POST, 6/19/02] Germany authorities had identified Zammar as a militant a decade earlier. [NEW YORK TIMES, 1/18/03] From 1995–2000, he makes frequent trips to Afghanistan. [NEW YORK TIMES, 1/18/03; STERN, 8/13/03] German intelligence is aware that he was personally invited to Afghanistan by bin Laden. [FRANKFURTER ALLGEMEINE ZEITUNG, 2/2/03] Spanish investigators later say Zammar is a longtime associate of Barakat Yarkas, the alleged boss of the al-Qaeda cell in Madrid, Spain. In 1998, Germany is given more evidence of his terrorist ties, and surveillance intensifies. He is periodically trailed, and all his calls are recorded. [STERN, 8/13/03] It is not clear if or when the investigation ends, but it continues until at least September 1999. [ASSOCIATED PRESS, 6/22/02]

**Mohammed
Haydar
Zammar**

August 1998: Germany Investigates Hamburg al-Qaeda Cell Member

A German inquiry into Mounir el Motassadeq, an alleged member of the Hamburg al-Qaeda cell with Mohamed Atta, begins by this time. Although Germany will not reveal details, documents show that by August 1998, Motassadeq is under surveillance. "The trail soon [leads] to most of the main [Hamburg] participants" in 9/11. Surveillance records Motassadeq and Mohammed Haydar Zammar, who had already been identified by police as a suspected extremist, as they meet at the Hamburg home of Said Bahaji, who is also under surveillance that same year. (Bahaji will soon move into an apartment with Atta and other al-Qaeda members.) German police monitor several other meetings between Motassadeq and Zammar in the following months. [NEW YORK TIMES, 1/18/03] Motassadeq is later convicted in August 2002 in Germany for participation in the 9/11 attacks, but his conviction is later overturned.

**Mounir el
Motassadeq**

September 20, 1998: Terrorist Arrested in
Germany Leads to New German Suspects

Mamdouh Mahmud Salim, an al-Qaeda terrorist from the United Arab Emirates connected to the 1998 U.S. embassy bombings, is arrested near Munich, Germany. [PBS NEWSHOUR, 9/30/98] In retrospect, it appears he was making one of many visits to the al-Qaeda cells in Germany. [THE BASE, BY JANE CORBIN, 8/02, P. 147] U.S. investigators later call him bin Laden's "right hand man." [NEW YORK TIMES, 9/29/01] However, the FBI is unwilling to brief their German counterparts on what they know about Salim and al-Qaeda, despite learning much that could have been useful as part of their investigation into the U.S. embassy bombings. By the end of the year, German investigators learn that Salim has a Hamburg bank account. [NEW YORK TIMES, 9/29/01] The cosignatory on the account is businessman Mamoun Darkazanli, whose home number had been programmed into Salim's cell phone. [CHICAGO TRIBUNE, 11/17/02] U.S. intelligence had first investigated Darkazanli in 1993, when a suspect was found with his telephone number [9/11 CONGRESSIONAL INQUIRY, 7/24/03], and German authorities had begun to suspect Darkazanli of terrorist money laundering in 1996. Wadih El-Hage, a former personal secretary to bin Laden, is also arrested in the wake of the embassy bombings. El-Hage had created a number of shell companies as fronts for al-Qaeda terrorist activities, and one of these uses the address of Darkazanli's apartment. [CHICAGO TRIBUNE, 11/17/02] Darkazanli's phone number and Deutsche Bank account number are also found in El-Hage's address book. [CNN, 10/16/01] The FBI also discovers that Darkazanli had power of attorney over a bank account of Hajer, a person on al-Qaeda's supreme council. [9/11 CONGRESSIONAL INQUIRY, 7/24/03] Based on these new connections, investigators ask a federal prosecutor for permission to open a formal investigation against Darkazanli. An investigation reportedly begins, at the insistence of the U.S., though Germany has claimed the request for the investigation was rejected. [AGENCE FRANCE-PRESSE, 10/28/01; NEW YORK TIMES, 1/18/03] German investigators also learn of a connection between Salim and Mohammed Haydar Zammar, who is already identified by police in 1997 as a suspected extremist. [ASSOCIATED PRESS, 6/22/02; NEW YORK TIMES, 1/18/03]

November 1, 1998–February 2001: Atta and Other Terrorists Are
Monitored by U.S., Germany in Hamburg Apartment

Mohamed Atta and al-Qaeda terrorists Said Bahaji and Ramzi bin al-Shibh move into a four bedroom apartment at 54 Marienstrasse, in Hamburg, Germany, and stay there until February 2001 (Atta is already living primarily in the U.S. well before this time). Investigators believe this move marks the formation of their Hamburg al-Qaeda terrorist cell. [LOS ANGELES TIMES, 1/27/02; NEW YORK TIMES, 9/10/02] Up to six men at a time live at the apartment, including other al-Qaeda agents such as hijacker Marwan Alshehhi and cell member Zakariya Essabar. [NEW YORK TIMES, 9/15/01 (F)] During the 28 months Atta's name is on the apartment lease, 29 ethnically Middle Eastern or North African men register the apartment as their home address. From the very beginning, the apartment was officially under surveillance by German intelligence, because of investigations into businessman Mamoun Darkazanli that connect to Said Bahaji. [WASHINGTON POST, 10/23/01] The Germans also suspect connections between Bahaji and al-Qaeda operative Mohammed Haydar Zammar. [LOS ANGELES TIMES, 9/1/02] German intelligence monitors the apartment off and on for months, and wiretaps Mounir el Motassadeq, an associate of the apartment-mates who is later put on trial in August 2002 for assisting the 9/11 plot, but apparently do not find any indication of suspicious activity. [CHICAGO TRIBUNE, 9/5/02] Bahaji is directly monitored at least for part of 1998, but German officials have not disclosed when the probe began or ended. That investigation is dropped for lack of evidence. [ASSOCIATED PRESS, 6/22/02; LOS ANGELES TIMES, 9/1/02] It is now clear that investigators would have found evidence if they looked more thoroughly. For instance, Zammar, a talkative man who has trouble keeping secrets, is a frequent visitor to the many late night meetings there. [CHICAGO TRIBUNE, 9/5/02; LOS ANGELES TIMES, 9/1/02; THE CELL, BY JOHN MILLER, MICHAEL STONE, AND CHRIS MITCHELL, 8/02, PP. 259–60] Another visitor later

Said Bahaji

recalls Atta and others discussing attacking the U.S. [KNIGHT RIDDER, 9/9/02] 9/11 mastermind Khalid Shaikh Mohammed is in Hamburg several times in 1999, and comes to the apartment. However, although there was a $2 million reward for Mohammed since 1998, the U.S. apparently fails to tell Germany what it knows about him. [NEW YORK TIMES, 11/4/02; NEWSWEEK, 9/4/02] Hijacker Waleed Alshehri also apparently stays at the apartment "at times." [WASHINGTON POST, 9/14/01; WASHINGTON POST, 9/16/01 (B)] The CIA also starts monitoring Atta while he is living at this apartment, and does not tell Germany of the surveillance. Remarkably, the German government will claim it knew little about the Hamburg al-Qaeda cell before 9/11, and nothing directed it them towards the Marienstrasse apartment. [DAILY TELEGRAPH, 11/24/01]

January 31, 1999: Germany Monitors Hijacker's Calls, Shares Information with CIA

German intelligence is tapping the telephone of al-Qaeda terrorist Mohammed Haydar Zammar, and on this date, Zammar gets a call from a "Marwan." This is later found to be hijacker Marwan Alshehhi. Marwan talks about mundane things, like his studies in Bonn, Germany, and promises to come to Hamburg in a few months. German investigators trace the telephone number and determine the call came from a mobile phone registered in the United Arab Emirates. [9/11 CONGRESSIONAL INQUIRY, 7/24/03; DEUTCHE PRESSE-AGENTEUR, 8/13/03; NEW YORK TIMES, 2/24/04] German intelligence will pass this information to the CIA about one month later, but the CIA apparently fails to capitalize on it.

February 17, 1999: Germans Intercept al-Qaeda Calls, One Mentions Atta's Name

German intelligence is periodically tapping suspected al-Qaeda terrorist Mohammed Haydar Zammar's telephone. On this day, investigators hear a caller being told Zammar is at a meeting with "Mohamed, Ramzi, and Said," and can be reached at the phone number of the Marienstrasse apartment where all three of them live. This refers to Mohamed Atta, Ramzi bin al-Shibh, and Said Bahaji, all members of the Hamburg al-Qaeda cell. However, apparently the German police fail to grasp its importance of these names, even though Said Bahaji is also under investigation. [ASSOCIATED PRESS, 6/22/02; NEW YORK TIMES, 1/18/03] Atta's last name is given as well. Agents check the phone number and confirm the street address, but it is not known what they make of the information. [DER SPIEGEL, 2/3/03]

March 1999: Germany Provides CIA Hijacker's
Name and Telephone Number

German intelligence gives the CIA the first name of hijacker Marwan Alshehhi and his telephone number in the United Arab Emirates. The Germans learned the information from surveillance of suspected terrorists. They tell the CIA that Alshehhi has been in contact with suspected al-Qaeda members Mohammed Haydar Zammar and Mamoun Darkazanli. He is described as a United Arab Emirates student who has spent some time studying in Germany. [9/11 CONGRESSIONAL INQUIRY, 7/24/03; DEUTCHE PRESSE-AGENTEUR, 8/13/03; *NEW YORK TIMES*, 2/24/04] The Germans consider this information "particularly valuable" and ask the CIA to track Alshehhi, but the CIA never responds until after the 9/11 attacks. The CIA decides at the time that this "Marwan" is probably an associate of bin Laden but never track him down. It is not clear why the CIA fails to act, or if they learn his last name before 9/11. [*NEW YORK TIMES*, 2/24/04] The Germans monitor others calls between Alshehhi and Zammar, but it isn't clear if the CIA is also told of these or not.

Friends of Ziad Jarrah at a wedding on April 1, 1999. Third from left in back row is Abdelghani Mzoudi, fifth is Mounir El Motassadeq, seventh is Ramzi bin al-Shibh. Far right in middle row is Mohamed Atta; Atta rests his hands on Mohammed Raji, now a wanted man.

April 1, 1999: Hijacker Jarrah
Photographs Wedding Party

Ziad Jarrah has an unofficial wedding with his girlfriend, Aisel Senguen. Interestingly, a photo apparently taken by Jarrah at the wedding is found by German intelligence several days after 9/11. An undercover agent is able immediately to identify ten of the 18 men in the photo, as well as where it was taken: the prayer room of Hamburg's Al-Quds mosque. He is also able to identify which of them attended Mohamed Atta's study group. He knows even "seemingly trivial details" of some of the men, showing that "probably almost all members of the Hamburg terror cell" has been watched by German state intelligence since this time, if not before. The head of the state intelligence had previously maintained that they knew nothing of any of these men. [*FRANKFURTER ALLGEMEINE ZEITUNG*, 2/2/03]

Summer 1999: U.S. Intelligence Links Zammar to
Senior bin Laden Operatives, Fails to Share Information

Around this time, U.S. intelligence notes that a man in Hamburg, Germany named Mohammed Haydar Zammar is in direct contact with one of bin Laden's senior operational coordinators. Zammar is an al-Qaeda recruiter with links to Mohamed Atta and the rest of the Hamburg terror cell. The U.S. had noted Zammar's terror links on "numerous occasions" before 9/11. [9/11 CONGRESSIONAL INQUIRY 7/24/03] However, apparently the U.S. does not share their information on Zammar with German intelligence.

Instead, the Germans are given evidence from Turkey that Zammar is running a travel agency as a terror front in Hamburg. In 1998, they get information from Italy confirming he is a real terrorist. However, his behavior is so suspicious that they have already started monitoring him closely. [STERN, 8/13/03; 9/11 CONGRESSIONAL INQUIRY, 7/24/03]

September 21, 1999: German Intelligence Records Calls Between Hijacker and Others Linked to al-Qaeda

German intelligence is periodically tapping suspected al-Qaeda terrorist Mohammed Haydar Zammar's telephone, and on this day investigators hear Zammar call hijacker Marwan Alshehhi. Officials initially claim that the call also mentions hijacker Mohamed Atta, but only his first name. [DAILY TELEGRAPH, 11/24/01; NEW YORK TIMES, 1/18/03] However, his full name, "Mohamed Atta Al Amir," is mentioned in this call and in another recorded call. [FRANKFURTER ALLGEMEINE ZEITUNG, 2/2/03] Alshehhi makes veiled references to plans to travel to Afghanistan. He also hands the phone over to Said Bahaji (another member of the Hamburg cell under investigation at the time), so he can talk to Zammar. [STERN, 8/13/03] German investigators still do not know Alshehhi's full name, but they recognize this "Marwan" also called Zammar in January, and they told the CIA about that call. Alshehhi, living in the United Arab Emirates at the time, calls Zammar frequently. German intelligence asks the United Arab Emirates to identify the number and the caller, but the request is not answered. [DER SPIEGEL, 2/3/03]

Late November 1999: Hamburg Cell Members Arrive in Afghanistan for Training

Investigators believe hijackers Mohamed Atta, Marwan Alshehhi, and Ziad Jarrah and associates Ramzi bin al-Shibh and Said Bahaji (all members of the same Hamburg, Germany, cell) arrive separately in Afghanistan around this time. They meet with bin Laden and train for several months. [CBS NEWS, 10/9/02; NEW YORK TIMES, 9/10/02] In a 2002 interview with Al Jazeera, bin al-Shibh says, "We had a meeting attended by all four pilots including Nawaf Alhazmi, Atta's right-hand man." The *Guardian* interprets this to mean that Alhazmi flew Flight 77, not Hani Hanjour as popularly believed. [GUARDIAN, 9/9/02]

December 1999: CIA Attempts to Recruit Man with Links to Atta and Hamburg Cell

The CIA begins "persistent" efforts to recruit German businessman Mamoun Darkazanli as an informant. Darkazanli knows Mohamed Atta and the other members of the Hamburg al-Qaeda cell. U.S. and German intelligence had previously opened investigations into Darkazanli in September 1998. Agents occasionally followed him, but Darkazanli obviously noticed the trail on him at least once. More costly and time-consuming electronic surveillance is not done however, and by the end of 1999, the investigation has produced little of value. German law does not allow foreign governments to have informants in Germany. So this month, Thomas Volz, the undercover CIA representative in Hamburg appears at the headquarters of the Hamburg state domestic intelligence agency, the LfV, responsible for tracking terrorists

Mamoun Darkazanli

and domestic extremists. He tells them the CIA believes Darkazanli has knowledge of an unspecified terrorist plot and encourages that he be "turned" against his al-Qaeda comrades. A source later recalls he says, "Darkazanli knows a lot." Efforts to recruit him will continue in the spring next year. The CIA has not admitted this interest in Darkazanli. [CHICAGO TRIBUNE, 11/17/02; STERN, 8/13/03]

2000: German Intelligence Issues Report on al-Qaeda Connections in Germany

The BKA, the German counterpart to the FBI, prepares an extensive report on al-Qaeda's connections in Germany. The BKA warns that "unknown structures" are preparing to stage attacks abroad. However, the German federal prosecutor's office rejects a proposed follow-up investigation. One of the persons named in the BKA report supposedly had contacts with the Hamburg terror cell. [BERLINER ZEITUNG, 9/24/01]

January–May 2000: CIA Has Atta Under Surveillance

Hijacker Mohamed Atta is put under surveillance by the CIA while living in Germany. [AGENCE FRANCE-PRESSE, 9/22/01; FOCUS, 9/24/01; BERLINER ZEITUNG, 9/24/01] He is "reportedly observed buying large quantities of chemicals in Frankfurt, apparently for the production of explosives [and/or] for biological warfare." "The U.S. agents reported to have trailed Atta are said to have failed to inform the German authorities about their investigation," even as the Germans are investigating many of his associates. "The disclosure that Atta was being trailed by police long before 11 September raises the question why the attacks could not have been prevented with the man's arrest." [OBSERVER, 9/30/01] A German newspaper adds that Atta is able to get a visa into the U.S. on May 18. According to some reports, the surveillance stops when he leaves for the U.S. at the start of June. However, "experts believe that the suspect [remains] under surveillance in the United States." [BERLINER ZEITUNG, 9/24/01] A German intelligence official also states, "We can no longer exclude the possibility that the Americans wanted to keep an eye on Atta after his entry in the U.S." [FOCUS, 9/24/01] This correlates with a *Newsweek* claim that U.S. officials knew Atta was a "known [associate] of Islamic terrorists well before [9/11]." [NEWSWEEK, 9/20/01] However, a congressional inquiry later reports that the U.S. "intelligence community possessed no intelligence or law enforcement information linking 16 of the 19 hijackers [including Atta] to terrorism or terrorist groups." [9/11 CONGRESSIONAL INQUIRY, 9/20/02]

January–March 2000: Hijacker Atta Begins E-mailing U.S. Flight Schools, Seeking Pilot Training

Hijacker Mohamed Atta, who is under CIA surveillance at this time, begins sending e-mails to U.S. flight schools, inquiring about their pilot training programs. One e-mail states, "We are a small group (2–3) of young men from different arab [sic] countries." "Now we are living in Germany since a while for study purposes. We would like to start training for the career of airline professional pilots. In this field we haven't yet any knowledge, but we are ready to undergo an intensive training program." Apparently, multiple e-mails are sent from the same Hotmail account. Some e-mails are signed "M. Atta," while others are signed "Mahmoud Ben Hamad." [CHICAGO TRIBUNE, 2/26/04]

Spring 2000: CIA Continues Efforts to Recruit Man Close to Hamburg Cell

German investigators finally agree to the CIA's request to recruit businessman Mamoun Darkazanli as an informant. An agent of the LfV, the Hamburg state intelligence agency, casually approaches Darkazanli and asks him whether he is interested in becoming a spy. Darkazanli replies that he is just a businessman who knows nothing about al-Qaeda or terrorism. The Germans inform the local CIA representative that the approach failed. The CIA agent persists, asking the German agent to continue to try. However, when German agents ask for more information to show Darkazanli they know of his terrorist ties, the CIA fails to give them any information. As it happens, at the end of January 2000, Darkazanli had just met with terrorist Barakat Yarkas in Madrid, Spain. [CHICAGO TRIBUNE, 11/17/02] Darkazanli is a longtime friend and business partner of Yarkas, the most prominent al-Qaeda agent in Spain. [LOS ANGELES TIMES, 1/14/03] The meeting included other suspected al-Qaeda figures, and it was monitored by Spanish police. If the CIA is aware of the Madrid meeting, they do not tell the Germans. [CHICAGO TRIBUNE, 11/17/02] A second LfV attempt to recruit Darkazanli also fails. The CIA then attempts to work with federal German intelligence officials in Berlin to "turn" Darkazanli. Results of that effort are not known. [CHICAGO TRIBUNE, 11/17/02]

May 22, 2000: German Intelligence Place Two Hijacker Associates on Watch List

By early 2000, German intelligence monitoring al-Qaeda suspect Mohammed Haydar Zammar notice that Mounir el Motassadeq and Said Bahaji, members of al-Qaeda's Hamburg cell with Mohamed Atta, regularly meet with Zammar. [9/11 CONGRESSIONAL INQUIRY, 7/24/03] In March 2000, Germany's internal intelligence service had placed Motassadeq and Bahaji on a border patrol watch list. Their international arrivals and departures are to be reported immediately. On this day, Motassadeq flies to Istanbul, Turkey, and from there goes to an al-Qaeda training camp in Afghanistan. Unfortunately, the border patrol only notes his destination of Istanbul. Bahaji does not travel, and when he finally does the week before 9/11, it isn't noted. [DER SPIEGEL, 2/3/03]

June or July 2001: Hijackers Plan Attacks from German University

Mohamed Atta, Marwan Alshehhi, and an unknown third person are seen in the ground-floor workshops of the architecture department at this time, according to at least two witnesses from the Hamburg university where Atta had studied. They are seen on at least two occasions with a white, three-foot scale model of the Pentagon. Between 60 and 80 slides of the Sears building in Chicago and the WTC are found to be missing from the technical library after 9/11. [SUNDAY TIMES, 2/3/02]

August 14, 2001: Atta's Hamburg Associates Purchase Tickets to Pakistan

Two apparent associates of Mohamed Atta's Hamburg al-Qaeda cell, Ismail Ben Mrabete and Ahmed Taleb, purchase tickets to fly to Pakistan on September 3, 2001. They will be joined on that flight by cell member Said Bahaji. All three will disappear into Afghanistan thereafter. It is later discovered that Taleb had been in e-mail contact with al-Qaeda leader Abu Zubaida. [CHICAGO TRIBUNE, 2/25/03]

September 3–5, 2001: Members of Hamburg's Terrorist Cell Leave for Pakistan

Members of Mohamed Atta's Hamburg terrorist cell leave Germany for Pakistan. Said Bahaji flies out of Hamburg on September 3. [CHICAGO TRIBUNE, 2/25/03] German intelligence already has Bahaji under surveillance, and German border guards are under orders to report if he leaves the country, yet the border guards fail to note his departure. [FRANKFURTER ALLGEMEINE ZEITUNG, 2/2/03] German agents later discover two other passengers on the same flight traveling with false passports who stay in the same room with Bahaji when they arrive in Karachi, Pakistan. [LOS ANGELES TIMES, 9/1/02] Investigators now believe his flight companions were Ismail Ben Mrabete and Ahmed Taleb, both Algerians in their late 40s. Three more associates—Mohammed Belfatmi, an Algerian extremist from the Tarragona region of Spain, and two brothers with the last name Joya—also travel on the same plane. To date none of these people have been located. [CHICAGO TRIBUNE, 2/25/03; CHICAGO TRIBUNE, 2/26/03] Ramzi bin al-Shibh flies out of Germany on September 5 and stays in Spain a few days before presumably heading for Pakistan. [LOS ANGELES TIMES 9/1/02]

September 24, 2001: Bank Accounts of Suspected al-Qaeda Supporters Frozen by U.S.

The U.S. freezes the accounts of 27 individuals and organizations, alleging that they had channeled money to al-Qaeda. Included in the list is Mamoun Darkazanli. U.S. officials say Darkazanli took part in a 1996 attack on government troops in Saudi Arabia. According to German investigators, Darkazanli attended Said Bahaji's wedding several years earlier. [NEW YORK TIMES, 9/29/01] The German government also freezes accounts connected to Darkazanli on October 2, 2001. Both governments suspect Darkazanli of providing financial and logistical support to the Hamburg al-Qaeda cell. [AGENCE FRANCE-PRESSE, 10/28/01] Shortly thereafter, Spanish police listening in to Barakat Yarkas' telephone hear Yarkas warn the leader of a Syrian extremist organization that Darkazanli has caught the "flu" going around. [CHICAGO TRIBUNE, 11/17/02]

October 7, 2001: Stolen 9/11 Documents Appear in Mysterious Circumstances

On this day, Zeljko E., a Kosovar Serb, enters a Hamburg, Germany police station and says he wants to turn himself in. He tells the police that he has robbed a business and stolen piles of paper written in Arabic, with the hopes of selling them. A friend of his told him that they relate to the 9/11 attacks. The 44 pounds of papers are translated and they prove to be a "treasure trove." The documents come from Mamoun Darkazanli's files, which were not in Darkazanli's apartment when police raided it two days after 9/11. "It makes for a great story. A petty thief pilfers files containing critical information about the largest terrorist attack in history and dutifully turns them over to the police. [But German] agents do not buy this story for a minute; they suspect that some other secret service was trying to find a way of getting evidence into [their] hands. The question is, whose secret service?" Some German investigators later suggest that the CIA was responsible; there are also reports that the FBI illegally monitored Darkazanli after 9/11. [DER SPIEGEL, 10/27/01; CHICAGO TRIBUNE, 11/17/02; INSIDE 9-11: WHAT REALLY HAPPENED, BY DER SPIEGEL MAGAZINE, 2/02, PP. 166–67]

October 27, 2001: Zammar Arrested, Detained by U.S. in Syria

Suspected terrorist Mohammed Haydar Zammar travels from Germany to Morocco. Not long after, perhaps in December, he is arrested by Moroccan police with U.S. assistance. Although he is a German citizen and under investigation by Germany, German intelligence remain unaware of his arrest, and only learn about it from the newspapers in June 2002. He is sent to Syria, where there are formal charges against him. Zammar reportedly now claims he recruited Mohamed Atta and others into the al-Qaeda Hamburg cell. [WASHINGTON POST, 6/19/02 (B)] It is widely suspected that the U.S. arranged for Zammar to be sent to Syria so that he could be more thoroughly interrogated using torture. The Germans are angry that the U.S. has been submitting questions for Zammar and learning answers from Syria, but have not informed Germany of what they have learned. [CHRISTIAN SCIENCE MONITOR, 7/26/02; DAILY TELEGRAPH, 6/20/02]

April 23, 2002: Spain Arrests al-Qaeda Financier

Spanish authorities arrest Syrian-born Spanish businessman Mohammed Galeb Kalaje Zouaydi, alleging that he is a key al-Qaeda financier. [CHICAGO TRIBUNE, 5/6/02] An accountant, Zouaydi is considered to be the "big financier" behind the al-Qaeda network in Europe, according to French investigator Jean-Charles Brisard. From 1996 to 2001, Zouaydi lived in Saudi Arabia and funnelled money into a series of companies set up to accept donations. (The source of the donations is unknown.) Around $1 million was then forwarded to al-Qaeda agents throughout Europe, especially to Germany. Mohamed Atta's Hamburg apartment telephone number was saved in the cell phone memory of one of Zouaydi's associates. [AGENCE FRANCE-PRESSE, 9/20/02] Zouaydi also allegedly sent money to Mamoun Darkazanli, a Syrian-born businessman who has admitted knowing Atta and others in the Hamburg al-Qaeda cell. [CHICAGO TRIBUNE, 5/6/02] One of Zouaydi's employees in Spain visited the WTC in 1997. While there, he extensively videotaped the buildings. Perhaps only coincidentally, while in Saudi Arabia, Zouaydi "was an accountant for the al-Faisal branch of the Saudi royal family, including Prince Mohammed al-Faisal al-Saud and Prince Turki al-Faisal." [AGENCE FRANCE-PRESSE, 9/20/02]

August 29, 2002: Germany Charges Moroccan with Complicity in 9/11

German authorities charge Mounir el Motassadeq with complicity in the 9/11 attacks. He was arrested in Germany two months after 9/11. He is only the second person in the world to be charged with any crime related to the 9/11 attacks (after Zacarias Moussaoui). He is charged with helping finance Mohamed Atta and others in the Hamburg terrorist cell. [AGENCE FRANCE-PRESSE, 8/29/02; NEW YORK TIMES, 8/29/02]

October 25, 2002: German-U.S. Breakdown in
Communications Hampers Anti-Terrorism Measures

PBS *Newshour* reports, "[German authorities] say they're not getting the cooperation they need from the authorities in the [U.S.], and they're worried that a political dispute between Washington and Berlin is hampering their ability to protect the public . . . In Hamburg, the police say that breakdown in communications between the U.S. and German governments has also led to a dramatic reduction in the amount of investigative help they're getting from the [U.S.]" The Bush administration has not spoken to the German government since it won re-election four months earlier while openly opposing Bush's planned war on Iraq. Germans say existing prosecutions of 9/11 suspects are now threatened by the information breakdown. [ONLINE NEWSHOUR, 10/25/02] The Germans helped capture terrorist Mohamed Heidar Zammar and turned him over to a third country, yet now they're learning very little from his interrogations, even though he has admitted to being involved in a plot to attack a consulate in Germany. A U.S. State Department official denies there is any problem, aside from a few "bumps in the road." [NEW YORK TIMES, 11/4/02]

February 18, 2003: Al-Qaeda Member Convicted in Germany

Mounir el Motassadeq, an alleged member of Mohamed Atta's Hamburg al-Qaeda cell, is convicted in Germany of accessory to murder in the 9/11 attacks. His is given the maximum sentence of 15 years. [ASSOCIATED PRESS, 2/19/03] Motassadeq admitted varying degrees of contact with Atta, Marwan Alshehhi, Ramzi bin al-Shibh, Said Bahaji, Ziad Jarrah, and Zakariya Essabar; admitted he had been given power of attorney over Alshehhi's bank account; and admitted attending an al-Qaeda training camp in Afghanistan from May to August 2000; but he claimed he had nothing to do with 9/11. [NEW YORK TIMES, 10/24/02] The conviction is the first one related to 9/11, but as the *Independent* puts it, "there are doubts whether there will ever be a second." This is because intelligence agencies have been reluctant to turn over evidence, or give access to requested witnesses. In Motassadeq's case, his lawyers tried several times unsuccessfully to obtain testimony by two of his friends, bin al-Shibh and Mohammed Haydar Zammar—a lack of evidence that will later become grounds for overturning his conviction. [INDEPENDENT, 2/20/03]

February 25, 2003: More 9/11 Terrorist Associates Believed to Be in Germany

The *Chicago Tribune* reveals that there appear to be many more members of Mohamed Atta's Hamburg cell than previously reported. While many members of the cell died in the attacks or fled Germany just prior to 9/11, up to a dozen suspected of belonging to the Hamburg cell stayed behind, apparently hoping to avoid government scrutiny. Many of their names have not yet been revealed. In some cases, investigators still do not know the names. For instance, phone records show that someone using the alias Karl Herweg was in close communication with the Hamburg cell and Zacarias Moussaoui, but Herweg's real identity is not known. [CHICAGO TRIBUNE, 2/25/03]

May 9, 2003: Mzoudi Is Charged for Role in 9/11 Attacks

Abdelghani Mzoudi is charged in Germany for an alleged role in the 9/11 plot. The 30-year-old electrical engineering student from Morocco is accused of accessory to murder and membership of a terrorist organization. He is alleged to have trained in Afghanistan, transferred money, and provided other logistical support to his fellow cell members involved in the 9/11 attacks. Mzoudi had known lead hijacker Mohamed Atta since 1996 and had roomed with Mounir Motassadeq, another Moroccan who

Abdelghani Mzoudi

was convicted of the same charges. Mzoudi denies any involvement in the hijacking plans. [ASSOCIATED PRESS, 5/9/03; *WASHINGTON POST*, 5/10/03; *WASHINGTON POST*, 8/15/03] In Mzoudi's trial, which begins in August 2003, his lawyers say they may explore theories during the trial about how the 9/11 attacks suspiciously served the foreign policy goals of U.S. conservatives. One defense attorney says, "As I take a close look at the results of the investigations through my glasses, I find anomalies that are immediately apparent. They begin with passenger lists that include the Arabic names of people who are still very much alive today." [*WASHINGTON POST*, 8/15/03; *DER SPIEGEL*, 9/8/03]

September 17, 2003: Spain Charges Darkazanli, 34 Others with Involvement in 9/11 Plot

A Spanish judge issues an indictment against Mamoun Darkanzali and 34 others, alleging that they belonged to or supported the al-Qaeda cell in Madrid, which assisted the 9/11 hijackers in planning the attack. Darkazanli's name appears 177 times in the 690-page indictment. He is accused of acting as bin Laden's "financier in Europe." "The list of those with whom Darkazanli has done business or otherwise exchanged money reads like a *Who's Who* of al-Qaeda: Wadih El-Hage, bin Laden's one-time personal secretary; Seedi al-Tayyib, the husband of bin Laden's niece and, before 9/11, al-Qaeda's chief financial officer; and Mustafa Setmariam Nasr, the head of a training camp for al-Qaeda in Afghanistan who journeyed to Hamburg to visit Darkazanli in 1996." [*CHICAGO TRIBUNE*, 10/5/03] The CIA had been monitoring Darkanzali sometime before December 1999 and had tried to convince Germany to "turn" him into an al-Qaeda informant. However, the CIA refused Germany's request to share information regarding Darkanzali's terrorist ties in the spring of 2000. [*CHICAGO TRIBUNE*, 11/17/02]

December 11, 2003: Secret Testimony Leads to Release of Mzoudi

Abdelghani Mzoudi, charged by the German government in assisting the 9/11 plot, is released from custody, pending completion of his trial. Mzoudi is released on bail following evidence submitted by Germany's federal criminal office of secret testimony from an unnamed informant who says that Mzoudi was not involved with the planning for the attacks. The presiding judge in the case identifies captured al-Qaeda operative Ramzi bin al-Shibh as the likely source of the testimony. [*GUARDIAN*, 12/12/03; BBC 1/21/04; REUTERS, 1/22/04; *CHICAGO TRIBUNE*, 1/22/04] Presumably, this information comes from the U.S. government. However, U.S. authorities have repeatedly rejected German attempts to have bin al-Shibh appear in court for his testimony to be examined. [ASSOCIATED PRESS, 10/23/03; AGENCE FRANCE-PRESSE, 1/22/04]

January 22, 2004: Iranian Spy Gives Evidence at Mzoudi Trial; Is Quickly Discounted

The prosecution in the trial of Abdelghani Mzoudi presents a witness who claims to be a defector from an Iranian intelligence agency. [BBC NEWS, 1/21/04] The witness, Hamid Reza Zakeri, does not appear in court himself, but instead Judge Klaus Ruehle reads out his testimony. [REUTERS, 1/22/04] According to Zakeri, the Iranian intelligence service was really behind the 9/11 attacks and had employed al-Qaeda to carry them out. Zakeri's claims are widely publicized. However, these claims are quickly discounted, and German intelligence notes that, "he presents himself as a witness on any theme which can bring him benefit." [DEUTSCHE PRESSE-AGENTEUR, 1/22/04; *CHICAGO TRIBUNE*, 1/22/04; REUTERS, 1/22/04; ASSOCIATED PRESS, 1/30/04]

January 29, 2004: Extent of German Government Knowledge of Hijackers Is Reported Missing

A Hamburg, Germany newspaper reports that a former senior official in the Hamburg state administration named Walter Wellinghausen has taken a "politically explosive" file from the government offices. "The file is said to contain an exact chronology of the knowledge of the [Hamburg] intelligence agency before September 11, 2001 about the people living in Hamburg who should later become the terrorists." He claims to have not been charged or even questioned about this matter and the file remains missing. [*HAMBURGER ABENDBLATT*, 1/29/04]

March 3, 2004: U.S. Secrecy Leads to Overturning of Motassadeq Conviction

A German appeals court overturns the conviction of Mounir el Motassadeq after finding that German and U.S. authorities withheld evidence. He had been sentenced to 15 years in prison for involvement in the 9/11 plot. According to the court, a key suspect in U.S. custody, Ramzi bin al-Shibh, had not been allowed to testify. European commentators blame U.S. secrecy, complaining that "the German justice system [is] suffering 'from the weaknesses of the way America is dealing with 9/11,' and 'absolute secrecy leads absolutely certainly to flawed trials.'" [AGENCE FRANCE PRESSE, 3/5/04] The court orders a new trial scheduled to begin later in the year. [ASSOCIATED PRESS, 3/4/04; *HOUSTON CHRONICLE*, 4/24/04] The release of Motassadeq (and the acquittal of Mzoudi earlier in the year) means that there is not a single person who has ever been successfully prosecuted for the events of 9/11.

7.

Hijackers Nawaf Alhazmi and Khalid Almihdhar

Of all the identified 9/11 hijackers, Nawaf Alhazmi and Khalid Almihdhar have the longest records of involvement with al-Qaeda. Of all the hijackers, they were the only ones known to have attended the pivotal al-Qaeda meeting in Malaysia in January 2000.

It appears that U.S. intelligence had more information about Alhazmi and Almihdhar prior to 9/11 than about any other identified hijackers. It also appears that the U.S. had the opportunity to learn even more, but declined to do so. The U.S. intelligence community had learned of Alhazmi and Almihdhar at least as early as 1999. As was too often the case in similar circumstances, vital information regarding these hijackers obtained by the CIA apparently was not shared with the FBI or NSA. For example, even after Alhazmi and Almihdhar were identified to the CIA as having attended a key January 2000 al-Qaeda strategy meeting in Malaysia, the CIA failed to add them to the terrorist watch list, and apparently failed to inform either the FBI or the NSA of their al-Qaeda connection. In other situations, the NSA obtained information but failed to share it with the CIA or FBI. Each agency held several pieces of the puzzle with respect to these two hijackers.

One week after the January 2000 al-Qaeda strategy meeting in Malaysia, Alhazmi and Almihdhar arrived in the U.S. and settled openly in San Diego. The CIA quickly learned that these men with clear al-Qaeda connections had arrived in the U.S., but it apparently did nothing, nor did it inform the FBI.

While in the U.S., the hijackers communicated with several other individuals known to have terrorist connections who were also under investigation.

Not until mid-August 2001 were the hijackers added to the U.S. terrorism watch list—but this was only for international flights, and they weren't added to a much smaller domestic U.S. terrorism watch list. So they were able to obtain plane tickets for the flights that they would later hijack on 9/11.

The case of Alhazmi and Almihdhar provides a compelling picture of how completely dysfunctional the relationships between U.S. intelligence agencies had become in failing to share key information—and the tragic consequences of this failure.

Interestingly, in July 2004, as this book was going to press, it was revealed that the Justice Department's Office of Inspector General (OIG) had completed a report examining how the FBI handled information obtained on Alhazmi and Almihdhar prior to 9/11. (The OIG report also addresses the FBI's handling of the "Phoenix Memo" submitted by Arizona Field Agent Ken Williams, addressed in chapter 1, and the Minnesota Field Office's request to obtain a search warrant for Zacarias Moussaoui, addressed in chapter 10.) Unfortunately the report remains classified, although senior Senate Judiciary Committee members Patrick Leahy (D) and Charles Grassley (R) have called for its release. (The senators also called for the release of two other reports addressing the accusations of Sibel Edmonds relating to the FBI's translation program, addressed in chapters 1 and 20.) According to the senators, "While the needs of national security must be weighed seriously, we fear the designation of information as classified, in some cases, serves to protect the executive branch against embarrassing revelations and full accountability. We hope that is not the case here." [WASHINGTON TIMES, 7/12/04]

1993–1999: Alhazmi and Almihdhar Fight for al-Qaeda

Of all the 9/11 hijackers, Nawaf Alhazmi and Khalid Almihdhar have the longest records of involvement with al-Qaeda. CIA Director Tenet calls them al-Qaeda veterans. According to the CIA, Alhazmi first travels to Afghanistan in 1993 as a teenager. In 1995, he travels with Almihdhar to Bosnia and fights against the Serbs. Almihdhar makes his first visit to Afghanistan training camps in 1996, and then fights in Chechnya in 1997. Both swear loyalty to bin Laden around 1998. Alhazmi fights in Afghanistan against the Northern Alliance with his brother, Salem Alhazmi. He fights in Chechnya, probably in 1998. He returns to Saudi Arabia in early 1999 and shares information about the 1998 U.S. embassy bombings. [OBSERVER, 9/23/01; ABC NEWS, 1/9/02; LOS ANGELES TIMES, 9/1/02; 9/11 CONGRESSIONAL INQUIRY, 7/24/03; CIA DIRECTOR TENET TESTIMONY, 6/18/02]

Nawaf Alhazmi (left) and Khalid Almihdhar (right)

August 1994–July 2001: Possible Terrorist Front Man with Saudi Backing Settles in San Diego

A Saudi named Omar al-Bayoumi arrives in San Diego, California. He will later become well known for his suspicious connections to both some 9/11 hijackers and the Saudi government, although the 9/11 Commission asserts that it received no evidence that he was involved in terrorism or the 9/11 attacks. [9/11 COMMISSION REPORT, 6/16/04] Acquaintances in San Diego long suspect he is a Saudi government spy reporting on the activities of Saudi-born college students. [SAN DIEGO UNION-TRIBUNE, 9/14/02; NEWSWEEK, 11/22/02; SAN DIEGO MAGAZINE, 9/03] Says one witness, "He was always watching [young Saudi college students], always checking up on them, literally following them around and then apparently reporting their activities back to Saudi Arabia." [NEWSWEEK, 11/24/02] Just prior to moving to the U.S., he worked for the Saudi Ministry of Defense and Aviation, headed by Prince Sultan. His salary in this job is approved by Hamid al-Rashid, a Saudi government official whose son, Saud al-Rashid, is strongly suspected of al-Qaeda ties. [9/11 CONGRESSIONAL INQUIRY, 7/24/03] Once in San Diego, al-Bayoumi tells people that he's a student or a pilot, and even claims to be receiving monthly payments from "family in India" (despite being Saudi). However, he is none of those things. [SUNDAY MERCURY, 10/21/01; WALL STREET JOURNAL, 8/11/03] In fact, as he tells some people, he

Omar al-Bayoumi

receives a monthly stipend from Dallah Avco, a Saudi aviation company that has extensive ties to the same Saudi Ministry of Defense and Aviation. [LOS ANGELES TIMES, 9/1/02; NEWSWEEK, 11/24/02] From early 1995 until 2002, he is paid about $3,000 a month for a project in Saudi Arabia even though he is living in the U.S. According to the *New York Times*, Congressional officials believe he is a "ghost employee" doing no actual work. The classified section of the 9/11 Congressional Inquiry report notes that his payments increase significantly just after he comes into contact with two hijackers in early 2000. [NEW YORK TIMES, 8/2/03] The FBI is investigating possible ties between Dallah Avco and al-Qaeda. [NEWSWEEK, 10/29/02] The firm's owner, Saudi billionaire Saleh Abdullah Kamel, has denied the accusation. [NEWSWEEK, 7/28/03]

April 1998: Hijacker Associate Receives Saudi Money; FBI Fails to Investigate

Osama Basnan, a Saudi living in California, claims to write a letter to Saudi Arabian Prince Bandar bin Sultan and his wife, Princess Haifa bint Faisal, asking for financial help because his wife needs thyroid surgery. The Saudi embassy sends Basnan $15,000 and pays the surgical bill. However, according to University of California at San Diego hospital records, Basnan's wife, Majeda Dweikat, is not treated until April 2000. [LOS ANGELES TIMES, 11/24/02] Basnan will later come under investigation for possibly using some of this money to support two of the 9/11 hijackers who arrive in San Diego, although the 9/11 Commission has concluded that evidence does not support these charges. [9/11 COMMISSION REPORT, 6/16/04] Prior to this time, the FBI had several chances to investigate Basnan, but failed to do so. In 1992, they received information suggesting a connection between him and a terror group later associated with bin Laden. In 1993, they received reports that Basnan hosted a party for terrorist leader Sheikh Omar Abdul-Rahman the year before, but again they failed to investigate. [9/11 CONGRESSIONAL INQUIRY, 7/24/03] According to one U.S. official, Basnan later "celebrate[s] the heroes of September 11" and talks about "what a wonderful, glorious day it had been" at a party shortly after 9/11. [NEWSWEEK, 12/24/02; SAN DIEGO MAGAZINE, 9/03]

June 1998: Saudi Benefactor Offers Funding for Mosque
if Suspected Advance Man Is Retained as Building Manager

An unknown Saudi benefactor pays a Saudi, Saad Al-Habeeb, to buy a building in San Diego, California, for a new Kurdish community mosque. However, the approximately $500,000 will only be given on the condition that Omar al-Bayoumi be installed as the building's maintenance manager with a private office at the mosque. After taking the job, al-Bayoumi rarely shows up for work. [NEWSWEEK, 11/24/02; SAN DIEGO MAGAZINE, 9/03] This means he has two jobs at once. The people in the mosque eventually begin a move to replace al-Bayoumi, but he moves to Britain in July 2001 before this can happen. [NEWSWEEK, 11/24/02] An anonymous federal investigator states, "Al-Bayoumi came here, set everything up financially, set up the San Diego [terrorist] cell and set up the mosque." An international tax attorney notes that anyone handling business for a mosque or a church could set it up as a tax-exempt charitable organization "and it can easily be used for money laundering." [SAN DIEGO UNION-TRIBUNE, 10/27/01; SAN DIEGO UNION-TRIBUNE, 10/22/02]

Late August 1998: Captured Terrorist
Leads U.S. to Safe House Phone Number

An al-Qaeda operative involved in the bombing of the U.S. embassy in Nairobi is captured and interrogated by the FBI. The FBI learns the telephone number of a safe house in Yemen, owned by bin Laden associate Ahmed Al-Hada, hijacker Khalid Almihdhar's father-in-law. [NEWSWEEK, 6/2/02; DIE ZEIT, 10/1/02] U.S. intelligence also learns that the safe house is an al-Qaeda "logistics center," used by agents around the world to communicate with each other and plan attacks. [NEWSWEEK, 6/2/02] It is later revealed that bin Laden called the safe house dozens of times from 1996 to 1998 (the two years he had a traced satellite phone). [SUNDAY TIMES, 3/24/02; LOS ANGELES TIMES, 9/1/02] The NSA and CIA jointly plant bugs inside the house, tap the phones, and monitor visitors with spy satellites. [MIRROR, 6/9/02] The NSA later records Khalid Almihdhar and other hijackers calling this house, including calls from the U.S. In late 1999, the phone line will lead the CIA to an important al-Qaeda meeting in Malaysia. [NEWSWEEK, 6/2/02] It appears al-Qaeda continues to use this phone line until the safe house is raided by the Yemeni government in February 2002. [CBS NEWS, 2/13/02]

September 1998–July 1999: FBI Conducts
Inquiry of Suspected al-Qaeda Advance Man

The FBI conducts a counterterrorism inquiry on Omar al-Bayoumi, suspected al-Qaeda advance man, and possible Saudi agent. The FBI discovers he has been in contact with several people also under investigation. [9/11 CONGRESSIONAL INQUIRY, 7/24/03] The FBI is given a tip that he was sent a suspicious package filled with wire from the Middle East, and that large numbers of Arab men routinely meet in his apartment. His landlord notices that he switches from driving a beat up old car to a new Mercedes. [NEWSWEEK, 7/28/03] According to the 9/11 Congressional Inquiry, the FBI notes that al-Bayoumi has "access to seemingly unlimited funding from Saudi Arabia." For instance, an FBI source identifies him as a person who has delivered about $500,000 from Saudi Arabia to buy a mosque in June 1998. However, the FBI closes the inquiry "for reasons that remain unclear." [9/11 CONGRESSIONAL INQUIRY, 7/24/03] Also in 1999, al-Bayoumi is working as an employee of the Saudi company Dallah Avco but apparently is doing no work. Someone in the company tries to fire him and sends a note to the Saudi government about this, since the company is so closely tied to the government. However, Mohammed Ahmed al-Salmi, the Director General of Civil Aviation, replies that it is "extremely urgent" his job is renewed "as quickly as possible," and so he keeps his job. [WALL STREET JOURNAL, 8/11/03]

Early 1999: NSA Monitoring Hears 9/11 Hijacker Names,
This Information Is Not Shared with CIA or FBI

As the NSA continues to monitor an al-Qaeda safe house in Yemen owned by hijacker Khalid Almihdhar's father-in-law, they find references to Almihdhar and the hijacker brothers, Salem and Nawaf Alhazmi. According to analysts, around late 1999 these men are among their very best sources on al-Qaeda, with "notorious" links to the organization and direct links to the 1998 U.S. embassy bombings. [9/11 COM-

Salem Alhazmi

MISSION REPORT, 1/26/04] In early 1999, the NSA intercepts communications mentioning the full name "Nawaf Alhazmi." More communications involving the names Nawaf, Salem, and Khalid together and alone are intercepted during 1999. However, this information is not shared with the CIA or FBI. As a result, as an important al-Qaeda meeting approaches in Malaysia (to be held January 5–8, 2000), the CIA does not know the last name of the "Nawaf" attending the meeting. [CONGRESSIONAL INQUIRY, 7/24/03; ASSOCIATED PRESS, 9/25/02; NSA DIRECTOR CONGRESSIONAL TESTIMONY, 10/17/02] In mid-January, after the Malaysian meeting is over, the CIA reports to the NSA what it has learned about Almihdhar, and asks the NSA for information about him. Some information about him is given back and some is not. The NSA still fails to report Alhazmi's last name. [9/11 CONGRESSIONAL INQUIRY, 7/24/03] The NSA continues to monitor calls from these hijackers to this safe house after they move to the U.S. in the first half of 2000.

April 3–7, 1999: Three Hijackers Obtain U.S. Visas

Hijackers Nawaf Alhazmi, Salem Alhazmi, and Khalid Almihdhar obtain U.S. visas through the U.S. Consulate in Jeddah, Saudi Arabia. [9/11 CONGRESSIONAL INQUIRY, 7/24/03 (B)] All three are already "al-Qaeda veterans" and battle-hardened killers. Almihdhar's visa is issued on April 7, and he thereafter leaves and returns to the U.S. multiple times until April 6, 2000. [STERN, 8/13/03] Nawaf Alhazmi gets the same kind of visa; details about Salem are unknown. The CIA claims the hijackers then travel to Afghanistan to participate in "special training" with at least one other suicide bomber on a different mission. The training is led by Khallad bin Attash. The U.S. learns about Almihdhar's visa in January 2000. The Jeddah Consulate keeps in its records the fact that Nawaf and Alhazmi obtain U.S. visas several days before Almihdhar, but apparently these records are never searched before 9/11. [9/11 CONGRESSIONAL INQUIRY, 7/24/03]

June 1999–March 2000: FBI Investigates Muslim Leader

The FBI conducts a counterterrorism inquiry into imam Anwar Al Aulaqi. From about February 2000, he serves as the "spiritual leader" to several of the hijackers while they live in San Diego, and again in March 2001 when he and they move to the East Coast. During the investigation, the FBI discovers he is in contact with a number of other people being investigated. For instance, in early 2000 he is visited by an associate of Sheikh Omar Abdul-Rahman. However, as the 9/11 Congressional Inquiry later notes, the investigation is closed "despite the imam's contacts with other subjects of counterterrorism investigations and reports concerning the imam's connection to suspect organizations." [9/11 CONGRESSIONAL INQUIRY 7/24/03]

November 1999: Hijackers Said to Lease Apartment in
San Diego, Two Months Before Alleged Arrival in U.S.

The *Washington Post* refers to hijackers Nawaf Alhazmi and Khalid Almihdhar when it later reports, "In November 1999, two Saudi Arabian men moved into a ground-floor apartment at the Parkwood Apartments, a town house complex near a busy commercial strip in San Diego." [WASHINGTON POST, 9/30/01]

Alhazmi's name is on the apartment lease beginning in November 1999. [WASHINGTON POST, 10/01] Some reports even have them visiting the U.S. as early as 1996. [WALL STREET JOURNAL, 9/17/01; LAS VEGAS REVIEW JOURNAL, 10/26/01] However, FBI Director Mueller has stated the two hijackers did not arrive in the U.S. until the middle of January 2000, after an important meeting in Malaysia. Some news reports mention that the hijackers first arrive in late 1999 [LOS ANGELES TIMES, 9/1/02 AND 11/24/02], but most reports concur with the FBI.

December 1999: U.S. Intelligence Learns of Planned al-Qaeda Meeting Involving Future Hijackers

A Yemeni safe house telephone monitored by the FBI and CIA reveals that there will be an important al-Qaeda meeting in Malaysia in January 2000, and that a "Khalid," a "Nawaf," and a "Salem" will attend. One intelligence agent notes at the time, "Salem may be Nawaf's younger brother." It turns out they are brothers—and the future hijackers Khalid Almihdhar, and Nawaf and Salem Alhazmi. U.S. intelligence is already referring to both of them as "terrorist operatives" because the safe house is known to be such a hotbed of al-Qaeda activity. Their last names are not yet known to the CIA, even though the NSA has learned Nawaf's last name. However, Khalid Almihdhar is at the safe house at the time, and as he travels to Malaysia, he is watched and followed. [9/11 CONGRESSIONAL INQUIRY, 7/24/03]

December 4, 1999: Saudi Ambassador's Wife Gives Funds That Are Possibly Passed to Hijackers

Princess Haifa bint Faisal, the wife of Prince Bandar, the Saudi Ambassador to the U.S., begins sending monthly cashier's checks of between $2,000 and $3,500 (accounts differ) to Majeda Dweikat, the Jordanian wife of Osama Basnan, a Saudi living in San Diego. Accounts also differ over when the checks were first sent (between November 1999 and about March 2000; a Saudi government representative has stated December 4, 1999 [FOX NEWS, 11/23/02]). Basnan's wife signs many of the checks over to her friend Manal Bajadr, the wife of Omar al-Bayoumi. [WASHINGTON TIMES, 11/26/02; NEWSWEEK, 11/22/02; NEWSWEEK, 11/24/02; GUARDIAN, 11/25/02] Some later suggest that the money from the wife of the Saudi ambassador passes through the al-Bayoumi and Basnan families as intermediaries and ends up in the hands of the two hijackers. The payments from Princess Haifa continue until May 2002 and may total $51,000, or as much as $73,000. [NEWSWEEK, 11/22/02; MSNBC, 11/27/02] While living in the San Diego area, al-Bayoumi and Basnan are heavily involved in helping with the relocation of, and offering financial support to, Saudi immigrants in the community. [LOS ANGELES TIMES, 11/24/02] In late 2002, al-Bayoumi claims he did not pass any money along to the hijackers. [WASHINGTON TIMES, 12/04/02] Basnan has variously claimed to know al-Bayoumi, not to know him at all, or to know him only vaguely. [ARAB NEWS, 11/26/02; ABC NEWS, 11/26/02; ABC NEWS, 11/25/02; MSNBC, 11/27/02] However, earlier reports say Basnan and his wife were "very good friends" of al-Bayoumi and his wife. Both couples lived at the Parkwood Apartments at the same time as the two hijackers; prior to that, the couples lived together in a different apartment complex. In addition, the two wives were arrested together in April 2001 for shoplifting. [SAN DIEGO UNION-TRIBUNE, 10/22/02]

December 11, 1999: Watch List Importance Is
Stressed but Procedures Are Not Followed

The CIA's Counter Terrorism Center sends a cable reminding all personnel about various reporting obligations. The cable clearly states that it is important to share information so terrorists can be placed on watch lists. The U.S. keeps a number of watch lists; the most important one, TIPOFF, contains about 61,000 names of suspected terrorists by 9/11. [LOS ANGELES TIMES, 9/22/02; KNIGHT RIDDER, 1/27/04] The list is checked whenever someone enters or leaves the U.S. "The threshold for adding a name to TIPOFF is low," and even a "reasonable suspicion" that a person is connected with a terrorist group, warrants being added to the database. [9/11 CONGRESSIONAL INQUIRY, 9/20/02] Within a month, two future hijackers, Nawaf Alhazmi and Khalid Almihdhar, are identified, but the cable's instructions are not followed for them. The CIA initially tells the 9/11 Congressional Inquiry that no such guidelines existed, and CIA Director Tenet fails to mention the cable in his testimony. [9/11 CONGRESSIONAL INQUIRY, 7/24/03 (B); NEW YORK TIMES, 5/15/03]

Late 1999: Saudis Claim to Add Two Hijackers to Watch List and Inform CIA

Former Saudi Intelligence Minister Prince Turki al Faisal later claims that around this time his intelligence agency tells the CIA that hijackers Nawaf Alhazmi and Khalid Almihdhar have been put on a Saudi terror watch list. Turki says, "What we told [the CIA] was these people were on our watch list from previous activities of al-Qaeda, in both the [1998] embassy bombings and attempts to smuggle arms into the kingdom in 1997." However, the CIA strongly denies any such warning. [ASSOCIATED PRESS, 10/16/03] Turki admits no documents concerning this were sent to the U.S., but claims the information was passed via word of mouth. [SALON, 10/18/03] The U.S. does not put these two on their watch list until August 2001.

January 2–5, 2000: CIA Tracks Alhazmi and Almihdhar to
al-Qaeda Meeting; Fails to Place Them on Terror Watch List

Hijackers Nawaf Alhazmi and Khalid Almihdhar travel to an important al-Qaeda meeting in Malaysia. Alhazmi is in Pakistan with a ticket to Malaysia for January 2. CIA and Pakistani officials plan to have his passport scrutinized as he passes through the airport, but he changes his ticket departure date twice. Officials get confused and are not there when he leaves the country, so they still don't learn his last name. Meanwhile, Almihdhar is watched when he leaves a safe house in Yemen. Agents from eight CIA offices and six friendly foreign intelligence services are all asked to help track him, in the hopes he will lead them to bigger al-Qaeda figures. [STERN, 8/13/03] United Arab Emirates officials secretly make copies of his passport as he is passing through the Dubai airport on his way to Malaysia and immediately report this to the CIA. Therefore, by the time he reaches Malaysia, the CIA knows his full name, and the fact that he has a multiple entry visa to the U.S. that is valid from April 1999 to April 2000. Even though the CIA now knows Almihdhar has a one-year visa to the U.S. and presumably plans to travel there, they do not place him on a terror watch list, despite receiving guidelines the previous month on the importance of doing so. The FBI also is not informed that a known terrorist has a valid U.S. visa. One FBI official familiar with the case complains, "[The CIA] purposely hid [him] from the FBI, purposely refused to tell the bureau. . . . The thing was, they didn't want John O'Neill and the FBI running over their case. And that's why September 11 happened. . . . They have blood on their hands." [A PRETEXT FOR WAR, BY JAMES BAMFORD, 6/04, P. 224] The FBI is merely told, "The operation is still going on. Until now, many suspicious activities were watched. But no evidence was found that indicated a coming attack or criminal acts." The FBI also should watch list Almihdhar merely from his description as a "terrorist operative," but they fail to do so. [9/11 COMMISSION REPORT, 1/26/04; STERN, 8/13/03; 9/11 CONGRESSIONAL INQUIRY, 7/24/03]

January 5–8, 2000: Al-Qaeda Summit in Malaysia Monitored by Authorities; Information Passed to U.S.

About a dozen of bin Laden's trusted followers hold a secret, "top-level al-Qaeda summit" in the city of Kuala Lumpur, Malaysia. [CNN, 8/30/02; SAN DIEGO UNION-TRIBUNE, 9/27/02] Plans for the October 2000 bombing of the USS *Cole* and the 9/11 attacks are discussed. [USA TODAY, 2/12/02; CNN, 8/30/02] At the request of the CIA, the Malaysian secret service monitors the meeting and then passes the information on to the U.S. Attendees of the meeting include:

- Hijackers Nawaf Alhazmi and Khalid Almihdhar. The CIA and FBI will later miss many opportunities to foil the 9/11 plot through Alhazmi and Almihdhar and the knowledge of their presence at this meeting. The CIA already knows many details about these two by the time the meeting begins.

- Khalid Shaikh Mohammed, a top al-Qaeda leader and the alleged "mastermind" of the 9/11 attacks. The U.S. has known Mohammed is a major terrorist since the exposure of Operation Bojinka in January 1995 (see chapter 1), and knows what he looks like. U.S. officials have stated that they only realized the meeting was important in the summer of 2001, but the presence of Mohammed should have proved the meeting's importance. [LOS ANGELES TIMES, 2/2/02] Although the possible presence of Mohammed at this meeting is highly disputed by U.S. officials, one terrorism expert testifies before the 9/11 Commission in 2003 that he has access to transcripts of Mohammed's interrogations since his capture, and that Mohammed admits leading this meeting. [NEWSWEEK, 7/9/03; NEW YORK POST, 7/10/03] Many media reports identify him there as well (for instance, [INDEPENDENT, 6/6/02; CNN, 8/30/02; CNN, 11/7/02; CANADIAN BROADCASTING CORP., 10/29/03]). According to *Newsweek*, "Mohammed's presence would make the intelligence failure of the CIA even greater. It would mean the agency literally watched as the 9/11 scheme was hatched—and had photographs of the attack's mastermind . . . doing the plotting." [NEWSWEEK, 7/9/03]

**Riduan
Isamuddin
("Hambali")**

- Riduan Isamuddin, an Indonesian terrorist better known as Hambali. [BBC 8/15/03] He was the main financier of Operation Bojinka. [CNN, 8/30/02; CNN, 3/14/02] Philippine intelligence officials learned of Hambali's importance in 1995, but did not track him down or share information about him. [CNN, 3/14/02] He will be arrested by Thai authorities in August 2003. [CNN, 8/14/03; CBS NEWS, 8/15/03]

- Yazid Sufaat, a Malaysian man who owned the condominium where the meeting was held. [NEW YORK TIMES, 1/31/02; NEWSWEEK, 6/2/02] A possibility to expose the 9/11 plot through Sufaat's presence at this meeting is later missed in September. Sufaat will travel to Afghanistan in June 2001; and be arrested by Malaysian authorities when he returns to Malaysia in late 2001. [AUSTRALIAN, 12/24/02]

- Fahad al-Quso, a top al-Qaeda operative. [NEWSWEEK, 9/20/01] Al-Quso will be arrested by Yemeni authorities in December 2000, but the FBI is not given a chance to interrogate him before 9/11. He will escape from prison in 2003. [CNN, 5/15/03]

- Tawifiq bin Attash, better known by his alias "Khallad." Bin Attash, a "trusted member of bin Laden's inner circle," was in charge of bin Laden's bodyguards, and served as bin Laden's personal intermediary at least for the USS *Cole* attack. [NEWSWEEK, 9/20/01] He is also thought to be a "mastermind" of that attack. Attash is reportedly planning to be one of the hijackers, but will be unable to get a U.S. visa. [9/11 COMMISSION REPORT, 6/16/04 (B)] A possibility to expose the 9/11 plot through bin Attash's presence at this meeting is later missed in January 2001. Bin Attash had been previously arrested in Yemen for suspected terror ties, but let go. [CONTEMPORARY SOUTHEAST ASIA, 12/1/02] He will be captured in Pakistan by the U.S. in April 2003. [NEW YORK TIMES, 5/1/03]

- Ramzi bin al-Shibh, who investigators believe wanted to be the twentieth hijacker. His presence at the meeting may not have been realized until after 9/11, despite the fact that U.S. intelligence had a picture of him next to bin Attash, and had video footage of him. [WASHINGTON POST, 7/14/02; TIME, 9/15/02; DIE ZEIT, 10/1/02; NEWSWEEK, 11/26/01; CNN, 11/7/02] German police have credit card receipts indicating bin al-Shibh is in Malaysia at the same time. [LOS ANGELES TIMES, 9/1/02] Anonymous Malaysian officials claim he is there, but U.S. officials deny it. [ASSOCIATED PRESS, 9/20/02] One account says he is recognized at the time of the meeting, which makes it hard to understand why he is not tracked back to Germany and the Hamburg cell with Mohamed Atta and other hijackers. [DER SPIEGEL, 10/1/02] Another opportunity to expose the 9/11 plot through bin al-Shibh's presence at this meeting will be missed in June. It appears bin al-Shibh and Almihdhar are directly involved in the attack on the USS *Cole* in October 2000 [NEWSWEEK, 9/4/02; WASHINGTON POST, 7/14/02; GUARDIAN, 10/15/01], so better surveillance or follow-up from this meeting could have prevented that attack as well.

- Ahmad Hikmat Shakir, an al-Qaeda agent of Iraqi nationality, may have attended this meeting, according to some documents [NEWSWEEK, 10/7/02; AUSTRALIAN, 12/24/02], but his presence at the meeting is uncertain. [ASSOCIATED PRESS, 10/2/02]

- Salem Alhazmi may have attended the meeting, although very few accounts mention it. [AUSTRALIAN, 12/24/02] U.S. intelligence intercepts from before the meeting indicate that he had plans to attend the meeting. [9/11 CONGRESSIONAL INQUIRY, 7/24/03; 11]

- Abu Bara al Taizi, a Yemeni al-Qaeda agent, is also said to attend. He is reportedly meant to be one of the hijackers, but will be unable to enter the U.S. due to greater scutiny for Yemenis. [9/11 COMMISSION REPORT, 6/16/04 (B)]

- More? Unnamed members of the Egyptian-based Islamic Jihad are also known to have been at the meeting. [COX NEWS SERVICE, 10/21/01] (The Islamic Jihad had merged with al-Qaeda in February 1998. [ABC NEWS, 11/17/01])

January 6–9, 2000: Malaysia Provides CIA with Information on al-Qaeda Summit and Attendees

At the CIA's request, the Malaysian secret service is monitoring an important al-Qaeda summit in Kuala Lumpur, Malaysia, and begins passing what it knows to the CIA even before the meeting is over. Media accounts are consistent that the terrorists at the meeting are photographed and even videotaped, but there is no wiretapping or other recording of their conversations. [NEWSWEEK, 6/2/02; OTTAWA CITIZEN, 9/17/01; OBSERVER, 10/7/01; CNN, 3/14/02; NEW YORKER, 1/14/02; CANADIAN BROADCASTING CORP., 10/29/03; STERN, 8/13/03] However, Malaysian officials are not informed what to look for, and focus more on monitoring the local Malaysian and Indonesian hosts who serve as drivers than the visitors attending the meeting. [ASSOCIATED PRESS, 9/20/02] Authorities find out what hotel Khalid Almihdhar is staying at and he and his associates are photographed there [NEWSWEEK, 9/20/01; OBSERVER, 10/7/01], as well as coming and going from the condo where the meeting is held. [LOS ANGELES TIMES, 9/1/02] On January 6, the CIA office in Malaysia begins passing details of the meeting to the CIA Counter Terrorism Center (CTC). Cofer Black, head of the CTC, orders that he be continually informed about the meeting, and CIA Director Tenet is frequently informed as well. [STERN, 8/13/03] National Security Adviser Sandy Berger, FBI Director Louis Freeh, and other top officials are briefed, but apparently President Clinton is not. [A PRETEXT FOR WAR, BY JAMES BAMFORD, 6/04, PP. 225-26] On January 7, Khalid Almihdhar and others go shopping, giving Malaysian security ample opportunity to collect information about them. They spend hours at Internet cafes, and after they leave, Malaysian intelligence searches the hard drives of the computers they used. [AUSTRALIAN, 12/24/02; STERN, 8/13/03] However, no photos or video and few details from any of this surveillance have been publicly released. It is known that some photos show Khallad bin Attash with Almihdhar, some show Fahad al-Quso next to Almihdhar, and that some photos are of Ramzi bin al-Shibh. By January 9, all the data and footage the Malaysians have collected are in the hands of the CIA. [STERN, 8/13/03; NEWSWEEK, 9/20/01]

January 8, 2000: Al-Qaeda Summit Ends; CIA Fails to Add Attendees to Watch List

The al-Qaeda terror summit in Malaysia ends and the participants leave. Hijackers Nawaf Alhazmi and Khalid Almihdhar fly to Bangkok, Thailand, traveling under their real names. Al-Qaeda leader Khallad bin Attash also travels with them and the three sit side by side in the airplane, but bin Attash travels under the false name "Salah Said." [ASSOCIATED PRESS, 9/20/02; 9/11 CONGRESSIONAL INQUIRY, 7/24/03] The CIA knows that a "Nawaf" has attended the meeting, but does not know his last name. Shortly afterwards, the CIA is told of this airplane flight, and the fact that the person sitting next to Almihdhar on the plane is named "Nawaf Alhazmi." CIA Headquarters asks the NSA to put Almihdhar on their watch list. [9/11 COMMISSION REPORT, 1/26/04] However, neither Alhazmi nor Almihdhar are placed on a terror watch list. The CIA still fails to tell the FBI that Almihdhar has a valid U.S. visa, and fails to give them Alhazmi's last name. [STERN, 8/13/03; 9/11 CONGRESSIONAL INQUIRY, 7/24/03] The CIA searches for the names in their databases but get no "hits." Yet they don't search the much larger NSA databases, which had vital information on them. [9/11 COMMISSION REPORT, 1/26/04]

January 8–15, 2000: CIA Fails to Apprehend Hijackers in Thailand

Hijackers Nawaf Alhazmi and Khalid Almihdhar stay in Thailand. Khallad bin Attash, who flew there with them, is met in Bangkok by two al-Qaeda operatives who give him money. Some of this money is reportedly passed on to Alhazmi and Almihdhar for their upcoming work in the U.S. [9/11 COMMISSION REPORT, 1/26/04] The CIA is trying to find Almihdhar, since they now know his full name and the fact that he has just come from an important al-Qaeda meeting. For six weeks, they look for him in Thailand. However, the search is unsuccessful, because, as one official puts it, "when they arrived we were unable to mobilize what we needed to mobilize." The CIA only alerts Thailand-based CIA officers to look for these men hours after they had already arrived and disappeared into the city. However, a few days later a CIA official notifies superiors that surveillance of the men is continuing, even though the men's location is unknown. [9/11 COMMISSION REPORT, 1/26/04] Nevertheless, in February, the CIA rejects a request from foreign authorities to give assistance. The CIA gives up the search in early March when an unnamed foreign government tells the CIA that Nawaf Alhazmi has already flown to the U.S. [9/11 CONGRESSIONAL INQUIRY 7/24/03]

January 15, 2000: Hijackers Alhazmi and Almihdhar Travel to U.S. Undetected

A week after the meeting in Malaysia, hijackers Nawaf Alhazmi and Khalid Almihdhar fly together from Bangkok, Thailand, to Los Angeles, California. [MSNBC, 12/11/01] Because the CIA has lost track of them in Thailand, no one in the U.S. government realizes they are coming to the U.S. The CIA will learn this information in early March, but still will take no action. [9/11 COMMISSION REPORT, 1/26/04]

January 15–Early February 2000: Suspected Advance Man Helps 9/11 Hijackers Settle in San Diego

Hijackers Nawaf Alhazmi and Khalid Almihdhar arrive in Los Angeles and stay there for two weeks. Omar al-Bayoumi, a suspected al-Qaeda advance man and possible Saudi agent, arrives in Los Angeles and visits the Saudi Consulate there. According to *Newsweek*, "Law-enforcement officials believe al-Bayoumi may [have] a closed-door meeting with Fahad al Thumairy, a member of the consulate's Islamic and Culture Affairs Section." [NEWSWEEK, 7/28/03] (In March 2003, al Thumairy is stripped of his diplomatic visa and barred from entry to the U.S., reportedly because of suspected links to terrorism. [WASHINGTON POST 11/23/03]) Later that same day, al-Bayoumi goes to a restaurant and meets Alhazmi and Almihdhar. Al-Bayoumi later claims that this first contact with the hijackers is accidental. However, one FBI source later recalls that before he drives to Los Angeles that day he says he is going "to pick up visitors." [NEWSWEEK, 7/28/03; 9/11 CONGRESSIONAL INQUIRY, 7/24/03] Al-Bayoumi returns to San Diego after inviting the two hijackers to move there; Alhazmi and Almihdhar follow him there shortly thereafter. [9/11 CONGRESSIONAL INQUIRY 7/24/03] The FBI's "best source" in San Diego says that al-Bayoumi "must be an intelligence officer for Saudi Arabia or another foreign power." A former top FBI official working on the al-Bayoumi investigation claims: "We firmly believed that he had knowledge [of the 9/11 plot], and that his meeting with them that day was more than coincidence." [NEWSWEEK 7/28/03] Al-Bayoumi helps Alhazmi and Almihdhar settle in the U.S. After meeting them in Los Angeles and after bringing them to San Diego, he finds them a place to live. Al-Bayoumi lives at the Villa Balboa apartments with a wife and children, and the two hijackers move into the Parkwood apartments directly across the street. [SUNDAY MERCURY, 10/21/01; LOS ANGELES TIMES, 9/1/02] It appears the lease was actually signed by Alhazmi a few months earlier. Al-Bayoumi cosigns the lease and pays $1,500 cash for their first month's rent and security deposit. Some FBI officials claim the hijackers immediately pay him back, others claim they do not. [NEWSWEEK, 11/24/02; 9/11 CONGRESSIONAL INQUIRY, 7/24/03] Within days of bringing them from Los Angeles, al-Bayoumi throws a welcoming party that introduces them to the local Muslim community. [WASHINGTON POST, 12/29/01] One associate later says an al-Bayoumi party "was a big deal . . . it meant that everyone accepted them without question." [SAN DIEGO UNION-TRIBUNE, 10/25/01] He also introduces hijacker Hani Hanjour to the community a short time later. [SAN DIEGO UNION-TRIBUNE, 9/14/02] He tasks an acquaintance, Modhar Abdallah, to serve as their translator and help them get driver's licenses, Social Security cards, information on flight schools, and more. [SAN DIEGO UNION-TRIBUNE, 9/8/02; 9/11 CONGRESSIONAL INQUIRY, 7/24/03]

Early February–Summer 2000: Hijackers
Alhazmi and Almihdhar Live Openly in San Diego

Hijackers Nawaf Alhazmi and Khalid Almihdhar move to San Diego and live there openly. [9/11 CONGRESSIONAL INQUIRY, 7/24/03] Hijacker Hani Hanjour joins them as a roommate in February 2000 but apparently does not stay long. [SAN DIEGO UNION-TRIBUNE, 9/21/01; SAN DIEGO CHANNEL 10, 9/18/01] The hijackers use their real names on their rental agreement [9/11 CONGRESSIONAL INQUIRY, 9/20/02], driver's licenses, Social Security cards, credit cards [NEWSWEEK, 6/2/02], car purchase, and bank account. Alhazmi is even listed in the 2000–2001 San Diego phone book. [SOUTH FLORIDA SUN-SENTINEL, 9/28/01; NEWSWEEK, 6/2/02] Neighbors notice odd behavior: They have no furniture, they are constantly using cell phones on the balcony, constantly playing flight simulator games, keep to themselves, and strange cars and limousines pick them up for short rides in the middle of the night. [WASHINGTON POST, 9/30/01; TIME, 9/24/01; TIME, 9/24/01 (B)]

March 5, 2000: CIA Learn Hijackers Have Entered U.S.; FBI Not Informed

Thailand tells the CIA that hijacker Nawaf Alhazmi had flown from the January meeting in Malaysia to Los Angeles. Thai intelligence actually knew this the same day they flew to the U.S., but they didn't share the information until the CIA finally asked them about it around this time. [9/11 COMMISSION REPORT, 1/26/04; NEW YORK TIMES, 10/17/02] According to a senior FBI official, the CIA also learns about hijacker Khalid Almihdhar: "In March 2000, the CIA received information concerning the entry of Almihdhar and Alhazmi into the United States." [MICHAEL ROLINCE TESTIMONY, 9/20/02] The CIA disputes this, however. [9/11 CONGRESSIONAL INQUIRY, 7/24/03] A cable is immediately sent to CIA Headquarters noting (at least) that Nawaf Alhazmi has traveled to Los Angeles. The cable is marked "Action Required: None, FYI [For Your Information]." CIA Director Tenet later claims, "Nobody read that cable in the March timeframe." [NEW YORK TIMES, 10/17/02] Yet the day after the cable is received, "another overseas CIA station note[s], in a cable to the bin Laden unit at CIA headquarters, that it had 'read with interest' the March cable, 'particularly the information that a member of this group traveled to the U.S. . . .'" [9/11 CONGRESSIONAL INQUIRY, 9/20/02] Yet again, CIA fails to put their names on a watch list, and again fails to alert the FBI so they can be tracked. [9/11 CONGRESSIONAL INQUIRY, 9/20/02] Senior CIA counterterrorism official Cofer Black later says, "I think that month we watch listed about 150 people. [The watch listing] should have been done. It wasn't." [9/11 CONGRESSIONAL INQUIRY, 7/24/03]

Spring 2000: Payments to Suspected Hijacker Associate Increase Significantly

According to leaks from the still-classified part of the 9/11 Congressional Inquiry, monthly payments to Omar al-Bayoumi increase significantly at this time. Al-Bayoumi has been receiving a salary from the Saudi civil authority of about $500 a month. However, shortly after hijackers Nawaf Alhazmi and Khalid Almihdhar move to San Diego, al-Bayoumi's salary increases to about $3,000 to $3,500 a month. [NEW YORK TIMES, 7/29/03 (B)] It is not clear whether this pay spike is from his Dallah Avco job, or an additional payment by the Saudi government [NEW YORK TIMES, 7/29/03 (B), NEW YORK TIMES, 8/2/03], but the pay spike appears to be a separate stream of money, because another report indicates his Dallah Avco job started with $3,000 a month payments and remained consistent. [WALL STREET JOURNAL, 8/11/03] It also fits in with his claims to acquaintances at the time that he is receiving a regular government scholarship. [LOS ANGELES TIMES, 9/1/02]

Spring–Summer 2000: NSA Intercepts Calls
Made by Hijacker Almihdhar in U.S., Fails to Trace Location

Hijacker Khalid Almihdhar, while living in San Diego, telephones an al-Qaeda safe house in Yemen owned by his father-in-law (note that the facility is not named, but references in the 9/11 Congressional Inquiry report are consistent with other mentions of this safe house). This safe house has been closely monitored since 1998, as even bin Laden himself makes calls to it. The NSA intercepts these calls but doesn't realize the "Khalid" calling the safe house is calling from the U.S. (This is only determined by an analysis of phone toll records obtained after 9/11.) The NSA had been aware of a "Khalid," "Nawaf Alhazmi," and his brother "Salem" having communications with this safe house in 1999. In summer 2000 there are additional communications to the safe house from "Khalid" and "Salem," but again the NSA does not realize the meaning or importance of these calls. There may have been more communications—the section of the 9/11 Congressional Inquiry dealing with these calls is heavily censored. Some, but not all, of the information about certain calls is passed on to the FBI and CIA. [9/11 CONGRESSIONAL INQUIRY, 7/24/03]

June 10, 2000: Almihdhar Flies from San Diego to Germany; Return Date Unclear

Hijacker Khalid Almihdhar flies from San Diego to Frankfurt, Germany. [9/11 CONGRESSIONAL INQUIRY, 9/20/02] Authorities later believe that Almihdhar visits his cousin-in-law Ramzi bin al-Shibh and bin al-Shibh's roommate Mohamed Atta and other al-Qaeda members in bin al-Shibh's terrorist cell. However, since the CIA fails to notify Germany about their suspicions of either Almihdhar or bin al-Shibh, both of whom were seen attending the al-Qaeda summit in Malaysia in January, German police fail to monitor them and another chance to uncover the 9/11 plot is missed. [DIE ZEIT. 10/1/02; 9/11 CONGRESSIONAL INQUIRY, 7/24/03] FBI Director Mueller and the congressional inquiry into 9/11 will claim that Almihdhar does not return to the U.S. for over a year [9/11 CONGRESSIONAL INQUIRY, 9/20/02; 9/11 CONGRESSIONAL INQUIRY, 9/26/02], despite obvious evidence to the contrary. For instance, an FBI agent is told Khalid Almihdhar is in the room when he calls Almihdhar's landlord in autumn 2000 and there are indications Almihdhar attends a flight school in Arizona in early 2001. [ARIZONA REPUBLIC. 9/28/01]

Summer 2000: San Diego Hijackers Meet Atta and Al-Bayoumi

Anonymous government sources later claim that Mohamed Atta visits fellow hijackers Nawaf Alhazmi, Khalid Almihdhar, and Omar al-Bayoumi. These same sources claim al-Bayoumi is identified after September 11 as an "advance man" for al-Qaeda. [WASHINGTON TIMES, 11/26/02] Other reports have suggested Atta visited Alhazmi and Almihdhar in San Diego, but the FBI has not confirmed this. [9/11 CONGRESSIONAL INQUIRY, 7/24/03]

Summer–December 2000: FBI Asset Withholds
Information About 9/11 Hijackers Living in His San Diego House

Hijackers Nawaf Alhazmi and Khalid Almihdhar move to the house of Abdussattar Shaikh in San Diego. [SAN DIEGO UNION-TRIBUNE, 9/16/01] Shaikh, a local Muslim leader, is later revealed to be a "tested" undercover "asset" working with the local FBI. [NEWSWEEK, 9/9/02] Shaikh inexplicably fails to tell his FBI handler important details about the hijackers and appears to be lying about many matters concerning them. In early media reports, the two are said to have moved in around September [SOUTH FLORIDA SUN-SENTINEL. 9/28/01; SAN DIEGO UNION-TRIBUNE, 9/16/01; WALL STREET JOURNAL, 9/17/01] but the 9/11 Congressional Inquiry implies that Shaikh lied about this, and they moved in much earlier. Alhazmi stays until December; Almihdhar appears to be mostly out of the U.S. after June. [9/11 CONGRESSIONAL INQUIRY, 7/24/03] Neighbors claim that Mohamed Atta is a frequent visitor, and Hani Hanjour visits as well. [CHICAGO TRIBUNE, 9/30/01; ASSOCIATED PRESS. 9/29/01; LAS VEGAS REVIEW JOURNAL. 10/26/01; SAN DIEGO CHANNEL 10. 9/27/01; SAN DIEGO CHANNEL 10. 10/11/01] However, Shaikh denies Atta's visits, the FBI never mentions them, and the media appears to have forgotten about them. [ASSOCIATED PRESS, 9/29/01] Echoing reports from their first apartment, neighbors witness strange late night visits with Alhazmi and Almihdhar. [ASSOCIATED PRESS. 9/16/01 (D)] For instance, one neighbor says, "There was always a series of cars driving up to the house late at night. Sometimes they were nice cars. Sometimes they had darkened windows. They'd stay about 10 minutes." [TIME, 9/24/01 (B)]

Autumn 2000: Hijackers Live and Work in San Diego; Connected with Other Potential Terrorists

Hijacker Nawaf Alhazmi works at a gas station while living in San Diego. This is the only apparent instance of any of the hijackers having a job while in the U.S. He and hijacker Khalid Almihdhar also frequently socialize at the gas station and Alhazmi works there on and off for about a month at some point after Almihdhar has gone overseas. [9/11 CONGRESSIONAL INQUIRY, 7/24/03; WASHINGTON POST, 12/29/01; LOS ANGELES TIMES, 9/1/02] The station, Sam's Star Mart, is owned by Osama "Sam" Mustafa. [SAN DIEGO UNION-TRIBUNE, 7/25/03] Mustafa was first investigated by the FBI in 1991 after he tells a police officer that the U.S. needs another Pan Am 103 attack and that he could be the one to carry out the attack. He also says all Americans should be killed because of the 1991 Iraq War. In 1994, he was investigated for being a member of the Palestinian terror groups PFLP and PLO and for threatening to kill an Israeli intelligence officer living in San Diego. The investigation was closed, but reopened again in 1997 when he was tied to a possible terror plot in North Carolina. Apparently, it is closed again before 9/11. He also associates with Osama Basnan and others who have contacts with the hijackers. Witnesses later claim he cheers when first told of the 9/11 attacks. [9/11 CONGRESSIONAL INQUIRY, 7/24/03] The gas station is managed by Ed Salamah. [SAN DIEGO UNION-TRIBUNE, 7/25/03; WASHINGTON POST, 12/29/01] In January 2000, the brother of a known al-Qaeda operative is under surveillance and is seen chatting with Salamah. The Los Angeles FBI office is investigating this operative, and it calls Salamah about it. Salamah refuses to come to Los Angeles for an interview, and refuses to give his home address to be interviewed there. Faced with a reluctant witness, the FBI drops the matter. [9/11 CONGRESSIONAL INQUIRY, 7/24/03; NEWSWEEK, 7/28/03] The hijackers are living with an FBI informant who is aware of their contact with at least Mustafa, and that informant has given reports about Mustafa to the FBI in the past. However, the informant fails to tell the FBI about their contacts with him. The 9/11 Congressional Inquiry strongly implies that Salamah and Mustafa assisted the hijackers with the 9/11 plot, but the FBI appears uninterested in them and maintains that the hijackers received no assistance from anyone. [9/11 CONGRESSIONAL INQUIRY, 7/24/03]

Autumn 2000: FBI Informant Fails to Share Valuable Information on Hijackers

While hijackers Nawaf Alhazmi and Khalid Almihdhar live in the house of an FBI informant, Abdussattar Shaikh, the informant continues to have contact with his FBI handler. The handler, Steven Butler, later claims that during summer Shaikh mentions the names "Nawaf" and "Khalid" in passing and that they are renting rooms from him. [NEWSWEEK, 9/9/02; ASSOCIATED PRESS, 7/25/03 (B), 9/11

CONGRESSIONAL INQUIRY, 7/24/03] On one occasion, Shaikh tells Butler on the phone he cannot talk because Khalid is in the room. [NEWSWEEK, 9/9/02] Shaikh tells Butler they are good, religious Muslims who are legally in the U.S. to visit and attend school. Butler asks Shaikh for their last names, but Shaikh refuses to provide them. Butler is not told that they are pursuing flight training. Shaikh tells Butler that they are apolitical and have done nothing to arouse suspicion. However, according to the 9/11 Congressional Inquiry, he later admits that Alhazmi has "contacts with at least four individuals [he] knew were of interest to the FBI and about whom [he] had previously reported to the FBI." Three of these four people are being actively investigated at the time the hijackers are there. [9/11 CONGRESSIONAL INQUIRY, 7/24/03] The report mentions Osama Mustafa as one, and Shaikh admits that suspected Saudi agent Omar al-Bayoumi was a friend. [LOS ANGELES TIMES, 7/25/03; 9/11 CONGRESSIONAL INQUIRY, 7/24/03] The FBI later concludes Shaikh is not involved in the 9/11 plot, but they have serious doubts about his credibility. After 9/11 he gives inaccurate information and has an "inconclusive" polygraph examination about his foreknowledge of the 9/11 attack. The FBI believes he has contact with hijacker Hani Hanjour, but he claims not to recognize him. There are other "significant inconsistencies" in Shaikh's statements about the hijackers, including when he first met them and later meetings with them. The 9/11 Congressional Inquiry later concludes that had the informant's contacts with the hijackers been capitalized upon, it "would have given the San Diego FBI field office perhaps the Intelligence Community's best chance to unravel the September 11 plot." [9/11 CONGRESSIONAL INQUIRY 7/24/03] The FBI later tries to prevent Butler and Shaikh from testifying before the 9/11 Congressional Inquiry in October 2002. Butler ends up testifying but Shaikh does not. [WASHINTON POST, 10/11/01]

Early December 2000: USS *Cole* Terrorist al-Quso Linked to Hijackers by CIA; Information Withheld from FBI

Terrorist Fahad al-Quso is arrested by the government of Yemen. [PBS FRONTLINE, 10/3/02; PBS FRONTLINE, 10/3/02] In addition to being involved in the USS *Cole* bombing, al-Quso was at the January 2000 Malaysian meeting with al-Qaeda agents Khallad bin Attash and hijackers Nawaf Alhazmi and Khalid Almihdhar. Al-Quso tells Yemeni investigators that he flew from Yemen to Bangkok in January 2000 for a secret meeting where he turned over $36,000 in cash to bin Attash. The FBI asks the CIA for more information about bin Attash and the Malaysian meeting, but later the FBI claims that the CIA does not provide the requested information that could have led them to Alhazmi and Almihdhar as well. [NEW YORK TIMES, 4/11/04 (B)] For instance, there are pictures from the Malaysian meeting of al-Quso next to hijacker Khalid Almihdhar, but the CIA does not share the pictures with the FBI before 9/11. [NEWSWEEK, 9/20/01] Meanwhile, FBI head investigator John O'Neill believes that al-Quso is holding back important information from his Yemeni captors and wants him interrogated by the FBI. However, O'Neill had been kicked out of Yemen by his superiors a week or two before, and without his influential presence, the Yemeni government will not allow an interrogation. Al-Quso is finally interrogated days after 9/11,

Fahad al-Quso

and he admits to meeting with Alhazmi and Almihdhar in January 2000. One investigator calls the missed opportunity of exposing the 9/11 plot through al-Quso's connections "mind-boggling." [PBS FRONT-LINE, 10/3/02]

January 4, 2001: FBI, CIA Miss Connection Between USS *Cole* Bomber and Hijackers

The FBI's investigation into the USS *Cole* bombing reveals that terrorist Khallad bin Attash had been a principal planner of the bombing [ASSOCIATED PRESS, 9/21/02 (B)], and that two other participants in the bombing had delivered money to bin Attash at the time of the January 2000 al-Qaeda meeting in Malaysia. The FBI shares this information with the CIA. Based on a description of bin Attash from an informant,

Tawfiq "Khallad" bin Attash

CIA analysts reexamine pictures from the Malaysian meeting and identify bin Attash with both hijackers Nawaf Alhazmi and Khalid Almihdhar. CIA Director Tenet later testifies that the presence of bin Attash, a known, important al-Qaeda operative, gives the Malaysian meetings "greater significance." [9/11 COMMISSION REPORT, 1/26/04] The CIA has already been informed that Alhazmi has entered the U.S. in March 2000, yet once again they fail to watch list either Alhazmi or Almihdhar. [9/11 CONGRESSIONAL INQUIRY, 7/24/03] CNN later notes that at this point the CIA, at the very least, "could have put Alhazmi and Almihdhar and all others who attended the meeting in Malaysia on a watch list to be kept out of this country. It was not done." [CNN, 6/4/02] More incredibly, bin Attash is not placed on the watch list at this time, despite being labeled as the principal planner of the *Cole* bombing. (He is finally placed on the watch list in August 2001.) [LOS ANGELES TIMES, 9/22/02] CIA headquarters is told what these CIA analysts have learned, but it appears the FBI is not told. [9/11 CONGRESSIONAL INQUIRY, 7/24/03]

March 2001: Hijackers Continue to Associate with Suspicious Imam

After living together in Phoenix since December 2000, hijackers Hani Hanjour and Nawaf Alhazmi move to Falls Church, Virginia. [9/11 COMMISSION REPORT, 1/26/04; WASHINGTON POST, 9/10/02 (B)] They live only a few blocks from where two nephews of bin Laden with ties to terrorism go to work. They continue to live there off and on until around August. They begin attending the Dar al Hijrah mosque. [WASHINGTON POST, 9/10/02 (B)] When they and Khalid Almihdhar lived in San Diego in early 2000, they attended a mosque there led by the imam Anwar Al Aulaqi. This imam moved to Falls Church in January 2001, and now the hijackers attend his sermons at the Dar al Hijrah mosque. Some later suspect that Aulaqi is part of the 9/11 plot because of their similar moves, and other reasons:

- The FBI says Aulaqi had closed door meetings with hijackers Nawaf Alhazmi and Khalid Almihdhar in 2000 while all three of them were living in San Diego. [9/11 CONGRESSIONAL INQUIRY, 7/24/03]

- Police later find the phone number of Aulaqi's mosque when they search "would-be twenti-eth hijacker" Ramzi bin al-Shibh's apartment in Germany. [9/11 CONGRESSIONAL INQUIRY, 7/24/03]

- The FBI was investigating Aulaqi for terrorist ties in early 2000.

- A neighbor of Aulaqi later claims that, in the first week of August 2001, Aulaqi knocks on his door and tells him he is leaving for Kuwait: "He came over before he left and told me that something very big was going to happen, and that he had to be out of the country when it happened." [NEWSWEEK, 7/28/03]

- Aulaqi is apparently in the country in late September, 2001, and claims not to recognize any of the hijackers. [COPLEY NEWS, 10/1/01]

- A week after 9/11, Aulaqi says the hijackers were framed, and suggests Israel was behind 9/11. [WASHINGTON POST, 7/23/03]

- Aulaqi leaves the U.S. in early 2002. [TIME, 8/11/03]

- In December 2002, Aulaqi briefly returns and is temporarily detained as part of the Green Quest money laundering investigation. However, he is let go. [WORLDNETDAILY, 8/16/03]

By late 2003, the U.S. is looking for him in Yemen. [NEW REPUBLIC, 8/21/03] The FBI appears to be divided about him, with some thinking he is part of the 9/11 plot and some disagreeing. [9/11 CONGRESSIONAL INQUIRY, 7/24/03; TIME, 8/11/03] The 9/11 Commission later reports that Aulaqi gave substantial help to the two hijackers, that his relationship with them is "suspicious," and it cannot be discounted that he knew of the plot in advance. [ASSOCIATED PRESS, 6/27/04]

April 1, 2001: Hijacker Gets Speeding Ticket, but His Illegal Status Is Not Noticed

Hijacker Nawaf Alhazmi is stopped by an Oklahoma police officer for speeding. His license information is run through a computer to determine whether there are any warrants for his arrest. There are none, so he is issued a ticket and sent on his way. The CIA has known that Alhazmi is a terrorist possibly living in the U.S. since March 2000, but has failed to share this knowledge with other agencies. [DAILY OKLAHOMAN, 1/20/02; NEWSWEEK, 6/2/02] He also has been in the country illegally since January 2001, but this also does not raise any flags. [CONGRESSIONAL INTELLIGENCE COMMITTEE, 9/20/02]

May 15, 2001: CIA Hides al-Qaeda Meeting Information from FBI

A supervisor at the CIA's Counter Terrorism Center sends a request to CIA headquarters for the surveillance photos of the January 2000 al-Qaeda meeting in Malaysia. Three days later, the supervisor explains the reason for his interest in an e-mail to a CIA analyst: "I'm interested because Khalid Almihdhar's two companions also were couriers of a sort, who traveled between [the Far East] and Los Angeles at the same time ([H]azmi and [S]alah)." Hazmi refers to hijacker Nawaf Alhazmi, and Salah Said is the alias al-Qaeda leader Khallad bin Attash traveled under during the meeting. Apparently, the supervisor receives the photos. Toward the end of May, a CIA analyst contacts a specialist working at FBI headquarters about the photographs. The CIA wants the FBI analyst to review the photographs and determine if a person who had carried money to Southeast Asia for bin Attash in January 2000 could be identified. The CIA fails to tell the FBI analyst anything about Almihdhar or Alhazmi. Around the same time, the CIA analyst receives an e-mail mentioning Alhazmi's travel to the U.S. These two analysts travel to New York the next month and again the CIA analyst fails to divulge what he knows. [9/11 CONGRESSIONAL INQUIRY, 7/24/03]

June 11, 2001: FBI and CIA Hold "Shouting Match" over Terrorist Information; CIA Still Withholds Information

A CIA analyst and FBI analyst travel to New York and meet with FBI officials at FBI headquarters about the USS *Cole* investigation. The CIA analyst has already shown photographs from the al-Qaeda Malaysia meeting attended by hijackers Nawaf Alhazmi and Khalid Almihdhar to an FBI analyst, but failed to explain what he knows about them. The CIA analyst now shows the same photos to the additional FBI agents. He wants to know if they can identify anyone in the photos for a different case he is working on. "The FBI agents recognized the men from the *Cole* investigation, but when they asked the CIA what they knew about the men, they were told that they did not have clearance to share that information. It ended up in a shouting match. " [ABC NEWS, 8/16/02] The CIA analyst later admits that at the time, he knows Almihdhar had a U.S. visa, that Alhazmi had traveled to the U.S. in March, that al-Qaeda leader Khallad bin Attash had been recognized in one of the photos, and that Alhazmi was known to be an experienced terrorist. However, he does not tell any of this to any FBI agent. He does not let them keep copies of the photos either. [9/11 CONGRESSIONAL INQUIRY, 7/24/03] He promises them more information later, but the FBI agents do not receive more information until after 9/11. [9/11 CONGRESSIONAL INQUIRY, 9/20/02] Two days after this meeting, Almihdhar has no trouble getting a new, multiple reentry U.S. visa. [U.S. NEWS AND WORLD REPORT, 12/12/01; 9/11 CONGRESSIONAL INQUIRY, 9/20/02] CIA Director Tenet later claims, "Almihdhar was not who they were talking about in this meeting." When Senator Carl Levin (D) reads the following to Tenet—"The CIA analyst who attended the New York meeting acknowledged to the joint inquiry staff that he had seen the information regarding Almihdhar's U.S. visa and Alhazmi's travel to the United States but he stated that he would not share information outside of the CIA unless he had authority to do so."—Tenet claims that he talked to the same analyst, who told him something completely different. [NEW YORK TIMES, 10/17/02]

July 4, 2001: Hijacker Who Should Have Been on Watch List Re-enters U.S. Without Difficulty

Hijacker Khalid Almihdhar reenters the U.S. The CIA and FBI have recently been showing interest in him, but have still failed to place him on a terrorist watch list. Had he been placed on a watch list by this date, he would have been stopped and possibly detained as he tried to enter the U.S. He enters on a new U.S. visa obtained in Jeddah, Saudi Arabia on June 13, 2001. [9/11 CONGRESSIONAL INQUIRY, 7/24/03] The

FBI notes that he returns just days after the last of the hijacker "muscle" has entered the U.S., and speculates that he returns because his job in bringing them over is finished. [9/11 CONGRESSIONAL INQUIRY, 7/24/03]

July 13, 2001: CIA Reexamines Malaysia Meeting but "Major League Killer" Is Not Put on Watch List

The same supervisor of the CIA's Counter Terrorism Center (CTC) who expressed interest two months earlier in surveillance photos from the al-Qaeda Malaysia meeting now finds a cable he had been looking for regarding that same meeting. The cable, from January 2001, discusses al-Qaeda leader Khallad bin Attash's presence at the meeting. The supervisor explains later that bin Attash's presence at the meeting had been troubling him. He writes an e-mail to the CTC, stating, "[Bin Attash] is a major league killer, who orchestrated the *Cole* attack and possibly the Africa bombings." Yet bin Attash is still not put on a terrorist watch list. An FBI analyst assigned to the CTC is given the task of reviewing all other CIA cables about the Malaysian meeting. It takes this analyst until August 21—over five weeks later—to put together that Khalid Almihdhar had a U.S. visa and that Nawaf Alhazmi had traveled to the U.S. Yet other CIA agents are already very aware of these facts but are not sharing the information. Working with immigration officials, this analyst then learns that Almihdhar entered and left the U.S. in 2000, and entered again on July 4, 2001, and that Alhazmi appears to still be in the U.S. [9/11 CONGRESSIONAL INQUIRY, 7/24/03]

August 23, 2001: Alhazmi and Almihdhar Are Finally Added to Terrorism Watch List

Thanks to the request of an unnamed FBI analyst assigned to the CIA's Counter Terrorism Center, the CIA sends a cable to the State Department, INS, Customs Service, and FBI requesting that "bin Laden-related individuals" Nawaf Alhazmi, Khalid Almihdhar, and two others be put on the terrorism watch list. Since March 2000, if not earlier, the CIA has had good reason to believe these two were al-Qaeda terrorists living in the U.S., but apparently did nothing and told no other agency about it until now. The hijackers are not located in time, and both die in the 9/11 attacks. FBI agents later state that if they been told about Alhazmi and Almihdhar sooner, "There's no question we could have tied all 19 hijackers together" given the frequent contact between these two and the other hijackers. [NEWSWEEK, 6/2/02] However, in what the *Washington Post* calls a "critical omission," the FAA, the Treasury Department's Financial Crimes Enforcement Network, and the FBI's Financial Review Group are not notified. The two latter groups have the power to tap into private credit card and bank data, and claim they could have readily found Alhazmi and Almihdhar, given the frequency the two used credit cards. [WASHINGTON POST, 7/25/03 (C)] Furthermore, counterterrorism "tsar" Richard Clarke and his Counterterrorism and Security Group are not told about these two terrorists before 9/11 either. [NEWSWEEK, 3/24/04] At the same time, the CIA requests that Khallad bin Attash be added to the watch list—eight months after he was known to have been the main planner of the USS *Cole* bombing on January 4, 2001. One other attendee of the

January 2000 Malaysian meeting is also put on the watch list, but that name remains confidential. [NEW YORK TIMES, 9/21/02] The CIA later claims the request was labeled "immediate," the second most urgent category (the highest is reserved for things like declarations of war). [LOS ANGELES TIMES, 10/28/01] The FBI denies that it was marked "immediate" and other agencies treated the request as a routine matter. [LOS ANGELES TIMES, 10/18/01; 9/11 CONGRESSIONAL INQUIRY, 9/20/02] The State Department places all four men on the watch list the next day. [9/11 CONGRESSIONAL INQUIRY, 7/24/03] However, this watch list, named TIPOFF, checks their names only if they use international flights. There is another watch list barring suspected terrorists from flying domestically. On 9/11, it contains about two dozen names, but it does not include any of these four men. [KNIGHT RIDDER, 1/27/04]

August 23, 2001 (B): FBI Begins Unhurried Search for Alhazmi and Almihdhar

The FBI begins a search for hijackers Nawaf Alhazmi and Khalid Almihdhar, in response to a CIA cable about them. The FBI later claims that they responded aggressively. An internal review after 9/11 found that "everything was done that could have been done" to find them. [LOS ANGELES TIMES, 10/28/01] However, even aside from a failed attempt to start a criminal investigation, the search is halfhearted at best. As the *Wall Street Journal* later explains, the search "consisted of little more than entering their names in a nationwide law enforcement database that would have triggered red flags if they were taken into custody for some other reason." [WALL STREET JOURNAL, 9/17/01] A national motor vehicle index is checked, but a speeding ticket issued to Alhazmi the previous April is not detected. [DAILY OKLAHOMAN, 1/20/02; 9/11 CONGRESSIONAL INQUIRY, 7/24/03] Nor is a recorded interaction between Alhazmi and local police in Fairfax, Virginia in May, which could have led investigators to Alhazmi's East Coast apartment. [SAN DIEGO UNION-TRIBUNE, 9/27/02] Even though the two were known to have entered the U.S. through Los Angeles, drivers' license records in California are not checked. The FBI also fails to check national credit card or bank account databases, and car registrations. All of these would have had positive results. Alhazmi's name was even in the 2000–2001 San Diego phone book, listing the address where he and Almihdhar were living off and on until about September 9, 2001. [NEWSWEEK, 6/2/02; SOUTH FLORIDA SUN-SENTINEL, 9/28/01; LOS ANGELES TIMES, 10/28/01]

August 24–25, 2001: Alhazmi and Almihdhar Buy 9/11 Plane Tickets

Hijacker Khalid Almihdhar buys his 9/11 plane ticket on-line using a credit card; Nawaf Alhazmi does the same the next day. [9/11 CONGRESSIONAL INQUIRY, 9/26/02] Both men are put on a terrorist watch list this same day, but the watch list only means they will be stopped if trying to enter or leave the U.S. Procedures are in place for law enforcement agencies to share watch list information with airlines and airports and such sharing is common, but the FAA and the airlines are not notified about this case, so the purchases raise no red flags. [LOS ANGELES TIMES, 9/20/01 (C)] An official later states that had the FAA been properly warned, "they should have been picked up in the reservation process." [WASHINGTON POST, 10/2/02]

August 27, 2001: INS Given Non-Urgent Request to
Determine Visa Status of Alhazmi and Almihdhar

The FBI contacts the State Department and the INS to determine the visa status of recently watch listed hijackers Nawaf Alhazmi and Khalid Almihdhar. Almihdhar's visa obtained in June is revoked the same day; Alhazmi's visa has already expired and he is in the country illegally. [9/11 CONGRESSIONAL INQUIRY, 7/24/03] However neither agency is asked "to assist in locating the individuals, nor was any other information provided [that] would have indicated either a high priority or imminent danger." An INS official later states, "if [the INS] had been asked to locate the two suspected terrorists . . . in late August on an urgent, emergency basis, it would have been able to run those names through its extensive database system and might have been able to locate them." The State Department says that "it might have been able to locate the two suspected terrorists if it had been asked to do so." [9/11 CONGRESSIONAL INQUIRY, 9/20/02]

August 28, 2001: FBI's New York Office Request to Open
Criminal Investigation on Hijacker Rejected by FBI Headquarters

A report is sent by the FBI's New York office recommending that an investigation be launched "to determine if [Khalid] Almihdhar is still in the United States." The New York office tries to convince FBI headquarters to open a criminal investigation, but it is immediately turned down. The reason given is a "wall" between criminal and intelligence work—Almihdhar could not be tied to the USS *Cole* investigation without the inclusion of sensitive intelligence information. [9/11 CONGRESSIONAL INQUIRY, 9/20/02] So instead of a criminal case, the New York office opens an "intelligence case," excluding all the "criminal case" investigators from the search. [FBI AGENT TESTIMONY, 9/20/02] One FBI agent expresses his frustration in an e-mail the next day, saying, "Whatever has happened to this—someday someone will die—and wall or not—the public will not understand why we were not more effective and throwing every resource we had at certain 'problems.' Let's hope the [FBI's] National Security Law Unit will stand behind their decisions then, especially since the biggest threat to us now, UBL [bin Laden], is getting the most 'protection.' " [NEW YORK TIMES, 9/21/02; FBI AGENT TESTIMONY, 9/20/02]

August 29, 2001: FBI Learns That Almihdhar Arrived in U.S.;
Fails to Notify San Diego FBI Office

The FBI learns that when hijacker Khalid Almihdhar arrived in the U.S. in July 2001, he indicated he would be staying at a Marriott hotel in New York City. By September 5, an investigation of all New York area Marriott hotels will turn up nothing. The FBI office in Los Angeles receives a request on the day of 9/11 to check Sheraton Hotels in Los Angeles, because that is where Almihdhar said he would be staying when he entered the country over a year and a half earlier. That search also turns up nothing. [9/11 CONGRESSIONAL INQUIRY, 9/20/02; 9/11 CONGRESSIONAL INQUIRY, 9/18/02] The San Diego FBI office is not notified about the need for a search until September 12, and even then, they are only provided with "sketchy" information. [LOS ANGELES TIMES, 9/16/01] The FBI handling agent in San Diego is certain they could have been located quickly had they known where to look. [9/11 CONGRESSIONAL INQUIRY, 7/24/03] There is some evidence from eyewitnesses that a few days before 9/11, Almihdhar and two other hijackers are living in the same San Diego apartment that they had been living in off and on for the past two years (see chapter 8).

September 21–28, 2001: Suspected Hijacker
Associate Is Arrested in Britain, Released

Omar al-Bayoumi, suspected al-Qaeda advance man and possible Saudi agent, is arrested, and held for one week in Britain. He moved from San Diego to Britain in July 2001 and is a studying at Aston University Business School in Birmingham when he is taken into custody by British authorities working with the FBI. [SAN DIEGO UNION-TRIBUNE, 10/27/01; WASHINGTON POST, 12/29/01; MSNBC, 11/27/02] During a search of al-Bayoumi's Birmingham apartment (which includes ripping up the floorboards), the FBI finds the names and phone numbers of two employees of the Saudi embassy's Islamic Affairs Department. [NEWSWEEK, 11/24/02] "There was a link there," a Justice Department official says, adding that the FBI interviewed the employees and "that was the end of that, in October or November of 2001." The official adds, "I don't know why he had those names." Nail al-Jubeir, chief spokesperson for the Saudi embassy in Washington, says al-Bayoumi "called [the numbers] constantly." [LOS ANGELES TIMES, 11/24/02] They also discover jihadist literature, and conclude he "has connections to terrorist elements" including al-Qaeda. [WASHINGTON POST, 7/25/03] However, he is released after a week. [LOS ANGELES TIMES, 11/24/02; NEWSWEEK, 11/24/02] British intelligence officials are frustrated that the FBI failed to give them information that would have enabled them to keep al-Bayoumi in custody longer than the seven days allowed under British anti-terrorism laws. [LONDON TIMES, 10/19/01; SAN DIEGO CHANNEL 10, 10/25/01] Even FBI officials in San Diego appear to have not been told of al-Bayoumi's arrest by FBI officials in Britain until after he is released. [SUNDAY MERCURY, 10/21/01] *Newsweek* claims that classified sections of the 9/11 Congressional Inquiry indicate the Saudi Embassy pushed for al-Bayoumi's release—"another possible indicator of his high-level [Saudi] connections." [NEWSWEEK, 7/28/03] A San Diego FBI agent later secretly testifies that supervisors fail to act on evidence connecting to a Saudi money trail. The FBI is said to conduct a massive investigation of al-Bayoumi within days of 9/11, which shows he has connections to foreign terrorists [9/11 CONGRESSIONAL INQUIRY, 7/24/03; SUNDAY MERCURY, 10/21/01; NEWSWEEK, 7/28/03], but two years later witnesses connecting him to Saudi money apparently are not interviewed by the FBI. Al-Bayoumi continues with his studies in Britain and is still there into 2002, and yet is still not rearrested. [WASHINGTON POST, 12/29/01; NEWSWEEK, 10/29/02] He disappears into Saudi Arabia by the time he reenters the news in November 2002. [SAN DIEGO MAGAZINE, 9/03]

April 25, 2002: Saudi Prince Said to Meet
Suspected Hijacker Associate While Visiting Bush

Osama Basnan, an alleged associate of 9/11 hijackers Nawaf Alhazmi and Khalid Almihdhar, reports his passport stolen to Houston police. [NEWSWEEK, 11/24/02] This confirms that Basnan is in Houston on the same day that Saudi Crown Prince Abdullah, Prince Saud Al-Faisal, and Saudi U.S. Ambassador Prince Bandar meet with President Bush, Vice President Cheney, Secretary of State Powell, and National Security Adviser Rice at Bush's ranch in nearby Crawford, Texas. [U.S.-SAUDI ARABIAN BUSINESS COUNCIL, 4/25/02] Abdullah's entourage passes through Houston that week en route to Bush's ranch. While in Texas, it is believed that Basnan "met with a high Saudi prince who has responsibilities for intelligence matters and is known to bring suitcases full of cash into the United States." [NEWSWEEK, 11/24/02; GUARDIAN, 11/25/02] The still-classified section of the 9/11 Congressional Inquiry is said to discuss the possibility of Basnan meeting this figure at this time. [ASSOCIATED PRESS, 8/2/03]

August 22–November 2002:
Possible Hijacker Associate Is Arrested, Then Deported

Osama Basnan, an alleged associate of 9/11 hijackers Nawaf Alhazmi and Khalid Almihdhar, and his wife are arrested for visa fraud. [LOS ANGELES TIMES, 11/24/02; NEWSWEEK, 11/22/02] One report says he is arrested for allegedly having links to Omar al-Bayoumi. [ARAB NEWS, 11/26/02] On October 22, Basnan and his wife, Majeda Dweikat, admit they used false immigration documents to stay in the U.S. [SAN DIEGO CHANNEL 10, 10/22/02] Possible financial connections between Basnan and al-Bayoumi, Alhazmi and Almihdhar, and the Saudi royal family are known to the 9/11 Congressional Inquiry (as well as the FBI and CIA) at this time. Remarkably, Basnan is deported to Saudi Arabia on November 17, 2002. His wife is deported to Jordan the same day. [WASHINGTON POST, 11/24/02] Less than a week after the deportations, new media reports make Basnan a widely known suspect. [NEWSWEEK, 11/22/02]

October 9, 2002: FBI Agent Handled Hijackers' Landlord

San Diego FBI agent Steven Butler reportedly gives "explosive" testimony to the 9/11 Congressional Inquiry. Butler, recently retired, has been unable to speak to the media, but he was the handler for Abdussattar Shaikh, an FBI informant who rented a room to hijackers Nawaf Alhazmi and Khalid Almihdhar. Butler claims he might have uncovered the 9/11 plot if the CIA had provided the FBI with more information earlier about Alhazmi and Almihdhar. [NEW YORK TIMES, 10/22/02] He says, "It would have made a huge difference." He suggests they would have quickly found the two hijackers because they were "very, very close." "We would have immediately opened . . . investigations. We would have given them the full court press. We would . . . have done everything—physical surveillance, technical surveillance, and other assets." [9/11 CONGRESSIONAL INQUIRY, 7/24/03; SAN DIEGO UNION-TRIBUNE, 7/25/03] Butler discloses that he had been monitoring a flow of Saudi Arabian money that wound up in the hands of two of the 9/11 hijackers, but his supervisors failed to take any action on the warnings. It is not known when Butler started investigating the money flow, or when he warned his supervisors. [U.S. NEWS AND WORLD REPORT, 11/29/02] The FBI unsuccessfully tries to prevent Butler from testifying. [WASHINGTON POST, 10/11/02] This testimony doesn't stop the U.S. government from deporting Basnan to Saudi Arabia several weeks later. [WASHINGTON POST, 11/24/02]

November 22, 2002: *Newsweek* Reports
Saudi Royals Sent Money to Hijackers' Associates

Newsweek reports that hijackers Nawaf Alhazmi and Khalid Almihdhar may have received money from Saudi Arabia's royal family through two Saudis, Omar al-Bayoumi and Osama Basnan. *Newsweek* bases its report on information leaked from the 9/11 Congressional Inquiry in October. [NEWSWEEK, 11/22/02; NEWSWEEK, 11/22/02; WASHINGTON POST, 11/23/02; NEW YORK TIMES, 11/23/02] Al-Bayoumi is in Saudi Arabia by this time. Basnan was deported to Saudi Arabia just five days earlier. Saudi officials and Princess Haifa immediately deny any terrorist connections. [LOS ANGELES TIMES, 11/24/02] *Newsweek* reports that while the money trail "could be perfectly innocent . . . it is nonetheless intriguing—and could ultimately expose the Saudi government to some of the blame for 9/11 . . . " [NEWSWEEK, 11/22/02] Some Saudi newspapers which usually reflect government thinking claim the leak is blackmail to pressure Saudi Arabia into supporting war with Iraq. [MSNBC, 11/27/02] Senior U.S. government officials claim the FBI and CIA failed to aggressively pursue leads that might have linked the two hijackers to Saudi Arabia. This causes a bitter dispute between FBI and CIA officials and the intelligence panel investigating the 9/11 attacks. [NEW YORK TIMES, 11/23/02] A number of senators, including Richard Shelby (R), John McCain (R), Mitch O'Connell (R), Joe Lieberman (D), Bob Graham (D), Joseph Biden (D), and Charles Schumer (D), express concern about the Bush administration's action (or non-action) regarding the Saudi royal family and its possible role in funding terrorists. [NEW YORK TIMES, 11/25/02; REUTERS, 11/24/02] Lieberman says, "I think it's time for the president to blow the whistle and remember what he said after September 11—you're either with us or you're with the terrorists." [ABC NEWS, 11/25/02] FBI officials strongly deny any deliberate connection between these two men and the Saudi government or the hijackers [TIME, 11/24/03], but later even more connections between them and both entities are revealed. [9/11 CONGRESSIONAL INQUIRY, 7/24/03]

March 24, 2004: FBI Says Saudi Associates of Hijackers Not Involved in Plot

It is reported that the FBI has closed down their investigation into Saudis Omar al-Bayoumi and Osama Basnan. The Associated Press reports, "The FBI concluded at most the two Saudi men occasionally provided information to their kingdom or helped Saudi visitors settle into the United States, but did so in compliance with Muslim custom of being kind to strangers rather than out of some relationship with Saudi intelligence." [ASSOCIATED PRESS 3/24/04 (B)]

8.

The Other Hijackers

This chapter does not attempt to chronicle all of the movements and actions of the 9/11 hijackers. It focuses on their behavior that suggests further U.S. intelligence failures and/or information that supplements—or contradicts—the mainstream and official narratives about them.

One of the mysteries surrounding the other 9/11 hijackers is simply *Who were they?* In the days following 9/11, the FBI released the names and photographs of the 19 alleged hijackers. Within weeks, reports surfaced that some of the individuals identified by the FBI were very much alive. Of the 19 hijacker names identified by the U.S. government, in at least eight cases individuals with the same name and similar appearance have come forward to prove that they were not the 9/11 hijackers. In other cases, family members and friends of the hijackers have reported that the FBI photographs bore no resemblance whatsoever to them. This book uses the popular names of the hijackers because there are no other names to use, but it is important to keep in mind that these are only the alleged names.

While mystery surrounds some of the hijackers, we do know a significant amount about others. Based on the public record, we know that U.S. intelligence missed many opportunities to identify and detain several of the hijackers.

While FBI agents in Arizona and Oklahoma raised concerns about men with potential terrorist connections attending flight schools in their areas, the hijackers were able to obtain pilot training at these and other schools. Between 1998 and 2001, the behavior of at least two hijackers, Hani Hanjour and Waleed Alshehri, was sufficiently suspicious that U.S. citizens reported them to intelligence agencies. In both cases, officials failed to follow up properly on the tips. In mid-2000, lead hijacker Mohamed Atta boldly sought a loan from the Department of Agriculture to obtain a crop-dusting airplane. Atta's loan request was refused, but his highly suspicious behavior was not reported to any other agency. This incident also took place before immigration records indicate Atta first entered the U.S. This is just one of many anomalies concerning where and when the hijackers traveled.

It's surprising how little we know about the hijackers and what they did, especially given how much time has passed. Some of the hijackers remain near-total mysteries. Important reports from just after 9/11, such as reports suggesting some of the hijackers trained at U.S. military bases, remain unanswered. What we do know is that, once again, the U.S. intelligence agencies missed multiple opportunities to investigate suspicious activities, and failed to act when they learned of potential terrorist connections to the hijackers, until it was too late.

Are the identities the FBI has ascribed to the hijackers in fact accurate? If not, who *were* they—and why has the FBI promulgated this mistaken information? Why does the location of the hijackers as reported by many witnesses in the Philippines, San Antonio, Texas, Spain, Florida, and other places not match the official narrative? Did any of the hijackers leave a trail suspicious enough that it should have led to their detection?

1990: First Hijacker Enters U.S.

Hani Hanjour, who will pilot Flight 77 on 9/11, enters the U.S. He takes an English course in Tucson, Arizona. [*TIME*, 9/24/01 (B); COX NEWS SERVICE, 10/15/01; *NEW YORK TIMES*, 6/19/02] Despite evidence to the contrary, the FBI claims Hanjour first arrived on October 3, 1991. [CONGRESSIONAL INTELLIGENCE COMMITTEE, 9/26/02] The 9/11 Commission does not take a position on this issue, finding only that "Hanjour came to the United States to attend school in three stints during the 1990s. His final arrival was in December 2000, through the Cincinnati/Northern Kentucky airport." [9/11 COMMISSION REPORT, 1/26/04]

March 1995–February 1996:
Hijacker Jarrah Living in New York or Lebanon?

A man named "Ziad Jarrah" rents an apartment in Brooklyn, New York. [*AMONG THE HEROES*, BY JERE LONGMAN, 2002, P. 90] The landlords later identify his photograph as being that of the 9/11 hijacker. A Brooklyn apartment lease bears Ziad Jarrah's name. [*BOSTON GLOBE*, 9/25/01] "Another man named Ihassan Jarrah lived with Ziad, drove a livery cab and paid the eight-hundred-dollar monthly rent. The men were quiet, well-mannered, said hello and good-bye. Ziad Jarrah carried a camera and told his landlords that he was a photographer. He would disappear for a few days on occasion, then reappear. Sometimes a woman who appeared to be a prostitute arrived with one of the men. 'Me and my brother used to crack jokes that they were terrorists,' said Jason Matos, a construction worker who lived in a basement there, and whose mother owned the house." However, another Ziad Jarrah is still in his home

Ziad Jarrah

country of Lebanon at this time. He is studying in a Catholic school in Beirut, and is in frequent contact with the rest of his family. His parents drive him home to be with the family nearly every weekend, and they are in frequent contact by telephone as well. [*LOS ANGELES TIMES*, 10/23/01] Not until April 1996 does this Ziad Jarrah leave Lebanon for the first time to study in Germany. [*BOSTON GLOBE*, 9/25/01] His family believes that the New York lease proves that there were two "Ziad Jarrahs." [CNN, 9/18/01 (B)] Evidence seems to indicate Jarrah was also in two places at the same time from November 2000 to January 2001.

1996–December 2000: Majority of Hijackers Disappear into Chechnya

At least 11 of the 9/11 hijackers travel to Chechnya between 1996 and 2000:

- Nawaf Alhazmi fights in Chechnya, Bosnia, and Afghanistan for several years, starting around 1995. [OBSERVER, 9/23/01; ABC NEWS, 1/9/02; 9/11 CONGRESSIONAL INQUIRY, 7/24/03; CIA DIRECTOR TENET TESTIMONY, 6/18/02]

- Khalid Almihdhar fights in Chechnya, Bosnia, and Afghanistan for several years, usually with Nawaf Alhazmi. [LOS ANGELES TIMES, 9/1/02; 9/11 CONGRESSIONAL INQUIRY, 7/24/03; CIA DIRECTOR TENET TESTIMONY, 6/18/02]

- Salem Alhazmi spends time in Chechnya with his brother Nawaf Alhazmi. [ABC NEWS, 1/9/02] He also possibly fights with his brother in Afghanistan. [9/11 CONGRESSIONAL INQUIRY, 7/24/03]

- Ahmed Alhaznawi leaves for Chechnya in 1999 [ABC NEWS, 1/9/02], and his family loses contact with him in late 2000. [ARAB NEWS, 9/22/01]

- Hamza Alghamdi leaves for Chechnya in early 2000 [INDEPENDENT, 9/27/01; [WASHINGTON POST, 9/25/01] or sometime around January 2001. He calls home several times until about June 2001, saying he is in Chechnya. [ARAB NEWS, 9/18/01]

- Mohand Alshehri leaves to fight in Chechnya in early 2000. [ARAB NEWS, 9/22/01]

- Ahmed Alnami leaves home in June 2000, and calls home once in June 2001 from an unnamed location. [ARAB NEWS, 9/19/01; WASHINGTON POST, 9/25/01]

- Fayez Ahmed Banihammad leaves home in July 2000 saying he wants to participate in a holy war or do relief work. [ST. PETERSBURG TIMES, 9/27/01; WASHINGTON POST, 9/25/01] He calls his parents one time since. [ARAB NEWS, 9/18/01]

- Ahmed Alghamdi leaves his studies to fight in Chechnya in 2000, and is last seen by his family in December 2000. He calls his parents for the last time in July 2001, but does not mention being in the U.S. [ARAB NEWS, 9/18/01; ARAB NEWS, 9/20/01]

- Waleed Alshehri disappears with Wail Alshehri in December 2000, after speaking of fighting in Chechnya. [WASHINGTON POST, 9/25/01; ARAB NEWS, 9/18/01]

- Wail Alshehri, who had psychological problems, went with his brother to Mecca to seek help and both disappear, after speaking of fighting in Chechnya. [WASHINGTON POST, 9/25/01]

- Majed Moqed is last seen by a friend in 2000 in Saudi Arabia, after communicating a "plan to visit the United States to learn English." [ARAB NEWS, 9/22/01]

Clearly, there is a pattern: eleven hijackers appear likely to have fought in Chechnya, and two others are known to have gone missing. It is possible that others have similar histories, but this is hard to confirm because "almost nothing [is] known about some." [NEW YORK TIMES, 9/21/01] Indeed, a colleague claims that hijackers Mohamed Atta, Marwan Alshehhi, Ziad Jarrah and would-be hijacker Ramzi bin al-Shibh wanted to fight in Chechnya but were told in early 2000 that they were needed elsewhere. [WASHINGTON POST, 10/23/02; REUTERS, 10/29/02] Reuters has reported, "Western diplomats play down any Chechen involvement by al-Qaeda." [REUTERS, 10/24/02] Many of the FBI hijackers' photos appear to be incorrect. If so, it might not be the first time this technique was used: former CIA director James Woolsey claims bin Laden agents murdered 12 men during the 1990 Iraqi invasion of Kuwait, stole their paperwork, then used their identities for later plots such as the WTC bombing in 1993. [MSNBC, 9/27/01]

1997 or 1998: Atta in Two Places at Once?

Spanish newspaper *El Mundo* later reports, "According to several professors at the Valencia School of Medicine, some of whom are forensic experts, [Atta] was a student there in 1997 or 1998. Although he used another name then, they remember his face among the students that attended anatomy classes." It is also suggested that "years before, as a student he went to Tarragona. That would explain his last visit to Salou [from July 8–19, 2001], where he could have made contact with dormant cells . . . " [EL MUNDO, 9/30/01] If this is true, it would contradict reports concerning Atta's presence as a student in Hamburg, Germany, during this entire period.

1997–July 2001: Hanjour Associate Freely Travels in and out of U.S.

Hijacker Hani Hanjour begins associating with an unnamed individual who is later mentioned in FBI agent Ken Williams' famous "Phoenix memo." Hanjour and this individual train at flight schools in Arizona. Several flight instructors later note the two were associates and may have carpooled together. They are known to share the same airplane on one occasion in 1999, and are at the school together on other occasions. The unnamed individual leaves the U.S. in April 2000. In May 2001, the FBI attempts to investigate this person, but after finding out that he has left the U.S., the FBI declines to open a formal investigation. The name of this person is not placed on a watch list, so the FBI is unaware that the person returns in June and stays in the U.S. for another month. By this time, this person is an experienced flight instructor who is certified to fly Boeing 737s. The FBI speculates the person may return to evaluate Hanjour's flying skills or provide final training before 9/11. There is considerable circumstantial evidence placing this person near Hanjour during this month. [9/11 CONGRESSIONAL INQUIRY, 7/24/03]

1998: By Some Accounts, al-Qaeda Begins Planning for 9/11

According to closed-session testimony by CIA, FBI and NSA heads, al-Qaeda begins planning the 9/11 attacks this year. [USA TODAY, 6/18/02] In a June 2002 interview, 9/11 mastermind Khalid Shaikh Mohammed also asserts that planning for the attacks begin at this time. [ASSOCIATED PRESS, 9/8/02] However, it appears the targeting of the WTC and pilot training began even earlier. An al-Qaeda operative in Spain will later be found with videos filmed in 1997 of major U.S. structures (including "innumerable takes from all distances and angles" of the WTC). There are numerous connections between Spain and the 9/11 hijackers, including an important meeting there in July 2001. [ASSOCIATED PRESS, 7/17/02] Hijacker Waleed Alshehri was living in Florida since 1995, started training for his commercial pilot training degree in 1996, and obtained his license in 1997. [SUNDAY HERALD, 9/16/01; BOSTON GLOBE, 9/14/01]

1998 (B): Does Atta Train at Base Conducting Pilotless Aircraft Exercises?

A military report released this year describes the "Joint Vision 2010" program, a series of "analyses, war games, studies, experiments, and exercises" which are "investigating new operational concepts, doc-

Mohamed Atta

trines, and organizational approaches that will enable U.S. forces to maintain full spectrum dominance of the battlespace well into the 21st century." "The Air Force has begun a series of war games entitled Global Engagement at the Air War College, Maxwell Air Force Base, Alabama." The same report mentions that the military is working on a "variety of new imaging and signals intelligence sensors, currently in advanced stages of development, deployed aboard the Global Hawk, DarkStar, and Predator unmanned aerial vehicles (UAVs) . . . " [DEPARTMENT OF DEFENSE ANNUAL REPORT, 1998]

Global Hawk is a technology that enables pilotless flight and has been functioning since at least early 1997. [DEPARTMENT OF DEFENSE, 2/20/97] While it may be mere coincidence, "Air Force spokesman Colonel Ken McClellan said a man named Mohamed Atta—which the FBI has identified as one of the five hijackers of American Airlines Flight 11—had once attended the International Officer's School at Maxwell/Gunter Air Force Base in Montgomery, Ala." [GANNETT NEWS SERVICE, 9/17/01]

1998 (C): Information on Hanjour Apparently Ignored by FBI

American Muslim Aukai Collins later says he reports to the FBI on hijacker Hani Hanjour for six months this year. [ASSOCIATED PRESS, 5/24/02] The FBI later acknowledges they paid Collins to monitor the Islamic and Arab communities in Phoenix between 1996 and 1999. [ASSOCIATED PRESS, 5/24/02; ABC NEWS, 5/23/02] Collins claims that he is a casual acquaintance of Hanjour while Hanjour is taking flying lessons. [ASSOCIATED PRESS, 5/24/02] Collins sees nothing suspicious about Hanjour as an individual, but he tells the FBI about him because Hanjour appears to be part of a larger, organized group of Arabs taking flying lessons. [FOX NEWS, 5/24/02] He says the FBI "knew everything about the guy," including his exact address, phone number and even what car he drove. The FBI denies Collins told them anything about Hanjour, and denies knowing about Hanjour before 9/11. [ABC NEWS, 5/23/02] Collins later calls Hanjour a "hanky panky" hijacker: "He wasn't even moderately religious, let alone fanatically religious. And I knew for a fact that he wasn't part of al-Qaeda or any other Islamic organization; he couldn't even spell jihad in Arabic." [MY JIHAD: THE TRUE STORY OF AN AMERICAN MUJAHID'S AMAZING JOURNEY FROM USAMA BIN LADEN'S TRAINING CAMPS TO COUNTERT-ERRORISM WITH THE FBI AND CIA, BY AUKAI COLLINS, 6/02, P. 248]

1998–2000: Atta and Alshehhi Are Seen Living in Philippines

Hijackers Mohamed Atta and Marwan Alshehhi live periodically in the town of Mabalacat, Philippines. They stay in the Woodland Resort hotel and apparently learn to fly planes at a nearby flight school. Although Philippine and U.S. investigators have not been able to confirm their presence in Mabalacat, locals are certain they saw them frequently partying, drinking alcohol, sleeping with local women, and consorting with many other, unknown Arabs (most of whom disappear shortly before 9/11). For instance, according to a former waitress at the hotel, Alshehhi throws a party in December 1999 with six or seven Arab friends: "They rented the open area by the swimming pool for 1,000 pesos. They drank Johnnie Walker Black Label whiskey . . . They came in big vehicles, and they had a lot of money. They all had girlfriends." Several employ-

Marwan Alshehhi

ees recall Atta staying at the hotel during the summer of 1999, acting unfriendly and cheap. One hotel employee claims that most of the guests were Arab, and many took flying lessons at the nearby school. These witnesses claim the two used aliases, but the other Arabs referred to Atta as "Mohamed." [MANILA TIMES, 10/2/01; INTERNATIONAL HERALD TRIBUNE, 10/5/01; ASSOCIATED PRESS, 9/28/01] Apparently, other hijackers and 9/11 mastermind Khalid Shaikh Mohammed visit the Philippines during this time. However, according to the official version of events, Atta and Alshehhi are in Hamburg, Germany during this time. Atta is still working on his thesis, which he completes in late 1999. [AUSTRALIAN BROADCASTING CORP., 11/12/01]

July 7, 1998: Stolen Passport Shows Ties Between Hijackers and Spanish Terrorist Cells

Thieves snatch a passport from a car driven by a U.S. tourist in Barcelona, Spain, which later finds its way into the hands of would-be hijacker Ramzi bin al-Shibh. Bin al-Shibh allegedly uses the name on the passport in the summer of 2001 as he wires money to pay flight school tuition for Zacarias Mous-saoui in Oklahoma. Investigators believe the movement of this passport shows connections between the 9/11 plotters in Germany and a support network in Spain, made up mostly by ethnic Syrians. "Investi-gators believe that the Syrians served as deep-cover mentors, recruiters, financiers and logistics providers for the hijackers—elite backup for an elite attack team." [LOS ANGELES TIMES, 1/14/03] Mohamed Atta twice travels to Spain in 2001, perhaps to make contact with members of this Spanish support team.

1999: Neighbors Report Suspicious Activities to CIA; No Apparent Response

Diane and John Albritton later say they call the CIA and police several times this year to report suspi-cious activity at a neighbor's home, but authorities fail to respond. [MSNBC, 9/23/01; NEW YORK DAILY NEWS, 9/15/01] Hijacker Waleed Alshehri is renting the house on Orrin Street in Vienna, Virginia at the time (three blocks from a CIA headquarters). [ASSOCIATED PRESS, 9/15/01 (B)] He makes his neighbors nervous. "There were always people coming and going," said Diane Albritton. "Arabic people. Some of them never uttered a word; I don't know if they spoke English. But they looked very focused. We thought they might be dealing drugs, or illegal immigrants." [NEW YORK TIMES, 9/15/01 (B)] Ahmed Alghamdi lived at the same address until July 2000. [FOX NEWS, 6/6/02; WORLD NET DAILY, 9/14/01] Waleed Alshehri lived with Ahmed Alghamdi in Florida for seven months in 1997. [DAILY TELEGRAPH, 9/20/01] Albritton says they observed a van parked outside the home at all hours of the day and night. A Middle Eastern man appeared to be monitoring a scanner or radio inside the van. Another neighbor says, "We thought it was a drug house. All the cars parked on the street were new BMWs, new Mercedes. People were always walking around out front with cell phones." There were frequent wild parties, numerous complaints to authorities, and even a police report about a woman shooting a gun into the air during a party. [WORLD NET DAILY, 9/14/01] Other neighbors also called the police about the house. [ASSOCIATED PRESS, 9/14/01 (B)] "Critics say [the case] could have made a difference [in stopping 9/11] had it been handled differently." Standard procedures require the CIA to notify the FBI of such domestic information. However, FBI officials have not been able to find any record that the CIA shared the information. [FOX NEWS, 6/6/02] FBI Director Mueller has said "the hijackers did all they could to stay below our radar." [SENATE JUDICIARY STATEMENT, 5/8/02]

Mid-June 1999: Hijackers Meet in Amsterdam and Get Saudi Cash

Hijackers Mohamed Atta and Marwan Alshehhi, plus would-be hijacker Ramzi Bin al-Shibh and associate Mounir el Motassadeq, hold a meeting in Amsterdam, Netherlands. All are living in Hamburg at the time, so it is not clear why they go to meet there, though some speculate that they are meeting someone else. Motassadeq also goes to the town of Eindhoven, Netherlands, on three occasions, in early 1999, late 1999, and 2001. [ASSOCIATED PRESS, 9/13/02] On at least one occasion, Motassadeq receives cash provided by unnamed "Saudi financiers" that is meant to fund a new Eindhoven mosque. Investigators believe he uses the money to help pay for some 9/11 hijacker flying lessons. [BALTIMORE SUN, 9/2/02]

July 1999: FBI Investigates Individual Linked to al-Qaeda and Nuclear Science

The FBI begins an investigation of an unnamed person for ties to important al-Qaeda figures and several organizations linked to al-Qaeda. The FBI is concerned that this person is in contact with several experts in nuclear sciences. After 9/11, the FBI determines that hijacker Marwan Alshehhi had contact with this person on the East Coast of the U.S. This person also may have ties to Mohamed Atta's sister. Most additional details about this person, including their name, when and how often Alshehhi had contact, and if the investigation was ever closed, remain classified. [9/11 CONGRESSIONAL INQUIRY, 7/24/03]

July 1999–November 2000: Hijacker Alshehhi Receives $100,000 from His Brother

Hijacker Marwan Alshehhi receives about $100,000 from an account in Sharjah, United Arab Emirates during this time. [FINANCIAL TIMES, 11/30/01; NEWSWEEK, 12/2/01; CONGRESSIONAL INTELLIGENCE COMMITTEE, 9/26/02] The money is apparently sent by Mohamed Yousef Mohamed Alqusaidi, believed to be Alshehhi's brother. Alqusaidi had been sending money to Alshehhi in Germany since at least March 1998. [CONGRESSIONAL INTELLIGENCE COMMITTEE, 9/26/02] It is not clear whether this money is innocently sent by Alshehhi's rich Arab family or deliberately sent for terrorism.

September 1999: Atta Obtains U.S. Store Membership Before Alleged Arrival in U.S.

BJ's Wholesale Club, a store in Hollywood, Florida, later tells the FBI that Mohamed Atta may have held a BJ's membership card since at least this time ("more than two years" before 9/11). Several cashiers at the store vaguely remember seeing Atta there. [MIAMI HERALD, 9/18/01] According to the official story, Atta does not arrive in the U.S. until June 3, 2000. [MIAMI HERALD, 9/22/01]

Late 1999: Hijackers Clear Their Passport Records

Hijackers Mohamed Atta and Marwan Alshehhi report their passports missing; Ziad Jarrah reports his missing in February 2000. [SOUTH FLORIDA SUN-SENTINEL, 9/28/01; INSIDE 9-11: WHAT REALLY HAPPENED, BY DER SPIEGEL, 2/02, PP. 257–58] Alshehhi receives a replacement passport on December 26, 1999. [LONDON TIMES, 9/20/01]

2000–September 10, 2001: Hijackers May Spend Time in Philippines

The names of four hijackers are later discovered in Philippines immigration records, according to Philippine Immigration Commissioner Andrea Domingo. However, whether these are the hijackers or just other Saudis with the same names has not been confirmed. Abdulaziz Alomari visits the Philippines once in 2000, then again in February 2001, leaving on February 12. [ASSOCIATED PRESS, 9/19/01; DAILY TELEGRAPH, 9/20/01; PHILIPPINES DAILY INQUIRER. 9/19/01] Ahmed Alghamdi visits Manila, Philippines more than 13 times in the two years before 9/11. He leaves the Philippines the day before the attacks. [ARIZONA DAILY STAR, 9/28/01; DAILY TELEGRAPH, 9/20/01] Fayez Ahmed Banihammad visits the Philippines on October 17–19, 2000. [ARIZONA DAILY STAR, 9/28/01; DAILY TELEGRAPH, 9/20/01] Saeed Alghamdi visits the Philippines on at least 15 occasions in 2001, entering as a tourist. The last visit ends on August 6, 2001. [DAILY TELEGRAPH, 9/20/01] Hijackers Mohamed Atta and Marwan Alshehhi may also have been living in the Philippines, and 9/11 mastermind Khalid Shaikh Mohammed occasionally stays there. While in the Philippines, it is possible the hijackers meet with associates of Mohammad Jamal Khalifa, bin Laden's brother-in-law. Khalifa has been closely linked with the Philippines chapter of the International Islamic Relief Organization, which gets much of its money from the Saudi Arabian government. The organization has recently been accused of being a front for al-Qaeda. Amongst other connections to terrorism, Khalifa helped fund the Islamic Army of Aden, a group that claimed responsibility for the bombing of the USS *Cole*. [BOSTON HERALD, 10/14/01] Khalifa has been connected through phone calls to Hambali, a major terrorist who attended a planning meeting for the 9/11 attacks in Malaysia (see chapter 7), also attended by two hijackers connected to the Islamic Army of Aden. [PBS FRONTLINE, 10/3/02 (C); COX NEWS SERVICE, 10/21/01] (The U.S. had Khalifa in custody for about six months in 1995.) Nothing more has been heard to confirm or deny the hijackers' Philippines connections since these reports.

April 2000: Atta in Portland Public Library Before Official Arrival Date

Spruce Whited, director of security for the Portland Public Library, later says Mohamed Atta and possibly a second hijacker are regulars at the library and frequently use public Internet terminals at this time.

He says four other employees recognize Atta as a library patron. "I remember seeing (Atta) in the spring of 2000," he says. "I have a vague memory of a second one who turned out to be (Atta's) cousin." Whited also says federal authorities have not inquired about the library sightings. [*BOSTON HERALD*, 10/5/01; *PORTLAND PRESS HERALD*, 10/5/01] According to the official story, Atta does not arrive in the U.S. until June 3, 2000. [*MIAMI HERALD*, 9/22/01; AUSTRALIAN BROADCASTING CORP., 11/12/01]

April–May 2000: Hijacker Tells Librarian About Major Attack in U.S.

Around this time hijacker Marwan Alshehhi boasts of planning an attack to a Hamburg librarian. He says, "There will be thousands of dead. You will think of me." He also specifically mentions the WTC. [AGENCE FRANCE-PRESSE, 8/29/02; *NEW YORK TIMES*, 8/29/02] "You will see," Alshehhi adds. "In America something is going to happen. There will be many people killed." [*NEW YORK TIMES*, 9/10/02] This "demonstrates that the members of the Hamburg cell were not quite as careful to keep secret their plans as had previously been thought. In addition, it appears to bury for good the theory that the pilots were informed of their targets only hours before they took off. Not least, though, Marwan Alshehhi's boast provides a key element for the reconstruction of the plot—a date by which the terrorists had decided on their target." [*GUARDIAN*, 8/30/02]

Late April–Mid-May 2000: Atta Leaves Numerous Clues While Seeking Crop-Dusting Airplane Loan

Mohamed Atta reportedly has a very strange meeting with Johnelle Bryant of the U.S. Department of Agriculture (incidentally, one month before the official story claims he arrived in the U.S. for the first time). According to Bryant, in the meeting Atta does all of the following:

- He initially refuses to speak with one who is "but a female."

- He asks her for a loan of $650,000 to buy and modify a crop-dusting plane.

- He mentions that he wants to "build a chemical tank that would fit inside the aircraft and take up every available square inch of the aircraft except for where the pilot would be sitting."

- He uses his real name even as she takes notes, and makes sure she spells it correctly.

- He says he has just arrived from Afghanistan.

- He tells about his travel plans to Spain and Germany.

- He expresses an interest in visiting New York.

- He asks her about security at the WTC and other U.S. landmarks.

- He discusses al-Qaeda and its need for American membership.

- He tells her bin Laden "would someday be known as the world's greatest leader."

- He asks to buy the aerial photograph of Washington hanging on her Florida office wall, throwing increasingly large "wads of cash" at her when she refuses to sell it. [ABC NEWS, 6/6/02]

- After Bryant points out one of the buildings in the Washington photograph as her former place of employment, he asks her, "How would you like it if somebody flew an airplane into your friends' building?"

- He asks her, "What would prevent [me] from going behind [your] desk and cutting [your] throat and making off with the millions of dollars" in the safe behind her.

- He asks, "How would America like it if another country destroyed [Washington] and some of the monuments in it like the cities in [my] country had been destroyed?"

- He gets "very agitated" when he isn't given the money in cash on the spot.

- Atta later tries to get the loan again from the same woman, this time "slightly disguised" by wearing glasses.

Three other terrorists also attempt to get the same loan from Bryant, but all of them fail. Bryant turns them down because they do not meet the loan requirements, and fails to notify anyone about these strange encounters until after 9/11. Government officials not only confirm the account and say that Bryant passed a lie detector test, but also elaborate that the account is consistent with other information they have received from interrogating prisoners. Supposedly, failing to get the loan, the terrorists switched plans from using crop dusters to hijacking aircraft. As of July 1, 2004; the 9/11 Commission has failed to mention any aspect of Johnelle Bryant's account. [ABC NEWS 6/6/02; LONDON TIMES 6/8/02] Compare Atta's meeting with FBI Director Mueller's later testimony about the hijackers: "There were no slip-ups. Discipline never broke down. They gave no hint to those around them what they were about." [CNN, 9/28/02]

June 2000: Hijackers Open Many Bank Accounts in Florida; Transactions Not Followed

Mohamed Atta and other hijackers begin to open bank accounts in Florida. At least 35 accounts are opened, 14 of them at SunTrust Bank. All are opened with fake social security numbers (some with randomly made up numbers), yet none of the accounts are checked or questioned by the banks. [NEW YORK TIMES, 7/10/02] One transfer from the United Arab Emirates three months later totaling $69,985 prompts the bank to make a "suspicious transaction report" to the U.S. Treasury's Financial Crimes Enforcement Network. Apparently, no investigation into this transaction occurs. [FINANCIAL TIMES 11/29/01]

June 3, 2000: Atta Supposedly Arrives in U.S. for
First Time, Despite Evidence of Prior Entries

Mohamed Atta supposedly arrives in the U.S. for the first time, flying from Prague to Newark on a tourist visa issued May 18 in Berlin. [MIAMI HERALD, 9/22/01; AUSTRALIAN BROADCASTING CORP., 11/12/01]

June 29, 2000–September 18, 2000: Hijackers Receive $100,000 in Funding from United Arab Emirates Location

Someone using the aliases "Isam Mansour," "Mustafa Ahmed Al-Hisawi," "Mr. Ali" and "Hani (Fawaz Trdng)," sends a total of $109,910 to the 9/11 hijackers in a series of transfers between these dates. [MSNBC, 12/11/01; NEWSWEEK, 12/2/01; NEW YORK TIMES, 12/10/01; FINANCIAL TIMES, 11/30/01; CONGRESSIONAL INTELLIGENCE COMMITTEE, 9/26/02] The money is sent from Sharjah, an emirate in the United Arab Emirates that is allegedly a center for al-Qaeda's illegal financial dealings. The identity of this moneyman "Mustafa Ahmed al-Hisawi" is in dispute. It has been claimed that the name "Mustafa Ahmed" is an alias used by Saeed Sheikh, a known ISI and al-Qaeda agent who sends the hijackers money in August 2001. [CNN, 10/6/01] India claims that Pakistani ISI Director Mahmood orders Saeed to send the hijackers the money at this time. [FRONT-LINE, 10/6/01; DAILY EXCELSIOR 10/18/01] FBI Director Mueller's theory is that this money is sent by "Ali Abdul Aziz Ali." However, of the four aliases used in the different transactions, Mueller connects this man only to three, and not to the alias "Mustafa Ahmed Al-Hisawi." [NEW YORK TIMES, 12/10/01; CONGRESSIONAL INTELLIGENCE COMMITTEE, 9/26/02; ASSOCIATED PRESS 9/26/02] It appears that most of the money is sent to an account shared by Marwan Alshehhi and Mohamed Atta, who would obtain money orders and distribute the money to the other hijackers. [CNN, 10/1/01; MSNBC, 12/11/01; CONGRESSIONAL INTELLIGENCE COMMITTEE, 9/26/02] The *New York Times* later suggests that the amount passed from "Mustafa Ahmed" to the Florida bank accounts right until the day before the attack is around $325,000. The rest of the $500,000–$600,000 they receive for U.S. expenses comes from another, still unknown source. [NEW YORK TIMES, 7/10/02]

July 2000: Atta and Alshehhi Move to Florida and Enroll in Pilot Classes

Mohamed Atta and Marwan Alshehhi move to Venice, Florida, and enroll in pilot classes at Huffman Aviation. [CHICAGO SUN-TIMES, 9/16/01]

Late November 2000–January 30, 2001: Conflicting Accounts of Jarrah's Location

When Ziad Jarrah is questioned at Dubai, United Arab Emirates on January 30, 2001, he reveals that he has been in Pakistan and Afghanistan for the previous two months and five days, and that he is returning to Florida. [CHICAGO TRIBUNE, 12/13/01] Investigators also later confirm that "Jarrah had spent at least three weeks in January 2001 at an al-Qaeda training camp in Afghanistan." [CNN, 8/1/02] However, the Florida Flight Training Center where Jarrah has been studying for the previous six months, later says he is in school there until January 15, 2001. His family later reports he arrives in Lebanon to visit them on January 26, five days before he supposedly passes through Dubai. His father had just undergone open-heart surgery, and Jarrah visits him every day in the hospital until after January 30. Pointing out this incident, his uncle

Ziad Jarrah (from document, left) and Ziad Jarrah (from singed passport recovered in Flight 93 wreckage)

Jamal Jarrah later asks, "How could he be in two places at one time?" [AMONG THE HEROES, JERE LONGMAN, 2002, P. 101–02] This is not the only example of Jarrah being in two places at the same time—there is also evidence he was in different places at once from March 1995–February 1996.

December 26, 2000: Hijackers Abandon
Stalled Plane on Florida Runway; No Investigation Ensues

Hijackers Mohamed Atta and Marwan Alshehhi, while learning to fly in Florida, stall a small plane on a Miami International Airport runway. Unable to start the plane, they simply walk away. Flight controllers have to guide the waiting passenger airliners around the stalled aircraft until it is towed away 35 minutes later. They weren't supposed to be using that airport in the first place. The FAA threatens to investigate the two students and the flight school they are attending. The flight school sends records to the FAA, but no more is heard of the investigation. [NEW YORK TIMES, 10/17/01] "Students do stupid things during their flight course, but this is quite stupid," says the owner of the flight school. Nothing was wrong with the plane. [CNN, 10/17/01]

January 2001: Hijackers Rent Post Office Box in Florida
Months Before They Officially Arrive

Hijackers Hamza Alghamdi and Mohand Alshehri rent a post office box in Delray Beach, Florida, according to the *Washington Post*. Yet FBI Director Mueller later claims they do not enter the country until May 28, 2001. [WASHINGTON POST 9/30/01; CONGRESSIONAL INTELLIGENCE COMMITTEE 9/26/02]

January–February 2001: Flight School's
Repeated Warnings About Hijacker Hanjour Ignored by FAA

In January, the Arizona flight school JetTech alerts the FAA about hijacker Hani Hanjour. No one at the school suspects Hanjour of terrorist intent, but they tell the FAA he lacks both the English and flying skills necessary for the commercial pilot's license he has already obtained. The flight school manager "couldn't believe he had a commercial license of any kind with the skills that he had." A former employee says, "I'm still to this day amazed that he could have flown into the Pentagon. He could not fly at all." They also note he is an exceptionally poor student who does not seem to care about passing

Hani Hanjour

his courses. [NEW YORK TIMES, 5/4/02 (B)] An FAA official named John Anthony actually sits next to Hanjour in class and observes his skills. He suggests the use of a translator to help Hanjour pass, but the flight school points out that goes "against the rules that require a pilot to be able to write and speak English fluently before they even get their license." [ASSOCIATED PRESS, 5/10/02] The FAA verifies that Hanjour's pilot's license is legitimate, but takes no other action. However, since 9/11, the FBI appears to have questions about how Hanjour got his license in 1999. They have questioned and polygraphed the Arab-American instructor who signed off on his flying skills. [CBS NEWS, 5/10/02] His license had in fact already expired in late 1999. [ASSOCIATED PRESS, 9/15/01 (B)] In February, Hanjour begins advanced simulator training, "a far more complicated task than he had faced in earning a commercial license." [NEW YORK TIMES, 6/19/02] The flight school again alerts the FAA about this and gives a total of five alerts about Hanjour, but no further action on him is taken. The FBI is not told about Hanjour. [CBS NEWS. 5/10/02] Ironically, the following July, Arizona FBI agent Ken Williams recommends in a memo that the FBI liaison with local flight schools and keep track of suspicious activity by Middle Eastern students (see chapter 1).

January–June 2001: Hijackers Pass Through Britain for Training or Fundraising

Eleven of the 9/11 hijackers stay in or pass through Britain, according to the British Home Secretary and top investigators. Most are in Britain between April and June, just passing through from Dubai, United Arab Emirates. However, investigators suspect some stay in Britain for training and fundraising. Not all 11 names are given, but one can deduce from the press accounts that Ahmed Alghamdi, Salem Alhazmi, Ahmed Alhaznawi, Ahmed Alnami, and Saeed Alghamdi were definitely in Britain. Ahmed Alghamdi was one of several that should have been "instantly 'red-flagged' by British intelligence," because of his links to Raed Hijazi, a suspected ally of bin Laden being held in Jordan on charges of conspiring to destroy holy sites. Two of the following three also were in Britain: Wail Alshehri, Fayez Banihammad, and Abdulaziz Alomari. Apparently, the investigation concludes that the "muscle," and leaders like Mohamed Atta and Marwan Alshehhi did not pass through Britain at this time. [LONDON TIMES, 9/26/01; WASHINGTON POST, 9/27/01; BBC, 9/28/01; SUNDAY HERALD, 9/30/01] However, police are investigating whether Atta visited Britain in 1999 and 2000, together with some Algerians. [DAILY TELEGRAPH, 9/30/01] The London Times also writes, "Officials hope that the inquiries in Britain will disclose the true identities of the suicide team. Some are known to have arrived in Britain using false passports and fake identities that they kept for the hijack."

January 4, 2001: Atta Moves Between U.S. and Spain

Mohamed Atta flies from Miami, Florida to Madrid, Spain. He has allegedly been in the U.S. since June 3, 2000, learning to fly in Florida with Marwan Alshehhi. [MIAMI HERALD, 9/22/01] He returns to the U.S. on January 10. He makes a second trip to Spain in July of the same year.

January 10, 2001: Two Attas Enter the U.S. on the Same Day?

"INS documents, matched against an FBI alert given to German police, show two men named Mohamed Atta [arrive] in Miami on January 10, each offering different destination addresses to INS agents, one in Nokomis, near Venice, the other at a Coral Springs condo. He (they?) is admitted, despite having overstayed his previous visa by a month. The double entry could be a paperwork error, or confusion over a visa extension. It could be Atta arrived in Miami, flew to another country like the Bahamas, and returned the same day. Or it could be that two men somehow cleared immigration with the same name using the same passport number." [MIAMI HERALD, 9/22/01] Officials later call this a bureaucratic snafu, and insist that only one Atta entered the U.S. on this date. [ASSOCIATED PRESS, 10/28/01] In addition, while Atta arrives on a tourist visa, he tells immigration inspectors that he is taking flying lessons in the U.S., which requires an M-1 student visa. [WASHINGTON POST, 10/28/01] The fact that he had overstayed his visa over a month on a previous visit also does not cause a problem. [LOS ANGELES TIMES, 9/27/01] The INS later defends its decision, but "immigration experts outside the agency dispute the INS position vigorously." For instance Stephen Yale-Loehr, co-author of a 20-volume treatise on immigration law, asserts, "They just don't want to tell you they blew it. They should just admit they made a mistake." [WASHINGTON POST, 10/28/01]

January 11–18, 2001: Overstaying Visa No Obstacle for Alshehhi

Hijacker Marwan Alshehhi flies from the U.S. to Casablanca, Morocco and back, for reasons unknown. He is able to reenter the U.S. without trouble, despite having overstayed his previous visa by about five weeks. [DEPARTMENT OF JUSTICE, 5/20/02; LOS ANGELES TIMES, 9/27/01]

January 30, 2001: Hijacker Questioned at Request of CIA, Then Released

Hijacker Ziad Jarrah is questioned for several hours at the Dubai International Airport, United Arab Emirates, at the request of the CIA for "suspected involvement in terrorist activities," then let go. This is according to United Arab Emirates, U.S., and European officials, but the CIA denies the story. The CIA notified local officials that he would be arriving from Pakistan on his way back to Europe, and they wanted to know where he had been in Afghanistan and how long he had been there. U.S. officials were informed of the results of the interrogation before Jarrah left the airport. Jarrah had already been in the U.S. for six months learning to fly. "UAE and European intelligence sources told CNN that the questioning of Jarrah fits a pattern of a CIA operation begun in 1999 to track suspected al-Qaeda operatives who were traveling through the United Arab Emirates." He was then permitted to leave, eventually going to the U.S. [CNN, 8/1/02; CHICAGO TRIBUNE, 12/13/01] Some accounts place this in January 2000.

February 2001: Two Hijackers Seen Living in
San Antonio with Swapped Identities

At least six people with no connections to one another later claim they recognize hijackers Satam Al Suqami and Salem Alhazmi living in San Antonio, Texas, until this month. The management of an apartment building says the two men abandoned their leases at about this time, and some apartment residents recognize them. However, all the witnesses say that Suqami was going by Alhazmi's name, and vice versa! [SAN ANTONIO, TEXAS, KENS 5 EYEWITNESS NEWS, 10/1/01] One pilot shop employee recognizes Alhazmi as a frequent visitor to the store and interested in a 757 or 767 handbook, though he also says Alhazmi used Suqami's name. [KENS 5 EYEWITNESS NEWS, 10/3/01] The apartment-leasing agent also recalls a Ziad Jarrah who once lived there in June 2001 and looked the same as the hijacker. [SAN ANTONIO EXPRESS-NEWS, 9/22/01; ASSOCIATED PRESS, 9/22/01 (B)] Local FBI confirm that a Salem Alhazmi attended the nearby Alpha Tango Flight School and lived in that apartment building, but they say he is a different Salem Alhazmi who is still alive and living in Saudi Arabia. [KENS 5 EYEWITNESS NEWS, 10/4/01] However, that "Salem Alhazmi" says he has never been to the U.S. and has proven to the authorities that he did not leave Saudi Arabia in the two years prior to 9/11. [WASHINGTON POST, 9/20/01] The FBI does not explain Satam Al Suqami's presence. Neither hijacker is supposed to have arrived in the U.S. before April 2001.

March 2001: Hijackers Pledge Martyrdom in Videos

Supposedly, all 13 of the "muscle" hijackers record a farewell video before leaving training in Kandahar, Afghanistan around this time. [CBS NEWS, 10/9/02] Several have been released. A video of Ahmed Alhaznawi is shown by the Al Jazeera television network in April 2002. In it, he pledges to give his life to "martyr-

Ahmed Alhaznawi recording from before 9/11 (background footage was added in after 9/11)

dom" and swears to send a "bloodied message" to Americans by attacking them in their "heartland." [GUARDIAN, 4/16/02] In September 2002, Al Jazeera shows a similar farewell video of Abdulaziz Alomari made around the same time. [ASSOCIATED PRESS, 9/9/02] Alomari states, "I am writing this with my full conscience and I am writing this in expectation of the end, which is near . . . God praise everybody who trained and helped me, namely the leader Sheik Osama bin Laden." [WASHINGTON POST, 9/11/02] Al Jazeera also shows Ahmed Alnami, Hamza Alghamdi, Saeed Alghamdi, and Wail Alshehri in Kandahar studying maps and flight manuals. [FINANCIAL TIMES, 9/11/02]

March–August 2001: Atta Familiarizes Himself with Flying Crop-Duster Planes

In March and August, Mohamed Atta visits a small airport in South Florida and asks detailed questions about how to start and fly a crop-duster plane. People there easily recall him because he was so persistent. After explaining his abilities, Atta is told he is not skilled enough to fly a crop-duster. [MIAMI HERALD, 9/24/01] Employees at South Florida Crop Care in Belle Glade, Florida later tell the FBI that Atta was among the men who in groups of two or three visited the crop dusting firm nearly every weekend for six or eight weeks before the attacks. Employee James Lester says, "I recognized him because he stayed on my feet all the time. I just about had to push him away from me." [ASSOCIATED PRESS, 9/15/01] Yet, according to U.S. investigators, Atta and the other hijackers gave up on the crop-duster idea back around May 2000.

Mid-March 2001: Hijackers Meet with ID Forger

Hijackers Ahmed Alghamdi, Majed Moqed, Hani Hanjour, and Nawaf Alhazmi stay for four days in the Fairfield Motor Inn, Fairfield, Connecticut. They meet with Eyad M. Alrababah, a Jordanian living in Bridgeport who has been charged with providing false identification to at least 50 illegal aliens. This meeting takes place about six weeks before the FBI says Moqed and Alghamdi enter the U.S. [ASSOCIATED PRESS, 3/6/02; CONGRESSIONAL INTELLIGENCE COMMITTEE, 9/26/02]

April 12–September 7, 2001: Hijackers Collect Multiple Drivers' License Copies

At least six hijackers get more than one Florida driver's license. They get the second license simply by filling out change of address forms:

- Waleed Alshehri—first license May 4, duplicate May 5;

- Marwan Alshehhi—first license, April 12, duplicate in June;

- Ziad Jarrah—first license May 2, duplicate July 10;

- Ahmed Alhaznawi—first license July 10, duplicate September 7;

- Hamza Alghamdi—first license June 27, two duplicates, the second in August; and

- "A sixth man" with a Florida duplicate is not named. [SOUTH FLORIDA SUN-SENTINEL, 9/28/01]

Additionally, some hijackers obtained licenses from multiple states. For instance, Nawaf Alhazmi had licenses from California, New York, and Florida at the same time, apparently all in the same name. [SOUTH FLORIDA SUN-SENTINEL, 9/28/01; NEWSDAY, 9/21/01; DAILY OKLAHOMAN, 1/20/02] [SOUTH FLORIDA SUN-SENTINEL, 9/28/01]

April 18, 2001: Hijacker Flies Between Miami and Amsterdam

Hijacker Marwan Alshehhi flies from Miami, Florida, to Amsterdam, Netherlands. He returns on May 2. Investigators have not divulged where he went or what he did while in Europe. [JUSTICE DEPARTMENT, 5/20/02]

April 23–June 29, 2001: Terrorist "Muscle" Team Arrives in U.S. at This Time—or Earlier

The 13 hijackers commonly known as the "muscle" allegedly first arrive in the U.S. The muscle provides the brute force meant to control the hijacked passengers and protect the pilots. [WASHINGTON POST, 9/30/01] According to FBI Director Mueller, they all pass through Dubai, United Arab Emirates, and their travel was probably coordinated from abroad by Khalid Almihdhar. [CONGRESSIONAL INTELLIGENCE COMMITTEE, 9/26/02] However, some information contradicts their official arrival dates:

- April 23: Waleed Alshehri and Satam Al Suqami arrive in Orlando, Florida. Suqami in fact arrived before February 2001. Alshehri was leasing a house near Washington in 1999 and 2000 with Ahmed Alghamdi. He also lived with Ahmed Alghamdi in Florida for seven months in 1997. [DAILY TELEGRAPH, 9/20/01] Alshehri appears quite Americanized in the summer of 2001, frequently talking with an apartment mate about football and baseball, even identifying himself a fan of the Florida Marlins baseball team. [ASSOCIATED PRESS, 9/21/01]

- May 2: Majed Moqed and Ahmed Alghamdi arrive in Washington. Both actually arrived by mid-March 2001. Ahmed Alghamdi was living with Waleed Alshehri near Washington until July 2000. He also lived with Waleed Alshehri in Florida for seven months in 1997. [DAILY TELEGRAPH, 9/20/01]

- May 28: Mohand Alshehri, Hamza Alghamdi, and Ahmed Alnami allegedly arrive in Miami, Florida. According to other reports, however, both Mohand Alshehri and Hamza Alghamdi arrived by January 2001.

- June 8: Ahmed Alhaznawi and Wail Alshehri arrive in Miami, Florida.

- June 27: Fayez Banihammad and Saeed Alghamdi arrive in Orlando, Florida.

- June 29: Salem Alhazmi and Abdulaziz Alomari allegedly arrive in New York. According to other reports, however, Alhazmi arrived before February 2001.

After entering the U.S. (or, perhaps, reentering), the hijackers arriving at Miami and Orlando airports settle in the Fort Lauderdale, Florida, area along with Mohamed Atta, Marwan Alshehhi, and Ziad Jarrah. The hijackers, arriving in New York and Virginia, settle in the Paterson, New Jersey, area along with Nawaf Alhazmi and Hani Hanjour. [CONGRESSIONAL INTELLIGENCE COMMITTEE, 9/26/02] Note the FBI's early conclusion that 11 of these muscle men "did not know they were on a suicide mission." [OBSERVER, 10/14/01] CIA Director Tenet's later claim that they "probably were told little more than that they were headed for a suicide mission inside the United States" [CIA DIRECTOR TENET TESTIMONY, 6/18/02] and reports that they did not know the exact details of the 9/11 plot until shortly after the attack [CBS NEWS, 10/9/02] are contradicted by video confessions made by all of them in March 2001.

April 26, 2001: Arrest Warrant Issued for Hijacker

Mohamed Atta is stopped at a random inspection near Fort Lauderdale, Florida, and given a citation for having no driver's license. He fails to show up for his May 28 court hearing, and a warrant is issued for his arrest on June 4. After this, he flies all over the U.S. using his real name, and even flies to Spain and back in July, but is never stopped or questioned. The police apparently never try to find him. [WALL STREET JOURNAL, 10/16/01; AUSTRALIAN BROADCASTING CORP., 11/12/01]

May 2001: Hijackers Take Advantage of New, Anonymous Visa Express Procedure

The U.S. introduces the "Visa Express" program in Saudi Arabia, which allows any Saudi Arabian to obtain a visa through his or her travel agent instead of appearing at a consulate in person. An official later states, "The issuing officer has no idea whether the person applying for the visa is actually the person in the documents and application." [U.S. NEWS AND WORLD REPORT, 12/12/01; CONGRESSIONAL INTELLIGENCE COMMITTEE, 9/20/02] At the time, warnings of an attack against the U.S. led by the Saudi Osama bin Laden are higher than they had ever been before—"off the charts" as one senator later puts it. [LOS ANGELES TIMES, 5/18/02; 9/11 CONGRESSIONAL INQUIRY, 9/18/02] A terrorism conference had recently concluded that Saudi Arabia was one of four top nationalities in al-Qaeda. [MINNEAPOLIS STAR-TRIBUNE, 5/19/02] Five hijackers—Khalid Almihdhar, Abdulaziz Alomari, Salem Alhazmi, Saeed Alghamdi, and Fayez Ahmed Banihammad—use Visa

Express over the next month to enter the U.S. [CONGRESSIONAL INTELLIGENCE COMMITTEE, 9/20/02] The widely criticized program is finally canceled in July 2002.

May–August 2001: Hijackers Take Practice Flights and Enjoy Las Vegas Diversions

A number of the hijackers make at least six trips to Las Vegas. It is probable they met here after doing practice runs on cross-country flights. At least Mohamed Atta, Marwan Alshehhi, Nawaf Alhazmi, Ziad Jarrah, Khalid Almihdhar and Hani Hanjour were involved. All of these "fundamentalist" Muslims drink alcohol, gamble, and frequent strip clubs. They even have strippers perform lap dances for them. [SAN FRANCISCO CHRONICLE, 10/4/01; NEWSWEEK, 10/15/01]

May 6–September 6, 2001: Some Hijackers Work Out at Gyms, Some Merely Hang Out

The hijackers work out at various gyms, presumably getting in shape for the hijacking. Ziad Jarrah appears to have trained intensively from May to August, and Mohamed Atta and Marwan Alshehhi also took exercising very seriously. [NEW YORK TIMES, 9/23/01; LOS ANGELES TIMES, 9/20/01] However, these three are presumably pilots who would need the training the least. For instance, Jarrah's trainer says "If he wasn't one of the pilots, he would have done quite well in thwarting the passengers from attacking." [LOS ANGELES TIMES, 9/20/01] For instance, Hani Hanjour, Majed Moqed, Khalid Almihdhar, Nawaf Alhazmi, and Salem Alhazmi work out for four days in early September. [ASSOCIATED PRESS, 9/21/01] Three others—Waleed Alshehri, Wail Alshehri and Satam al-Suqami—"simply clustered around a small circuit of machines, never asking for help and, according to a trainer, never pushing any weights. 'You know, I don't actually remember them ever doing anything . . . They would just stand around and watch people.' " [NEW YORK TIMES, 9/23/01] Those three also had a one month membership in Florida—whether they ever actually worked out there is unknown. [LOS ANGELES TIMES, 9/20/01]

June 2001: Hijackers Meet in London

British investigators believe that at least five of the hijackers have a "vital planning meeting" held in a safe house in north London, Britain. [LONDON TIMES, 9/26/01] Authorities suspect that Mustapha Labsi, an Algerian now in British custody, trains the hijackers in this safe house, as well as previously training the hijackers in Afghanistan. [DAILY TELEGRAPH, 9/30/01]

June or July 2001: Hijackers Plan Attacks from German University

Mohamed Atta, Marwan Alshehhi, and an unknown third person are seen in the ground-floor workshops of the architecture department at this time, according to at least two witnesses from the Hamburg university where Atta had studied. They are seen on at least two occasions with a white, three-foot scale model of the Pentagon. Between 60 and 80 slides of the Sears building in Chicago and the World Trade Center are found to be missing from the technical library after 9/11. [SUNDAY TIMES, 2/3/02]

June 25, 2001: Hijacker Opens Bank Account in Dubai

Hijacker Fayez Banihammad opens a bank account in Dubai, United Arab Emirates (UAE), with 9/11 paymaster "Mustafa Ahmed al-Hawsawi." That name is a likely alias for Saeed Sheikh, who is known to visit Dubai frequently in this time period. [MSNBC, 12/11/01] Banihammad flies to the U.S. the next day. Banihammad gives power of attorney to "al-Hawsawi" on July 18, and then "al-Hawsawi" sends Banihammad Visa and ATM cards in Florida. Banihammad uses the Visa card to buy his airplane ticket for 9/11. [WASHINGTON POST, 12/13/01; MSNBC, 12/11/01] The same pattern of events occurs for some other hijackers, though the timing is not fully known. [CONGRESSIONAL INTELLIGENCE COMMITTEE, 9/26/02] Visa cards are given to several other hijackers in Dubai. [LONDON TIMES, 12/1/01] Other hijackers, including Hani Hanjour, Abdulaziz Alomari, and Khalid Almihdhar, open foreign bank and credit card accounts in the UAE and in Saudi Arabia. Majed Moqed, Saeed Alghamdi, Hamza Alghamdi, Ahmed Alnami, Ahmed Alhaznawi, Wail Alshehri, and possibly others purchase travelers checks in the UAE, presumably with funds given to them when they pass through Dubai. It is believed that "al-Hawsawi" is in Dubai every time the hijackers pass through. [CONGRESSIONAL INTELLIGENCE COMMITTEE, 9/26/02]

July 8–19, 2001: Atta, Bin Al-Shibh, Alshehhi, and Others Meet in Spain to Finalize Attack Plans

Mohamed Atta travels to Spain again (his first trip was in January). Three others cross the Atlantic with him but their names are not known, as they apparently use false identities. [EL MUNDO, 9/30/01] Ramzi bin al-Shibh, a member of his Hamburg terrorist cell, arrives in Spain on July 9, and stays until July 16. [NEW YORK TIMES, 5/1/02] Hijacker Marwan Alshehhi also comes to Spain at about the same time and leaves on July 17. [ASSOCIATED PRESS, 6/30/02] Alshehhi must have traveled under another name, because U.S. immigration has no records of his departure or return. [DEPARTMENT OF JUSTICE, 5/20/02] Investigators believe Atta, Alshehhi, and bin al-Shibh meet with at least three unknown others in a secret safe house near Tarragona. [LOS ANGELES TIMES, 9/1/02; ASSOCIATED PRESS, 6/30/02] It is theorized that the final details of the 9/11 attacks are set at this meeting. [LOS ANGELES TIMES, 9/1/02] Atta probably meets with, and is hosted by, Barakat Yarkas and other Spanish al-Qaeda members. [INTERNATIONAL HERALD TRIBUNE, 11/21/01] One of the unknowns at the meeting could be Yarkas's friend Mamoun Darkazanli, a German with connections to the Hamburg al-Qaeda cell. Darkazanli travels to Spain and meets with Yarkas during the time Atta is there. He travels with an unnamed Syrian Spanish suspect, who lived in Afghanistan and had access there to al-Qaeda leaders. [LOS ANGELES TIMES, 1/14/03] The Spanish newspaper La Vanguardia later reports that Atta also meets with fellow hijackers, Waleed Alshehri and Wail Alshehri, on July 16. [ASSOCIATED PRESS, 9/27/01] Strangely enough, on July 16, Atta stayed in the same hotel in the town of Salou that had hosted FBI counterterrorist expert John O'Neill a few days earlier, when he made a speech to other counterterrorism experts on the need for greater international cooperation by police agencies to combat terrorism. Bin al-Shibh arrived in Salou on July 9, which means he would have been there when the counter-terrorist meeting took place. [THE CELL, BY JOHN MILLER, MICHAEL STONE, AND CHRIS MITCHELL, 8/02, P. 135]

August 2001: Six Hijackers Live Near Entrance to NSA

At least six 9/11 hijackers, including all of those who boarded Flight 77, live in Laurel, Maryland from about this time. They reportedly include Hani Hanjour, Majed Moqed, Khalid Almihdhar, Nawaf Alhazmi, and Salem Alhazmi. Laurel, Maryland is home to a Muslim cleric named Moataz Al-Hallak who teaches at a local Islamic school and has been linked to bin Laden. He has testified three times before a grand jury investigating bin Laden. NSA expert James Bamford later states, "the terrorist cell that eventually took over the airliner that crashed into the Pentagon ended up living, working, planning and developing all their activities in Laurel, Maryland, which happens to be the home of the NSA. So they were actually living alongside NSA employees as they were plotting all these things." [WASHINGTON POST, 9/19/01; BBC, 6/21/02]

August 1–2, 2001: Hijackers Illegally Purchase Virginia Identity Cards

Hijackers Hani Hanjour and Khalid Almihdhar meet Luis Martinez-Flores, an illegal immigrant from El Salvador, in a 7–Eleven parking lot in Falls Church, Virginia. Martinez-Flores is paid $100 cash to accompany the two to a local Department of Motor Vehicles office and sign forms attesting to their permanent residence in Virginia. Given new state identity cards, the cards are used the next day to get Virginia identity cards for several (five to seven) additional hijackers, including Abdulaziz Alomari, Ahmed Alghamdi, Majed Moqed, and Salem Alhazmi. [ARIZONA DAILY STAR, 9/28/01; WASHINGTON POST, 9/30/01; WALL STREET JOURNAL, 10/16/01]

August 20, 2001: Atta Announces Approximate Date
of Attack in E-mail to bin Al-Shibh

In a later interview, would-be hijacker Ramzi bin al-Shibh claims that roughly around this day, he receives a coded e-mail about the 9/11 plot from Mohamed Atta. It reads, "The first term starts in three weeks. . . . There are 19 certificates for private studies and four exams." Bin al-Shibh learns the exact day of the attack on August 29. [GUARDIAN, 9/9/02] Hijacker Hani Hanjour also makes surveillance test flights near the Pentagon and WTC around this time, showing the targets have been confirmed as well. [CBS NEWS, 10/9/02 (B)] Information in a notebook later found in Afghanistan suggests the 9/11 attack was planned for later, but was moved up at the last minute. [MSNBC, 1/30/02] The FBI later notices spikes in cell phone use between the hijackers just after the arrest of Zacarias Moussaoui and just before the hijackers begin to buy tickets for the flights they would hijack. [NEW YORK TIMES, 9/10/02] CIA Director Tenet has hinted that Zacarias Moussaoui's arrest a few days earlier (on August 15; see chapter 10) may be connected to when the date of the attack was picked. [CIA, 6/18/02] On the other hand, some terrorists appear to have made plans to flee Germany in advance of the 9/11 attacks one day before Moussaoui's arrest.

August 24–29, 2001: Hijackers Buy 9/11
Plane Tickets Using Their Apparent Real Names

All of the hijackers book their flights for 9/11, using their apparent real names. Most pay using credit cards on the Internet. [MIAMI HERALD, 9/22/01] At least five tickets are one way only. [LOS ANGELES TIMES, 9/18/01]

August 28, 2001: Atta Buys Flight Ticket Despite Being Wanted by Police

Hijacker Mohamed Atta is able to buy his flight ticket, despite having an arrest warrant for driving without a license and also having violated visa regulations. He should have been wanted for abandoning a stalled aircraft in December 2000 as well. [AUSTRALIAN BROADCASTING CORP., 11/12/01]

August 29, 2001: Atta Tells bin Al-Shibh Exact Date of Attack

In a later interview, would-be hijacker Ramzi bin al-Shibh claims that on this day Mohamed Atta calls him (he is in Germany at the time) from the U.S. Atta asks him what is "two sticks, a dash and a cake with a stick down?" The answer, which bin al-Shibh figures out, is "11–9"—the European and Arabic way of writing 9/11. [KNIGHT RIDDER, 9/9/02; CBS NEWS, 10/9/02] Now knowing the date of the attack, bin al-Shibh later claims that he orders active cells in Europe, the U.S., and elsewhere to evacuate.

Early September 2001: Accounts Place Three Hijackers on East and West Coasts at the Same Time

The standard accounts place hijackers Hani Hanjour, Nawaf Alhazmi, and Khalid Almihdhar on the East Coast for the entire time in the weeks before the attacks. [NEW YORK TIMES, 11/6/01; CNN, 9/26/01; NEW YORK TIMES, 9/21/01; SOUTH FLORIDA SUN-SENTINEL, 9/28/01; ST. PETERSBURG TIMES, 9/27/01; ASSOCIATED PRESS, 9/21/01; NEWSDAY, 9/23/01 (B); CONGRESSIONAL INTELLIGENCE COMMITTEE, 9/26/02] However, neighbors at the San Diego apartment complex where the three lived are clear in their assertions that all three were there until days before 9/11. For instance, one article states, "Authorities believe Almihdhar, Hanjour and Alhazmi . . . moved out a couple of days before the East Coast attacks." [SAN DIEGO CHANNEL 10, 11/1/01] Ed Murray, a resident at the complex, said that all three "started moving out Saturday night–and Sunday [September 9] they were gone." [SAN DIEGO CHANNEL 10, 9/14/01; SAN DIEGO CHANNEL 10, 9/20/01] This is the same day that Alhazmi is reportedly seen in an East Coast shopping mall. [CNN, 9/26/01] As with previous reports, neighbors also see them getting into strange cars late at night. A neighbor interviewed shortly after 9/11 said, "A week ago, I was coming home between 12:00 and 1:00 A.M. from a club. I saw a limo pick them up. It was not the first time. In this neighborhood you notice stuff like that. In the past couple of months, I have seen this happen at least two or three times." [TIME, 9/24/01] To add to the confusion, there have been reports that investigators think Almihdhar is still alive and the *Chicago Tribune* says of Alhazmi, Almihdhar, and Hanjour: "The most basic of facts—the very names of the men—are uncertain. The FBI has said each used at least three aliases. 'It's not going to be a terrible surprise down the line if these are not their true names,' said Jeff Thurman, an FBI spokesman in San Diego." [CHICAGO TRIBUNE, 9/30/01]

September 7, 2001: Bush's Plan to Visit Sarasota on 9/11 Is Publicly Announced; Atta and Marwan Alshehhi Are Seen in Sarasota That Evening

Bush's plan to visit a Sarasota, Florida elementary school on September 11 is publicly announced. According to a later news article, numerous eyewitnesses see hijackers Mohamed Atta and Marwan

Alshehhi in Sarasota later that evening. They appear to have stayed at a Holiday Inn very close to the place Bush will later stay. [LONGBOAT OBSERVER, 11/21/01]

September 7, 2001 (B): Story of Hijackers Drinking Alcohol Changes Over Time

One of the first and most frequently told stories about the hijackers is their visit to Shuckums, a sports bar in Hollywood, Florida, on this day. What is particularly interesting about this story is how it has changed over time. In the original story, first reported on September 12 [ASSOCIATED PRESS, 9/12/01 (E)], Mohamed Atta, Marwan Alshehhi, and an unidentified man come into the restaurant already drunk. "They were wasted," says bartender Patricia Idrissi, who directs them to a nearby Chinese restaurant. [ST. PETERSBURG TIMES, 9/27/01] Later they return and drink—Atta orders five vodka and orange juices, while Alshehhi orders five rum and Cokes. [TIME, 9/24/01] According to manager Tony Amos, "The guy Mohamed was drunk, his voice was slurred and he had a thick accent." Idrissi says they argue about the bill, and when she asks if there was a problem, "Mohamed said he worked for American Airlines and he could pay his bill." [ASSOCIATED PRESS, 9/12/01 (E)] This story was widely reported through much of September. [NEW YORK TIMES, 9/13/01; SUNDAY HERALD, 9/16/01; SOUTH FLORIDA SUN-SENTINEL, 9/15/01; MIAMI HERALD, 9/22/01; NEWSWEEK, 9/17/01; TIME, 9/24/01] However, beginning on September 15, a second story appears. [TORONTO STAR, 9/15/01] This story is similar to the first, except that here, Atta is playing video games and drinking cranberry juice instead of vodka, and Alshehhi is the one who argues over the bill and pays. After some coexistence, the second story seems to have become predominant in later September. [WASHINGTON POST, 9/16/01; WASHINGTON POST, 9/22/01; LOS ANGELES TIMES, 9/27/01; ST. PETERSBURG TIMES, 9/27/01; AUSTRALIAN BROADCASTING CORP., 11/12/01; SUNDAY TIMES, 2/3/02]

September 8–11, 2001: Some Hijackers Sleep with Prostitutes

In the days before the attacks, some of the hijackers (including Waleed Alshehri and/or Wail Alshehri) apparently sleep with prostitutes in Boston hotel rooms, or at least try to do so. A driver working at an "escort service" used by the hijackers claims he regularly drove prostitutes to a relative of bin Laden about once a week until 9/11, when the relative disappeared. Bin Laden has several relatives in the Boston area, most or all of whom returned to Saudi Arabia right after 9/11. [BOSTON HERALD, 10/10/01] On September 10, four other hijackers in Boston (Marwan Alshehhi, Fayez Ahmed Banihammad, Mohand Alshehri, and Satam Al Suqami) call around to find prostitutes to sleep with on their last night alive, but in the end decline. Says one official, "It was going to be really expensive and they couldn't come to a consensus on price, so that was the end of it . . . Either they thought it was too extravagant [over $400] or they did not have enough money left." [BOSTON GLOBE, 10/10/01]

September 9, 2001: Hijacker Jarrah Is Stopped for Speeding but No Red Flag in Computer Records

Hijacker Ziad Jarrah is stopped in Maryland for speeding, ticketed, and released. No red flags show up when his name is run through the computer by the state police, even though he already had been questioned in

January 2001 in the United Arab Emirates [UAE] at the request of the CIA for "suspected involvement in terrorist activities." Baltimore's mayor has criticized the CIA for not informing them that Jarrah was on the CIA's watch list. [CHICAGO TRIBUNE, 12/14/01; ASSOCIATED PRESS, 12/14/01] The CIA calls the whole story (of their alleged request that the UAE question Jarrah) "flatly untrue." [CNN, 8/1/02]

Before September 11, 2001: Hijackers Drink, Watch Strip Shows on Eve of Attacks

A number of the hijackers apparently drink alcohol heavily in bars and watch strip shows. On September 10, three terrorists spend $200 to $300 apiece on lap dances and drinks in the Pink Pony, a Daytona Beach, Florida strip club. While the hijackers had left Florida by this time, Mohamed Atta is reported to have visited the same strip club, and these men appear to have had foreknowledge of the 9/11 attacks. [BOSTON HERALD, 10/10/01] FBI agents have also reportedly questioned the owners of Nardone's Go-Go Bar in Elizabeth, New Jersey. Several of the terrorists spent time in nearby Paterson and Newark and reportedly patronized the club, even on the weekend before 9/11. [BOSTON HERALD, 10/10/01; WALL STREET JOURNAL, 10/16/01] Majed Moqed visits a porn shop and rents a porn video. The major of Paterson, New Jersey, says of the six hijackers who stayed there: "Nobody ever saw them at mosques, but they liked the go-go clubs." [NEWSWEEK, 10/15/01] Nawaf Alhazmi and Khalid Almihdhar often frequented Cheetah's, a nude bar in San Diego. [LOS ANGELES TIMES, 9/1/02] Hamza Alghamdi watched a porn video on September 10. [WALL STREET JOURNAL, 10/16/01] University of Florida religion professor Richard Foltz states, "It is incomprehensible that a person could drink and go to a strip bar one night, then kill themselves the next day in the name of Islam . . . People who would kill themselves for their faith would come from very strict Islamic ideology. Something here does not add up." [SOUTH FLORIDA SUN-SENTINEL, 9/16/01]

9.

Khalid Shaikh Mohammed

Khalid Shaikh Mohammed is widely considered the "mastermind" of 9/11. He was born into a Pakistani family, but his family lived in Kuwait for most of his childhood. He went to college in the U.S. and spent time in Afghanistan in the late 1980s, where he apparently met bin Laden and other future al-Qaeda leaders. [FINANCIAL TIMES, 2/15/03]

Mohammed was involved in the 1993 WTC bombing, working with his nephew Ramzi Yousef. While the 9/11 Commission denies Yousef and Mohammed had ties to al-Qaeda at this time, there are, in fact, many ties. For instance, when Yousef fled from the bombing, he stayed at al-Qaeda's House of the Martyrs in Peshawar, Pakistan. [FINANCIAL TIMES, 2/15/03]

Mohammed moved to the Philippines, and in early 1995, he plotted (with nephew Yousef) to destroy twelve passenger airplanes in what became known as Operation Bojinka. He reportedly lived a very extravagant lifestyle there. The funding for this lifestyle is unknown. Some suggest that he was supported by the Pakistani ISI.

In 1993, U.S. agents uncovered photographs showing Mohammed with close associates of future Pakistani Prime Minister Nawaz Sharif. The *Financial Times* has noted that Mohammed and his allies "must have felt confident that their ties to senior Pakistani Islamists, whose power had been cemented within the country's intelligence service [the ISI], would prove invaluable." [FINANCIAL TIMES, 2/15/03] In that same year, he was involved in an operation to assassinate Benazir Bhutto, then prime minister of Pakistan (and the opponent of Sharif and the ISI). [SLATE, 9/21/01; GUARDIAN, 3/3/03]

The *Los Angeles Times* reports that Mohammed "spent most of the 1990s in Pakistan. Pakistani leadership through the 1990s sympathized with Osama bin Laden's fundamentalist rhetoric. This sympathy allowed Mohammed to operate as he pleased in Pakistan. . . ." [LOS ANGELES TIMES, 6/24/02] If Mohammed had some relationship with the ISI, when—if ever—did this relationship end? When he was plotting the various al-Qaeda attacks in the 1990s, did he do so with ISI backing?

By 1996, U.S. investigators discovered Mohammed's ties to both the 1993 WTC bombing and the 1995 Operation Bojinka plot. These were the two most significant terror attack plots on U.S. targets in years, and the FBI launched a worldwide manhunt for the man they knew to be Mohammed's nephew, Ramzi Yousef, but they failed to launch a similar search for Mohammed himself. Why not?

There is conflicting evidence on just how serious a threat the U.S. considered Mohammed to be. The FBI announced a $2 million dollar reward for Mohammed in 1998. However, in July 2001 he obtained a visa to enter the U.S. (using a Saudi passport and alias, but his own photograph). [LOS ANGELES TIMES, 1/27/04] Yousef's picture had been distributed throughout the world on free matchboxes, but the same was not done for Mohammed. During the same period in which he applied for the visa, U.S. intelligence had determined he was "sending terrorists to the United States" and planning to assist their activities once they arrive. U.S. intelligence intercepted communications between Mohammed and Mohamed Atta, including a call apparently providing a final go ahead for the 9/11 attacks. Given the knowledge of his participation in previous major terrorist plots, why did these intercepts not lead to the detection of the 9/11 plot?

Even after 9/11, the government's response to Mohammed is clouded in mystery. Throughout 2002, there were numerous reports that he'd been arrested or killed. He was allegedly finally arrested in March 2003, but there are numerous problems with the accounts of his arrest. As the Asia Times put it a few days after the arrest, "Clearly, no one has the final word on whether [he] is dead, was captured earlier, or is still free." [ASIA TIMES, 3/6/03] At the time of his alleged March 2003 arrest, he was staying in a neighborhood filled with ISI officials, just a short distance from ISI headquarters, leading to suspicions that he'd been doing so with ISI approval. [AUSTRALIAN BROADCASTING CORP., 3/3/03] One expert noted that after his arrest, "Those who think they have ISI protection will stop feeling that comfort level." [AUSTRALIAN BROADCASTING CORP., 3/2/03] Journalist Robert Fisk reports, "Mohammed was an ISI asset; indeed, anyone who is 'handed over' by the ISI these days is almost certainly a former (or present) employee of the Pakistani agency whose control of Taliban operatives amazed even the Pakistani government during the years before 2001." [TORONTO STAR, 3/3/03]

Assuming that Mohammed is actually alive and in U.S. custody, questions regarding the veracity of any alleged testimony must be considered highly suspect. Some reports indicate that he has been tortured while in U.S. custody. Indeed, even the CIA appears to doubt testimony provided by Mohammed. One CIA report of his interrogations is called, "Khalid Shaikh Mohammed's Threat Reporting—Precious Truths, Surrounded by a Bodyguard of Lies." [LOS ANGELES TIMES, 7/23/04] Why does the 9/11 Commission largely trust his account of the 9/11 plot, and fail to mention any of his alleged ties to the ISI?

Early 1994–January 1995: Khalid Shaikh Mohammed
Receives Financial Support in the Philippines

9/11 mastermind Khalid Shaikh Mohammed lives in the Philippines for a year, planning Operation Bojinka until the plot is exposed in January 1995 and he has to flee (see chapter 7). Police later say he lives a very expensive and non-religious lifestyle. He goes to karaoke bars and go-go clubs, dates go-go dancers, stays in four-star hotels, and takes scuba diving lessons. Once he rents a helicopter just to fly it past the window of a girlfriend's office in an attempt to impress her. This appears to be a pattern; for instance, he has a big drinking party in 1998. [LOS ANGELES TIMES, 6/24/02] Offi- cials believe his obvious access to large sums of money indicate that some larger network is backing him by this time. [LOS ANGELES TIMES 6/6/02] It has been suggested that Mohammed, a Pakistani, is able "to operate as he please[s] in Pakistan" during the 1990s [LOS ANGELES TIMES, 6/24/02], and he is linked to the Pakistani ISI. His hedonistic time in the

Khalid Shaikh Mohammed

Philippines resembles reports of hijackers Mohamed Atta and Marwan Alshehhi in the Philippines in 1998–2000 (see chapter 8). Mohammed returns to the Philippines occasionally, and he is even spotted there after 9/11. [KNIGHT RIDDER, 9/9/02] He is almost caught while visiting an old girlfriend there in 1999, and fails in a second plot to kill the Pope when the Pope cancels his visit to the Philippines that year. [LOS ANGELES TIMES, 9/1/02; LONDON TIMES, 11/10/02]

January–May 1996: U.S. Fails to Capture Mohammed Living Openly in Qatar

Since Operation Bojinka was uncovered in the Philippines (in January 1995), nearly all of the plot's major planners, including Ramzi Yousef, are found and arrested. The one exception is 9/11 mastermind Khalid Shaikh Mohammed. He flees to Qatar in the Persian Gulf, where he lives openly using his real name, enjoying the patronage of Abdallah bin Khalid al-Thani, Qatar's Interior Minister and a member of the royal family. [ABC NEWS, 2/7/03] In January 1996, he is indicted in the U.S. for his role in the 1993 WTC bombing, and in the same month, the U.S. determines his location in Qatar. FBI Director Louis Freeh sends a letter to the Qatari government asking for permission to send a team after him. [LOS ANGELES TIMES, 12/22/02] One of Freeh's diplomatic notes states that Mohammed was involved in a conspiracy to "bomb U.S. airliners" and is believed to be "in the process of manufacturing an explosive device." [NEW

YORKER, 5/27/02] Qatar confirms that Mohammed is there and is making explosives, but they delay handing him over. After waiting several months, a high-level meeting takes place in Washington to consider a commando raid to seize him. However, the raid is deemed too risky, and another letter is sent to the Qatari government instead. One person at the meeting later states, "If we had gone in and nabbed this guy, or just cut his head off, the Qatari government would not have complained a bit. Everyone around the table for their own reasons refused to go after someone who fundamentally threatened American interests. . . ." *[LOS ANGELES TIMES, 12/22/02]* Around May 1996, Mohammed's patron, Abdallah bin Khalid al-Thani, makes sure that Mohammed and four others are given blank passports and a chance to escape. Qatar's police chief later says the other men include Ayman al-Zawahiri and Mohammed Atef, al-Qaeda's number two and number three leaders respectively. *[LOS ANGELES TIMES, 9/1/02; ABC NEWS, 2/7/03]* Bin Laden twice visits al-Thani in Qatar. *[NEW YORK TIMES, 6/8/02; ABC NEWS, 2/7/03]* It appears that bin Laden visits al-Thani in Qatar between the years 1996 and 2000. *[NEW YORK TIMES, 6/8/02; ABC NEWS, 2/7/03]* Al-Thani continues to support al-Qaeda, providing Qatari passports and more than $1 million in funds to al-Qaeda. Even after 9/11, Mohammed is provided shelter in Qatar for two weeks in late 2001. *[NEW YORK TIMES, 2/6/03]* Yet the U.S. apparently still has not frozen al-Thani's assets or taken other action.

Mid-1996–September 11, 2001: Mohammed Travels World; Involved in Many Terrorist Activities

After fleeing Qatar, Khalid Shaikh Mohammed travels the world and plans many terrorist acts. He previously was involved in the 1993 WTC bombing, and the Operation Bojinka plot. *[TIME, 1/20/03]*

Khalid Shaikh Mohammed (note that the picture on the right is an artist rendering of Mohammed in a business suit, based on the picture on the left)

And now, he is apparently involved in the 1998 U.S. embassy bombings, the 2000 USS *Cole* bombing, and other attacks. One U.S. official says, "There is a clear operational link between him and the execution of most, if not all, of the al-Qaeda plots over the past five years." *[LOS ANGELES TIMES, 12/22/02]* He lives in Prague, Czech Republic, through much of 1997. *[LOS ANGELES TIMES, 9/1/02]* By 1999, he is living in Germany and visiting with the hijackers there. *[NEW YORK TIMES, 9/22/02]* Using 60 aliases and as many passports, he travels through Europe, Africa, the Persian Gulf, Southeast Asia and South America, personally setting up al-Qaeda cells. *[LOS ANGELES TIMES, 12/22/02; TIME, 1/20/03]*

December 1997: CIA Ignores Ex-Agent's Warnings About Mohammed

CIA agent Robert Baer, newly retired from the CIA and working as a terrorism consultant, meets a former police chief from the Persian Gulf nation of Qatar. He learns that Khalid Shaikh Mohammed was sheltered from the FBI by then Qatari Interior Minister Abdallah bin Khalid al-Thani. He passes this information to the CIA in early 1998, but the CIA takes no action against Qatar's al-Qaeda patrons. The ex-police chief also tells Baer that Mohammed is a key aide to bin Laden, and that based on Qatari intelligence, Mohammed "is going to hijack some planes." Baer passes this information to the CIA as well, but again the CIA does not seem interested, even when he tries yet again after 9/11. [UPI, 9/30/02; VANITY FAIR, 2/02; SEE NO EVIL, BY ROBERT BAER, 2/02 PP. 270–71] Baer also tries to interest reporter Daniel Pearl in a story about Mohammed before 9/11, but Pearl is still working on it when he is kidnapped and later murdered in early 2002. [UPI, 9/30/02] Baer's source later disappears, presumably kidnapped in Qatar. It has been speculated that the CIA turned on the source to protect its relationship with the Qatari government. [BREAKDOWN, BY BILL GERTZ, 8/02, PP. 55–58] It appears bin Laden visits al-Thani in Qatar between the years 1996 and 2000. [ABC NEWS, 2/7/03] Al-Thani continues to support al-Qaeda, providing Qatari passports and more than $1 million in funds to al-Qaeda. Even after 9/11, Mohammed is provided shelter in Qatar for two weeks in late 2001. [NEW YORK TIMES, 2/6/03] Yet the U.S. still has not frozen al-Thani's assets or taken other action.

January 8, 1998: Mohammed Revealed as Major Terrorist at Yousef Sentencing

Terrorist Ramzi Yousef is sentenced to 240 years for his role in the 1993 WTC bombing. At the same time, prosecutors unseal an indictment against Khalid Shaikh Mohammed for participating with Yousef in the 1995 Operation Bojinka plot. In unsealing this, U.S. Attorney Mary Jo White calls Mohammed a "major player" and says he is believed to be a relative of Yousef. [WASHINGTON POST, 1/9/98] The U.S. announces a $2 million reward for his capture in 1998 and wanted posters with his picture are distributed. [NEW YORK TIMES, 6/5/02] This contradicts the FBI's claim after 9/11 that they did not realize he was a major terrorist before 9/11. [COMMITTEE FINDINGS, 12/11/02] For instance, a senior FBI official later says, "He was under everybody's radar. We don't know how he did it. We wish we knew. . . . He's the guy nobody ever heard of." [LOS ANGELES TIMES, 12/22/02] However, another official says, "We have been after him for years, and to say that we weren't is just wrong. We had identified him as a major al-Qaeda operative before September 11." [NEW YORK TIMES, 9/22/02] Yet strangely, despite knowing Mohammed is a major terrorist plotter and putting out a large reward for his capture at this time, there is no worldwide public manhunt for him as there successfully was for his nephew Ramzi Yousef. Mohammed's name remains obscure and he isn't even put on the FBI's Most Wanted Terrorists list until one month after 9/11. [1000 YEARS FOR REVENGE, BY PETER LANCE, 9/03, PP. 327–30]

FBI Reward Notice for Khalid Shaikh Mohammed

1999: Mohammed Repeatedly Visits Atta, Others in al-Qaeda's Hamburg Cell

Khalid Shaikh Mohammed "repeatedly" visits Mohamed Atta and others in the Hamburg al-Qaeda cell. [ASSOCIATED PRESS, 8/24/02] U.S. and German officials say a number of sources place Mohammed at Atta's Hamburg apartment, although when he visits, or who he visits while he is there, is unclear. [LOS ANGELES TIMES, 6/6/02; NEW YORK TIMES, 11/4/02] However, it would be logical to conclude that he visits Atta's housemate Ramzi bin al-Shibh, since investigators believe he is the "key contact between the pilots" and

Mohammed. [LOS ANGELES TIMES, 1/27/03] Mohammed is living elsewhere in Germany at the time. [NEW YORK TIMES, 9/22/02] German intelligence monitors the apartment in 1999 but apparently does not notice Mohammed. U.S. investigators have been searching for Mohammed since 1996, but apparently never tell the Germans what they know about him. [NEW YORK TIMES, 11/4/02] Even after 9/11, German investigators complain that U.S. investigators do not tell them what they know about Mohammed living in Germany until they read it in the newspapers in June 2002. [NEW YORK TIMES, 6/11/02]

June 2001: Intercepts Indicate Mohammed Wants to Send Terrorists to U.S.

U.S. intelligence learns that Khalid Shaikh Mohammed is interested in "sending terrorists to the United States" and planning to assist their activities once they arrive. The 9/11 Congressional Inquiry will note that the significance of this is not understood at the time, and data collection efforts are not subsequently "targeted on information about [Mohammed] that might have helped understand al-Qaeda's plans and intentions." [COMMITTEE FINDINGS, 12/11/02; LOS ANGELES TIMES, 12/12/02; USA TODAY, 12/12/02]

Summer 2001: NSA Fails to Share Intercepted Information About Calls Between Atta and Mohammed

Around this time, the NSA intercepts telephone conversations between Khalid Shaikh Mohammed and Mohamed Atta, but apparently does not share the information with any other agencies. The FBI has a $2 million reward for Mohammed at the time, while Atta is in charge of operations inside the U.S. [KNIGHT RIDDER, 6/6/02; INDEPENDENT, 6/6/02] The NSA either fails to translate these messages in a timely fashion or fails to understand the significance of what was translated. [KNIGHT RIDDER, 6/6/02]

September 10, 2001: NSA Monitors Call as Mohammed Gives Final Approval to Launch Attacks

Mohamed Atta calls Khalid Shaikh Mohammed in Afghanistan. Mohammed gives final approval to Atta to launch the attacks. This call is monitored and translated by the U.S., although it is not known how quickly the call is translated, and the specifics of the conversation haven't been released. [INDEPENDENT, 9/15/02]

April 11, 2002: Khalid Shaikh Mohammed Connected to New Bombing in Tunisia

A truck bomb kills 19 people, mostly German tourists, at a synagogue in Djerba, Tunisia. It is later claimed that al-Qaeda is behind the attack, and that the suspected bomber speaks with Khalid Shaikh Mohammed by phone about three hours before the attack. [ASSOCIATED PRESS, 8/24/02]

June 4, 2002: 9/11 Mastermind Publicly Identified

Khalid Shaikh Mohammed is publicly identified as the "mastermind" behind the 9/11 attacks. He is believed to have arranged the logistics while on the run in Germany, Pakistan, and Afghanistan. In 1996, he had been secretly indicted in the U.S. for his role in Operation Bojinka, and the U.S. began offering a $2 million reward for his capture in 1998, which increased to $25 million in December 2001. [AP, 6/4/02 (B); *NEW YORK TIMES*, 6/5/02] There are conflicting accounts on how much U.S. investigators knew about Mohammed before 9/11. Mohammed is Pakistani (though born in Kuwait [CBS NEWS, 6/5/02]) and a relative of Ramzi Yousef, the bomber of the WTC in 1993. [*NEW YORK TIMES*, 6/5/02] Though not widely reported, Yossef Bodansky, Director of the Congressional Task Force on Terrorism and Unconventional Warfare, says Mohammed also has ties to the ISI, and they had acted to shield him in the past. [UPI, 9/30/02]

June 16, 2002: 9/11 Mastermind Reported Captured

In September 2002, articles appear in the Pakistani and Indian press suggesting that 9/11 mastermind Khalid Shaikh Mohammed is actually captured on this day. Supposedly he has been sent to the U.S., though the U.S. and Pakistan deny the story and say Mohammed has not been captured at all. [*DAILY TIMES*, 9/9/02; *TIMES OF INDIA*, 9/9/02; *ECONOMIC TIMES*, 9/10/02] If it happened, Mohammed may have been captured *before* a reported interview with Al Jazeera reporter Yosri Fouda.

September 11, 2002: 9/11 Planner Ramzi bin Al-Shibh Captured in Pakistan; Khalid Shaikh Mohammed Possibly Killed as Well

Would-be hijacker Ramzi bin al-Shibh is arrested after a huge gunfight in Karachi, Pakistan, involving thousands of police. [*OBSERVER*, 9/15/02] He is considered "a high-ranking operative for al-Qaeda and one of the few people still alive who know the inside details of the 9/11 plot." [*NEW YORK TIMES*, 9/13/02] Khalid Shaikh Mohammed called bin al-Shibh "the coordinator of the Holy Tuesday [9/11] operation" in an interview aired days before. Captured with him are approximately nine associates, as well as numerous computers, phones and other evidence. [*TIME*, 9/15/02; *NEW YORK TIMES*, 9/13/02] There are conflicting claims that Mohammed is killed in the raid [ASIA TIMES, 10/30/02; *DAILY TELEGRAPH*, 3/4/03; ASIA TIMES, 3/6/03]; shot while escaping [AUSTRALIAN BROADCASTING CORP., 3/2/03]; someone who looks like him is killed, leading to initial misidentification [*TIME*, 1/20/03]; someone matching his general appearance is captured [ASSOCIATED PRESS, 9/16/02]; or that he narrowly escapes capture but his young children are captured. [*LOS ANGELES TIMES*, 12/22/02]

Ramzi bin al-Shibh arrested in Pakistan

March 1, 2003: Mohammed Reportedly Arrested in Pakistan, Doubts Persist

Khalid Shaikh Mohammed is reportedly arrested in Rawalpindi, Pakistan. [ASSOCIATED PRESS, 3/1/03] Officials claim that he is arrested in a late-night joint Pakistani and FBI raid, in which they also arrest Mustafa Ahmed Al-Hawsawi, the purported main financer of the 9/11 attacks. [MSNBC, 3/3/03] However, some journalists immediately cast serious doubts about this arrest. For instance, MSNBC reports, "Some analysts questioned whether Mohammed was actually arrested

Photo of Mohammed's Alleged Arrest in Pakistan

Saturday, speculating that he may have been held for some time and that the news was made public when it was in the interests of the United States and Pakistan" [MSNBC, 3/3/03] There are numerous problems surrounding the U.S.-alleged arrest of Mohammed:

- Witnesses say Mohammed is not present when the raid occurs. [GUARDIAN, 3/3/03 (B), ASSOCIATED PRESS, 3/2/03; ASSOCIATED PRESS, 3/2/03 (B), AUSTRALIAN BROADCASTING CORP., 3/2/03; NEW YORK TIMES, 3/3/03];

- There are differing accounts about which house he is arrested in. [LOS ANGELES TIMES, 3/2/03; LOS ANGELES TIMES, 3/3/03; ASSOCIATED PRESS, 3/1/03]

- There are differing accounts about where he was before the arrest and how authorities found him. [WASHINGTON POST, 3/2/03; TIME, 3/1/03; NEW YORK TIMES, 3/4/03; NEW YORK TIMES, 3/3/03; WASHINGTON POST, 3/2/03]

- Some accounts have him sleeping when the arrest occurs [NEW YORK TIMES, 3/3/03; LOS ANGELES TIMES, 3/2/03; DAILY TELEGRAPH, 3/4/03; REUTERS, 3/2/03]

- Accounts differ on who arrests him—Pakistanis, Americans, or both. [CNN, 3/2/03; LOS ANGELES TIMES, 3/2/03; NEW YORK TIMES, 3/2/03; DAILY TELEGRAPH, 3/3/03; LONDON TIMES, 3/3/03; ASSOCIATED PRESS, 3/3/03 (C)]

- There are previously published accounts that Mohammed may have been killed in September 2002. [DAILY TELEGRAPH, 9/16/02; CHRISTIAN SCIENCE MONITOR, 10/29/02; ASIA TIMES, 10/30/02; LOS ANGELES TIMES, 12/22/02; DAILY TELEGRAPH, 3/4/03; ASIA TIMES, 3/6/03]

- There are accounts that he was captured the year before. [DAILY TIMES, 9/9/02; TIMES OF INDIA, 9/9/02; ASSOCIATED PRESS, 9/16/02; AUSTRALIAN BROADCASTING CORP., 3/2/03]

- A few days later the ISI show what they claim is a video of the capture. It is openly mocked as a bad forgery by the few reporters allowed to see it. [ABC NEWS, 3/11/03; REUTERS, 3/11/03; PAKNEWS, 3/11/03; DAILY TIMES, 3/13/03] For instance, a Fox News reporter says, "Foreign journalists looking at it laughed and said this is baloney, this is a reconstruction." [FOX NEWS, 3/10/03]

These are just some of the difficulties with the arrest story. There are so many problems with it that one *Guardian* reporter says, "The story appears to be almost entirely fictional." [GUARDIAN, 3/6/03]

10.

Zacarias Moussaoui

Zacarias Moussaoui, a Muslim Arab from France, is a self-confessed member of al-Qaeda. In mid-August 2001, he was arrested when the FBI became highly suspicious of his behavior at a Minnesota flight-training school.

Deeply concerned by the information obtained through a preliminary investigation, the Minnesota FBI desperately tried to learn more about him. Because he was a French national, they sought information from the French intelligence service. The French had been tracking Moussaoui since 1995 and provided a substantial dossier on him, including information demonstrating that he had links to al-Qaeda, had journeyed to Afghanistan several times, and had trained at a terrorism camp. This information was passed to Minnesota FBI Agent Coleen Rowley, and to the counterterrorism section at FBI headquarters. [SEATTLE TIMES, 7/7/02]

During this same period, the Defense Department sought to point the FBI's attention to information it had obtained from Milan, Italy. In a newspaper interview, Sheik Omar Bakri Muhmmad, an al-Qaeda spokesman, bragged that al-Qaeda had recruited "kamikaze bombers ready to die for Palestine." The earliest the FBI would grant the Department of Defense an audience to present their information was September 12, 2001. [INTELLIGENCE FAILURE, BY DAVID BOSSIE, 5/04, PP. 44-45]

Yet FBI headquarters continued to thwart the Minnesota FBI agents' attempts to obtain a search warrant to search Moussaoui's home and computer. According to agent Coleen Rowley, her FBI supervisors "deliberately . . . undercut" their attempt to obtain a search warrant by withholding available relevant information provided by the field office, and by watering down the information that headquarters did retain. FBI Headquarters' resistance was so bad that agents in the Minnesota office joked that some FBI officials "had to be spies or moles . . . working for Osama bin Laden." [TIME, 5/21/02]

Why did the Minnesota office face so much resistance from FBI headquarters? Why did CIA Director Tenet, already frantic with worry about a surge in attack warnings, fail to warn the President? Was Moussaoui, in fact, part of the 9/11 plot? If so, who should be held accountable for failing to learn the extent of his involvement in time to prevent 9/11? If not, why is he still being held on that charge, rather than some other terrorism charge? Could part or all of the 9/11 plot been thwarted if the Minnesota FBI was quickly given access to Moussaoui's belongings?

As with other aspects of 9/11, the current public record raises more questions than answers. And, although there was an internal governmental investigation examining the FBI's failures with respect to the Moussaoui case, the report remained classified as this book went to press. [WASHINGTON TIMES, 7/12/04]

1999: French Observe Moussaoui Traveling
Between London, Pakistan, and Afghanistan

Zacarias Moussaoui, living in London, is observed by French intelligence making several trips to Pakistan and Afghanistan. French investigators later claim the British spy agency MI5 was alerted and requested to place Moussaoui under surveillance. The request appears to have been ignored. [INDEPENDENT, 12/11/01]

September–October 2000: Moussaoui Visits
Malaysia After CIA Stops Surveillance There

Zacarias Moussaoui visits Malaysia twice, and stays at the very same condominium where the January al-Qaeda meeting was held. [CNN, 8/30/02; LOS ANGELES TIMES, 2/2/02; WASHINGTON POST, 2/3/02] After that meeting, Malaysian intelligence keeps watch on the condominium at the request of the CIA. However, the CIA stops the surveillance before Moussaoui arrives, spoiling a chance to expose the 9/11 plot by monitoring Moussaoui's later travels. The Malaysians later say they were surprised by the CIA's lack of interest. "We couldn't fathom it, really," Rais Yatim, Malaysia's Legal Affairs minister, told *Newsweek*. "There was no show of concern." [NEWSWEEK, 6/2/02] While Moussaoui is in Malaysia, Yazid Sufaat, the owner of the condominium, signs letters falsely identifying Moussaoui as a representative of his wife's company. [REUTERS, 9/20/02;

Yazid Sufaat

WASHINGTON POST, 2/3/02] When Moussaoui is later arrested in the U.S. about one month before the 9/11 attacks, this letter in his possession could have led investigators back to the condominium and the connections with the January 2000 meeting attended by two of the hijackers. [USA TODAY, 1/30/02] Moussaoui's belongings also contained phone numbers that could have linked him to Ramzi bin al-Shibh (and his roommate, Mohamed Atta), another participant in the Malaysian meeting. [ASSOCIATED PRESS, 12/12/01 (B)]

February 23–June 2001: Moussaoui Takes
Lessons at Flight School Previously Used by al-Qaeda

Zacarias Moussaoui flies to the U.S. Three days later, he starts flight training at the Airman Flight School in Norman, Oklahoma. (Other terrorists had trained at the same flight school or other schools in the area in 1998 and 1999. See chapter 1.) He trains there until May, but does not do well and drops out before getting a pilot's license. His visa expires on May 22, but he does not attempt to renew it or get another one. He stays in Norman, arranging to change flight schools, and frequently exercising in a gym. [9/11 CONGRESSIONAL INQUIRY, 10/17/02; MSNBC, 12/11/01] According to U.S. investigators, would-be hijacker Ramzi bin al-Shibh later says he meets Moussaoui in Karachi (Pakistan) in June 2001. [WASHINGTON POST, 11/20/02]

August 1, 2001: Atta, Alshehhi, and Moussaoui Possibly Meet in Oklahoma City Motel

A motel owner in Oklahoma City later claims that Zacarias Moussaoui and hijackers Mohamed Atta and Marwan Alshehhi all come to his motel on this day. Although the FBI has investigated this lead, they have not commented on it, and prosecutors have not attempted to use the incident as evidence in their case against Moussaoui. It is widely admitted that the case against Moussaoui is not strong (for instance, *Newsweek* states: "there's nothing that shows Moussaoui ever had contact with any of the 9/11 hijackers" [NEWSWEEK, 8/5/02]). The *LA Weekly* speculates the FBI may want to ignore this lead because it "could force the FBI to reopen its investigation of Middle Eastern connections to the 1995 Oklahoma City blast, because convicted bombers Timothy McVeigh and Terry Nichols reportedly stayed at the same motel, interacting with a group of Iraqis during the weeks before the bombing." [LA WEEKLY, 8/2/02 (B)]

Early August 2001: Moussaoui Moves to Minnesota

Zacarias Moussaoui moves from Oklahoma to Minnesota some time in early August, in order to attend flight school training there. [9/11 CONGRESSIONAL INQUIRY, 10/17/02; MSNBC, 12/11/01]

August 13–15, 2001: Moussaoui Immediately Raises Suspicions at Flight School

Zacarias Moussaoui

Zacarias Moussaoui trains at the Pan Am International Flight School in Minneapolis, Minnesota, where he pays $8,300 ($1500 by credit card and the remainder in cash) to use a Boeing 474 Model 400 aircraft simulator. After just one day of training most of the staff is suspicious that he is a terrorist, especially after they discuss with him "how much fuel [is] on board a 747–400 and how much damage that could cause if it hit[s] anything." Staff members call the FBI with their concerns later that day. [NEW YORK TIMES, 2/8/02; 9/11 CONGRESSIONAL INQUIRY, 10/17/02] They are suspicious because:

- In contrast to all the other students at this high-level flight school, he has no aviation background, little previous training, and no pilot's license. [9/11 CONGRESSIONAL INQUIRY, 10/17/02]

- He wants to fly a 747 not because he plans to be a pilot, but as an "ego boosting thing." [NEW YORK TIMES, 10/18/02] Yet within hours of his arrival, it is clear he "was not some affluent joyrider." [NEW YORK TIMES, 2/8/02]

- He is "extremely" interested in the operation of the plane's doors and control panel. [9/11 CONGRESSIONAL INQUIRY, 10/17/02] He also is very keen to learn the protocol for communicating with the flight tower, despite having no plans to become an actual pilot. [NEW YORK TIMES, 2/8/02]

- He is evasive and belligerent when asked about his background. When an instructor, who notes from his records that Moussaoui is from France, attempts to greet him in French, Moussaoui appears not to understand, saying that he had spent very little time in France and that he is from the Middle East. The instructor considers it odd that Moussaoui did not specify the Middle Eastern country. [MINNEAPOLIS STAR-TRIBUNE, 12/21/01; WASHINGTON POST, 1/2/02]

- He tells a flight instructor he is not a Muslim, but the instructor senses he is lying about this. [NEW YORKER, 9/30/02]

- He says he would "love" to fly a simulated flight from London to New York, raising fears he has plans to hijack such a flight. [9/11 CONGRESSIONAL INQUIRY, 10/17/02] His original e-mail to the flight school similarly stated he wanted to be good enough to fly from London to New York. [NEW YORK TIMES, 2/8/02]

- He pays for thousands of dollars in expenses from a large wad of cash. [NEW YORK TIMES, 2/8/02]

- He seems determined to pack a large amount of training in a short period for no apparent reason. [NEW YORK TIMES, 2/8/02]

- He mostly practices flying in the air, not taking off or landing (although that reports claiming he did not want to take off or land at all appear to be an exaggeration). [NEW YORK TIMES, 2/8/02; SLATE, 5/21/02; MINNEAPOLIS STAR-TRIBUNE, 12/21/01; NEW YORK TIMES, 5/22/02]

Failing to get much initial interest from the FBI, the flight instructor tells the FBI agents, "Do you realize how serious this is? This man wants training on a 747. A 747 fully loaded with fuel could be used as a weapon!" [NEW YORK TIMES, 2/8/02]

August 15, 2001: Moussaoui Is Arrested; FBI Headquarters Uninterested

Based on the concerns of flight school staff, Zacarias Moussaoui is arrested and detained in Minnesota on the excuse of an immigration violation. [TIME, 5/27/02] The FBI confiscates his possessions, including a computer laptop, but does not have a search warrant to search through them. When arresting him, they note that he possesses two knives, fighting gloves and shin guards, and has prepared "through physical training for violent confrontation." An FBI interview of him adds more concerns. For example, he states that he is in the U.S. working as a "marketing consultant" for a computer company, but is unable to provide any details of his employment. Nor can he convincingly explain his $32,000 bank balance. [MSNBC, 12/11/01; 9/11

CONGRESSIONAL INQUIRY, 10/17/02] An FBI report states that when asked about his trips to Pakistan, "the questioning caused him to become extremely agitated, and he refused to discuss the matter further." The report also notes, "Moussaoui was extremely evasive in many of his answers." [CNN, 9/28/02] His roommate is interviewed on the same day, and tells agents that Moussaoui believes it is "acceptable to kill civilians who harm Muslims," that Moussaoui approves of Muslims who die as "martyrs, and that Moussaoui might be willing to act on his beliefs. [WASHINGTON POST, 5/24/02] But Minnesota FBI agents quickly become frustrated at the lack of interest in the case from higher ups, and grow increasingly concerned. [NEW YORK TIMES, 2/8/02]

August 17 and 31, 2001: Tenet Briefs President Bush; Fails to Mention Moussaoui

CIA records show that Tenet briefed the president twice in August—once in Crawford, Texas, on August 17, and once in Washington, on August 31. By the time of the second briefing, Tenet is aware of Zacarias Moussaoui's arrest, but, apparently, he fails to tell Bush about it. [WASHINGTON POST, 4/15/04 (B)] In April 2004, Tenet will testify under oath before the 9/11 Commission that he had no direct communication with President Bush during the month of August. [NEW YORK TIMES, 4/15/04] This is quickly discovered to be untrue. A CIA spokesperson will then claim, "He momentarily forgot [about the briefings]." [WASHINGTON POST, 4/15/04 (B)]

August 21, 2001: Local FBI Pleads with Headquarters to Warn Secret Service About Moussaoui

The Minnesota FBI office e-mails FBI headquarters on this day, saying it is "imperative" that the Secret Service be warned of the danger that a plot involving Zacarias Moussaoui might pose to the President's safety. However, no such warning is ever sent. [9/11 CONGRESSIONAL INQUIRY, 10/17/02; NEW YORK TIMES, 10/18/02]

August 22, 2001: France Gives FBI Information on Moussaoui; FBI Headquarters Still Refuses Search Warrant

Responding to the request of the FBI's Minnesota field office, the French provide intelligence information they have compiled over the past several years relating to Zacarias Moussaoui. [9/11 CONGRESSIONAL INQUIRY, 10/17/02] The French say Moussaoui has ties with radical Islamic groups and recruits men to fight in Chechnya. They believe he spent time in Afghanistan in 1999. He had been on a French watch list for several years, preventing him from entering France. A French justice official later says that "the government gave the FBI 'everything we had'" on Moussaoui, "enough to make you want to check this guy out every way you can. Anyone paying attention would have seen he was not only operational in the militant Islamist world but had some autonomy and authority as well." [TIME, 5/27/02] A senior French investigator later says, "Even a neophyte working in some remote corner of Florida, would have understood the threat based on what was sent." [TIME, 8/4/02] The French Interior Minister also emphasizes, "We did not hold back any information." [ABC NEWS, 9/5/02] However, senior officials at FBI headquarters still maintain that the information "was too sketchy to justify a search warrant for his computer." [TIME, 8/4/02]

August 23, 2001: FBI Agents Visit Moussaoui's
Former Flight School; Fail to Make Connections

Two agents from the Oklahoma City FBI office visit Airman Flight School in Norman, Oklahoma, to learn about Zacarias Moussaoui's training there earlier in the year. One of these agents had visited the same school in September 1999 to learn more about Ihab Ali, an al-Qaeda agent who trained there in 1993. Apparently, this agent forgets the connection when he visits the school to look into Moussaoui. He later admits he should have connected the two cases. [9/11 CONGRESSIONAL INQUIRY, 7/24/03 (B), *BOSTON GLOBE*, 9/18/01] The staff director of the 9/11 Congressional Inquiry later states, "No one will ever know whether a greater focus on the connection between these events would have led to the unraveling of the September 11 plot." [*NEW YORK DAILY NEWS*, 9/25/02] The Oklahoma City office also does not connect Moussaoui to a memo that had come from its office in May 1998 warning that "large numbers of Middle Eastern males" were receiving flight training in Oklahoma and could be planning terrorist attacks. Furthermore, Moussaoui's Oklahoma roommate Hussein Attas is also under suspicion at this time. The person who attempted to post bond for Attas had previously been the subject of an extensive investigation by the same Oklahoma City FBI office. That person had numerous terror ties, and was involved in recruiting for a Palestinian terror group. This connection is also not noticed. [9/11 CONGRESSIONAL INQUIRY, 7/24/03 (B)]

August 23 or 24, 2001: CIA Senior Staff Sits on Moussaoui Memo

CIA Director Tenet and CIA senior staff are briefed about the arrest of Zacarias Moussaoui in a briefing entitled "Islamic Extremist Learns to Fly." However, apparently others such as President Bush and the White House counterterrorism group are not told about Moussaoui until after the 9/11 attacks begin. Even the acting director of the FBI is not told, despite the fact that lower level FBI officials who made the arrest tried to pass on the information. Tenet later maintains that there was no reason to alert President Bush or to share information about Moussaoui during an early September 2001 Cabinet-level meeting on terrorism, saying, "All I can tell you is, it wasn't the appropriate place. I just can't take you any farther than that." [*WASHINGTON POST*, 4/17/04]

August 23–27, 2001: Minnesota FBI Agents Convinced Moussaoui Plans
"to Do Something with a Plane," Undermined by FBI Headquarters

In the wake of the French intelligence report on Zacarias Moussaoui, FBI agents in Minnesota are "in a frenzy" and "absolutely convinced he [is] planning to do something with a plane." One agent writes notes speculating Moussaoui might "fly something into the World Trade Center." [*NEWSWEEK*, 5/20/02] Minnesota FBI agents become "desperate to search the computer lap top" and "conduct a more thorough search of his personal effects," especially since Moussaoui acted as if he was hiding something important in the laptop when arrested. [*TIME*, 5/21/02; *TIME*, 5/27/02] They decide to apply for a search warrant under the Foreign Intelligence Surveillance Act (FISA). "FISA allows the FBI to carry out wiretaps and searches that would otherwise be unconstitutional" because "the goal is to gather intelligence, not evidence." [WASHINGTON POST, 11/4/01] Standards to get a warrant through FISA are so low that out of 10,000 requests over more than 20 years, not a single one was turned down. Previously, when the FBI did not have a strong enough case, it allegedly simply lied to FISA. In May 2002, the FISA court complained that the FBI had lied in at least 75 warrant cases during the Clinton administration, once even by the FBI Director. [NEW YORK TIMES, 8/27/02] However, as FBI agent Coleen Rowley later puts it, FBI headquarters "almost inexplicably, throw[s] up roadblocks" and undermines their efforts. Headquarters personnel bring up "almost ridiculous questions in their apparent efforts to undermine the probable cause." One Minneapolis agent's e-mail says FBI headquarters is "setting this up for failure." That turns out to be correct. [TIME, 5/21/02; TIME, 5/27/02]

August 24, 2001: Frustrated Minnesota FBI
Asks CIA for Help with Moussaoui Case

Frustrated with the lack of response from FBI headquarters about Zacarias Moussaoui, the Minnesota FBI contacts an FBI agent working with the CIA's Counter Terrorism Center, and asks for help. [9/11 CONGRESSIONAL INQUIRY, 10/17/02] On this day, the CIA sends messages to stations and bases overseas requesting information about Moussaoui. The message says that the FBI is investigating Moussaoui for possible involvement in the planning of a terrorist attack and mentions his efforts to obtain flight training. It also suggests he might be "involved in a larger plot to target airlines traveling from Europe to the U.S." [9/11 CONGRESSIONAL INQUIRY, 9/18/02] It calls him a "suspect 747 airline attacker" and a "suspect airline suicide hijacker"—showing that the form of the 9/11 attack is not a surprise, at least to the CIA. [9/11 CONGRESSIONAL INQUIRY, 10/17/02] FBI headquarters responds by chastising the Minnesota FBI for notifying the CIA without approval. [TIME, 5/21/02]

August 27, 2001: Minnesota FBI Suffers Further
Resistance and Does Not Receive Phoenix Memo

An agent at the FBI headquarters' Radical Fundamentalist Unit (RFU) tells the FBI Minnesota office supervisor that the supervisor is getting people "spun up" over Zacarias Moussaoui. The supervisor replies that he is trying to get people at FBI headquarters "spun up" because he is trying to make sure that Moussaoui does "not take control of a plane and fly it into the World Trade Center." He later alleges the headquarters agent replies, "[T]hat's not going to happen. We don't know he's a terrorist. You don't have enough to show he is a terrorist. You have a guy interested in this type of aircraft—that is it." [9/11 CONGRESSIONAL INQUIRY, 10/17/02] Three weeks earlier, Dave Frasca, the head of the RFU unit, had received Ken Williams' memo expressing concern about terrorists training in U.S. flight schools and he also knew all about the Moussaoui case, but he apparently was not "spun up" enough to connect the two cases. [TIME, 5/27/02] Neither he nor anyone else at FBI headquarters who saw Williams's memo informed anyone at the FBI Minnesota office about it before 9/11. [TIME, 5/21/02]

August 28, 2001: Minnesota FBI's Moussaoui Warrant Request Is Edited,
Then Dropped by FBI Deputy General Counsel

A previously mentioned unnamed RFU (Radical Fundamentalism Unit) agent edits the Minnesota FBI's request for a FISA search warrant to search Zacarias Moussaoui's possessions. Minnesota is trying to prove that Moussaoui is connected to al-Qaeda through a rebel group in Chechnya, but the RFU agent removes information connecting the Chechnya rebels to al-Qaeda. Not surprisingly, the FBI Deputy General Counsel who receives the edited request decides on this day that the connection to al-Qaeda is insufficient to allow an application for a search warrant through FISA, so FISA is never even sought. [9/11 CONGRESSIONAL INQUIRY, 10/17/02] According to a later memo written by Minneapolis FBI legal officer Coleen Rowley, FBI headquarters is to blame for not getting the FISA warrant because of this rewrite of the request. She states: "I feel that certain facts . . . have, up to now, been omitted, downplayed, glossed over and/or mischaracterized in an effort to avoid or minimize personal and/or institutional embarrassment on the part of the FBI and/or perhaps even for improper political reasons." She asks, "Why would an FBI agent deliberately sabotage a case?" The superiors acted so strangely that some agents in the Minneapolis office openly joked that these higher-ups "had to be spies or moles . . . working for Osama bin Laden." FBI headquarters also refuses to contact the Justice Department to try to get a search warrant through ordinary means. Rowley and others are unable to search Moussaoui's computer until after the 9/11 attacks. Rowley later notes that the headquarters agents who blocked the Minnesota FBI were promoted after 9/11. [SYDNEY MORNING HERALD, 5/28/02; TIME, 5/21/02]

September 4, 2001: FBI Dispatches Vague Message to
U.S. Intelligence Community About Moussaoui Investigation

FBI headquarters dispatches a message to the entire U.S. intelligence community about the Zacarias Moussaoui investigation. According to a later Congressional inquiry, the message notes "that Moussaoui was being held in custody but [it does not] describe any particular threat that the FBI thought he posed, for example, whether he might be connected to a larger plot. [It also does] not recommend that the addressees take any action or look for any additional indicators of a terrorist attack, nor [does] it provide any analysis of a possible hijacking threat or provide any specific warnings." [9/11 CONGRESSIONAL INQUIRY, 9/24/02] The FAA is also given the warning, but the FAA decides not to issue a security alert to the nation's airports. An FAA representative says, "He was in jail and there was no evidence he was connected to other people." [NEW YORK POST, 5/21/02] This is in sharp contrast to an internal CIA warning sent out on August 24 based on even less information, which stated Moussaoui might be "involved in a larger plot to target airlines traveling from Europe to the U.S." [9/11 CONGRESSIONAL INQUIRY, 9/18/02] It turns out that prior to this time, terrorist Ahmed Ressam had started cooperating with investigators. He had trained with Moussaoui in Afghanistan and willingly shared this information after 9/11. The FBI dispatch, with its notable lack of urgency and details, failed to prompt the agents in Seattle holding Ressam to ask him about Moussaoui. Had the connection between these two been learned before 9/11, presumably the search warrant for Moussaoui would have been approved and the 9/11 plot might have unraveled. [SUNDAY TIMES, 2/3/02]

September 5–6, 2001: French Again Warn U.S. About Moussaoui

French and U.S. intelligence officials hold meetings in Paris on combating terrorism. The French newspaper *Le Monde* claims that the French try again to warn their U.S. counterparts about Moussaoui, "but the American delegation . . . paid no attention . . . basically concluding that they were going to take no one's advice, and that an attack on American soil was inconceivable." The U.S. participants also say Moussaoui's case is in the hands of the immigration authorities and is not a matter for the FBI. [INDEPENDENT, 12/11/01; VILLAGE VOICE, 5/28/02] The FBI arranges to deport Moussaoui to France on September 17, so the French can search his belongings and tell the FBI the results. Due to the 9/11 attacks, the deportation never happens. [9/11 CONGRESSIONAL INQUIRY, 10/17/02]

September 11, 2001: FBI Agents Obtain Warrant for Moussaoui—Too Late

Zacarias Moussaoui reportedly cheers as he watches the 9/11 attack on television inside a prison, where he is being held on immigration charges. [BBC, 12/12/01] Within an hour of the attacks, the Minnesota FBI uses a memo written to FBI headquarters shortly after Moussaoui's arrest to ask permission from a judge for the search warrant they have been desperately seeking. Even after the attacks, FBI headquarters is still attempting to block the search of Moussaoui's computer, characterizing the similarities between the actual attack and the fears expressed by the local FBI agents before 9/11 as a mere coincidence. [TIME, 5/21/02] However, a federal judge approves the warrant that afternoon. [NEW YORKER, 9/30/02] Minnesota FBI agent Coleen Rowley notes that this very memo was previously deemed insufficient by FBI headquarters to get a search warrant, and the fact that they are immediately granted one when finally allowed to ask shows "the missing piece of probable cause was only the [FBI headquarters'] failure to appreciate that such an event could occur." [TIME, 5/21/02] The search uncovers information suggesting Moussaoui may have been planning an attack using crop dusters, but it does not reveal any direct connection to the 9/11 hijackers. However, investigators find some German telephone numbers and the name "Ahad Sabet." The numbers allow them to determine the name is an alias for Ramzi bin al-Shibh, Mohamed Atta's former roommate, and they find he wired Moussaoui money. They also find a document connecting Moussaoui with the Malaysian Yazid Sufaat, a lead that could have led to hijackers Nawaf Alhazmi and Khalid Almihdhar. [NEW YORKER, 9/30/02; MSNBC, 12/11/01] Rowley later suggests that if they would had received the search warrant sooner, "There is at least some chance that . . . may have limited the September 11th attacks and resulting loss of life." [TIME, 5/27/02]

December 11, 2001: Moussaoui Indicted, Could Face Death Penalty

Zacarias Moussaoui is criminally indicted for his role in the 9/11 attacks. If he is found guilty, he could be sentenced to death. [MSNBC, 12/11/01; ASSOCIATED PRESS, 12/12/01] Moussaoui has admitted to being a member of al-Qaeda, but while he has been involved in terrorist activity, many have expressed doubts that he was involved in the 9/11 plot.

May 21, 2002: FBI Whistleblower Reveals Slip-Ups in Moussaoui Arrest Before 9/11

Minnesota FBI agent Coleen Rowley, upset with what she considers lying from FBI Director Mueller and others in the FBI about the handling of the Zacarias Moussaoui case, releases a long memo she wrote about the case two weeks before 9/11. [TIME, 5/21/02] She also applies for whistleblower protection. *Time* magazine calls the memo a "colossal indictment of our chief law-enforcement agency's neglect" and says it "raises serious doubts about whether the FBI is capable of protecting the public—and whether it still deserves the public's trust." [TIME, 5/27/02]

Coleen Rowley

Three days after 9/11, Mueller made statements such as "There were no warning signs that I'm aware of that would indicate this type of operation in the country." Coleen Rowley and other Minnesota FBI agents "immediately sought to reach [Mueller's] office through an assortment of higher-level FBI [headquarters] contacts, in order to quickly make [him] aware of the background of the Moussaoui investigation and forewarn [him] so that [his] public statements could be accordingly modified." Yet Mueller continued to make similar comments, including in a Senate hearing on May 8, 2002. [TIME, 5/21/02; NEW YORK TIMES, 5/30/02] Finally, after Rowley's memo becomes public, Mueller states, "I cannot say for sure that there wasn't a possibility we could have come across some lead that would have led us to the hijackers." He also admits: "I have made mistakes occasionally in my public comments based on information or a lack of information that I subsequently got." [NEW YORK TIMES, 5/30/02] *Time* magazine later names Rowley one of three "Persons of the Year" for 2002, along with fellow whistleblowers Cynthia Cooper of WorldCom and Sherron Watkins of Enron. [TIME, 12/22/02; TIME, 12/22/02]

June 3, 2002: FBI Downplays Significance of Moussaoui's E-mails

Former FBI Deputy Director Weldon Kennedy states: "Even in the [Zacarias] Moussaoui case, there's lots of uproar over the fact that the—there was a failure to obtain a warrant to search his computer. Well, the facts now are that warrant was ultimately obtained. The computer was searched and guess what? There was nothing significant on there pertaining to 9/11." [CNN, 6/3/02] Three days later, The *Washington Post* reports: "Amid the latest revelations about FBI and CIA lapses prior to the September 11 attacks, congressional investigators say it is now clear that the evidence that lay unexamined in Zacarias Moussaoui's possession was even more valuable than previously believed. A notebook and correspondence of Moussaoui's not only appears to link him to the main hijacking cell in Hamburg, Germany, but also to an al-Qaeda associate in Malaysia whose activities were monitored by the CIA more than a year before the terror attacks on New York and Washington." [WASHINGTON POST, 6/6/02] Slate magazine later gives Kennedy the "Whopper of the Week" award for his comment. [SLATE, 6/7/02]

September 5, 2002: French Believe Moussaoui
Was Prepared for Second Wave of Attacks

Based on the recent interrogations of terrorist Zacarias Moussaoui's al-Qaeda associates, including his alleged handler, French intelligence believes Moussaoui was not part of the 9/11 attacks, but was being readied for a second wave of attacks. Says one French official: "Moussaoui was going to be a foot soldier in a second wave of attacks that was supposed to culminate in early 2002 with simultaneous bombings against U.S. embassies in Europe, the Middle East and Asia, as well as several hijackings in the United States." However, the U.S. has charged him with being the "20th hijacker" who planned to be on Flight 77 in the 9/11 attack. [ABC NEWS, 9/5/02]

September 24, 2002: Discovered Business Card Helps Case Against Moussaoui

Federal prosecutors say a business card found in the wreckage of Flight 93 provides a link between alleged conspirator Zacarias Moussaoui and hijacker Ziad Jarrah. Supposedly a business card belonging to Jarrah has a phone number written on it, and Moussaoui had once called that number. It was not explained what the number is, whose phone number it was, when Moussaoui called it, when the card was found, or how investigators know the card belonged to Jarrah. [MSNBC, 9/24/02] Interestingly, this find comes just as the case against Moussaoui is facing trouble. For instance, one month earlier, *USA Today* reported that investigators had found no link between Moussaoui and the other hijackers. [USA TODAY, 8/29/02] Prosecutors have been trying to get permission to play the Flight 93 cockpit voice recordings to the jury, but on September 13, the judge said, "the recordings appear to have marginal evidentiary value while posing unfair prejudice to the defendant." [WASHINGTON POST, 9/25/02]

September 25, 2002: FBI Director Denies
Moussaoui Leads Could Have Prevented 9/11

In an interview with CBS, FBI Director Mueller states, "I can tell you there are things I wish we had done differently. That there are things we should have followed up on. But the bottom line is I do not believe that we would have been able to prevent 9/11." Speaking about the Zacarias Moussaoui case, he says, "That took us several months, to follow that lead, and it also required the full support of the German authorities, and it would have been very, I think impossible to have followed that particular lead in the days between the time in which Moussaoui was detained and September 11th." [CBS NEWS, 9/25/02] This negativism is in sharp contrast to a previous statement he made on May 21, 2002, as well as the opinion of many rank and file FBI officers, some of whom have made a chart showing how all the hijackers could have been caught if certain leads had been followed. [NEWSWEEK, 6/2/02] Mueller's opinion on the Moussaoui case is contradicted by many, including FBI agents working on that case. [TIME, 5/21/02] The media also does not agree. For instance the *Independent* stated information on Moussaoui's computer "might have been enough to expose the Hamburg cell, which investigators believe was the key planning unit for 11 September." [INDEPENDENT, 12/11/01]

September 30, 2002: No Plea Bargain
Sought in Case Against Moussaoui

Seymour Hersh of *New Yorker* magazine reveals that, despite a weak case against Zacarias Moussaoui, no federal prosecutor has discussed a plea bargain with him since he was indicted in November 2001. Hersh reports that "Moussaoui's lawyers, and some FBI officials, remain bewildered at the government's failure to pursue a plea bargain." Says a federal public defender, "I've never been in a conspiracy case where the government wasn't interested in knowing if the defendant had any information—to see if there wasn't more to the conspiracy." Apparently a plea bargain isn't being considered because Attorney General Ashcroft wants nothing less than the death penalty for Moussaoui. One former CIA official claims, "They cast a wide net and [Moussaoui] happened to be a little fish who got caught up in it. They know it now. And nobody will back off." A legal expert says, "It appears that Moussaoui is not competent to represent himself, because he doesn't seem to understand the fundamentals of the charges against him, but I am starting to feel that the rest of us are crazier . . . we may let this man talk himself to death to soothe our sense of vulnerability." [NEW YORKER, 9/30/02]

November 20, 2002: Moussaoui Served as Backup Only

The U.S. claims that captured would-be hijacker Ramzi bin al-Shibh has said that Zacarias Moussaoui met 9/11 mastermind Khalid Shaikh Mohammed in Afghanistan during the winter of 2000–01 and that Mohammed gave Moussaoui the names of U.S. contacts. [WASHINGTON POST, 11/20/02] Bin al-Shibh and Mohammed agreed Moussaoui should be nothing more than a backup figure in the 9/11 plot because he could not keep a secret and was too volatile and untrustworthy. Supposedly, bin al-Shibh wired Moussaoui money intended for other terrorist activities, not 9/11. [USA TODAY, 11/20/02] There have been suggestions that the U.S. may move Moussaoui's case from a civilian court to a military tribunal, which would prevent bin al-Shibh from testifying, but the issue remains undecided. [USA TODAY, 11/20/02]

February 26, 2003: Whistleblower Believes FBI
Not Prepared for New Terrorist Threats

Coleen Rowley, the FBI whistleblower who was proclaimed *Time* magazine's Person of the Year in 2002, sends another public letter to FBI Director Mueller. She believes the FBI is not prepared for new terrorist attacks likely to result from the upcoming Iraq war. She also says counterterrorism cases are being mishandled. She claims the FBI and the Justice Department have not questioned captured al-Qaeda suspects Zacarias Moussaoui and Richard Reid about their al-Qaeda contacts, choosing instead to focus entirely on prosecution. She writes, "Lack of follow-through with regard to Moussaoui and Reid gives a hollow ring to our 'top priority'—i.e. preventing another terrorist attack. Moussaoui almost certainly would know of other al-Qaeda contacts, possibly in the U.S., and would also be able to alert us to the motive behind his and Mohamed Atta's interest in crop-dusting." Moussaoui's lawyer also says the government has not attempted to talk to Moussaoui since 9/11. [NEW YORK TIMES, 3/5/03 (C), NEW YORK TIMES, 3/6/03 (B)]

January 30, 2003: Government Reveals Fifth-Jet Theory

The government reveals in a closed-door court hearing that recent interrogations of top al-Qaeda prisoners indicate that Moussaoui may have been part of a plot to hijack a fifth plane on the day of 9/11, perhaps with the White House as its target. This is in contrast to the government's original accusation that Moussaoui was to be the "20th hijacker" on Flight 93. Because Moussaoui does not have a security clearance, he cannot see the classified evidence against him, but he later learns of this "fifth-jet theory" while reading an inadequately blacked-out transcript of the hearing. [CNN, 8/8/03; *TIME*, 10/19/03]

March 27, 2003: Khalid Shaikh Mohammed
Says Moussaoui Not Involved in 9/11

The *Washington Post* reports that information obtained from interrogations of 9/11 mastermind Khalid Shaikh Mohammed further undermines the government's case against Zacarias Moussaoui for his alleged involvement in the 9/11 attacks. Apparently, Mohammed told his interrogators that Moussaoui was not part of the 9/11 hijacker group, but was in the U.S. for a second wave of attacks that were planned for early 2002. Details of any such plan have not been revealed. Legal experts agree that at the very least, "on the death penalty, [this information] is quite helpful to Moussaoui." In spite of Mohammed's revelations, the government still feels that it can convict Moussaoui of being involved in a conspiracy with al-Qaeda. [*WASHINGTON POST*, 3/28/03]

May 14, 2003: Judge Rules That Moussaoui Should
Have Same Access to Top al-Qaeda Prisoners as Prosecution

In January 2003, Judge Leonie Brinkema ruled that Zacarias Moussaoui must be allowed to conduct a videotaped deposition of bin al-Shibh. However, the government still refuses to allow Moussaoui access to bin al-Shibh, stating that even its own lawyers do not have access to question al-Qaeda captives. But on May 12, the government revealed that lawyers have been submitting questions to al-Qaeda detainees about Moussaoui's role in the 9/11 plot. Two days later, Judge Brinkema demands to know, "If circumstances have changed such that submission of written questions is now possible, when did the circumstances change and why was neither this court nor the district court so informed at the time?" She also suggests that since the prosecution can submit questions to al-Qaeda operatives in custody, Moussaoui should also be allowed to do the same. [*NEW YORK TIMES*, 5/15/04]

April 22, 2004: Death Penalty Allowed by Appeals Court

In spite of multiple rulings beginning in 2002 that Zacarias Moussaoui must be allowed to question witnesses, including Ramzi bin al-Shibh, the government has continued to refuse any access to high-level al-Qaeda prisoners. Because of this, Judge Brinkema sanctions the government by ruling in October 2003 that the prosecution could not seek the death penalty. [*TIME*, 10/19/03] Prosecutors have appealed the decision and, on this day, a federal appeals panel restores the government's right to seek the death penalty. However, the same ruling hands a partial victory to Moussaoui, ordering prosecutors to work out a method that would permit Moussaoui to question three high-level prisoners. CBS News reports that the judge ruled, "Moussaoui could have access to information from three al-Qaeda prisoners [Khalid Shaikh Mohammed, Ramzi bin al-Shibh, and Mustafa al-Hawsawi] who may be able to exonerate him." [CBS NEWS, 4/23/04] As a result of the appeals decision, the government will file a motion in July 2004, seeking to conduct a psychiatric evaluation of Moussaoui. The motion explains that the evaluation would only be used to counter any defense strategy to spare Moussaoui the death penalty by citing his mental condition. The motion states, "Like most capital cases, the mental condition of the defendant is likely to play a significant rule during the penalty phase." [CBS NEWS, 4/23/04; *GUARDIAN*, 7/7/04]

11.

Nabil al-Marabh

Few have heard of Nabil al-Marabh. Fewer still understand his possible role in 9/11. Al-Marabh's story is either a major chapter in the 9/11 story, or a major chapter in the Bush administration's handling of innocent Muslims following 9/11. This chapter details the long and complex alleged ties between al-Marabh and other terrorists, including some of the 9/11 hijackers.

Author Steven Brill has reported that following the 9/11 attacks, Attorney General John Ashcroft directed FBI and INS agents to question anyone they could find with a Muslim-sounding name. Some agents simply looked for names in the phone book. Anyone who could be held was detained, even if based on a minor violation of law or immigration rules. [VILLAGE VOICE, 6/27/03] Not surprisingly, given such a scattershot approach, virtually none of these detainees turned out to have any ties to terrorism.

Yet there are many like al-Marabh who seem to have clear, long-standing terrorist ties, but who have been deported and are now outside U.S. jurisdiction. For example, the 9/11 Commission reported that Mohdar Abdullah, an associate of hijackers Nawaf Alhazmi and Khalid Almihdhar while living in San Diego, may have had advance knowledge of the 9/11 attacks, and even boasted about it while in custody. Yet he was deported a few weeks before the commission report was released to the public. [ASSOCIATED PRESS, 6/27/04] Abduallah's deportation appears to be a part of a pattern. In another instance, Osama Basnan was deported to Saudi Arabia on November 17, 2002. [WASHINGTON POST, 11/24/02] Five days later, a *Newsweek* story made headlines around the world with the report that Basnan may have assisted Alhazmi and Almihdhar. [NEWSWEEK, 11/22/02]

There are two conclusions that one can reasonably draw based on a review of the public record: Either al-Marabh is an innocent man who has been unfairly tarred by both the media and the U.S. government, or he is a terrorist who quite possibly played a major role in the 9/11 plot, and who was knowingly released from U.S. custody and allowed to leave the country. Either way, this is a remarkable story, one that apparently has never been comprehensively covered by any media outlet.

The U.S. government has publicly maintained that there were no U.S.-based accomplices who aided the hijackers (with the possible exception of Zacarias Moussaoui). In June 2002, FBI Director Mueller testified, "While here, the hijackers effectively operated without suspicion. . . . As far as we know, they contacted no known terrorist sympathizers in the United States." [ASSOCIATED PRESS, 9/26/02] Privately, however, the FBI had already concluded the opposite. In a November 2001 internal FBI document, they concluded that the lead hijackers "maintained a web of contacts both in the United States and abroad. . . . [Some] contacts provided legal, logistical, or financial assistance, facilitated U.S. entry and flight school enrolment, or were known from [al-Qaeda]-related activities or training." [9/11 CONGRESSIONAL INQUIRY, 7/24/03] The 9/11 Congressional Inquiry's final report concluded that some of the hijackers received "substantial assistance" from associates in the U.S., though it is "not known to what extent any of these contacts in the United States were aware of the plot." The Inquiry reported that the hijackers came into contact with at least 14 people who had been investigated by the FBI for terrorist ties prior to 9/11. [9/11 CONGRESSIONAL INQUIRY, 7/24/03] Evidence of other accomplices exists. For example, a few hours after the attacks, German intelligence intercepted a conversation between two known al-Qaeda followers discussing "the 30 people traveling for the operation." [NEW YORK TIMES, 9/29/01]

Once again, too many questions remain unanswered. Who were the hijackers' associates and where are they now? Are some still operating in the U.S.? Has the U.S. government allowed some to disappear back into the protection of terrorist-friendly countries? If so, why? In 2002, the FBI suggested there were about 5,000 al-Qaeda agents in the U.S., and then in 2003 reduced the number to "several hundred." [NEW YORK TIMES, 2/16/03] Who are they, and why have they not been arrested? Why did many Muslims languish in detention, for months apparently, simply because they had a Muslim-sounding name, while individuals such as Nabil al-Marabh have been released from custody and U.S. jurisdiction?

Al-Marabh's case may be considered a minor mystery, but it is symptomatic of a larger mystery: the failure of the U.S. government to criminally charge or even acknowledge the existence of any U.S. accomplices to the 9/11 plot.

1989–May 2000: Hijacker Associate al-Marabh Busy Inside U.S.

Nabil al-Marabh moves to Boston in 1989 and apparently lives there as a taxi driver and al-Qaeda sleeper agent for the next ten years. [NEW YORK TIMES, 9/18/01; BOSTON HERALD, 9/19/01] In 1992, he learns to use weapons in an Afghan al-Qaeda training camp with a terrorist named Raed Hijazi. [CHICAGO SUN-TIMES, 9/5/02] He and Hijazi live together and drive taxis at the same company in Boston for several years. [LOS ANGELES TIMES, 9/21/01] A mutual friend at the same taxi company is later killed participating in a 1999 al-Qaeda terrorist attack. [BOSTON HERALD, 9/19/01] Hijazi participates in a failed attempt to bomb a hotel in Jordan in November 1999, and helps plan the USS Cole bombing in October 2000. In May 1999, the FBI approaches al-Marabh looking for Hijazi, but al-Marabh lies and says he does not know Hijazi. [WASHINGTON POST, 9/4/02] Hijazi is arrested in Syria in October 2000 and imprisoned in Jordan for his bomb attempt there. [TORONTO SUN, 10/16/01] He begins to cooperate with investigators and identi-

Nabil al-Marabh

fies al-Marabh as a U.S. al-Qaeda operative. [NEW YORK TIMES, 9/18/01] Terrorist Ahmed Ressam gives evidence helping to prove that al-Marabh sent money to Hijazi for the Jordan bombing. [TORONTO SUN, 11/16/01; ABC NEWS, 1/31/02] By February 1999, al-Marabh is driving taxis in Tampa, Florida while maintaining a cover of living in Boston. [TORONTO STAR, 10/26/01; ABC NEWS, 1/31/02] [NEW YORK TIMES, 9/18/01] He apparently lives in Tampa at least part time until February 2000; investigators later wonder if he is an advance man for the Florida-based hijackers. [NEW YORK TIMES, 9/18/01; ABC NEWS, 1/31/02] Al-Marabh is living in Detroit by May 2000, though he maintains a Boston address until September 2000. [BOSTON HERALD, 9/19/01]

May 30, 2000–September 11, 2001: Possible al-Qaeda Sleeper Agent Repeatedly Evades Authorities

Nabil al-Marabh engages in many suspicious activities. He stabs his Detroit roommate in the knee during an argument on May 30, 2000. He pleads guilty in December 2000 to assault and battery with a dangerous weapon. [BOSTON HERALD, 9/20/01] He is given a six-month suspended sentence, but; he fails to report in to his probation officer. An arrest warrant is issued for him in March 2001. [LOS ANGELES TIMES, 9/21/01; OTTAWA CITIZEN, 10/29/01] Al-Marabh lives in Detroit with an al-Qaeda agent named Yousef Hmimssa. [BOSTON HERALD, 9/20/01; ABC NEWS, 1/31/02] Al-Marabh receives five drivers' licenses in Michigan over a period of 13 months in addition to carrying driver's licenses for Massachusetts, Illinois, Florida, and Ontario, Canada. [TORONTO STAR, 10/26/01] On September 11, 2000, he obtains a Michigan license permitting him to drive semi-trucks containing hazardous materials, including explosives and caustic materials. He is still unsuccessfully trying to find a tractor-trailer driving job one month before 9/11. [LOS ANGELES TIMES, 9/21/01; ABC NEWS, 1/31/02] In early 2001, he mostly lives in Toronto, Canada, with Hassan Almrei, a man running some al-Qaeda front businesses. [ABC NEWS, 1/31/02] Many witnesses see al-Marabh with two 9/11 hijackers at his uncle's Toronto photocopy store. On June 27, 2001, al-Marabh is arrested while trying to enter the U.S. from Canada in the back of a tractor-trailer, carrying a false Canadian passport and citizenship card. [ST. CATHERINES STANDARD, 9/28/01; ST. CATHERINES STANDARD, 10/2/01] He had been illegally crossing the U.S.-Canadian border for years. [OTTAWA CITIZEN, 10/29/01] Despite suspicions that he is connected to al-Qaeda, the U.S. immediately deports him to Canada. [NEW YORK TIMES, 7/13/02] He spends two weeks in a Canadian prison, where he boasts to other prisoners that he is in contact with the FBI. He is ordered to live with his uncle in Toronto. These prisoners are puzzled that the FBI does not try to interview them about al-Marabh after 9/11. Al-Marabh fails to show up for a deportation hearing in August and for a court date in September. [ST. CATHERINES STANDARD, 10/2/01] "Had Canadian security agents investigated Mr. al-Marabh when they had the chance back in June, when he was jailed by immigration authorities, they may have discovered any number of his worldwide links to convicted and suspected terrorists, including two of the [9/11 hijackers]." [OTTAWA CITIZEN, 10/29/01]

January 2001–September 11, 2001: Hijackers Witnessed Preparing False IDs in Toronto Photocopy Shop with al-Marabh

Numerous witnesses later recall seeing hijackers Mohamed Atta and/or Marwan Alshehhi in Nabil al-Marabh's Toronto apartment building and photocopy shop owned by al-Marabh's uncle at various times during the year. [TORONTO SUN, 9/28/01; ABC NEWS, 1/31/02] Some of the dozens of eyewitness accounts say Atta sporadically works in the photocopy shop. [TORONTO SUN, 10/21/01] There is a large picture of bin Laden hanging in the store. [TORONTO SUN, 10/21/01] Partially completed fake IDs are found in the store and at al-Marabh's apartment. [TORONTO SUN, 9/28/01; TORONTO SUN, 10/16/01] "Forensic officers said there are similarities in the paper stock, laminates and ink seized from the downtown store and that which was used in identification left behind by the [9/11 hijackers]." [TORONTO SUN, 10/16/01] U.S. and Canadian police later determine that there is a flurry of phone calls and financial transactions between al-Marabh, Atta, and Alshehhi days before the attacks. [TORONTO SUN, 11/16/01] U.S. intelligence also intercepts al-Marabh's associates making post-9/11 phone calls praising attacks. [OTTAWA CITIZEN, 10/29/01] U.S. police say Al-Marabh is one of their top five suspects in the 9/11 attacks. Canadian police say he is linked through financial and phone records to some of the suicide pilots, and believe he heads a Toronto al-Qaeda cell. [TORONTO SUN, 11/23/01]

Spring 2001: U.S. Customs Investigate Two Hijackers Before 9/11

A U.S. Customs Service investigation finds evidence that Nabil al-Marabh has funneled money to hijackers Ahmed Alghamdi and Satam Al Suqami. [COX NEWS SERVICE, 10/16/01; ABC NEWS, 1/31/02] By summer,

Customs also uncovers a series of financial transactions between al-Marabh and al-Qaeda agent Raed Hijazi. [NEW YORK TIMES, 9/21/01; ASSOCIATED PRESS, 11/17/01]

September 17, 2001: Associates of al-Marabh Arrested on Conspiracy Charges

Federal agents looking for al-Marabh fail to find him at an old address, but they accidentally discover three other potential terrorists there. Karim Koubriti, Ahmed Hannan, and Farouk Ali-Haimoud are arrested. They worked as dishwashers at the Detroit airport. Investigators believe they were casing the airport for possible security breaches. [BOSTON GLOBE, 11/15/02] In the apartment, the FBI discovers a day planner that includes notes about the "American base in Turkey," the "American Foreign Minister," and "Alia Airport" in Jordan. [WASHINGTON POST, 9/20/01] They believe the men were planning to assassinate ex-Defense Secretary William Cohen during a visit to Turkey. [ASSOCIATED PRESS, 11/17/01] A stash of false documents is also found, and all three have false passports, Social Security cards, and immigration papers. [BOSTON HERALD, 9/20/01; BOSTON GLOBE, 11/15/02] Fake documents linking al-Marabh and another suspected terrorist named Yousef Hmimssa are also found [ABC NEWS, 1/31/02], as is videotaped surveillance of major tourist spots like Disneyland and the MGM Grand Hotel in Las Vegas. [BOSTON GLOBE, 11/15/02] Abel-Ilah Elmardoudi, the apparent ringleader of this group, is arrested in North Carolina in November 2002. All are tried on terrorist charges in 2003. [BOSTON GLOBE, 11/15/02] Two of the four are convicted of being part of a terrorism conspiracy. [ASSOCIATED PRESS, 6/3/04]

September 19, 2001: Al-Marabh Arrested, Given Light Prison Sentence on Non-Terrorism Charges

Nabil al-Marabh is arrested on September 19, 2001 at an Illinois convenience store. [LOS ANGELES TIMES, 9/21/01] FBI investigators claim al-Marabh helped the hijackers get false IDs, and helped launder money for al-Qaeda. [ABC NEWS, 1/31/02] The FBI decides not to charge al-Marabh on any terrorism related charge. Instead, on September 3, 2002, Nabil al-Marabh pleads guilty to illegally entering the U.S., and he is sentenced to only eight months in prison. [CHICAGO SUN-TIMES, 9/5/02] Federal prosecutors claim that "at this time" there is no evidence "of any involvement by [al-Marabh] in any terrorist organization," even though he has admitted to getting weapons training in Afghanistan. [WASHINGTON POST, 9/4/02] The judge states he cannot say "in good conscience" that he approves of the plea bargain worked out between the prosecution and defense, but he seems unable to stop it. He says, "Something about this case makes me feel uncomfortable. I just don't have a lot of information." The judge has a number of unanswered questions, such as how al-Marabh had $22,000 in cash and $25,000 worth of amber jewels on his possession when he was arrested, despite holding only a sporadic series of low-paying jobs. "These are the things that kind of bother me. It's kind of unusual, isn't it?" says the judge. [NATIONAL POST, 9/4/02]

Late 2002: Informant Details Even More of al-Marabh's Terrorist Ties

Nabil al-Marabh is serving an eight month prison sentence for illegally entering the U.S. A Jordanian in prison with al-Marabh at this time later informs against him, claiming that al-Marabh tells him many details of his terrorism ties. The informant, who shows "a highly detailed knowledge of his former cell-mate's associations and movements" [GLOBE AND MAIL, 6/4/04] claims that al-Marabh:

Raed Hijazi

- admitted he sent money to a former roommate, Raed Hijazi, who is later convicted of trying to blow up a hotel in Jordan, and that he aided Hijazi's flight from authorities. [ASSOCIATED PRESS, 6/3/04]

- planned to die a martyr by stealing a gasoline truck, driving it into the Lincoln or Holland tunnels in New York City, turning it sideways, opening its fuel valves and having an al-Qaeda operative shoot a flare to ignite a massive explosion (the plan is foiled when Hijazi is arrested in Jordan in October 2000). [TORONTO SUN, 10/16/01; ASSOCIATED PRESS, 6/3/04]

- trained on rifles and rocket propelled grenades at militant camps in Afghanistan. [ASSOCIATED PRESS, 6/3/04]

- boasted about getting drunk with two 9/11 hijackers. [GLOBE AND MAIL, 6/4/04]

- asked his uncle to hide an important CD from Canadian police. [GLOBE AND MAIL, 6/4/04]

- claimed he took instructions from a mysterious figure in Chicago known as "al Mosul" which means "boss" in Arabic. [ASSOCIATED PRESS, 6/3/04]

- acknowledged he distributed as much as $200,000 a month to training camps in Afghanistan in the early 1990's. [ASSOCIATED PRESS, 6/3/04]

FBI agents are able to confirm portions of the informant's claims. In Chicago, U.S. Attorney Patrick Fitzgerald drafts an indictment against al-Marabh on multiple counts of making false statements in his interviews with FBI agents. However, Justice headquarters declines prosecution in the name of protecting intelligence sources. Prosecutors in Detroit also develop evidence against al-Marabh but are not allowed to charge him for the same reason. [ASSOCIATED PRESS, 6/3/04]

January 2004: Terror Suspect al-Marabh Mysteriously Returned to Syria

After Nabil al-Marabh's eight-month prison sentence is complete in 2003, he remains in a Chicago prison awaiting deportation. However, deportation proceedings are put on hold because federal prosecutors have lodged a material witness warrant against him. When the warrant is dropped, al-Marabh is cleared to be deported to Syria. [ASSOCIATED PRESS, 1/29/03; ASSOCIATED PRESS, 6/3/04] A footnote in his deportation ruling states, "The FBI has been unable to rule out the possibility that al-Marabh has engaged in terrorist activity or will do so if he is not removed from the United States." [ASSOCIATED PRESS, 6/3/04] Senator Charles Schumer says, "It's hard to believe that the best way to deal with the FBI's 27th most wanted terrorist is to send him back to a terrorist-sponsoring country," and claims al-Marabh could have been tried in a military tribunal or a classified criminal trial. [ASSOCIATED PRESS 6/3/04]

June 2004: Several Senators Demand Ashcroft
Explain al-Marabh's Deportation Decision

The *Associated Press* reports that both Republicans and Democrats have expressed outrage that al-Marabh was deported in January. Several senators have written letters to Attorney General Ashcroft, demanding an explanation. Charles Grassley (R) states that the circumstances of al-Marabh's deportation—who was "at one time No. 27 on the [FBI] list of Most Wanted Terrorists—are "of deep concern and appear to be a departure from an aggressive, proactive approach to the war on terrorism." Patrick Leahy (D) wrote to Aschroft that "The odd handling of this case raises questions that deserve answers from the Justice Department. . . . Why was a suspected terrorist returned to a country that sponsors terrorism? We need to know that the safety of the American people and our strategic goals in countering terrorism are paramount factors when decisions like this are made." [ASSOCIATED PRESS, 6/30/04]

PART III
GEOPOLITICS AND 9/11

12.

The Pakistani ISI

For more than 20 years, Pakistan's spy agency, the ISI—Inter-Services Intelligence—and the U.S. intelligence agencies have engaged in a troubled, dangerous relationship. In the 1980s, the U.S. gave billions of dollars in aid to the ISI, and the ISI in turn passed most, but not all, of the money to the Afghans fighting the Soviet Union. Flush from this windfall of aid money, and from involvement in a growing regional illegal drug trade, the ISI became a self-funding operation. Soon it was called a "state within a state," and wielded more power and independence than possibly any other country's intelligence agency.

It is widely accepted that the Taliban was created by the ISI, who then facilitated the Taliban's successful conquest of Afghanistan. Bin Laden regularly met with ISI officials, and ISI officials allegedly passed on U.S. military intelligence and attack plans to bin Laden (and others), enabling him (and others) to evade U.S. efforts to capture or kill him on more than one occasion. And yet, for some reason, U.S. officials continued to work with the ISI both before and after 9/11.

Although generally underreported in the Western media and not acknowledged by Bush administration officials, disturbing evidence suggests that some ISI leaders were directly involved in facilitating the 9/11 attacks. A key figure to understanding the connections between the ISI, al-Qaeda, and 9/11 is Saeed Sheikh. This chapter goes into considerable detail about Sheikh and his terrorist career, because he is a central link between the hijackers and both ISI and al-Qaeda leaders.

Shortly after 9/11, Sheikh was revealed as a key financier of the 9/11 attacks. A few days later, the link between Sheikh and Mahmood was revealed, and the reporting about Sheikh and his connection to 9/11 stopped. A bewildering series of alternatives was put forth, each with names vaguely similar to

Sheikh's or his aliases. In early 2002, Sheikh was captured in Pakistan for involvement in the murder of reporter Daniel Pearl. This murder once again showed the links between al-Qaeda and the ISI, and even between Sheikh and the 9/11 mastermind, Khalid Shaikh Mohammed. But few articles identified Sheikh as both an al-Qaeda and ISI agent. To date, the author has not uncovered a single mainstream U.S. media outlet that has seriously explored the possibility of Pakistani involvement in the 9/11 plot.

Most importantly, Sheikh directly leads to Mahmood Ahmed, the director of the ISI at the time of 9/11. It has been alleged that Mahmood directed Saeed Sheikh to provide funds to hijacker Mohamed Atta. Mahmood's possible role in 9/11 is certainly intriguing, but not definitively proven. Yet it is remarkable how little the U.S. media and government seems to care about Mahmood or ISI ties to the Taliban, al-Qaeda, or to the 9/11 attacks. One can only imagine the outcry had there been evidence of links between 9/11 and the head of Iraq's intelligence agency.

Pakistan seems incapable of upsetting or provoking the Bush administration. In early 2004, it was revealed that Dr. Abdul Qadeer Khan, the father of Pakistan's nuclear weapons program, participated in the illicit trade in nuclear technology that gave vital bomb making secrets to "rogue" states such as North Korea, Iran, and Libya. Pakistani President Pervez Musharraf was implicated in this trade, a trade that continued at least into 2002. Dr. Khan confessed to his role in all this and was pardoned. [NEW YORK TIMES, 1/29/04; GUARDIAN, 1/31/04; REUTERS. 2/10/04; NEW YORKER, 3/1/04] He and ten other senior Pakistani nuclear engineers are also suspected of assisting al-Qaeda in developing its own program. [OSAMA'S REVENGE, BY PAUL WILLIAMS. 6/04, PP. 129-41] The U.S. went to war with Iraq based on far more tenuous evidence of participation in development and proliferation of weapons of mass destruction. Yet, not only has Pakistan suffered no negative consequences, but the U.S. has canceled and/or rescheduled billions of dollars of Pakistani debt. [DAILY TIMES, 7/20/04]

Pakistan is considered a close U.S. ally. Furthermore, thanks to its nuclear weapons, no one in the U.S. would look forward to war with Pakistan even if it were not considered an ally. But is Pakistan a genuine ally? Did the ISI have a role in the 9/11 attacks? Did the ISI and/or Musharraf have foreknowledge of the attacks and fail to warn the U.S.? Have the pro-Taliban and pro-al-Qaeda elements actually been purged from the Pakistani government? And, why does the Bush administration seemingly care so little about the Pakistani ISI's ties to drugs and terrorism?

1982–1991: Afghan Opium Production Skyrockets

Afghan opium production rises from 250 tons in 1982 to 2,000 tons in 1991, coinciding with CIA support and funding of the mujahedeen. Alfred McCoy, a professor of Southeast Asian history at the University of Wisconsin, says U.S. and Pakistani intelligence officials sanctioned the rebels' drug trafficking because of their fierce opposition to the Soviets: "If their local allies were involved in narcotics trafficking, it didn't trouble [the] CIA. They were willing to keep working with people who were heavily involved in narcotics." For instance, Gulbuddin Hekmatyar, a rebel leader who received about half of all the CIA's covert weapons, was known to be a major heroin trafficker. The director of the CIA in Afghanistan claims later to be oblivious about the drug trade: "We found out about it later on." [MIN-NEAPOLIS STAR-TRIBUNE, 9/30/01; ATLANTIC MONTHLY, 5/96]

1984: Bin Laden Develops Ties with Pakistani ISI and Afghan Warlord

Bin Laden moves to Peshawar, a Pakistani town bordering Afghanistan, and helps run a front organization for the mujahedeen known as Maktab al-Khidamar (MAK), which funnels money, arms, and fighters from the outside world into the Afghan war. [NEW YORKER, 1/24/00] "MAK was nurtured by Pakistan's state security services, the Inter-Services Intelligence agency, or ISI, the CIA's primary conduit for conducting the covert war against Moscow's occupation." [MSNBC, 8/24/98] Bin Laden becomes closely tied to the warlord Gulbuddin Hekmatyar, and greatly strengthens Hekmatyar's opium smuggling operations. [LE MONDE, 9/14/01] Hekmatyar, who also has ties with bin Laden, the CIA, and drug running, has been called "an ISI stooge and creation." [ASIA TIMES, 11/15/01]

Mid-1980s: Pakistani ISI and CIA Gain from Drug Production

The Pakistani ISI starts a special cell of agents who use profits from heroin production for covert actions "at the insistence of the CIA." "This cell promotes the cultivation of opium, the extraction of heroin in Pakistani and Afghan territories under mujahedeen control. The heroin is then smuggled into the Soviet controlled areas, in an attempt to turn the Soviet troops into heroin addicts. After the withdrawal of the Soviet troops, the ISI's heroin cell started using its network of refineries and smugglers for smuggling heroin to the Western countries and using the money as a supplement to its legitimate economy. But for these heroin dollars, Pakistan's legitimate economy must have collapsed many years ago." [FINANCIAL TIMES (ASIAN EDITION), 8/10/01] The ISI grows so powerful on this money, that "even by the shadowy standards of spy agencies, the ISI is notorious. It is commonly branded 'a state within the state', or Pakistan's 'invisible government'." [TIME, 5/6/02]

Mid 1980s: ISI Head Regularly Meets with bin Laden

According to controversial author Gerald Posner, ex-CIA officials claim that General Akhtar Abdul Rahman, Pakistani ISI's head from 1980 to 1987, regularly meets bin Laden in Peshawar, Pakistan. The ISI and bin Laden form a partnership that forces Afghan tribal warlords to pay a "tax" on the opium trade. By 1985, bin Laden and the ISI are splitting annual profits of up to $100 million a year. [WHY AMER-ICA SLEPT, BY GERALD POSNER, 9/03, P. 29]

July 5, 1991: "Outlaw" BCCI Bank Is Shut Down

The Bank of England shuts down Bank of Credit and Commerce International (BCCI), the largest Muslim bank in the world. Based in Pakistan, this bank financed numerous Muslim terrorist organizations and laundered money generated by illicit drug trafficking and other illegal activities, including arms trafficking. Bin Laden and many other terrorists had accounts there. [DETROIT NEWS, 9/30/01] One money-laundering expert claims, "BCCI did dirty work for every major terrorist service in the world." [LOS ANGELES TIMES, 1/20/02] American and British governments knew about all this yet kept the bank open for years. The Pakistani ISI had major connections to the bank. However, as later State Department reports indicate, Pakistan remains a major drug trafficking and money-laundering center despite the bank's closing. [DETROIT NEWS, 9/30/01] "The CIA used BCCI to funnel millions of dollars to the fighters battling the Soviet occupation of Afghanistan" according to the *Washington Post*. A French intelligence report in 2001 suggests the BCCI network has been largely rebuilt by bin Laden. [WASHINGTON POST, 2/17/02] The ruling family of Abu Dhabi, the dominant emirate in the United Arab Emirates, owned 77 percent of the bank. [LOS ANGELES TIMES 1/20/02]

June 1993–October 1994: 9/11 Funder
Saeed Sheikh Becomes a Terrorist

Saeed Sheikh, a brilliant British student at the London School of Economics, drops out of school and moves to his homeland of Pakistan to become a terrorist. Two months later, he begins training in Afghanistan at camps run by al-Qaeda and the Pakistani ISI. By mid-1994, he has become a terrorist instructor. In June 1994, he begins kidnapping Western tourists in India. In October 1994, he is captured after kidnapping three Britons and an American, and is put in an Indian maximum-security prison, where he remain for five years. The ISI pays a lawyer to defend him. [LOS ANGELES TIMES, 2/9/02; DAILY MAIL, 7/16/02; VANITY FAIR, 8/02] His supervisor for his terror work is Ijaz Shah, an ISI officer. [TIMES OF INDIA, 3/12/02; GUARDIAN, 7/16/02]

September 1994: ISI Creates the Taliban, Helps Them Begin Afghanistan Conquest

Starting as Afghani exiles in Pakistan religious schools, the Taliban begin their conquest of Afghanistan. [MSNBC, 10/2/01] "The Taliban are widely alleged to be the creation of Pakistan's military intelligence [the ISI], which, according to experts, explains the Taliban's swift military successes." [CNN, 10/5/96] Richard Clarke, a counterterrorism official during the Reagan and George H. W. Bush administrations and the counterterrorism "tsar" by 9/11, later claims that not only does the ISI create the Taliban, but they also facilitate connections between the Taliban and al-Qaeda to help the Taliban achieve victory. [AGAINST ALL ENEMIES, BY RICHARD CLARKE, 3/04, P. 53] Less often reported is that the CIA worked with the ISI to create the Taliban. A long-time regional expert with extensive CIA ties says, "I warned them that we were creating a monster." He adds that even years later, "The Taliban are not just recruits from 'madrassas' (Muslim theological schools) but are on the payroll of the ISI." [TIMES OF INDIA, 3/7/01] The same claim is made on CNN in February 2002. [CNN, 2/27/02] An edition of the *Wall Street Journal* will state in November 2001, "Despite their clean chins and pressed uniforms, the ISI men are as deeply fundamentalist as any bearded fanatic; the ISI created the Taliban as their own instrument and still support it." [ASIA TIMES, 11/15/01]

November 1994–December 1999: 9/11 Funder Saeed Sheikh Is Captured; Makes Connections in Indian Prison

Saeed Sheikh is imprisoned in India for kidnapping Westerners. While there, he meets Aftab Ansari, another prisoner, an Indian gangster who will be released from prison near the end of 1999. [INDIA TODAY, 2/25/02] Saeed also meets another prisoner named Asif Raza Khan, who also is released in 1999. [REDIFF, 11/17/01] After Saeed is rescued from prison at the end of 1999, he works with Ansari and Khan to kidnap Indians and then uses some of the profits to fund the 9/11 attacks. [FRONTLINE, 2/2/02; INDIA TODAY, 2/14/02] Saeed also becomes good friends with prisoner Maulana Masood Azhar, a terrorist with al-Qaeda connections. [SUNDAY TIMES, 4/21/02] Saeed will later commit further terrorist acts together with Azhar's group, Jaish-e-Mohammed. [INDEPENDENT, 2/26/02]

June 1996: Bin Laden Meets with Pakistani Military Leaders

Controversial author Gerald Posner claims bin Laden and al-Qaeda leader Abu Zubaida meet with senior members of Pakistan's military, including Mushaf Ali Mir, who becomes chief of Pakistan's air force in 2000. Bin Laden had moved to Afghanistan the month before, and the Pakistanis offer bin Laden protection if he allies with the Taliban. The alliance proves successful, and bin Laden calls it "blessed by the Saudis," who are already giving money to both the Taliban and al-Qaeda. [WHY AMERICA SLEPT, BY GERALD POSNER, 9/03, PP. 105–06; TIME, 8/31/03] Perhaps not coincidentally, this meeting comes only one month after a deal is reportedly made that reaffirms Saudi support for al-Qaeda. Bin Laden is initially based in Jalalabad, which is free of Taliban control, but after the deal he moves his base to Kandahar, which is the center of Taliban power. [ASIA TIMES, 9/17/03]

Mid-1996–October 2001: Ariana Airlines
Becomes Transport Arm of al-Qaeda

In 1996, al-Qaeda assumes control of Ariana Airlines, Afghanistan's national airline, for use in its illegal trade network. Passenger flights become few and erratic, as planes are used to fly drugs, weapons, gold, and personnel, primarily between Afghanistan, the United Arab Emirates (UAE), and Pakistan. The Emirate of Sharjah, in the UAE, becomes a hub for al-Qaeda drug and arms smuggling. Typically, "large quantities of drugs" are flown from Kandahar, Afghanistan, to Sharjah, and large quantities of

Victor Bout

weapons are flown back to Afghanistan. [LOS ANGELES TIMES, 11/18/01] About three to four flights run the route each day. Many weapons come from Victor Bout, a notorious Russian arms dealer based in Sharjah. [LOS ANGELES TIMES, 1/20/02] Afghan taxes on opium production are paid in gold, and then the gold bullion is flown to Dubai, UAE, and laundered into cash. [WASHINGTON POST, 2/17/02] Taliban officials regularly provide terrorists with false papers identifying them as Ariana Airlines employees so they can move freely around the world. A former National Security Council official later claims the U.S. is well aware at the time that al-Qaeda agents regularly fly on Ariana Airlines, but the U.S. fails to act for several years. The U.S. does press the UAE for tighter banking controls, but moves "delicately, not wanting to offend an ally in an already complicated relationship," and little changes by 9/11. [LOS ANGELES TIMES, 11/18/01] Much of the money for the 9/11 hijackers flows though these Sharjah, UAE, channels. There also are reports suggesting that Ariana Airlines might have been used to train Islamic militants as pilots. The illegal use of Ariana Airlines helps convince the United Nations to impose sanctions against Afghanistan in 1999, but the sanctions lack teeth and do not stop the airline. A second round of sanctions finally stops foreign Ariana Airlines flights, but its charter flights and other charter services keep the illegal network running. [LOS ANGELES TIMES, 11/18/01]

October 1996: Arms Dealer Aligns with Taliban and ISI

Russian arms merchant Victor Bout, who has been selling weapons to Afghanistan's Northern Alliance since 1992, switches sides, and begins selling weapons to the Taliban and al-Qaeda instead. [GUARDIAN, 4/17/02; LOS ANGELES TIMES, 1/20/02; LOS ANGELES TIMES, 5/17/02] The deal comes immediately after the Taliban captures Kabul in late October 1996 and gains the upper hand in Afghanistan's civil war. In one trade in 1996, Bout's company delivers at least 40 tons of Russian weapons to the Taliban, earning about $50 million. [GUARDIAN, 2/16/02] Two intelligence agencies later confirm that Bout trades with the Taliban "on behalf of the Pakistan government." In late 2000, several Ukrainians sell 150 to 200 T-55 and T-62 tanks to the Taliban in a deal conducted by the ISI, and Bout helps fly the tanks to Afghanistan. [MONTREAL GAZETTE, 2/5/02] Bout formerly worked for the Russian KGB, and now operates the world's largest private weapons transport network. Based in the United Arab Emirates (UAE), Bout operates freely there until well after 9/11. The U.S. becomes aware of Bout's widespread illegal weapons trading in Africa in 1995, and of his ties to the Taliban in 1996, but they fail to take effective action against him for years. [LOS ANGE-

LES TIMES, 5/17/02] U.S. pressure on the UAE in November 2000 to close down Bout's operations there is ignored. Press reports calling him "the merchant of death" also fail to pressure the UAE. [FINANCIAL TIMES, 6/10/00; GUARDIAN, 12/23/00] After President Bush is elected, it appears the U.S. gives up trying to get Bout, until after 9/11. [WASHINGTON POST, 2/26/02; GUARDIAN, 4/17/02] Bout moves to Russia in 2002. He is seemingly protected from prosecution by the Russian government, which in early 2002 will claim, "There are no grounds for believing that this Russian citizen has committed illegal acts." [GUARDIAN, 4/17/02] The _Guardian_ suggests that Bout may have worked with the CIA when he traded with the Northern Alliance, and this fact may be hampering current international efforts to catch him. [GUARDIAN, 4/17/02]

Late 1996: Bin Laden Becomes Active in Opium Trade

Bin Laden establishes and maintains a major role in opium drug trade, soon after moving the base of his operations to Afghanistan. Opium money is vital to keeping the Taliban in power and funding bin Laden's al-Qaeda network. Yossef Bodansky, Director of the Congressional Task Force on Terrorism and Unconventional Warfare and author of a 1999 biography on bin Laden, says bin Laden takes a 15 percent cut of the drug trade money in exchange for protecting smugglers and laundering their profits. [MINNEAPOLIS STAR-TRIBUNE, 9/30/01] Another report estimates that bin Laden takes up to 10 percent of Afghanistan's drug trade by early 1999. This would give him a yearly income of up to $1 billion out of $6.5 to $10 billion in annual drug profits from within Afghanistan. [FINANCIAL TIMES, 11/28/01]

May 28, 1998: Pakistan Tests Nuclear Bomb

Pakistan conducts a successful nuclear test. Former Clinton official Karl Inderfurth later notes that concerns about an Indian-Pakistani conflict, or even nuclear confrontation, compete with efforts to press Pakistan on terrorism. [9/11 CONGRESSIONAL INQUIRY, 7/24/03 (B)]

1999: 9/11 Funder Returns to Britain Without Being Arrested

The _London Times_ later claims that British intelligence secretly offers 9/11 paymaster Saeed Sheikh (imprisoned in India from 1994 to December 1999 for kidnapping Britons and Americans) amnesty and the ability to "live in London a free man" if he will reveal his links to al-Qaeda. He apparently refuses. [DAILY MAIL, 7/16/02; LONDON TIMES, 7/16/02] Yet after he is rescued in a hostage swap deal in December, the press reports that he, in fact, is freely able to return to Britain. [PRESS TRUST OF INDIA, 1/3/00] He visits his parents there in 2000 and again in early 2001. [VANITY FAIR, 8/02; BBC, 7/16/02 (B); DAILY TELEGRAPH, 7/16/02] He is not charged with kidnapping until well after 9/11. Saeed's kidnap victims call the government's decision not to try him a "disgrace" and "scandalous." [PRESS TRUST OF INDIA, 1/3/00] The _Pittsburgh Tribune-Review_ later suggests that not only is Saeed closely tied to both the ISI and al-Qaeda, but may also have been working for the CIA: "There are many in Musharraf's government who believe that Saeed Sheikh's power comes not from the ISI, but from his connections with our own CIA. The theory is that . . . Saeed Sheikh was bought and paid for." [PITTSBURGH TRIBUNE-REVIEW, 3/3/02]

Spring 1999: New Jersey HMO Is Possibly Funding al-Qaeda

Randy Glass is a con artist turned government informant participating in a sting called Operation Diamondback. [PALM BEACH POST, 9/29/01] He discusses an illegal weapons deal with an Egyptian-American named Mohamed el Amir. In wiretapped conversations, Mohamed discusses the need to get false papers to disguise a shipment of illegal weapons. His brother, Dr. Magdy el Amir, has been a wealthy neurologist in Jersey City for the past twenty years. Two other weapons dealers later convicted in a sting operation involving Glass also lived in Jersey City, and both el Amirs admit knowing one of them, Diaa Mohsen. Mohsen has been paid at least once by Dr. el Amir. In 1998, Congressman Ben Gilman was given a foreign intelligence report suggesting that Dr. el Amir owns an HMO that is secretly funded by bin Laden, and that money is being skimmed from the HMO to fund terrorist activities. The state of New Jersey later buys the HMO and determines that $15 million were unaccounted for and much of that has been diverted into hard-to-trace offshore bank accounts. However, investigators working with Glass are never given the report about Dr. el Amir. Neither el Amir has been charged with any crime. Mohamed now lives in Egypt and Magdy continues to practice medicine in New Jersey. Glass's sting, which began in late 1998, will uncover many interesting leads before ending in June 2001. [MSNBC, 8/2/02]

July 1999: Ex-ISI Head Is Providing Taliban
Information on U.S. Missile Launches

The U.S. gains information that former ISI head Hamid Gul contacts Taliban leaders at this time and advises them that the U.S. is not planning to attack Afghanistan to get bin Laden. He assures them that he will provide them three or four hours warning of any future U.S. missile launch, as he did "last time." Counterterrorism "tsar" Richard Clarke later suggests Gul gave al-Qaeda warning about the missile strike in August 1998. [NEW YORKER, 7/28/03]

July 14, 1999: Pakistani ISI Agent Promises
Attack on WTC in Recorded Conversation

U.S. government informant Randy Glass records a conversation at a dinner attended by him, illegal arms dealers Diaa Mohsen, Mohammed Malik, a former Egyptian judge named Shireen Shawky, and ISI agent Rajaa Gulum Abbas, held at a restaurant within view of the WTC. FBI agents pretending to be restaurant customers sit at nearby tables. [WPBF CHANNEL 25, 8/5/02; MSNBC, 8/2/02] Abbas says he wants to buy a whole shipload of weapons stolen from the U.S. military to give to bin Laden. [COX NEWS SERVICE, 8/2/02] Abbas points to the WTC and says, "Those towers are coming down." This ISI agent later makes two other references to an attack on the WTC. [WPBF CHANNEL 25, 8/5/02; COX NEWS SERVICE, 8/2/02; PALM BEACH POST, 10/17/02] Abbas also says, "Americans [are] the enemy," and, "We would have no problem with blowing up this entire restaurant because it is full of Americans." [MSNBC, 3/18/03] The meeting is secretly recorded, and parts are shown on television in 2003. [MSNBC, 3/18/03 (R)]

Randy Glass with Stinger missile

August 17, 1999: Arms Merchants Seek Nuclear Materials in U.S.; Report Is Sanitized

A group of illegal arms merchants, including an ISI agent with foreknowledge of 9/11, had met in a New York restaurant the month before. This same group meets at this time in a West Palm Beach, Florida, warehouse, and it is shown Stinger missiles as part of a sting operation. [SOUTH FLORIDA SUN-SENTINEL, 3/20/03] U.S. intelligence soon discovers connections between two in the group, Rajaa Gulum Abbas and Mohammed Malik, terrorist groups in Kashmir (where the ISI assists terrorists fighting against India), and the Taliban. Mohamed Malik suggests in this meeting that the Stingers will be used in Kashmir or Afghanistan. His colleague Diaa Mohsen also says Abbas has direct connections to "dignitaries" and bin Laden. Abbas also wants heavy water for a "dirty bomb" or other material to make a nuclear weapon. He says he will bring a Pakistani nuclear scientist to the U.S. to inspect the material. [MSNBC, 8/2/02; MSNBC, 3/18/03] Government informant Randy Glass passes these warnings on before 9/11, but he claims, "The complaints were ordered sanitized by the highest levels of government." [WPBF CHANNEL 25, 8/5/02] In June 2002, the U.S. secretly indicts Abbas, but apparently they aren't trying very hard to find him: In August 2002, MSNBC is easily able to contact Abbas in Pakistan and speak to him by telephone. [MSNBC, 8/2/02]

October 12, 1999: General Musharraf Takes Control of Pakistan; Ousted ISI Leader Has Curious Finances

General Pervez Musharraf becomes leader of Pakistan in a coup. One major reason for the coup is the ISI felt the previous ruler had to go "out of fear that he might buckle to American pressure and reverse Pakistan's policy [of supporting] the Taliban." [NEW YORK TIMES, 12/8/01] Shortly thereafter Musharraf replaces the leader of the ISI, Brig Imtiaz, because of his close ties to the previous leader. Imtiaz is arrested and convicted of "having assets disproportionate to his known sources of income." It comes out that he was keeping tens of millions of dollars earned from heroin smuggling in a Deutschebank account. This is interesting because insider trading just prior to 9/11 will later connect to a branch of Deutschebank recently run by "Buzzy" Krongard, now Executive Director of the CIA. [FINANCIAL TIMES (ASIAN EDITION), 8/10/01] The new Director of the ISI is Lieutenant General Mahmood Ahmed, a close ally of Musharraf who is instrumental in the success of the coup. [GUARDIAN, 10/9/01]

General
Musharraf

November 29, 1999: UN Says ISI Makes Billions from Drugs

The United Nations Drug Control Programme determines that the ISI makes around $2.5 billion annually from the sale of illegal drugs. [TIMES OF INDIA, 11/29/99]

December 24–31, 1999: Hijacked Flight Leads to Freeing of Future 9/11 Funder

An Indian Airlines flight is hijacked and flown to Afghanistan where 155 passengers are held hostage for eight days. They are freed in return for the release of three militants held in Indian prisons. One of the hostages is killed. One of the men freed in the exchange is 9/11 paymaster Saeed Sheikh. [BBC, 12/31/99]

Another freed militant is terrorist leader Maulana Masood Azhar. Azhar emerges in Pakistan a few days later, and tells a crowd of 10,000, "I have come here because this is my duty to tell you that Muslims should not rest in peace until we have destroyed America and India." [ASSOCIATED PRESS, 1/5/00] He then tours Pakistan for weeks under the protection of the ISI. [VANITY FAIR, 8/02] The ISI and Saeed helps Azhar form a new terrorist group called Jaish-e-Mohammed, and Azhar is soon plotting terrorist acts again. [PITTSBURGH TRIBUNE-REVIEW, 3/3/02; GUARDIAN, 7/16/02; WASHINGTON POST, 2/8/03]

January 1, 2000–September 11, 2001: Following Release from Prison, Saeed Sheikh Lives Openly; Supports Future 9/11 Hijackers

After being released from prison at the end of 1999, Saeed Sheikh stays in Kandahar, Afghanistan, for several days and meets with Taliban leader Mullah Omar. He also meets with bin Laden, who is said to call Saeed "my special son." He then travels to Pakistan and is given a house by the ISI. [VANITY FAIR, 8/02] He lives openly and opulently in Pakistan, even attending "swanky parties attended by senior Pakistani government officials." U.S. authorities conclude he is an asset of the ISI. [NEWSWEEK, 3/13/02] Amazingly, he is allowed to travel freely to Britain, and visits family there at least twice. [VANITY FAIR, 8/02] He works with Ijaz Shah, a former ISI official in charge of handling two terrorist groups, Lieutenant-General Mohammed Aziz Khan, former deputy chief of the ISI in charge of relations with Jaish-e-Mohammed, and Brigadier Abdullah, a former ISI officer. He is well known to other senior ISI officers. He regularly travels to Afghanistan and helps train new terrorist recruits in training camps there. [NEW YORK TIMES, 2/25/02; NATIONAL POST, 2/26/02; GUARDIAN, 7/16/02; INDIA TODAY, 2/25/02] Saeed helps train the 9/11 hijackers also, presumably in Afghanistan. [DAILY TELEGRAPH, 9/30/01] He also helps al-Qaeda develop a secure web-based communications system, and there is talk that he could one day succeed bin Laden. [VANITY FAIR, 8/02; DAILY TELEGRAPH, 7/16/02] He wires money to the 9/11 hijackers in August 2001 and possibly several other times. Presumably, he sends the money from the United Arab Emirates during his many trips there. [GUARDIAN, 2/9/02]

March 17, 2000: Bin Laden Reportedly Ill

Reports suggest bin Laden appears weak and gaunt at an important meeting of supporters. He may be very ill with liver ailments, and is seeking a kidney dialysis machine. [ASSOCIATED PRESS, 3/25/00] It is believed he gets the dialysis machine in early 2001. [LONDON TIMES, 11/01/01] He is able to talk, walk with a cane, and hold meetings, but little else. [DEUTSCHE PRESSE-AGENTUR, 3/16/00; ASIAWEEK, 3/24/00] The ISI is said to help facilitate his medical treatment. [CBS NEWS, 1/28/02]

March 25, 2000: Clinton Visits Pakistan Despite Fears of ISI Terror Ties

President Clinton visits Pakistan. It is later revealed that the U.S. Secret Service believed that the ISI was so deeply infiltrated by terrorist organizations, that it begged Clinton to cancel his visit. Specifically, the U.S. government determined that the ISI had long-standing ties with al-Qaeda. When Clinton decided to go over the Secret Service's protestations, his security took extraordinary and unprecedented precautions.

For instance, an empty Air Force One was flown into the country, and the president made the trip in a small, unmarked plane. [*NEW YORK TIMES*, 10/29/01]

April 4, 2000: ISI Director Visits Washington and Is Told to Give Warning to Taliban

ISI Director and "leading Taliban supporter" Lieutenant General Mahmood Ahmed visits Washington. In a message meant for both Pakistan and the Taliban, U.S. officials tell him that al-Qaeda has killed Americans and "people who support those people will be treated as our enemies." However, no actual action, military or otherwise, is taken against either the Taliban or Pakistan. [*WASHINGTON POST*, 12/19/01]

July 2000: Taliban Bans Poppy Growing

In response to Western pressure, the Taliban bans poppy growing in Afghanistan. As a result, the opium yield drops dramatically in 2001, from 3,656 tons to 185 tons. Of that, 83 percent is from Northern Alliance controlled lands. However, the effect isn't that great because there is a surplus in the West, and it is believed the Taliban have a large stockpile as well. [*GUARDIAN*, 2/21/02; REUTERS, 3/3/02; *OBSERVER*, 11/25/01]

January 19, 2001: UN Sanctions on Taliban Do Not Stop Illegal Trade Network

New United Nations sanctions against Afghanistan take effect, adding to those from November 1999. The sanctions limit travel by senior Taliban authorities, freeze bin Laden's and the Taliban's assets, and order the closure of Ariana Airlines offices abroad. The sanctions also impose an arms embargo against the Taliban, but not against Northern Alliance forces battling the Taliban. [ASSOCIATED PRESS, 12/19/00] The arms embargo has no visible effect because the sanctions fail to stop Pakistani military assistance. [9/11 COMMISSION REPORT, 3/24/04] The sanctions also fail to stop the illegal trade network that the Taliban is secretly running through Ariana. Two companies, Air Cess and Flying Dolphin, take over most of Ariana's traffic. Air Cess is owned by the Russian arms dealer Victor Bout, and Flying Dolphin is owned by the United Arab Emirates' former ambassador to the U.S., who is also an associate of Bout. In late 2000, despite reports linking Flying Dolphin to arms smuggling, the United Nations gives Flying Dolphin permission to take over Ariana's closed routes, which it does until the new sanctions take effect. Bout's operations are still functioning and he has not been arrested. [*LOS ANGELES TIMES*, 1/20/02; *MONTREAL GAZETTE*, 2/5/02] Ariana is essentially destroyed in the October 2001 U.S. bombing of Afghanistan. [*LOS ANGELES TIMES*, 11/18/01]

May 2001: Tenet Visits Pakistan; Armitage Calls on India

Deputy Secretary of State Richard Armitage, a former covert operative and Navy Seal, travels to India on a publicized tour while CIA Director Tenet makes a quiet visit to Pakistan to meet with President General Musharraf. Armitage has long and deep Pakistani intelligence connections (as well as a role in the Iran-Contra affair). It would be reasonable to assume that while in Islamabad, Tenet, in what was described as "an unusually long meeting," also meets with his Pakistani counterpart, ISI Director

Richard Armitage, Deputy Secretary of State

Mahmood. A long-time regional expert with extensive CIA ties publicly says, "The CIA still has close links with the ISI." [SAPRA, 5/22/01; *TIMES OF INDIA*, 3/7/01]

Early June 2001: Extensive ISI Support for Taliban Continues

UPI reporters visiting Taliban leader Mullah Omar note, "Saudi Arabia and the [United Arab Emirates] secretly fund the Taliban government by paying Pakistan for its logistical support to Afghanistan. Despite Pakistan's official denials, the Taliban is entirely dependent on Pakistani aid. This was verified on the ground by UPI. Everything from bottled water to oil, gasoline and aviation fuel, and from telephone equipment to military supplies, comes from Pakistan." [UPI, 6/14/01]

June 12, 2001: Sting Operation Exposes al-Qaeda, ISI, and Drug Connections; Investigators Face Obstacles to Learn More

Operation Diamondback, a sting operation uncovering an attempt to buy weapons illegally for the Taliban, bin Laden, and others, ends with a number of arrests. An Egyptian named Diaa Mohsen and a Pakistani named Mohammed Malik are arrested and accused of attempting to buy Stinger missiles, nuclear weapon components, and other sophisticated military weaponry for the Pakistani ISI. [SOUTH FLORIDA SUN-SENTINEL, 8/23/01; WASHINGTON POST, 8/2/02 (B)] Malik appears to have had links to important Pakistani officials and Kashmiri terrorists, and Mohsen claims a connection to a man "who is very connected to the Taliban" and funded by bin Laden. [WASHINGTON POST, 8/2/02 (B), MSNBC, 8/2/02] Some other ISI agents came to Florida on several occasions to negotiate, but they escaped being arrested. They wanted to pay partially in heroin. One mentioned that the WTC would be destroyed. These ISI agents said some of their purchases would go to the Taliban in Afghanistan and/or terrorists associated with bin Laden. [NEW YORK TIMES, 6/16/01; WASHINGTON POST, 8/2/02 (B); MSNBC, 8/2/02] Both Malik and Mohsen lived in Jersey City, New Jersey. [JERSEY JOURNAL, 6/20/01] Mohsen pleads guilty after 9/11, "but remarkably, even though [he was] apparently willing to supply America's enemies with sophisticated weapons, even nuclear weapons technology, Mohsen was sentenced to just 30 months in prison." [MSNBC, 8/2/02] Malik's case appears to have been dropped, and reporters find him working in a store in Florida less than a year after the trial ended. [MSNBC, 8/2/02] Malik's court files remain completely sealed, and in Mohsen's court case, prosecutors "removed references to Pakistan from public filings because of diplomatic concerns." [WASHINGTON POST, 8/2/02 (B)] Also arrested are Kevin Ingram and Walter Kapij. Ingram pleads guilty to laundering $350,000 and he is sentenced to 18 months in prison. [ASSOCIATED PRESS, 12/1/01] Ingram was a former senior investment banker with Deutschebank, but resigned in January 1999 after his division suffered costly losses. [JERSEY JOURNAL, 6/20/01] Walter Kapij, a pilot with a minor role in the plot, is given the longest sentence, 33 months in prison. [PALM BEACH POST, 1/12/02] Informant Randy Glass plays a key role in the sting, and has thirteen felony fraud charges against him reduced as a result, serving only seven months in prison. Federal agents involved in the case later express puzzlement that Washington higher-ups did not make the case a higher priority, pointing out that bin Laden could have gotten a nuclear bomb if the deal was for real. Agents on the case complain that the FBI did not make the case a counterterrorism matter, which would have improved bureaucratic backing and opened access to FBI information and U.S. intelligence from around the world. [WASHINGTON POST, 8/2/02 (B), MSNBC, 8/2/02] Federal agents frequently couldn't get prosecutors to approve wiretaps. [COX NEWS SERVICE, 8/2/02] Glass says, "Wouldn't you think that there should have been a wire tap on Diaa [Mohsen]'s phone and Malik's phone?" [WPBF CHANNEL 25, 8/5/02] An FBI supervisor in Miami refused to front money for the sting, forcing agents to use money from U.S. Customs and even Glass's own money to help keep the sting going. [COX NEWS SERVICE, 8/2/02]

Summer 2001: Pakistani Intelligence Rescues bin Laden Associate

Egyptian investigators track down a close associate of bin Laden named Ahmed al-Khadir, wanted for bombing the Egyptian embassy in Islamabad in 1995. Egyptians surround the safe house in Pakistan where al-Khadir is hiding. They notify the ISI to help arrest him, and the ISI promises swift action. Instead, a car sent by the ISI filled with Taliban and having diplomatic plates arrives, grabs al-Khadir and drives him to safety in Afghanistan. *Time* magazine later reports the incident as demonstrating the strong ties between the ISI and both the Taliban and al-Qaeda. [TIME, 5/6/02]

July 2, 2001: Osama bin Laden Periodically Undergoes Dialysis with Approval of the ISI

Indian sources claim that "bin Laden, who suffers from renal deficiency, has been periodically undergoing dialysis in a Peshawar military hospital with the knowledge and approval of the Inter-Services Intelligence (ISI), if not of [Pakistani President] Musharraf himself." [SARPA, 7/2/01] While one might question the bias of an Indian newspaper on this issue, highly respected intelligence newsletter Jane's later reports the story, and adds, "None of [these details] will be unfamiliar to U.S. intelligence operatives who have been compiling extensive reports on these alleged activities." [JANE'S INTELLIGENCE DIGEST, 9/20/01] CBS will later report bin Laden had emergency medical care in Pakistan the day before 9/11. [CBS NEWS, 01/28/02] If these stories are true, it appears Pakistan could have captured bin Laden for the U.S. at any time. The *Jane's Intelligence Digest* article adds, "It is becoming clear that both the Taliban and al-Qaeda would have found it difficult to have continued functioning—including the latter group's terrorist activities—without substantial aid and support from Islamabad [Pakistan]." [JANE'S INTELLIGENCE DIGEST, 9/20/01]

August–October 2001: Britain Seeks Indian Assistance in Catching Saeed Sheikh

British intelligence asks India for legal assistance in catching Saeed Sheikh sometime during August 2001. Saeed has been openly living in Pakistan since 1999 and has even traveled to Britain at least twice during that time, despite having kidnapped Britons and Americans in 1993 and 1994. [LONDON TIMES, 4/21/02; VANITY FAIR, 8/02] According to the Indian media, informants in Germany tell the internal security service there that Saeed helped fund hijacker Mohamed Atta. [FRONTLINE, 10/6/01] On September 23, it is revealed, without explanation, that the British have asked India for help in finding Saeed. [LONDON TIMES,

9/23/01] Saeed Sheikh's role in training the hijackers and financing the 9/11 attacks soon becomes public knowledge, though some elements are disputed. [DAILY TELEGRAPH, 9/30/01; CNN, 10/6/01; CNN, 10/8/01] The Gulf News claims that the U.S. freezes the assets of Pakistani terrorist group Jaish-e-Mohammed on October 12, 2001 because it has established links between Saeed Sheikh and 9/11. [GULF NEWS, 10/11/01] However, in October, an Indian magazine notes, "Curiously, there seems to have been little international pressure on Pakistan to hand [Saeed] over" [FRONTLINE, 10/6/01], and the U.S. does not formally ask Pakistan for help to find Saeed until January 2002.

Early August 2001: Saeed Sheikh Receives
Ransom Money; Sends $100,000 to Atta

The ransom for a wealthy Indian shoe manufacturer kidnapped in Calcutta, India, two weeks earlier is paid to an Indian gangster named Aftab Ansari. Ansari is based in Dubai, United Arab Emirates and has ties to the ISI and Saeed Sheikh. Ansari gives about $100,000 of the about $830,000 in ransom money to Saeed, who sends it to hijacker Mohamed Atta. [LOS ANGELES TIMES, 1/23/02; INDEPENDENT, 1/24/02] A series of recovered e-mails shows the money is sent just after August 11. This appears to be one of a series of Indian kidnappings this gang carries out in 2001. [INDIA TODAY, 2/14/02; TIMES OF INDIA, 2/14/02] Saeed provides training and weapons to the kidnappers in return for a percentage of the profits. [FRONTLINE, 2/2/02; INDIA TODAY, 2/25/02]

August 22, 2001: U.S. and Pakistan Negotiate to Capture or Kill bin Laden

The Asia Times reports that the U.S. is engaged in "intense negotiations" with Pakistan for assistance in an operation to capture or kill bin Laden. However, despite promised rewards, there is a "very strong lobby within the [Pakistani] army not to assist in any U.S. moves to apprehend bin Laden." [ASIA TIMES, 8/22/01]

August 25, 2001: 9/11 Paymaster Has Ties to Khalid Shaikh Mohammed

A supplemental Visa credit card on a "Mustafa Al-Hawsawi" bank account is issued in the name of Abdulrahman A. A. Al-Ghamdi, who the FBI says is an alias for Khalid Shaikh Mohammed. The FBI believes this helps prove Mohammed is a superior to the 9/11 paymaster. [9/11 CONGRESSIONAL INQUIRY, 9/26/02; HOUSTON CHRONICLE, 6/5/02] The identity of "Mustafa Al-Hawsawi" is highly contested, but may well be Saeed Sheikh. Mohammed and Sheikh appear to work together in the kidnapping of reporter Daniel Pearl in January 2002.

August 28–30, 2001: U.S. Politicians Visit Pakistan and Discuss bin Laden

Senator Bob Graham (D), Representative Porter Goss (R) and Senator Jon Kyl (R) travel to Pakistan and meet with President Musharraf. They reportedly discuss various security issues, including the possible extradition of bin Laden. They also meet with Abdul Salam Zaeef, the Taliban ambassador to

Pakistan. Zaeef apparently tells them that the Taliban wants to solve the issue of bin Laden through negotiations with the U.S. Pakistan says it wants to stay out of the bin Laden issue. [AGENCE FRANCE-PRESSE, 8/28/01; SALON, 9/14/01]

September 4–11, 2001: ISI Director Visits Washington for "Mysterious" Meetings

ISI Director Mahmood visits Washington for the second time. On September 10, a Pakistani newspaper reports on his trip so far. It says his visit has "triggered speculation about the agenda of his mysterious meetings at the Pentagon and National Security Council" as well as meetings with CIA Director Tenet, unspecified officials at the White House and the Pentagon, and his "most important meeting" with Mark Grossman, U.S. Under Secretary of State for Political Affairs. The article suggests, "[O]f course, Osama bin Laden" could be the focus of some discussions. Prophetically, the article adds, "What added interest to his visit is the history of such visits. Last time [his] predecessor was [in Washington], the domestic [Pakistani] politics turned topsy-turvy within days." [THE NEWS, 9/10/01] This is a reference to the Musharraf coup just after an ISI Director's visit on October 12, 1999.

September 8–11, 2001: Last-Minute Money Transfers
Between Hijackers and United Arab Emirates

The hijackers send money to and receive money from a man in the United Arab Emirates (UAE) who uses the aliases "Mustafa Ahmed," "Mustafa Ahmad," and "Ahamad Mustafa." [MSNBC, 12/11/01] This "Mustafa" transfers money to Mohamed Atta in Florida on September 8 and 9 from a branch of the Al Ansari Exchange in Sharjah, UAE, a center of al-Qaeda financial dealings. [FINANCIAL TIMES, 11/30/01] On September 9, three hijackers, Atta, Waleed Alshehri, and Marwan Alshehhi, transfer about $15,000 back to "Mustafa's" account. [TIME, 10/1/01; LOS ANGELES TIMES, 10/20/01] Apparently the hijackers are returning money meant for the 9/11 attacks that they have not needed. "Mustafa" then transfers $40,000 to his Visa card and then, using a Saudi passport, flies from the UAE to Karachi, Pakistan, on 9/11. He makes six ATM withdrawals there two days later, and then disappears into Pakistan. [MSNBC, 12/11/01] In early October 2001, it is reported that the financier "Mustafa Ahmed" is an alias used by Saeed Sheikh. [CNN, 10/6/01] It will later be reported that Saeed wired money to Atta the month before. These last-minute transfers are touted as the "smoking gun" proving al-Qaeda involvement in the 9/11 attacks, since Saeed is a known financial manager for bin Laden. [GUARDIAN, 10/1/01]

September 10, 2001: Pakistan Guards Osama
as He Receives Medical Treatment

CBS later reports that on this day, bin Laden is admitted to a military hospital in Rawalpindi, Pakistan, for kidney dialysis treatment. Pakistani military forces guard bin Laden. They also move out all the regular staff in the urology department and send in a secret team to replace them. It is not known how long he stays there. [CBS NEWS, 01/28/02]

September 11, 2001: Intelligence Committee Chairs Meet with ISI Head and Possible 9/11 Attack Funder as the Attack Occurs

At the time of the attacks, ISI Director Mahmood is at a breakfast meeting at the Capitol with the chairmen of the House and Senate Intelligence Committees, Senator Bob Graham (D) and Representative Porter Goss (R) (Goss is a 10-year veteran of the CIA's clandestine operations wing). The meeting is said to last at least until the second plane hits the WTC. [WASHINGTON POST, 5/18/02] Graham and Goss later co-head the joint House-Senate investigation into the 9/11 attacks, which has made headlines for saying there was no "smoking gun" of Bush knowledge before 9/11. [WASHINGTON POST, 7/11/02] Note that Senator Graham should have been aware of a report made to his staff the previous month that one of Mahmood's

Senator Bob Graham (D), Senator Jon Kyl (R), and Representative Porter Goss (R)

subordinates had told a U.S. undercover agent that the WTC would be destroyed. Evidence suggests that attendee Mahmood ordered that $100,000 be sent to hijacker Mohamed Atta. Also present at the meeting were Senator Jon Kyl (R) and the Pakistani ambassador to the U.S., Maleeha Lodhi. (All or virtually all of the people in this meeting had previously met in Pakistan just a few weeks earlier.) Senator Graham says of the meeting: "We were talking about terrorism, specifically terrorism generated from Afghanistan." The *New York Times* reports that bin Laden was specifically discussed. [VERO BEACH PRESS JOURNAL, 9/12/01; SALON, 9/14/01; NEW YORK TIMES, 6/3/02]

September 11–16, 2001: Pakistan Threatened; Promises to Support U.S.

ISI Director Mahmood, extending his Washington visit because of the 9/11 attacks [JAPAN ECONOMIC NEWSWIRE, 9/17/01], meets with U.S. officials and negotiates Pakistan's cooperation with the U.S. against al-Qaeda. It is rumored that later in the day of 9/11 and again the next day, Deputy Secretary of State Richard Armitage visits Mahmood and offers him the choice: "Help us and breathe in the 21st century along with the international community or be prepared to live in the Stone Age." [DEUTSCHE PRESSE-AGENTUR, 9/12/01; LA WEEKLY, 11/9/01] Secretary of State Powell presents Mahmood seven demands as an ultimatum and Pakistan supposedly agrees to all seven. [WASHINGTON POST, 1/29/02] Mahmood also has meetings with Senator Joseph Biden (D), Chairman of the Senate Foreign Relations Committee, and Secretary of State Powell, regarding Pakistan's position. [MIAMI HERALD, 9/16/01; NEW YORK TIMES, 9/13/01; REUTERS, 9/13/01; ASSOCIATED PRESS, 9/13/01] On September 13, the airport in Islamabad, the capital of Pakistan, is shut down for the day. A government official later says the airport had been closed because of threats made against Pakistan's "strategic assets," but does not elaborate. The next day, Pakistan declares "unstinting" support for the U.S., and the airport is reopened. It is later suggested that Israel and India threatened to attack Pakistan and take control of its nuclear weapons if Pakistan did not side with the U.S. [LA WEEKLY, 11/9/01] It is later reported that Mahmood's presence in Washington was a lucky blessing; one Western diplomat saying it "must have helped in a crisis situation when the U.S. was clearly very, very angry." [FINANCIAL TIMES, 9/18/01]

September 11, 2001–January 2002:
Saeed Sheikh Lives Openly in Pakistan

After probably completing last-minute financial transactions with some 9/11 hijackers, Saeed Sheikh flies to Pakistan. [KNIGHT RIDDER, 10/7/01] He meets with bin Laden in Afghanistan a few days later. [WASHINGTON POST, 2/18/02; LONDON TIMES, 2/25/02; GUARDIAN, 7/16/02] The U.S. government claims Saeed fights for the Taliban in Afghanistan in September and October 2001. [CNN, 3/14/02] Some reports indicate that after the defeat of the Taliban in Afghanistan, Saeed acts as a go-between with bin Laden and the ISI seeking to hide bin Laden. [PITTSBURGH TRIBUNE-REVIEW, 3/3/02] He also helps produce a video of a bin Laden interview. [PITTSBURGH TRIBUNE-REVIEW, 3/3/02] Sometime in October 2001 [GUARDIAN, 7/16/02], Saeed moves back to his home in Lahore, Pakistan, and lives there openly. He is frequently seen at local parties hosted by government leaders. In January 2002, he hosts a party to celebrate the birth of his newborn baby. [USA TODAY, 2/25/02; PITTSBURGH TRIBUNE-REVIEW, 3/3/02] He stays in his well-known Lahore house with his new wife and baby until January 19, 2002—four days before reporter Daniel Pearl is kidnapped. [BBC, 7/16/02]

Ahmed Omar Saeed Sheikh is seen partying in Pakistan after 9/11

Mid-September 2001: Pakistani Leaders Side with Taliban

The *Guardian* later claims that Pakistani President Musharraf has a meeting with his 12 or 13 most senior officers. Musharraf proposes to support the U.S. in the imminent war against the Taliban and bin Laden. Supposedly, four of his most senior generals oppose him outright in "a stunning display of disloyalty." The four are ISI Director Mahmood, Lieutenant General Muzaffar Usmani, Lieutenant General Jamshaid Gulzar Kiani, and Lieutenant General Mohammed Aziz Khan. All four are removed from power over the next month. If this meeting took place, it is hard to see when it could have happened, since the article states it happened "within days" of 9/11, but Mahmood was in the U.S. until late September 16, then flew to Afghanistan for two days, then possibly to Saudi Arabia. [GUARDIAN, 5/25/02]

Mid-September 2001 (B): Israel and U.S. Plan Contingency
to Steal Nuclear Weapons from Pakistan

According to Seymour Hersh of the *New Yorker*, a few days after 9/11 members of the elite Israeli counterterrorism unit Sayeret Matkal arrive in the U.S. and begin training with U.S. Special Forces in a secret location. The two groups are developing contingency plans to attack Pakistan's military bases and remove its nuclear weapons if the Pakistani government or the nuclear weapons fall into the wrong hands. [NEW YORKER, 10/29/01] There may have been threats to enact this plan on September 13, 2001. The *Japan Times* later notes that this "threat to divest Pakistan of its 'crown jewels' was cleverly used by the U.S., first to force Musharraf to support its military campaign in Afghanistan, and then to warn would-be coup plotters against Musharraf." [JAPAN TIMES, 11/10/01] Note the curious connection between Sayeret Matkal and Daniel Lewin, one of the 9/11 passengers on Flight 11 (see chapter 17).

September 17–18 and 28, 2001: Taliban Refuses to Extradite bin Laden

On September 17, ISI Director Mahmood heads a six-man delegation that visits Mullah Omar in Kandahar, Afghanistan. It is reported he is trying to convince Omar to extradite bin Laden or face an immediate U.S. attack. [PRESS TRUST OF INDIA, 9/17/01; FINANCIAL TIMES, 9/18/01; LONDON TIMES, 9/18/01] Also in the delegation is Lieutenant General Mohammed Aziz Khan, an ex-ISI official who appears to be one of Saeed Sheikh's contacts in the ISI. [PRESS TRUST OF INDIA, 9/17/01] On September 28, Ahmed returns to Afghanistan with a group of about ten religious leaders. He talks with Omar, who again says he will not hand over bin Laden. [AGENCE FRANCE-PRESSE, 9/28/01] A senior Taliban official later claims that on these trips Mahmood in fact urges Omar not to extradite bin Laden, but instead urges him to resist the U.S. [ASSOCIATED PRESS, 2/21/02; TIME, 5/6/02] Another account claims Mahmood does "nothing as the visitors [pour] praise on Omar and [fails] to raise the issue" of bin Laden's extradition. [KNIGHT RIDDER, 11/3/01] Two Pakistani brigadier generals connected to the ISI also accompany Mahmood, and advise al-Qaeda to counter the coming U.S. attack on Afghanistan by resorting to mountain guerrilla war. The advice is not followed. [ASIA TIMES, 9/11/02] Other ISI officers also stay in Afghanistan to advise the Taliban.

September 19, 2001: Rumored Meeting
Between Saudi Fundamentalists and ISI

According to the private intelligence service Intelligence Online, a secret meeting between fundamentalist supporters in Saudi Arabia and the ISI takes place in Riyadh, Saudi Arabia, on this day. Crown Prince Abdullah, the de facto ruler of Saudi Arabia, and Nawaf bin Abdul Aziz, the new head of Saudi intelligence, meet with General Mohamed Youssef, head of the ISI's Afghanistan Section, and ISI Director Mahmood (just returning from discussions in Afghanistan). They agree "to the principle of trying to neutralize Osama bin Laden in order to spare the Taliban regime and allow it to keep its hold on Afghanistan." There has been no confirmation that this meeting in fact took place, but if it did, its goals were unsuccessful. [INTELLIGENCE ONLINE, 10/4/01] There may have been a similar meeting before 9/11 in the summer of 2001.

September 23, 2001: Experts Say Muslim
Terrorist Groups Linked to Organized Crime

European law enforcement experts claim that numerous links tie major Muslim terrorist organizations, including al-Qaeda, with international organized crime groups. For approximately the last decade, mutually benefiting strong ties have developed between the two groups. Organized crime launders an estimated $900 billion a year, some of it from terrorist groups. France's chief financial crime prosecutor: "The nerve center of war is money . . . Without money, terrorist networks do not exist. They can't finance their operations overseas or purchase arms." Terrorist groups are also deeply involved in the international narcotics trade. [SAN FRANCISCO CHRONICLE, 9/23/01]

September 24, 2001–December 26, 2002:
Identity of 9/11 Financier Constantly Changes

In 2000, the 9/11 hijackers receive money from a man using "Mustafa Ahmed Al-Hisawi" and other aliases. On September 8–11, 2001, the hijackers send money to a man in the United Arab Emirates who uses the aliases "Mustafa Ahmed," "Mustafa Ahmad," and "Ahamad Mustafa." Soon the media begins reporting on who this 9/11 "paymaster" is, but his reported names and identities will continually change. The media has sometimes made the obvious connection that the paymaster is Saeed Sheikh, a British financial expert who studied at the London School of Economics, undisputedly sent hijacker Mohamed Atta money the month before, made frequent trips to Dubai (where the money is sent), and is known to have trained the hijackers. However, the FBI consistently deflects attention onto other possible explanations, with a highly confusing series of names vaguely similar to Mustafa Ahmed or Saeed Sheikh:

- September 24, 2001: *Newsweek* reports that the paymaster for the 9/11 attacks is someone named "Mustafa Ahmed." [NEWSWEEK, 9/24/01] This refers to Mustafa Mahmoud Said Ahmed, an Egyptian al-Qaeda banker who was captured in Tanzania in 1998 then later released. [SYDNEY MORNING HERALD, 9/28/01; NEWSDAY, 10/3/01]

- October 1, 2001: The *Guardian* reports that the real name of "Mustafa Mohamed Ahmad" is "Sheikh Saeed." [GUARDIAN, 10/1/01] A few days later, CNN confirms that this "Sheik Syed" is the British man Ahmed Omar Saeed Sheikh rescued in from an Indian prison in 1999. [CNN, 10/6/01; CNN, 10/8/01] However, starting on October 8, the story that ISI Director Mahmood ordered Saeed to give Mohamed Atta $100,000 begins to break. References to the 9/11 paymaster being the British Saeed Sheikh (and the connections to Ahmed) suddenly disappear from the Western media (with one exception [CNN, 10/28/01]).

- October 2001: Other articles continue to use "Mustafa Mohammed Ahmad" or "Shaykh Saiid" with no details of his identity, except for suggestions that he is Egyptian. There are numerous spelling variations and conflicting accounts over which name is the alias. [EVENING STANDARD, 10/1/01; BBC, 10/1/01; NEWSDAY, 10/3/01; ASSOCIATED PRESS, 10/6/01; WASHINGTON POST, 10/7/01; SUNDAY TIMES, 10/7/01; KNIGHT RIDDER, 10/9/01; NEW YORK TIMES, 10/15/01; LOS ANGELES TIMES, 10/20/01]

- October 16, 2001: CNN reports that the 9/11 paymaster "Sheik Sayid" is mentioned in a May 2001 trial of al-Qaeda members. However, this turns out to be a Kenyan named Sheik Sayyid el Masry. [CNN, 10/16/01; TRIAL TRANSCRIPT, 2/20/01; TRIAL TRANSCRIPT, 2/21/01]

- November 11, 2001: The identity of 9/11 paymaster "Mustafa Ahmed" is suddenly no longer Egyptian, but is now a Saudi named Sa'd Al-Sharif who is said to be bin Laden's brother-in-law. [NEWSWEEK, 11/11/01; UNITED NATIONS, 3/8/01; ASSOCIATED PRESS, 12/18/01]

- December 11, 2001: The federal indictment of Zacarias Moussaoui calls the 9/11 paymaster "Mustafa Ahmed al-Hawsawi a/k/a 'Mustafa Ahmed,'" and gives him Sa'd's nationality and birth date. [MSNBC, 12/11/01] Many articles begin adding "al-Hawsawi" to the Mustafa Ahmed name. [WASHINGTON POST, 12/13/01; WASHINGTON POST, 1/7/02; LOS ANGELES TIMES, 1/20/02]

- January 23, 2002: As new information is reported in India, the media returns to the British Saeed Sheikh as the 9/11 paymaster. [LOS ANGELES TIMES, 1/23/02; DAILY TELEGRAPH, 1/24/02; INDEPENDENT, 1/24/02; DAILY TELEGRAPH, 1/27/02] While his role in the kidnapping of Daniel Pearl is revealed on February 6, many articles connect him to 9/11, but many more do not. Coverage of Saeed's 9/11 connections generally dies out by the time of his trial in July 2002.

- June 4, 2002: Without explanation, the name "Shaikh Saiid al-Sharif" begins to be used for the 9/11 paymaster, presumably a combination of Saeed Sheikh and S'ad al-Sharif. [ASSOCIATED PRESS, 6/5/02; INDEPENDENT, 9/15/02; ASSOCIATED PRESS, 9/26/02; SAN FRANCISCO CHRONICLE, 11/15/02] Many of the old names continue to be used, however. [NEW YORK TIMES, 7/10/02; CHICAGO TRIBUNE, 9/5/02; WASHINGTON POST, 9/11/02; LOS ANGELES TIMES, 12/24/02 (B), LOS ANGELES TIMES, 9/1/02; KNIGHT RIDDER, 9/8/02; KNIGHT RIDDER, 9/9/02; TIME, 8/4/02 (B)]

- June 18, 2002: FBI Director Mueller testifies that the money sent in 2000 is sent by someone named "Ali Abdul Aziz Ali" but the money in 2001 is sent by "Shaikh Saiid al-Sharif." The "Aziz Ali" name has not been mentioned again by the press or FBI (outside of coverage of this testimony in September 2002). [9/11 CONGRESSIONAL INQUIRY, 9/26/02]

- September 4, 2002: *Newsweek* says "Mustafa Ahmad Adin Al-Husawi," presumably Saudi, is a deputy to the Egyptian "Sayyid Shaikh Al-Sharif." However, it adds he "remains almost a total mystery," and they are unsure of his name. [NEWSWEEK, 9/4/02]

- December 26, 2002: U.S. officials now say there is no such person as Shaikh Saiid al-Sharif. Instead, he is probably a composite of three different people: "[Mustafa Ahmed] Al-Hisawi; Shaikh Saiid al-Masri, al-Qaeda's finance chief, and Saad al-Sharif, bin Laden's brother-in-law and a midlevel al-Qaeda financier." [ASSOCIATED PRESS, 12/27/02] Shaikh Saiid al-Masri is likely a reference the Kenyan Sheik Sayyid el Masry. Note that, now, al-Hisawi is the assistant to Shaikh Saiid, a flip from a few months before.

Saiid and/or al-Hisawi still haven't been added to the FBI's official most wanted lists. [LONDON TIMES, 12/1/01; WALL STREET JOURNAL, 6/17/02; FBI MOST WANTED TERRORISTS, 2002] Despite the confusion, the FBI isn't even seeking information about them. [FBI SEEKING INFORMATION, 2002] Mustafa Ahmed Al-Hawsawi is said to be arrested with Khalid Shaikh Mohammed in Pakistan in 2003, but no photos of him are released, and witnesses of the supposed arrest did not see Al-Hawsawi or Mohammed there (see chapter 9). [REUTERS, 3/3/03 (C)]

September 27, 2001: ISI Has Connections to Taliban, Drug Trade, CIA

The *Sydney Morning Herald* discusses the connections between the CIA and Pakistan's ISI, and the ISI's long-standing control over the Taliban. Drugs are a big part of their operation: "opium cultivation and heroin production in Pakistan's northern tribal belt and adjoining Afghanistan were a vital offshoot of the ISI-CIA cooperation. It succeeded in turning some of the Soviet troops into addicts. Heroin sales in Europe and the U.S., carried out through an elaborate web of deception, transport networks, couriers, and payoffs, offset the cost of the decade-long war in Afghanistan." [SYDNEY MORNING HERALD, 9/27/01]

Late September–November 2001: Pakistani ISI Aids Taliban Against U.S.

The ISI secretly assists the Taliban in its defense against a U.S.-led attack. Between three and five ISI officers give military advice to the Taliban in late September. [DAILY TELEGRAPH, 10/10/01] At least five key ISI operatives help the Taliban prepare defenses in Kandahar, yet none are punished for their activities. [TIME, 5/6/02] Secret advisers begin to withdraw in early October, but some stay on into November. [KNIGHT RIDDER, 11/3/01] Large convoys of rifles, ammunition, and rocket-propelled grenade launchers for Taliban fighters cross the border from Pakistan into Afghanistan on October 8 and 12, just after U.S. bombing of Afghanistan begins and after a supposed crackdown on ISI fundamentalists. The Pakistani ISI secretly gives safe passage to these convoys, despite having promised the U.S. in September that such assistance would immediately stop. [NEW YORK TIMES, 12/8/01] Secret ISI convoys of weapons and nonlethal supplies continue into November. [UPI, 11/1/01; TIME, 5/6/02] An anonymous Western diplomat later states, "We did not fully understand the significance of Pakistan's role in propping up the Taliban until their guys withdrew and things went to hell fast for the Talibs." [NEW YORK TIMES, 12/8/01]

October 2001: Report: Bin Laden's Financial Network Is Successor to the BCCI Bank

A 70-page French intelligence report claims: "The financial network of bin Laden, as well as his network of investments, is similar to the network put in place in the 1980s by BCCI for its fraudulent operations, often with the same people (former directors and cadres of the bank and its affiliates, arms merchants, oil merchants, Saudi investors). The dominant trait of bin Laden's operations is that of a terrorist network backed up by a vast financial structure." The BCCI was the largest Muslim bank in the world before it collapsed in July 1991. A senior U.S. investigator later says U.S. agencies are looking into the ties outlined by the French because "they just make so much sense, and so few people from BCCI ever went to jail. BCCI was the mother and father of terrorist financing operations." The report identifies dozens of companies and individuals who were involved with BCCI and were found to be dealing with bin Laden after the bank collapsed. Many went on to work in banks and charities identified by the U.S. and others as supporting al-Qaeda. The role of Saudi billionaire Khalid bin Mahfouz in supporting bin Laden is emphasized in the report. In 1995, bin Mahfouz paid a $225 million fine in a settlement with U.S. prosecutors for his role in the BCCI scandal. [WASHINGTON POST, 2/17/02] Representatives of bin Mahfouz later argue that this report was in fact prepared by Jean-Charles Brisard and not the French intelligence service. Bin Mahfouz has begun libel proceedings against Mr. Brisard, claiming that he has made unfounded and defamatory allegations. [KENDALL FREEMAN, 5/13/2004]

October 1, 2001: Some Officials Question If
Intelligence Service Helped bin Laden in Plot

The *New Yorker* reports that "a number of intelligence officials have raised questions about bin Laden's capabilities. 'This guy sits in a cave in Afghanistan and he's running this operation?' one CIA official asked. 'It's so huge. He couldn't have done it alone.' A senior military officer told me that because of the visas and other documentation needed to infiltrate team members into the United States a major foreign intelligence service might also have been involved." [NEW YORKER, 10/1/01] No specific service is named, but the ISI would be one likely candidate.

October 1, 2001 (B): Kashmir Suicide Attack Involves 9/11 Funder Saeed Sheikh

A suicide truck-bomb attack on the provincial parliamentary assembly in Indian-controlled Kashmir leaves 36 dead. It appears that Saeed Sheikh and Aftab Ansari, working with the ISI, are behind the attacks. [VANITY FAIR, 8/02; PITTSBURGH TRIBUNE-REVIEW, 3/3/02] Indian intelligence claims that Pakistani President Musharraf is later given a recording of a phone call between Jaish-e-Mohammed leader Maulana Masood Azhar and ISI Director Mahmood in which Azhar allegedly reports the bombing is a "success." [UPI, 10/10/01]

October 7, 2001: ISI Director Replaced at U.S. Urging;
Role in Funding 9/11 Plot Is One Explanation

ISI Director Mahmood is replaced in the face of U.S. pressure after links are discovered between him, Saeed Sheikh, and the funding of the 9/11 attacks. Mahmood instructed Saeed to transfer $100,000 into hijacker Mohamed Atta's bank account prior to 9/11. This is according to Indian intelligence, which claims the FBI has privately confirmed the story. [PRESS TRUST OF INDIA, 10/8/01; TIMES OF INDIA, 10/9/01; INDIA TODAY, 10/15/01; DAILY EXCELSIOR, 10/18/01] The story is not widely reported in Western countries, though it makes the *Wall Street Journal.* [AUSTRALIAN, 10/10/01; AGENCE FRANCE-PRESSE, 10/10/01; WALL STREET JOURNAL, 10/10/01] It is reported in Pakistan as well. [DAWN, 10/8/01] The Northern Alliance also repeats the claim in late October. [FEDERAL NEWS SERVICE, 10/31/01] In Western countries, the usual explanation is that Mahmood is fired for being too close to the Taliban. [LONDON TIMES, 10/9/01; GUARDIAN, 10/9/01] The *Times of India* reports that Indian intelligence helped the FBI discover the link, and says, "A direct link between the ISI and the WTC attack could have enormous repercussions. The U.S. cannot but suspect whether or not there were other senior Pakistani Army commanders who were in the know of things. Evidence of a larger conspiracy could shake U.S. confidence in Pakistan's ability to participate in the anti-terrorism

Lt. Gen.
Mahmood
Ahmed

THE PAKISTANI ISI   261

coalition." [TIMES OF INDIA, 10/9/01] There is evidence some ISI officers may have known of a plan to destroy the WTC as early as July 1999. Two other ISI leaders, Lieutenant General Mohammed Aziz Khan and Lieutenant General Muzaffar Usmani, are sidelined on the same day as Mahmood. [FOX NEWS, 10/8/01] Saeed had been working under Khan. The firings are said to have purged the ISI of its fundamentalists. However, according to one diplomat, "To remove the top two or three doesn't matter at all. The philosophy remains . . . . [The ISI is] a parallel government of its own. If you go through the officer list, almost all of the ISI regulars would say, of the Taliban, 'They are my boys.'" [NEW YORKER, 10/29/01] It is believed Mahmood has been living under virtual house arrest in Pakistan ever since (which would seem to imply more than just a difference of opinion over the Taliban), but no charges have been brought against him, and there is no evidence the U.S. has asked to question him. [ASIA TIMES, 1/5/02] He also has refused to speak to reporters since being fired [ASSOCIATED PRESS, 2/21/02], and outside India and Pakistan, the story has only been mentioned a couple times in the media since. [SUNDAY HERALD, 2/24/02; LONDON TIMES, 4/21/02]

November 2001–February 5, 2002: Saeed Sheikh Indicted for Role in 1994 Kidnapping

A U.S. grand jury secretly indicts Saeed Sheikh for his role in the 1994 kidnapping of an American. The indictment is revealed in late February 2002. The U.S. later claims it begins asking Pakistan for help in arresting and extraditing Saeed in late November. [ASSOCIATED PRESS, 2/26/02; NEWSWEEK, 3/13/02] However, it is not until January 9, 2002 that Wendy Chamberlin, the U.S. ambassador to Pakistan, officially asks the Pakistani government for assistance. [ASSOCIATED PRESS, 2/24/02; CNN, 2/24/02; LOS ANGELES TIMES, 2/25/02 (B)] Saeed is seen partying with Pakistani government officials well into January 2002. The *Los Angeles Times* later reports that Saeed "move[s] about Pakistan without apparent impediments from authorities" up until February 5, when he is identified as a suspect in the Daniel Pearl kidnapping. [LOS ANGELES TIMES, 2/13/02] The *London Times* reports: "It is inconceivable that the Pakistani authorities did not know where he was" before then. [LONDON TIMES, 4/21/02]

November 10, 2001: Reporter Investigating ISI-Taliban Ties Is Expelled from Pakistan

Daily Telegraph reporter Christina Lamb is arrested and expelled from Pakistan by the ISI. She had been investigating the connections between the ISI and the Taliban. [DAILY TELEGRAPH 11/11/01]

December 7, 2001: Indian Police Shoot Terrorist Gangster Dead

Indian gangster Asif Raza Khan, terrorist associate of Saeed Sheikh and Aftab Ansari, is shot dead by Indian police. Police claim he was trying to escape. [LOS ANGELES TIMES, 1/23/02] A month or two before he died, Indian investigators recorded a confession of his involvement in a plot with Ansari and Saeed to send kidnapping profits to hijacker Mohamed Atta. This information becomes public just before Saeed is suspected in the kidnapping and murder of reporter Daniel Pearl. [INDEPENDENT, 2/24/02; INDIA TODAY, 2/25/02]

Many in Ansari's Indian criminal network are arrested in October and November 2001, and they confirm Khan's money connection to Atta. [INDIA TODAY, 2/14/02]

December 13, 2001: ISI-Connected
Terrorists Attack Indian Parliament

The Indian Parliament building in New Delhi is attacked by terrorists. Fourteen people, including the five attackers, are killed. India blames the Pakistani terrorist group Jaish-e-Mohammed for the attacks. Twelve days later, Maulana Masood Azhar, head of Jaish-e-Mohammed, is arrested by Pakistan and his group is banned. He is freed one year later. [AGENCE FRANCE-PRESSE, 12/25/01; CHRISTIAN SCIENCE MONITOR, 12/16/02] The Parliament attack leads to talk of war, even nuclear war, between Pakistan and India, until President Musharraf cracks down on terrorist groups in early January. [DAILY TELEGRAPH, 12/28/01; WALL STREET JOURNAL, 1/3/02; GUARDIAN, 5/25/02] It appears that Saeed Sheikh and Aftab Ansari, working with the ISI, are also involved in the attacks. [VANITY FAIR, 8/02; PITTSBURGH TRIBUNE-REVIEW, 3/3/02]

December 24, 2001–January 23, 2002:
Reporter Daniel Pearl Investigates Sensitive Topics in Pakistan

Wall Street Journal reporter Daniel Pearl writes stories about the ISI that will lead to his kidnapping and murder. On December 24, 2001, he reports about ties between the ISI and a Pakistani organization that was working on giving bin Laden nuclear secrets before 9/11. A few days later, he reports that the ISI-supported terrorist organization Jaish-e-Mohammed still has its office running and bank accounts working, even though President Musharraf claims to have banned the group. "If [Pearl] hadn't been on the ISI's radarscope before, he was now." [VANITY FAIR, 8/02; GUARDIAN, 7/16/02] He begins investigating links between shoe bomber Richard Reid and Pakistani militants, and comes across connections to the ISI and a mysterious religious group called Al-Fuqra. [WASHINGTON POST, 2/23/02] He also may be looking into the U.S. training and backing of the ISI. [GULF NEWS, 3/25/02] He is writing another story on Dawood Ibrahim, a powerful terrorist and gangster protected by the ISI, and other Pakistani organized crime figures. [NEWSWEEK, 2/4/02; VANITY FAIR, 8/02] Former CIA agent Robert Baer later claims to be working with Pearl on an investigation of 9/11 mastermind Khalid Shaikh Mohammed. [UPI, 9/30/02] It is later suggested that Mohammed masterminds both Reid's shoe bomb attempt and the Pearl kidnapping, and has connections to Pakistani gangsters and the ISI, so some of these explanations could fit together. [UPI, 9/30/02; ASIA TIMES, 10/30/02; CNN, 1/30/03] Kidnapper Saeed will later say of Pearl, "because of his hyperactivity he caught our interest." [THE NEWS, 2/15/02] Pearl is kidnapped on January 23, 2002, and his murder is confirmed on February 22, 2002. [CNN, 2/22/02]

December 30, 2001: Afghan Minister Claims ISI Supports bin Laden

The new Afghan Interior Minister Younis Qanooni claims that the ISI helped bin Laden escape from Afghanistan: "Undoubtedly they (ISI) knew what was going on." He claims that the ISI is still supporting bin Laden even if Pakistani President Musharraf isn't. [BBC, 12/30/01]

January 5, 2002: FBI Interested in Captured Pakistani Terrorist Leader

The FBI has asked Pakistan for permission to question Maulana Masood Azhar, the leader of Jaish-e-Mohammed, according to reports. Pakistan arrested him on December 25, 2001 after U.S. pressure to do so. One Pakistani official says, "The Americans are aware Azhar met bin Laden often, and are convinced he can give important information about bin Laden's present whereabouts and even the September 11 attacks." But the "primary reason" for U.S. interest is the link between Azhar and Saeed Sheikh. They hope to learn about Saeed's involvement in financing the 9/11 attacks. Whether Pakistan gives permission to question Azhar is unclear. Four days later, the U.S. officially asks Pakistan for help in finding and extraditing Saeed. [GULF NEWS, 1/5/02]

January 6, 2002 (B): Shoe Bomber Is Believed to Be Involved with Pakistani Jihadists

The *Boston Globe* reports that shoe bomber Richard Reid may have had ties with an obscure Pakistani group called Al-Fuqra. Reid apparently visited the Lahore, Pakistan home of Ali Gilani, the leader of Al-Fuqra. [BOSTON GLOBE, 1/6/02] Reporter Daniel Pearl reads the article, and decides to investigate. [VANITY FAIR, 8/02] Pearl believes he is on his way to interview Gilani when he is kidnapped. [PITTSBURGH TRIBUNE-REVIEW, 3/3/02] A 1995 State Department report said Al-Fuqra's main goal is "purifying Islam through violence." [VANITY FAIR, 8/02] Intelligence experts now say Al-Fuqra is a splinter group of Jaish-e-Mohammed, with ties to al-Qaeda. [UPI, 1/29/02] Al-Fuqra claims close ties with the Muslims of the Americas, a U.S. tax-exempt group claiming about 3,000 members living in rural compounds in 19 states, the Caribbean, and Europe. Members of Al-Fuqra are suspected of at least 13 fire bombings and 17 murders, as well as theft and credit-card fraud. Gilani, who had links to people involved in the 1993 WTC bombing, fled the U.S. after the bombing, and admitted he works with the ISI, now lives freely in Pakistan. [BOSTON GLOBE, 1/6/02; THE NEWS, 2/15/02; PITTSBURGH TRIBUNE-REVIEW, 3/3/02; VANITY FAIR, 8/02] Saeed Sheikh "has long had close contacts" with the group, and praises Gilani for his "unexplained services to Pakistan and Islam." [THE NEWS, 2/18/02; PITTSBURGH TRIBUNE-REVIEW, 3/3/02]

January 12, 2002: Pakistan Takes Half-Hearted Anti-Terrorist Measures

Pakistan President Musharraf makes "a forceful speech . . . condemning Islamic extremism." [WASHINGTON POST, 3/28/02] Around this time, he also arrests about 2,000 people he calls extremists. He is hailed in the Western media as redirecting the ISI to support the U.S. agenda. Yet, by the end of the month, at least 800 of the arrested are set free [WASHINGTON POST, 3/28/02] including "most of their firebrand leaders." [TIME, 5/6/02] Within one year, "almost all" of those arrested have been quietly released. Even the most prominent leaders, such as Maulana Masood Azhar have been released. Their terrorist organizations are running again, most under new names. [WASHINGTON POST, 2/8/03]

January 22, 2002: Saeed Sheikh and ISI Terrorize India

A crowd of mostly unarmed Indian police near the U.S. Information Service building in Calcutta, India, is attacked by gunmen; four policemen are killed and 21 people injured. The gunmen escape.

India claims that Aftab Ansari immediately calls to take credit, and India charges that the gunmen belong to Ansari's kidnapping ring also connected to funding the 9/11 attacks in August 2001. [DAILY TELE-GRAPH, 1/24/02; ASSOCIATED PRESS, 2/10/02] Saeed Sheikh and the ISI assist Ansari in the attack. [VANITY FAIR, 8/02; PITTSBURGH TRIBUNE-REVIEW, 3/3/02] This is the fourth terrorist attack in which they have cooperated, including the 9/11 attacks, and attacks in October and December 2001.

January 22–25, 2002: India Tells FBI Director
About Saeed Sheikh Connection to 9/11

FBI Director Mueller visits India, and is told by Indian investigators that Saeed Sheikh sent ransom money to hijacker Mohamed Atta in the U.S. In the next few days, Saeed is publicly blamed for his role with gangster Aftab Ansari in financing Atta and organizing the Calcutta terrorist attack. [PRESS TRUST OF INDIA, 1/22/02; LOS ANGELES TIMES, 1/23/02; INDEPENDENT, 2/24/02; AGENCE FRANCE-PRESSE, 1/27/02; DAILY TELEGRAPH, 1/27/02] Meanwhile, on January 23, Saeed helps kidnap reporter Daniel Pearl and is later arrested. Also on January 23, Ansari is placed under surveillance after flying to Dubai, United Arab Emirates. On January 24, Mueller and U.S. Ambassador to Pakistan Wendy Chamberlin discuss Saeed at a previously scheduled meeting with Pakistani President Musharraf. Apparently Saeed's role in Pearl's kidnapping is not yet known. [ASSOCIATED PRESS, 2/24/02] Mueller then flies to Dubai on his way back to the U.S. to pressure the government there to arrest Ansari and deport him to India. Ansari is arrested on February 5 and deported four days later. [ASSOCIATED PRESS, 2/10/02; FRONTLINE, 2/16/02; INDIA TODAY, 2/25/02]

January 23, 2002: Reporter Daniel Pearl Is Kidnapped While Investigating the ISI

Wall Street Journal report Daniel Pearl is kidnapped while investigating the ISI's connection to terrorism. [GUARDIAN, 1/25/02; BBC, 7/5/02] Saeed Sheikh is later convicted as the mastermind of the kidnap, and though it appears he lured Pearl into being kidnapped beginning January 11, the actual kidnapping is perpetrated by others who remain at large. [VANITY FAIR, 8/02; WALL STREET JOURNAL, 1/23/03] Both al-Qaeda and the ISI appear to be behind the kidnapping. The overall mastermind behind the kidnapping seems to be Khalid Shaikh Mohammed, also mastermind of the 9/11 attacks. [TIME, 1/26/03; CNN, 1/30/03]

January 28, 2002: Daniel Pearl's Kidnappers Make Odd Demands for His Release

Reporter Daniel Pearl's kidnappers e-mail the media a picture of Pearl and a list of very strange demands. [BBC, 7/5/02] The kidnappers call themselves "The National Movement for the Restoration of Pakistani Sovereignty," a previously unheard of name. [VANITY FAIR, 8/02] Their demands include the return of U.S.-held Pakistani prisoners and the departure of U.S. journalists from Pakistan. [ABC NEWS, 2/7/02] Most unusually, they demand that the U.S. sell F-16 fighters to Pakistan. No terrorist group had ever shown interest in the F-16's, but this demand and the others reflect the desires of Pakistan's military and the ISI to obtain the fighters. [LONDON TIMES, 4/21/02; GUARDIAN, 7/16/02] On January 29, "a senior Pakistani official" presumably from the ISI leaks the fact that Pearl is Jewish to the Pakistani press. This may have been an attempt to ensure the kidnappers would want to murder him, which they do shortly thereafter. [VANITY FAIR, 8/02] On the same day, it is reported that U.S. intelligence believes the kidnappers are connected to the ISI. [UPI, 1/29/02]

January 31, 2002: Reporter Daniel Pearl
Is Murdered by Pakistani Kidnappers

Wall Street Journal reporter Daniel Pearl is murdered. He is reported dead on February 21, but his mutilated body is found months later. Police investigators say "there were at least eight to ten people present on the [murder] scene" and at least 15 who participated in his kidnapping and murder. "Despite issuing a series of political demands shortly after Pearl's abduction four weeks ago, it now seems clear that the kidnappers planned to kill Pearl all along." [WASHINGTON POST, 2/23/02] Some captured participants later claim 9/11 mastermind Khalid Shaikh Mohammed is the one who cuts Pearl's throat. [MSNBC, 9/17/02; TIME, 1/26/03]

A photo taken by Daniel Pearl's kidnappers taken shortly before his death

February 5, 2002: Pakistan Apprehends Saeed Sheikh

Pakistani police, with the help of the FBI, determine Saeed Sheikh is behind the kidnapping of Daniel Pearl, but are unable to find him. They round up about ten of his relatives and threaten to harm them unless he turns himself in. Saeed Sheikh does turn himself in, but to Ijaz Shah, his former ISI boss. [BOSTON GLOBE, 2/7/02; VANITY FAIR, 8/02] The ISI holds Saeed for a week, but fails to tell Pakistani police or anyone else that they have him. This "missing week" is the cause of much speculation. The ISI never tells Pakistani police any details about this week. [NEWSWEEK, 3/11/02] Saeed also later refuses to discuss the week or his connection to the ISI, only saying, "I will not discuss this subject. I do not want my family to be killed." He adds, "I know people in the government and they know me and my work." [NEWSWEEK, 3/13/02; VANITY FAIR, 8/02] It is suggested Saeed is held for this week to make sure that Pearl was killed. Saeed later says that during this week he got a coded message from the kidnappers that Pearl had been murdered. Also, the time might have been spent working out a deal with the ISI over what Saeed would tell police and the public. [NEWSWEEK, 3/11/02] Several others with both extensive ISI and al-Qaeda ties wanted for the kidnapping are arrested around this time. [WASHINGTON POST, 2/23/02; LONDON TIMES, 2/25/02] One of these men, Khalid Khawaja, "has never hidden his links with Osama bin Laden. At one time he used to fly Osama's personal plane." [PAKNEWS, 2/11/02]

February 6, 2002: Western Media Largely Ignores
Links Between Saeed Sheikh, ISI, and 9/11

Pakistani police publicly name Saeed Sheikh and a terrorist group he belongs to, Jaish-e-Mohammed, as those responsible for reporter Daniel Pearl's murder. [OBSERVER, 2/24/02] In the next several months, at least 12 Western news articles mention Saeed's links to al-Qaeda [ABC NEWS, 2/7/02; BOSTON GLOBE, 2/7/02;

ASSOCIATED PRESS, 2/24/02; *LOS ANGELES TIMES*, 3/15/02], including his financing of 9/11 [*NEW YORK DAILY NEWS*, 2/7/02; CNN, 2/8/02; ASSOCIATED PRESS, 2/9/02; *GUARDIAN*, 2/9/02; *INDEPENDENT*, 2/10/02; *TIME*, 2/10/02; *NEW YORK POST*, 2/10/02; *EVENING STANDARD*, 2/12/02; *LOS ANGELES TIMES*, 2/13/02; *NEW YORK POST*, 2/22/02; *SUNDAY HERALD*, 2/24/02; *USA TODAY*, 3/8/02], and at least 16 articles mention his links to the ISI. [COX NEWS SERVICE, 2/21/02; *OBSERVER*, 2/24/02; *DAILY TELEGRAPH*, 2/24/02; *NEWSWEEK*, 2/25/02; *NEW YORK TIMES*, 2/25/02; *USA TODAY*, 2/25/02; *NATIONAL POST*, 2/26/02; *BOSTON GLOBE*, 2/28/02; *NEWSWEEK*, 3/11/02; *NEWSWEEK*, 3/13/02; *GUARDIAN*, 4/5/02; MSNBC, 4/5/02] However, many other articles fail to mention either link. Only a few articles consider that Saeed could have been connected to both groups at the same time [*LONDON TIMES*, 2/25/02; *LONDON TIMES*, 4/21/02; *PITTSBURGH TRIBUNE-REVIEW*, 3/3/02], and apparently, only one of these mentions he could be involved in the ISI, al-Qaeda and financing 9/11. [*LONDON TIMES*, 4/21/02] By the time Saeed is convicted of Pearl's murder in July 2002, Saeed's possible connections to al-Qaeda and/or the ISI are virtually unreported in U.S. newspapers, while many British newspapers are still making one or the other connection.

February 9, 2002: Pakistani Gangster Admits Ties to ISI, Saeed Sheikh, and Terrorism

Aftab Ansari

Gangster Aftab Ansari is deported to India. He was arrested in Dubai, United Arab Emirates, on February 5. [*INDEPENDENT*, 2/10/02] He admits funding terrorist attacks through kidnapping ransoms, and building a network of arms and drug smuggling. [DEUTSCHE PRESSE-AGENTUR, 2/11/02] He later also admits to close ties with the ISI and Saeed Sheikh, whom he befriended in prison. [PRESS TRUST OF INDIA, 5/13/02]

February 12, 2002: ISI Deliver Saeed Sheikh to Pakistani Police

Saeed Sheikh, already in ISI custody for a week, is handed over to Pakistani police. Shortly afterwards, he publicly confesses to his involvement in reporter Daniel Pearl's murder. Later he will recant this confession. It appears that initially he thought he would get a light sentence. *Newsweek* describes him initially "confident, even cocky," saying he would only serve three to four years if convicted, and would never be extradited. [*NEWSWEEK*, 3/11/02] He is sentenced in July 2002 to hang instead. Pakistani militants respond to his arrest with three suicide attacks that kill more than 30 people. [*GUARDIAN*, 7/16/02]

February 18, 2002: Pakistan Applies Censorship on Link Between ISI and Saeed Sheikh

The Pakistani government unsuccessfully tries to stop Pakistani newspaper *The News* from publishing a story revealing Saeed Sheikh's connections to the ISI, based on leaks from Pakistani police interrogations. [*WASHINGTON POST*, 3/10/02; *LONDON TIMES*, 4/21/02; *GUARDIAN*, 7/16/02] According to the article, Saeed admits his involvement in recent attacks on the Indian parliament in Delhi and in Kashmir, and says the ISI helped him finance, plan, and execute them. [*THE NEWS*, 2/18/02] On March 1, the ISI pressures *The News* to fire the four journalists who worked on the story. The ISI also demands an apology from the newspaper's editor, who flees the country instead. [WASHINGTON POST, 3/10/02; LONDON TIMES, 4/21/02; GUARDIAN, 7/16/02]

March 1, 2002: ISI Maintains Huge Drug Economy

Vanity Fair suggests the ISI is still deeply involved in the drug trade in Central Asia. It estimates that Pakistan has a parallel drug economy worth $15 billion a year. Pakistan's official economy is worth about $60 billion. The article notes that the U.S. has not tied its billions of dollars in aid to Pakistan to assurances that Pakistan will stop its involvement in drugs. [VANITY FAIR, 3/1/02]

March 3, 2002: Powell Denies ISI Links to Daniel Pearl Murder

Secretary of State Powell rules out any links between "elements of the ISI" and the murderers of reporter Daniel Pearl. [DAWN, 3/3/02] The *Guardian* later calls Powell's comment "shocking," given the overwhelming evidence that the main suspect, Saeed Sheikh, worked for the ISI. [GUARDIAN, 4/5/02] Defense Secretary Rumsfeld called Saeed a possible "asset" for the ISI a week earlier. [LONDON TIMES, 2/25/02] The *Washington Post* says, "The [ISI] is a house of horrors waiting to break open. Saeed has tales to tell." [WASHINGTON POST, 3/28/02] The *Guardian* says Saeed "is widely believed in Pakistan to be an experienced ISI 'asset.'" [GUARDIAN, 4/5/02]

March 7, 2002: Pakistan's President Wants Saeed Hanged

Pakistani President Musharraf says Saeed Sheikh, chief suspect in the killing of reporter Daniel Pearl, will not be extradited to the U.S., at least not until after he is tried by Pakistan. [GUARDIAN, 3/15/02] The U.S. ambassador later reports to Washington that Musharraf privately said, "I'd rather hang him myself" than extradite Saeed. [WASHINGTON POST, 3/28/02] Musharraf even brazenly states, "Perhaps Daniel Pearl was over-intrusive. A mediaperson should be aware of the dangers of getting into dangerous areas. Unfortunately, he got over-involved." [THE HINDU, 3/8/02] He also says Pearl was caught up in "intelligence games." [WASHINGTON POST, 5/3/02] In early April, Musharraf apparently says he wants to see Saeed sentenced to death. Defense lawyers are appalled, saying Musharraf is effectively telling the courts what to do. [BBC, 4/12/02] The *Washington Post* reports in early March that Pakistani "police alternately fabricate and destroy evidence, depending on pressure from above" [WASHINGTON POST, 3/10/02], and in fact Saeed's trial will be plagued with problems.

March 14, 2002: U.S. Indicts Saeed for Murder of Daniel Pearl

Attorney General Ashcroft announces a second U.S. criminal indictment of Saeed Sheikh, this time for his role in the kidnapping and murder of Daniel Pearl. The amount of background information given about Saeed is very brief, and of all his many terrorist acts since he was released from prison in 1999, the only one mentioned is that he fought in Afghanistan with al-Qaeda in September and October 2001. The indictment and Ashcroft fail to mention Saeed's financing of the 9/11 attacks, and no reporters ask Ashcroft about this either. [CNN, 3/14/02; LOS ANGELES TIMES, 3/15/02]

April 5, 2002: Saeed Shaikh Tried in Secret

The Pakistani trial of Saeed Sheikh and three others begins. [BBC, 7/5/02] NBC reports that death sentences are expected for the four accused killers of Daniel Pearl, despite a lack of evidence. The case will be decided in top secret by handpicked judges in Pakistan's anti-terrorism courts. "Some in Pakistan's government also are very concerned about what [the defendant] Saeed might say in court. His organization and other militant groups here have ties to Pakistan's secret intelligence agency [the ISI]. There are concerns he could try to implicate that government agency in the Pearl case, or other questionable dealings that could be at the very least embarrassing, or worse." [MSNBC, 4/5/02] Later in the month the *London Times* says that the real truth about Saeed will not come out in the trial because, "Sheikh is no ordinary terrorist but a man who has connections that reach high into Pakistan's military and intelligence elite and into the innermost circles of Osama Bin Laden and the al-Qaeda organization." [LONDON TIMES, 4/21/02]

June 25, 2002: Suspicions of ISI-al-Qaeda Links Continue

Although the Western media continues to report that the ISI has reformed itself, "few in Pakistan believe it." The *Independent* later reports rumors that on this day ISI officers hide three al-Qaeda members after a gun battle in which ten soldiers were killed. This follows several other betrayals—now the FBI and the other Pakistani law enforcement authorities no longer tell the ISI about their raids in advance. Other Pakistani investigators are forced to build terrorist files from scratch. because the ISI will not share what they know. [INDEPENDENT, 7/21/02]

July 2002: Terrorists Tied to al-Qaeda and ISI Indicted by U.S., but Little Effort Is Made to Find Them

The U.S. secretly indicts Rajaa Gulum Abbas and Abdul Malik for attempting to buy $32 million in Stinger missiles and other military weaponry in an undercover arms-dealing investigation. However, a U.S. official states that Abbas is an alleged member of the ISI, and is thought to have ties to Middle East terrorist groups and arms-trafficking operations. He also appears to have foreknowledge of the 9/11 attacks. Abdul Malik is said to be Abbas's money man. Abdul Malik is not related to Mohammed Malik, another Pakistani targeted by the undercover operation. The chief U.S. informant in the case, Randy Glass, says that both men also have clear ties to al-Qaeda, and the arms were going to be funnelled to al-Qaeda and used against American targets. [PALM BEACH POST, 3/20/03; SOUTH FLORIDA SUN-SENTINEL, 3/20/03] The indictment is not revealed until March 2003; both men still remain missing and are presumed to be in Pakistan. The U.S. says it is still working on capturing and extraditing Abbas and Malik. [MSNBC, 3/18/03] NBC seems to have no trouble reaching Abbas in Pakistan by telephone. [MSNBC, 8/2/02; MSNBC, 3/18/03] The indictment "makes no mention of Pakistan, any ties to Afghanistan's former Taliban regime or the ultimate destination of the weapons." [PALM BEACH POST, 3/20/03]

July 15, 2002: U.S. Media Ignore ISI
Link in Reports on Saeed's Conviction

Saeed Sheikh and three co-defendants are judged guilty for the murder of reporter Daniel Pearl. Saeed, the supposed mastermind of the murder, is sentenced to death by hanging, and the others are given 25-year terms. Saeed threatens the judge with retribution. As if to confirm that his death covers up unpleasant truths, in the stories of his sentencing every major U.S. media story fails to report Saeed's connections to 9/11 or even to the ISI. [ASSOCIATED PRESS, 7/15/02; ASSOCIATED PRESS, 7/15/02 (B); CBS NEWS, 7/15/02; CNN, 7/15/02; LOS ANGELES TIMES, 7/15/02; MSNBC, 7/15/02; NEW YORK TIMES, 7/15/02; REUTERS, 7/15/02; DAILY TELEGRAPH, 7/16/02; WALL STREET JOURNAL, 7/15/02; WASHINGTON POST, 7/15/02] In contrast, the British media connects Saeed to the ISI [GUARDIAN, 7/16/02; GUARDIAN, 7/16/02 (B); DAILY MAIL, 7/16/02];, al-Qaeda [INDEPENDENT, 7/16/02], the 9/11 attacks [SCOTSMAN, 7/16/02], or some combination of the three [LONDON TIMES, 7/16/02; DAILY MAIL, 7/16/02; DAILY TELEGRAPH, 7/16/02] (with one exception: [BBC, 7/16/02; BBC, 7/16/02 (B)]). The U.S. and British governments both approve of the verdict. [WALL STREET JOURNAL, 7/15/02; BBC, 7/15/02] In the U.S., only the *Washington Post* questions the justice of the verdict. [WASHINGTON POST, 7/15/02; WASHINGTON POST, 7/16/02] By contrast, all British newspapers question the verdict, and subsequently raise additional questions about it. Saeed has appealed the decision but a second trial has yet to begin. [ASSOCIATED PRESS, 8/18/02]

July 16–21, 2002: More Controversy
Surrounding Saeed Trial

More questions emerge in British newspapers about the conviction of Saeed Sheikh for reporter Daniel Pearl's murder in the days immediately after the verdict. Pakistani police have secretly arrested two men who many believe are the real masterminds of Pearl's murder, and official confirmation of these crucial arrests could have ended Saeed's trial. [GUARDIAN, 7/18/02] On May 16, Pearl's body was found and identified, but the FBI does not officially release the DNA results because official confirmation of the body would also have meant a new trial. [INDEPENDENT, 7/16/02] Pakistani officials admit they waited to release the results until after the verdict. [GUARDIAN, 7/18/02] After the trial ends, Pakistani officials admit that the key testimony of a taxi driver is doubtful. The "taxi driver" turns out to be a head constable policeman. [GUARDIAN, 7/18/02] One of the co-defendants turns out to be working for the Special Branch. [INDEPENDENT, 7/21/02] The law states the trial needs to finish in a week, but in fact it took three months. The trial judge and the venue were changed three times. [BBC, 7/16/02] The trial was held in a bunker underneath a prison, and no reporters were allowed to attend. When all the appeals are done, it is doubtful that Saeed will be extradited to the U.S., "because Mr. Sheikh might tell the Americans about the links between al-Qaeda and Pakistan's own intelligence organization." [INDEPENDENT, 7/16/02] Meanwhile, at least seven more suspects remain at large. All have ties to the ISI, and as one investigator remarks, "It seems inconceivable that there isn't someone in the ISI who knows where they're hiding." [TIME, 5/6/02]

July 19, 2002: Why Is U.S. Not Interrogating Saeed, Indian Paper Wonders

An editorial in an Indian newspaper wonders why the U.S. is still not interrogating Saeed Sheikh, recently convicted of murdering Daniel Pearl. Saeed was briefly interrogated by the FBI in February, but they were unable to ask about his links to al-Qaeda, and no known U.S. contact has taken place since. [INDEPENDENT, 7/16/02; INDIAN EXPRESS, 7/19/02] The editorial suggests that if the U.S. pressures its close ally Pakistan to allow Saeed to be interrogated in his Pakistani prison, they could learn more about his financing of the 9/11 attacks and the criminal underworld that Saeed was connected to. Also, U.S. attempts to find al-Qaeda cells in Pakistan could be strongly boosted with new information. [INDIAN EXPRESS, 7/19/02]

August 2, 2002: ISI Tried to Buy Nuclear Material for bin Laden

MSNBC airs recordings informant Randy Glass made of arms dealers and Pakistani ISI agents attempting to buy nuclear material and other illegal weapons for bin Laden. [MSNBC, 8/2/02] Meanwhile, it is reported that federal investigators are re-examining the arms smuggling case involving Glass "to determine whether agents of the Pakistani government tried to buy missiles and nuclear weapons components in the United States last year for use by terrorists or Pakistan's military." [WASHINGTON POST, 8/2/02 (B)] Two such ISI agents, Rajaa Gulum Abbas and Abdul Malik, are already secretly indicted by this time. But Glass still says, "The government knows about those involved in my case who were never charged, never deported, who actively took part in bringing terrorists into our country to meet with me and undercover agents." [COX NEWS SERVICE, 8/2/02] One such person may be a former Egyptian judge named Shireen Shawky, who was interested in buying weapons for the Taliban and attended a meeting in July 1999 in which ISI agent Rajaa Gulum Abbas said the WTC would be destroyed. [WPBF CHANNEL 25, 8/5/02; MSNBC, 8/2/02] Others not charged may include Mohamed Amir and Dr. Magdy el Amir.

December 14, 2002: Al-Qaeda Associate Freed by Pakistani Court

A Pakistani court frees Maulana Masood Azhar, leader of the terrorist group Jaish-e-Mohammed, from prison. [CHRISTIAN SCIENCE MONITOR, 12/16/02] Two weeks later, he is freed from house arrest. He was held for exactly one year without charge, the maximum allowed in Pakistan. [ASSOCIATED PRESS, 12/29/02] He was arrested shortly after an attack on the Indian Parliament that was blamed on his terrorist group. In December 1999 he and Saeed Sheikh were rescued by al-Qaeda from an Indian prison, and he has ties to al-Qaeda and possibly the 9/11 attacks. Pakistan frees several other top terrorist leaders in the same month. It is believed they are doing this so these terrorists can fight in a secret proxy war with India over Kashmir. U.S. officials have remained silent about the release of Azhar and others Pakistani terrorist leaders. [CHRISTIAN SCIENCE MONITOR, 12/16/02] The U.S. froze the funds of Jaish-e-Mohammed in October 2001, but the group simply changed its name to al-Furqan, and the U.S. has not frozen the funds of the "new" group. [FINANCIAL TIMES, 2/8/03; WASHINGTON POST, 2/8/03]

January 18, 2003: Pakistan's President
Warns of Imminent Western Attack

Pakistani President Pervez Musharraf warns of an "impending danger" that Pakistan will become a target of war for "Western forces" after the Iraq crisis. "We will have to work on our own to stave off the danger. Nobody will come to our rescue, not even the Islamic world. We will have to depend on our muscle." [PRESS TRUST OF INDIA, 1/19/03; FINANCIAL TIMES, 2/8/03] Pointing to "a number of recent 'background briefings' and 'leaks'" from the U.S. government, "Pakistani officials fear the Bush administration is planning to change its tune dramatically once the war against Iraq is out of the way." [FINANCIAL TIMES, 2/8/03] Despite evidence that the head of Pakistan's intelligence agency, the ISI, ordered money given to the hijackers, so far only one partisan newspaper has suggested Pakistan was involved in 9/11. [WORLDNETDAILY, 1/3/02]

January 22, 2003: Still No Clarity in Pearl Murder Case

One year after reporter Daniel Pearl's kidnapping and murder, the investigation is mired in controversy. "Mysteries still abound. . . . Suspects disappear or are found dead. Crucial dates are confused. Confessions are offered and then recanted. . . . Nobody who physically carried out the killing has been convicted. None of the four men sentenced is even believed to have ever been at the shed where Pearl was held" and killed. The government arrested three suspects in May 2002 but hasn't charged them and still will not admit to holding them, because acknowledging their testimony would ruin the case against Saeed Sheikh. [ASSOCIATED PRESS, 8/18/02; ASSOCIATED PRESS, 1/22/03] Two of the three claim that 9/11 mastermind Khalid Shaikh Mohammed cut Pearl's throat with a knife. [MSNBC, 9/17/02; TIME, 1/26/03]

July 31, 2003: FBI Claims 9/11 Money Came from Pakistan

John S. Pistole, Deputy Assistant Director of the FBI's Counterterrorism Division, testifies before a Congressional committee. He states, the 9/11 investigation "has traced the origin of the funding of 9/11 back to financial accounts in Pakistan, where high-ranking and well-known al-Qaeda operatives played a major role in moving the money forward, eventually into the hands of the hijackers located in the U.S." [SENATE TESTIMONY, 7/31/03] Pistole does not reveal any further details, but in India it is noted that this is consistent with previous reports that Saeed Sheikh and ISI Director Mahmood Ahmed were behind the funding of 9/11. [TIMES OF INDIA, 8/1/03; PIONEER, 8/7/03]

October 2003: Members of 9/11 Commission Meet with ISI

9/11 Commission staff director Philip Zelikow and several members of his staff embark on a fact-finding mission to Pakistan, Afghanistan, and other countries. While in Pakistan, they interview at least two senior members of the ISI. Whether they are investigating a possible ISI role in the 9/11 plot remains unclear. [UPI, 11/5/03]

June 16, 2004: 9/11 Commission Figure Says Pakistan
Was "Up to Their Eyeballs" with Taliban and al-Qaeda

"An unnamed senior staff member" on the 9/11 Commission tells the *Los Angeles Times* that, before 9/11, Pakistani officials were "up to their eyeballs" in collaboration with the Taliban and al-Qaeda. As an example, he says of bin Laden moving to Afghanistan in 1996, "He wouldn't go back there without Pakistan's approval and support, and had to comply with their rules and regulations." From "day one," the ISI helped al-Qaeda set up an infrastructure, and jointly operated terrorist training camps. However, these findings have only been lightly hinted at in publicly-released 9/11 Commission statements. [LOS ANGELES TIMES, 6/14/04]

13.

Saudi Arabia and the bin Laden Family

Osama bin Laden has been a sworn enemy of the Saudi ruling family, and has long advocated the overthrow of the Saudi government. Therefore, considering any Saudi governmental role in the 9/11 plot would appear patently absurd.

Yet the situation is more complicated than those few facts might suggest. Bin Laden and 15 of the 19 hijackers came from Saudi Arabia. Bin Laden and the Saudi royals share a common belief in Wahhabism, a fundamentalist and expansionist version of Islam. Saudi contributions to support and promote Wahhabism outside of Saudi Arabia have been used for terrorist causes, accidentally or otherwise. Additionally, it has long been suspected that bin Laden effectively blackmailed the Saudi royal family, threatening to bring terrorism to Saudi soil if they did not provide financial support for al-Qaeda operations elsewhere. Finally, the Saudi royal family now includes thousands of wealthy princes, some of whom may closely sympathize with al-Qaeda for various reasons.

So, while few argue that the Saudi government, *as* a government, would have wanted the 9/11 plot to succeed, some rich and/or powerful Saudi individuals may have provided support—and, however unintentionally, the Saudi royal family's significant donations to Wahhabism expansion efforts may well have contributed to the funding of 9/11.

Yet again, there are many questions and unknowns. Did some prominent Saudis have foreknowledge of the attack? Were any such Saudis involved, financially or otherwise, directly or indirectly, in the plot? Are there any connections between Saudis who have supported terror, including the 9/11 plot and President Bush, or other U.S. government officials? Have financial ties between prominent Americans and Saudis hindered counterterrorism efforts? Are some simply too quick to blame Saudi Arabia, perhaps as part of an effort to gain control of the country's vast oil wealth?

Osama bin Laden has dozens of brothers and sisters, and, undoubtedly the vast majority of his hundreds of relatives are eager to distance themselves from terrorism. But has he really been ostracized by all his family members? In 2004, Yeslam Binladin, one of Osama's half-brothers, was asked if he would turn in Osama if given the chance. He replied, "What do you think? Would you turn in your brother?" [MSNBC, 7/10/04] Yeslam's ex-wife, Carmen Binladin, commented, "From what I have seen and what I have read, I cannot believe that [the rest of the bin Laden family] have cut off Osama completely. . . . And I cannot believe that some of the sisters (don't support him.) They are very close to Osama." [SALON, 7/10/4]

What, if anything, did anyone in Osama bin Laden's extended family know prior to 9/11? Did any member provide support to his operation? Has the fact that bin Laden comes from one of the richest and most powerful families in Saudi Arabia assisted him in maintaining his organization and avoiding capture?

October 1980: Osama's Brother Allegedly Involved in "October Surprise"

Salem bin Laden, Osama's oldest brother, described by a French secret intelligence report as one of two closest friends of Saudi Arabia's King Fahd who often performs important missions for Saudi Arabia, is involved in secret Paris meetings between U.S. and Iranian emissaries this month, according to a French report. Frontline, which published the French report, notes that such meetings have never been confirmed. Rumors of these meetings have been called the "October Surprise" and some have speculated that in these meetings, George H. W. Bush negotiated a delay to the release of the U.S. hostages in Iran, thus helping Ronald Reagan and Bush win the 1980 Presidential election. All of this is highly speculative, but if the French report is correct, it points to a long-standing connection of highly improper behavior between the Bush and bin Laden families. [PBS *FRONTLINE*, 2001 (B)]

Mid-1980s: Osama's Brother Is Allegedly Involved in the Iran-Contra Affair

Quoting a French intelligence report posted by PBS Frontline, the *New Yorker* reports, "During the nineteen-eighties, when the Reagan administration secretly arranged for an estimated $34 million to be funneled through Saudi Arabia to the Contras in Nicaragua, [Osama's eldest brother] Salem bin Laden aided in this cause." [*NEW YORKER*, 11/5/01; CITING PBS *FRONTLINE* 2001 (B)]

1988: Bin Ladens Bail Out George W. Bush?

Prior to this year, President George W. Bush is a failed oilman. Three times, friends and investors have bailed him out to keep his business from going bankrupt. However, in 1988, the same year his father becomes President, some Saudis buy a portion of his small company, Harken, which has never performed work outside of Texas. Later in the year, Harken wins a contract in the Persian Gulf and starts doing well financially. These transactions seem so suspicious that the *Wall Street Journal* in 1991 states it "raises the question of . . . an effort to cozy up to a presidential son." Two major investors in Bush's company during this time are Salem bin Laden and Khalid bin Mahfouz. [SALON, 11/19/01; INTELLIGENCE NEWSLETTER, 3/2/00] Salem bin Laden is Osama's oldest brother; Khalid bin Mahfouz is a Saudi banker with a 20 percent stake in BCCI. The bank will be shut down a few years later and bin Mahfouz will have to pay a $225 million fine (while admitting no wrongdoing) (see chapter 12). [*FORBES*, 3/18/02]

August 1990–March 1991: The First Gulf War

Iraq invades and conquers Kuwait in August 1990. Bin Laden, newly returned to Saudi Arabia, offers the Saudi government the use of his thousands of mujahedeen fighters to defend the country in case Iraq attacks it. The Saudi government turns him down, and allows 300,000 U.S. soldiers on Saudi soil instead. Bin Laden is incensed, and immediately goes from ally to enemy of the Saudis. [GHOST WARS, STEVE COLL, 2/04, PP. 221–24, 270–71] After a slow buildup, the U.S. invades Iraq in March 1991 and reestablishes Kuwait. [WHY AMERICA SLEPT, BY GERALD POSNER, 9/03, PP. 40–41] Bin Laden soon leaves Saudi Arabia and starts his career as terrorist (see chapter 4).

March 1991: U.S. Military Remains in Saudi Arabia

As the Gulf War against Iraq ends, the U.S. stations some 15,000–20,000 soldiers in Saudi Arabia permanently. [NATION, 2/15/99] President George H. W. Bush falsely claims that all U.S. troops have withdrawn. [GUARDIAN, 12/21/01] The U.S. troop's presence is not admitted until 1995, and there has never been an official explanation as to why they remained. The *Nation* postulates that they are stationed there to prevent a coup. Saudi Arabia has an incredible array of high-tech weaponry, but lacks the expertise to use it and it is feared that Saudi soldiers may have conflicting loyalties. In 1998, bin Laden will release a statement: "For more than seven years the United States has been occupying the lands of Islam in the holiest of places, the Arabian peninsula, plundering its riches, dictating to its rulers, humiliating its people, terrorizing its neighbors, and turning its bases in the peninsula into a spearhead through which to fight the neighboring Muslim peoples." [NATION, 2/15/99]

Summer 1991: Bin Laden Leaves Saudi Arabia

Bin Laden, recently returned to Saudi Arabia, has been placed under house arrest for his opposition to the continued presence of U.S. soldiers on Saudi soil. [PBS FRONTLINE, 9/01] Controversial author Gerald Posner claims that a classified U.S. intelligence report describes a secret deal between bin Laden and Saudi intelligence minister Prince Turki al-Faisal at this time. Although bin Laden has become an enemy of the Saudi state, he is nonetheless too popular for his role with the mujahedeen in Afghanistan to be easily imprisoned or killed. According to Posner, bin Laden is allowed to leave Saudi Arabia with his

money and supporters, but the Saudi government will publicly disown him. Privately, the Saudis will continue to fund his supporters with the understanding that they will never be used against Saudi Arabia. The wrath of the fundamentalist movement is thus directed away from the vulnerable Saudis. [WHY AMERICA SLEPT, BY GERALD POSNER, 9/03, PP. 40–42] Posner alleges the Saudis "effectively had [bin Laden] on their payroll since the start of the decade." [TIME, 8/31/03] This deal is reaffirmed in 1996 and 1998. Bin Laden leaves Saudi Arabia in the summer of 1991, returning first to Afghanistan. [GHOST WARS, BY STEVE

Prince Turki al-Faisal

COLL, 2/04, PP. 229–31, 601–02] After staying there a few months, he moves again, settling into Sudan with hundreds of ex-Mujahedeen supporters. [PBS FRONTLINE, 9/01]

June 4, 1992: FBI Investigates Ties Between George W. Bush and Saudi Money

The FBI investigates connections between James Bath and George W. Bush, according to published reports. Bath is Salem bin Laden's official representative in the U.S. Bath's business partner contends that "Documents indicate that the Saudis were using Bath and their huge financial resources to influence U.S. policy," since George W. Bush's father is president. George W. Bush denies any connections to Saudi money. What became of this investigation is unclear. [HOUSTON CHRONICLE, 6/4/92]

1994: U.S. Declines to Accept Documents Exposing
Saudi Scandals and Terrorist Ties

Mohammed al-Khilewi, the First Secretary at the Saudi Mission to the United Nations, defects and seeks political asylum in the U.S. He brings with him 14,000 internal government documents depicting the Saudi royal family's corruption, human-rights abuses, and financial support for terrorists. He meets with two FBI agents and an Assistant U.S. Attorney. "We gave them a sampling of the documents and put them on the table," says his lawyer, "but the agents refused to accept them." [NEW YORKER, 10/16/01] The documents include "details of the $7 billion the Saudis gave to [Iraq leader] Saddam Hussein for his nuclear program—the first attempt to build an Islamic Bomb." However, FBI agents were "ordered not to accept evidence of Saudi criminal activity, even on U.S. soil." [BEST DEMOCRACY MONEY CAN BUY, BY GREG PALAST, 2/03, PP. 101]

April 9, 1994: Saudi Government Publicly Breaks with Osama bin Laden

The Saudi government revokes bin Laden's citizenship and moves to freeze his assets in Saudi Arabia because of his support for Muslim fundamentalist movements. [NEW YORK TIMES, 4/10/94; PBS FRONTLINE, 9/01] However, allegedly, this is only a public front and they privately continue to support him as part of a secret deal allegedly made in 1991.

December 14, 1994: Osama's Brother-in-Law Held in U.S.,
Then Let Go Despite Terror Ties

Mohammed Jamal Khalifa, a brother-in-law to bin Laden, is arrested in the U.S. Khalifa, who financed the Abu Sayyaf terror group in the Philippines, has recently been sentenced to death in Jordan for funding a group that staged a series of bombings in that country. The FBI finds and quickly translates literature in his luggage advocating training in assassination, explosives, and weapons, bombing churches, and murdering Catholic priests. At the time, he is linked to funding bin Laden's terrorism efforts, as well as to terrorist Ramzi Yousef and other Operation Bojinka plotters. Bin Laden could be connected to many terrorist activities through Khalifa's connections. However, Secretary of State Warren Christopher argues that Khalifa should be released to Jordan. Khalifa is sent to Jordan in May 1995, where his conviction has already been overturned. In a later retrial there, a witness recants and Khalifa is set free. Says one expert working at the CIA's Counter Terrorism Center at the time, "I remember people at the CIA who were ripshit at the time. Not even speaking in retrospect, but contemporaneous with what the intelligence community knew about bin Laden, Khalifa's deportation was unreal." [1000 YEARS FOR REVENGE, BY PETER LANCE, 9/03, PP. 233–35; NEW YORK TIMES, 5/2/02 (B); SAN FRANCISCO CHRONICLE, 4/18/95; ASSOCIATED PRESS, 4/26/95]

1995–2001: Persian Gulf Elite Go Hunting with bin Laden in Afghanistan

After the Taliban takes control of the area around Kandahar, Afghanistan in September 1994, prominent Persian Gulf state officials and businessmen, including high-ranking United Arab Emirates and Saudi government ministers, such as Saudi intelligence minister Prince Turki al-Faisal, frequently secretly fly into Kandahar on state and private jets for hunting expeditions. [LOS ANGELES TIMES, 11/18/01] General Wayne Downing, Bush's former national director for combating terrorism, says: "They would go out and see Osama, spend some time with him, talk with him, you know, live out in the tents, eat the simple food, engage in falconing, some other pursuits, ride horses." [MSNBC, 9/5/03] While there, some develop ties to the Taliban and al-Qaeda and give them money. Both bin Laden and Taliban leader Mullah Omar sometimes participate in these hunting trips. Former U.S. and Afghan officials suspect that the dignitaries' outbound jets may also have smuggled out al-Qaeda and Taliban. [LOS ANGELES TIMES, 11/18/01]

Late 1995: Illness of Saudi King Generates Long-Term Power Struggle

King Fahd of Saudi Arabia suffers a severe stroke. Afterwards, he is able to sit in a chair and open his eyes, but little more. The resulting lack of leadership begins a behind-the-scenes struggle for power and leads to increased corruption. Crown Prince Abdullah has been urging his fellow princes to address the problem of corruption in the kingdom—so far unsuccessfully. A former White House adviser says: "The only reason Fahd's being kept alive is so Abdullah can't become king." [NEW YORKER, 10/16/01] This internal power struggle persists on 9/11, and continues through at least mid-2004.

1996: Saudi Regime Has "Gone to the Dark Side"

The Saudi Arabian government, which allegedly initiated payments to al-Qaeda in 1991, increases its payments in 1996, becoming al-Qaeda's largest financial backer. It also gives money to other extremist groups throughout Asia, vastly increasing al-Qaeda's capabilities. [NEW YORKER, 10/16/01] Presumably, two meetings in early summer bring about the change. Says one U.S. official, "[19]96 is the key year . . . Bin Laden hooked up to all the bad guys—it's like the Grand Alliance—and had a capability for conducting large-scale operations." The Saudi regime, he says, had "gone to the dark side." Electronic intercepts by the NSA "depict a regime increasingly corrupt, alienated from the country's religious rank and file, and so weakened and frightened that it has brokered its future by channeling hundreds of millions of dollars in what amounts to protection money to fundamentalist groups that wish to overthrow it." U.S. officials later privately complain "that the Bush administration, like the Clinton administration, is refusing to confront this reality, even in the aftermath of the September 11th terrorist attacks." [NEW YORKER, 10/16/01]

May 1996: Saudis and al-Qaeda Allegedly Strike a Secret Deal

French intelligence secretly monitors a meeting of Saudi billionaires at the Hotel Royale Monceau in Paris this month with the financial representative of al-Qaeda. "The Saudis, including a key Saudi prince joined by Muslim and non-Muslim gun traffickers, [meet] to determine who would pay how much to Osama. This [is] not so much an act of support but of protection—a payoff to keep the mad bomber away from Saudi Arabia." [BEST DEMOCRACY MONEY CAN BUY, BY GREG PALAST, 2/03, P. 100] Participants also agree that bin Laden should be rewarded for promoting Wahhabism (an austere form of Islam that requires literal interpretation of the Koran) in Chechnya, Kashmir, Bosnia, and other places. [CANADIAN BROADCASTING CORP., 10/29/03 (C)] This extends an alleged secret deal first made between the Saudi government and bin Laden in 1991. Later, 9/11 victims' relatives will rely on the "nonpublished French intelligence report" of this meeting in their lawsuit against important Saudis. [MINNEAPOLIS STAR-TRIBUNE, 8/16/02] According to French terrorism expert Jean-Charles Brisard and/or reporter Greg Palast, there are about 20 people at the meeting, including an unnamed brother of bin Laden and an unnamed representative from the Saudi Defense Ministry. [CBC, 10/29/03 (C)] Palast, noting that the French monitored the meeting, asks, "Since U.S. intelligence was thus likely informed, the question becomes why didn't the government immediately move against the Saudis?" [BEST DEMOCRACY MONEY CAN BUY, BY GREG PALAST, P. 100]

September 11, 1996: Investigation of
bin Laden Family Members Is Closed

The FBI investigation into two relatives of bin Laden, begun in February 1996, is closed. The FBI wanted to learn more about Abdullah bin Laden, "because of his relationship with the World Assembly of Muslim Youth [WAMY]—a suspected terrorist organization." [GUARDIAN, 11/7/01] Abdullah was the U.S. director of WAMY and lived with his brother Omar in Falls Church, Virginia, a suburb of Washington. The coding on the document, marked secret, indicates the case involved espionage, murder, and national security. WAMY's office address is 5613 Leesburg Pike. It is later determined that four of the 9/11 hijackers lived at 5913 Leesburg Pike at the same time the two bin Laden brothers were there. WAMY has been banned in Pakistan by this time. [BBC NEWSNIGHT, 11/6/01; GUARDIAN, 11/7/01] The Indian and Philippine governments have also cited WAMY for funding terrorism. The 9/11 Commission heard testimony that WAMY "has openly supported Islamic terrorism. There are ties between WAMY and 9/11 hijackers. It is a group that has openly endorsed the notion that Jews must be killed. . . . [It] has consistently portrayed the United States, Jews, Christians, and other infidels as enemies who have to be defeated or killed. And there is no doubt, according to U.S. intelligence, that WAMY has been tied directly to terrorist attacks." [TESTIMONY OF STEVE EMERSON, 9/11 COMMISSION HEARING, 7/9/03, P. 66] "WAMY was involved in terrorist-support activity," says a security official who served under President Bush. "There's no doubt about it." [VANITY FAIR, 10/03] A high-placed intelligence official tells the *Guardian*: "There were always constraints on

WAMY logo

investigating the Saudis. There were particular investigations that were effectively killed." [GUARDIAN, 11/7/01] An unnamed U.S. source says to the BBC, "There is a hidden agenda at the very highest levels of our government." [BBC NEWSNIGHT, 11/6/01] The Bosnian government says in September 2002 that a charity with Abdullah bin Laden on its board had channeled money to Chechen guerrillas, something that "is only possible because the Clinton CIA gave the wink and nod to WAMY and other groups who were aiding Bosnian guerrillas when they were fighting Serbia, a U.S.-approved enemy." The investigation into WAMY is only restarted a few days after 9/11, around the same time these bin Ladens leave the U.S. [BEST DEMOCRACY MONEY CAN BUY, BY GREG PALAST, 2/03, PP. 96–99] (Note that this Abdullah bin Laden is possibly bin Laden's cousin, not brother, and is not the same as the Abdullah bin Laden who serves as the bin Laden family spokesperson. [BEST DEMOCRACY MONEY CAN BUY BY GREG PALAST 2/03 PP. 98–99]) Interestingly, as of mid-2004, WAMY is still not listed by the U.S. as a terrorist organization.

Spring 1998: Bin Laden's Stepmother Visits Afghanistan

Sources who know bin Laden claim his stepmother has the first of two meetings with her son in Afghanistan during this period. This trip was arranged by Prince Turki al-Faisal, then the head of Saudi intelligence. Turki was in charge of the "Afghanistan file" for Saudi Arabia, and had long-standing ties to bin Laden and the Taliban since 1980. [NEW YORKER, 11/5/01]

June 1998: Taliban and Saudis Discuss bin Laden

Relations between Taliban head Mullah Omar and bin Laden grow tense, and Omar discusses a secret deal with the Saudis, who have urged the Taliban to expel bin Laden from Afghanistan. Head of Saudi intelligence Prince Turki al-Faisal travels to Kandahar, Afghanistan and brokers the deal. According to Prince Turki, he seeks to have the Taliban turn bin Laden over to Saudi custody. Omar agrees in principle, but requests that the parties establish a joint commission to work out how bin Laden would be dealt with in accordance with Islamic law. [GHOST WARS, BY STEVE COLL, 2/04, P. 400–02] Note that some reports of this meeting—and the deal discussed—vary significantly from Prince Turki's version. However, before a deal can be reached, the U.S. strikes Afghanistan in August in retaliation for the U.S. African embassy bombings (see chapter 1), driving Omar and bin Laden back together. Prince Turki later states that "the Taliban attitude changed 180 degrees," and that Omar is "absolutely rude" to him when he visit[s] again in September. [GUARDIAN, 11/5/01; LONDON TIMES, 8/3/02]

July 1998: Taliban and Saudis Meet and Purportedly Make a Deal

Taliban officials allegedly meet with Prince Turki, head of Saudi intelligence, to continue talks concerning the Taliban's ouster of bin Laden from Afghanistan. Reports on the location of this meeting, and the deal under discussion differ. According to some reports, including documents exposed in a later lawsuit, this meeting takes place in Kandahar. Those present include Prince Turki al-Faisal, head of Saudi Arabian intelligence, Taliban leaders, senior officers from the ISI, and bin Laden. According to these reports, Saudi Arabia agrees to give the Taliban and Pakistan "several hundred millions" of dollars, and in return, bin Laden promises no attacks against Saudi Arabia. The Saudis also agree to ensure that requests for the extradition of al-Qaeda members will be blocked and promise to block demands by other countries to close down bin Laden's Afghan training camps. Saudi Arabia had previously given money to the Taliban and bribe money to bin Laden, but this ups the ante. [SUNDAY TIMES, 8/25/02] A few weeks after the meeting, Prince Turki sends 400 new pickup trucks to Afghanistan. At least $200 million follow. [HONOLULU STAR-BULLETIN, 9/23/01; NEW YORK POST, 8/25/02] Controversial author Gerald Posner gives a similar account said to come from high U.S. government officials, and adds that al-Qaeda leader Abu Zubaida also attends the meeting. [WHY AMERICA SLEPT, BY GERALD POSNER, 9/03, PP. 189–90] Note that reports of this meeting seemingly contradict reports of a meeting the month before between Turki and the Taliban, in which the Taliban agreed to get rid of bin Laden.

October 1998: Chicago FBI Investigation of Terrorist Cell Thwarted

FBI agents Robert Wright and John Vincent, tracking a terrorist cell in Chicago, are told to simply follow suspects around town and file reports. The two agents believe some of the money used to finance the 1998 U.S. embassy bombing in Africa leads back to Chicago and Saudi multimillionaire businessman Yassin al-Qadi. Supervisors try, but temporarily fail, to halt the investigation into al-Qadi's possible terrorist connections. However, at this time, a supervisor prohibits Wright and Vincent from making any arrests connected to the bombings, or opening new criminal investigations. Even though they believe their case is growing stronger, in January 2001, Wright is told that the Chicago case is being closed and that "it's just better to let sleeping dogs lie." Wright tells ABC: "Those dogs weren't sleeping, they were training, they were getting ready. . . . September the 11th is a direct result of the incompetence of the FBI's International Terrorism Unit. . . . Absolutely no doubt about that." Chicago federal prosecutor Mark Flessner, also working on the case, says there "were powers bigger than I was in the Justice Department and within the FBI that simply were not going to let [the building of a criminal case] happen." Wright will write an internal FBI memo harshly criticizing their decisions regarding the case in June 2001. Al-Qadi is named in a March 2000 affidavit as a source of terrorist funds in Chicago. [ABC NEWS, 11/26/02; ABC NEWS, 12/19/02; ABC NEWS, 12/19/02 (B)] He is also on secret U.S. and UN lists of major al-Qaeda financiers. His charity, the Muwafaq Foundation, is allegedly an al-Qaeda front that transferred $820,000 to the Palestinian group Hamas through a Muslim foundation called the Quranic Literacy Institute. (The date of the transfer has not been released.) [CNN, 10/15/01 (B)] Al-Qadi says he shut down Muwafaq in 1996, but the charity has received money from the United Nations since then. [BBC, 10/20/01; CNN, 10/15/01 (B)]

November 1998: Former President George H. W. Bush Meets with bin Laden Family

Former President George H. W. Bush meets with the bin Laden family on behalf of the Carlyle Group. The meeting takes place in Jeddah, Saudi Arabia. [SUNDAY HERALD, 10/7/01]

Late 1998: Key Embassy Bombing Witnesses
Are Beheaded Before They Can Talk to FBI

FBI counterterrorism expert John O'Neill and his team investigating the 1998 U.S. embassy bombings are repeatedly frustrated by the Saudi government. Guillaume Dasquié, one of the authors of *The Forbidden Truth*, later tells the *Village Voice*: "We uncovered incredible things . . . Investigators would arrive to find that key witnesses they were about to interrogate had been beheaded the day before." [VILLAGE VOICE 1/2/02; THE FORBIDDEN TRUTH, BY JEAN-CHARLES BRISARD, GUILLAUME DASQUIÉ, AND WAYNE MADSEN, 7/02, PP. XXIX]

January 2000: Former President Bush Meets
with bin Laden Family on Behalf of Carlyle Group

Former President George H. W. Bush meets with the bin Laden family on behalf of the Carlyle Group. He had also met with them in November 1998, but it is not known if he meets with them again after this. Bush denies this meeting took place until a thank you note is found confirming that it took place. [WALL STREET JOURNAL, 9/27/01; GUARDIAN, 10/31/01]

March 21, 2000: Complaints About FBI Agent Are Ignored,
Dubious Agent Promoted to Head FBI Investigations in Saudi Arabia

FBI agent Robert Wright, having been accused of tarnishing the reputation of fellow agent Gamal Abdel-Hafiz, makes a formal internal complaint about Abdel-Hafiz. FBI agent Barry Carmody seconds Wright's complaint. Wright and Carmody accuse Abdel-Hafiz, a Muslim, of hindering investigations by openly refusing to record other Muslims. The FBI was investigating if BMI Inc., a New Jersey based company with connections to Saudi financier Yassin al-Qadi, had helped fund the 1998 U.S. embassy bombings. [WALL STREET JOURNAL, 11/26/02; ABC NEWS, 12/19/02] Federal prosecutor Mark Flessner and other FBI agents back up the allegations against Abdel-Hafiz. [ABC NEWS, 12/19/02] Carmody also claims that Abdel-Hafiz hindered an inquiry into the possible terrorist ties of fired University of South Florida Professor Sami Al-Arian by refusing to record a conversation with the professor in 1998. [TAMPA TRIBUNE, 3/4/03] Complaints to superiors and headquarters about this never get a response. [FOX NEWS, 3/6/03] Furthermore, "Far from being reprimanded, Abdel-Hafiz [is] promoted to one of the FBI's most important anti-terrorism posts, the American Embassy in Saudi Arabia, to handle investigations for the FBI in that Muslim country." [ABC NEWS, 12/19/02] Abdel-Hafiz is finally suspended in February 2003, after his scandal is widely reported in the press. [TAMPA TRIBUNE, 3/4/03] Bill O'Reilly of Fox News claims that on March 4, 2003, the FBI threatens to fire Wright if he speaks publicly about this, one hour before Wright is scheduled to appear on Fox News. [FOX NEWS, 3/4/03]

Spring 2000: Saudi Suggestion to Track bin Laden's
Stepmother in Planned Meeting with bin Laden Is Rejected

Sources who know bin Laden later claim that his stepmother has a second meeting with her son in Afghanistan (her first visit took place in the spring of 1998). The trip is approved by the Saudi royal family. The Saudis pass the message to him that "'they wouldn't crack down on his followers in Saudi Arabia' as long as he set his sights on targets outside the desert kingdom." In late 1999, the Saudi government had told the CIA about the upcoming trip, and suggested placing a homing beacon on her luggage. This does not happen—Saudis later claim they weren't taken seriously, and Americans claim they never received specific information on her travel plans. [NEW YORKER, 11/5/01; WASHINGTON POST, 12/19/01]

Late January 2001: U.S. Intelligence Told to Back Off from bin Laden and Saudis

The BBC later reports, "After the elections, [U.S. intelligence] agencies [are] told to 'back off' investigating the bin Ladens and Saudi royals, and that anger[s] agents." This follows previous orders to abandon an investigation of bin Laden relatives in 1996, and difficulties in investigating Saudi royalty. [BBC, 11/6/01] An unnamed "top-level CIA operative" says there is a "major policy shift" at the National Security Agency at this time. Bin Laden could still be investigated, but agents could not look too closely at how he got his money. [BEST DEMOCRACY MONEY CAN BUY, BY GREG PALAST, 2/03, P. 99] Presumably one such investigation canceled is an investigation by the Chicago FBI into ties between Saudi multimillionaire Yassin al-Qadi and the U.S. embassy bombings in August 1998, because during this month an FBI agent is told that the case is being closed and that "it's just better to let sleeping dogs lie." Reporter Greg Palast notes that President Clinton was already hindering investigations by protecting Saudi interests. However, as he puts it, "Where Clinton said, 'Go slow,' Bush policymakers said, 'No go.' The difference is between closing one eye and closing them both." [BEST DEMOCRACY MONEY CAN BUY, BY GREG PALAST, 2/03, PP. 102]

February 2001: Bin Laden's Sisters Seen Handing Money to al-Qaeda Member

A former CIA anti-terror expert later claims that an allied intelligence agency sees "two of Osama's sisters apparently taking cash to an airport in Abu Dhabi [United Arab Emirates], where they are suspected of handing it to a member of bin Laden's al-Qaeda organization." This is cited as one of many incidents showing an "interconnectedness" between bin Laden and the rest of his family. [NEW YORKER, 11/5/01]

February 26, 2001: Osama Attends Son's Wedding
with Other bin Laden Family Members

Bin Laden attends the wedding of his son Mohammed in Kandahar, Afghanistan. Although bin Laden is supposedly long estranged from his family, bin Laden's stepmother, two brothers, and sister are also said to attend, according to the only journalist who was invited. [REUTERS, 3/1/01; SUNDAY HERALD, 10/7/01]

Summer 2001: Saudi and Taliban Leaders Reportedly Discuss bin Laden

An Asia Times article published just prior to 9/11 claims that Crown Prince Abdullah, the de facto ruler of Saudi Arabia, makes a clandestine visit to Pakistan around this time. After meeting with senior army officials, he visits Afghanistan with ISI Director Mahmood. They meet Taliban leader Mullah Omar and try to convince him that the U.S. is likely to launch an attack on Afghanistan. They insist bin

Laden be sent to Saudi Arabia, where he would be held in custody and not handed over to any third country. If bin Laden were to be tried in Saudi Arabia, Abdullah would help make sure he is acquitted. Mullah Omar apparently rejects the proposal. The article suggests that Abdullah is secretly a supporter of bin Laden and is trying to protect him from harm. [ASIA TIMES, 8/22/01] A similar meeting may also take place on September 19, 2001.

Mid-July 2001: John O'Neill Rails Against White House and Saudi Obstructionism

FBI counterterrorism expert John O'Neill privately discusses White House obstruction in his bin Laden investigation. O'Neill says, "The main obstacles to investigate Islamic terrorism were U.S. oil corporate interests and the role played by Saudi Arabia in it." He adds, "All the answers, everything needed to dismantle Osama bin Laden's organization, can be found in Saudi Arabia." O'Neill also believes the White House is obstructing his investigation of bin Laden because they are still keeping the idea of a pipeline deal with the Taliban open (see chapter 5). [CNN, 1/8/02; CNN, 1/9/02; *IRISH TIMES*, 11/19/01; *THE FORBIDDEN TRUTH*, BY JEAN-CHARLES BRISARD, GUILLAUME DASQUIÉ, AND WAYNE MADSEN, 7/02, PP. XXIX]

August 31, 2001: Head of Saudi Arabia's Intelligence Service Is Replaced

Prince Turki al-Faisal, head of Saudi Arabia's intelligence service for 24 years, is replaced. No explanation is given. He is replaced by Nawaf bin Abdul Aziz, his nephew and the king's brother, who has "no background in intelligence whatsoever." [AGENCE FRANCE-PRESSE, 8/31/01; *SEATTLE TIMES*, 10/29/01; *WALL STREET JOURNAL*, 10/22/01] The *Wall Street Journal* later reports: "The timing of Turki's removal—August 31—and his Taliban connection raise the question: Did the Saudi regime know that bin Laden was planning his attack against the U.S.? The current view among Saudi-watchers is probably not, but that the House of Saud might have heard rumors that something was planned, although they did not know what or when. (An interesting and possibly significant detail: Prince Sultan, the defense minister, had been due to visit Japan in early September, but canceled his trip for no apparent reason less than two days before an alleged planned departure.)" [WALL STREET JOURNAL, 10/22/01] Turki is later sued in August 2002 for his role in 9/11, and is appointed ambassador to Britain in October 2002.

September 5–8, 2001: Raid on Arab Web Hosting Company Precedes 9/11 Attacks

The Joint Terrorism Task Force conducts a three-day raid of the offices of InfoCom Corporation, a Texas-based company that hosts about 500 mostly Arab websites, including Al Jazeera, the Arab world's leading news channel. [GUARDIAN, 9/10/01; WEB HOST INDUSTRY REVIEW, 9/10/01] The task force includes agents from the FBI, Secret Service, and Diplomatic Security, as well as tax inspectors, immigration officials, customs officials, department of commerce officials, and computer experts. [GUARDIAN, 9/10/01] The FBI declines to give a reason for the raid [NEWSFACTOR NETWORK, 9/7/01], but a spokeswoman said it is not aimed at InfoCom's clients. [BBC, 9/7/01] The reasons for the raid "may never be known, because a judge ordered the warrant to be sealed." [WEB HOST INDUSTRY REVIEW, 9/10/01] Three days after the initial raid, the task force is "still busy inside the building, reportedly copying every hard disc they could find." [GUARDIAN, 9/10/01] Info-Com's offices are located in Richardson, a suburb of Dallas. Two charities in Richardson, The Global Relief Foundation Inc. and The Holy Land Foundation for Relief and Development, have been investigated for possible ties to Palestinian terrorist organizations. [BOSTON HERALD, 12/11/01] Five or six years earlier, counterterrorism "tsar" Richard Clarke had wanted to raid the Holy Land offices, but was prevented from doing so by the FBI and Treasury Department. [AGAINST ALL ENEMIES, BY RICHARD CLARKE, 3/04, P. 98] Not only are InfoCom and Holy Land across the road from each other, they are intimately connected through two brothers: Ghassan and Bayan Elashi. [GUARDIAN, 9/10/01] Ghassan Elashi is both the vice president of InfoCom and chairman of Holy Land. [NEW YORK TIMES, 12/20/02] These two and others are later arrested on a variety of serious charges. Approximately one week before 9/11, Bank One closes Holy Land's checking accounts totaling about $13 million, possibly because of an investigation begun by the New York State Attorney General. [DALLAS BUSINESS JOURNAL, 9/7/01] The U.S. freezes Holy Land's assets two months later for suspected terrorist associations. Holy Land is represented by Akin, Gump, Strauss, Hauer, and Feld, a Washington, D.C., law firm with unusually close ties to the Bush White House. [WASHINGTON POST, 12/17/01] In the Garland suburb adjoining Richardson, a fifth-grade boy apparently has foreknowledge of 9/11. [HOUSTON CHRONICLE, 9/19/01]

September 10, 2001: Three Hijackers at Same Hotel with Senior Saudi Official

Three hijackers, Hani Hanjour, Khalid Almihdhar, and Nawaf Alhazmi, check into the same hotel as a prominent Saudi government official. [WASHINGTON POST, 10/2/03] Investigators have not found any evidence that the hijackers met with the official, and stress it could be a coincidence. [DAILY TELEGRAPH, 3/10/03] However, one prosecutor working on a related case asserts, "I continue to believe it can't be a coincidence." [WALL STREET JOURNAL, 10/2/03] The official, Saleh Ibn Abdul Rahman Hussayen, is interviewed by the FBI shortly after 9/11, but according to testimony from an FBI agent, the interview is cut short when Hussayen "feign[s] a seizure, prompting the agents to take him to a hospital, where the attending physicians [find] nothing wrong with him." The agent recommends that Hussayen "should not be allowed to leave until a follow-up interview could occur." However, that "recommendation, for whatever reason, [is] not complied with." Hussayen returns to Saudi Arabia a few days later, as soon as the U.S. ban on international flights has ended. [WASHINGTON POST, 10/2/03] For most of the 1990s, Hussayen was director of the SAAR Foundation, a Saudi charity that is being investigated for terrorism ties. A few months after 9/11 he is named a minister of the Saudi government and put in charge of its two holy mosques. Hussayen had arrived in the U.S. in late August 2001 planning to visit some Saudi-sponsored charities. Many of the charities on his itinerary, including the Global Relief Foundation, World Muslim League, IIRO (International Islamic Relief Organization), IANA (Islamic Assembly of North America), WAMY, have since been shut down or investigated for alleged ties to terrorism. [WASHINGTON POST, 10/2/03] His nephew, Sami Omar Hussayen, is indicted in early 2004 for using his computer expertise to assist terrorist groups, and is charged with administering a website associated with IANA, which expressly advocated suicide attacks and using airliners as weapons in the months before 9/11. Investigators also claim the nephew was in contact with important al-Qaeda figures. [WASHINGTON POST, 10/2/03; SEATTLE POST-INTELLIGENCER, 1/10/04] IANA is under investigation, as well as the flow of money from the uncle to nephew. [DAILY TELEGRAPH, 3/10/03] The uncle has not been charged with any crime. [WALL STREET JOURNAL, 10/2/03]

September 11, 2001: Bin Laden Brother Attends Carlyle Group Conference

The Carlyle Group is a company closely associated with officials of the Bush and Reagan administrations, and has considerable ties to Saudi oil money, including ties to the bin Laden family. Those ties are well illustrated by the fact that on this day the Carlyle Group is hosting a conference at a Washington hotel. Among the guests of honor is investor Shafig bin Laden, brother to bin Laden. [OBSERVER, 6/16/02]

September 13, 2001: Saudi Royals Fly to Kentucky in Violation of Domestic Flight Ban

After a complete air flight ban in the U.S. begun during the 9/11 attacks, some commercial flights begin resuming this day. However, all private flights are still banned from flying. Nonetheless, at least one private flight carrying Saudi royalty takes place on this day. And in subsequent days, other flights carry royalty and bin Laden family members. These flights take place even as fighters escort down three other private planes attempting to fly. Most of the Saudi royals and bin Ladens in the U.S. at the time are high school or college students and young professionals. [NEW YORK TIMES, 9/30/01; VANITY FAIR, 10/03] The first flight is a Lear Jet that leaves from a private Raytheon hangar in Tampa, Florida, and takes three Saudis to Lexington, Kentucky. [TAMPA TRIBUNE, 10/5/01] Prince Bandar, the Saudi ambassador to the U.S. who is so close to the Bush family that he is nicknamed "Bandar Bush," pushes for and helps arrange the flights at the request of frightened Saudis. [VANITY FAIR, 10/03; CANADIAN BROADCASTING CORP., 10/29/03 (D)] For two years, a violation of the air ban is denied by the FAA, FBI, and White House, and decried as an urban legend except for one article detailing them in a Tampa newspaper. [TAMPA TRIBUNE, 10/5/01] Finally in 2003; Richard Clarke, National Security Council Chief of Counterterrorism, confirms the existence of these flights, and Secretary of State Powell confirms them as well. [VANITY FAIR, 10/03; MSNBC, 9/7/03] However, the White House remains silent on the matter. [NEW YORK TIMES, 9/4/03] Officials at the Tampa International Airport finally confirm this first flight in 2004. But whether the flight violated the air ban or not rests on some technicalities that remain unresolved. [LEXINGTON HERALD-LEADER, 6/10/04] The Saudis are evacuated to Saudi Arabia over the next several days.

September 14, 2001: Revealed: Saudi Students May Attend Florida Flight Schools Without Background Checks

In interviews with the *Boston Globe*, flight instructors in Florida say that it was common for students with Saudi affiliations to enter the U.S. with only cursory background checks and sometimes none. Some flight schools, including some of those attended by the hijackers, have exemptions that allow the schools to unilaterally issue paperwork that students can present at U.S. embassies and consulates so they can obtain visas. Saudi Arabia is possibly the only Arab country with such an exemption. [BOSTON GLOBE, 9/14/01]

September 14–19, 2001: Bin Laden Family Members, Saudi Royals Quietly Leave U.S.

Following a secret flight inside the U.S. that is in violation of a national private airplane flight ban, members of the bin Laden family and Saudi royalty quietly depart the U.S. The flights are only publicly acknowledged after all the Saudis have left. [BOSTON GLOBE, 9/21/01; NEW YORK TIMES, 9/30/01] About 140 Saudis, including around 24 members of the bin Laden family, are passengers in these flights. The identity of most of these passengers is not known. However, some of the passengers include:

- The son of the Saudi Defense Minister Prince Sultan. Sultan is sued in August 2002 for alleged complicity in the 9/11 plot. [TAMPA TRIBUNE, 10/5/01] He is alleged to have contributed at least $6 million since 1994 to four charities that finance al-Qaeda. [VANITY FAIR, 10/03];

- Khalil bin Laden. He has been investigated by the Brazilian government for possible terrorist connections. [VANITY FAIR, 10/03];

- Abdullah bin Laden and Omar bin Laden, cousins of bin Laden. Abdullah was the U.S. director of the Muslim charity World Assembly of Muslim Youth (WAMY). The governments of India, Pakistan, Philippines, and Bosnia have all accused WAMY of funding terrorism. These two relatives were investigated by the FBI in 1996 in a case involving espionage, murder, and national security. Their case is reopened on September 19, right after they leave the country. [VANITY FAIR, 10/03] Remarkably, four of the 9/11 hijackers briefly live in the town of Falls Church, Virginia, three blocks from the WAMY office headed by Abdullah bin Laden. [BBC NEWSNIGHT, 11/6/01]

- Saleh Ibn Abdul Rahman Hussayen. He is a prominent Saudi official who is in the same hotel as three of the hijackers the night before 9/11. He leaves on one of the first flights to Saudi Arabia before the FBI can properly interview him about this. [WASHINGTON POST, 10/2/03]

There is a later dispute regarding how thoroughly the Saudis are interviewed before they leave and who approves the flights. Richard Clarke, National Security Council Chief of Counterterrorism, says he agrees to the flights after the FBI assures him none of those on board has connections to terrorism and that it is "a conscious decision with complete review at the highest levels of the State Department and the FBI and the White House." [CONGRESSIONAL TESTIMONY, 9/3/03] Clarke says the decision to approve the flights "didn't get any higher than me." However, the question of who made the request of Clarke is still unknown. [THE HILL, 05/18/04] According to *Vanity Fair*, both the FBI and the State Department "deny playing any role whatsoever in the episode." However, Dale Watson, the FBI's former head of counterterrorism, says the Saudis on the planes "[are] identified, but they [are] not subject to serious interviews or interrogations" before they leave. [VANITY FAIR, 10/03] An FBI spokesperson says the bin Laden relatives are

only interviewed by the FBI "at the airport, as they [are] about to leave." [NATIONAL REVIEW, 9/11/02] There are claims that some passengers are not interviewed by the FBI at all. [VANITY FAIR, 10/03] Abdullah bin Laden, who stays in the U.S., says that even a month after 9/11, his only contact with the FBI is a brief phone call. [BOSTON GLOBE, 9/21/01; NEW YORKER, 11/5/01] Numerous experts are surprised that the bin Ladens are not interviewed more extensively before leaving, pointing out that interviewing the relatives of suspects is standard investigative procedure. [NATIONAL REVIEW, 9/11/02; VANITY FAIR, 10/03] MSNBC claims that "members of the Saudi royal family met frequently with bin Laden—both before and after 9/11" [MSNBC, 9/5/03], and many Saudi royals and bin Laden relatives are being sued for their alleged role in 9/11. The *Boston Globe* opines that the flights occur "too soon after 9/11 for the FBI even to know what questions to ask, much less to decide conclusively that each Saudi [royal] and bin Laden relative [deserve] an 'all clear,' never to be available for questions again." [BOSTON GLOBE, 9/30/03] Senator Charles Schumer (D) says of the secret flights, "This is just another example of our country coddling the Saudis and giving them special privileges that others would never get. It's almost as if we didn't want to find out what links existed." [NEW YORK TIMES, 9/4/03]

Abdullah bin Laden, bin Laden family spokesman (not the Abdullah connected to WAMY)

September 20, 2001: Saudi Arabia Uncooperative in 9/11 Investigation

President Bush states: "Either you are with us, or you are against us. From this day forward, any nation that continues to harbor or support terrorism will be regarded by the United States as a hostile regime." [WHITE HOUSE, 9/20/01] Shortly thereafter, Bush says, "As far as the Saudi Arabians go, they've been nothing but cooperative," and "[Am] I pleased with the actions of Saudi Arabia? I am." However, several experts continue to claim Saudi Arabia is being "completely unsupportive" and is giving "zero cooperation" to the 9/11 investigation. Saudi Arabia refuses to help the U.S. trace the names and other background information on the 15 Saudi hijackers. One former U.S. official says, "They knew that once we started asking for a few traces the list would grow. . . . It's better to shut it down right away." [LOS ANGELES TIMES, 10/13/01; NEW YORKER, 10/16/01]

October 7, 2001: Bin Laden Brother's Ties to Terrorism Revealed

It is reported that Mahrous bin Laden, brother to Osama bin Laden, is currently manager of the Medina, Saudi Arabia branch of the bin Laden family company, the Binladin Group. In 1979, Binladin company trucks were used by 500 dissidents who seized the Grand Mosque in Mecca, Islam's holiest city. All the men who took part were later beheaded except Mahrous, who is eventually released from prison apparently because of the close ties between the bin Ladens and the Saudi royal family. The bin Laden family claims that no family members have any ties to terrorism except Osama. [SUNDAY HERALD, 10/7/01]

October 12, 2001: Additional Suspected Terrorist Supporters' Assets Frozen

The U.S. freezes the assets of 39 additional individuals and organizations connected to terrorism. Five of the names are al-Qaeda leaders on a list the United Nations published in March 8, 2001, with a recommendation that all nations freeze their assets. Other countries froze assets of those on that list, but the U.S. did not. "Members of Congress want to know why treasury officials charged with disrupting the finances of terrorists did not follow their lead." [ASSOCIATED PRESS,10/12/01; *GUARDIAN*, 10/13/01 (B)] The most detailed case is laid out against Saudi multimillionaire businessman Yassin al-Qadi. [*CHICAGO TRIBUNE*, 10/14/01; *CHICAGO TRIBUNE*, 10/29/01] Al-Qadi is "horrified and shocked" and offers to open his financial books to U.S. investigators. [*CHICAGO TRIBUNE*, 10/16/01] There have been several accusations that al-Qadi laundered money to fund Hamas [*CHICAGO TRIBUNE*, 10/16/01; *CHICAGO TRIBUNE*, 10/29/01], and an investigation into his al-Qaeda connections was canceled by higher-ups in the FBI in October 1998. Saudi Arabia also later freezes al-Qadi's accounts, an action the Saudis have taken against only three people, but he has yet to be charged or arrested by the Saudis or the U.S. [*NEWSWEEK*, 12/6/02]

October 14, 2001: Bin Laden Reportedly
"Has Supporters at All Levels of Saudi Arabia"

The *Boston Herald* reports: "Three banks allegedly used by Osama bin Laden to distribute money to his global terrorism network have well-established ties to a prince in Saudi Arabia's royal family, several billionaire Saudi bankers, and the governments of Kuwait and Dubai. One of the banks, Al-Shamal Islamic Bank in the Sudan, was controlled directly by bin Laden, according to a 1996 U.S. State Department report." A regional expert states, "I think we underestimate bin Laden. He comes from the highest levels of Saudi society and he has supporters at all levels of Saudi Arabia." [*BOSTON HERALD*, 10/14/01]

November 5, 2001: Has bin Laden Family Really Disowned Osama?

The *New Yorker* points to evidence that the bin Laden family has generally not ostracized itself from bin Laden as is popularly believed, but retains close ties in some cases. The large bin Laden family owns and runs a $5 billion a year global corporation that includes the largest construction firm in the Islamic world. One counterterrorism expert says, "There's obviously a lot of spin by the Saudi Binladin Group [the family corporation] to distinguish itself from Osama. I've been following the bin Ladens for years, and it's easy to say, 'We disown him.' Many in the family have. But blood is usually thicker than water." The article notes that neither the bin Laden family nor the Saudi royal family have publicly denounced bin Laden since 9/11. [*NEW YORKER*, 11/5/01]

November 28, 2001: Bin Laden Family Business Valued at $36 Billion

The *Financial Times* estimates that the bin Laden family's business, the Saudi Binladin Group, is worth about $36 billion. Osama bin Laden inherited about $300 million at the age of ten on the death of his father, but he may be worth much more today. While he spends large amounts each month supporting terror, he reportedly gets large amounts from rich Saudis every month to make up for the losses. [FINANCIAL TIMES, 11/28/01] The 9/11 Commission later disputes these figures and claims that bin Laden only gets about $1 million a year for about two decades until around 1994. [9/11 COMMISSION REPORT, 6/16/04 (B)]

Early December 2001: Bush Officials Again Look for Saudi Cooperation

Bush administration officials go to Saudi Arabia in a second attempt to obtain Saudi government cooperation in the 9/11 investigation. The Saudis have balked at freezing assets of organizations linked to bin Laden. Shortly thereafter, the *Boston Herald* runs a series of articles on the Saudis, citing an expert who says, "If there weren't all these other arrangements—arms deals and oil deals and consultancies—I don't think the U.S. would stand for this lack of cooperation." Another expert states that "it's good old fashioned 'I'll scratch your back, you scratch mine.' You have former U.S. officials, former presidents, aides to the current president, a long line of people who are tight with the Saudis . . . We are willing to basically ignore inconvenient truths that might otherwise cause our blood to boil." These deals are worth an incredible amount of money; one *Washington Post* reporter claims that prior to 1993, U.S. companies spent $200 billion on Saudi Arabia's defenses alone. [BOSTON HERALD, 12/10/01; BOSTON HERALD, 12/11/01 (B); PBS FRONTLINE, 2/16/93]

December 4, 2001: Largest U.S. Islamic Charity Assets Are Frozen

Holy Land Foundation for Relief and Development, the largest Islamic charity in the U.S., has its assets frozen by the Treasury Department. [CNN, 12/4/01; JERUSALEM POST, 12/5/01] Foundation offices in San Diego, California; Paterson, New Jersey; and Bridgeview, Illinois are also raided. [CNN, 12/4/01 (B)] Holy Land is represented by the powerful law firm of Akin, Gump, Strauss, Hauer & Feld. Three partners at Akin, Gump are very close to President Bush: George R. Salem chaired Bush's 2000 campaign outreach to Arab-Americans; Barnett A. "Sandy" Kress was appointed by Bush as an "unpaid consultant" on education reform, and has an office in the White House; and James C. Langdon is one of Bush's closest Texas friends. The firm has also represented Khalid bin Mahfouz. [BOSTON HERALD, 12/11/01; WASHINGTON POST, 12/17/01]

March 2002: Al-Qaeda "Golden Chain" Financiers Revealed

Authorities in Bosnia, Sarajevo, raid the offices of the Benevolence International Foundation due to suspected funding of al-Qaeda. The raid uncovers a handwritten list of twenty wealthy donors sympathetic to al-Qaeda. The list, referred to as "The Golden Chain," reveals both the names of the donors, and the names of the recipients. Seven of the payments were made to bin Laden, and at least one donation to him was made by the "bin Laden brothers." UPI points out that "the discovery of this document in Sarajevo calls into question whether al-Qaeda has received support from one of Osama's scores of wealthy brothers." Adel Batterjee, a wealth businessman from Jeddah, Saudi Arabia, who is also the founder of both the charity BFI and its predecessor, Lajnatt Al-Birr Al-Islamiah, appears to be mentioned as a recipient three times. Batterjee has been named as a defendant in a 9/11 lawsuit against wealthy Saudis, but he has been missing in Saudi Arabia for over a year by the time this report comes out. When the document was written is not revealed, but it may date from the early 1990s. [UPI, 2/11/03]

March 28, 2002: Al-Qaeda Leader Zubaida Is Captured

FBI agents and Pakistani police commandos raid a house in the city of Faisalabad, Pakistan, and capture al-Qaeda leader Abu Zubaida. He's shot three times but survives. [NEW YORK TIMES, 4/14/02] Many documents are found that lead to the indictment of 100 more people. [NEWSWEEK, 9/4/02] U.S. intelligence found his location by tracing his phone calls. [NEW YORK TIMES, 4/14/02] He has since given the U.S. useful information on 9/11 and other al-Qaeda plans. [NEWSWEEK, 9/4/02] Zubaida is considered one of the highest in al-Qaeda's leadership and one of the highest ranking prisoners captured by the U.S. [NEW YORK TIMES, 4/14/02] It is believed that 9/11 mastermind Khalid Shaikh Mohammed takes over Zubaida's tasks. [ASIA TIMES, 9/11/02] The claim is later made that during interrogation Zubaida claims ties to people high in the Saudi and Pakistani governments.

March 29, 2002: Bin Laden Family Denies Terrorist Connections

Abdullah bin Laden, spokesman for the bin Laden family and one of Osama's many brothers, speaks directly to the press for the first time since 9/11. He says that the family cut all personal and financial ties to Osama in 1993 and that no family member has contact with him or provides any kind of support for him. "We went through a tough time. It was difficult. We felt we are a victim as well." [ABC NEWS, 3/29/02]

March 31, 2002: Al-Qaeda Leader Zubaida
Allegedly Incriminates Saudi Princes

Author Gerald Posner, controversial for his books dismissing JFK assassination and other conspiracy theories, claims a remarkable interrogation of al-Qaeda prisoner Abu Zubaida begins at this time. Zubaida, arrested three days earlier, is tricked into thinking the U.S. has handed him to the Saudis for a more brutal interrogation, but in fact the Saudis are still American agents. Zubaida expresses great relief at this, and under the influence of the "truth serum" sodium pentothal, tells his interrogators to call Prince Ahmed bin Abdul-Aziz, a nephew of the Saudi king. He provides telephone numbers from memory and says, "He will tell you what to do." He proceeds to give more information and phone numbers, claiming ties with higher ups in both the Saudi and Pakistani governments. He also names:

- Pakistani air force chief Mushaf Ali Mir, said to be closely tied to the fundamentalists in the ISI

- Saudi Intelligence Minister Prince Turki al-Faisal

- Prince Sultan bin Faisal, another nephew of the Saudi King

- Prince Fahd bin Turki, another member of the Saudi royalty

According to Posner, Zubaida claims that all of these people were intermediaries he dealt with in the frequent transfer of money to al-Qaeda. The phone numbers and other details are consistent with information already known by U.S. intelligence. According to Zubaida, he was present in a meeting in 1996 where the Pakistanis and the Saudis struck a deal with bin Laden, promising him protection, arms, and supplies in exchange for not being the targets of future terror attacks. He claimed both governments were told the U.S. would be attacked on 9/11, but not given the details of how the attack would work. Within months, all of the people named by Zubaida die mysteriously except for Prince Turki, who is made an ambassador, giving him diplomatic immunity. [WHY AMERICA SLEPT, BY GERALD POSNER, 9/03, PP. 186–94] Posner says he learned this story from two unnamed U.S. government sources who gave similar, independent accounts. One is from the CIA and the other is a senior Bush administration official "inside the executive branch." [SALON, 10/18/03] With the notable exception of a prominent *Time* magazine article [TIME, 8/31/03], few news outlets cover the story [SALON, 10/18/03; MSNBC, 9/5/03; ASIA TIMES, 9/17/03 (B)], and some that do cover it only in the form of book reviews. [WASHINGTON POST, 9/10/03 (C); NEW YORK TIMES, 10/12/03; NEW YORK TIMES, 10/29/03 (B)] Some experts put forth the theory that the story could be made up by neoconservatives interested in starting a war with Saudi Arabia. It is also possible Zubaida mixed facts with lies, as he was found to lie to interrogators on many other occasions. [SALON, 10/18/03] There is also speculation that the gist of the story may be true, but that Zubaida's Saudi and Pakistani contacts may have been pinned on dead men to protect the actual guilty parties. [ASIA TIMES, 9/17/03; SALON, 10/18/03]

June 13, 2002: In Arrest of al-Qaeda Strongman, Sudan Is Helpful; Saudi Arabia Is Not

Sudan arrests an al-Qaeda leader who has confessed to firing a missile at a U.S. plane taking off from Prince Sultan Air Base, Saudi Arabia, in May. Saudi Arabia had failed to arrest him. This is just the latest in a series of events where "some countries long deemed key U.S. allies—such as Saudi Arabia—are considered less than helpful in the war against terror, while other states remaining on the U.S. State Department's blacklist of terrorist sponsors, such as Syria and Sudan, are apparently proving more cooperative than their pariah status would suggest." The U.S. hasn't been given access to al-Qaeda members arrested by Saudi Arabia, and "concerns over the Saudi authorities' 'unhelpful' stance are increasing." [JANE'S INTELLIGENCE REVIEW, 7/5/02]

July 10, 2002: U.S. Military Alleges Saudis Are Heavily Involved in Terrorism

A briefing given to a top Pentagon advisory group states, "The Saudis are active at every level of the terror chain, from planners to financiers, from cadre to foot-soldier, from ideologist to cheerleader . . . Saudi Arabia supports our enemies and attacks our allies." They are called "the kernel of evil, the prime mover, the most dangerous opponent." This position still runs counter to official U.S. policy, but the *Washington Post* says it "represents a point of view that has growing currency within the Bush administration." The briefing suggests that the Saudis be given an ultimatum to stop backing terrorism or face seizure of its oil fields and its financial assets invested in the United States. The group, the Defense Policy Board, is headed by Richard Perle. [WASHINGTON POST, 8/6/02] An international controversy follows the public reports of the briefing in August 2002 (for instance, [SCOTSMAN, 8/12/02]). In an abrupt change, the media starts calling the Saudis enemies, not allies of the U.S. Slate reports details of the briefing the Post failed to mention. The briefing states, "There is an 'Arabia,' but it needs not be 'Saudi.'" The conclusion of the briefing: "Grand strategy for the Middle East: Iraq is the tactical pivot. Saudi Arabia the strategic pivot. Egypt the prize." [SLATE, 8/7/02] Note that a similar meeting of the Defense Policy Board appears to have preceded and affected the U.S.'s decision to take a warlike stance against Iraq (see chapter 15).

July 19, 2002: U.S. Finally Ends Controversial Visa Program in Saudi Arabia

Faced with growing criticism of its Visa Express program, the State Department decides merely to change the name of the program in early July 2002. When that fails to satisfy critics, the program is abandoned altogether on July 19. The Visa Express program allowed anyone in Saudi Arabia to apply for U.S. visas through their travel agents instead of having to show up at a consulate in person. [WASHINGTON POST, 7/11/02 (B)] Mary Ryan, the head of the State Department's consular service that was responsible for letting most of the hijackers into the U.S., is also forced to retire. It has been pointed out that Ryan deceived Congress by testifying that "there was nothing State could have done to prevent the terrorists from obtaining visas." However, after all this, Ryan and the other authors of the Visa Express program are given "outstanding performance" awards of $15,000 each. The reporter who wrote most of the stories critical of Visa Express is briefly detained and pressured by the State Department. [WASHINGTON TIMES, 10/23/02; PHILADELPHIA DAILY NEWS, 12/30/02]

July 22, 2002–February 20, 2003: Saudi Princes and
Pakistani General Later Implicated in 9/11 Plot Die Mysteriously

Three prominent members of the Saudi royal family die in mysterious circumstances. Prince Ahmed bin Abdul-Aziz, a nephew of the Saudi king, prominent businessman, and owner of the winning 2002 Kentucky Derby horse, is said to die of a heart attack at the age of 43. The next day, Prince Sultan bin Faisal, another nephew of the king, dies driving to Prince Ahmed's funeral. A week later, Prince Sultan bin Faisal supposedly "dies of thirst" in the Arabian desert. Seven months later, on February 20, 2003, Pakistan's air force chief Mushaf Ali Mir, dies in a plane crash in clear weather, along with his wife and closest confidants. Controversial author Gerald Posner implies that all of these events are linked together and the deaths not accidental, but have occurred because of the testimony of captured al-Qaeda leader Abu Zubaida in March 2002. The deaths all occurred not long after the respective gov-

Mushaf Ali Mir

ernments were told of Zubaida's confessions. Only one other key figure named by Zubaida remains alive: Saudi Intelligence Minister Prince Turki al-Faisal. Posner says, "He's the J. Edgar Hoover of Saudi Arabia," too powerful and aware of too many secrets to be killed off. Prince Turki lost his Intelligence

Minister job ten days before 9/11, and is later made Saudi ambassador to Britain, giving him diplomatic immunity from any criminal prosecution. [WHY AMERICA SLEPT, BY GERALD POSNER, 9/03, PP. 190–94; TIME, 8/31/03]

August 15, 2002: Relatives File Lawsuit Against Alleged Saudi al-Qaeda Financiers

More than 600 relatives of victims of the 9/11 attacks file a 15-count, $1 trillion lawsuit against various parties they accuse of financing al-Qaeda and Afghanistan's former Taliban regime. (The number of plaintiffs will increase to 2,500. Up to 10,000 were eligible to join this suit. [NEWSWEEK, 9/13/02]) The defendants include the Binladin Group (the company run by Osama bin Laden's family), seven international banks, eight Islamic foundations and charities, individual terrorist financiers, three Saudi princes, and the government of Sudan. [CNN, 8/15/02; WASHINGTON POST, 8/16/02] Individuals named include Saudi Defense Minister Prince Sultan, former Saudi intelligence chief Prince Turki al-Faisal, Yassin al-Qadi, and Khalid bin Mahfouz. [ASSOCIATED PRESS, 8/15/02; MSNBC, 8/25/02] "The attorneys and investigators were able to obtain, through French intelligence, the translation of a secretly recorded meeting between representatives of bin Laden and three Saudi princes in which they sought to pay him hush money to keep him from attacking their enterprises in Saudi Arabia." [CNN, 8/15/02] The plaintiffs also accuse the U.S. government of failing to pursue such institutions thoroughly enough because of lucrative oil interests. [BBC, 8/15/02] Ron Motley, the lead lawyer in the suit, says the case is being aided by intelligence services from France and four other foreign governments, but no help has come from the Justice Department. [MINNEAPOLIS STAR-TRIBUNE, 8/16/02] The plaintiffs acknowledge the chance of ever winning any money is slim, but hope the lawsuit will help bring to light the role of Saudi Arabia in the 9/11 attacks. [BBC, 8/15/02] A number of rich Saudis respond by threatening to withdraw hundreds of billions of dollars in U.S. investments if the lawsuit goes forward. [DAILY TELEGRAPH, 8/20/02]

August 15, 2002 (B): Son of Saudi Official Associated with Hijackers

The picture of a young Saudi man named Saud al-Rashid is discovered on a CD-ROM that also contains the pictures of four 9/11 hijackers in an al-Qaeda safe house in Pakistan. A senior U.S. official says that investigators "were able to take this piece of information and it showed clear signals or lines that he was connected to 9/11." [ASSOCIATED PRESS, 8/21/02 (B)] Rashid was in Afghanistan in 2000 and 2001. [NEW YORK TIMES, 7/29/03 (B)] Six days later, the U.S. issues a worldwide dragnet to find him. [ASSOCIATED PRESS, 8/21/02 (B)] But they are unable to catch him because a few days later, he flees from Egypt to Saudi Arabia and turns himself in to the Saudi authorities. The Saudis apparently will not try him for any crime or allow the FBI to interview him. [CNN, 8/26/02; CNN, 8/31/02] Intriguingly, Al-Rashid's father is Hamid al-Rashid, a Saudi government official who paid a salary to Omar al-Bayoumi, an associate of two hijackers who is later suspected of being a Saudi agent. [NEW YORK TIMES, 7/29/03 (B)]

August 20, 2002: Saudis Retract Billions from U.S. in Response to 9/11 Lawsuit

The *Financial Times* reports that "disgruntled Saudis have pulled tens of billions of dollars out of the U.S., signaling a deep alienation from America." Estimates range from $100 billion to over $200 billion.

Part of the anger is in response to reports that the U.S. might attack Saudi Arabia and freeze Saudi assets unless Saudi Arabia fights terrorism more effectively. It is also in response to a lawsuit against many Saudi Arabians that also may lead to a freeze of Saudi assets. Estimates of total Saudi investments in the U.S. range from $400 billion to $600 billion. [FINANCIAL TIMES, 8/20/02]

August 25, 2002: CIA Collects No Intelligence About Saudi Arabia

Former CIA agent Bob Baer says the U.S. collects virtually no intelligence about Saudi Arabia nor are they given any intelligence collected by the Saudis. He says this is because there are implicit orders from the White House, "Do not collect information on Saudi Arabia because we're going to risk annoying the royal family." In the same television program, despite being on a U.S. terrorist list since October 2001, Saudi millionaire Yassin al-Qadi says, "I'm living my life here in Saudi Arabia without any problem" because he is being protected by the Saudi government. Al-Qadi admits to giving bin Laden money for his "humanitarian" work, but says this is different from bin Laden's terrorist work. Presented with this information, the U.S. Treasury Department only says that the U.S. "is pleased with and appreciates the actions taken by the Saudis" in the war on terror. The Saudi government still has not given U.S. intelligence permission to talk to any family members of the hijackers, even though some U.S. journalists have had limited contact with a few. [MSNBC, 8/25/02]

August 27, 2002: Close Relationship Between
Saudi Ambassador and Bush Raise Questions

Prince Bandar, Saudi ambassador to the U.S., meets privately for more than an hour with President Bush and National Security Adviser Rice in Crawford, Texas. [DAILY TELEGRAPH, 8/28/02] Press Secretary Ari Fleischer characterizes it as a warm meeting of old friends. Bandar, his wife (Princess Haifa), and seven of their eight children stay for lunch. [FOX NEWS, 8/27/02] Prince Bandar, a long-time friend of the Bush family, donated $1 million to the George W. Bush Presidential Library in College Station, Texas. [BOSTON HERALD, 12/11/01 (B), BUSH LIBRARY] This relationship later becomes news when it is learned that Princess Haifa gave between $51,000 and $73,000 to two Saudi families in California who may have financed two of the 9/11 hijackers (see chapter 7). [NEW YORK TIMES, 11/23/02; MSNBC, 11/27/02]

**President Bush and
Prince Bandar meeting
in Crawford, Texas**

September 17, 2002: Tie Between
9/11 Arrest and bin Laden Family Members

CBS reports that in the days after the arrest of Ramzi bin al-Shibh and four other al-Qaeda operatives in Pakistan on September 11, 2002, "a search of the home of the five al-Qaeda suspects turned up passports belonging to members of the family of Osama bin Laden." No more details, such as which family members, or why bin al-Shibh's group had these passports, is given. [CBS NEWS, 9/17/02]

September 20, 2002: Saudi Charity in Bosnia Linked to al-Qaeda

A Bosnian government probe connects the Saudi charity Talibah International Aid Association to terrorist funding and an al-Qaeda front group. Talibah has been under investigation since shortly after 9/11 due to a foiled terror attack in Bosnia that has been connected to Talibah and al-Qaeda. Abdullah bin Laden (his cousin and not the bin Laden family spokesman of the same name), is a Talibah officer in its Virginia office. An investigation into Abdullah bin Laden was cancelled in September 1996. The U.S. has been criticized for failing to list Talibah as a sponsor of terrorism and for not freezing its assets. [WALL STREET JOURNAL, 9/20/02]

October 18, 2002: Saudi Acquaintance of
bin Laden Given Immunity by Becoming Ambassador

Saudi Arabia announces that Turki al-Faisal will be its next ambassador to Britain. Turki is a controversial figure because of his long-standing relationship to bin Laden. He has also been named in a lawsuit by 9/11 victims' relatives against Saudi Arabians for their support of al-Qaeda before 9/11. It is later noted that his ambassador position could give him diplomatic immunity from the lawsuit. [NEW YORK TIMES, 12/30/02] Turki's predecessor as ambassador was recalled after it was revealed he had written poems praising suicide bombers. [OBSERVER, 3/2/03 (C)] Articles reporting on his new posting suggest that Turki last met bin Laden in the early 1990s, before bin Laden became a wanted terrorist. [LONDON TIMES, 10/18/02; GUARDIAN, 10/19/02] However, these reports fail to mention other reported contacts with bin Laden, including a possible secret meeting in July 2001.

November 1, 2002: 9/11 Victims' Relatives
Protest Against Opposition to Sue Saudis

Some of the 9/11 victims' relatives hold a rally at the U.S. Capitol to protest what they fear are plans by the Bush administration to delay or block their lawsuit against prominent Saudi individuals for an alleged role in financing al-Qaeda. [WASHINGTON POST, 11/1/02] U.S. officials say they have not decided whether to submit a motion seeking to block or restrict the lawsuit, but they are concerned about the "diplomatic sensitivities" of the suit. Saudis have withdrawn hundreds of billions of dollars from the U.S. in response to the suit. The *Guardian* previously reported that "some plaintiffs in the case say the Bush administration is pressuring them to pull out of the lawsuit in order to avoid damaging U.S.-Saudi relations, threatening them with the prospect of being denied any money from the government's own compensation scheme if they continue to pursue it. Bereaved relatives who apply to the federal compensation scheme must, in any case, sign away their rights to sue the government, air carriers in the U.S., and other domestic bodies—a condition that has prompted some of them to call the government compensation 'hush money.' The fund is expected, in the end, to pay out $4 billion. They remain, however, free to sue those they accuse of being directly responsible for the attacks, such as bin Laden, and—so they thought—the alleged financers of terrorism." [GUARDIAN, 9/20/02]

November 22, 2002: More Defendants Added to Saudi Lawsuit

9/11 victims' relatives add 50 defendants to the 100 defendants previously named in their $1 trillion lawsuit against Saudi citizens and organizations. New defendants include Saudi Interior Minister Prince Naif and the Saudi American Bank, that nation's second largest financial institution. The suit alleges that Naif and Minister of Defense and Aviation Prince Sultan bin Abdulaziz al-Saud engaged in payoffs to al-Qaeda. The suit further alleges that the Saudi American Bank, partly owned and managed by Citibank, financed development projects in Sudan benefiting bin Laden in the early 1990s when he was living there. [WALL STREET JOURNAL, 11/22/02]

November 26, 2002: Secret List of Saudi Terror Financiers Is Revealed

In the wake of news that two Saudis living in San Diego, California may have helped two of the 9/11 hijackers, reports surface that the U.S. has a secret, short list of wealthy individuals who are the alleged key financiers of al-Qaeda and other terrorist groups. The *Washington Post* claims there are nine names on the list: seven Saudi, plus one from Egypt, and one from Pakistan. [WASHINGTON POST, 11/26/02] ABC News claims the list consists of 12 names, all Saudis, and says they were financing al-Qaeda through accounts in Cyprus, Switzerland, and Malaysia, among other countries. [ABC NEWS, 11/25/02] They also claim the Saudi government has a copy of the list. U.S. officials privately say all the people listed have close personal and business ties with the Saudi royal family. [ABC NEWS, 11/26/02] French investigator Jean-Charles Brisard names seven prominent Saudi financiers of terror; the number matches the seven Saudis mentioned in the *Washington Post* article, though it's not known if all the names are the same. The Saudis mentioned by Brisard include Yassin al-Qadi and Wa'el Hamza Julaidan (who has had his assets frozen by the U.S. [STATE DEPARTMENT, 9/6/02]). Brisard says al-Qaeda has received between $300 million and $500 million over the last ten years from wealthy businessmen and bankers. He claims that the combined fortunes of these men equal about 20 percent of Saudi Arabia's GDP (gross domestic product). [LOS ANGELES TIMES, 12/24/02; UN REPORT, 12/19/02 OR UN REPORT, 12/19/02 (B)] However, Brisard's study has been mistakenly described as a United Nations report. While he submitted the study to the UN, the UN didn't request it. [MONEY LAUNDERING ALERT, 10/03] It is also reported that a National Security Council task force recommends that the U.S. demand that Saudi Arabia crack down on terrorist financiers within 90 days of receiving evidence of misdeeds and if they do not, the U.S. should take unilateral action to bring the suspects to justice. [WASHINGTON POST, 11/26/02] However, the U.S. denies this, calling Saudi Arabia a "good partner in the war on terrorism." [WASHINGTON POST, 11/26/02] Press Secretary Ari Fleischer says: "I think the fact that many of the hijackers came from that nation [Saudi Arabia] cannot and should not be read as an indictment of the country." [RADIO FREE EUROPE, 11/27/02]

December 3, 2002: Trial of Motassadeq
Reveals Connections with Saudi Embassy Official

During the German trial of Mounir Motassadeq, accused of participation in the 9/11 attacks, a German police officer testifies the business card of Muhammad J. Fakihi, the chief of Islamic affairs at the Saudi Embassy in Berlin, was found in a raid on Motassadeq's apartment. The raid also turned up a credit card belonging to Mohamed Atta and the password to Atta's e-mail account. [INTERNATIONAL HERALD TRIBUNE, 12/4/02; SYDNEY MORNING HERALD, 12/5/02; NEW YORK TIMES, 12/8/02; NEWSWEEK 12/9/02; DER SPIEGEL, 3/26/03; NEW YORK TIMES, 4/25/03] Saudi officials deny that Fakihi had ever met Motassadeq. Fakihi is recalled to Saudi Arabia three months later, following demands by Germany that he leaves. [NEW YORK TIMES, 4/25/03]

December 5, 2002: Software Company with Access to Government Secrets May Have al-Qaeda Ties

Federal agents search the offices of Ptech, Inc., a computer software company in Quincy, Massachusetts, looking for evidence of links to bin Laden. A senior Ptech official confirms that Yassin al-Qadi, one of 12 Saudi businessmen on a secret CIA list suspected of funnelling millions of dollars to al-Qaeda, was an investor in the company, beginning in 1994. [NEWSWEEK, 12/6/02; WBZ4, 12/9/02] Some of Ptech's customers include the White House, Congress, Army, Navy, Air Force, NATO, FAA, FBI, and the IRS. [BOSTON GLOBE, 12/7/02] A former FBI counterterrorism official states, "For someone like [al-Qadi] to be involved in a capacity, in an organization, a company that has access to classified information, that has access to government open or classified computer systems, would be of grave concern." Yacub Mirza—"a senior official of major radical Islamic organizations that have been linked by the U.S. government to terrorism"—has recently been on Ptech's board of directors. Hussein Ibrahim, the Vice President and Chief Scientist of Ptech, was vice chairman of a now defunct investment group called BMI. An FBI affidavit names BMI as a conduit to launder money from al-Qadi to Hamas terrorists. [WBZ4, 12/9/02] The search into Ptech is part of Operation Green Quest, which has served 114 search warrants in the past 14 months involving suspected terrorist financing. Fifty arrests have been made and $27.4 million seized. [FORBES, 12/6/02] Al-Qadi's assets were frozen by the FBI in October 2001. [ARAB NEWS, 9/26/02] That same month, a number of Ptech employees told the Boston FBI that Ptech was financially backed by al-Qadi, but the FBI did little more than take their statements. A high level government source claims the FBI did not convey the information to a treasury department investigation of al-Qadi, and none of the government agencies using Ptech software were warned. Indira Singh, a second whistleblower, spoke with the FBI in June 2002 and was "shocked" and "frustrated" when she learned the agency had done nothing. [BOSTON GLOBE, 12/7/02; WBZ4, 12/9/02] Beginning in mid-June 2002, WBZ-TV Boston had prepared an lengthy investigative report on Ptech, but withheld it for more than three months after receiving "calls from federal law enforcement agencies, some at the highest levels." The station claims the government launched its Ptech probe in August 2002, after they "got wind of our investigation" and "asked us to hold the story so they could come out and do their raid and look like they're ahead of the game." [BOSTON GLOBE, 12/7/02 (B), WBZ4, 12/9/02]

December 18, 2002: Family Members Arrested on Suspicion of Sponsoring Terrorism

Four brothers in Texas—Ghassan Elashi, Bayan Elashi, Hazim Elashi, and Basman Elashi—are arrested on charges of conspiracy, money laundering, dealing in the property of a designated terrorist, illegal export and making false statements. [ASSOCIATED PRESS, 12/18/02 (B), WASHINGTON POST, 12/19/02] Ghassan Elashi is the vice president of InfoCom Corporation, which was raided on September 5, 2001 by 80 members of a Joint Terrorism Task Force. [GUARDIAN, 9/10/01] He is also chairman of Holy Land Foundation for Relief and Development, which had its assets frozen by the FBI in December 2001. [ASSOCIATED PRESS, 12/23/02] The 33-count indictment names a fifth brother, Ihsan Elashi, who is already in custody, as well as Mousa Abu Marzook, whom the U.S. deported in 1997, and his wife Nadia Elashi (both believed to be in the Middle East). [BBC, 12/18/01] On December 20, a judge rules that Bayan and Hazim Elashi must remain in federal detention, but frees Basman Elashi on $15,000 bail. [ASSOCIATED PRESS, 12/23/02] Ghassan Elashi is released without bond, but will wear an electronic monitor. [ASSOCIATED PRESS, 12/23/02]

May 12, 2003: Saudi Arabia Bombing
Changes Saudi Stance Toward al-Qaeda

Saudi Arabia is attacked by three suicide bombings in the capital of Riyadh. At least 34 people are killed. The Saudi royal family had taken very little action against al-Qaeda prior to this. However, it appears to more aggressively combat al-Qaeda afterward. [LOS ANGELES TIMES, 6/14/04]

August 2003: FBI and CIA Reopen
Investigation Into Saudi Link to 9/11

In the wake of the 9/11 Congressional Inquiry report, and "under intense pressure from Congress," as the *Boston Globe* puts it, the FBI and CIA reopen an investigation into whether Saudi Arabian officials aided the 9/11 plot. [BOSTON GLOBE, 8/3/03] In early August, Saudi Arabia allows the FBI to interview Omar al-Bayoumi. However, the interview takes place in Saudi Arabia, and apparently on his terms, with Saudi government handlers present. [NEW YORK TIMES, 8/5/03; ASSOCIATED PRESS, 8/6/03] Says one anonymous government terrorism consultant, "They are revisiting everybody. The [FBI] did not do a very good job of unraveling the conspiracy behind the hijackers." [BOSTON GLOBE, 8/3/03] But by September, the *Washington Post* reports that the FBI has concluded that the idea al-Bayoumi was a Saudi government agent is "without merit and has largely abandoned further investigation . . . The bureau's September 11 investigative team, which is still tracking down details of the plot, has reached similar conclusions about other associates named or referred to in the congressional inquiry report." [WASHINGTON POST, 9/10/03] Yet another article claims that by late August, some key people who interacted with al-Bayoumi have yet to be interviewed by the FBI. "Countless intelligence leads that might help solve" the mystery of a Saudi connection to the hijackers "appear to have been underinvestigated or completely overlooked by the FBI, particularly in San Diego." [SAN DIEGO MAGAZINE, 9/03] Not only were they never interviewed when the investigation was supposedly reopened, they were not interviewed in the months after 9/11 either, when the FBI supposedly opened an "intense investigation" of al-Bayoumi, visiting "every place he was known to have gone, and [compiling] 4,000 pages of documents detailing his activities." [NEWSWEEK, 7/28/03]

August 1–3, 2003: Leaks Hint at Saudi Involvement in 9/11

In the wake of the release of the 9/11 Congressional Inquiry's full report, anonymous officials leak some details from a controversial, completely censored 28-page section that focuses on possible Saudi support for 9/11. According to leaks given to the *New York Times*, the section says that Omar al-Bayoumi and/or Osama Basnan "had at least indirect links with two hijackers [who] were probably Saudi intelligence agents and may have reported to Saudi government officials." It also says that Anwar Al Aulaqi "was a central figure in a support network that aided the same two hijackers." Most connections drawn in the report between the men, Saudi intelligence, and 9/11 is said to be circumstantial. [NEW YORK TIMES, 8/2/03] One key section is said to read, "On the one hand, it is possible that these kinds of connections could suggest, as indicated in a CIA memorandum, 'incontrovertible evidence that there is support for these terrorists . . . On the other hand, it is also possible that further investigation of these allegations could reveal legitimate, and innocent, explanations for these associations.'" Some of the most sensitive information involves what U.S. agencies are doing currently to investigate Saudi business figures and organizations. [ASSOCIATED PRESS, 8/2/03] According to the *New Republic*, the section outlines "connections between the hijacking plot and the very top levels of the Saudi royal family." An anonymous official is quoted as saying, "There's a lot more in the 28 pages than money. Everyone's chasing the charities. They should be chasing direct links to high levels of the Saudi government. We're not talking about rogue elements. We're talking about a coordinated network that reaches right from the hijackers to multiple places in the Saudi government. . . . If the people in the administration trying to link Iraq to al-Qaeda had one–one-thousandth of the stuff that the 28 pages has linking a foreign government to al-Qaeda, they would have been in good shape. . . . If the 28 pages were to be made public, I have no question that the entire relationship with Saudi Arabia would change overnight." [NEW REPUBLIC, 8/1/03] The section also is critical that the issue of foreign government support remains unresolved. One section reads, "In their testimony, neither CIA or FBI officials were able to address definitely the extent of such support for the hijackers, globally or within the United States, or the extent to which such support, if it exists, is knowing or inadvertent in nature. This gap in intelligence community coverage is unacceptable." [BOSTON GLOBE, 8/3/03]

November 14, 2003: Two Saudi
Princes Given Immunity from Lawsuit

A federal judge rules that two Saudi Arabian princes have immunity from a 9/11 lawsuit filed on August 15, 2002 because they are foreign leaders. The two Saudis are Prince Turki al-Faisal, former head of Saudi intelligence and now the Saudi ambassador to Great Britain, and Prince Sultan bin Abdul Aziz, the Saudi defense minister. The judge writes in his ruling, "whatever their actions, they were performed in their official (government) capacities." [ASSOCIATED PRESS, 11/16/03; CHARLESTON POST AND COURIER, 11/18/03] Prince Sultan is represented in the case by three lawyers from prestigious Dallas-based law firm, Baker and Botts. James Baker, former Secretary of State and close associate of the Bush family, is one of the senior partners of the law firm. [NEWSWEEK, 4/16/03] However, another ruling in July 2003 has already allowed charges against other prominent Saudis and Saudi organizations to proceed. [CHARLESTON POST AND COURIER, 8/7/03; STATEN ISLAND ADVANCE, 2/20/04]

March 2004: German Intelligence Links Suspected
Hamburg al-Qaeda Associates to Saudi Arabian Intelligence Agency

According to German intelligence, two private Saudi companies, The Twaik Group and Rawasin Media Productions, linked with suspected al-Qaeda cells in Germany, have connections to the Saudi Arabian intelligence agency and its longtime chief, Prince Turki bin Faisal. Both companies, based in Riyadh, have served as fronts for the Saudi General Intelligence Directorate. The Twaik Group deposited more than $250,000 in bank accounts controlled by Mamoun Darkazanli, who had been under CIA surveillance for nearly a decade while living in Hamburg, Germany, and who is suspected of being a longtime al-Qaeda member with close ties to the 9/11 hijackers who were members of the Hamburg al-Qaeda cell (see chapter 6). Darkazanli remains free in Germany, although he is reportedly being kept under surveillance and continues to be the target of a German investigation into the suspected laundering of al-Qaeda funds. [CHICAGO TRIBUNE, 3/31/04]

June 14, 2004: Some 9/11 Commissioners and Investigators
See Deep Collaboration Between al-Qaeda, Saudis, and Pakistanis

The *Los Angeles Times* reports that, according to some 9/11 Commission members and U.S. counter-terrorism officials, Pakistan and Saudi Arabia helped set the stage for the 9/11 attacks by cutting secret deals with the Taliban and bin Laden. These deals date to at least 1996, and appear to have shielded both countries from al-Qaeda attacks until long after 9/11. "Saudi Arabia provided funds and equipment to the Taliban and probably directly to Bin Laden, and didn't interfere with al-Qaeda's efforts to raise money, recruit and train operatives, and establish cells throughout the kingdom, commission and U.S. officials said. Pakistan provided even more direct assistance, its military and intelligence agencies often coordinating efforts with the Taliban and al-Qaeda, they said." The two countries become victims of terrorist attacks only after they launch comprehensive efforts to eliminate domestic al-Qaeda cells. In Saudi Arabia, such efforts didn't begin until late 2003. 9/11 Commission Bob Kerrey says, "There's no question the Taliban was getting money from the Saudis . . . and there's no question they got much more than that from the Pakistani government. Their motive is a secondary issue for us." He adds that the commission's findings are based almost entirely on information known to the U.S. government before 9/11. "All we're doing is looking at classified documents from our own government, not from some magical source. So we knew what was going on, but we did nothing." The *Los Angeles Times* comments the publicly-released 9/11 Commission staff statement of this matter only touches the "tip of the iceberg." [LOS ANGELES TIMES, 6/14/04] In fact, the commission reports, "There is no convincing evidence that any government financially supported al-Qaeda before 9/11." [9/11 COMMISSION REPORT, 6/16/04 (B)]

July 11, 2004: Saudi-Terrorist Link Continues?

Senator Charles Schumer (D-NY) claims that Saudi leaders and members of the Saudi royal family continue to fund terrorist schools and groups in the U.S. He calls on the Bush administration to cut U.S. ties with Saudi Arabia, and says, "There's been much too close a relationship between Saudi royal family, the White House, and big oil. We have to be much tougher with the Saudis." [ASSOCIATED PRESS, 7/11/04]

14.

Israel

There is a large amount of evidence that fundamentalist Muslims played important roles in the 9/11 attacks, such as the many warnings and examples of foreknowledge coming out of Afghanistan. The U.S. and Israel have been al-Qaeda's prime targets. Indeed, U.S. support of Israel is a primary reason al-Qaeda purports to have targeted the U.S. So any theories indicating an active Israeli role in 9/11 seem utterly preposterous.

Yet there may be *some* connection between Israel and 9/11. Multiple reports and U.S. government documents have indicated that, at a minimum, in the months before and after 9/11, a peculiar group of Israelis were living in cities and towns across the U.S. Posing as art students, these individuals were observed conducting surveillance of multiple federal buildings in the U.S., and seeking contacts with agents from multiple federal agencies, including the DEA and FBI. Many of the "students" were active or former Israeli military personnel with intelligence expertise. At least some of these "students" lived down the street from hijacker Mohamed Atta and three of his associates in Florida prior to 9/11. More than 100 of these "students" were deported prior to 9/11, and up to 60 were deported afterward. Although the U.S. government has consistently maintained that these individuals were deported for routine visa violations, internal documents leaked to the press indicate that at least some in the government suspected the students were part of an Israeli spy ring. [SALON, 5/7/02]

The most common theory offered to explain this story is that the spies were assisting the U.S.'s counterterrorism efforts, and that they actually warned the U.S. of the pending attack based on the intelligence they obtained, including names of some of the hijackers, in the weeks prior to 9/11.

Another theory is that the spies were in the U.S. for entirely different purposes, wholly unrelated to 9/11. [SALON, 5/7/02] Most controversially, some suggest that Israel knew about the 9/11 attacks but failed to warn the U.S., or even that Israel itself was behind the attacks. [REUTERS, 9/11/03; *NEW JERSEY STAR-LEDGER*, 12/29/03]

Although some news outlets have reported on this alleged Israeli "art student spy ring," such stories have generally failed to garner front-page coverage. The case of one Israeli spy in the U.S., Jonathan Pollard, probably generated more news coverage when he was arrested about 15 years earlier than the news coverage hundreds of "art student" spies combined have generated. Fox News gave into pressure and pulled its ground-breaking story on the spy ring from its website, and anonymous U.S. officials have warned that to speak of the topic could endanger their careers.

The actions of the Mossad, Israel's spy agency, are shrouded in secrecy and often hard to fathom. Ephraim Halevy, head of the Mossad at the time of 9/11, claims, "Not one big success of the Mossad has ever been made public." [CBS NEWS, 2/5/03] As with other issues related to 9/11, the purpose of this book is not to theorize, or speculate, but to compile the facts as they are known to date, based on what has been reported on this issue. The "art student spy ring" story remains one of the most baffling of the many 9/11 mysteries.

January 2000: Israeli Spy Ring Begins Penetrating U.S.

A DEA government document later leaked to the press [DEA REPORT, 6/01] suggests that a large Israeli spy ring starts penetrating the U.S. from at least this time, if not earlier. This ring, which will later become popularly known as the "art student spy ring," is later shown to have strange connections to the events of 9/11. [INSIGHT, 3/11/02]

February 6, 2000: Apparent Mossad Attempt to Infiltrate al-Qaeda Thwarted

India's largest newsweekly reports that it appears a recent Mossad attempt to infiltrate al-Qaeda failed when undercover agents were stopped on their way to Bangladesh by Indian customs officials. These 11 men appeared to be from Afghanistan, but had Israeli passports. One expert states, "It is not unlikely for Mossad to recruit 11 Afghans in Iran and grant them Israeli citizenship to penetrate a network such as bin Laden's. They would begin by infiltrating them into an Islamic radical group in an unlikely place like Bangladesh." [THE WEEK, 2/6/00]

**Mossad Director
Ephraim Halevy**

April 19, 2000: Reports Indicate Israeli Organized Crime Units Dominate Ecstasy Distribution

USA Today reports that "Israeli crime groups . . . dominate distribution" of Ecstasy. [USA TODAY, 4/19/00] The DEA also states that most of the Ecstasy sold in the U.S. is "controlled by organized crime figures in Western Europe, Russia, and Israel." [UPI, 10/25/01] According to DEA documents, the Israeli "art student spy ring" "has been linked to several ongoing DEA [Ecstasy] investigations in Florida, California, Texas, and New York now being closely coordinated by DEA headquarters." [INSIGHT, 3/11/02]

December 2000–April 2001: Israeli Investigators Deported After Identifying Two Hijackers

According to later German reports, "a whole horde of Israeli counter-terror investigators, posing as students, [follow] the trails of Arab terrorists and their cells in the United States . . . In the town of Hollywood, Florida, they [identify] . . . Atta and Marwan Alshehhi as possible terrorists. Agents [live] in the vicinity of the apartment of the two seemingly normal flight school students, observing them around the clock." Supposedly, around April, the Israeli agents are discovered and deported, terminating the investigation. [DER SPIEGEL, 10/1/02]

March 23, 2001: DEA Issues Alert to Look Out for Israeli Spies

The Office of National Drug Control Policy issues a National Security Alert describing "apparent attempts by Israeli nationals to learn about government personnel and office layouts." This later becomes known through a leaked DEA document called "Suspicious Activities Involving Israeli Art Students at DEA Facilities." A crackdown ensues and by June, around 120 Israelis are apprehended. More are apprehended later. [DEA REPORT, 6/01]

June 2001: DEA Draws Up Report on Israeli Spies

The DEA's Office of Security Programs prepares a 60-page internal memo on the Israeli "art student spy ring." [DEA REPORT, 6/01] The memo is a compilation of dozens of field reports, and was meant only for the eyes of senior officials at the Justice Department (of which the DEA is adjunct), but it is leaked to the press around December 2001. The report connects the spies to efforts to foil investigations into Israeli organized crime activity involving the importation of the drug Ecstasy. The spies also appear to be snooping on top-secret military bases. For instance, on April 30, 2001, an Air Force alert was issued from Tinker Air Force Base in Oklahoma City concerning "possible intelligence collection being conducted by Israeli art students." Tinker AFB houses AWACS surveillance craft and Stealth bombers. By the time of the report, the U.S. has "apprehended or expelled close to 120 Israeli nationals" but many remain at large. [LE MONDE, 3/5/02; SALON, 5/7/02] An additional 20 or so Israeli spies are apprehended between June and 9/11. [FOX NEWS, 12/12/01]

August 8–15, 2001: Israel Reportedly Warns of "Major Assault on the U.S."

At some point between these dates, Israel warns the U.S. that an al-Qaeda attack is imminent. [FOX NEWS, 5/17/02] Reportedly, two high-ranking agents from the Mossad come to Washington and warn the FBI and CIA that from 50 to 200 terrorists have slipped into the U.S. and are planning "a major assault on the United States." They say indications point to a "large scale target," and that Americans would be "very vulnerable." They add there could be Iraqi connections to the al-Qaeda attack. [DAILY TELEGRAPH, 9/16/01; LOS ANGELES TIMES, 9/20/01; OTTAWA CITIZEN, 9/17/01] The Los Angeles Times later retracts its story after a CIA spokesperson says, "There was no such warning. Allegations that there was are complete and utter nonsense." [LOS ANGELES TIMES, 9/21/01 (B)] Other newspapers do not retract it.

August 23, 2001: Mossad Reportedly Gives CIA List of Terrorist Living in U.S.; at Least Four 9/11 Hijackers Named

According to German newspapers, the Mossad gives the CIA a list of 19 terrorists living in the U.S. and say that they appear to be planning to carry out an attack in the near future. It is unknown if these are the

19 9/11 hijackers or if the number is a coincidence. However, four names on the list are known, and these four will be 9/11 hijackers: Nawaf Alhazmi, Khalid Almihdhar, Marwan Alshehhi, and Mohamed Atta. [DIE ZEIT, 10/1/02; DER SPIEGEL, 10/1/02; BBC, 10/2/02; HA'ARETZ, 10/3/02] The Mossad appears to have learned about this through its "art student spy ring." Yet apparently, this warning and list are not treated as particularly urgent by the CIA and the information is not passed on to the FBI. It is unclear whether this warning influenced the decision to add Alhazmi and Almihdhar to a terrorism watch list on this same day, and if so, why only those two. [DER SPIEGEL, 10/1/02] Israel has denied that there were any Mossad agents in the U.S. [HA'ARETZ, 10/3/02]

September 4, 2001: Mossad Gives Another Warning of Major, Imminent Attack

"On or around" this day, the Mossad give their "latest" warning to the U.S. of a major, imminent terrorist attack, according to sources close to Mossad. One former Mossad agent says, "My understanding is that the warning was not specific. No target was identified. But it should have resulted in an increased state of security." U.S. intelligence claims this never happened. [SUNDAY MAIL, 9/16/01]

September 4, 2001 (B): Israeli Company Moves Out of WTC

The Zim-American Israeli Shipping Co. moves their North American headquarters from the 16th floor of the WTC to Norfolk, Virginia, one week before the 9/11 attacks. The Israeli government owns 49 percent of the company. [VIRGINIAN-PILOT, 9/4/01] Zim announced the move and its date six months earlier. [VIRGINIAN-PILOT, 4/3/01] More than 200 workers had just been moved out; about ten are still in the building making final moving arrangements on 9/11, but escape alive. [JERUSALEM POST, 9/13/01; JOURNAL OF COMMERCE, 10/18/01] The move leaves only one Israeli company, ClearForest, with 18 employees, in the WTC on 9/11. The four or five employees in the building at the time manage to escape. [JERUSALEM POST, 9/13/01] One year later, a Zim ship is impounded while attempting to ship Israeli military equipment to Iran; it is speculated that this is done with the knowledge of Israel. [AGENCE FRANCE-PRESSE, 8/29/02 (B)]

September 11, 2001: Two Hours Before Attacks, Israeli Company Employees Receive Warnings

Two employees of Odigo, Inc., an Israeli company, receive warnings of an imminent attack in New York City around two hours before the first plane hits the WTC. Odigo, one of the world's largest instant messaging companies, has its headquarters two blocks from the WTC. The Odigo Research and Development offices where the warnings were received are located in Herzliyya, a suburb of Tel Aviv. Israeli security and the FBI were notified immediately after the 9/11 attacks began. The two employees claim not to know who sent the warnings. "Odigo service includes a feature called People Finder that allows users to seek out and contact others based on certain interests or demographics. [Alex] Diamandis [Odigo vice president of sales and marketing] said it was possible that the attack warning was broadcast to other Odigo members, but the company has not received reports of other recipients of the message." [HA'ARETZ]

Odigo logo

9/26/01; WASHINGTON POST, 9/27/01 (C)] Odigo claims the warning did not specifically mention the WTC, but the company refuses to divulge what was specified, claiming, "Providing more details would only lead to more conjecture." *[WASHINGTON POST, 9/28/01]* Odigo gave the FBI the Internet address of the message's sender so the name of the sender could be found. *[DEUTSCHE PRESSE-AGENTUR, 9/26/01]* Two months later, it is reported that the FBI is still investigating the matter, but there have been no reports since. *[COURIER MAIL, 11/20/01]*

September 11, 2001 (B): Israeli Special-Ops Passenger Shot or Stabbed by Hijackers?

An FAA memo written on the evening of 9/11, and later leaked, suggests that a man on Flight 11 was shot and killed by a gun before the plane crashed into the WTC. The "Executive Summary," based on information relayed by a flight attendant to the American Airlines Operation Center, stated "that a passenger located in seat 10B [Satam Al Suqami] shot and killed a passenger in seat 9B [Daniel Lewin] at 9:20 A.M." (Note that since Flight 11 crashed at 8:46, the time must be a typo, probably meaning 8:20). A report in Israeli newspaper *Ha'aretz* on September 17 identifies Lewin as a former member of the Israel Defense Force Sayeret Matkal, Israel's most successful special-operations unit. *[UPI, 3/6/02]* Sayeret Matkal is a deep-penetration unit that has been involved in assassinations, the theft of foreign signals-intelligence materials, and the theft and destruction of foreign nuclear weaponry. Sayeret Matkal is best known for the 1976 rescue of 106 passengers at Entebbe Airport in Uganda. *[NEW YORKER, 10/29/01]* Note that Lewin founded Akamai, a successful computer company, and his connections to Sayeret Matkal remained hidden until the gun story became known. *[GUARDIAN, 9/15/01]* FAA and American Airline officials later deny the gun story and suggest that Lewin was probably stabbed to death instead. *[UPI, 3/6/02; WASHINGTON POST, 3/2/02 (B)]* Officials assert that the leaked document was a "first draft," and subsequently corrected, but declines to release the final draft, calling it "protected information." However, an FAA official present when the memo was drafted will dispute the FAA's claim, asserting that "[t]he document was reviewed for accuracy by a number of people in the room, including myself and a couple of managers of the operations center." *[WORLDNETDAILY, 3/7/02]*

September 11, 2001 (C): Five Israelis Arrested for "Puzzling Behavior" near WTC

Five Israelis are arrested around 4:30 P.M. for "puzzling behavior" related to the WTC attacks: filming the burning WTC from the roof of their company's building near Liberty State Park, New Jersey, then shouting in what was interpreted as cries of joy and mockery. A neighbor spotted them and called the police and the FBI. The police tracked them down in a van with the words "Urban Moving Systems" written on the side. *[BERGEN RECORD, 9/12/01; HA'ARETZ, 9/17/01]* One man was found with $4,700 in cash hidden in his sock, another had two passports on him, and a box cutter was found in the van. *[ABC NEWS, 6/21/02]* Investigators say that "[t]here are maps of the city in the car with certain places highlighted . . . It looked like they're hooked in with this. It looked like they knew what was going to happen." *[BERGEN RECORD, 9/12/01]* One of these Israelis later says, "our purpose was to document the event." *[ABC NEWS, 6/21/02]* The FBI later concludes at least two are Mossad agents and that all were on a Mossad surveillance mission.

The FBI interrogates them for weeks. [FORWARD, 3/15/02] They are held on immigration violation charges, but are released 71 days later. [ABC NEWS, 6/21/02] Their names are later identified as Sivan and Paul Kurzberg, Oded Ellner, Omer Marmari, and Yaron Shmuel. [FORWARD, 3/15/02]

September 11, 2001 (D): Former Israeli Prime Minister: 9/11 "Very Good" for Israeli-U.S. Relations

Former Israeli Prime Minister Benjamin Netanyahu, when asked what the 9/11 attacks means for relations between the U.S. and Israel, replies, "It's very good." Then he edits himself: "Well, not very good, but it will generate immediate sympathy." [NEW YORK TIMES, 9/12/01 (C)] A week later, the *Village Voice* states, "From national networks to small-town newspapers, the view that America's terrible taste of terrorism will finally do away with even modest calls for the restraint of Israel's military attacks on Palestinian towns has become an instant, unshakable axiom. . . . Now, support for Israel in America is officially absolute, and Palestinians are cast once again as players in a global terrorist conspiracy." [VILLAGE VOICE, 9/19/01]

September 14, 2001: Head of Shadowy Company Flees U.S.

Dominick Suter, owner of the company Urban Moving Systems, flees the country to Israel. The FBI later tells ABC News, "Urban Moving may have been providing cover for an Israeli intelligence operation." Suter has been tied to the five Israeli agents caught filming the WTC attack. The FBI had questioned Suter around September 12, removing boxes of documents and a dozen computer hard drives. However, when the FBI returns a few days later, he is gone. [FORWARD, 3/15/02; NEW JERSEY DEPARTMENT OF LAW AND PUBLIC SAFETY, 12/13/01; ABC NEWS, 6/21/01]

September 29–30, 2001: Suspected Mossad Agents Detained, Released

Police in the Midwest stop six men carrying suspicious documents. They possess photos and descriptions of a nuclear power plant in Florida and the Trans-Alaska pipeline, and have "box cutters and other equipment." All six have Israeli passports. They are released the same day after their passports are shown to be valid, but before anyone interviews them. The FBI is reportedly furious about their release. [MIAMI HERALD, 10/3/01; KNIGHT RIDDER, 10/31/01; LONDON TIMES, 11/2/01] The six men may have been Mossad agents. In addition to snooping on the DEA and Muslim terrorists, some Mossad agents in the "art student spy ring" have been caught trying to break into military bases and other top-secret facilities. [SALON, 5/7/02]

October 16, 2001: Several Arrested for Curious Sears Tower Surveillance

Two men, Moshe Elmakias and Ron Katar, are arrested after being found with a detailed video of the Sears Tower in Chicago. In addition, a woman named Ayelet Reisler is found with them, carrying conflicting identification information. They are arrested for illegal dumping, using a van with the name Moving Systems Incorporated. The video contains extensive zoom in shots of the Sears Tower; it is not known when the video was filmed. [PHILADELPHIA MERCURY, 10/18/01]

November 20, 2001: Israelis Who Videotaped
WTC Attack Are Released, Deported

The five Israelis arrested on 9/11 for videotaping the WTC attack and then cheering about it are released and deported to Israel. Some of the men's names had appeared in a U.S. national intelligence database, and the FBI has concluded that at least two of the men were working for the Mossad, according to ABC News. However, the FBI says that none of the Israelis had any advanced knowledge of the 9/11 attacks, and they were released as part of a deal between the U.S. and the Israeli government. After their release, they claim to have been tortured. [FORWARD, 3/15/02; ABC NEWS, 6/21/02]

November 23, 2001: More Israelis Arrested in Wake of 9/11

The *Washington Post* reports that "[a]t least 60 young Israeli Jews have been arrested and detained around the country on immigration charges since the September 11 attacks, many of them held on U.S. government officials' invocation of national security." An INS official who requested anonymity says the use of the term "special interest" for Israelis being held in Cleveland, St. Louis, and other places means the case in question is "related to the investigation of September 11th." [WASHINGTON POST, 11/23/01]

December 12–15, 2001: News Reports Raises Israeli Spying Questions

Fox News reports, "Investigators within the DEA, INS, and FBI have all told Fox News that to pursue or even suggest Israeli spying . . . is considered career suicide." "A highly placed investigator says there are 'tie-ins' between the spy ring and 9/11. However, when asked for details, he flatly refuses to describe them, saying, 'evidence linking these Israelis to 9/11 is classified. I cannot tell you about evidence that has been gathered. It's classified information.'" The report also reveals that Amdocs, an Israeli company, is recording virtually every phone call in the U.S. and could be passing information on to the Israeli government (similar claims were first raised in 2000 [INSIGHT, 5/29/00]). Fox News suggests that the position of this company might impede the 9/11 investigation. [FOX NEWS 12/12/01]

December 16, 2001: Fox News Removes Controversial Story
from Website, but Story Nonetheless Makes an Impact

Fox News removes its series on the "art student spy ring" from its website after only two days, in response to pressure from the Jewish Institute for National Security Affairs (JINSA), the Anti-Defamation League (ADL), the Committee for Accuracy in Middle East Reporting in America (CAMERA) and others. CAMERA suggests the reporter "has something, personally, about Israel . . . Maybe he's very sympathetic to the Arab side." [SALON, 5/7/02] The head of the ADL calls the report "sinister dangerous innuendo which fuels anti-Semitism." [FORWARD, 12/21/01] Yet there does not appear to be any substance to these personal attacks (and *Forward* later reverses its stance on the spy ring [FORWARD, 3/15/02]). Fox News also never makes a formal repudiation or correction about the series. The contents of the series continues to be generally ignored by the mainstream media, but it makes a big impact inside the U.S. government: An internal DEA communiqué from December 18 mentions the Fox report by name, and warns of security breaches in telecommunications as described in the Fox report. [SALON, 5/7/02]

March 5, 2002: Israeli Spies Reportedly Tracked 9/11 Hijackers

It is reported that many spies in the uncovered Israeli spy ring seemed to have been trailing the 9/11 hijackers. At the least, they were in close proximity. For instance, five Israeli spies are intercepted in the tiny town of Hollywood, Florida, and four 9/11 hijackers are known to have spent time in Hollywood, Florida. [LE MONDE, 3/5/02; REUTERS, 3/5/02; JANE'S INTELLIGENCE DIGEST. 3/15/02] In one case, some Israeli spies lived at 4220 Sheridan Street, only a few hundred feet from where Mohamed Atta was living at 3389 Sheridan Street. Israeli spies appear to have been close to at least ten of the 19 9/11 hijackers. [SALON, 5/7/02]

March 6, 2002: U.S. Officials Deny Existence of Israeli Spy Ring

A *Washington Post* article relying on U.S. officials, denies the existence of any Israeli spy ring. A "wide array of U.S. officials" supposedly deny it, and Justice Department spokeswoman Susan Dryden says: "This seems to be an urban myth that has been circulating for months. The department has no information at this time to substantiate these widespread reports about Israeli art students involved in espionage." [WASHINGTON POST, 3/6/02] The *New York Times* fails to cover the story at all, even months later. [SALON, 5/7/02] By mid-March, *Jane's Intelligence Digest,* the respected British intelligence and military analysis service, notes: "It is rather strange that the U.S. media seems to be ignoring what may well be the most explosive story since the 11 September attacks—the alleged breakup of a major Israeli espionage operation in the U.S.A." [JANE'S INTELLIGENCE DIGEST, 3/15/02]

March 11, 2002: Suspected Israeli Students Reportedly Served in Military

A newspaper reports that the DEA study on Israeli "art students" determined the "students" all had "recently served in the Israeli military, the majority in intelligence, electronic signal intercept, or explosive ordnance units." [PALM BEACH POST. 3/11/02]

March 15, 2002: Jewish Magazine Says Israelis Spied in U.S. on Radical Muslims

Forward, a U.S. publication with a large Jewish audience, admits that there has been an Israeli spy ring in the U.S. [FORWARD, 3/15/02] This is a reversal of their earlier stance. [FORWARD, 12/21/01] But, "far from pointing to Israeli spying against U.S. government and military facilities, as reported in Europe last week, the incidents in question appear to represent a case of Israelis in the United States spying on a common enemy, radical Islamic networks suspected of links to Middle East terrorism." [FORWARD, 3/15/02]

May 7, 2002: Explosives Detected on Illegal Israelis

A moving truck is pulled over for speeding in the middle of the night in Oak Harbor, Washington, near the Whidbey Island Naval Air Station. The base is the home of the advanced electronic warfare Prowler jets. A bomb-sniffing dog detects explosives on one of the men and inside the truck. High-tech equipment is then used to confirm the presence of TNT on the gearshift and RDX plastic explosive on the steering wheel. Both men turn out to be Israeli (one with an altered passport) and in the country illegally. [FOX NEWS, 5/13/02] However, the FBI later clears the two men, saying both the dog and the tests just detected false positives from "residue left by a cigarette lighter." [SEATTLE POST-INTELLIGENCER, 5/14/02; JERUSALEM POST, 5/14/02] The "art student spy ring" frequently uses moving vans as cover, and has been caught spying on the most top secret military bases. [SALON, 5/7/02] In a possibly related story, the Seattle FBI office that handled this case will be broken into a few weeks later, and even a room containing evidence will be penetrated. [SEATTLE POST-INTELLIGENCER, 7/29/02]

May 7, 2002 (B): Sloppy Israeli Spy Ring Could Have Been Smoke Screen

Salon reports on the Israeli "art student spy ring." All the "students" claim to have come from either Bezalel Academy, or the University of Jerusalem. A look in the Bezalel database shows that not a single "art student" appears to have attended school there. There is no such thing as the University of Jerusalem. In fact, the article points out that the sheer sloppiness and brazenness of the spy operation appears to be a great mystery, especially since the Mossad is renowned as one of the best spy agencies in the world. One government source suggests a theory to Salon that the "art students" were actually a smoke screen. They were meant to be caught and connected to DEA surveillance so that a smaller number of spies also posing as art students could complete other missions. One such mission could have been the monitoring of al-Qaeda terrorists. [SALON, 5/7/02] Shortly afterwards, a major Israeli newspaper publishes a story about the spy ring, but does not come to any conclusions. [HA'ARETZ, 5/14/02]

November 2002: Saudi Interior Minister Blames Jews for 9/11 Attacks

Saudi Interior Minister Prince Nayef blames Zionists and Jews for the 9/11 attacks. He tells journalists, "Who has benefited from September 11 attacks? I think [the Jews] were the protagonists of such attacks." Nayef is in charge of the Saudi investigation into the attacks, and some U.S. congresspeople respond to the comments by questioning how strongly Saudi Arabia is investigating the involvement of the 15 Saudi 9/11 hijackers. [ASSOCIATED PRESS, 12/5/02]

May 2004: More Israelis Are Arrested in Suspicious Circumstances, Again Traveling in Moving Vans

On May 9, two Israelis are arrested after a high-speed chase in Tennessee. They are found with false documents. [WYCB, 5/9/04] On May 27, two others are arrested after trying to enter an Atlanta military base. Explosives are possibly discovered in their van. [KINGS BAY PERISCOPE, 5/27/04]

15.

Iraq

Polls taken since 2003 show that most Americans believe Iraq was involved in the 9/11 attacks, and even that some of the hijackers were Iraqi. These mistaken assumptions have effectively been disproved, but there is a connection of sorts between 9/11 and Iraq—the way the Bush administration used 9/11 to justify a war in Iraq.

This chapter chronicles events relating to three basic questions about connections between Iraq and 9/11.

First, did the Bush administration's seeming obsession with Iraq divert its attention from the terrorist threat posed by al-Qaeda and bin Laden? The Bush administration was highly focused on Iraq from the first moments of Bush's tenure. The first National Security Council meeting focused not on terrorism, as expected by officials from the prior administration, but on Iraq. Throughout the first eight months of the Bush administration, when counterterrorism officials sought greater focus on bin Laden, other officials rejected this plea, choosing instead to focus on Iraq. This chapter chronicles events in which key Bush administration officials indicated that regime change in Iraq was a higher priority than dealing with the growing threat of terrorism from bin Laden and al-Qaeda prior to 9/11.

Second, how focused was the Bush administration on Iraq just after 9/11, and when was the decision made to go to war with Iraq? Just hours after the 9/11 attacks, while the ruins still burned, senior Bush administration officials sought to leverage the tragedy to justify its agenda of regime change in Iraq. Counterterrorism officials were ordered to search for an Iraqi connection. Even as President Bush approved the Afghanistan war plan, he directed the military to develop a war plan for Iraq. In December

2001 "victory" was declared in Afghanistan, even though many Taliban officials continued to live openly throughout the country, bin Laden had been neither captured nor killed, and al-Qaeda continued to exist. A short time later, in early 2002, the decision for war in Iraq was made with the transfer of resources out of Afghanistan and into the Iraq region, so this focus concludes at that time.

Third, did the Bush administration mislead the public about connections between Iraq and 9/11 to garner public support for its war with Iraq? Just days after the attacks, Bush administration officials began its public effort to connect the 9/11 attacks to Saddam Hussein, even though no credible evidence indicated such a connection existed. This chapter illustrates how it came to be that so many people saw an Iraq-9/11 connection, despite the slim to nonexistent evidence of one.

(Chapter 16 addresses the issue of why the Bush administration was determined to bring regime change to Iraq.)

January 30, 2001: First National Security Council
Meeting Focuses on Iraq, Not Terrorism

The Bush administration holds its first National Security Council meeting, ten days after Bush's inauguration. According to Treasury Secretary Paul O'Neill's account in a 2004 book, the first and most important topic discussed is Iraq. O'Neill states that, "From the very beginning, there was a conviction that Saddam Hussein was a bad person and that he needed to go." There was no dissent amongst top officials to this idea. O'Neill states, "It was all about finding a way to do it. That was the tone of it. The president saying 'Go find me a way to do this.' For me, the notion of pre-emption, that the U.S. has the unilateral right to do whatever we decide to do, is a really huge leap." [CBS NEWS, 1/11/04; *THE PRICE OF LOYALTY*, BY RON SUSKIND, 1/04] O'Neill also claims, "In the 23 months I was [Treasury Secretary], I never saw anything that I would characterize as evidence of weapons of mass destruction." [*TIME*, 1/10/04] President Clinton had publicly advocated regime change in Iraq in 1998, and Bush later claims that his Iraq policy before 9/11 was no different from Clinton's. [*NEW YORK TIMES*, 1/12/04] However, an anonymous official also at this meeting later says the tone is much more aggressive. "The president told his Pentagon officials to explore the military options, including use of ground forces. That went beyond the Clinton administration's halfhearted attempts to overthrow Hussein without force." [ABC NEWS, 1/13/04]

February 1, 2001: Rumsfeld Envisions Post-Saddam Iraq

The Bush White House holds its second National Security Council meeting, and again regime change in Iraq is a central topic. According to Treasury Secretary Paul O'Neill, who attends the meeting, Defense Secretary Rumsfeld argues in favor of removing Saddam Hussein. He talks about post-Saddam Iraq in some detail and the problems likely to arise. He says removing Hussein would "demonstrate what U.S. policy is all about," and help transform the Middle East. [*NEW YORK TIMES*, 1/12/04]

March 5, 2001: Treasury Secretary Sees Bush
Administration Is Preoccupied with Iraq Regime Change

Paul O'Neill, Bush's Treasury Secretary at this time, later recalls that the most important topic of the Bush administration in its early months is regime change in Iraq. Planning at this time envisions peacekeeping troops, war crimes tribunals, and even divvying up Iraq's oil wealth. One document from around February 2001 is titled, "Plan for post-Saddam Iraq." Another Pentagon document from this date is titled, "Foreign Suitors for Iraqi Oilfield contracts." It includes a map of potential areas for exploration in Iraq. [CBS NEWS, 1/11/04]

Spring 2001: Securing Energy Supplies Becomes U.S. Military Doctrine

The *Sydney Morning Herald* later reports, "The months preceding September 11 [see] a shifting of the U.S. military's focus . . . Over several months beginning in April [2001] a series of military and governmental policy documents [are] released that [seek] to legitimize the use of U.S. military force . . . in the pursuit of oil and gas." Michael Klare, an international security expert and author of *Resource Wars,* says the military has increasingly come to "define resource security as their primary mission." An article in the Army War College's journal by Jeffrey Record, a former staff member of the Senate armed services committee, argues for the legitimacy of "shooting in the Persian Gulf on behalf of lower gas prices." He also "advocate[s] the acceptability of presidential subterfuge in the promotion of a conflict" and "explicitly urge[s] painting over the U.S.'s actual reasons for warfare with a nobly high-minded veneer, seeing such as a necessity for mobilizing public support for a conflict." In April, Tommy Franks, the commander of U.S. forces in the Persian Gulf/South Asia area, testifies to Congress that his command's key mission is "access to [the region's] energy resources." The next month, U.S. Central Command begins planning for war with Afghanistan, plans that are later used after 9/11. [SYDNEY MORNING HERALD, 12/26/02]

April 2001: Report Argues U.S. Should Take Control of Iraqi Oil

A report commissioned by former U.S. Secretary of State James Baker and the Council on Foreign Relations entitled "Strategic Energy Policy Challenges For The 21st Century" is submitted to Vice President Cheney this month. "The report is linked to a veritable who's who of U.S. hawks, oilmen, and corporate bigwigs." The report says the "central dilemma" for the U.S. administration is that "the American people continue to demand plentiful and cheap energy without sacrifice or inconvenience." It warns that the U.S. is running out of oil, with a painful end to cheap fuel already in sight. It argues that "the United States remains a prisoner of its energy dilemma," and that one of the "consequences" of this is a "need for military intervention" to secure its oil supply. It argues that Iraq needs to be overthrown so the U.S. can control its oil. [SUNDAY HERALD, 10/5/02; SYDNEY MORNING HERALD, 12/26/02] In what may be a reference to a pipeline through Afghanistan, the report suggests the U.S. should "investigate whether any changes to U.S. policy would quickly facilitate higher exports of oil from the Caspian Basin region . . . the exports from some oil discoveries in the Caspian Basin could be hastened if a secure, economical export route could be identified swiftly." [STRATEGIC ENERGY POLICY CHALLENGES FOR THE 21ST CENTURY, 4/01]

April 8, 2001: Claims That Atta Meets Iraqi Spy at this Time Later Debunked

According to later highly disputed and ultimately discredited reports, Mohamed Atta flies from the U.S. to Prague, Czech Republic, and meets with Ahmed Khalil Ibrahim Samir al-Ani, an Iraqi spy, then returns on April 9 or 10. [NEW YORK TIMES, 10/27/01] However, as the chapter details further below, the meeting is confirmed and denied repeatedly for several years. It is eventually generally refuted.

September 11, 2001: Hours After 9/11, Rumsfeld Seeks Iraq Ties

Hours after the 9/11 attacks, Defense Secretary Rumsfeld is given information indicating that three of the names on the airplane passenger manifests are suspected al-Qaeda operatives. Notes composed by aides who were with Rumsfeld in the National Military Command Center on 9/11 are leaked nearly a year later. According to the notes, Rumsfeld wants the "best info fast. Judge whether good enough hit S.H. [Saddam Hussein] at same time. Not only UBL. [bin Laden]" "Go massive," he notes. "Sweep it all up. Things related, and not." [CBS NEWS, 9/4/02; A PRETEXT FOR WAR, BY JAMES BAMFORD, 6/04, P. 285] He presents the idea to Bush the next day. It is later revealed that shortly after 9/11, Rumsfeld sets up "a small team of defense officials outside regular intelligence channels to focus on unearthing details about Iraqi ties with al-Qaeda and other terrorist networks." The team sifts "through much of the same databases available to government intelligence analysts but with the aim of spotlighting information the spy agencies have either overlooked or played down." [WASHINGTON POST, 10/25/02] *Time* magazine will report in May 2002 that Defense Secretary "Rumsfeld has been so determined to find a rationale for an attack that on ten separate occasions he asked the CIA to find evidence linking Iraq to the terror attacks of September 11. The intelligence agency repeatedly came back empty-handed." [TIME, 5/6/02 (B)] However, while the CIA hasn't been helpful to Rumsfeld, one former senior official later says, "If it became known that [Rumsfeld] wanted [the Defense Intelligence Agency] to link the government of Tonga to 9/11, within a few months they would come up with sources who'd do it." [NEW YORKER, 12/16/02]

September 12, 2001: Bush Meeting Raises Iraq Attack Possibility

Top officials meeting with President Bush discuss attacking Iraq. Counterterrorism "tsar" Richard Clarke later recalls, "At first I was incredulous that we were talking about something other than getting al-Qaeda. Then I realized with almost a sharp physical pain that [Defense Secretary] Rumsfeld and [Deputy Defense Secretary] Wolfowitz were going to try to take advantage of this national tragedy to promote their agenda about Iraq. Since the beginning of the administration, indeed well before, they had been pressing for a war with Iraq. My friends in the Pentagon had been telling me that the word was we would be invading Iraq sometime in 2002." Wolfowitz tries to argue that al-Qaeda must have had

**National Security Council Meeting,
September 12, 2001**

Iraqi backing. Following his notes from the day before seeking to blame 9/11 on Iraq and not just al-Qaeda, Defense Secretary Rumsfeld proposes that Iraq should be "a principal target of the first round in the war against terrorism." Rumsfeld complains that there are no decent targets to bomb in Afghanistan and that it would be better to bomb Iraq because they have better targets. Secretary of State Powell agrees with Clarke that the immediate focus should be on al-Qaeda. However, Powell also seems open to a later attack, saying, "Public opinion has to be prepared before a move against Iraq is possible." Clarke complains to him, "Having been attacked by al-Qaeda, for us now to go bombing Iraq in response would be like our invading Mexico after the Japanese attacked us at Pearl Harbor." President Bush notes the goal should be replacing the Iraqi government, not just bombing it, but the military warns an invasion would need a large force and many months to assemble. [WASHINGTON POST, 1/28/02; LOS ANGE-LES TIMES, 1/12/03; AGAINST ALL ENEMIES, BY RICHARD CLARKE, 3/04, PP. 30–31]

September 12, 2001 (B): Bush to Clarke: "Look into Iraq"

Counterterrorism "tsar" Richard Clarke later claims that on the evening of this day, President Bush speaks to Clarke and others, saying, "Look, I know you have a lot to do and all . . . but I want you, as soon as you can, to go back over everything, everything. See if Saddam did this. See if he's linked in any way." Clarke later recalls, "Now he never said, 'Make it up.' But the entire conversation left me in absolutely no doubt that George Bush wanted me to come back with a report that said Iraq did this." Clarke complains to Bush that the link between al-Qaeda and Iraq has already been investigated several times, and always come up empty. Bush responds in a testy and intimidating manner, "Look into Iraq, Saddam." Note that Deputy National Security Adviser Steve Hadley denies that this exchange ever occurred, but *60 Minutes* finds two sources to confirm Clarke's account independently. [CBS NEWS, 3/20/04] Others also emerge to back up Clarke's account, though some dispute his description of Bush as "intimidating." [NEW YORK TIMES, 3/24/04; GUARDIAN, 3/26/04] White House aides eventually concede that the meeting "probably" occurred. [NEW YORK DAILY NEWS, 3/27/04] Clarke delegates an investigation to look again at any link between Iraq and al-Qaeda. He later writes, "All agencies and departments agreed, there was no cooperation between the two (i.e., al-Qaeda and Iraq). A memorandum to that effect was sent up to the President, but there never was any indication that it reached him." The agencies are told to prepare another report on the same subject, with the implication that they should come up with a different answer. However, the second time, they reach the same conclusion, and again Clarke doubts if Bush will ever see it. [AGAINST ALL ENEMIES, BY RICHARD CLARKE, 3/04, PP. 30–31, CBS NEWS, 3/20/04]

September 13, 2001: White House Decides Afghanistan Is First, Then Iraq

According to counterterrorism "tsar" Richard Clarke, top Bush officials come to a consensus that the first response to the 9/11 attacks will be war in Afghanistan against al-Qaeda and the Taliban. However, it is clear there will be a second stage in a larger war on terrorism, and the implication is that will involve invading Iraq. [AGAINST ALL ENEMIES, BY RICHARD CLARKE, 3/04, PP. 32–33]

September 15, 2001: Wolfowitz Again Pushes for Iraq War

Bush meets at Camp David with his senior advisers to discuss the Afghanistan war plans. Deputy Defense Secretary Paul Wolfowitz pushes for regime change in Iraq, claiming there is a 10 to 50 percent chance that Iraq was involved in the 9/11 attacks. [VANITY FAIR, 5/04] The Defense Department submits a paper depicting Iraq, the Taliban, and al-Qaeda as priority targets in the first stage of action. Wolfowitz and Undersecretary of Defense Douglas Feith argue in three memos in subsequent days why Iraq should be included. One memo dated September 18 is titled, "Were We Asleep?" and suggests links between Iraq and al-Qaeda. But these links, such as the theory that 1993 WTC bomber Ramzi Yousef was an Iraqi agent, are considered highly dubious by most at the time. [WASHINGTON POST, 7/23/04 (B)]

Deputy Defense Secretary Paul Wolfowitz

Mid-September 2001: Propaganda Campaign to Tie 9/11 to Iraq Is Said to Begin

Retired General Wesley Clark later claims that "immediately after 9/11" there is a "concerted effort . . . to pin 9/11 and the terrorism problem on Saddam Hussein" and use the attacks as an excuse to go after the Iraqi dictator. When asked who is behind the concerted effort, Clark will respond, "Well, it came from the White House, it came from people around the White House. It came from all over." He claims that he personally receives a phone call at this time from an overseas think tank urging him to push an Iraq connection on his television appearances. [MSNBC, 6/15/03; NEW YORK TIMES, 7/18/03]

Mid-September 2001 (B): Atta-Iraq Spy Meeting Story Begins with Dubious Tip

The *New York Times* reports that a claim that Mohamed Atta met with an Iraqi spy in the Czech Republic prior to the 9/11 attacks, on April 8, 2001, has its origins at this time and not any time before 9/11. A single informant in the local Prague Arab community apparently claims to know about the meeting. Czech intelligence treats the claim skeptically because it comes only after Atta's picture has been broadcast on television and after the Czech press reported that records showed Atta had traveled to Prague. However, Czech Prime Minister Milos Zeman gives the story to the media before Czech intelligence can properly investigate it. [NEW YORK TIMES, 10/21/02] The claim is widely publicized, but is substantially discredited by mid-2003.

September 15, 2001–April 6, 2002: Bush Shifts Public Focus from bin Laden to Iraq

On September 15, 2001, President Bush says of bin Laden: "If he thinks he can hide and run from the United States and our allies, he will be sorely mistaken." [LOS ANGELES TIMES, 9/16/01 (B)] Two days later, he says, "I want justice. And there's an old poster out West, I recall, that says, 'Wanted: Dead or Alive.'" [ABC NEWS, 9/17/01] On December 28, 2001, even as the U.S. was declaring victory in Afghanistan, Bush says, "Our objective is more than bin Laden." [ASSOCIATED PRESS, 8/19/02 (B)] Bush's January 2002 State of the

Union speech describes Iraq as part of an "axis of evil" and fails to mention bin Laden at all. On March 8, 2002, Bush still vows: "We're going to find him." [WASHINGTON POST 10/1/02] Yet, only a few days later on March 13, Bush says, "He's a person who's now been marginalized. . . . I just don't spend that much time on him. . . . I truly am not that concerned about him." Instead, Bush is "deeply concerned about Iraq." [WHITE HOUSE, 3/13/02] The rhetoric shift is complete when Joint Chiefs of Staff Chairman Richard Myers states on April 6, "The goal has never been to get bin Laden." [DEPARTMENT OF DEFENSE, 4/6/02] In October 2002, the *Washington Post* notes that since March 2002, Bush has avoided mentioning bin Laden's name, even when asked about him directly. Bush sometimes uses questions about bin Laden to talk about Saddam Hussein instead. In late 2001, nearly two-thirds of Americans say the war on terrorism could not be called a success without bin Laden's death or capture. That number falls to 44 percent in a March 2002 poll, and the question has since been dropped. [WASHINGTON POST, 10/1/02] Charles Heyman, editor of *Jane's World Armies,* later points out: "There appears to be a real disconnect" between the U.S. military's conquest of Afghanistan and "the earlier rhetoric of President Bush, which had focused on getting bin Laden." [CHRISTIAN SCIENCE MONITOR, 3/4/02 (B)]

September 14, 2001: Congress to Bush: Use All Necessary Military Force

Congress authorizes Bush to use all necessary military force against the perpetrators of the 9/11 attacks, their sponsors, and those who protected them. [STATE DEPARTMENT, 12/26/01]

September 17, 2001: Bush Signs Afghanistan War Plan—But Is Focus on Iraq?

President Bush signs a document marked "TOP SECRET" that outlines a plan for going to war in Afghanistan. The document also directs the Pentagon to begin planning military options for an invasion of Iraq. Iraq secretly becomes a "central focus" of the U.S.'s counterterrorism effort over the next nine months, without much in the way of internal debate, public pronouncements, or paper trail. [WASHINGTON POST, 1/12/03] The document further orders the military to be ready to occupy Iraq's oil fields if the country acts against U.S. interests. [WASHINGTON POST, 7/23/04]

September 19, 2001: Rumsfeld and His Advisers Push for Iraq War

The Defense Policy Board, an advisory group to the U.S. military, secretly meets in the Pentagon and has a 19-hour discussion about military action against Iraq. [NEW YORK TIMES, 10/12/01] Among those attending are board chairman Richard Perle, Deputy Defense Secretary Paul Wolfowitz, Defense Secretary Rumsfeld, and Iraqi exile Ahmed Chalabi. Secretary of State Powell is not invited. [NEW YORK TIMES 10/12/01; VANITY FAIR, 5/2004] The attendees write a letter to President Bush calling for the overthrow of Saddam Hussein. The letter is published the next day in the name of the Project for a New American Century (PNAC), the neoconservative think tank established in 1997, since many members of this board are also members of that think tank. [PROJECT FOR A NEW AMERICAN CENTURY, 9/20/01; MANILA TIMES, 7/19/03]

September 19, 2001–Present: Claims of an Atta-Iraqi
Spy Meeting Are Repeatedly Asserted and Denied

Media coverage relating to an alleged meeting between hijacker Mohamed Atta and an Iraqi spy named Ahmed al-Ani took place in Prague, Czech Republic has changed repeatedly over time:

- September 19, 2001: It is first reported that an April 8, 2001, meeting took place; Atta is named later. [LOS ANGELES TIMES, 9/19/01; CNN, 10/11/01]

- October 20, 2001: The story is denied. [NEW YORK TIMES, 10/20/01]

- October 27, 2001: The story is confirmed. [NEW YORK TIMES, 10/27/01]

- October 27, 2001: It is claimed Atta met with Iraqi agents four times in Prague, plus in Germany, Spain, and Italy. [LONDON TIMES, 10/27/01]

- November 12, 2001: Conservative columnist William Safire calls the meeting an "undisputed fact." [NEW YORK TIMES, 11/12/01]

- December 9, 2001: Vice President Cheney asserts that the existence of the meeting is "pretty well confirmed." [WASHINGTON POST, 12/9/01]

- December 16, 2001: The identities of both al-Ani and Atta, alleged to have been at the meetings, are disputed. [NEW YORK TIMES, 12/16/01]

- January 12, 2002: It is claimed at least two meetings took place, including one a year earlier. [DAILY TELEGRAPH, 1/12/02]

- February 6, 2002: It is reported that the meeting probably took place, but was not connected to the 9/11 attacks. [NEW YORK TIMES, 2/6/02]

- March 15, 2002: Evidence that the meeting took place is considered between "slim" and "none." [WASHINGTON POST, 3/15/02]

- March 18, 2002: William Safire again strongly asserts that the meeting took place. [NEW YORK TIMES, 3/18/02]

- April 28–May 2, 2002: The meeting is largely discredited. For example, the *Washington Post* quotes FBI Director Mueller stating that, "We ran down literally hundreds of thousands of leads and checked every record we could get our hands on, from flight reservations to car rentals to bank accounts," yet no evidence that he left the country was found. According to the *Post*, "[a]fter months of investigation, the Czechs [say] they [are] no longer certain that Atta was the person who met al-Ani, saying 'he may be different from Atta.'" [WASHINGTON POST, 5/1/02] *Newsweek* cites a U.S. official who contends that, "Neither we nor the Czechs nor anybody else has any information [Atta] was coming or going [to Prague] at that time." [*NEWSWEEK*, 4/28/02; *WASHINGTON POST*, 5/1/02; *NEW YORK TIMES*, 5/2/02]

- May 8: Some Czech officials continue to affirm the meeting took place. [*PRAGUE POST*, 5/8/02]

- May 9, 2002: William Safire refuses to give up the story, claiming a "protect-Saddam cabal" in the high levels of the U.S. government is burying the story. [*NEW YORK TIMES*, 5/9/02]

- July 15, 2002: The head of Czech foreign intelligence states that reports of the meeting are unproved and implausible. [*PRAGUE POST*, 7/15/02]

- August 2, 2002: With a war against Iraq growing more likely, Press Secretary Ari Fleischer suggests the meeting did happen, "despite deep doubts by the CIA and FBI." [*LOS ANGELES TIMES*, 8/2/02]

- August 19, 2002: *Newsweek* states: "The sole evidence for the alleged meeting is the uncorroborated claim of a Czech informant." According to *Newsweek*, Deputy Defense Secretary Paul Wolfowitz is nonetheless pushing the FBI to have the meeting accepted as fact. [*NEWSWEEK*, 8/19/02]

- September 10, 2002: The Bush administration is no longer actively asserting that the meeting took place. [*WASHINGTON POST*, 9/10/02]

- September 17, 2002: Vice President Cheney and Defense Secretary Rumsfeld "accept reports from Czech diplomats" that the meeting took place. [*USA TODAY*, 9/17/02]

- September 23, 2002: *Newsweek* reports that the CIA is resisting Pentagon demands to obtain pictures of the alleged meeting from Iraqi exiles. One official says, "We do not shy away from evidence. But we also don't make it up." [*NEWSWEEK*, 9/23/02]

- October 20, 2002: Czech officials, including President Vaclav Havel, emphatically deny that the meeting ever took place. It now appears Atta was not even in the Czech Republic during the month the meeting was supposed to have taken place. President Havel told Bush "quietly some time earlier this year" that the meeting did not happen. [UPI, 10/20/02; *NEW YORK TIMES*, 10/21/02]

- December 8, 2002: Bush adviser Richard Perle continues to push the story, stating, "To the best of my knowledge that meeting took place." [CBS NEWS, 12/8/02]

- July 9, 2003: Iraqi intelligence officer Ahmed al-Ani is captured by U.S. forces in Iraq. [*WASHINGTON POST*, 7/9/03]

- July 10, 2003: In a story confirming al-Ani's capture, ABC News cites U.S. and British intelligence officials who have seen surveillance photos of al-Ani's meetings in Prague, and who say that there is a man who looks somewhat like Atta, but is not Atta. [ABC NEWS, 7/10/03]

- September 14, 2003: Vice President Cheney repeats the claims that Atta met with al-Ani in Prague on NBC's "Meet the Press." [WASHINGTON POST, 9/15/03]

- December 13, 2003: It is reported that al-Ani told interrogators he did not meet Atta in Prague. [WASHINGTON POST, 9/29/03; REUTERS, 12/13/03]

- June 16, 2004: The 9/11 Commission concludes that the meeting never happened. They claim cell phone records and other records show Atta never left Florida during the time in question. [9/11 COMMISSION REPORT, 6/16/04 (B)]

- July 17, 2004: Vice President Cheney says no one has "been able to confirm" the Atta meeting in Prague or to "to knock it down." [CNN, 6/18/04]

September 20, 2001: PNAC Think Tank Pushes for Iraq War

The Project for the New American Century (PNAC), an influential neoconservative think tank, publicly publishes a letter to President Bush, advising him to conquer Iraq quickly. "Failure to undertake such an effort will constitute an early and perhaps decisive surrender in the war on international terrorism." According to PNAC, the U.S. should demand that Iran and Syria cease all support of Hezbollah, and if they fail to do so, the U.S. should "retaliate" against those two countries as well. PNAC also praises Israel as "America's staunchest ally against international terrorism." [PNAC, 9/20/01] The next day, the *Los Angeles Times* notes that there is an internal battle inside the Bush administration about launching a war against Iraq. On one side are Secretary of State Powell and his allies, who argue that al-Qaeda needs to be defeated first. On the other side is the "string of Perles"—Richard Perle, Deputy Secretary of Defense Paul Wolfowitz, and their allies, who argue that Iraq should not wait. [LOS ANGELES TIMES, 9/21/01 (C)] Powell successfully delays an invasion, arguing there is no evidence implicating Iraq in the 9/11 attacks. [NEW YORK TIMES, 10/12/01]

September 20, 2001 (B): Bush to Blair:
After Afghanistan, "We Must Come Back to Iraq"

President Bush meets with British Prime Minister Tony Blair. [STATE DEPARTMENT, 12/26/01] Sir Christopher Meyer, the British ambassador to Washington at this time, later claims to overhear a discussion between the two leaders. Blair tells Bush that he wants to concentrate on ousting the Taliban in Afghanistan. Bush replies, "I agree with you Tony. We must deal with this first. But when we have dealt with Afghanistan, we must come back to Iraq." Blair says nothing to disagree. The *Independent* notes that this incident "presents a new challenge to Mr. Blair's assertion that no decision was taken on the invasion of Iraq until just days before operations began, in March 2003. It implies regime change in Iraq was U.S. policy immediately after 11 September." [BBC, 4/3/03; INDEPENDENT, 4/4/04 (B), OBSERVER, 4/4/04; VANITY FAIR, 5/04]

Late September 2001: Neoconservatives Look to Tie Iraq to 9/11

At the behest of Deputy Defense Secretary Paul Wolfowitz, former CIA Director and fellow neoconservative James Woolsey flies to London to look for evidence that would support the Bush administration goal of regime change in Iraq (i.e. war), including whether there were any ties between Saddam Hussein and the 1993 WTC bombing; or whether hijacker Mohamed Atta had worked with Iraqi intelligence to plan the 9/11 attacks. Woolsey is advised to meet with Iraqi exiles and others who may have useful intelligence. [*OBSERVER*, 10/14/01; *VILLAGE VOICE*, 11/21/01; *OBSERVER*, 12/2/01] Secretary of State Powell and CIA Director Tenet are not informed of the intelligence gathering mission. [*VILLAGE VOICE*, 11/21/01]

October 2001: "Secret Agenda" to Target Iraq Is Advanced by New Military Office Despite CIA Evidence

Douglas Feith, Under Secretary of Defense for Policy, creates the Counter Terrorism Evaluation Group (CTEG). It begins analyzing raw data, searching for evidence of links between terrorist groups and host countries. By the end of 2001, the unit concludes that there is an increasingly unified threat between different Muslim terrorist groups and that there are links between al-Qaeda and Iraq, and it reports

Douglas Feith, Under Secretary of Defense for Policy

these conclusions directly to high government officials. These conclusions are at odds with years of CIA analysis. Feith and the unit's few employees are closely connected to Richard Perle, a leading neoconservative long in favor of toppling Saddam Hussein and a vocal critic of the CIA. Note that the CTEG is not the same as the Office of Special Plans, although Feith sets both up, and both see ties between al-Qaeda and Iraq. The *New York Times* reports, "Some intelligence experts charge that the unit had a secret agenda to justify a war with Iraq and was staffed with people who were handpicked by conservative Pentagon policy makers to arrive at preordained conclusions about Iraq and al-Qaeda." [*NEW YORK TIMES*, 4/28/04]

November 21, 2001: Bush Wants Iraq Invasion Plan

President Bush privately directs Defense Secretary Rumsfeld to devise a plan to attack Iraq. Bush asks the plan be kept secret, and even some other top officials are not fully informed. [*WASHINGTON POST*, 1/18/04; *NEW YORK TIMES*, 4/17/04]

February 16, 2002: Bush Makes a Secret Decision to Prepare for Iraq War; Diverts Resources from War on Terror

Bush signs a secret National Security Council directive establishing the goals and objectives for going to war with Iraq, according to documents obtained by conservative author Rowan Scarborough. He details this in his book *Rumsfeld's War*. The *Guardian* notes, "The revelation casts doubt on the public insistence by U.S. and British officials throughout 2002 that no decision had been taken to go to war, pending negotiations at the United Nations." [*GUARDIAN*, 2/24/04] An unnamed former White House official says that around this time, "The Bush administration [takes] many intelligence operations that had been aimed at al-Qaeda and other terrorist groups around the world and redirect[s] them to the Persian Gulf. Linguists and special operatives [are] abruptly reassigned, and several ongoing anti-terrorism intelligence programs [are] curtailed." [NEW YORKER, 10/20/03] A senior military commander tells Senator Bob Graham (D) about this reassignment. [COUNCIL ON FOREIGN RELATIONS, 3/26/04]

September 4, 2002: Iraq Sued for Conspiring with al-Qaeda in 9/11 Attacks

Over 1,400 relatives of 9/11 attack victims sue Iraq for more than $1 trillion, claiming there is evidence Iraq conspired with al-Qaeda on the 9/11 attacks. [CBS NEWS, 9/5/02] One of the key pieces of evidence cited is an article in a small town Iraqi newspaper written by Naeem Abd Muhalhal on July 21, 2001. He describes bin Laden thinking "seriously, with the seriousness of the Bedouin of the desert, about the way he will try to bomb the Pentagon after he destroys the White House." He adds that bin Laden is "insisting very convincingly that he will strike America on the arm that is already hurting," which has been interpreted as a possible reference to the 1993 bombing of the WTC. Iraqi leader Saddam Hussein apparently praised this writer on September 1, 2001. The lawsuit is based largely on the idea that "Iraqi officials were aware of plans to attack American landmarks," yet did not warn their archenemy, the U.S. [ASSOCIATED PRESS, 9/4/02] Former CIA agent and terrorist consultant Robert Baer is hired by the prosecuting legal team to find evidence of a meeting between Mohamed Atta and Iraqi agents on April 8, 2001, but despite the help of the CIA, he is unable find any evidence of such a meeting. [CBS NEWS, 12/8/02]

October 3–11, 2002: French and British Deny Link Between Iraq and al-Qaeda

French and British officials deny that there is any link between al-Qaeda and Iraq. The British specifically deny any meeting between Mohamed Atta and Iraqi agents in the Czech Republic. They state that Iraq has purposely distanced itself from al-Qaeda, not embraced it. [FINANCIAL TIMES, 10/4/02; GUARDIAN, 10/10/02; GUARDIAN, 10/10/02 (B)] Meanwhile, Vincent Cannistraro, the CIA's former head of counterintelligence, says, "Basically, cooked information is working its way into high-level pronouncements and there's a lot of unhappiness about it in intelligence, especially among analysts at the CIA." A source connected to the 9/11 investigation says, "The FBI has been pounded on to make this link." [SYDNEY MORNING HERALD, 10/10/02] The Los Angeles Times also reports an escalating "war" between the Pentagon and the CIA over tying Iraq to al-Qaeda. [LOS ANGELES TIMES, 10/11/02]

March 20, 2003: U.S. and Partners Invade Iraq

The U.S., Britain, Australia, and Poland invade Iraq. [ASSOCIATED PRESS, 3/19/03] Bush sends a letter to Congress giving two reasons for the war. First, he has determined that further diplomacy will not protect the U.S. Second, he is "continuing to take the necessary actions against international terrorists and terrorist organizations, including those nations, organizations, or persons who planned, authorized, committed, or aided the terrorist attacks that occurred on September 11, 2001." [WHITE HOUSE, 3/18/03] This mimics language from a bill passed by Congress in October 2002, which granted Bush the power to declare war against Iraq if a link with the 9/11 attacks is shown and several other conditions are met. [WHITE HOUSE, 10/2/02] Yet on January 31, 2003, when a reporter asked both Bush and British Prime Minister Blair, "Do you believe that there is a link between Saddam Hussein, a direct link, and the men who attacked on September the 11th?" Bush replied, "I can't make that claim." Blair then replied, "That answers your question." [WHITE HOUSE, 1/31/03] A *New York Times*/CBS poll from earlier in the month indicates that 45 percent of Americans believe Saddam Hussein was "personally involved" in the 9/11 attacks. [NEW YORK TIMES, 3/11/03 (B)] The *Christian Science Monitor* notes, "Sources knowledgeable about U.S. intelligence say there is no evidence that Hussein played a role in the September 11 attacks, nor that he has been or is currently aiding al-Qaeda. Yet the White House appears to be encouraging this false impression . . . ." [CHRISTIAN SCIENCE MONITOR, 3/14/03] For instance, Bush claims Hussein has supported "al-Qaeda-type organizations," and "al-Qaeda types." [NEW YORK TIMES, 3/9/03]

June 25, 2003: Top al-Qaeda Prisoners Deny al-Qaeda-Iraq Link

U.S. officials admit that Khalid Sheikh Mohammed and Abu Zubaida have said in interrogations that bin Laden vetoed a long-term relationship with Saddam because he did not want to be in Hussein's debt. [NEWSWEEK, 6/25/03; ASSOCIATED PRESS, 7/13/03]

August 2003: Seventy Percent of Americans Misled by Bush Administration Insinuations of Iraq–9/11 Links

Several polls taken since 9/11 have shown that a majority of Americans believe Iraq was deeply involved in the attacks. A poll taken October 2002 showed 66 percent of Americans believe Iraq was involved in the 9/11 attacks. [REUTERS, 10/10/02] This month, polling by the *Washington Post* finds the number slightly rising to 69 percent. In a January 2003 Knight-Ridder poll, 50 percent of respondents even believed one or more of the 9/11 hijackers came from Iraq; only 17 percent said none came from Iraq. [WASHINGTON POST, 9/6/03] President Bush and Vice President Cheney have made carefully worded statements implying Iraq 9/11 links. For instance, in a September 2002 speech, President Bush claimed, "You can't distinguish between al-Qaeda and Saddam." [WHITE HOUSE, 09/25/02] A year later, he states, "there's no question that Saddam Hussein had al-Qaeda ties." [WHITE HOUSE, 09/17/03] Vice President Cheney also pushes these claims, claiming in January 2003, "There's overwhelming evidence there was a connection between al-Qaeda and the Iraqi government. I am very confident that there was an established relationship there." [SAN FRANCISCO CHRONICLE, 01/23/04]

September 14–17, 2003: Cheney Links Iraq to 9/11; Bush, Rumsfeld, and Rice All Disavow Cheney's Claim

Vice President Cheney says, "I think it's not surprising that people make [the] connection" between Iraq and 9/11. He adds, "If we're successful in Iraq . . . then we will have struck a major blow right at the heart of the base, if you will, the geographic base of the terrorists who had us under assault now for many years, but most especially on 9/11." However, two days later, Defense Secretary Rumsfeld states that he sees no Iraq-9/11 link. [ASSOCIATED PRESS, 9/16/03] National Security Adviser Rice says the administration has never accused Hussein of directing the 9/11 attacks. [REUTERS, 9/16/03] The next day, Bush also disavows the Cheney statement, stating, "We've had no evidence that Saddam Hussein was involved with September the 11th . . . [but] there's no question that Saddam Hussein has al-Qaeda ties." [ASSOCIATED PRESS, 9/17/03; *WASHINGTON POST*, 9/18/03; *WASHINGTON POST*, 9/29/03]

December 14, 2003: Dubious Document Links Atta to Saddam Hussein's Government

Future Iraqi interim Prime Minister Iyad Allawi backs the validity of a document purporting to show that terrorist Abu Nidal trained Mohamed Atta in Baghdad a few months before 9/11. [*DAILY TELEGRAPH*, 12/14/03] *Newsweek* reports that the document is probably a fabrication, citing both the FBI's detailed Atta timeline and a document expert who, amongst other things, distrusts an unrelated second "item" on the same document, which supports a discredited claim that Iraq sought uranium from Niger. [*NEWSWEEK*, 12/17/03]

January 11, 2004: Former Treasury Secretary Criticizes Bush Administration Focus on Iraq Before 9/11

Paul O'Neill, Bush's treasury secretary from inauguration until early 2003, appears on CBS's *60 Minutes* and on the cover of *Time* magazine as a new book containing his criticisms of Bush is released. [CBS NEWS, 1/10/04; CBS NEWS, 1/11/04; *TIME*, 1/10/04] Amongst his many critical charges reported in the book *The Price of Loyalty,* perhaps the most controversial is the claim, as CBS puts it, that "The Bush administration began making plans for an invasion of Iraq, including the use of American troops, within days of President Bush's inauguration in January of 2001—not eight months later after the 9/11 attacks, as has been previously reported." [CBS NEWS, 1/10/04] O'Neill's book, written by Ron Suskind, is based not only on O'Neill's account, but also 19,000 government documents, including transcripts of private, high-level National Security Council meetings. [CBS NEWS, 1/11/04] The Bush administration angrily reacts to O'Neill's charges, admitting they were targeting Iraq from the first days in office, but claiming they were merely considering different options. They open a probe into whether O'Neill was authorized to disclose the documents he released. O'Neill is later cleared. [*WASHINGTON POST*, 1/13/04]

Treasury Secretary Paul O'Neill

June 14–18, 2004: Bush Administration and 9/11 Commission Clash over Iraq Connections to al-Qaeda

On June 14, Vice President Dick Cheney repeats his insistence that Hussein "had long-established ties with al-Qaeda." [CNN, 6/14/04] However, three days later the 9/11 Commission states it found "no credible evidence" of collaboration by Iraq with the 9/11 attacks, and "no collaborative relationship" between

Iraq and al-Qaeda. [CNN, 6/17/04] President Bush forcefully disputes the commission's findings: "The reason I keep insisting that there was a connection between Iraq and Saddam and al-Qaeda is because there was a relationship between Iraq and al-Qaeda." [ABC NEWS, 6/18/04] Cheney says reporters who doubt the connection are "lazy." He asserts "we don't know" if Iraq was involved in 9/11, says no one has "been able to confirm," or "to knock it down" the claims regarding Atta's alleged meeting in Prague. He also says that he "probably" knows information the commission does not. [CNN 6/18/04] A few days later, the commission says that after asking Cheney for any additional evidence he might have, they stand by their position. Cheney maintains his position as well. [LOS ANGELES TIMES, 7/2/04]

U.S. Dominance in Central Asia and the World

Chapter 15 chronicled the Bush administration's focus on "regime change" in Iraq, and how it used the events of 9/11 to justify its war on Iraq. This chapter chronicles events that address the obvious question: Why? Why was the administration so focused on Iraq, even from the earliest moments of the Bush presidency? Did new (or old) intelligence indicate that the increased terrorist threat originated from Iraq?

While President Bush was new to national politics, his vice president and many of those he selected for senior administration posts were not. Vice President Dick Cheney, Defense Secretary Donald Rumsfeld, and Deputy Defense Secretary Paul Wolfowitz, among others, had all held positions under President George H. W. Bush's administration between 1989 and 1993, when the Soviet Union collapsed and the U.S. became the world's sole superpower.

Beginning in 1992, Cheney, Rumsfeld, Wolfowitz, and others began to articulate their strategy for a new world order in which the U.S. maintained its position of world dominance. The preemptive doctrine of attacking countries that posed a threat before they acted became public policy after 9/11, but in fact it was formed in the waning days of President George H. W. Bush's administration.

Throughout the Clinton presidency, the 1992 plan was further developed at the Project for the New American Century (PNAC), a neoconservative think tank made up mostly of Republicans strategizing and anticipating their return to power. PNAC publicly advocated for regime change in Iraq in order to protect the world's oil supply, and they were ready when commissioned to draft a strategy document in preparation for George W. Bush's 2000 presidential campaign.

PNAC outlined its grand blueprint for global dominance while noting that implementation of the plan could take years "absent some catastrophic and catalyzing event—like a new Pearl Harbor." They repeated many of the same ideas first presented in 1992.

When Bush became president in January 2001, these strategic architects became top ranking government officials. In late 2003, 9/11 Commissioner and former Senator Max Cleland said, "This administration had a point of view the day that [9/11] happened. If you look at 9/11 separately you realize it had nothing to do with Saddam Hussein. Except Cheney and Wolfowitz put a plan together in '92 to try to convince [George H. W.] Bush to invade Iraq. . . . Now, this administration bought the Cheney-Wolfowitz plan from '92 hook line and sinker." [SALON, 11/21/03]

The global dominance plan was centered not just on Iraq and the Middle East, but also on Central Asia. (Because there is a separate chapter on Iraq, this chapter focuses more on the latter region.) The Caspian Sea basin of Central Asia has huge natural resource potential and is also very strategic in a dominance struggle between the U.S., Russia, and China. Former National Security Adviser Zbigniew Brzezinski wrote in 1997 that controlling Central Asia was the key to controlling all of Eurasia. Just as the PNAC did, Brzezinski gave the example of how Pearl Harbor catalyzed public support for war, and noted his Central Asian strategy could not be implemented "except in the circumstance of a truly massive and widely perceived direct external threat." [THE GRAND CHESSBOARD, BY ZBIGNIEW BRZEZINSKI, 10/97, PP. 24–25, 210–11]

Zalmay Khalilzad typifies all these threads. He helped write the 1992 plan; he was a member of the PNAC; he worked with Unocal in the late 1990s to bring a gas pipeline across Afghanistan; and he worked under Brzezinski. [ASIA TIMES, 12/23/03; NEW YORK TIMES, 3/22/03] Today he is the de facto U.S. ambassador to Afghanistan.

Sandy Berger, Clinton's National Security Adviser, stated, "You show me one reporter, one commentator, one member of Congress who thought we should invade Afghanistan before September 11 and I'll buy you dinner in the best restaurant in New York City." [THE CELL, BY JOHN MILLER, MICHAEL STONE, AND CHRIS MITCHELL, 8/02, P. 219] In July 2002, British Prime Minister Tony Blair stated: "To be truthful about it, there was no way we could have got the public consent to have suddenly launched a campaign on Afghanistan but for what happened on September 11." [LONDON TIMES, 7/17/02] So it is certainly interesting how these Iraqi and Central Asian ambitions were able to be fulfilled in the wake of 9/11.

Did the PNAC group anticipate a 9/11-like event to implement a predetermined agenda? How much was the U.S. increasing its influence in Central Asia before 9/11? Did the Bush administration take advantage of the Afghan war to cement U.S. power and influence throughout Central Asia?

January 20, 1989: George H. W. Bush Is Inaugurated U.S. President

George H. W. Bush replaces Ronald Reagan and remains president until 1993. Many of the key members in his government will have important positions again when his son George W. Bush becomes president in 2001. For instance, Joint Chiefs of Staff Chairman Colin Powell later becomes Secretary of State, and Defense Secretary Dick Cheney later becomes Vice President.

George
H.W. Bush

November 9, 1989: Cold War Ends; U.S. Asserts World Dominance

The Berlin Wall begins to fall in East Germany, signifying the end of the Soviet Union as a superpower. Just six days later, Joint Chiefs of Staff Chairman Colin Powell presents a new strategy document to President George H. W. Bush, proposing that the U.S. shift its strategic focus from countering Soviet attempts at world dominance to ensuring U.S. world dominance. George H. W. Bush accepts this plan in a public speech, with slight modifications, on August 2, 1990, the same day Iraq invades Kuwait. In early 1992, Powell, counter to his usual public dove persona, tells congresspersons that the U.S. requires "sufficient power" to "deter any challenger from ever dreaming of challenging us on the world stage." He says, "I want to be the bully on the block." Powell's early ideas of global hegemony will be formalized by others in a 1992 policy document and finally realized as policy when George W. Bush becomes president in 2001. [HARPER'S, 10/02]

Secretary of
State Colin
Powell

March 8, 1992: Raw U.S. World Dominance Plan Is Leaked to the Media

The Defense Planning Guidance, "a blueprint for the department's spending priorities in the aftermath of the first Gulf War and the collapse of the Soviet Union," is leaked to the *New York Times*. [NEW YORK TIMES, 3/8/92; NEWSDAY, 3/16/03] The document causes controversy, because it hadn't yet been "scrubbed" to replace candid language with euphemisms. [NEW YORK TIMES, 3/10/92; NEW YORK TIMES, 3/11/92; OBSERVER, 4/7/02] The document argues that the U.S. dominates the world as sole superpower, and to maintain that role, it "must maintain the mechanisms for deterring potential competitors from even aspiring to a larger regional or global role." [NEW YORK TIMES, 3/8/92; NEW YORK TIMES, 3/8/92 (B)] As the *Observer* summarizes it, "America's friends are potential enemies. They must be in a state of dependence and seek solutions to their problems in Washington." [OBSERVER, 4/7/02] The document is mainly written by Paul Wolfowitz and I. Lewis "Scooter" Libby, who hold relatively low posts at the time, but become Deputy Defense Secretary and Vice President Cheney's Chief of Staff, respectively, under George W. Bush. [NEWSDAY, 3/16/03] The authors conspicuously avoid mention of collective security arrangements through the United Nations, instead suggesting the U.S. "should expect future coalitions to be ad hoc assemblies, often not lasting beyond the crisis being confronted." [NEW YORK TIMES, 3/8/92] They call for "punishing" or "threatening punishment" against regional aggressors before they act. Interests to be defended preemptively include "access to vital raw materials, primarily Persian Gulf oil, proliferation of weapons of mass destruction and ballistic missiles, [and] threats to U.S. citizens from terrorism." [HARPER'S, 10/02] Senator Lincoln Chafee (R) later says, "It is my opinion that [George W. Bush's] plan for preemptive strikes was formed back at the end of the first Bush administration with that 1992 report." [NEWSDAY, 3/16/03] In response to the controversy, U.S. releases an updated version of the document in May 1992, which stresses that the U.S. will work with the United Nations and its allies. [WASHINGTON POST 5/24/92; HARPER'S 10/02]

January 1993: Cheney Releases New Global Domination Strategy

While still serving as Defense Secretary, Dick Cheney releases a documented titled "Defense Strategy for the 1990s," in which he reasserts the plans for U.S. global domination outlined in the Defense Policy Guide leaked to the press in March 1992. [HARPER'S, 10/02] Clinton's inauguration as president later in the month precludes Cheney from actually implementing his plans.

January 20, 1993: Bill Clinton Inaugurated

Bill Clinton replaces George H. W. Bush as U.S. President and remains president until 2001.

July 7, 1996: "A Clean Break" Outlines New Middle East Strategy for Israel

The Institute for Advanced Strategic and Political Studies, an Israeli think tank, publishes a paper entitled "A Clean Break: A New Strategy for Securing the Realm." [CHICAGO SUN-TIMES, 3/6/03] Lead author Richard Perle will later become chairman of President Bush's influential Defense Policy Board. Several other co-authors will hold key positions in Washington after Bush's election. In the paper, Perle and his co-authors advise the new, right wing Israeli leader Binyamin Netanyahu to make a complete break with the past by adopting a strategy "based on an entirely new intellectual foundation, one that restores strategic initiative and provides the nation the room to engage every possible energy on rebuilding Zionism. . . ." The first step is to be the removal of Saddam Hussein in Iraq. A war with Iraq will destabilize the entire Middle East, allowing governments in Syria, Iran, Lebanon, and other countries to be replaced. "Israel will not only contain its foes; it will transcend them," the paper concludes [GUARDIAN, 9/3/02], citing the original paper at [THE INSTITUTE FOR ADVANCED STRATEGIC AND POLITICAL STUDIES, 7/8/96]. Perle will be instrumental is moving Bush's U.S. policy toward war with Iraq after the 9/11 attacks.

June 3, 1997: PNAC Think Tank Issues Statement of Principles

The Project for the New American Century (PNAC), a neoconservative think tank formed in the spring of 1997, issues its statement of principles. PNAC's stated aims are:

- to "shape a new century favorable to American principles and interests"

- to achieve "a foreign policy that boldly and purposefully promotes American principles abroad"

- to "increase defense spending significantly"

- to challenge "regimes hostile to U.S. interests and values"

Project for the New American Century

- to "accept America's unique role in preserving and extending an international order friendly to our security, our prosperity, and our principles." [PNAC PRINCIPLES, 6/3/97]

The Statement of Principles is significant, because it is signed by a group who will become "a roll call of today's Bush inner circle." [GUARDIAN, 2/26/03] ABC's Ted Koppel will later say PNAC's ideas have "been called a secret blueprint for U.S. global domination." [ABC NEWS, 3/5/03 (B)]

October 1997: Brzezinski Highlights the Importance
of Central Asia to Achieving World Domination

Former National Security Adviser Zbigniew Brzezinski publishes a book in which he portrays the Eurasian landmass as the key to world power, and Central Asia with its vast oil reserves as the key to domination of Eurasia. He states that for the U.S. to maintain its global primacy, it must prevent any possible adversary from controlling that region. He notes, "The attitude of the American public toward the external projection of American power has been much more ambivalent. The public supported America's engagement in World War II largely because of the shock effect of the Japanese attack on Pearl Harbor." He predicts that because of popular resistance to U.S. military expansionism, his ambitious Central Asian strategy can not be implemented "except in the circumstance of a truly massive and widely perceived direct external threat." [THE GRAND

Zbigniew Brzezinski

CHESSBOARD: AMERICAN PRIMACY AND ITS GEOSTRATEGIC IMPERATIVES, BY ZBIGNIEW BRZEZINSKI, 10/97, PP. 24–25, 210–11]

January 26, 1998: PNAC Think Tank Urges War Against Iraq

The PNAC think tank publishes a letter to President Clinton, urging war against Iraq and the removal of Saddam Hussein because he is a "hazard" to "a significant portion of the world's supply of oil." Providing a foretaste of what is to come in 2003, PNAC calls for the U.S. to go to war alone, attacks the United Nations, and asserts that the U.S. should not be "crippled by a misguided insistence on unanimity in the UN Security Council." The letter is signed by many who will later lead the 2003 Iraq war: Ten of the 18 signatories later join the Bush administration, including (future) Defense Secretary

Donald Rumsfeld, Deputy Defense Secretary Paul Wolfowitz, Deputy Secretary of State Richard Armitage, Undersecretaries of State John Bolton and Paula Dobriansky, Presidential Adviser for the Middle East Elliott Abrams, and Special Iraq Envoy Zalmay Khalilzad. [SUNDAY HERALD, 3/16/03; PNAC LETTER, 1/26/98] Clinton heavily bombs Iraq in late 1998, but the bombing does not last long. Some argue that the bombing's long-term effect is the halt in United Nations weapons inspections. [NEW YORK TIMES, 3/22/03]

March 3, 1999: New "Pearl Harbor" Needed
to Change U.S. Military Policies, Says Expert

Andrew Krepinevich, Executive Director of the Center for Strategic and Budgetary Assessments, testifies before the Senate Armed Services Subcommittee on Emerging Threats and Capabilities: "There appears to be general agreement concerning the need to transform the U.S. military into a significantly different kind of force from that which emerged victorious from the Cold and Gulf Wars. Yet this verbal support has not been translated into a defense program supporting transformation . . . the 'critical mass' needed to effect it has not yet been achieved. One may conclude that, in the absence of a strong external shock to the United States—a latter-day 'Pearl Harbor' of sorts—surmounting the barriers to transformation will likely prove a long, arduous process." [CSBA, 3/5/99] This comment echoes other strategists at PNAC who wait for a second Pearl Harbor to fulfill their visions.

Early 2000: U.S. Builds Up Influence in Central Asia

By the start of this year, the U.S. has already begun "to quietly build influence" in Central Asia. The U.S. has established significant military-to-military relationships with Kyrgyzstan, Uzbekistan, and Kazakhstan. Americans have trained soldiers from those countries. The militaries of all three have an ongoing relationship with the National Guard of a U.S. state—Kazakhstan with Arizona, Kyrgyzstan with Montana, and Uzbekistan with Louisiana. The countries also participate in NATO's Partnership for Peace program. [WASHINGTON POST, 8/27/02]

April 2000: U.S. Granted Permission to Expand Qatar Military Base

The U.S. obtains permission to expand greatly a military base in the Persian Gulf nation of Qatar, and construction begins shortly thereafter. The justification for expanding Al Adid, a billion-dollar base, is preparedness for renewed action against Iraq. [LOS ANGELES TIMES, 1/6/02] Dozens of other U.S. military bases had sprung up in the region in the 1990s. [VILLAGE VOICE, 11/13/02]

September 2000: PNAC Report Recommends Policies That
Need "New Pearl Harbor" for Quick Implementation

PNAC drafts a strategy document, "Rebuilding America's Defenses: Strategies, Forces and Resources for a New Century," for George W. Bush's team before the 2000 Presidential election. The document was commissioned by future Vice President Cheney, future Defense Secretary Rumsfeld, future Deputy

Defense Secretary Paul Wolfowitz, Florida Governor Jeb Bush (Bush's brother), and future Vice President Cheney's Chief of Staff Lewis Libby. The document outlines:

- A "blueprint for maintaining global U.S. preeminence, precluding the rise of a great power rival, and shaping the international security order in line with American principles and interests."

- PNAC states further, "The United States has for decades sought to play a more permanent role in Gulf regional security. While the unresolved conflict with Iraq provides the immediate justification, the need for a substantial American force presence in the Gulf transcends the issue of the regime of Saddam Hussein."

- PNAC calls for the control of space through a new "U.S. Space Forces," the political control of the Internet, and the subversion of any growth in political power of even close allies, and advocates "regime change" in China, North Korea, Libya, Syria, Iran, and other countries.

- It also mentions that "advanced forms of biological warfare that can 'target' specific genotypes may transform biological warfare from the realm of terror to a politically useful tool."

- However, PNAC complains that the desired changes are likely to take a long time, "absent some catastrophic and catalyzing event—like a new Pearl Harbor." [LOS ANGELES TIMES, 1/12/03]

This strategy document clearly demonstrates that Bush and/or the leading members of his administration planned to attain military control of Persian Gulf oil whether or not Saddam Hussein was in power and sought to retain control of the region even if there is no threat. Notably, while Cheney commissioned this plan (along with other future key leaders of the Bush administration), he defends Bush's position of maintaining Clinton's policy not to attack Iraq, during an NBC interview in the midst of the 2000 presidential campaign, asserting that the U.S. should not act as though "we were an imperialist power, willy-nilly moving into capitals in that part of the world, taking down governments." [WASHINGTON POST, 1/12/02] A British Member of Parliament will later say of the report: "This is a blueprint for U.S. world domination—a new world order of their making. These are the thought processes of fantasist Americans who want to control the world." [SUNDAY HERALD, 9/7/02] Both PNAC and its strategy plan for Bush are almost virtually ignored by the media until a few weeks before the start of the Iraq war.

September 2000: General Tommy Franks Tours Central Asia to Build Military Aid Relationships

U.S. General Tommy Franks tours Central Asia in an attempt to build military aid relationships with nations there, but finds no takers. Russia's power in the region appears to be on the upswing instead. Russian Defense Minister Igor Sergeyev writes, "The actions of Islamic extremists in Central Asia give Russia the chance to strengthen its position in the region." However, shortly after 9/11, Russia and China agree to allow the U.S. to establish temporary U.S. military bases in Central Asia to prosecute the

Afghanistan war. The bases become permanent, and the *Guardian* will write in early 2002, "Both countries increasingly have good reasons to regret their accommodating stand. Having pushed, cajoled, and bribed its way into their Central Asian backyard, the U.S. clearly has no intention of leaving any time soon." [GUARDIAN, 1/10/02]

January 21, 2001: George W. Bush Inaugurated

George W. Bush is inaugurated as President, replacing Bill Clinton.

May 16, 2001: U.S. Strengthens Military Relations with Central Asian Republics

U.S. General Tommy Franks, later to head the U.S. occupation of Afghanistan, visits the capital of Tajikistan. He says the Bush administration considers Tajikistan "a strategically significant country" and offers military aid. This follows a visit by a Department of Defense official earlier in the year. The *Guardian* later asserts that by this time, "U.S. Rangers were also training special troops in Kyrgyzstan. There were unconfirmed reports that Tajik and Uzbek special troops were training in Alaska and Montana." [GUARDIAN, 9/26/01]

June 2001: New Organization Aims to Counter U.S. Dominance in Central Asia

China, Russia, and four Central Asian countries create the Shanghai Cooperation Organization. Its explicit purpose is to oppose U.S. dominance, especially in Central Asia. [GUARDIAN, 10/23/01] Russian defense minister Igor Sergeyev writes, "The actions of Islamic extremists in Central Asia give Russia the chance to strengthen its position in the region." [GUARDIAN, 1/16/02] In March 2003, the *Guardian* will note that the new ring of U.S. military bases built in the Afghan war "has, in effect, destroyed the Shanghai Cooperation Organization which Russia and China had established in an attempt to develop a regional alternative to U.S. power." [GUARDIAN, 3/11/03]

August 21, 2001: PNAC Think Tank Leader States
U.S. Should Embrace Imperialist Hegemon Role

Thomas Donnelly, deputy executive director of the PNAC, explains to the *Washington Post* that the U.S. should embrace its role as imperialist hegemon over the world. He says many important politicians privately agree with him. "There's not all that many people who will talk about it openly," he says. "It's discomforting to a lot of Americans. So they use code phrases like 'America is the sole superpower.'" He also says, "I think Americans have become used to running the world and would be very reluctant to give it up, if they realized there were a serious challenge to it." [WASHINGTON POST, 8/21/01] Such statements of policy had been publicly denounced by Bush prior to his election, and some claim that the Bush administration only changes its mind toward a more aggressive policy after 9/11. However, this claim is inconsistent with the roles of senior Bush officials such as Vice President Cheney, Defense Secretary Rumsfeld, and Deputy Defense Secretary Wolfowitz in formulating the preemptive doctrine in 1992 then pushing for it in PNAC during the Clinton administration. In the summer of 2001, Defense Secretary Rumsfeld's office "sponsored a study of ancient empires—Macedonia, Rome, the Mongols—to figure out how they maintained dominance." [NEW YORK TIMES, 3/5/03]

September 15, 2001: Bush Approves the CIA's "Worldwide Attack Matrix" Action Plan

CIA Director Tenet briefs Bush "with a briefcase stuffed with top-secret documents and plans, in many respects the culmination of more than four years of work on bin Laden, the al-Qaeda network and worldwide terrorism." In his briefing, Tenet advocates "a strategy to create 'a northern front, closing the safe haven [of Afghanistan].' His idea [is] that Afghan opposition forces, aided by the United States, would move first against the northern city of Mazar-i-Sharif, try to break the Taliban's grip on that city and open up the border with Uzbekistan. From there the campaign could move to other cities in the north . . ." Tenet also explains that the CIA had begun working with a number of tribal leaders to stir up resistance in the south the previous year. Tenet then turns to a top secret document called the "Worldwide Attack Matrix," which describes covert operations in 80 countries that are either underway or now recommended. The actions range from routine propaganda to lethal covert action in preparation for military attacks. The military, which typically plans such military campaigns, is caught relatively unprepared and so it defers to the CIA plans. [WASHINGTON POST, 1/31/02]

September 22, 2001–December 2001: U.S. Secretly Increases Military Presence in Central Asia

Witnesses begin to report U.S. military planes secretly landing at night in the Central Asian nations of Uzbekistan and Tajikistan. The U.S., Tajik, and Uzbek governments initially deny that any U.S. troops have been sent there. [DAILY TELEGRAPH, 9/23/01 (D); ASSOCIATED PRESS, 9/25/01 (D)] By October 5, witnesses say a "huge military buildup" has already occurred. [DAILY TELEGRAPH, 10/4/01] On October 7, the U.S. and Uzbekistan sign a secret agreement that reportedly is "a long-term commitment to advance security and regional stability." [FINANCIAL TIMES, 10/13/01] It is later reported that the U.S. military bases here, "originally agreed as temporary and emergency expedients, are now permanent." [GUARDIAN, 1/16/02] The U.S. begins building a military base in the nearby country of Kyrgyzstan in December 2001. "There are no restrictions" in the agreement on what the U.S. can do with this base, and it will be a "transportation hub" for the whole region. [NEW YORK TIMES, 1/9/02] The base is only 200 miles from China. [CHRISTIAN SCIENCE MONITOR, 1/17/02] The building of these bases is the culmination of the strategy first proposed in 1992 by the men now in power.

Early October 2001: U.S. Launches Attacks on Afghanistan from Pakistani Bases

The U.S. begins using the Shahbaz air force base and other bases in Pakistan in their attacks against Afghanistan. [LONDON TIMES, 10/15/01] However, because of public Pakistani opposition to U.S. support, the two governments claim the U.S. is there for purely logistical and defensive purposes. Even six months later, the U.S. refuses to confirm it is using the base for offensive operations. [LOS ANGELES TIMES, 3/6/03] Such bases in Pakistan become a link in a chain of U.S. military outposts in Central Asia. Other countries also falsely maintain that such bases are not being used for military operations in Afghanistan despite clear evidence to the contrary. [REUTERS, 12/28/01]

November 21, 2001: Bush Says "Afghanistan Is Just the Beginning"

Bush states that "Afghanistan is just the beginning on the war against terror. There are other terrorists who threaten America and our friends, and there are other nations willing to sponsor them. We will not be secure as a nation until all of these threats are defeated. Across the world and across the years, we will fight these evil ones, and we will win." [WHITE HOUSE, 11/21/01] A short time later, it is reported that "the U.S. has honed a hit list of countries to target for military action in rogue regions across the globe where it believes terror cells flourish," including Iraq. [GUARDIAN, 12/10/01]

December 19, 2001: U.S. Official Proclaims "We Will Not Leave Central Asia"

Speaking in Kazakhstan, U.S. Deputy Secretary of State Elizabeth Jones states: "We will not leave Central Asia after resolving the conflict [in Afghanistan]. We want to support the Central Asian countries in their desire to reform their societies as they supported us in the war against terrorism. These are not only new but long-term relations." [BBC, 12/19/01]

January 2002: Central Asian Countries See U.S. Military Bases Expand

Reportedly, the U.S. is improving bases in "13 locations in nine countries in the Central Asian region" [CHRISTIAN SCIENCE MONITOR, 1/17/02] U.S. military personnel strength in bases surrounding Afghanistan has increased to 60,000. [LOS ANGELES TIMES, 1/6/02] "Of the five ex-Soviet states of Central Asia, Turkmenistan alone is resisting pressure to allow the deployment of U.S. or other Western forces on its soil . . . ." [GUARDIAN, 1/10/02] "The task of the encircling U.S. bases now shooting up on Afghanistan's periphery is only partly to contain the threat of political regression or Taliban resurgence in Kabul. Their bigger, longer-term role is to project U.S. power and U.S. interests into countries previously beyond its reach. . . . The potential benefits for the U.S. are enormous: growing military hegemony in one of the few parts of the world not already under Washington's sway, expanded strategic influence at Russia and China's expense, pivotal political clout and—grail of holy grails—access to the fabulous, non-OPEC oil and gas wealth of central Asia." [GUARDIAN, 1/16/02] On January 9, the speaker of the Russian parliament states, "Russia would not approve of the appearance of permanent U.S. bases in Central Asia," but Russia seems helpless to stop what a Russian newspaper calls "the inexorable growth" of the U.S. military presence in Central Asia. [GUARDIAN, 1/10/02]

March 30, 2002: U.S. Wants Our Energy, Claims Kazakhstan official

With U.S. troops already in many Central Asian countries, U.S. Special Forces soldiers are now reportedly training Kazakhstan troops in a secret location. [LONDON TIMES, 3/30/02] An anonymous source in the Kazakh government previously stated, "It is clear that the continuing war in Afghanistan is no more than a veil for the U.S. to establish political dominance in the region. The war on terrorism is only a pretext for extending influence over our energy resources." [OBSERVER, 1/20/02]

April 30, 2002: U.S. Military Plans Long-Term Presence in Central Asia

It is reported that the U.S. military is drawing up a plan for a long-term military "footprint" in Central Asia. The U.S. says it plans no permanent bases, but the leaders of Central Asia speak of the U.S. being there for decades, and the temporary structures that had been hastily constructed over the past several months are being replaced by permanent buildings. [ASSOCIATED PRESS, 4/30/02; WASHINGTON POST, 8/27/02; LOS ANGELES TIMES, 4/4/02] All of the countries are encumbered by corrupt dictatorships, and many experts say their serious social and economic problems are growing worse. Some experts wonder if the U.S. is increasing Muslim resentment and the risk of terrorism by closely associating with such regimes. [WASHINGTON POST, 8/27/02]

June 1, 2002: Bush Launches Doctrine of Preemptive Attack

In a speech, President Bush announces a "new" U.S. policy of preemptive attacks: "If we wait for threats to fully materialize we will have waited too long. We must take the battle to the enemy, disrupt his plans and confront the worst threats before they emerge." [NEW YORK TIMES, 6/2/02] This preemptive strategy is included in a defensive strategic paper the next month, and formally announced in September 2002. Despite the obvious parallels, the mainstream media generally fails to report that this "new" antiterrorism strategy was first proposed by his key administration officials in 1992 and continually advocated by the same people ever since. [NEW YORK TIMES, 9/20/02; WASHINGTON POST, 9/21/02; GUARDIAN, 9/21/02]

July 13, 2002: U.S. Military Plans for Global Dominance

The U.S. military releases a new Defense Planning Guidance strategic vision. It "contains all the key elements" of a similar document written ten years earlier by largely the same people now in power. Like the original, the centerpiece of this vision is preventing any other powers from challenging U.S. world dominance. Some new tactics are proposed, such as using nuclear weapons for a preemptive strike, but the basic plan remains the same. [LOS ANGELES TIMES, 7/13/02; LOS ANGELES TIMES, 7/16/02; HARPER'S, 10/02] David Armstrong notes in *Harper's* magazine, "[In 1992] the goal was global dominance, and it met with bad reviews. Now it is the answer to terrorism. The emphasis is on preemption, and the reviews are generally enthusiastic. Through all of this, the dominance motif remains, though largely undetected." [HARPER'S, 10/02]

August 11, 2002: Bush's Advisers Advocate Attacking Saudi Arabia, Iran, Egypt, and Other Countries

A *Newsweek* article suggests that some of Bush's advisers advocate not only attacking Iraq, but also Saudi Arabia, Iran, North Korea, Syria, Egypt, and Burma, shocking many. One senior British official says: "Everyone wants to go to Baghdad. Real men want to go to Tehran." [NEWSWEEK, 8/11/02; NEWSWEEK, 8/11/02 (B)] Later in the year, Bush's influential adviser Richard Perle states, "No stages. This is total war. We are fighting a variety of enemies. There are lots of them out there. All this talk about first we are going to do Afghanistan, then we will do Iraq . . . this is entirely the wrong way to go about it. If we just let our vision of the world go forth, and we embrace it entirely and we don't try to piece together clever diplomacy, but just wage a total war . . . our children will sing great songs about us years from now." [NEW STATESMAN, 12/16/02] In February 2003, U.S. Undersecretary of State John Bolton says in meetings with Israeli officials that he has no doubt America will attack Iraq, and that it will be necessary to deal with threats from Syria, Iran, and North Korea afterward. This is not reported in the U.S. media. [HA'ARETZ, 2/17/03]

August 27, 2002: U.S. Establishes Military Presence in Uzbekistan

The Central Asian nation of Uzbekistan has recently signed a treaty committing the U.S. to respond to "any external threat" to the country. Uzbekistan's foreign minister explains: "The logic of the situation suggests that the United States has come here with a serious purpose, and for a long time." According to a *Washington Post* report, the other Central Asian nations—Kazakhstan, Kyrgyzstan, Tajikistan, and Turkmenistan—have similar agreements with the U.S. The U.S. claims it is supporting democracy in these nations, but experts say authoritarianism has been on the rise since 9/11. A new U.S. military base in Uzbekistan currently holds about 1,000 U.S. soldiers, but is being greatly enlarged. The article makes the general point that the U.S. is replacing Russia as the dominant power in Central Asia. [WASHINGTON POST, 8/27/02]

September 26, 2002: New U.S. Military Organization Will "Stimulate" Terrorist Attacks Instead of Stopping Them

A leaked August 16, 2002 report from Defense Secretary Rumsfeld's influential Defense Science Board 2002 is exposed. [UPI, 9/26/02] The board "recommends creation of a super-intelligence support [agency], an organization it dubs the Proactive, Preemptive Operations Group, (P2OG), to bring together CIA and military covert action, information warfare, intelligence, and cover and deception. Among other things, this body would launch secret operations aimed at 'stimulating reactions' among terrorists and states possessing weapons of mass destruction—that is, for instance, prodding terrorist cells into action and exposing themselves to 'quick-response' attacks by U.S. forces. Such tactics would hold 'states/sub-state actors accountable' and 'signal to harboring states that their sovereignty will be at risk.'" [LOS ANGELES TIMES, 10/27/02; ASIA TIMES, 11/5/02] An editorial in the *Moscow Times* comments: "In other words—and let's say this plainly, clearly and soberly, so that no one can mistake the intention of Rumsfeld's plan—the United States government is planning to use 'cover and deception' and secret military operations to provoke murderous terrorist attacks on innocent people." It is further suggested terrorists could be instigated in countries the U.S. wants to gain control over.

Defense Secretary Donald Rumsfeld

[MOSCOW TIMES, 11/1/02]

February–March 20, 2003: Stories About PNAC
Global Domination Agenda Gets Some Media Coverage

With war against Iraq imminent, numerous media outlets finally begin reporting on PNAC's role in influencing Iraq policy specifically, and U.S. foreign policy generally. PNAC's plans for global domination had been noted before 9/11 [WASHINGTON POST, 8/21/01], and PNAC's 2000 report recommending the conquest of Iraq even if Saddam Hussein is not in power was first reported on in September 2002 [SUNDAY HERALD, 9/7/02], but there are few follow-up mentions until February 2003. (Exceptions: [ATLANTA JOURNAL AND CONSTITUTION, 9/29/02; BANGOR DAILY NEWS, 10/18/02; NEW STATESMAN, 12/16/02; LOS ANGELES TIMES, 1/12/03]) Many of these articles use PNAC to suggest that global and regional domination is the real reason for the Iraq war. Coverage increases as war gets nearer, but many media outlets still fail to do any reporting on this, and some of the reporting that is done is not prominently placed (a *New York Times* article on the topic is buried in the Arts section! [NEW YORK TIMES, 3/11/03]). One *Newsweek* editorial notes that "not until the last few days" before war have many reasons against the war been brought up. It calls this "too little, too late" to make an impact. [NEWSWEEK, 3/18/03] (Articles that discuss PNAC before war begins: [PHILADELPHIA DAILY NEWS, 1/27/03; NEW YORK TIMES, 2/1/03; PBS FRONTLINE, 2/20/03; OBSERVER, 2/23/03; BERGEN RECORD, 2/23/03; GUARDIAN, 2/26/03; MOTHER JONES, 3/03; BBC, 3/2/03; OBSERVER, 3/2/03 (B); DER SPIEGEL, 3/4/03; ABC NEWS, 3/5/03; SALON, 3/5/03; INDEPENDENT, 3/8/03; TORONTO STAR, 3/9/03; ABC NEWS, 3/10/03; AUSTRALIAN BROADCASTING CORP., 3/10/03; CNN, 3/10/03; GUARDIAN, 3/11/03; NEW YORK TIMES, 3/11/03; AMERICAN PROSPECT, 3/12/03; CHICAGO TRIBUNE, 3/12/03; GLOBE AND MAIL, 3/14/03; JAPAN TIMES, 3/14/03; SYDNEY MORNING HERALD, 3/15/03; SALT LAKE TRIBUNE, 3/15/03; MINNEAPOLIS STAR-TRIBUNE, 3/16/03; OBSERVER, 3/16/03; SUNDAY HERALD, 3/16/03; TORONTO STAR, 3/16/03; CANADIAN BROADCASTING CORP., 3/17/03; GLOBE AND MAIL, 3/19/03; ASIA TIMES, 3/20/03; THE AGE, 3/20/03].)

April 3, 2003: Ex-CIA Director Foresees Many More Wars in Middle East

Former CIA Director James Woolsey says the U.S. is engaged in a world war, and that it could continue for years: "As we move toward a new Middle East, over the years and, I think, over the decades to come . . . we will make a lot of people very nervous." He calls it World War IV (World War III being the Cold War according to neoconservatives like himself), and says it will be fought against the religious rulers of Iran, the "fascists" of Iraq and Syria, and Islamic extremists like al-Qaeda. He singles out the leaders of Egypt and Saudi Arabia, saying, "We want you nervous." This echoes the rhetoric of the PNAC, of which Woolsey is a supporter, and the singling out of Egypt and Saudi Arabia echoes the rhetoric of the Defense Policy Board, of which he is a member. In July 2002, a presentation to that board concluded, "Grand strategy for the Middle East: Iraq is the tactical pivot. Saudi Arabia the strategic pivot. Egypt the prize." [CNN, 4/3/03; CNN, 4/3/03 (B)]

PART IV
THE ATTACKS

ALLEGED 9/11 HIJACKERS

Waleed Alshehri

American Airlines Flight 11

Wail Alshehri   Mohamed Atta   Abdulaziz Alomari   Satam Al Suqami

United Airlines Flight 93

  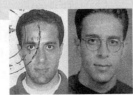 

Ahmed Alhaznawi   Ahmed Alnami   Ziad Jarrah   Saeed Alghamdi

United Airlines Flight 175

Marwan Alshehhi   Fayez Ahmed Banihammad

Mohand Alshehri   Hamza Alghamdi   Ahmed Alghamdi

Khalid Almihdhar   Majed Moqed

American Airlines Flight 77

Nawaf Alhazmi   Salem Alhazmi   Hani Hanjour

17.

The Day of 9/11

Regardless of what may or may not have been known prior to 9/11, no analysis of the terrorist threat and the U.S. response would be complete without a close examination of exactly what took place on that day. That morning, every minute mattered. A response delayed by a single minute could—and quite possibly did—affect countless lives. Could one or more of the plane hijackings have been prevented? Could one or more of the planes have been prevented from crashing? To answer these questions, one must analyze *exactly* who knew what, when they learned it, and how they responded to what they may have known.

The failures began even before the hijackers boarded the airplanes. All the hijackers managed to proceed successfully through airport security, although several of them were identified for increased security screening and apparently had dubious passports or forged U.S. identity cards; at least three of the hijackers were either listed on the U.S. terrorism watch list or had outstanding arrest warrants.

Once the hijackings were confirmed by flight controllers, it is difficult to understand why all planes in the air were not alerted to the crisis more quickly. For example, Flight 93 was apparently hijacked around 9:30 A.M., long after it was clear that there was a crisis involving multiple planes. But the flight dispatcher who tried to warn Flight 93, just moments before it was hijacked, was apparently instructed not to give the details of other hijackings to the pilots he wanted to warn. American Airlines headquarters was notified that its Flight 11 had been hijacked at 8:21 A.M., nearly 30 minutes before its Flight 77 was hijacked, yet failed to provide potentially life-saving information to its pilots (or, apparently, to anyone else). Why was such information withheld?

Although airport security failed to prevent the hijackers from boarding, and although withholding key information resulted in potentially avoidable additional hijackings, there remained the possibility that the loss of life could have been diminished. And yet U.S. defense forces failed to seize that opportunity as well.

The most mysterious failure concerns the slow response of U.S. fighters to the hijacked planes. The North American Aerospace Defense Command (NORAD) is the military agency charged with defending America's skies. According to NORAD leaders, fighters on alert were generally prepared to scramble (i.e. launch) within 15 minutes of being notified. Because military exercises were in process on 9/11, fighters were actually able to scramble in only 6–8 minutes [9/11 COMMISSION, 6/17/04 (B)] and then could have reached target cities in another 6–10 minutes had they traveled close to full speed.

How, then, were the fighters unable to intercept even the last hijacked plane, Flight 93, an hour and forty-five minutes after the U.S. government knew that the first plane had been hijacked? Was NORAD so completely unprepared for such an event? It cannot be said that such an event was wholly unexpected, because NORAD had been conducting contingency exercises that anticipated attacks inside the U.S. startlingly similar to the 9/11 attack. For example, in the two years before 9/11, one NORAD war game simulated a hijacker airliner crashing into the World Trade Center. [USA TODAY, 4/18/04] A live fighter exercise conducted prior to 9/11 simulated a foreign airplane crashing into a famous U.S. building. [CNN, 4/19/04] An exercise in the planning stages on the day of 9/11 hypothesized simultaneous hijackings, including one from Salt Lake City—deep inside the U.S. [CNN, 6/4/02 (B); AMERICAN FORCES PRESS SERVICE, 6/4/02]

In 2000, fighters were scrambled 129 times when problems (e.g., pilots who did not file flight plans, diverted from filed flight plans, or used the wrong radio frequency) were not immediately resolved. [CALGARY HERALD, 10/13/01] Between September 2000 and June 2001, fighters scrambled 67 times. [ASSOCIATED PRESS, 8/13/02] The standard fighter-response time in these cases unfortunately has not been made available to the public.

A NORAD spokesperson explains what typically happens when a fighter scrambles: "When planes are intercepted, they typically are handled with a graduated response. The approaching fighter may rock its wingtips to attract the pilot's attention, or make a pass in front of the aircraft. Eventually, it can fire tracer rounds in the airplane's path, or, under certain circumstances, down it with a missile." [BOSTON GLOBE, 9/15/01] Standard procedures to determine when flight controllers should consider a situation an emergency were also in place. The regulations governing flight controllers make clear that an aircraft emergency exists when "there is unexpected loss of radar contact and radio communications with any . . . aircraft." [FAA ORDER 7110.65P] Another regulation provided further guidance: "If . . . you are in doubt that a situation constitutes an emergency or potential emergency, handle it as though it were an emergency." [FAA FLIGHT SERVICES MANUAL]

One would think, then, that NORAD and the flight controllers had the technology, procedures, and training necessary to respond quickly to an emerging crisis. Even in the case of the first attack, flight controllers responded quickly enough to attempt to contact fighters in the air. Had the flight controllers been able to reach them, the fighters presumably could have reached Flight 11 before it crashed into the World Trade Center. While these fighters were not armed with missiles, they could have initiated the procedural steps outlined above to determine whether it was, in fact, hijacked, and whether the crash was accidental or intentional. Presumably, had such information been immediately available, the rest of the morning could have unfolded quite differently.

Could the loss have been less? How many of the planes could have been stopped before wreaking so much devastation in Pennsylvania, New York, and Washington?

One newspaper has called the "restrained—even failed—standard U.S. military air defense protocols while the attacks were occurring" a "real mystery" that deserves a serious investigation. [SARASOTA HERALD-TRIBUNE, 5/20/03] A few days after 9/11, the *Village Voice* commented, "In the wake of the terrorist attacks, questions are mounting about America's airline safety system and, beyond that, the military's air-defense screen." [VILLAGE VOICE, 9/13/01] The *Miami Herald* quoted one flight controller who asked, "What the hell went on here? Was anyone doing anything about it? Just as a national defense thing, how [were the hijackers] able to fly around and no one go after them?" [MIAMI HERALD, 9/14/01] Yet the mainstream U.S. media has devoted little attention to these questions in the years since.

As this chapter details, there are many conflicting accounts as to what happened. Finding the truth is no easy task. It appears that accounts of the organizations and individuals involved have often tended to minimize their own failures. For example, Flight 77 was hijacked around 8:55 A.M., yet fighters were not launched toward Washington until 9:30 A.M., and even then they may not have been launched with the knowledge that Flight 77 was headed in that direction. Where did the breakdown in communication and or procedure occur? Not surprisingly, the FAA has blamed NORAD, and NORAD has blamed the FAA. Both have given accounts that dramatically conflict with each other with respect to some aspects, and refused to provide any information whatsoever with respect to other aspects. As a result, there remain many conflicting accounts of exactly what occurred on that day.

The closest approximation to a full investigation conducted to date has been conducted by the 9/11 Commission. However, the commission is rife with conflicts of interest, staffed with employees from the FAA, the military, the FBI, the White House, and other governmental agencies—i.e., the very institutions that responded to the crisis on 9/11. The creation of the commission itself was a political compromise, and partisan politics and political pressures are inherent in the way it was established and has operated. Indeed, *Newsweek* has reported that at least one committee conclusion was watered down prior to public release due to intense pressure from the White House. [NEWSWEEK, 6/20/04] As can be seen in the detailed account below, some of the commission's conclusions seem to fly in the face of known facts and testimonies, and conflicting accounts and details have not yet been resolved or explained.

Nonetheless, the 9/11 Commission's June 2004 interim staff report explored the day of 9/11 in some detail, and reported many failures on the part of multiple governmental agencies. Yet, the report concluded, "We do not believe that an accurate understanding of the events of that morning reflects discredit on the operational personnel from NEADS or FAA facilities." [9/11 COMMISSION REPORT, 6/17/04] No specific individuals were blamed. Several airline managers learned of a vital early clue that could have saved lives if shared, but instead they were recorded saying, "Don't spread this around," and "Keep it quiet." [NEW YORK OBSERVER, 6/17/04] But not even these managers have been held accountable in any way. In fact, the 9/11 Commission often seemed to go out of its way to exonerate people rather than to assess shortcomings that could be avoided in the future.

The 9/11 Commission has taken some important first steps in the investigation of the events leading up to, and occurring on, that tragic day. However, it should not—indeed cannot—be considered the final arbiter of facts on the events of 9/11. Even if the commission were a wholly apolitical institution subject to no outside pressures, its conclusions should not be considered the definitive account simply because several other investigations into events relating to 9/11, such as the FBI's, are still ongoing. Additionally, the Bush administration continues to withhold much information in the name of "national security" or executive privilege.

In short, the full truth is not known at this time—and may never be known. There remain simply too many unanswered questions, and too many conflicting accounts, including those of the most senior administration officials, all the way up to President Bush himself. However, one can review what we *do* know, judge the conflicting information based on what we *do* know, and draw at least some of our own conclusions.

And we can continue searching for the truth. Until we can fully understand what happened—and why—we will not be able to fully learn from the failures of that day in order to prevent another, even more devastating attack on America.

Responding Organizations and Key Players

NORAD

The North American Aerospace Defense Command is in charge of defending America's airspace.

Key NORAD players:

General Ralph Eberhart: NORAD's Commander in Chief. He is at Tyndall Air Force Base, Florida, on 9/11.

Major General Larry Arnold: Commander of NORAD's continental U.S. region on 9/11. He is also in Florida.

Major General Rick Findley: A Canadian Air Force officer, he is in charge of battle stations at NORAD's Cheyenne Mountain, Colorado headquarters due to a joint U.S.–Canada exercise known as Vigilant Guardian.

Captain Mike Jellinek: A Canadian, Jellinek is overseeing NORAD's Colorado headquarters on 9/11.

NEADS

NORAD's "Northeast Air Defense Sector." Based in upstate New York, the NEADS sector covers much of the eastern U.S., including the Washington, D.C., New York City, and Chicago areas. Out of NORAD's 14 fighters on a permanent state of alert, only four are based in the NEADS sector. Two are at Langley Air Force Base, Virginia (south of Washington), and two are at Otis Air National Guard Base, Massachusetts (east of New York City). [BBC, 8/29/02] NORAD claims that on 9/11 these fighters are guaranteed to get airborne within 15 minutes. [NORAD TESTIMONY, 5/23/03]

Key NEADS players:

Colonel Robert Marr: head of NEADS.

FAA

The Federal Aviation Administration is responsible for the safety of civil aviation in the U.S.

Key FAA players:

Jane Garvey: FAA administrator.

Ben Sliney: FAA's National Operations Manager. 9/11 is his first day serving as Manager.

Mike McCormick: manager at the New York flight control center.

NMCC

The National Military Command Center is the Pentagon's emergency response center. It is located inside the Pentagon (on the opposite side from where the Pentagon is attacked on 9/11).

Key NMCC players:

Donald Rumsfeld: Secretary of Defense. He enters the NMCC on 9/11 at 10:30 A.M.

Brigadier General Montague Winfield: Commander of the NMCC.

Captain Charles Leidig: Deputy for Command Center Operations at the NMCC. He takes over temporarily from Brigadier General Montague Winfield, and he is effectively in charge of the NMCC during the 9/11 crisis.

PEOC

The Presidential Emergency Operation Center is an underground bunker in the East Wing of the White House.

Key PEOC players:

Richard "Dick" Cheney: Vice President. Present in the PEOC on 9/11.

Condoleezza Rice: National Security Adviser. Present in the PEOC on 9/11.

Norman Mineta: Secretary of Transportation. Present in the PEOC on 9/11.

Other key players:

General Henry Shelton: Chairman of the Joint Chiefs of Staff, the head of the U.S. military. On 9/11, he is flying across the Atlantic on the way to Europe and he is effectively out of the picture during the crisis.

General Richard Myers: Vice Chairman of the Joint Chiefs of Staff. Myers is the acting Joint Chiefs of Staff Chairman on 9/11 because General Shelton is unavailable. Myers enters the NMCC around 10:30 A.M.

Richard Clarke: National Coordinator for Counterterrorism, commonly known as the counterterrorism "tsar." On 9/11, he is in the Secure Video Conferencing Center (West Wing of the White House). He directs the response to the 9/11 attacks, at the request of National Security Adviser Rice. He stays in contact with the PEOC and other key players (including Myers, Rumsfeld, and Garvey) through video links.

The Hijackers and the Flights

American Airlines Flight 11
Boeing 767–223ER flying from Boston to Los Angeles with 23,980 gallons of fuel. 81 passengers, nine flight attendants, and two pilots.

Alleged Hijackers
Waleed Alshehri (Passenger 2, Seat 2B [1])
Wail Alshehri (Passenger 1, Seat 2A [1])
Mohamed Atta (the likely pilot) (Seat 8A [2] or 8D [1, 7])
Abdulaziz Alomari (Passenger 14, Seat 8B [2] or 8G [1, 7])
Satam Al Suqami (Passenger 20, Seat 10B [1])

United Airlines Flight 93
A Boeing 757 flying from Newark to San Francisco. 38 passengers (out of 182 seats), five flight attendants, and two pilots.

Alleged Hijackers
Ahmed Alhaznawi (Seat 6B [7])
Ahmed Alnami (Seat 3C [7])
Ziad Jarrah (the likely pilot) (Seat 1B [7])
Saeed Alghamdi (had flight training) (Seat 3D [7])

United Airlines Flight 175
Boeing 767–222 flying from Boston to Los Angeles with 23,980 gallons of fuel. 56 passengers, seven flight attendants, and two pilots.

Alleged Hijackers
Marwan Alshehhi (the likely pilot) (Passenger 4, Seat 6C [4])
Fayez Ahmed Banihammad (had flight training) (Passenger 6, Seat 2A [3])
Mohand Alshehri (had flight training) (Passenger 5, Seat 2B [7])
Hamza Alghamdi (Passenger 3, Seat 9C [3])
Ahmed Alghamdi (Passenger 2, Seat 9D [3])

American Airlines Flight 77

Boeing 757–223 flying from Dulles Airport outside Washington to Los Angeles with 11,489 gallons of fuel. 58 passengers, four flight attendants, and two pilots.

Alleged Hijackers

Khalid Almihdhar (Passenger 20, Seat 12B [1])

Majed Moqed (Passenger 19, Seat 12A [1])

Nawaf Alhazmi (Passenger 12, Seat 5C [6] or 5E [7])

Salem Alhazmi (Passenger 13, Seat 5F [1])

Hani Hanjour (the likely pilot) (Seat 1B [5])

[1]=[ABC NEWS, 9/15/01]; [2]=[*CHICAGO SUN–TIMES*, 9/16/01]; [3]=[*NEW YORK TIMES*, 9/15/01]; [4]=[*NEW YORK TIMES*, 11/4/01]; [5]=[*WALL STREET JOURNAL*, 10/16/01]; [6]=[CNN, 9/28/01]; [7] = [*SAN FRANCISCO CHRONICLE*, 7/23/04]; ALL PASSENGER NUMBERS FROM [ABC NEWS, 9/15/01] MAXIMUM PLANE CAPACITIES FROM [*TIME*, 9/14/01]

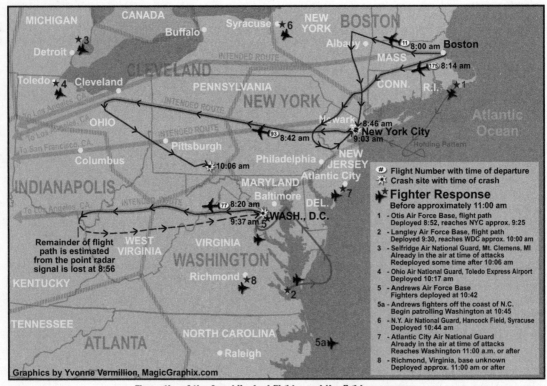

The paths of the four hijacked flights, and the fighter response

Note: All times given are Eastern Standard Time. Where time is reported in seconds, the time noted has been rounded to the nearest minute. An asterisk indicates approximate times. This occurs when different times are mentioned in different sources, and/or sources indicate times that are not precise. When accounts of exact times conflict, the time noted reflects an average of the reported times.

Abdulaziz Alomari and Mohamed Atta (in dark shirt) go through security in the Portland, Maine, airport

5:53 A.M.*: Two Hijackers Caught on Video as They Board a Flight to Boston

Hijackers Mohamed Atta and Abdulaziz Alomari board a Colgan Air flight from Portland, Maine, to Boston, Massachusetts. A security camera captures Atta and Alomari going through security at the Portland airport. [TIME, 9/24/01; FBI, 10/4/01; MIAMI HERALD, 9/22/01; NEW YORK DAILY NEWS, 5/22/02]

6:00 A.M.*: Two Hijackers Fly to Boston

Mohamed Atta and Abdulaziz Alomari's flight from Portland to Boston takes off. [FBI, 10/4/01] Two passengers later say Atta and Alomari board separately, keep quiet, and do not draw attention to themselves. [WASHINGTON POST, 9/16/01; CHICAGO SUN-TIMES, 9/16/01]

6:00 A.M.*: Bush Interview or Assassination Attempt?

President Bush has just spent the night at Colony Beach and Tennis Resort on Longboat Key, Florida. Surface-to-air missiles have been placed on the roof of the resort (it is not known if this was typical of presidential security before 9/11, or if this was related to increased terror warnings). [SARASOTA HERALD-TRIBUNE, 9/10/02] Bush wakes up around 6:00 A.M. and is preparing for his morning jog. [NEW YORK TIMES, 9/16/01 (B); DAILY TELEGRAPH, 12/16/01; MSNBC, 10/27/02] A van occupied by men of Middle Eastern descent arrives at the Colony Beach Resort, stating they have a "poolside" interview with the president. They do not have an appointment and they are turned away. [LONGBOAT OBSERVER, 9/26/01] Some question whether this was an assassination attempt modeled on the one used on Afghani leader Ahmed Massoud two days earlier. [TIME, 8/4/02 (B)]

**The Colony Beach and Tennis Resort
on Longboat Key, Florida**

Longboat Key Fire Marshal Carroll Mooneyhan was reported to have overheard the conversation between the men and the Secret Service, but he later denies the report. The newspaper that reported this, the *Longboat Observer,* say they stand by the story. [*ST. PETERSBURG TIMES,* 7/4/04] Witnesses recall seeing Mohamed Atta in the Longboat Key Holiday Inn a short distance from where Bush was staying as recently as September 7, the day Bush's Sarasota appearance was publicly announced. [*LONGBOAT OBSERVER,* 11/21/01; *ST. PETERSBURG TIMES,* 7/4/04]

6:30 A.M.*: NORAD on Alert for Emergency Exercises

Lieutenant Colonel Dawne Deskins and other day shift employees at NEADS start their workday. NORAD is unusually prepared on 9/11 because it is conducting a weeklong, semi-annual exercise called Vigilant Guardian. [NEWHOUSE NEWS SERVICE, 1/25/02] Deskins is regional Mission Crew Chief for the Vigilant Guardian exercise. [ABC NEWS, 9/11/02] The exercise poses "an imaginary crisis to North American Air Defense outposts nationwide." [NEWHOUSE NEWS SERVICE, 1/25/02] Accounts by participants vary on whether 9/11 was the second, third, or fourth day of the exercise. [NEWHOUSE NEWS SERVICE, 1/25/02; *OTTAWA CITIZEN,* 9/11/02; *CODE ONE MAGAZINE,* 1/02] NORAD is also running another fighter exercise named Operation Northern Vigilance. NORAD is thus fully staffed and alert, and senior officers are manning stations throughout the U.S. The entire chain of command is in place and ready when the first hijacking is reported. An article later says, "In retrospect, the exercise would prove to be a serendipitous enabler of a rapid military response to terrorist attacks on September 11." [*AVIATION WEEK AND SPACE TECHNOLOGY,* 6/3/02; *BERGEN RECORD,* 12/5/03] ABC News later reports that because NORAD is "conducting training exercises [it has] extra fighter planes on alert." [ABC NEWS, 9/14/02] Colonel Robert Marr, in charge of NEADS, says, "We had the fighters with a little more gas on board. A few more weapons on board." [ABC NEWS, 9/11/02] However, Deskins and other NORAD officials later are initially confused about whether the 9/11 attacks are real or part of the exercise. There is a National Reconnaissance Office exercise planned to occur as well, involving a scenario of an airplane as a flying weapon. [ASSOCIATED PRESS, 8/21/02; UPI, 8/22/02]

6:30 A.M.*: Hijackers Cause Scene in Airport, Have Pass to Restricted Airport Areas

A man has an argument with five Middle Eastern men over a parking space in the parking lot of Boston's Logan Airport. He reports the event later in the day and officials discover that the car was rented by Mohamed Atta. Inside the car, police find a ramp pass, which allows access to restricted airport areas. [*NEWS OF THE WORLD,* 9/16/01; *MIAMI HERALD,* 9/22/01]

6:31 A.M.*: Bush Goes Jogging

President Bush goes for a four-mile jog around the golf course at the Colony Beach and Tennis Resort.
[*WASHINGTON POST,* 1/27/02; MSNBC, 10/27/02; *WASHINGTON TIMES,* 10/7/02]

6:45 A.M.*: Israeli Company Given Two Hours' Notice of Attack

"Approximately two hours prior to the first attack," at least two workers at Odigo, an Israeli-owned instant messaging company, receive messages warning of the attack. Odigo's U.S. headquarters are located two blocks from the WTC. The source of the warning is unknown. [WASHINGTON POST, 9/28/01; HA'ARETZ, 9/26/01]

6:47 A.M.: WTC Building 7 Alarm Not Operating

According to later reports, the alarm system in WTC 7 is placed on "TEST" status for a period due to last eight hours. This ordinarily happens during maintenance or other testing, and any alarms received from the building are generally ignored. [NIST PROGRESS REPORT, 6/04, CH.1, P. 28]

6:50 A.M.: Hijacker's Connecting Flight Arrives in Boston

Mohamed Atta and Abdulaziz Alomar's Portland-Boston flight arrives on time at Boston's Logan Airport. [INSIDE 9–11: WHAT REALLY HAPPENED, BY DER SPIEGEL, 2/02]

7:00 A.M.–7:45 A.M.*: Computer Screening Program Selects Some Hijackers; Fails to Stop Them

Sometime during this period, the hijackers pass through airport security checkpoints at the various airports. The FAA has a screening program in place called the Computer Assisted Passenger Prescreening System (CAPPS). CAPPS automatically targets passengers for additional screening based on suspicious behavior such as buying one-way tickets or paying with cash. If a passenger is selected, their bags are thoroughly screened for explosives, but their bodies are not searched. [WASHINGTON POST, 1/28/04] CAPPS selects three of the five Flight 11 hijackers. Since Waleed Alshehri checked no bags, his selection had no consequences. Wail Alshehri and Satam Al Suqami have their bags scanned for explosives, but are not stopped. No Flight 175 hijackers are selected. Only Ahmad Alhaznawi is selected from Flight 93. His bag is screened for explosives, but he is not stopped. The 9/11 Commission later concludes that Alhaznawi and Ahmed Alnami, also headed to Flight 93, have suspicious indicators and that they could have been linked to al-Qaeda upon inspection, but it has not been explained why or how. [9/11 COMMISSION REPORT, 1/27/04; BALTIMORE SUN, 1/27/04] Screening of the Flight 77 hijackers is described below.

7:18 A.M.: Hijack Suspects Set Off Airport Alarms; Allowed to Board Anyway

According to a security video apparently viewed by the 9/11 Commission, Flight 77 hijackers Majed Moqed and Khalid Almihdhar pass through a security checkpoint at Washington's Dulles International Airport. They are selected by the CAPPS program for closer inspection. While their carry-on bags fail to set off any alarms, both set off alarms when passing through the magnetometer. They are directed to a second magnetometer. Almihdhar passes, but Moqed fails again. He is subjected to a personal screening with a metal detection hand wand. This time he is cleared and he is permitted to pass through the checkpoint. The 9/11 Commission later concludes that Almihdhar's passport was "suspicious" and could have been linked to al-Qaeda upon inspection, but it has not been explained why or how. [9/11 COMMISSION, 1/27/04; *BALTIMORE SUN* 1/27/04]

7:35 A.M.: More Hijackers Have Checkpoint Problems; Allowed to Board Anyway

According to the 9/11 Commission's review of airport security footage, the remaining three Flight 77 hijackers pass through a security checkpoint at Dulles Airport. Hani Hanjour is selected for further inspection by the CAPPS program. His two carry-on bags fail to set off any alarms. One minute later,

Hijacker brothers Salem (white shirt) and Nawaf Alhazmi (dark shirt) pass through security at Dulles Airport in Washington

Nawaf Alhazmi and Salem Alhazmi enter the same checkpoint. They are selected for further inspection because one of them does not have photo identification nor is able to understand English and a security agent finds both of them suspicious. [*SAN FRANCISCO CHRONICLE*, 7/23/04] Salem Alhazmi successfully clears the magnetometer, and he is permitted through the checkpoint. Nawaf Alhazmi sets off the alarms for both the first and second magnetometers. He is subsequently subjected to a personal screening with a metal detection hand wand. He finally passes. His shoulder strap carry-on bag is swiped by an explosive trace detector and returned without further inspection. The 9/11 Commission later concludes that the Alhazmi brothers' passports had suspicious indicators, and that they could have been linked to al-Qaeda upon closer inspection. [9/11 COMMISSION, 1/27/04; *BALTIMORE SUN* 1/27/04] To date, video footage of the Flight 77 hijackers has been released to the public, but none of the footage of other hijackers going through security has been.

7:45 A.M.*: Hijack Suspects' Bags Contain Airline Uniforms

Mohamed Atta and Abdulaziz Alomari board Flight 11. Atta's bags are not loaded onto the plane in time and are later found by investigators. Investigators later find airline uniforms and many other remarkable items. [*BOSTON GLOBE*, 9/18/01] It is later reported that at least two other hijackers on Flight 11 use stolen uniforms and IDs to board the plane. [*SUNDAY HERALD*, 9/16/01]

Before 7:59 A.M.*: Inter Flight Phone Call Between Hijackers

Hijacker Mohamed Atta on Flight 11 calls hijacker Marwan Alshehhi in Flight 175 as both planes sit on the runway. They presumably confirm the plot is on. [*TIME*, 8/4/02 (B)]

7:59 A.M.*: Flight 11 Is Late Taking Off

Flight 11 takes off from Boston's Logan Airport, 14 minutes after its scheduled 7:45 departure time. [LOS ANGELES TIMES, 9/20/01; ABC NEWS, 7/18/02; CNN, 9/17/01; WASHINGTON POST, 9/12/01; GUARDIAN, 10/17/01; ASSOCIATED PRESS, 8/19/02 (B); NEWSDAY, 9/10/02; 9/11 COMMISSION REPORT, 6/17/04]

8:00 A.M.*: Bush Receives Daily Intelligence Briefing

President Bush sits down for his daily intelligence briefing. "The President's briefing appears to have included some reference to the heightened terrorist risk reported throughout the summer," but it contained nothing serious enough to cause Bush to call National Security Adviser Rice. The briefing ends around 8:20 A.M. [DAILY TELEGRAPH, 12/16/01]

8:01 A.M.: Flight 93 Is Delayed for 41 Minutes

Flight 93 is delayed for 41 minutes on the runway in Newark, New Jersey. It will take off at 8:42 A.M. The *Boston Globe* credits this delay as a major reason why this was the only one of the four flights not to succeed in its mission. [BOSTON GLOBE, 11/23/01; NEWSWEEK, 9/22/01; PITTSBURGH POST-GAZETTE, 10/28/01 (B)] Apparently, Flight 93 has to wait in a line of about a dozen planes before it can take off. [USA TODAY, 8/12/02]

8:13 A.M.*: Flight 11 Hijacked, but Pilot Makes No Distress Call

The last routine communication takes place between ground control and the pilots of Flight 11 around this time. Flight controller Pete Zalewski is handling the flight. The pilot responds when told to turn right, but immediately afterwards fails to respond to a command to climb. Zalewski repeatedly tries to reach the pilot, even using the emergency frequency, but gets no response. [BOSTON GLOBE, 11/23/01; NEW YORK TIMES, 10/16/01; MSNBC, 9/11/02 (B); 9/11 COMMISSION REPORT, 6/17/04] Flight 11 is apparently hijacked around this time. One flight controller says the plane is hijacked over Gardner, Massachusetts, less than 50 miles west of Boston. [NASHUA TELEGRAPH, 9/13/01] The *Boston Globe* notes, "It appears that the hijackers' entry was surprising enough that the pilots did not have a chance to broadcast a traditional distress call." It would only have taken a few seconds to press the right buttons. [BOSTON GLOBE, 11/23/01] Yet flight attendant Amy Sweeney appears to witness three of the hijackers storming the cockpit around 8:20 A.M. [LOS ANGELES TIMES, 9/20/01 (C)] This would imply that, at most, one or two hijackers enter the cockpit at this time, before the others do.

Between 8:13–8:21 A.M.*: Flight 11 Transponder Turned Off

Shortly after flight controllers ask Flight 11 to climb to 35,000 feet, the transponder stops transmitting. A transponder is an electronic device that identifies a plane on a controller's screen and gives its exact location and altitude. Among other vital functions, it is also used to transmit a four-digit emergency hijack code. Flight control manager Glenn Michael later says, "We considered it at that time to be a

Flight controller Matt McCluskey stands in the Boston tower where the Flight 11 hijack was first detected

possible hijacking." [ASSOCIATED PRESS, 8/12/02; *CHRISTIAN SCIENCE MONITOR*, 9/13/01; MSNBC, 9/15/01] Initial stories after 9/11 suggest the transponder is turned off around 8:13 A.M., but Pete Zalewski, the flight controller handling the flight, later says the transponder is turned off at 8:20 A.M. [MSNBC, 9/11/02 (B)] The 9/11 Commission places it at 8:21 A.M. [9/11 COMMISSION REPORT, 6/17/04] Colonel Robert Marr, head of NEADS, claims the transponder is turned off some time after 8:30 A.M. [ABC NEWS, 9/11/02]

8:14 A.M.: Flight 175 Takes Off 16 Minutes Late

Flight 175 takes off from Boston's Logan Airport, 16 minutes after its scheduled 7:58 departure time.
[*WASHINGTON POST*, 9/12/01; CNN, 9/17/01; *GUARDIAN*, 10/17/01; ASSOCIATED PRESS, 8/19/02 (B); *NEWSDAY*, 9/10/02]

After 8:14 A.M.–8:38 A.M.*: Flight 11 Pilot Repeatedly Pushes Talk Back Button

At some unknown point after the hijacking begins, the talkback button is activated, which enables Boston flight controllers to hear what is being said in the cockpit. It is unclear whether John Ogonowski, the pilot of Flight 11, activates the talkback button, or whether a hijacker accidentally does so when he takes over the cockpit. A controller later says, "The button [is] being pushed intermittently most of the way to New York." An article later notes that "his ability to do so also indicates that he [is] in the driver's seat much of the way" to the WTC. Such transmissions continue until about 8:38 A.M.
[*CHRISTIAN SCIENCE MONITOR*, 9/13/01; MSNBC, 9/15/01]

8:15 A.M.*: Flight Controllers Cannot Contact Flight 11

Two Boston flight controllers, Pete Zalewski and Lino Martins, discuss the fact that Flight 11 cannot be contacted. Zalewski says to Martins, "He won't answer you. He's nordo [no radio] roger thanks." [CNN, 9/17/01; *GUARDIAN*, 10/17/01; *NEW YORK TIMES*, 10/16/01 (C); MSNBC, 9/11/02 (B)]

Before 8:20 A.M.*: Hijackers Attack Passenger on Flight 77

Four hijackers get up from their seats and stab or shoot passenger Daniel Lewin, a multimillionaire who once belonged to the Israel Defense Force's Sayeret Matkal, a top-secret counterterrorist unit. Lewin is sitting in front of one of the three hijackers in business class. An initial FAA memo regarding the flight states that Satam Al Suqami shoots Lewin at 9:20 A.M. The time is certainly a typo; perhaps 8:20 A.M. is meant? The killing is mentioned in a phone call from the flight that starts at 8:20 A.M. [ABC NEWS, 7/18/02; UPI, 3/6/02; *WASHINGTON POST*, 3/2/02 (B)]

8:20 A.M.: Flight 11 IFF Signal Transmission Stops

Flight 11 stops transmitting its IFF (identify friend or foe) beacon signal. [CNN, 9/17/01]

8:20 A.M.*: Flight 11 Veers Off Course

Flight 11 starts to veer dramatically off course. [MSNBC, 9/11/02 (B)]

8:20 A.M.*: Boston Flight Control Thinks Flight 11 May Be Hijacked?

According to some reports, Boston flight control decides that Flight 11 has probably been hijacked, but apparently, it does not notify other flight control centers for another five minutes, and does not notify NORAD for approximately 20 minutes. [NEWSDAY, 9/23/01; NEW YORK TIMES, 9/15/01 (C)] ABC News will later say, "There doesn't seem to have been alarm bells going off, [flight] controllers getting on with law enforcement or the military. There's a gap there that will have to be investigated." [ABC NEWS, 9/14/01] (Note the conflicting account at 8:21 A.M.)

8:20 A.M.*: Flight 11 Attendant Sweeney Phones in Hijacking Details

Flight 11 attendant Amy (Madeline) Sweeney borrows a calling card from flight attendant Sara Low and uses an Airfone to call Boston's Logan Airport. She speaks to Michael Woodward, an American Airlines flight service manager. Because Woodward and Sweeney are friends, he does not have to verify the call is not a hoax. The call is not recorded, but Woodward takes detailed notes. [NEW YORK OBSERVER, 2/11/04; ABC NEWS, 7/18/02] She calmly tells Woodward, "Listen, and listen to me very carefully. I'm on Flight 11. The airplane has been hijacked." [ABC NEWS, 7/18/02] According to one account, she gives him the seat locations of three hijackers: 9D, 9G, and 10B. She says they are all of Middle Eastern descent, and one speaks English very well. [NEW YORK OBSERVER, 2/11/04] Another account states that she identifies four hijackers (but still not the five said to be on the plane), and notes that not all the seats she gave matched up with the seats assigned to the hijackers on their tickets. [ABC NEWS, 7/18/02; LOS ANGELES TIMES, 9/20/01 (C)] She says she cannot contact the cockpit, and does not believe the pilots are flying the plane any longer. [NEW YORK OBSERVER, 2/11/04] According

Amy
(Madeline)
Sweeney

to a later *Los Angeles Times* report, "Even as she was relating details about the hijackers, the men were storming the front of the plane and 'had just gained access to the cockpit,'" (Note that Sweeney witnesses the storming of the cockpit at least seven minutes after radio contact from Flight 11 stops and at least one of the hijackers begins taking control of the cockpit.) [LOS ANGELES TIMES, 9/20/01 (C)] She says the hijackers have stabbed the two first-class flight attendants, Barbara Arestegui and Karen Martin. She adds, "A hijacker cut the throat of a business-class passenger [later identified as Daniel Lewin], and he appears to be dead." She also says the hijackers have brought a bomb into the cockpit. Woodward asks Sweeney, "How do you know it's a bomb?" She answers, "Because the hijackers showed me a bomb." She describes its yellow and red wires. Sweeney continues talking with Woodward until Flight 11 crashes. [BOSTON GLOBE, 11/23/01; NEW YORK OBSERVER, 2/11/04]

8:20 A.M.*: Flight 77 Takes Off 10 Minutes Late

Flight 77 departs Dulles International Airport near Washington, ten minutes after its 8:10 scheduled departure time. [CNN, 9/17/01; WASHINGTON POST, 9/12/01; GUARDIAN, 10/17/01; 9/11 COMMISSION REPORT, 6/17/04; ASSOCIATED PRESS, 8/19/02 (B)]

8:21 A.M.: Sweeney's Call Reaches American
Headquarters, but Managers Cover Up the News

American Airlines flight service manager Michael Woodward is listening to Flight 11 attendant Amy Sweeney on the telephone, and he wants to pass on the information he is hearing from her. Since there is no tape recorder, he calls Nancy Wyatt, the supervisor of pursers at Logan Airport. Holding telephones in both hands, he repeats to Wyatt everything that Sweeney is saying to him. Wyatt in turn simultaneously transmits his account to the airline's Fort Worth, Texas, headquarters. The conversation between Wyatt and managers at headquarters is recorded. All vital details from Sweeney's call reach American Airlines' top management almost instantly. However, according to victims' relatives who later hear this recording, the two managers at headquarters immediately begin discussing a cover-up of the hijacking details. They say, "Don't spread this around. Keep it close," "Keep it quiet," and "Let's keep this among ourselves. What else can we find out from our own sources about what's going on?" One former American Airlines employee who has also heard this recording recalls, "In Fort Worth, two managers in SOC [Systems Operations Control] were sitting beside each other and hearing it. They were both saying, "Do not pass this along. Let's keep it right here. Keep it among the five of us." Apparently, this decision prevents early and clear evidence of a hijacking from being shared during the crisis. Gerard Arpey, American Airlines' Executive Vice President for Operations, soon hears details of the hijacking from flight attendant Betty Ong's phone call at 8:30 A.M., but apparently, he does not learn of Sweeney's call until much later. Victims' relatives will later question whether lives could have been saved if only this information had been quickly shared with other airplanes. [NEW YORK OBSERVER, 6/17/04]

8:21 A.M.*: Flight 11 Attendant Ong Phones in
Hijack Report, Officials Doubt Validity

Flight 11 attendant Betty Ong calls Vanessa Minter, an American Airlines reservations agent in North Carolina, using a seatback Airfone from the back of the plane. Ong speaks to Minter and an unidentified man for about two minutes. Then supervisor Nydia Gonzales is patched in to the conference call as well. Ong says, "The cockpit's not answering. Somebody's stabbed in business class and . . . I think there's mace . . . that we can't breathe. I don't know, I think we're getting hijacked." A minute later, she continues, "And the cockpit is not answering their phone. And there's somebody stabbed in business class. And there's . . . we can't breathe in business class. Somebody's got mace or something . . . I'm sitting in the back. Somebody's coming back from business. If you can hold on for one second, they're coming back." As this quote shows, other flight attendants relay information from the front of the airplane to Ong sitting in the back, and she periodically waits for updates. She goes on, "I think the guys are up there [in the cockpit]. They might have gone there—jammed the way up there, or

Betty Ong

something. Nobody can call the cockpit. We can't even get inside." The first four and a half minutes of the call is later played in a public 9/11 Commission hearing. Ong apparently continues speaking to Gonzales and Minter until the plane crashes. [NEW YORK OBSERVER, 2/11/04; 9/11 COMMISSION. 1/27/04] 9/11 Commissioner Bob Kerrey, who has heard more recordings than have been made public, says that some officials on the ground greet her account skeptically: "They did not believe her. They said, 'Are you sure?' They asked her to confirm that it wasn't air-rage. Our people on the ground were not prepared for a hijacking." [NEW YORK TIMES, 4/18/04]

8:21 A.M.*: Boston Controller Suspects Something "Seriously Wrong" with Flight 11, but NORAD Not Notified

Boston flight controller Pete Zalewski, handling Flight 11, sees that the flight is off course and that the plane has turned off both transponder and radio. Zalewski later claims he turns to his supervisor and says, "Would you please come over here? I think something is seriously wrong with this plane. I don't know what. It's either mechanical, electrical, I think, but I'm not sure." When asked if he suspected a hijacking at this point, he replies, "Absolutely not. No way." According to the 9/11 Commission, "the supervisor instructed the controller [presumably Zalewski] to follow standard operating procedures for handling a 'no radio' aircraft" once the controller told the supervisor the transponder had been turned off." Another flight controller, Tom Roberts, has another nearby American Airlines flight try to contact Flight 11. There is still no response. The flight is now "drastically off course" but NORAD is still not notified. [MSNBC, 9/11/02 (B); 9/11 COMMISSION REPORT, 6/17/04] Note that this response contradicts flight control manager Glenn Michael's assertion that Flight 11 was considered a possible hijacking as soon as the transponder was discovered turned off.

8:23 A.M.*: Flight 11 Attendant Ong's Hijacking Account Forwarded to American Airlines Headquarters

Nydia Gonzalez, an American Airlines supervisor with expertise on security matters, is patched in to a call with flight attendant Betty Ong on Flight 11. [9/11 COMMISSION, 1/27/04] At 8:27 A.M., Gonzalez calls Craig Marquis, a manager at American Airlines' headquarters. Gonzalez holds the phone to Ong to one ear, and the phone to Marquis to the other. [WALL STREET JOURNAL, 10/15/01; NEW YORK OBSERVER, 2/11/04] Gonzalez talks to Marquis continuously until Flight 11 crashes. The first four minutes of this call are later played before the 9/11 Commission. Marquis quickly says, "I'm assuming they've declared an emergency. Let me get ATC [air traffic control] on here. Stand by. . . . Okay, we're contacting the flight crew now and we're . . . we're also contacting ATC." In the four, recorded minutes, Gonzalez relays that Ong is saying the hijackers from seats 2A and 2B are in the cockpit with the pilots. There are no doctors on board. All the first class passengers have been moved to the coach section. The airplane is flying very erratically. [9/11 COMMISSION REPORT, 1/27/04]

8:24 A.M.*: Flight 11 Turns, Many Watch It on Primary Radar

Boston flight control radar sees Flight 11 making an unplanned 100-degree turn to the south (the plane is already way off course). Flight controllers never lose sight of the flight, though they can no longer determine altitude once the transponder is turned off. [NEWHOUSE NEWS SERVICE, 1/25/02; MSNBC, 9/11/02 (B); CHRIST-IAN SCIENCE MONITOR, 9/13/01] Before this turn, the FAA had tagged Flight 11's radar dot for easy visibility and, at American Airlines headquarters at least, "All eyes watched as the plane headed south. On the screen, the plane showed a squiggly line after its turn near Albany, then it straightened." [WALL STREET JOURNAL, 10/15/01] Boston flight controller Mark Hodgkins later says, "I watched the target of American 11 the whole way down." [ABC NEWS. 9/6/02] However, apparently, NEADS has different radar. When they are finally told about the flight, they cannot find it. Boston has to update NEADS on Flight 11's position periodically by telephone until NEADS finally finds it a few minutes before it crashes into the WTC. [AVIATION WEEK AND SPACE TECHNOLOGY, 6/3/02; ABC NEWS, 9/11/02; NEWHOUSE NEWS SERVICE, 1/25/02]

8:24 A.M.*: Boston Flight Controllers
Hear Flight 11 Hijacker: "We Have Some Planes"

Because the talkback button on Flight 11 has been activated, Boston flight controllers can hear a hijacker on Flight 11 say to the passengers: "We have some planes. Just stay quiet and you will be OK. We are returning to the airport." Flight controller John Zalewski responds, "Who's trying to call me?" The hijacker continues, "Everything will be OK. If you try to make any moves you'll endanger yourself and the airplane. Just stay quiet." [GUARDIAN, 10/17/01; NEW YORK TIMES. 10/16/01; 9/11 COMMISSION REPORT, 6/17/04; BOSTON GLOBE, 11/23/01; MSNBC, 9/11/02 (B); NEW YORK TIMES, 9/12/01; CHANNEL 4 NEWS, 9/13/01] Immediately after hearing this voice, Flight controller John Zalewski "knew right then that he was working a hijack" and calls for his supervisor. The frequency of Flight 11 is played on speakers so everyone in Boston flight control can hear. [MSNBC, 9/11/02 (B); VILLAGE VOICE, 9/13/01]

8:25 A.M.: Boston Flight Control Tells Other Centers About Hijack, but Not NORAD

The *Guardian* reports that Boston flight control "notifies several air traffic control centers that a hijack is taking place," But it does not notify NORAD for another 6–15 minutes, depending on the account. [GUARDIAN, 10/17/01] However, the Indianapolis flight controller monitoring Flight 77 claims to not know about this or Flight 175's hijacking twenty minutes later at 8:56 A.M. Additionally, the flight controllers at New York City's La Guardia airport are never told about the hijacked planes and learn about them from watching the news. [BERGEN RECORD, 1/4/04]

Before 8:26 A.M.*: Hijackers Identified by Seat Locations

Having been told by flight attendant Amy Sweeney the seat locations of three hijackers, American Airlines flight service manager Michael Woodward orders a colleague at Boston's Logan Airport to look up those seat locations on the reservations computer. The names, addresses, phone numbers, and credit cards of these hijackers are quickly identified: Abdulaziz Alomari is in 9G, Mohamed Atta is in 9D, and Satam Al Suqami is in 10B. 9/11 Commissioner Bob Kerrey notes that from this information, American Airlines officials monitoring the call would probably have known or assumed right away that the hijacking was connected to al-Qaeda. [ABC NEWS, 7/18/02; *NEW YORK OBSERVER*, 2/11/04]

Between 8:27–8:30 A.M.*: Ong Gives Flight 11 Details; Seating Accounts Differ

Craig Marquis, listening to information coming from flight attendant Betty Ong on Flight 11, calls American Airlines' system operations control center in Fort Worth. He says, "She said two flight attendants had been stabbed, one was on oxygen. A passenger had his throat slashed and looked dead and they had gotten into the cockpit." He relays that Ong said the four hijackers had come from first-class seats: 2A, 2B, 9A, and 9B. She said the wounded passenger was in seat 10B. [*BOSTON GLOBE*, 11/23/01] Note that this conflicts with the seats flight attendant Amy Sweeney gives for the hijackers at about the same time: 9D, 9G, and 10B. By 8:27 A.M., this information is passed to Gerard Arpey, the effective head of American Airlines that morning. By 9:59 A.M., counterterrorism "tsar" Richard Clarke and other top officials receive the information. [*AGAINST ALL ENEMIES*, BY RICHARD CLARKE, 3/04, PP. 13–14]

8:28 A.M.: FAA Centers Have Hijacking Conference Call; NORAD Not Notified

Boston flight control calls the FAA Command Center and tells them that they believe Flight 11 has been hijacked and it is heading toward New York airspace. At 8:32 A.M., the Command Center shares this with the Operations Center at FAA headquarters. Headquarters replies that they have just begun discussing the hijack situation with the main FAA New England office. The Command Center immediately establishes a teleconference between the Boston, New York, and Cleveland flight control centers so that Boston can help the others understand what is happening. Even though by 8:24 A.M. Boston is fairly certain that Flight 11 has been hijacked, they do not contact NORAD. [9/11 COMMISSION REPORT, 6/17/04]

8:28 A.M.*: Flight 11 Is a Confirmed Hijacking; NORAD Still Not Notified

American Airlines manager Craig Marquis is talking to Nydia Gonzalez, who in turn is talking to flight attendant Betty Ong on Flight 11. Marquis says, "We contacted air traffic control, they are going to handle this as a confirmed hijacking. So they're moving all the traffic out of this aircraft's way. . . . He turned his transponder off, so we don't have a definitive altitude for him. We're just going by . . . They seem to think that they have him on a primary radar. They seem to think that he is descending." This transmission further indicates that Boston flight control believes that Flight 11 has been hijacked by this time. [9/11 COMMISSION, 1/27/04]

8:30 A.M.: FAA Command Center Informed of Hijacking; NORAD Still Not Notified

The FAA's Command Center in Herndon, Virginia begins their usual daily senior staff meeting. National Operations Manager Ben Sliney interrupts the meeting to report a possible hijacking in progress, as the Center had been told about the Flight 11 hijacking two minutes earlier. Later, a supervisor interrupts the meeting to report that a flight attendant on the hijacked aircraft may have been stabbed. The meeting breaks up before the first WTC crash at 8:46 A.M. Apparently, no one in the meeting contacts NORAD. [AVIATION WEEK AND SPACE TECHNOLOGY, 12/17/01]

8:30 A.M.: American Airlines Vice President Informed of Hijacking

Gerard Arpey (American Airlines' Executive Vice President for Operations) learns from manager Joe Burdepelly that Flight 11 may have been hijacked. Burdepelly tells Arpey that he has been told that another manager, Craig Marquis, is in contact with flight attendant Betty Ong on the hijacked flight. Arpey learns that Ong has said two other attendants have been stabbed, that two or three passengers are in the cockpit, and more. Arpey is the effective head of American Airlines during the early phase of the crisis, because the company's president is still at home and out of contact. [9/11 COMMISSION REPORT, 1/27/04] At some point before Flight 11 crashes, Arpey is told about the "We have some planes" comment made by the hijackers. [USA TODAY, 8/13/02]

8:30 A.M.: Rookie in Command of the NMCC

**Captain
Charles Leidig**

Captain Charles Leidig, the Deputy for Command Center Operations at the NMCC, takes over temporarily from Brigadier General Montague Winfield and is effectively in charge of NMCC during the 9/11 crisis. Winfield had requested the previous day that Leidig stand in for him on September 11. Leidig had started his role as Deputy for Command Center Operations two months earlier and had qualified to stand in for Winfield just the previous month. Leidig remains in charge from a few minutes before the 9/11 crisis begins until about 10:30 A.M., after the last hijacked plane crashes. He presides over an important crisis response teleconference that has a very slow start, not even beginning until 9:39 A.M. [LEIDIG TESTIMONY, 6/17/04; 9/11 COMMISSION, 6/17/04 (B)]

8:30 A.M.*: Some U.S. Leaders Are Scattered; Others in D.C.

Just prior to learning about the 9/11 attacks, top U.S. leaders are scattered across the country and overseas:

- President Bush is in Sarasota, Florida. [WASHINGTON POST, 1/27/02]

- Secretary of State Powell is in Lima, Peru. [WASHINGTON POST, 1/27/02]

- General Henry Shelton, Chairman of the Joint Chiefs of Staff, is flying across the Atlantic on the way to Europe. [WASHINGTON POST, 1/27/02]

- Attorney General Ashcroft is flying to Milwaukee, Wisconsin. [WASHINGTON POST, 1/27/02]

- Federal Emergency Management Agency Director Joe Allbaugh is at a conference in Montana. [ABC NEWS, 9/14/02 (B)]

Others are in Washington:

- Vice President Cheney and National Security Adviser Rice are at their offices in the White House. [WASHINGTON POST, 1/27/02]

- Defense Secretary Rumsfeld is at his office in the Pentagon, meeting with a delegation from Capitol Hill. [WASHINGTON POST, 1/27/02]

- CIA Director Tenet is at breakfast with his old friend and mentor, former senator David Boren (D), at the St. Regis Hotel, three blocks from the White House. [WASHINGTON POST, 1/27/02]

- FBI Director Mueller is in his office at FBI headquarters on Pennsylvania Avenue. [WASHINGTON POST, 1/27/02]

- Transportation Secretary Norman Mineta is at his office at the Department of Transportation. [SENATE COMMERCE COMMITTEE, 9/20/01]

- Counterterrorism "tsar" Richard Clarke is at a conference in the Ronald Reagan Building three blocks from the White House. [AGAINST ALL ENEMIES, BY RICHARD CLARKE, 3/04, P. 1]

8:34 A.M.: Boston Flight Control Attempts to Contact Air Base Directly; Result Unknown

Boston flight controllers attempt to contact the military through the FAA's Cape Cod, Massachusetts facility. Two fighters are on twenty-four hour alert at the Otis Air National Guard Base, at Cape Cod. Boston tries reaching this base so the fighters there can scramble after Flight 11. Apparently, they do this before going through the usual NORAD channels. The 9/11 Commission is vague about the outcome of this call. [9/11 COMMISSION REPORT, 6/17/04] However, the lead pilot at the Otis base, Lieutenant Colonel Timothy Duffy (codenamed Duff), later claims he is given an advance warning to get ready to scramble before the official notification, thanks to a call from Boston flight control at 8:40 A.M. [AVIATION WEEK AND SPACE TECHNOLOGY, 6/3/02]

8:34 A.M.*: Atlantic City Fighters Not Reached; Not Redeployed Until Much Later

Around this time, Boston flight control attempts to contact an Atlantic City, New Jersey, air base, to send fighters after Flight 11. For decades, the air base had two fighters on 24-hour alert status, but this changed in 1998 due to budget cutbacks. The flight controllers do not realize this, and apparently try in vain to reach someone. Two F-16s from this base are practicing bombing runs over an empty stretch of the Pine Barrens near Atlantic City. Only eight minutes away from New York City, they are not alerted to the emerging crisis. Shortly after the second WTC crash at 9:03 A.M., the two F-16s are ordered to land and are refitted with air-to-air missiles, then sent aloft. However, the pilots re-launch more than an hour after the second crash. They are apparently sent to Washington, but do not reach there until almost 11:00 A.M. After 9/11, one newspaper questions why NORAD "left what seems to be a yawning gap in the midsection of its air defenses on the East Coast—a gap with New York City at the center." [BERGEN RECORD, 12/5/03; 9/11 COMMISSION REPORT, 6/17/04] Had these two fighters been notified at 8:37 A.M. or before, they could have reached New York City before Flight 11.

8:34 A.M.*: Boston Flight Control Hears Hijacker Announcement

Flight controllers hear a hijacker on Flight 11 say to the passengers: "Nobody move, please, we are going back to the airport. Don't try to make any stupid moves." [BOSTON GLOBE, 11/23/01; GUARDIAN, 10/17/01; NEW YORK TIMES, 10/16/01; 9/11 COMMISSION REPORT, 6/17/04] Apparently, shortly after this, the transmission tapes that are automatically recorded are played back to hear the words that were spoken by the hijackers a few minutes before. Everyone in the Boston flight control center hears the hijackers say, "We have some planes." [MSNBC, 9/11/02 (B)] Ben Sliney, the FAA's National Operations Manager, soon gets word of the "We have some planes," message and later says the phrase haunts him all morning. [USA TODAY, 8/13/02]

8:35 A.M.*: Sweeney Continues to Provide Flight 11 Updates

Flight attendant Amy Sweeney continues to describe what is happening onboard Flight 11 to Michael Woodward at Logan Airport. At some point prior to this, she explains that flight attendants are giving injured people oxygen. They have made an announcement over the PA system asking if there is a doctor or nurse on board. Sweeney is calling from the rear of the coach section. She explains that the passengers in coach, separated by curtains from the violence in first class, are calm, believing that there is some type of medical emergency at the front of the plane. Then, at this time, the plane suddenly lurches, tilting all the way to one side, then becomes horizontal again. Then she says it begins a rapid descent. She tries to contact the cockpit again, but still gets no response. [ABC NEWS, 7/18/02; NEW YORK OBSERVER, 2/11/04]

8:35 A.M.*: Bush Motorcade Leaves for Elementary School

President Bush's motorcade leaves for Emma E. Booker Elementary School in Sarasota, Florida. [WASHINGTON POST, 1/27/02; BBC, 9/1/02; SARASOTA MAGAZINE, 9/19/01; WASHINGTON TIMES, 10/7/02] His official schedule had him leaving at 8:30. [ST. PETERSBURG TIMES, 7/4/04] He said farewell to the management at the Colony Beach and Tennis Resort at 8:20 A.M. [DAILY TELEGRAPH, 12/16/01]

8:36 A.M.*: Flight 11 Attendants Ong and Sweeney Report Plane Maneuvers

On Flight 11, flight attendant Betty Ong reports that the plane tilts all the way on one side and then becomes horizontal again. Flight attendant Amy Sweeney reports that the plane has begun a rapid descent. [ABC NEWS, 7/18/02] Sweeney also says that the hijackers are Middle Easterners. [SAN FRANCISCO CHRONICLE, 7/23/04]

8:37 A.M.: Flight 11 Enters New York Control Space

Flight 11 passes from Boston flight control airspace into New York flight control airspace. Flight controller John Hartling takes over monitoring the plane. However, when a colleague tells him the flight is hijacked, he is incredulous: "I didn't believe him. Because I didn't think that that stuff would happen anymore, especially in this country." [MSNBC, 9/11/02 (B)]

8:37 A.M.: Flight 175 Pilots Asked to Look for Flight 11

Flight controllers ask the United Airlines Flight 175 pilots to look for a lost American Airlines plane 10 miles to the south—a reference to Flight 11. They respond that they can see it. They are told to keep away from it. [GUARDIAN, 10/17/01; BOSTON GLOBE, 11/23/01; 9/11 COMMISSION REPORT, 6/17/04] Apparently, Flight 175 is not told Flight 11 has been hijacked. Flight 175 itself is hijacked a few minutes later.

8:37 A.M.*: Boston Flight Control Notifies NORAD; Timing Disputed

According to the 9/11 Commission, Boston flight control contacts NEADS at this time. This is apparently the first successful notification to the military about the crisis. Tech. Sgt. Jeremy Powell, a member of the Air National Guard at NEADS, initially takes the call from Boston flight control. [AVIATION WEEK AND SPACE TECHNOLOGY, 6/3/02; NEWHOUSE NEWS SERVICE, 1/25/02] Boston flight control says, "Hi. Boston [flight control], we have a problem here. We have a hijacked aircraft headed toward New York, and we need you guys to, we need someone to scramble some F-16s or something up there, help us out." Powell replies, "Is this real-world or exercise?" Boston answers, "No, this is not an exercise, not a test." [9/11 COMMISSION REPORT, 6/17/04; BBC, 9/1/02] Powell gives the phone to Lieutenant Colonel Dawne Deskins, regional Mission Crew Chief for the Vigilant Guardian exercise. Deskins later says that initially she and "everybody" else at NEADS thinks the call is part of Vigilant Guardian. After the phone call, she has to clarify to everyone that it is not a drill. [NEWHOUSE NEWS SERVICE, 1/25/02] NORAD commander Major General Larry Arnold in Tyndall Air Force Base, Florida, also says that when he hears of the hijacking at this time, "The first thing that went through my mind was, is this part of the exercise? Is this some kind of a screw-up?" [ABC NEWS, 9/11/02] Deskins recalls, "I picked up the line and I identified myself to the Boston [flight] controller, and he said, we have a hijacked aircraft and I need to get you some sort of fighters out here to help us out." However, the timing of this vital notification is in some dispute. Deskins herself claimed the call occurred at 8:31 A.M. [ABC NEWS, 9/11/02] Another report later states, "Shortly after 8:30 A.M., behind the scenes, word of a possible hijacking [reaches] various stations of NORAD." [ABC NEWS, 9/14/02] FAA Administrator Jane Garvey testified that the FAA notified NORAD at 8:34 A.M. [NEW YORK TIMES, 12/30/03] NORAD on the other hand, originally claimed they were first notified at 8:40 A.M., and this was widely reported in the media prior to the 9/11 Commission's 2004 report. [NORAD, 9/18/01; ASSOCIATED PRESS, 8/19/02 (B); BBC, 9/1/02; NEWSDAY, 9/10/02] If the 8:37 A.M. time is accurate, then flight controllers failed to notify NORAD until approximately 13 minutes after the hijackers in the cockpit clearly stated that the plane had been hijacked at 8:24 A.M.; 17 minutes after the transponder signal was lost and the flight goes far off course; and 24 minutes after radio contact was lost at 8:13 A.M.

Lieutenant
Colonel Dawne
Deskins

After 8:37 A.M.*: NORAD Scramble Order
Moves Through Official and Unofficial Channels

NORAD gives the command to scramble fighters after Flight 11 after receiving Boston's call. Lieutenant Colonel Dawne Deskins at NEADS tells Colonel Robert Marr, head of NEADS, "I have FAA on the phone, the shout line, Boston [flight control]. They said they have a hijacked aircraft." Marr then

calls Major General Larry Arnold at NORAD's command Center in Tyndall Air Force Base, Florida, and says, "Boss, I need to scramble [fighters at] Otis [Air National Guard Base]." Arnold recalls, "I said go ahead and scramble them, and we'll get the authorities later." Arnold then calls NORAD headquarters to report. [ABC NEWS, 9/11/02; 9/11 COMMISSION REPORT, 6/17/04] Then, upon receiving proper authorization, NEADS calls Canadian Captain Mike Jellinek at NORAD's Colorado headquarters. Jellinek is sitting near Canadian

**General
Larry Arnold**

Air Force Major General Rick Findley, director of combat operations there. Findley's staff is "already on high alert" because of a Vigilant Guardian and Operation Northern Vigilance, two emergency training exercises that were currently in progress. Jellinek gets the thumbs up authorization from Findley to send fighters after Flight 11. Findley later states, "At that point all we thought was we've got an airplane hijacked and we were going to provide an escort as requested. We certainly didn't know it was going to play out as it did." Findley remains in charge of NORAD headquarters while his staff feeds information to NORAD Commander in Chief Ralph Eberhart, who is stationed in Florida. [CANADIAN BROADCASTING CORP., 11/27/01; TORONTO STAR, 12/9/01; OTTAWA CITIZEN, 9/11/02; AVIATION WEEK AND SPACE TECHNOLOGY, 6/3/02]

8:38 A.M.*: Flight 11 Pilot Stops Activating Talk Back Button

The talkback button on Flight 11, which has been periodically activated since around 8:14 A.M., stops around this time. Some have suggested that this indicates that the hijackers replace pilot John Ogonowski at this time. [CHRISTIAN SCIENCE MONITOR, 9/13/01; MSNBC, 9/15/01]

8:40 A.M.*: Fighter Pilots Unofficially Told to
"Get Ready to Scramble" After Flight 11

Major Daniel Nash (codenamed Nasty) and Lieutenant Colonel Timothy Duffy (codenamed Duff) are the two F-15 pilots who would scramble after Flight 11 and then Flight 175. Apparently, they get several informal calls warning to get ready. According to Nash, at this time, a colleague at the Otis Air National Guard Base tells him that a flight out of Boston has been hijacked, and that he should be on

alert. [CAPE COD TIMES, 8/21/02] NEADS senior technician Jeremy Powell (informed about the hijacking at 8:37 A.M.), says that he telephones Otis Air National Guard Base soon thereafter to tell it to upgrade its "readiness posture." [NEWHOUSE NEWS SERVICE, 1/25/02] Colonel Robert Marr, head of NEADS, also says that after being told of the hijacking at 8:37 A.M., he says, "I'll call First Air Force [at Otis] and let them know we've got a potential incident." [BBC, 9/1/02] Boston flight control had tried calling the Otis base

**Major
Daniel Nash**

directly at 8:34 A.M., although the result of that call remains unclear. Duffy recalls being warned: "I was just standing up by the ops desk and I was told I had a phone call. I asked who it was and they said the [Boston] tower calling and something about a hijacking. It was Flight American 11, a 767, out of Boston going to California. At the time we ran in and got suited up." [AVIATION WEEK AND SPACE TECHNOLOGY, 6/3/02; BBC, 9/1/02; CAPE COD TIMES, 8/21/02] Duffy says, "Halfway to the jets, we got 'battle stations' . . . which means to get ready for action." [AVIATION WEEK AND SPACE TECHNOLOGY, 6/3/02] The actual scramble order does not come until the pilots are already waiting in the fighters: "We went out, we hopped in the jets and we were ready to go—standby for a scramble order if we were going to get one." [BBC, 9/1/02] Duffy continues, "I briefed Nasty on the information I had about the American Airlines flight. About four–five minutes later, we got the scramble order and took off." [AVIATION WEEK AND SPACE TECHNOLOGY, 6/3/02] However, the official notification to scramble these fighters does not come until 8:46 A.M. The six-minute (or more) delay between unofficial and official notification has not been explained.

8:41 A.M.: Flight 175 Reports Suspicious Flight 11 Radio Transmission; Hijacked Moments Later

The pilots of Flight 175 tell ground control about Flight 11, "We figured we'd wait to go to your center. We heard a suspicious transmission on our departure out of Boston. Someone keyed the mic and said, 'Everyone stay in your seats.' It cut out." [GUARDIAN, 10/17/01; NEWSDAY, 9/10/02; NEW YORK TIMES, 10/16/01] An alternate version: "We heard a suspicious transmission on our departure from B-O-S [Boston's airport code]. Sounds like someone keyed the mic and said, 'Everyone, stay in your seats.'" [BOSTON GLOBE, 11/23/01] The last transmission from Flight 175, still discussing this message, comes a few seconds before 8:42 A.M. [NEW YORK TIMES, 10/16/01] Presumably Flight 175 is hijacked within the next minute.

8:41 A.M.: New York Flight Control Knows Flight 11 Has Been Hijacked

Flight 175 flies into New York flight control airspace. Dave Bottoglia takes over monitoring the flight. Bottoglia has just been told by the pilot of Flight 175 that he has heard threatening communications from Flight 11. Seconds later, a controller sitting next to Bottoglia gets up and points to a radar blip. He says, "You see this target here? This is American 11. Boston [flight control] thinks it's a hijack." John Hartling has been watching the hijacked Flight 11 since 8:37 A.M. Bottoglia joins Hartling in watching Flight 11's blip until it disappears over New York City. He does not pay attention to Flight 175 for several minutes. [MSNBC, 9/11/02 (B)] The New York flight control center was notified of the hijacking around 8:25 A.M.

8:42 A.M.*: Flight 93 Takes Off 41 Minutes Late

Flight 93 takes off from Newark International Airport, bound for San Francisco, California. It leaves 41 minutes late because of heavy runway traffic. [MSNBC, 9/3/02; NEWSWEEK, 9/22/01; ASSOCIATED PRESS, 8/19/02 (B); PITTSBURGH POST-GAZETTE, 10/28/01; 9/11 COMMISSION REPORT, 6/17/04]

8:43 A.M.: NORAD Notified That Flight 175 Has Been Hijacked

National Guard troops stationed at NORAD's Northeast Air Defense Sector (NEADS) in Rome, New York

After 9/11, NORAD and other sources claim that NORAD is notified at this time Flight 175 has been hijacked. [NORAD, 9/18/01; CNN, 9/17/01; WASHINGTON POST, 9/12/01; ASSOCIATED PRESS, 8/19/02; NEWSDAY, 9/10/02] The 9/11 Commission, however, later concludes that New York flight control gives NEADS its first notification that Flight 175 has been hijacked at 9:03 A.M. [9/11 COMMISSION REPORT, 6/17/04] If this earlier account is the accurate one, NEADS technicians learn of the hijacking at the exact same time the flight controllers do. They already have their headsets linked to Boston flight control to track Flight 11 at this time, and so they learn instantly about Flight 175. [NEWHOUSE NEWS SERVICE, 1/25/02]

8:44 A.M.: Other Pilots Notice Flight 175's Emergency Signal

The pilot of U.S. Airlines Flight 583 tells an unidentified flight controller, regarding Flight 175, "I just picked up an ELT [emergency locator transmitter] on 121.5. It was brief but it went off." The controller responds, "O.K. they said it's confirmed believe it or not as a thing, we're not sure yet. . . ." One minute later, another pilot says, "We picked up that ELT, too, but it's very faint." [NEW YORK TIMES, 10/16/01 (B)] Flight 175 appears to have been the only trigging of any emergency signal on 9/11. It is possible the ELT came from Flight 11 instead.

Before 8:45 A.M.*: American Airlines Tells Crisis Center, Leaders of Hijacking, but Not Other Pilots

At American Airlines' headquarters in Fort Worth, their crisis command center used in emergencies is activated. A page is sent to American's top executives and operations personnel: "Confirmed hijacking Flight 11." However, pilots on other American flights apparently are not notified. Top managers gather at the command center and watch the radar blip of Flight 11 until it disappears over New York City. [WALL STREET JOURNAL, 10/15/01]

8:45–8:46 A.M.*: Flight 11 Attendants Calm as End Approaches

Flight attendant Amy Sweeney is still on the phone with Michael Woodward, describing conditions on Flight 11. The plane is nearing New York City, but the coach section passengers are still quiet, apparently unaware a hijacking is in progress. Woodward asks Sweeney to look out of the window and see if she can tell where they are. She replies, "I see the water. I see the building. I see buildings." She tells him the plane is flying very low. Then she takes a slow, deep breath and slowly, calmly says, "Oh my God!" Woodward hears a loud click, and then silence. [LOS ANGELES TIMES, 9/20/01; ABC NEWS, 7/18/02] Flight attendant Betty Ong, on another phone, apparently does not realize what is about to happen. She is repeatedly saying, "Pray for us. Pray for us," before her phone call comes to a halt. [9/11 COMMISSION REPORT, 6/17/04]

Before 8:46 A.M.*: Rumsfeld Reportedly Predicts Terror Attacks

Defense Secretary Rumsfeld, Deputy Defense Secretary Paul Wolfowitz, and Representative Christopher Cox (R) are meeting in Rumsfeld's private Pentagon dining room, discussing missile defense. Rumsfeld later recalls, "I had said at an eight o'clock breakfast that sometime in the next two, four, six, eight, ten, twelve months there would be an event that would occur in the world that would be sufficiently shocking that it would remind people again how important it is to have a strong healthy defense department that contributes to—that underpins peace and stability in our world." [CNN, 12/5/01] Wolfowitz recalls, "And we commented to them that based on what Rumsfeld and I had both seen and worked on the Ballistic Missile Threat Commission, that we were probably in for some nasty surprises over the next ten years." [DEFENSE DEPARTMENT, 5/9/03] There are confused accounts that Rumsfeld says, "I've been around the block a few times. There will be another event," just before the Pentagon is hit by Flight 77, but such comments may have been made around this time instead. Rumsfeld says, "And someone walked in and handed a note that said that a plane had just hit the World Trade Center. And we adjourned the meeting, and I went in to get my CIA briefing . . . right next door here [in my office]." [CNN, 12/5/01]

8:46 A.M.: Flight 11 Hits the North Tower of the World Trade Center

Flight 11 slams into the WTC North Tower (Building 1). Seismic records pinpoint the crash at 26 seconds after 8:46 A.M. [CNN, 9/12/01; NEW YORK TIMES, 9/12/01; NORAD, 9/18/01; ASSOCIATED PRESS, 8/19/02 (B); USA TODAY, 8/13/02; NEWSDAY, 9/10/02; NEW YORK TIMES, 9/11/02; USA TODAY, 12/20/01] Investigators believe the plane still has about 10,000 gallons of fuel and is traveling approximately 470 mph. [NEW YORK TIMES, 9/11/02] The plane strikes the 93rd through 98th floors in the 110 story building. No one above the crash line survives; approximately 1,360 people die. Below the crash line, approximately 72 die and more than 4,000 survive. Both towers are slightly less than half full at the time of the attack, with between 5,000 to 7,000 people in each tower. This number is lower than expected. Many office workers have not yet shown up to work, and tourists to the observation deck opening at 9:30 A.M. have yet to arrive. [USA TODAY, 12/20/01]

Flight 11 crashes into the North Tower of the World Trade Center

8:46 A.M.: First WTC Attack Recorded on Video, but Not Broadcast Until Evening

Two French documentary filmmakers are filming a documentary on New York City firefighters about ten blocks from the WTC. One of them hears a roar, looks up, and captures a distant image of the first WTC crash. They continue shooting footage nonstop for many hours, and their footage is first shown that evening on CNN. [NEW YORK TIMES, 1/12/02] President Bush later claims that he sees the first attack live on television, but this is technically impossible, as there was no live news footage of the attack. [WALL STREET JOURNAL, 3/22/04]

8:46 A.M.: Fighters Are Training over North Carolina; Not Recalled to Washington Until Much Later

At the time of the first WTC crash, three F-16s assigned to Andrews Air Force Base, ten miles from Washington, are flying an air-to-ground training mission on a range in North Carolina, 207 miles away. Eventually they are recalled to Andrews, but they do not begin patrolling Washington until 10:45 A.M. [AVIATION WEEK AND SPACE TECHNOLOGY, 9/9/02] F-16s can travel a maximum speed of 1,500 mph. [ASSOCIATED PRESS, 6/16/00] Traveling even at 1,100 mph (the speed NORAD Major General Larry Arnold says two fighters from Massachusetts travel toward Flight 175 [MSNBC, 9/23/01 (C), SLATE, 1/16/02]), at least one of the F-16s could have returned to Washington within ten minutes and started patrolling the skies well before 9:00 A.M.

8:46 A.M.: Flight 175 Changes Transponder Signal but Remains Easily Traceable

Flight 175 stops transmitting its transponder signal. It is 50 miles north of New York City, headed toward Baltimore. [GUARDIAN, 10/17/01; NEWSDAY, 9/10/02; 9/11 COMMISSION REPORT, 6/17/04] However, the transponder is turned off for only about 30 seconds, and then changed to a signal that is not designated for any plane on that day. [NEWSDAY, 9/10/02] This "allow[s] controllers to track the intruder easily, though they couldn't identify it." [WASHINGTON POST, 9/17/01]

8:46 A.M.: New York Flight Control Suspects Flight 175 Hijacking

New York flight controller Dave Bottoglia is in charge of monitoring both Flights 11 and 175. He has just watched Flight 11's radar blip disappear over New York City, but does not yet realize the plane has crashed. "Within seconds" of losing Flight 11's blip, he realizes that Flight 175 is also missing. He has another controller take over all his other planes so he can focus on finding Flight 175. He tries contacting the planes several times unsuccessfully. Curt Applegate, sitting at the radar screen next to Bottoglia, sees a blip that might be the missing Flight 11. In fact, it is the missing Flight 175. Just as Bottoglia notices it, its transponder signal turns back on, but at a different signal than before. "There is no longer any question in Bottoglia's mind that he's looking at a second hijacked airliner," according to later MSNBC reports. Bottoglia then notices Flight 175 turn east and start descending. He keeps an eye on it and sees it head right toward Delta Flight 2315. He recalls saying to the Delta flight, "Traffic, 2:00, ten miles. I think he's been hijacked. I don't know his intentions. Take any evasive action necessary." Flight 2315 takes evasive action, missing Flight 175 by less than 200 feet. [MSNBC, 9/11/02 (B)] However, there is no claim that NORAD is notified about the hijacking at this time. On the other hand, according to a NORAD timeline from shortly after 9/11, NORAD is notified by Boston flight control three minutes earlier at 8:43 A.M. [NORAD, 9/18/01] The 9/11 Commission seems to ignore this account from Bottoglia completely, asserting that he notices the transponder change at 8:51 A.M. [9/11 COMMISSION REPORT, 6/17/04]

8:46 A.M.*: Fighters Ordered to Scramble to Flight 11
Nine Minutes After NORAD Notification

Two F-15 fighters are ordered to scramble from Otis Air National Guard Base in Massachusetts to find Flight 11, approximately 190 miles from the known location of the plane and 188 miles from New York City. [CHANNEL 4 NEWS, 9/13/01; CNN, 9/17/01; *WASHINGTON POST*, 9/15/01; *LOS ANGELES TIMES*, 9/17/01; NORAD, 9/18/01; 9/11 COMMISSION REPORT, 6/17/04] According to the 9/11 Commission, NORAD makes the decision to scramble after only one phone call, as the decision is made to act first and get clearances later. Yet there is a nine-minute gap between when the 9/11 Commission says NORAD is notified about the hijacking at 8:37 A.M., and when the fighters are ordered scrambled. This delay has not been explained. The pilots had already received several unofficial warnings before this order—possibly as early as 8:34 A.M., 12 minutes earlier. One of the pilots recalls sitting in the cockpit, ready and waiting for the scramble order to come. [9/11 COMMISSION REPORT, 6/17/04; BBC, 9/1/02] Yet, according to some reports, they do not take off for another six minutes, at 8:52 A.M. [9/11 COMMISSION REPORT, 6/17/04; NORAD, 9/18/01] The fighters' initial target, Flight 11, is already crashing into the WTC at this time. Unaware of this development, the fighter pilots scramble to New York City.

8:46 A.M.*: Bush, Some Aides Reportedly Still Unaware of Flight 11 Hijack

President Bush is traveling through Sarasota, Florida in a motorcade when the first WTC attack occurs. According to the 9/11 Commission, "no one in the White House or traveling with the President knew that [Flight 11] had been hijacked [at this time]. Immediately afterward, duty officers at the White House and Pentagon began notifying senior officials what had happened." However, according to reports, Bush is not notified about the crash until his motorcade reaches its destination, even though there is a secure phone in his vehicle for just this type of emergency, and even though others in the motorcade are notified. Reportedly, not even Jane Garvey, head of the FAA, nor her deputy have been told of a confirmed hijacking before they learn about the crash from the television. [9/11 COMMISSION REPORT, 6/17/04; *A PRETEXT FOR WAR*, BY JAMES BAMFORD, 6/04, P. 17]

8:46–8:50 A.M.*: New York and Boston Flight Control Conclude Flight 11 Has Hit WTC

Rick Tepper, a flight controller at the Newark, New Jersey tower, looks across the Hudson River at New York City in time to see the explosion caused by Flight 11. Another flight controller there tries to find out what caused it. He recalls that in the next few minutes, "We contacted La Guardia, Kennedy Tower, and Teterboro Tower to find out if they lost an airplane. And they all said they didn't know what it was. I got on the phone to the en route air traffic control's facility out in New York on Long Island, and I asked them if they'd lost any airplanes, and they said, 'No, but Boston [flight control] lost an airplane. They lost an American 767.'" New Jersey flight controller Bob Varcadapane says to the Long Island flight controller, "I have a burning building and you have a missing airplane. This is very coincidental." The assumption is quickly made at New York and Boston flight control centers that Flight 11 has hit the WTC. NBC later reports, "Word of the fate of Flight 11 quickly travels throughout the air traffic control world." [MSNBC, 9/11/02 (B)] However, the Indianapolis flight control center that handles Flight 77 reportedly does not learn of Flight 11's crash until around 9:20 A.M. [9/11 COMMISSION REPORT, 6/17/04]

Between 8:46–8:55 A.M.*: Bush's Motorcade Quickly Hears of Flight 11 Crash, but Bush Reportedly Still Unaware

When Flight 11 hits the WTC at 8:46 A.M., President Bush's motorcade is crossing the John Ringling Causeway on the way to Booker Elementary from the Colony Beach and Tennis Resort on Longboat Key. [WASHINGTON TIMES, 10/8/02] *Sarasota Magazine* claims that Bush is on Highway 301, just north of Main Street when he is told that a plane had crashed in New York City. [SARASOTA MAGAZINE, 9/19/01] Around the same time, Press Secretary Ari Fleischer, who riding in another car in the motorcade, is talking on his cell phone, when he blurts out: "Oh, my God, I don't believe it. A plane just hit the World Trade Center." (The person with whom Fleischer is speaking remains unknown.) Fleischer is told he will be needed on arrival to discuss reports of the crash. [CHRISTIAN SCIENCE MONITOR, 9/17/01; ALBUQUERQUE TRIBUNE, 9/10/02] This call takes place "just minutes" after the first news reports. [MSNBC, 10/29/02] Congressman Dan Miller also says he is told about the crash just before meeting Bush at Booker elementary school at 8:55 A.M. [SARASOTA MAGAZINE, 9/19/01] Some reporters waiting for him to arrive also learn of the crash just minutes after it happens. [CBS NEWS, 9/11/02 (B)] It would make sense that Bush is told about the crash immediately, at the same time that others hear about it. Yet the official

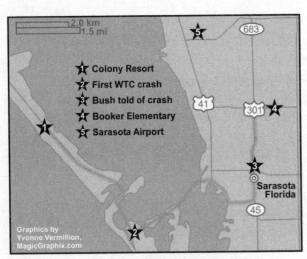

Bush's travels in the Sarasota, Florida, region, with key locations marked

- ☆ Colony Resort
- ☆ First WTC crash
- ☆ Bush told of crash
- ☆ Booker Elementary
- ☆ Sarasota Airport

story remains that Bush is not told about the crash until he arrives at the school. Author James Bamford comments, "Despite having a secure STU-III phone next to him in the presidential limousine and an entire national security staff at the White House, it appears that the President of the United States knew less than tens of millions of other people in every part of the country who were watching the attack as it unfolded." [A PRETEXT FOR WAR, BY JAMES BAMFORD, 6/04, P. 17]

After 8:46 A.M.*: Emergency Response Plans Activated by Officials, Not by Bush

President Bush will say in a speech later that evening, "Immediately following the first attack, I implemented our government's emergency response plans." [WHITE HOUSE, 9/11/01] However, the *Wall Street Journal* reports that lower level officials activate CONPLAN (Interagency Domestic Terrorism Concept of Operations Plan) in response to the emerging crisis. CONPLAN, created in response to a 1995 Presidential Decision Directive issued by Clinton and published in January 2001, details the responsibility of seven federal agencies if a terrorist attack occurs. It gives the FBI the responsibility for activating the plan and alerting other agencies. Bush in fact later states that he doesn't give any orders responding to the attack until after 9:55 A.M. [WALL STREET JOURNAL, 3/22/04; CONPLAN, 1/02]

After 8:46 A.M.*: United Flight Dispatcher Decides Flight 175 Is Hijacked

Ed Ballinger, a United Airlines flight dispatcher, is handling 16 United transcontinental flights, including Flights 175 and 93. Shortly after hearing about the WTC crash, he contacts all of his flights to warn them. However, Flight 175 is "not acting appropriately," and fails to respond. Ballinger concludes the flight has been hijacked. Whether he contacts anyone about his conclusions is unclear. [CHICAGO DAILY HERALD, 4/14/04]

After 8:46 A.M.*: FAA Establishes Open
Telephone Line with the Secret Service

Shortly after the WTC is hit, the FAA opens a telephone line with the Secret Service to keep the White House informed of all events. A few days later, Vice President Cheney will state, "The Secret Service has an arrangement with the FAA. They had open lines after the World Trade Center was . . ." (He stopped himself before finishing the sentence.) [MSNBC, 9/16/01]

8:48 A.M.: CNN First Major Network to Show WTC Crash Footage

CNN is the first major network to show the footage of the crash site. It breaks into a commercial and anchor Carol Lin says, "This just in. You are looking at . . . obviously a very disturbing live shot there—that is the World Trade Center, and we have unconfirmed reports this morning that a plane has crashed into one of the towers of the World Trade Center." CNN then switches to Sean Murtagh, the network's vice president of finance, who says in a live telephone interview, "I just witnessed a plane that appeared to be cruising at a slightly lower than normal altitude over New York City. And it appears to have crashed into—I don't know which tower it is—but it hit directly in the middle of one of the World Trade Center towers. It was a jet, maybe a two-engine jet, maybe a 737 . . . a large passenger commercial jet . . . It was teetering back and forth, wing-tip to wing-tip, and it looks like it has crashed into—probably, twenty stories from the top of the World Trade Center—maybe the eightieth to eighty-fifth floor. There is smoke billowing out of the World Trade Center." [CNN, 9/11/01; A PRETEXT FOR WAR, BY JAMES BAMFORD, 6/04, PP. 16–17] Many reports do not come until a few minutes later. For instance, ABC first breaks into regular programming with the story at 8:52 A.M. [ABC NEWS, 9/14/02] Incredibly, a NORAD timeline presented to the 9/11 Commission in 2003 erroneously claims that CNN doesn't begin its coverage of the attacks until 8:57. [9/11 COMMISSION, 5/23/03]

8:48 A.M.: New York Flight Control Center Manager Aware of Ong Phone Call, Unaware Flight 11 Has Crashed

A New York flight control center manager speaks in a teleconference between flight centers. The person says, "Okay. This is New York [flight control]. We're watching the airplane [Flight 11]. I also had conversation with American Airlines, and they've told us that they believe that one of their stewardesses was stabbed and that there are people in the cockpit that have control of the aircraft, and that's all the information they have right now." The manager is unaware Flight 11 has already crashed. [9/11 COMMISSION REPORT, 6/17/04] This appears to be a simplified version of flight attendant Betty Ong's phone call, given to American Airlines leader Gerard Arpey and others around 8:30 A.M.

8:48 A.M.*: NORAD's Colorado Headquarters Sees WTC Television Footage

Canadian Air Force Major General Rick Findley is in charge of battle stations at NORAD's Colorado Springs, Colorado headquarters. According to Findley, "As the phones were beginning to ring, someone said, 'Sir, you might want to look at that.' I looked up and there was the CNN image of the World Trade Center. There was a hole in the side of one of the buildings." CNN broadcasts this footage starting at 8:48 A.M. An as-yet unidentified person reportedly tells Findley that it was a small plane, who responded, "I said the hole's too big for a small airplane. . . . I asked if it was the hijacked aircraft. I was scratching my head, wondering if it was another aircraft altogether." [CALGARY HERALD, 10/1/01]

Canadian Air Force Major General Rick Findley

After 8:48 A.M.*: Joint Chiefs of Staff Vice Chairman Still Oblivious? Accounts Are Contradictory

Air Force General Richard Myers, Vice Chairman of the Joint Chiefs of Staff, sees the first WTC crash on television. Myers is acting Chairman of the U.S. military during the 9/11 crisis because Chairman of the Joint Chiefs of Staff Army General Henry Shelton is flying in a plane across the Atlantic. [ABC NEWS, 9/11/02] Myers sees the television in an outer office of Senator Max Cleland (D), but he says, "They thought it was a small plane or something like that," so he goes ahead and meets with Cleland. He says, "Nobody informed us" about the second WTC crash, and remains oblivious to the emergency until the meeting with Cleland ends, and as the Pentagon explosion takes place at 9:37 A.M. Then he speaks to General Ralph Eberhart. [ARMED FORCES PRESS SERVICE, 10/23/01] Yet, in testimony on September 13, 2001, he states, "after the second tower was hit, I spoke to the commander of NORAD, General Eberhart. And at that point, I think the decision was at that point to start launching aircraft." [MYERS CONFIRMATION TESTIMONY, 9/13/01] NORAD claims the first fighters are scrambled even before the first WTC hit. [NORAD, 9/18/01] In his 2004 testimony before the 9/11 Commission, Myers' account changes again. He says that he gets a call from

Air Force General Richard Myers, Vice Chairman of the Joint Chiefs of Staff, and acting Chairman on 9/11

Eberhart, and then "shortly thereafter that the Pentagon was hit as we were on our way back to the Pentagon." [9/11 COMMISSION, 6/17/04 (B)] Myers' claim that he is out of the loop contradicts not only his previous account but also counterterrorism "tsar" Richard Clarke's account of what Myers does that day. According to Clarke's recollection, Myers takes part in a video conference from about 9:10 A.M. until after 10:00 A.M. If Myers is not involved in this conference, then his whereabouts and actions remain unknown until he arrives at the NMCC around 10:30 A.M.

8:49 A.M.*: United Airlines Headquarters Learns Flight 175 Is Missing; NORAD Apparently Not Informed

Apparently, managers at United Airlines' headquarters in Chicago are unaware of any unfolding emergency until they watch CNN break the story at 8:48 A.M. "Within minutes," United headquarters gets a call from the FAA, stating that the plane that crashed into the WTC was an American Airlines passenger plane. At about the same time and before a call about the flight that will take place at about 8:50 A.M., a manager says to Jim Goodwin (United's chairman and chief executive), "Boss, we've lost contact with one of our airplanes [Flight 175]." [WALL STREET JOURNAL, 10/15/01] At around 9:00 A.M., a United dispatcher reports that Flight 175 has been lost (it is not clear whether this is a clarification of the earlier message or a change in the timing that one call occurred). [9/11 COMMISSION, 1/27/04]

Before 8:50 A.M.*: CIA Director Expresses Worry About al-Qaeda Attack

According to one report, CIA Director Tenet is eating breakfast with his mentor, former Senator David Boren (D). "Before the planes hit the World Trade Center, CIA Director George Tenet warned [Boren] . . . that he was worried about a possible attack by Osama bin Laden's al-Qaeda network." [CHICAGO SUN TIMES, 12/6/02]

8:50 A.M.*: CIA Director, Told of "Attack," Immediately Suspects bin Laden

CIA Director Tenet is told of the first WTC crash while he is eating breakfast with his mentor, former Senator David Boren (D). They are interrupted when CIA bodyguards converge on the table to hand Tenet a cell phone. Tenet is told that the WTC has been attacked by an airplane. Boren later says, "I was struck by the fact that [the messenger] used the word 'attacked.'" Tenet then hands a cell phone back to an aide and says to Boren, "You know, this has bin Laden's fingerprints all over it." "'He was very collected,' Boren recalls. 'He said he would be at the CIA in 15 minutes, what people he needed in the room and what he needed to talk about.'" [USA TODAY, 9/24/01; ABC NEWS, 9/14/02] According to other accounts, Tenet responds to the caller, "They steered the plane directly into the building?" Vice President Cheney then says to Boren, "That looks like bin Laden." Tenet muses aloud, "I wonder if this has something to do with the guy [Zacarias Moussaoui] who trained for a pilot's license." (Moussaoui had been arrested several weeks earlier.) [STERN, 8/13/03; SAINT PAUL PIONEER PRESS 5/29/02] According to another account, Tenet pauses while on the phone to tell Boren, "The World Trade Center has been hit. We're pretty sure it wasn't an accident. It looks like a terrorist act," then returns to the phone to identify who should be summoned to the CIA situation room. [TIME, 9/14/01] (Note that according to two accounts, Tenet was not informed of the developing crisis until after the second WTC tower had been struck. [WASHINGTON POST, 1/26/02; A PRETEXT FOR WAR, BY JAMES BAMFORD, 6/04, PP. 18–19] However, the majority of reports indicate that Tenet was informed of the crisis right after the first WTC tower was struck.)

8:50 A.M.: Last Radio Contact with Flight 77

The last radio contact with Flight 77 is made when a pilot asks for clearance to fly higher. However, six minutes later, the plane fails to respond to a routine instruction. Presumably, it is hijacked during that time. Indianapolis flight control center is handling the plane by this time [GUARDIAN, 10/17/01; BOSTON GLOBE, 11/23/01; NEW YORK TIMES, 10/16/01; 9/11 COMMISSION REPORT, 6/17/04]

8:50 A.M.: Flight 175 Heads for New York City

Flight 175, already off course, makes a near complete U-turn and starts heading north toward New York City. [CNN, 9/17/01]

8:50 A.M.*: Flight 175 Attendant Reports Plane Has Been Hijacked; United Headquarters Informed

Rich Miles, manager of United's Chicago system operations center, receives a call from a mechanic at an airline maintenance center in San Francisco. (This center takes in-flight calls from flight attendants about broken items.) The mechanic informs Miles that a female flight attendant from Flight 175 has just called to report "Oh my God. The crew has been killed, a flight attendant has been stabbed. We've been hijacked." Then the line goes dead. A dispatcher monitoring the flight then sends messages to the plane's cockpit computer but gets no response. [WALL STREET JOURNAL, 10/15/01; BOSTON GLOBE, 11/23/01; 9/11 COMMISSION, 1/27/04] This information is quickly relayed to United's headquarters. [9/11 COMMISSION, 1/27/04] There is no published record identifying this female flight attendant. According to published accounts, a male flight attendant, Robert Fangman, calls from this flight at some unknown time. [CNN, 5/28/04] It is unclear whether the mechanic (or Miles) confused the gender of the caller, or two different attendants call from this flight. [CNN, 5/28/04]

8:50 A.M.*: Boston Flight Control Informs NORAD That Flight 11 Has Hit WTC

As soon as Boston flight controllers hear news that a plane might have hit the WTC, they know it was Flight 11. They have been tracking it continually since it began behaving erratically. It takes "several minutes" for Boston to report to NORAD that Flight 11 is responsible. [NEW YORK TIMES, 9/13/01 (F); NEWHOUSE NEWS SERVICE, 1/25/02]

8:51–8:53 A.M.: Flight Controller Declares Flight 175 Hijacked

According to the 9/11 Commission, the flight controller handling Flight 175 (presumably Dave Bottoglia [MSNBC, 9/11/02 (B)]) only notices now that the flight's transponder signal has changed, although, according to other published reports, this happened around 8:46 A.M. The controller asks the plane to return to its proper transponder code. There is no response. Beginning at 8:52 A.M., the controller makes repeated attempts to contact the plane, but there is still no response. [9/11 COMMISSION REPORT, 6/17/04] He contacts another controller at 8:53 A.M., and says, "We may have a hijack. We have some problems over here right now." [GUARDIAN, 10/17/01; NEW YORK TIMES, 10/16/01; 9/11 COMMISSION REPORT, 6/17/04] This account conflicts with earlier accounts that NORAD is notified at 8:43 A.M. that Flight 175 has been hijacked. [NORAD, 9/18/01] It also conflicts with Bottoglia's own account of finding Flight 175 at 8:46 A.M. and realizing it is hijacked at that time. [MSNBC, 9/11/02]

8:52 A.M.*: Flight 175 Passenger Details Stabbing

Businessman Peter Hanson calls his father from Flight 175 and says, "Oh, my God! They just stabbed the airline hostess. I think the airline is being hijacked." Despite being cut off twice, he manages to report how men armed with knives are stabbing flight attendants, apparently in an attempt to force crewmembers to unlock the doors to the cockpit. He calls again a couple of minutes before the plane crashes. [DAILY TELEGRAPH, 9/16/01 (B); TORONTO SUN, 9/16/01; BBC, 9/13/01] Hanson's father immediately calls the local police department and relays what he heard. [SAN FRANCISCO CHRONICLE, 7/23/04]

8:52 A.M.: Fighters Ordered Toward the Crashed Flight 11, Head for Flight 175 Instead

Two F-15s take off from Otis Air National Guard Base. This occurs six minutes after being ordered to go after Flight 11(which has already crashed); 26 minutes after flight controllers were certain Flight 11 was hijacked; and 39 minutes after flight controllers lost contact with Flight 11. [NORAD, 9/18/01; CNN, 9/17/01; WASHINGTON POST, 9/15/01; ABC NEWS, 9/11/02; WASHINGTON POST, 9/12/01; 9/11 COMMISSION REPORT, 6/17/04] The fighters inadvertently head toward Flight 175 instead. According to one of the pilots, as soon as they strap in, the green light to launch goes on, and they're up in the air even before their fighters' radar kicks in. [CAPE COD TIMES. 8/21/02]

8:52 A.M.*: New York Flight Controller Tracks Flight 175 into New York; NORAD Not Warned?

Mike McCormick, head of New York flight control center, sees the first WTC attack on CNN. He assumes that Flight 175, which he is tracking on his radar screen, is also headed into the WTC. He says, "Probably one of the most difficult moments of my life was the 11 minutes from the point I watched that aircraft, when we first lost communications until the point that aircraft hit the World Trade Center. For those 11 minutes, I knew, we knew, what was going to happen, and that was difficult." [CNN, 8/12/02] Yet, according to the 9/11 Commission, this flight control center will not notify NORAD about Flight 175 until after it crashes at 9:03 A.M.

**Mike
McCormick**

8:52 A.M. (and After): Otis Fighters Scramble to New York; Conflicting Accounts of Urgency and Destination

The F-15 Fighters are scrambling to New York City. Later accounts concerning these fighters conflict significantly. According one account, pilot Lieutenant Colonel Timothy Duffy later recalls that they are in a hurry at this time: "We've been over the flight a thousand times in our minds and I don't know what we could have done to get there any quicker." However, though Duffy says he's been warned Flight 11 had been hijacked and appears headed toward New York City, he does not yet realize that his

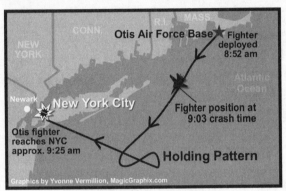

Route of the Otis Air National Guard fighters to New York City

flight is anything other than a routine exercise: "It's just peacetime. We're not thinking anything real bad is going to happen out there." [BBC, 9/1/02; CAPE COD TIMES, 8/21/02] But, in another account, Duffy claims that fellow officer tells him before takeoff, "This looks like the real thing." "It just seemed wrong. I just wanted to get there. I was in full-blower all the way." [AVIATION WEEK AND SPACE TECHNOLOGY, 6/3/02] Full-blower means the fighters are traveling at or near full speed. An F-15 can travel over 1,875 mph. [AIR FORCE NEWS, 7/30/97] A considerable amount of fuel is required to maintain such high speeds for long, but a NORAD commander notes that, coincidentally, these fighters are stocked with extra fuel. [AVIATION WEEK AND SPACE TECHNOLOGY, 6/3/02] Duffy later says, "As we're climbing out, we go supersonic on the way, which is kind of nonstandard for us." He says his target destination is over Kennedy airport in New York City. [ABC NEWS, 9/11/02] He says, "When we [take] off we [start] climbing a 280-heading, basically towards New York City. I [am] supersonic. . . . We [are] to proceed to Manhattan directly and set up a combat air patrol." [BBC, 9/1/02] There are different accounts as to just how quickly they travel. According to Major General Paul Weaver, director of the Air National Guard, "The pilots [fly] "like a scalded ape," topping 500 mph but [are] unable to catch up to the airliner." [DALLAS MORNING NEWS, 9/16/01] ABC News later says, "The fighters are hurtling toward New York at mach 1.2, nearly 900 miles per hour." [ABC NEWS, 9/11/02] NORAD commander Major General Larry Arnold later states that the fighters head straight for New York City at about 1,100 to 1,200 mph. [MSNBC, 9/23/01 (C); SLATE, 1/16/02] "An F-15 departing from Otis can reach New York City in ten to twelve minutes, according to an Otis spokeswoman." [CAPE COD TIMES, 9/16/01] At an average speed of 1,125 mph, the fighters would reach the city in ten minutes—9:02 A.M. If NORAD commander Arnold's recollection is correct, these fighters should reach Flight 175 just before it crashes. Yet according to a NORAD timeline developed just after 9/11, the fighters take about 19 minutes to reach New York City (arriving at about 9:11 A.M.), traveling below supersonic speeds at less than 600 mph. [NORAD, 9/18/01] The 9/11 Commission later concludes, in direct contrast to the recollections of the pilots and others involved that day, that the fighters are never directed toward New York City at all, but rather are ordered to head out over the Atlantic Ocean. According to the 9/11 Commission's conclusions, the fighters do not reach New York City until 9:25 A.M. [9/11 COMMISSION REPORT, 6/17/04]

After 8:52 A.M.*: NORAD Scramble Delayed?
Witness Casts Doubt on NORAD's Scramble Time

William Wibel, principal of a school inside Otis Air National Guard Base, is inside the Otis base preparing for a meeting when he learns that the WTC has been attacked and his meeting is canceled. He says, "As I drove away, and was listening to the news on the radio, the 102nd was scrambling into duty." [CAPE COD TIMES, 9/12/01] The WTC crash does not break on local news and radio until about 8:52 A.M. Even if he hears CNN's early reporting starting at 8:48 A.M., it still presumably takes time to learn the meeting is canceled, go back to his car and so forth. NORAD says the fighters took off from Otis at 8:52 A.M.

8:54 A.M.*: Flight 77 Veers Off Course

Flight 77 from Washington begins to go off course over southern Ohio, turning to the southwest. [WASHINGTON POST, 9/12/01; NEWSDAY, 9/23/01; 9/11 COMMISSION REPORT, 6/17/04]

8:55 A.M.*: Situation Room Director Informs Bush of WTC Crash

Captain Deborah Loewer, director of the White House Situation Room, is traveling in President Bush's motorcade toward Booker Elementary School, when she learns of the first WTC crash from her deputy in the White House Situation Room. According to some reports, as soon as the motorcade reaches the school, Loewer runs from her car to Bush's car, and informs Bush. [CATHOLIC TELEGRAPH, 12/7/01; ASSOCIATED PRESS, 11/26/01] Note that Bush maintains that he learns of the crash at a later time.

Captain Deborah Loewer

8:55 A.M.*: Bush Arrives at
Elementary School for Photo-Op

President Bush's motorcade arrives at Booker Elementary School

President Bush's motorcade arrives at Booker Elementary School for a photo-op to push his education policies. [ABC NEWS, 9/11/02; WASHINGTON TIMES, 10/7/02; SARASOTA MAGAZINE, 9/19/01; DAILY TELEGRAPH, 12/16/01; SARASOTA HERALD-TRIBUNE, 9/10/02; NEW YORK TIMES, 9/16/01 (B); ALBUQUERQUE TRIBUNE, 9/10/02] If he left the resort around 8:35 A.M. as reported, the timing of his arrival at 8:55 A.M. is consistent with the fact that the trip from the Colony Resort to the school is said to take 20 minutes. [NEW YORK TIMES, 9/16/01 (B); ST. PETERSBURG TIMES, 9/8/02 (B); MSNBC, 10/29/02]

8:55 A.M.*: New York Flight Control Believes Flight 175 Has Been Hijacked; NORAD Reportedly Not Informed

The head New York flight controller notifies a manager at the facility that she believes Flight 175 has been hijacked. The manager tries to notify regional managers about this, but cannot reach them because they are discussing the hijacking of Flight 11 and refuse to be disturbed. However, even though the controller managing Flight 175 said, "we may have a hijack" at 8:53 A.M., the 9/11 Commission concluded that NORAD is not notified by this time. [9/11 COMMISSION REPORT, 6/17/04] The commission's account conflicts with previous accounts stating that NORAD is notified of the Flight 175 hijacking at 8:43 A.M. The head of the flight control center, Mike McCormick, has already decided at 8:52 A.M. that Flight 175 has been hijacked and is on a suicide run to New York City. [CNN, 8/12/02]

Between 8:55–9:00 A.M.*: Bush First Told About WTC Crash?; Suggests Accident

President Bush's motorcade has arrived at Booker Elementary School and Bush enters the school with his entourage. The beepers of politicians' aides are going off with news of the first WTC crash as Bush arrives. According to one account, Bush learns of the crash when adviser Karl Rove takes Bush aside in a school corridor and tells him about the calamity. According to this account, Rove says the cause of the crash was unclear. Bush replies, "What a horrible accident!" Bush also suggests the pilot may have had a heart attack. This account is recalled by photographer Eric Draper, who was standing nearby at the time. [DAILY MAIL, 9/8/02] Dan Bartlett, White House Communications Director, also says he is there when Bush is told: "[Bush] being a former pilot, had kind of the same reaction, going, was it bad weather? And I said no, apparently not." [ABC NEWS, 9/11/02] One account states that Rove tells Bush the WTC has been hit by a large commercial airliner. [DAILY TELEGRAPH, 12/16/01] However, Bush later remembers Rove saying it appeared to be an accident involving a small, twin-engine plane. [WASHINGTON POST, 1/27/02] In a third version of the story, Bush later recalls that he first learns of the crash from chief of staff Andrew Card, who says, "Here's what you're going to be doing; you're going to meet so-and-so, such-and-such." And Andy Card says, "By the way, an aircraft flew into the World Trade Center." [WASHINGTON TIMES, 10/7/02] "From the demeanor of the President, grinning at the children, it appeared that the enormity of what he had been told was taking a while to sink in," according to a reporter standing nearby at the time. [DAILY MAIL, 9/8/02; DAILY TELEGRAPH, 12/16/01]

8:56 A.M.*: Flight 77 Transponder Signal Disappears; NORAD Not Informed

Flight 77's transponder signal is turned off at this time. [GUARDIAN, 10/17/01; BOSTON GLOBE, 11/23/01; NEWSDAY, 9/23/01] According to the 9/11 Commission, the Indianapolis flight controller in charge of the flight has watched it go off course and head southwest before the signal disappears. He looks for primary radar signals along its projected flight path as well as in the airspace where it has started to turn. He cannot find the plane. He tries contacting the airline but gets no answer. "'American 77, Indy,' the controller said, over and over. 'American 77, Indy, radio check. How do you read?' By 8:56 A.M., it was evident that Flight 77 was lost." [NEW YORK TIMES, 10/16/01] The controller has not been told about any other hijacked planes. (Other centers have been notified about the Flight 11 hijacking more than 20 minutes earlier at 8:25 A.M. [GUARDIAN, 10/17/01]) He assumes Flight 77 has experienced electrical or mechanical failure. [9/11 COMMISSION REPORT, 6/17/04] Apparently, American Airlines headquarters and the Pentagon's NMCC are notified that Flight 77 is off course with its radio and transponder not working, but NORAD is not notified at this time. [9/11 COMMISSION, 1/27/04]

8:56–9:05 A.M.*: Flight 77 Disappears from Radar Screens

According to the 9/11 Commission, "Radar reconstructions performed after 9/11 reveal that FAA radar equipment tracked [Flight 77] from the moment its transponder was turned off at 8:56 [A.M.]." However, for eight minutes and 13 seconds, this primary radar data is not displayed to Indianapolis flight controllers. "The reasons are technical, arising from the way the software processed radar information, as well as from poor primary radar coverage where American 77 was flying." [9/11 COMMISSION REPORT, 6/17/04] Apparently, a radar tower in West Virginia doesn't have primary radar. [WASHINGTON POST, 11/3/01] But the 9/11 Commission notes that other centers had primary radars that covered the missing areas, yet they weren't asked to do a primary radar search. [9/11 COMMISSION REPORT, 6/17/04]

After 8:56–9:24 A.M.*: Pentagon Emergency Center Knows Flight 77 Is Hijacked; NORAD Not Notified?

The *New York Times* reports, "During the hour or so that American Airlines Flight 77 [is] under the control of hijackers, up to the moment it struck the west side of the Pentagon, military officials in [the Pentagon's NMCC] [are] urgently talking to law enforcement and air traffic control officials about what to do." [NEW YORK TIMES, 9/15/01 (C)] Yet, although the Pentagon's NMCC reportedly knows of the hijacking, NORAD reportedly is not notified until 9:24 A.M. by some accounts, and not notified at all by others. [NORAD, 9/18/01; 9/11 COMMISSION, 6/17/04]

8:58 A.M.: Flight 175 Passenger Says Passengers Want to "Do Something" About Hijacking

Brian Sweeney on Flight 175 calls his wife, but can only leave a message. "We've been hijacked, and it doesn't look too good." Then he calls his mother and tells her what is happening onboard. [HYANNIS NEWS, 9/13/01; WASHINGTON POST, 9/21/01] She recalls him saying, "They might come back here. I might have to go. We are going to try to do something about this." She also recalls him identifying the hijackers as Middle Eastern. Then he tells his mother he loves her and hangs up the phone. The mother turns on the television and soon sees Flight 175 crash into the WTC. The 9/11 Commission later concludes that the Flight 175 passengers planned to storm the cockpit but did not have time before the plane crashed. [CNN, 3/10/04; NEW YORK DAILY NEWS, 3/9/04]

Before 9:00 A.M.*: Fire Department Advice to Evacuate
WTC Tower Fails to Reach People Inside

Shortly before 9:00 A.M., fire department commanders at the WTC Tower One advise Port Authority police and building personnel to evacuate Tower Two. However, there is no evidence that this advice is communicated effectively to the building personnel in Tower Two. When an announcement is made to evacuate at 9:02 A.M. (one minute before it is hit), it does not direct everyone to evacuate, and advises only that everyone may wish to start an orderly evacuation if warranted by conditions on their floor. [9/11 COMMISSION REPORT, 5/19/04]

Before 9:00 A.M.*: American Airlines Learns of Flight 77 Problems;
Cancels All Flight Take Offs in the Northeast; NORAD Not Notified

American Airlines headquarters in Forth Worth, Texas, learns that Flight 77 is not responding to radio calls, is not emitting a transponder signal, and flight control has lost its location since 8:56 A.M. [9/11 COMMISSION, 1/27/04] According to the *Wall Street Journal*, a call from the FAA roughly says that Flight 77 has "turned off its transponder and turned around. Controllers [have] lost radio communications with the plane. Without hearing from anyone on the plane, American [doesn't] know its location." Airline executive Gerard Arpey gives an order to stop all American flight take-offs in the Northeast. Within minutes, American gets word that United also has lost contact with a missing airliner (presumably Flight 175). When reports of the second WTC come through after 9:03 A.M., one manager recalls mistakenly shouting, "How did 77 get to New York and we didn't know it?" [WALL STREET JOURNAL, 10/15/01]

9:00 A.M.: Flight 175 Passenger Brian Sweeney Predicts Crash into Building

Flight 175 passenger Brian Sweeney calls his father a second time. He says, "It's getting bad, Dad—A stewardess was stabbed—They seem to have knives and Mace—They said they have a bomb—It's getting very bad on the plane—Passengers are throwing up and getting sick—The plane is making jerky movements—I don't think the pilot is flying the plane—I think we are going down—I think they intend to go to Chicago or someplace and fly into a building—Don't worry, Dad—If it happens, it'll be very fast—My God, my God." [SAN FRANCISCO CHRONICLE, 7/23/04]

9:00 A.M.: Pentagon Ups Alert Status

The Pentagon moves its alert status up one notch from normal to Alpha. After Flight 77 crashes into the Pentagon, it increases the alert to Delta, the highest level. The alert level will be reduced to Charlie on September 12th. [MSNBC, 9/11/01; AGENCE FRANCE-PRESSE, 9/12/01]

9:00 A.M.*: 9/11–Styled Simulation Cancelled

The National Reconnaissance Office plans a simulation of an airplane accidentally crashing into its headquarters. The office is located four miles from Washington's Dulles airport, where one of the real hijacked planes takes off. The NRO "operates many of the nation's spy satellites. It draws its personnel from the military and the CIA." The simulation is apparently run by John Fulton "and his team at the CIA." An agency spokesman says, "It was just an incredible coincidence that this happened to involve an aircraft crashing into our facility. As soon as the real world events began, we canceled the exercise." [ASSOCIATED PRESS, 8/21/02; UPI, 8/22/02]

NRO logo

9:00 A.M.*: Rice Informs Bush Flight 11 Has Accidentally Hit the WTC, but Knows Nothing Else

National Security Adviser Rice later claims she is in her White House office when she hears about the first WTC crash just before 9:00 A.M. She recalls, "I thought to myself, what an odd accident." She reportedly speaks to President Bush around 9:00 A.M. on the telephone, and tells him that a twin-engine plane has struck the WTC tower. She says, "That's all we know right now, Mr. President." [NEWSWEEK, 12/31/01] Rice later claims, "He said, what a terrible, it sounds like a terrible accident. Keep me informed." [ABC NEWS, 9/11/02] Despite her title of National Security Adviser, she is apparently unaware that NORAD has scrambled planes after learning of two hijackings in progress at least 15 minutes ago. She goes ahead with her usual national security staff meeting. [NEWSWEEK, 12/31/01] Author James Bamford comments, "Neither Rice nor Bush was aware that the United States had gone to 'battle stations' alert and had scrambled fighter jets into the air to intercept and possibly take hostile action against multiple hijacked airliners, something that was then known by hundreds of others within NORAD, the Federal Aviation Administration, and the Pentagon." [A PRETEXT FOR WAR, BY JAMES BAMFORD, 6/04, P. 17] Congressman Dan Miller, who is waiting in a receiving line to meet Bush, says he waits a few minutes for the call to end. Bush appears unbothered when he greets Miller after the call. Miller recalls, "It was nothing different from the normal, brief greeting with the president." [ST. PETERSBURG TIMES, 7/4/04]

9:00 A.M.*: Informed of the First Plane Crash, Bush Goes Ahead with Photo-Op

Sarasota elementary school principal Gwen Tose-Rigell is summoned to a room to talk with President Bush. She recalls, "He said a commercial plane has hit the World Trade Center, and we're going to go ahead and go on, we're going on to do the reading thing anyway." [ASSOCIATED PRESS, 8/19/02 (D)] One local reporter notes that at this point, "He could and arguably should have left Emma E. Booker Elementary School immediately, gotten onto Air Force One and left Sarasota without a moment's delay." [SARASOTA HERALD-TRIBUNE, 9/12/01 (B)]

9:00 A.M.*: WTC South Tower Announcement: OK to Return to Offices

A public announcement is broadcast inside the WTC Tower Two (the South Tower, which has yet to be hit), saying that the building is secure and people can return to their offices. [NEW YORK TIMES, 9/11/02] Such announcements continue until a few minutes before the building is hit, and "may [lead] to the deaths of hundreds of people." No one knows exactly what is said (though many later recall the phrase "the building is secure"), or who gives the authority to say it. [USA TODAY, 9/3/02] Additionally, security agents inside the building repeat similar messages to individuals in the tower. For instance, one survivor recounts hearing, "Our building is secure. You can go back to your floor. If you're a little winded, you can get a drink of water or coffee in the cafeteria." [NEW YORK TIMES, 9/13/01 (G)] Another survivor recalls an escaping crowd actually running over a man with a bullhorn encouraging them to return to their desks. [NEWSDAY, 9/12/01] Businessman Steve Miller recalls hearing a voice say over the building's loudspeaker something similar to: "There's a fire in Tower One. Tower Two in unaffected. If you want to leave, you can leave. If you want to return to your office, it's okay." [WASHINGTON POST, 9/16/01 (D)] British visitor Mike Shillaker recalls, "As we got to around floor 50, a message came over the [loudspeaker], telling us that there was an isolated fire in Tower One, and we did not need to evacuate Tower Two. Again, thank god we continued down, others didn't." [BBC, 9/1/02 (B)] Despite messages to the contrary, about two-thirds of the tower's occupants evacuate during the 17 minutes between the attacks. [USA TODAY, 12/20/01]

9:00 A.M.*: Northern Vigilance Exercise Canceled; False Radar Blips Purged from Radar Screens

Deep below Cheyenne Mountain at NORAD headquarters in Colorado, NORAD is at "full battle staff'" levels for a major annual exercise, Operation Northern Vigilance, which tests every facet of the organization. This military exercise, which began two days earlier, deploys fighters to Alaska and Northern Canada to monitor a Russian air force exercise in the Russian arctic. Canadian Captain Mike Jellinek is one hour into his shift, overseeing the Colorado command center, when he learns the FAA believes there is a hijacking in progress and is asking NORAD for support. Northern Vigilance is called off. As the *Toronto Star* reports, "Any simulated information, what's known as an 'inject,' is purged from the [radar] screens." [NORAD, 9/9/01; TORONTO STAR, 12/9/01] Therefore, many minutes into the real 9/11 attack, there may have been false radar blips causing confusion. Additional details, such as whose radar screens had false blips, or from when to when, are unknown. According to Jellinek, the Northern Vigilance is canceled just a minute or two before the second WTC crash at 9:03 A.M. The Russians, having seen the second WTC crash on television, quickly communicate that they are canceling their Russian arctic exercise. [TORONTO STAR, 12/9/01; NATIONAL POST, 10/19/02]

9:00 A.M.*: Cheney Perplexed over WTC Footage on Television

Vice President Cheney later says he is in his White House office watching the television images of the first WTC crash wreckage. According to his recollection, he was puzzled. "I was sitting there thinking about it. It was a clear day, there was no weather problem—how in hell could a plane hit the World Trade Center?" His staff members elsewhere in the White House are apparently unaware of the emerging crisis. For instance, his chief of staff, I. Lewis "Scooter" Libby, sees the television images briefly, but turns off the television so as not to be distracted from a conversation on another topic. [NEWSWEEK, 12/31/01]

9:00 A.M.*: Clarke Alerted to Crisis, Immediately Activates CSG

Counterterrorism "tsar" Richard Clarke is at a conference three blocks from the White House when a telephone call alerts him to the crisis. He runs to his car. He responds, "Activate the CSG on secure video. I'll be there in less than five." The CSG is the Counterterrorism and Security Group, comprising the leaders of the government's counterterrorism and security agencies. Clarke hurriedly drives to the White House. [AGAINST ALL ENEMIES, BY RICHARD CLARKE, 3/04, P. 1]

After 9:00 A.M.*: Indianapolis Flight Control Issues Alert to Look for Flight 77; FAA and NORAD Not Notified

According to the 9/11 Commission, shortly after 9:00 A.M., Indianapolis flight control begins to notify other government agencies that American 77 is missing and has possibly crashed. For instance, at 9:08 A.M., Indianapolis contacts Air Force Search and Rescue at Langley Air Force Base, Virginia, and tells them to look out for a downed aircraft. They also contact the West Virginia State Police, and ask whether they have any reports of a downed aircraft. However, they apparently do not notify the FAA or NORAD. [9/11 COMMISSION REPORT, 6/17/04]

After 9:00 A.M.*: Flight Dispatcher Sends Warning to All United Flights, Including Flight 93

Ed Ballinger, a flight dispatcher for United Airlines, is continuing to send messages one by one to the 16 transcontinental flights he is covering, warning them of the first WTC crash. He is handling both Flights 175 and 93, and 175 has failed to respond to his message. A few minutes after 9:00, he sends a message to Flight 93. The exact content of the message is not known, but apparently it doesn't advise the pilots to bar the cockpit door. [PITTSBURGH POST-GAZETTE, 10/28/01; NEW YORK OBSERVER, 6/17/04] Another flight controller at the Cleveland tower in charge of Flight 93 at the time later recalls, "I saw controllers step up to the plate and start warning flight crews. This was totally by the seat of their pants. It's not because they're directed to by anybody. It's just, OK, everybody's on alert right now." [MSNBC, 9/11/02 (B)] Ballinger later says, "One of the things that upset me was that [the FAA and United Airlines headquarters] knew, 45 minutes before [Flight 93 crashed], that American Airlines had a problem. I put the story together myself [from news accounts]. Perhaps if I had the information sooner, I might have gotten the message to [Flight] 93 to bar the door." [NEW YORK OBSERVER, 6/17/04] Ballinger will send Flight 93 a second, more detailed warning that does warn to bar the door. It will reach Flight 93 at 9:24 A.M., shortly before the flight is hijacked.

After 9:00 A.M.*: United Flight 23 Hijacking Averted?

Shortly after 9:00 A.M., United Airlines Flight 23 receives a warning message from flight dispatcher Ed Ballinger. Flight 23 is still on a Newark, New Jersey, runway, about to take off for Los Angeles. Apparently in response to Ballinger's message, the crew tells the passengers there has been a mechanical problem and returns to the departure gate. A number of Middle Eastern men (one account says three, others says six) argue with the flight crew and refuse to get off the plane. Security is called, but they flee before it arrives. [CBS NEWS, 9/14/01 (B); *CHICAGO DAILY HERALD*, 04/14/04] Later, authorities check their luggage and find copies of the Koran and al-Qaeda instruction sheets. Ballinger suspects they got away. "When all we have is a photo from a fake ID, the chances of finding [someone] in Afghanistan or Pakistan are rather slim." [*CHICAGO DAILY HERALD*, 04/14/04] A NORAD deputy commander later says, "From our perception, we think our reaction on that day was sufficiently quick that we may well have precluded at least one other hijacking. We may not have. We don't know for sure." [*GLOBE AND MAIL*, 6/13/02]

9:01 A.M.*: Bush Claims to See First WTC
Crash on Television While at Elementary School

President Bush later makes the following statement: "And I was sitting outside the classroom waiting to go in, and I saw an airplane hit the tower—the television was obviously on, and I use to fly myself, and I said, 'There's one terrible pilot.' And I said, 'It must have been a horrible accident.' But I was whisked off there—I didn't have much time to think about it." [CNN, 12/4/01] He has repeated the story on other occasions. [WHITE HOUSE, 1/5/02; CBS NEWS, 9/11/02] Notably, the first WTC Crash was not shown live on television. Further, Bush does not have access to a television until 15 or so minutes later. [*WASHINGTON TIMES*, 10/7/02] A *Boston Herald* article later notes, "Think about that. Bush's remark implies he saw the first plane hit the tower. But we all know that video of the first plane hitting did not surface until the next day. Could Bush have meant he saw the second plane hit—which many Americans witnessed? No, because he said that he was in the classroom when Andrew Card whispered in his ear that a second plane hit." The article, noting that Bush has repeated this story more than once, asks, "How could the commander in chief have seen the plane fly into the first building—as it happened?" [*BOSTON HERALD*, 10/22/02] A Bush spokesman later calls Bush's repeated comments "just a mistaken recollection." [*WALL STREET JOURNAL*, 3/22/04]

9:01 A.M.*: La Guardia Flight Controllers
and Port Authority Unaware of Hijackings

An unidentified woman in the La Guardia control tower speaks to a Port Authority police officer. La Guardia is one of two major New York City airports. The Port Authority patrols both the WTC and the city's airports. The woman asks the officer what has happened at the WTC, and the officer replies that he has learned from the news that a plane crashed into it. [*NEW YORK TIMES*, 12/30/03] Around the same time, one flight controller in the tower says to another, "But you don't know anything." The other responds, "We don't know. We're looking at it on Channel 5 right now." [*BERGEN RECORD*, 1/4/04] "Nothing on the [later released transcripts] shows that the La Guardia controllers knew that the planes flying into their airspace had been seized by terrorists, or that military aircraft were screaming in pursuit over the Hudson River." Port Authority officials appear to be equally oblivious. [*NEW YORK TIMES*, 12/30/03]

9:01 A.M.*: New York Flight Control Informs Low Altitude
NYC Controllers About Flight 175 Hijacking

New York flight control contacts New York terminal approach control and asks for help in locating Flight 175. Different flight controllers scan different altitudes, and terminal approach controllers only deal with low-flying planes. These low altitude flight controllers have remained uninformed about the fate of Flight 11 until about now: "We had 90 to 120 seconds; it wasn't any 18 minutes," says one controller, referring to the actual elapsed time between the two crashes. Another such controller says of both planes: "They dove into the airspace. By the time anybody saw anything, it was over." [NEW YORK TIMES, 9/13/01 (F); 9/11 COMMISSION REPORT, 6/17/04]

9:01 A.M.*: New York Flight Control Tells FAA
Command Center About Flight 175 Hijack; Wants NORAD Help

A manager from New York flight control tells the FAA Command Center in Herndon, Virginia, "We have several situations going on here. It's escalating big, big time. We need to get the military involved with us. . . . We're, we're involved with something else, we have other aircraft that may have a similar situation going on here. . . ." The 9/11 Commission calls this the first notification to FAA leadership of the second hijack, but NORAD is not yet notified. [9/11 COMMISSION REPORT, 6/17/04] If this is true, then it means United Airlines headquarters has not yet contacted the FAA, despite knowing Flight 175 has been hijacked since about 8:50 A.M.

Between 9:01–9:03 A.M.*: Flight Controllers, American
Headquarters Watch Flight 175 Head into New York City

Flight 175 is an unmarked blip to flight controllers in New York City. One controller stands up in horror. "No, he's not going to land. He's going in!" Another controller shouts, "Oh, my God! He's headed for the city. . . . Oh, my God! He's headed for Manhattan!" [WASHINGTON POST, 9/21/01] Managers at American Airlines' headquarters in Forth Worth, Texas also closely watch Flight 175 head into New York City on radar. [USA TODAY, 8/12/02] Yet, according to the 9/11 Commission, no one has notified NORAD about the flight.

Before 9:03 A.M.*: Special FAA-Military Link
Fails to Help Communication Problems

At some point before the second WTC crash, the FAA Command Center sets up a teleconference with FAA facilities in the New York area. Also on the same floor of the same building is "the military cell"— the Air Traffic Services Cell—created by the FAA and the Defense Department to coordinate priority aircraft movement during warfare or emergencies if needed. "The Pentagon staffs it only three days per month for refresher training, but September 11 happen[s] to be one of those days." [AVIATION WEEK AND SPACE TECHNOLOGY, 12/17/01] Additionally, just weeks earlier the cell had been given a secure terminal and other hardware "greatly enhancing the movement of vital information." [AVIATION WEEK AND SPACE TECHNOLOGY, 6/10/02] The 9/11 Commission later determines that communication between the FAA and the military is extremely poor. It is unclear why this connection, which the 9/11 Commission fails to mention, does not help.

9:03 A.M.: Flight 175 Crashes into WTC South Tower; Millions Watch Live on Television

Flight 175 crashes into the South Tower of the World Trade Center

Flight 175 hits the South Tower of the World Trade Center (Tower Two). Seismic records pinpoint the time at six seconds before 9:03 A.M. (rounded to 9:03 A.M.). [CNN, 9/17/01; NORAD, 9/18/01; NEW YORK TIMES, 9/12/01; CNN, 9/12/01; ASSOCIATED PRESS, 8/19/02 (B); USA TODAY, 9/3/02; NEW YORK TIMES, 9/11/02; USA TODAY, 12/20/01] Millions watch the crash live on television. The plane strikes the 78th through 84th floors in the 110-story building. Approximately 100 people are killed or injured in the initial impact; 600 people in the tower eventually die. All but four of those killed work above the crash point. The death toll is far lower than in the North Tower because about two-thirds of the South Tower's occupants have evacuated the building in the 17 minutes since the first tower was struck. [USA TODAY, 12/20/01] The combined death toll from the two towers is estimated at 2,819, not including the hijackers. [ASSOCIATED PRESS, 8/19/02 (B)]

9:03 A.M.: Bush's Security Agents Watch Second WTC Crash on Television; Bush Continues with Photo-Op

According to Sarasota County Sheriff Bill Balkwill, just after President Bush enters a Booker Elementary classroom, a Marine responsible for carrying Bush's phone walks up to Balkwill, who is standing in a nearby side room. While listening to someone talk to him in his earpiece, the Marine asks, "Can you get me to a television? We're not sure what's going on, but we need to see a television." Three Secret Service agents, a SWAT member, the Marine, and Balkwill turn on the television in a nearby front office just as Flight 175 crashes into the WTC. "We're out of here," the Marine tells Balkwill. "Can you get everyone ready?" [SARASOTA HERALD-TRIBUNE, 9/10/02] However, Bush stays at the school until after 9:30 A.M. Who makes the decision to stay—and why—remains unclear, and the Secret Service won't comment on the matter. Philip Melanson, author of a book on the Secret Service, comments, "With an unfolding terrorist attack, the procedure should have been to get the president to the closest secure location as quickly as possible, which clearly is not a school. You're safer in that presidential limo, which is bombproof and blastproof and bulletproof. . . . In the presidential limo, the communications system is almost duplicative of the White House—he can do almost anything from there but he can't do much sitting in a school." [ST. PETERSBURG TIMES, 7/4/04]

A large explosion immediately follows the crash of Flight 175 into the World Trade Center

9:03 A.M.: Newark Flight Controllers Watch Flight 175 Hit WTC

Flight controllers in Newark, New Jersey are on the phone with New York flight controllers and are asked to find Flight 175 from their windows. They see it and watch in horror as it drops the last five thousand feet and crashes into the WTC. Rick Tepper (who also saw the explosion of the first crash) recalls, "He was in a hard right bank, diving very steeply and very fast. And he—as he was coming up the Hudson River, he—he made another hard left turn and—just heading for downtown Manhattan. . . . You could see that he was trying to line himself up on the tower. Just before he hit the tower, he almost leveled it out and just—just hit the building." Newark immediately calls the Air Traffic Control System Command Center in Washington and tells them they will not land any more airplanes in Newark, in an effort to keep aircraft away from New York City. It is the first step in shutting down the national airspace system. [MSNBC, 9/11/02 (B)]

9:03 A.M.*: Contradictions over Otis Fighter Mission and Whereabouts

The minute Flight 175 hits the South Tower, pilot Major Daniel Nash says that clear visibility allows him to see smoke pour out of Manhattan, even though NORAD says he is 71 miles away. [CAPE COD TIMES, 8/21/02] The other Otis pilot, Lieutenant Colonel Timothy Duffy, recalls, "We're 60 miles out, and I could see the smoke from the towers." They call to NORAD right then for an update, and Duffy relates, "At that point, they said the second aircraft just hit the World Trade Center. That was news to me. I thought we were still chasing American [Airlines Flight] 11." [ABC NEWS, 9/14/02] In another account Duffy again relates, "It was right about then when they said the second aircraft had just hit the World Trade Center, which was quite a shock to both [Nash] and I, because we both thought there was only one aircraft out there. We were probably 70 miles or so out when the second one hit. So, we were just a matter of minutes away." [BBC, 9/1/02] He asks for clarification of their mission, but the request is met with "considerable confusion." [AVIATION WEEK AND SPACE TECHNOLOGY, 6/3/02] Bob Varcadapane, a Newark, New Jersey flight controller who sees the Flight 175 crash, claims, "I remember the two F-15s. They were there moments after the impact. And I was just—said to myself, 'If only they could have gotten there a couple minutes earlier.' They just missed it." [MSNBC, 9/11/02 (B)] However, the 9/11 Commission appears to believe that the pilots never get near New York City at this time. According to the commission's account, from 8:46 A.M. until 8:52 A.M., NORAD personnel are unable to find Flight 11. Shortly after 8:50 A.M., and just before the fighters take off, NORAD is given word that a plane has hit the WTC. Lacking a clear target, the fighters take off toward a military controlled airspace over the ocean, off the coast of Long Island. A map released by the 9/11 Commission indicates that at 9:03 the fighters are about 100 miles away and heading southwest instead of west to New York City. [9/11 COMMISSION REPORT, 6/17/04]

9:03 A.M.*: New York Flight Control Informs NORAD That Flight 175 Has Been Hijacked; Timing of Notice in Question

The 9/11 Commission later concludes that New York flight control tells NEADS that Flight 175 has been hijacked at this time. The commission refers to this as "the first indication that the NORAD air defenders had of the second hijacked aircraft." [9/11 COMMISSION REPORT, 6/17/04] Colonel Robert Marr, head of NEADS, claims that he first learns a flight other than Flight 11 has been hijacked when he sees Flight 175 crash into the WTC on television. [AVIATION WEEK AND SPACE TECHNOLOGY, 6/3/02] However, this account contradicts NORAD's conclusion reached shortly after 9/11 that it was first notified about Flight 175 at 8:43 A.M. [NORAD, 9/18/01] Additionally, as Flight 175 crashes into the WTC, Canadian Captain Mike Jellinek (who is overseeing the command center in NORAD's Colorado headquarters), is on the phone with NEADS. He sees the crash live on television and asks NEADS, "Was that the hijacked aircraft you were dealing with?" The reply is yes. [TORONTO STAR, 12/9/01] If the commission's account is correct, several questions remain unanswered. Flight 175 lost radio contact at 8:42 A.M. and changed transponder signals at 8:46 A.M.; a flight controller declared it possibly hijacked sometime between 8:46 A.M. and 8:53 A.M.; and a flight control manager called it hijacked at 8:55 A.M. The commission has not explained why New York flight control would wait 10–17 minutes before warning NORAD that Flight 175 is possibly hijacked. [9/11 COMMISSION REPORT, 6/17/04] It also would not explain why United Airlines headquarters would fail to notify NORAD National Guard after learning that the plane has been hijacked at about 8:50 A.M.

9:03 A.M.*: Fighters Do Not Have Shootdown Authority

A fighter pilot flying from Otis Air Base toward New York City later notes that it wouldn't have mattered if he caught up with Flight 175, because only President Bush could order a shoot down, and Bush is at a public event at the time. [CAPE COD TIMES, 8/21/02] "Only the president has the authority to order a civilian aircraft shot down," according to a 1999 CNN report. [CNN, 10/26/99] In fact, by 9/11, Defense Secretary Rumsfeld also has the authority to order a shootdown, but he is not responding to the crisis at this time. [NEW YORK OBSERVER, 6/17/04]

9:03 A.M.*: Boston Flight Control Tells FAA That Hijackers Said "We Have Planes," FAA Suggests Notifying NORAD

A manager at Boston flight control reports to the FAA's New England regional headquarters the "we have some planes" comment made by a Flight 11 hijacker at 8:24 A.M. The Boston controller says, "I'm gonna reconfirm with, with downstairs, but the, as far as the tape . . . seemed to think the guy said that 'we have planes.' Now, I don't know if it was because it was the accent, or if there's more than one [hijacked plane], but I'm gonna, I'm gonna reconfirm that for you, and I'll get back to you real quick. Okay?" Asked, "They have what?," this person clarifies, "Planes, as in plural. . . . It sounds like, we're talking to New York, that there's another one aimed at the World Trade Center. . . . A second one just hit the Trade Center." The person at New England headquarters replies, "Okay. Yeah, we gotta get—we gotta alert the military real quick on this." At 9:05 A.M., Boston confirms for this headquarters and the FAA Command Center in Herndon, Virginia that a hijacker said, "we have planes" (forgetting the "some"). [9/11 COMMISSION REPORT, 6/17/04] It appears Boston replays the recording of the hijacker saying this from about 30 minutes earlier. Other people, such as American Airlines leader Gerard Arpey at that airline's headquarters, apparently learned about this comment before the Flight 11 crash at 8:46 A.M.

9:03–9:06 A.M.*: Bush Enters Classroom Photo-Op, Still Claims to Think WTC Crash Is Accidental

President Bush enters Sandra Kay Daniels' second-grade class for a photo-op to promote Bush's education policies. [DAILY MAIL, 9/8/02] Numerous reporters who travel with the president, as well as members of the local media, watch from the back of the room. [ASSOCIATED PRESS, 8/19/02 (D)] Altogether, there about 150 people in the room, 16 of whom are children in the class. He is introduced to the children and poses for a number of staged pictures. The teacher then leads the students through some reading exercises (video footage shows this lasts about three minutes). [SALON, 9/12/01 (B)] Bush later claims that during this lesson, he is thinking what he will say about the WTC crash. "I was concentrating on the program at this point, thinking about what I was going to say. Obviously, I felt it was an accident. I was concerned about it, but there were no alarm bells." [WASHINGTON TIMES, 10/7/02] The children are just getting their books from under their seats to read a story together when Chief of Staff Andrew Card comes in to tell Bush of the second WTC crash. [WASHINGTON TIMES, 10/8/02; DAILY TELEGRAPH, 12/16/01] According to the *Washington Times*, Card comes in at the conclusion of the first half of the planned lesson, and "[seizes] a pause in the reading drill to walk up to Mr. Bush's seat." [WASHINGTON TIMES, 10/7/02; WASHINGTON TIMES, 10/8/02]

President Bush enters Sandra Kay Daniels' classroom

9:03–9:08 A.M.*: Flight Control Managers Ban Aircraft Around New York and Washington

In a series of stages, flight control managers ban aircraft from flying near the cities targeted by the hijackers. All takeoffs and landings in New York City are halted within a minute of the Flight 175 crash, without asking for permission from Washington. Boston and Newark flight control centers follow suit in the next few minutes. Around 9:08 A.M., departures nationwide heading to or through New York and Boston airspace are canceled. [ASSOCIATED PRESS, 8/12/02; NEWSDAY, 9/10/02; ASSOCIATED PRESS, 8/19/02 (B); USA TODAY, 8/13/02] Mike McCormick, head of a Long Island, New York air traffic control center, makes the decision without consulting any superiors. [ABC NEWS, 8/12/02] In addition, "a few minutes" after 9:03 A.M., all takeoffs from Washington are stopped. [USA TODAY, 8/12/02; USA TODAY, 8/13/02]

After 9:03 A.M.*: New York and Washington Flight Controllers Told to Watch for Suspicious Aircraft

New York flight controllers are told by the FAA to watch for airplanes whose speed indicates that they are jets, but which either are not responding to commands or have disabled their transponders. "Controllers in Washington [get] a similar briefing, which [help] them pick out hijacked planes more quickly." [NEW YORK TIMES, 9/13/01 (F)] Other centers are apparently not told the same, and Indianapolis flight control apparently remains unaware of any crisis. [9/11 COMMISSION REPORT, 6/17/04]

After 9:03 A.M.*: Rice Learns of Second Attack; Goes to Basement Bunker

National Security Adviser Rice has just started her daily national security staff meeting at 9:00 A.M. Shortly after 9:03 A.M., an aide hands her a note saying a second plane has hit the WTC. Rice later claims that she thinks, "This is a terrorist attack," and then leaves the meeting, quickly walking to the White House Situation Room. [NEWSWEEK, 12/31/01] However, according to counterterrorism "tsar" Richard Clarke, Rice leaves the meeting for Vice President Cheney's office. Clarke meets her there a few minutes later and only then does she go down to the basement bunker. [AGAINST ALL ENEMIES, BY RICHARD CLARKE, 3/04, PP. 1–2]

After 9:03 A.M.*: Boston Controllers Give Cockpit Security Alert to New England Planes and Asks FAA to Issue Nationwide Warning; FAA Fails to Do So

"Within minutes of the second impact," Boston flight control's Operations Manager instructs all flight controllers in his center to inform all aircraft in the New England region to monitor the events unfolding in New York and to advise aircraft to heighten cockpit security. Boston asks the FAA Command Center to issue a similar cockpit security alert to all aircraft nationwide. The 9/11 Commission concludes, "We have found no evidence to suggest that Command Center managers instructed any centers to issue a cockpit security alert." [9/11 COMMISSION REPORT, 6/17/04] United Airlines flight dispatchers give their pilots a cockpit warning about 20 minutes later.

After 9:03 A.M.*: Wolfowitz Continues Routine Meeting, Rumsfeld Stays in Office

Deputy Defense Secretary Paul Wolfowitz has recently left a meeting with Defense Secretary Rumsfeld around 8:46 A.M. Wolfowitz later recalls, "We were having a meeting in my office. Someone said a plane had hit the World Trade Center. Then we turned on the television and we started seeing the shots of the second plane hitting, and this is the way I remember it. It's a little fuzzy. . . . There didn't seem to be much to do about it immediately and we went on with whatever the meeting was." [DEFENSE DEPARTMENT, 5/9/03] Rumsfeld recalls, "I was in my office with a CIA briefer and I was told that a second plane had hit the other tower." [9/11 COMMISSION, 3/23/04] Deputy Defense Secretary Torie Clarke recalls, "A couple of us had gone into . . . Secretary Rumsfeld's office, to alert him to that, tell him that the crisis management process was starting up. He wanted to make a few phone calls. So a few of us headed across the hallway to an area called the National Military Command Center [around 200 feet away]. He stayed in his office." [DEFENSE DEPARTMENT, 9/15/01 (B)]

After 9:03 A.M.*: Air Base Commanders Offer to Help NORAD; Timing of Acceptance Unclear

Shortly after the second WTC crash, calls from fighter units begin "pouring into NORAD and sector operations centers, asking, "What can we do to help?" In Syracuse, New York, an Air National Guard commander tells NEADS commander Robert Marr, "Give me ten [minutes] and I can give you hot guns. Give me 30 [minutes] and I'll have heat-seeker [missiles]. Give me an hour and I can give you slammers [Amraams]." Marr replies, "I want it all." [*AVIATION WEEK AND SPACE TECHNOLOGY, 6/3/02*] Reportedly, Marr says, "Get to the phones. Call every Air National Guard unit in the land. Prepare to put jets in the air. The nation is under attack." [NEWHOUSE NEWS SERVICE, 1/25/02] Canadian Major General Eric Findley, based in Colorado and in charge of NORAD that day, reportedly has his staff immediately order as many fighters in the air as possible. [*OTTAWA CITIZEN, 9/11/02*] However, according to another account, NORAD does not accept the offers until about an hour later: "By 10:01 A.M., the command center began calling several bases across the country for help." [*TOLEDO BLADE, 12/9/01*] The 9/11 Commission later concludes that a command for other bases to prepare fighters to scramble is not given until 9:49 A.M. In fact, it appears the first fighters from other bases to take off are those from Syracuse at 10:44 A.M. This is over an hour and a half after Syracuse's initial offer to help at 9:26 A.M., and not long after a general ban on all flights, including military ones, is lifted at 10:31 A.M. These are apparently the fourth set of fighters scrambled from the ground. Previously, three fighters from Langley, two from Otis, Ohio, and two from Toledo, Ohio, were scrambled at 10:01 A.M., but did not launch until fifteen minutes later. [*TOLEDO BLADE, 12/9/01*]

After 9:03 A.M.[*]: NMCC Commander Not at the NMCC; Later Misleads Regarding His Role in Crisis Response

Brigadier General Montague Winfield, commander of the NMCC, the Pentagon's emergency response center, later says, "When the second aircraft flew into the second tower, it was at that point that we realized that the seemingly unrelated hijackings that the FAA was dealing with were in fact a part of a coordinated terrorist attack on the United States." [ABC NEWS, 9/14/02] However, despite the tenor of this and other media reports (for instance, [CNN, 9/4/02; ABC NEWS, 9/15/02]), Winfield isn't actually at the NMCC during the 9/11 crisis. [LEIDIG TESTIMONY, 6/17/04; 9/11 COMMISSION, 6/17/04 (B)]

Brig. Gen.
Montague
Winfield

After 9:03 A.M.[*]: Secret Service Wants Fighters Scrambled from Andrews; None Are Ready to Fly

A few minutes after 9:03 A.M., a squadron pilot at Andrews Air Force Base, Maryland (just ten miles from Washington), hears that two planes have crashed into the WTC. He calls a friend in the Secret Service to see what's going on. The Secret Service calls back, and asks whether Andrews can scramble fighters. Apparently anticipating the need, one commander has already started preparing weapons for the fighters. However, the weapons are located in a bunker on the other side of the base, and the process takes time. The fighters don't take off for about another hour and a half (10:42 A.M.). Meanwhile, there are also three unarmed F-16 fighters assigned to the Andrews base on a training mission 207 miles to the south in North Carolina. These are not recalled until much later, and don't reach Washington until 10:45 A.M. [*AVIATION WEEK AND SPACE TECHNOLOGY, 9/9/02*] NORAD commander Major General Larry Arnold has said, "We [didn't] have any aircraft on alert at Andrews." [MSNBC, 9/23/01 (C)] However, prior to 9/11, the District of Columbia Air National Guard [DCANG] based at Andrews had a publicly stated mission "to provide combat units in the highest possible state of readiness." Prior to 9/11, the mission statement was posted on the D.C. National Guard's public website. Shortly after 9/11, this mission statement is removed and replaced by a DCANG "vision" to "provide peacetime command and control and administrative mission oversight to support customers, DCANG units, and NGB in achieving the highest levels of readiness." [DCANG HOME PAGE (BEFORE AND AFTER THE CHANGE)]

After 9:03 A.M.: WTC Building 7 Evacuated; Exact Timing Unclear

According to a soldier at the scene, WTC Building 7 is evacuated before the second tower is hit. [D.C. MILITARY, 10/18/01] However, a firefighter who arrived there after the second tower is hit is told that the building is being evacuated due to reports of a third plane, indicating that two planes have already crashed. [JEMS AND FIRERESCUE SUPPLEMENT, MARCH 2002]

9:05 A.M.*: Clarke, Cheney, and Rice Talk, Clarke's Recommendation to Evacuate White House Is Ignored

Counterterrorism "tsar" Richard Clarke is driving up to a gate outside the White House when an aide calls and tells him, "The other tower was just hit." He responds, "Well, now we know who we're dealing with. I want the highest level person in Washington from each agency on-screen now, especially the FAA." He has already ordered this aide to set up a secure video conference, about five minutes earlier. A few minutes later, he finds Vice President Cheney and National Security Adviser Rice in Vice President Cheney's White House office. Cheney tells Clarke, "It's an al-Qaeda attack and they like simultaneous attacks. This may not be over." Rice asks Clarke for recommendations, and he says, "We're putting together a secure teleconference to manage the crisis." He also recommends evacuating the White House (However, evacuation does not begin until 9:45 A.M., after a critical 40 minutes has passed). Rice notes the Secret Service wants them to go to the bomb shelter below the White House, and as Clarke leaves the other two, he sees Rice and Cheney gathering papers and preparing to evacuate. [AGAINST ALL ENEMIES, BY RICHARD CLARKE, 3/04, PP. 1–2, AUSTRALIAN, 3/27/04]

9:05 A.M. (and After): Flight 77 Reappears on Radar, but Flight Controllers Do Not Notice

According to the 9/11 Commission, Flight 77's radar blip, missing for the last eight minutes, reappears on Indianapolis flight control's primary radar scope. It is east of its last known position. It remains in air space managed by Indianapolis until 9:10 A.M., and then passes into Washington air space. Two managers and one flight controller continue to look west and southwest for the flight, but don't look east. Managers don't instruct other Indianapolis controllers to join the search for the flight. Neither they nor

FAA headquarters issues an "all points bulletin" to surrounding centers to search for Flight 77. [9/11 COM-MISSION REPORT, 6/17/04] *Newsday* claims that rumors circulate the plane might have exploded in midair. [NEWS-DAY, 9/23/01] However, the 9/11 Commission's conclusion that Indianapolis flight controllers did not look east is contradicted by an account indicating that American Airlines headquarters was told that Flight 77 had turned around.

9:06 A.M.: Flight Controllers Nationwide Are Told Flight 11 Crash Caused by Hijacking

All flight control facilities nationwide are notified that the Flight 11 crash into the WTC was probably a hijacking. [HOUSE OF REPRESENTATIVES COMMITTEE, 9/21/01; NEWSDAY, 9/23/01]

9:06 A.M.*: Bush Told WTC Hit Again and "America's Under Attack"; He Continues Photo-Op

President Bush is in a Booker Elementary School second-grader classroom. His chief of staff, Andrew Card, enters the room and whispers into his ear, "A second plane hit the other tower, and America's under attack." [NEW YORK TIMES, 9/16/01 (B); NEW YORK TIMES, 9/16/01 (B); DAILY TELEGRAPH, 12/16/01; ALBUQUERQUE TRIBUNE, 9/10/02; WASHINGTON TIMES, 10/8/02; ABC NEWS, 9/11/02] Intelligence expert James Bamford describes Bush's reac-

tion: "Immediately [after Card speaks to Bush] an expression of befuddlement passe[s] across the President's face. Then, having just been told that the country was under attack, the commander in chief appear[s] uninterested in further details. He never ask[s] if there had been any additional threats, where the attacks were coming from, how to best protect the country from further attacks. . . . Instead, in the middle of a modern-day Pearl Harbor, he simply turn[s] back to the matter at hand: the day's photo-op." [BODY OF SECRETS, JAMES BAMFORD, 4/02 EDITION, P. 633] Bush continues listening to the goat story. Then, as one newspaper put it, "For some reason, Secret Service agents [do] not bustle him away." [GLOBE AND MAIL, 9/12/01] Bush later says of the experience, "I am very aware of the cameras. I'm trying to absorb that knowledge. I have nobody to talk to. I'm sitting

Andrew Card speaks to President Bush and tells him of the second World Trade Center attack

in the midst of a classroom with little kids, listening to a children's story and I realize I'm the commander in chief and the country has just come under attack." [DAILY TELEGRAPH, 12/16/01] Bush continues to listen to the goat story for about ten more minutes. The reason given is that, "Without all the facts at hand, George Bush ha[s] no intention of upsetting the schoolchildren who had come to read for him." [MSNBC, 10/29/02] Sarasota-Bradenton International Airport is only three and a half miles away. In fact, the elementary school was chosen for the photo-op partly because of its closeness to the airport. [SARASOTA HER-

ALD-TRIBUNE, 9/12/02] Why the Secret Service does not move Bush away from his publicized location that morning remains unclear.

9:06–9:16 A.M.*: Bush Reads Pet Goat Story
for Nearly Ten Minutes; Warned Not to Talk

President Bush, having just been told of the second WTC crash, stays in the Booker Elementary School Classroom, and listens as 16 Booker Elementary School second-graders take turns reading "The Pet Goat." It's a simple story about a girl's pet goat. [AGENCE FRANCE-PRESSE. 9/7/02; EDITOR AND PUBLISHER, 7/2/04] They are just about to begin reading when Bush is told of the attack. One account says that the classroom is

President Bush and Sandra Kay Daniels' students read the goat story while the media watches

then silent for about 30 seconds, maybe more. Bush then picks up the book and reads with the children "for eight or nine minutes." [TAMPA TRIBUNE, 9/1/02] In unison, the children read aloud, "The—Pet—Goat. A—girl—got—a—pet—goat. But—the—goat—did—some—things—that—made—the—girl's—dad—mad." And so on. Bush mostly listens, but does ask the children a few questions to encourage them. [WASHINGTON TIMES, 10/7/02] At one point he says, "Really good readers, whew! . . . These must be sixth-graders!" [TIME, 9/12/01]

In the back of the room, Press Secretary Ari Fleischer catches Bush's eye and holds up a pad of paper for him to read, with "DON'T SAY ANYTHING YET" written on it in big block letters. [WASHINGTON TIMES, 10/7/02] (Note that three articles claim that Bush leaves the classroom at 9:12 A.M.) [NEW YORK TIMES, 9/16/01 (B); DAILY TELEGRAPH, 12/16/01; DAILY MAIL, 9/8/02] However, a videotape of the event lasts for "at least seven additional minutes" and ends before Bush leaves. [WALL STREET JOURNAL, 3/22/04] (The timing of this entry is a rough approximation based mostly on the *Tampa Tribune* estimate. Much of this video footage is shown in Michael Moore's documentary *Fahrenheit 9/11.* [NEW YORK TIMES, 6/18/04 (C)])

9:08–9:13 A.M.*: Fighters Put in Holding Pattern
over Ocean Instead of Defending New York City

The two F-15s scrambled to find Flight 11 in New York are now ordered to circle in a 150-mile window of air space off the coast of Long Island. It is not clear whether they reach New York City before being directed over the ocean. Pilot Major Daniel Nash states, "Neither the civilian controller or the military controller knew what they wanted us to do." [CAPE COD TIMES, 8/21/02] At 9:09 A.M., the NEADS Mission Crew Commander learns of the second WTC crash, and decides to send the fighters to New York City. The 9/11 Commission says the fighters remain in a holding pattern over the ocean until 9:13 A.M. while the FAA clears the airspace. The fighters then establish a Combat Air Patrol over the city at 9:25 A.M. What the fighters do between 9:13 A.M. and 9:25 A.M. is unclear. The distance between the two locations is unknown but presumably not large. [9/11 COMMISSION REPORT, 6/17/04] These fighters remain over New York City for the next four hours. [CAPE COD TIMES, 8/21/02]

9:09 A.M.: NORAD Said to Order Langley Fighters to Battle Stations Alert; Pilots Say This Happens Much Later

According to some reports, NORAD orders F-16s at Langley Air Force Base, Virginia, on battle stations alert. Around this time, the FAA command center reports that 11 aircraft are either not in communication with FAA facilities, or flying unexpected routes. [AVIATION WEEK AND SPACE TECHNOLOGY, 6/3/02] The 9/11 Commission also later accepts this version, claiming that the intent of the alert was not to protect Washington, but because there is a concern that the fighters currently hovering over New York City will run low on fuel, and need to be replaced. [9/11 COMMISSION REPORT, 6/17/04] However, at least one pilot, Major Dean Eckmann, asserts that the battle stations alert does not occur until 9:21 A.M. Another pilot, code-named Honey (presumably Craig Borgstrom), asserts that this does not occur until 9:24 A.M. [BBC, 9/1/02]

9:09 A.M.: Indianapolis Flight Control Tells Local FAA Flight 77 Is Missing, but FAA Headquarters and NORAD Are Not Yet Told

Indianapolis fight control reports the loss of contact with Flight 77 to the FAA regional center. They describe it as a possible crash. The center waits 16 minutes before passing the information to FAA headquarters at 9:25 A.M. [WASHINGTON POST, 11/3/01; 9/11 COMMISSION REPORT, 6/17/04] However, American Airlines headquarters has been notified of the same information before 9:00 A.M.

9:10 A.M.: Port Authority Tells La Guardia Airport WTC Crashes Are Criminal Acts

According to released transcripts, a caller from the Port Authority police desk tells a La Guardia Airport control tower employee that, "they are considering [the crashes into the WTC] a criminal act." The control tower employee replies, "We believe that, and we are holding all aircraft on the ground." [ASSOCIATED PRESS, 12/29/03] La Guardia is one of two major New York City airports, and the Port Authority patrol both the WTC and the city's airports.

9:10 A.M.*: Rice and Cheney Apparently Go to White House Bunker; Other Accounts Have Cheney Moving Locations Later

According to counterterrorism "tsar" Richard Clarke and others, Vice President Cheney goes from his White House office to the Presidential Emergency Operations Center (PEOC), a bunker in the East Wing of the White House at about this time. National Security Adviser Rice, after initiating a video conference with Richard Clarke in the West Wing, goes to the PEOC to be with Cheney. There is no video link between response centers in the East and West Wings, but a secure telephone line is used instead. [AGAINST ALL ENEMIES, BY RICHARD CLARKE, 3/04, PP. 3–4, ABC NEWS, 9/14/02 (B); NEW YORK TIMES, 9/16/01 (B); DAILY TELEGRAPH, 12/16/01] One eyewitness account, David Bohrer, a White House photographer, says Cheney leaves for the PEOC just after 9:00 A.M. [ABC NEWS, 9/14/02 (B)] However, there is a second account claiming that Cheney doesn't leave until sometime after 9:30 A.M. In this account, Secret Service agents burst into Vice President Cheney's White House office. They carry him under his arms—nearly lifting him off the ground—and propel him down the steps into the White House basement and through a long tunnel toward an underground bunker. [WASHINGTON POST, 1/27/02; BBC, 9/1/02; NEWSWEEK, 12/31/01; NEW YORK TIMES, 10/16/01; MSNBC, 9/11/02 (B); 9/11 COMMISSION REPORT, 6/17/04] At about the same time, National Security Adviser Rice is told to go to the bunker as well. [ABC NEWS, 9/11/02] In addition to the eyewitness accounts of Clarke and Bohrer, ABC News claims that Cheney is in the bunker when he is told Flight 77 is 50 miles away from Washington at 9:27 A.M., suggesting that accounts of Cheney entering the bunker after 9:27 A.M. are likely incorrect.

9:10 A.M.*: Washington Flight Control Sees Unidentified Plane, Apparently Fails to Notify FAA or NORAD

Washington flight control notices a new eastbound plane entering its radar with no radio contact and no transponder identification. They do not realize it is Flight 77. They are aware of the hijackings and crashes of Flights 11 and 175, yet they apparently fail to notify anyone about the unidentified plane. [NEWSDAY, 9/23/01; 9/11 COMMISSION REPORT, 6/17/04] Another report says they never notice it, and it is only noticed when it enters radar coverage of Washington's Dulles International Airport at 9:24 A.M. [WASHINGTON POST, 11/3/01]

9:10 A.M.*: Clarke Directs Crisis Response Through Video Conference with Top Officials; 9/11 Commission and Others Barely Mention the Conference

Around this time, counterterrorism "tsar" Richard Clarke reaches the Secure Video Conferencing Center next to the Situation Room in the West Wing of the White House. From there, he directs the response to the 9/11 attacks and stays in contact with other top officials through video links. On video are Defense Secretary Rumsfeld, CIA Director Tenet, FBI Director Mueller, FAA Administrator Jane Garvey, Deputy Attorney General Larry Thompson (filling in for the traveling Attorney General Ashcroft), Deputy Secretary of State Richard Armitage (filling in for the traveling Secretary of State Powell), and Vice-Chairman of the Joint Chiefs of Staff Richard Myers (filling in for the traveling Chairman Henry Shelton). National Security Adviser Rice is with Clarke, but she lets Clarke run the crisis response, deferring to his longer experience on terrorism matters. Clarke is also told by an aide, "We're on the line with NORAD, on an air threat conference call." [AGAINST ALL ENEMIES, BY RICHARD CLARKE, 3/04, PP. 2–4; AUSTRALIAN, 3/27/04] The 9/11 Commission acknowledges the existence of this conference, but only gives it one sentence in a staff report about the day of 9/11: "The White House Situation Room initiated a video teleconference, chaired by Richard Clarke. While important, it had no immediate effect on the emergency defense efforts." [9/11 COMMISSION REPORT, 6/17/04] Yet, as the *Washington Post* puts it, "everyone seems to agree" Clarke is the chief crisis manager on 9/11. [WASHINGTON POST, 3/28/04 (B)] Even his later opponent, National Security Adviser Rice, calls him 9/11's "crisis management guy." [UPI, 4/10/04] The conference is where the government's emergency defense efforts are concentrated.

9:10 A.M.*: Cheney and Rice Establish Telephone
Connection Between Their White House Bunker and Clarke's

Around this time, Vice President Cheney goes from his White House office to the PEOC, the Presidential Emergency Operations Center, a bunker in the East Wing of the White House. National Security Adviser Rice, after initiating a video conference with counterterrorism "tsar" Richard Clarke in the West Wing, goes to the PEOC to be with Cheney. There is no video link between response centers in the East and West Wings, but a secure telephone line is used instead. [AGAINST ALL ENEMIES, BY RICHARD CLARKE, 3/04, PP. 3–4]

9:12 A.M.*: Flight 77 Attendant Has Confirmed Hijacking, American Airlines Learns

Renee May, a flight attendant on Flight 77, uses a cell phone to call her mother in Las Vegas. She tells her mother that the flight has been hijacked, and that everyone has been asked to move to the back of the plane. She asks her mother to call American Airlines and let them know Flight 77 has been hijacked. Her mother (Nancy May) calls the airline. [LAS VEGAS REVIEW JOURNAL, 9/13/01; LAS VEGAS REVIEW JOURNAL, 9/15/01; 9/11 COMMISSION, 1/27/04; SAN FRANCISCO CHRONICLE, 7/23/04] American Airlines headquarters is already aware that Flight 77 is hijacked, but supposedly Indianapolis flight control covering the flight still is not told.

9:13 A.M.: Port Authority Asks New York Airports
About Hijacked Planes, Airports Know Little

A Port Authority police officer calls a flight controller at La Guardia Airport in New York City. The officer asks, "They are inquiring whether or not you can call Kennedy's tower, because they can't get through, and inquire whether or not they had any contact with these aircrafts." The flight controller responds, "At this time, we do not think that anyone in the FAA had any contact with them." [NEW YORK TIMES, 12/30/03] "Kennedy" is a reference to John. F. Kennedy Airport, another major airport in New York City. Port Authority police, who patrol both the WTC and the airports, seek information from the controllers about the hijackers. However, the controllers are unable to offer any news. [NEW YORK TIMES, 12/30/03]

9:15 A.M.: American Airlines Orders No New
Take Offs in U.S.; United Airlines Follows Suit

American Airlines orders no new take-offs in the U.S.; United Airlines follows suit five minutes later. [WALL STREET JOURNAL, 10/15/01]

Between 9:15–9:25 A.M.*: FAA Head Tells Clarke
Videoconference That Flight 11 and 175 Were Hijacked

Counterterrorism "tsar" Richard Clarke begins a crisis response video conference by asking FAA Administrator Jane Garvey what she knows. Garvey replies, "The two aircraft that went in [to the WTC] were American flight 11, a 767, and United 175, also a 767. Hijacked." She says that she has put a hold on all takeoffs and landings in New York and Washington, then states, "We have reports of eleven aircraft off course or out of communications." Clarke and Garvey discuss the feasibility of canceling all takeoffs nationally, and grounding all planes in the air. Garvey says it is possible, but will take time. [AGAINST ALL ENEMIES, BY RICHARD CLARKE, 3/04, PP. 4–5]

9:16 A.M.*: NORAD's Original Claim Flight 93 Is Hijacked at This Time Is Apparently Wrong; One Hijacker May Have Snuck Into Cockpit Early

According to a NORAD timeline from a week after 9/11, NORAD claims that Flight 93 may have been hijacked at this time. The timeline inexplicably fails to say when the FAA told them about the hijack, the only flight for which they fail to provide this data. [CNN, 9/17/01; NORAD, 9/18/01] However, there may be one explanation: There are media reports that "investigators had determined from the cockpit voice recorder from United Airlines Flight 93 . . . that one of the four hijackers had been invited into the cockpit area before the flight took off from Newark, New Jersey." Cockpit voice recordings indicate that the pilots believed their guest was a colleague "and was thereby extended the typical airline courtesy of allowing any pilot from any airline to join a flight by sitting in the jumpseat, the folded over extra seat located inside the cockpit." [FOX NEWS, 9/24/01; HERALD SUN, 9/25/01] However, this account has not been confirmed. The 9/11 Commission asserts the hijacking begins around 9:28 A.M. [9/11 COMMISSION REPORT, 6/17/04] Note that during the 9/11 Commission hearings in May 2003, NORAD officials stated that the FAA informed NEADS at 9:16 A.M. that United Flight 93 was hijacked. According to a commission report in 2004, "this statement was incorrect." No further explanation is offered for NORAD's incorrect timeline. [9/11 COMMISSION REPORT, 6/17/04]

9:16 A.M.*: Bush Takes His Time Leaving Classroom Photo-Op

President Bush leaves the Sarasota classroom where he has been since about 9:03 A.M. The children finish their lessons and put away their readers. [SARASOTA MAGAZINE, 9/19/01] Bush advises the children to stay in school and be good citizens. [TAMPA TRIBUNE, 9/1/02; ST. PETERSBURG TIMES, 9/8/02 (B)] He also tells the children, "Thank you all so very much for showing me your reading skills." [ABC NEWS, 9/11/02] One student also asks Bush a question, and Bush gives a quick response on his education policy. [NEW YORK POST, 9/12/02] A reporter asks, "Mr. President, are you aware of the reports of the plane crash in New York? Is there any . . ." This question is interrupted by an aide who has come into the room, saying, "All right. Thank you. If everyone could please step outside." Bush then says, "We'll talk about it later." [CBS NEWS, 9/11/02 (B)] Bush then tells school principal Gwen Tose-Rigell, who is in the room, about the terror attacks and why he has to leave. [WASHINGTON TIMES, 10/7/02] He then goes into an empty classroom next door and meets with his staff there. [ABC NEWS, 9/11/02] Bush's program with the children was supposed to start at 9:00 A.M. and end 20 minutes later. [SARASOTA HERALD-TRIBUNE, 9/16/01] He leaves the classroom only a couple of minutes earlier than planned, if at all. The "goodbyes" and questions on the way out may have taken another minute or two.

Between 9:16–9:29 A.M.*: Bush Works on Speech with Staff; Makes No Decisions

President Bush works with his staff to prepare a speech he will deliver at 9:29 A.M. He intermittently watches the television coverage in the room. [ALBUQUERQUE TRIBUNE, 9/10/02] He also speaks on the phone to advisers, first calling National Security Adviser Rice, then Vice President Cheney, then New York Governor George Pataki. [DAILY MAIL, 9/8/02] Bush often turns to look at a television screen. He declares, "We're at war." [BBC, 9/1/02] Bush later claims he makes no major decisions about the crisis until after boarding Air Force One at 9:55 A.M.

Director of Communications Dan Bartlett points the television while, President Bush talks on the telephone. Also pictures are Deborah Loewer (directly behind Bush) and advisor Karl Rove (right).

9:17 A.M.: FAA Shuts Down All New York City Airports

The FAA shuts down all New York City area airports. [CNN. 9/12/01; NEW YORK TIMES. 9/12/01] A flight controller at La Guardia airport reports the taxiways, runways, and airspace completely clear at 9:37 A.M. [NEW YORK TIMES, 12/30/03]

9:18 A.M.*: FAA Command Center Warns Flight Controllers Nationwide to Watch for Suspicious Aircraft

The FAA Command Center finally issues a nationwide alert to flight controllers to watch for planes disappearing from radar or making unauthorized course changes. [WASHINGTON POST, 11/3/01]

9:20 A.M.*: FAA Command Center Notifies Field Facilities That Flight 77 Is Lost; Indianapolis Flight Control Reportedly Finally Learns of National Crisis

According to the 9/11 Commission, Indianapolis flight control learns that there are other hijacked aircraft by this time (presumably at least Flights 11 and 175). Millions of people have known about the crashes since CNN and all other media began broadcasting images from New York at 8:48 A.M., but Indianapolis is reportedly unaware until this time. The Indianapolis flight controllers begin to doubt their assumption that Flight 77 has crashed and consider that it might be hijacked. After a discussion between the Indianapolis manager and the FAA Command Center, the Command Center notifies some other FAA facilities that Flight 77 is lost. By 9:21 A.M., the Command Center, some FAA field facilities, and American Airlines join the search for Flight 77. [9/11 COMMISSION REPORT, 6/17/04]

9:20 A.M.*: FAA Sets Up Ineffectual Hijacking Teleconference

The FAA sets up a hijacking teleconference with several agencies, including the Defense Department. This is almost one hour after the FAA's Boston flight control notified other flight control centers about the first hijacking at 8:25 A.M. Yet even after this delay, FAA and Defense Department participants in the teleconference later claim it plays no role in coordinating the response to the hijackings. [9/11 COMMISSION REPORT, 6/17/04]

**Norman
Mineta**

9:20 A.M.*: Mineta Reaches Bunker, Meets Cheney

Transportation Secretary Mineta arrives at the White House bunker containing Vice President Cheney and others. In later testimony, he recalls that Cheney is already there when he arrives. [ST. PETERSBURG TIMES, 7/4/04] This supports accounts of Cheney reaching the bunker not long after the second WTC crash, but the 9/11 Commission concludes Cheney doesn't arrive until a few minutes before 10:00. [9/11 COMMISSION REPORT, 6/17/04]

9:20 A.M.*: Barbara Olson Said to Call from Flight 77,
but Account Is Full of Contradictions

A passenger on Flight 77, Barbara Olson, calls her husband, Theodore (Ted) Olson, who is Solicitor General at the Justice Department. [SAN FRANCISCO CHRONICLE, 7/23/04] Ted Olson is in his Justice Department office watching WTC news on television when his wife calls. A few days later, he says, "She told me that she had been herded to the back of the plane. She mentioned that they had used knives and box cutters to hijack the plane. She mentioned that the pilot had announced that the plane had been hijacked." [CNN, 9/14/01 (C)] He tells her that two planes have hit the WTC. [DAILY TELEGRAPH, 3/5/02] She feels nobody is taking charge. [CNN, 9/12/01] He doesn't know if she was near the pilots, but at one point she asks, "What shall I tell the pilot? What can I tell the pilot to do?" [CNN, 9/14/01 (C)] Then she is cut off without warning. [NEWSWEEK, 9/29/01] Ted Olson's recollection of the call's timing is extremely vague, saying it "must have been 9:15 [A.M.] or 9:30 [A.M.]. Someone would have to reconstruct the time for me." [CNN, 9/14/01 (C)] Other accounts place it around 9:25 A.M. [MIAMI HERALD, 9/14/01; NEW YORK TIMES, 9/15/01 (C); WASHINGTON POST, 9/21/01] The call is said to have lasted about a minute. [WASHINGTON POST, 9/12/01 (B)] By some accounts, his message that planes have hit the WTC comes later, in a second phone call. [WASHINGTON POST, 9/21/01] In one account, Barbara Olson calls from inside a bathroom. [EVENING STANDARD, 9/12/01] In another account, she is near a pilot, and in yet another she is near two pilots. [BOSTON GLOBE, 11/23/01] Ted Olson's account of how Barbara Olson made her calls is also conflicting. Three days after 9/11, he says, "I found out later that she was having, for some reason, to call collect and was having trouble getting through. You know how it is to get through to a government institution when you're calling collect." He says he doesn't know what kind of phone she used, but he has "assumed that it must have been on the airplane phone, and that she somehow didn't have access to her credit cards. Otherwise, she would have used her cell phone and called me." [FOX NEWS, 9/14/01] Why Barbara Olson would have needed access to her credit cards to call him on her cell phone is not explained. However, in another interview on the same day, he says that she used a cell phone and that she may have been cut off "because the signals from cell phones coming from airplanes don't work that well." [CNN, 9/14/01 (C)] Six months later, he claims she called collect "using the phone in the passengers' seats." [DAILY TELEGRAPH, 3/5/02] However, it is not possible to call on seatback phones, collect or otherwise, without a credit card, which would render making a collect call moot. Many other details are conflicting, and Olson faults his memory and says that he "tends to mix the two [calls] up because of the emotion of the events." [CNN, 9/14/01 (C)] The couple liked to joke that they were at the heart of what Hillary Clinton famously called a "vast, right-wing conspiracy." Ted Olson has been a controversial choice as Solicitor General since he argued on behalf of Bush before the Supreme Court in the 2000 presidential election controversy before being nominated for his current position.

9:21 A.M.: All New York City Bridges and Tunnels Are Closed

The New York City Port Authority closes all bridges and tunnels in New York City. [MSNBC, 9/22/01; CNN, 9/12/01; NEW YORK TIMES, 9/12/01; ASSOCIATED PRESS, 8/19/02 (B)]

9:21 A.M.: FAA Headquarters Mistakenly Tells
Boston Controller Flight 11 Is Still Airborne

According to the 9/11 Commission, NEADS is contacted by Boston flight control. A controller says, "I just had a report that American 11 is still in the air, and it's on its way towards—heading towards Washington. . . . That was another—it was evidently another aircraft that hit the tower. That's the latest report we have. . . . I'm going to try to confirm an ID for you, but I would assume he's somewhere over, uh, either New Jersey or somewhere further south." The NEADS official asks, "He—American 11 is a hijack? . . . And he's heading into Washington?" The Boston controller answers yes both times and adds, "This could be a third aircraft." Somehow Boston is told by FAA headquarters that Flight 11 is still airborne, but the commission hasn't been able to find where this mistaken information came from. [9/11 COMMISSION REPORT, 6/17/04]

9:21 A.M.: FAA Command Center Advises Dulles Airport Control to Be on Lookout

According to the 9/11 Commission, the FAA Command Center advises the Dulles Airport terminal control facility in Washington to look for primary targets. [9/11 COMMISSION REPORT, 6/17/04] By at least one account, Dulles notices Flight 77 a few minutes later.

9:21 A.M.: One Langley Pilot Claims to Be Put on
Battle Stations Now, Not 12 Minutes Earlier

Major Dean Eckmann, an F-16 fight pilot at Langley Air Force Base in Virginia, recalls, "The scramble horn goes off and we get the yellow light, which is our battle stations. So at that point I go running out to the airplanes—to my assigned alert airplane—get suited up and I get into the cockpit ready to start." [BBC, 9/1/02] A few minutes before the battle stations order, Eckmann is told that a plane has hit the WTC. He assumes it's some kind of accident. [ASSOCIATED PRESS, 8/19/02 (C)] However, another pilot, codenamed Honey (apparently Craig Borgstrom), claims the battle stations command happens at 9:24 A.M.; while the 9/11 Commission claims it happens at 9:09 A.M.

9:21–9:26 A.M.*: United Airlines Dispatchers Advise Pilots to Secure Cockpit Doors; Flight 93 Gets the Message

At 9:21 A.M., United Airlines dispatchers are told to advise their flights to secure cockpit doors. Flight dispatcher Ed Ballinger has apparently already started doing this on his own volition a couple of minutes earlier. Sending electronic messages one by one, at 9:24 he sends a message to Flight 93 reading: "Beware of cockpit intrusion. Two aircraft in New York hit Trade Center buildings." Ballinger claims that he was specifically instructed by superiors not to tell pilots why they needed to land (apparently he added the detail about the WTC against orders). [NEW YORK OBSERVER, 6/17/04] Flight 93 pilot Jason Dahl acknowledges the message two minutes later, replying, "Ed, confirm latest message please Jason." This is the last vocal contact from the cockpit of Flight 93. [9/11 COMMISSION, 1/27/04] Note that this formal warning is in addition to an informal one sent by Ballinger that reached Flight 93 around 9:00 A.M. In contrast to United Airlines, the 9/11 Commission finds no evidence that American Airlines sends such warnings to their pilots at any time during the hijackings.

9:23 A.M.: NEADS Wants Fighters to Track Phantom Flight 11

According to the 9/11 Commission, NEADS has just been told that the hijacked Flight 11 is still in the air and heading toward Washington. The NEADS Battle Commander says, "Okay, uh, American Airlines is still airborne. Eleven, the first guy, he's heading towards Washington. Okay? I think we need to scramble Langley right now. And I'm gonna take the fighters from Otis, try to chase this guy down if I can find him." The NEADS Mission Crew Commander issues the order, "Okay . . . scramble Langley. Head them towards the Washington area." The Langley, Virginia base gets the scramble order at 9:24 A.M. NEADS keeps their fighters from the Otis base over New York City. [9/11 COMMISSION REPORT, 6/17/04]

9:24 A.M.*: By Some Accounts, FAA Notifies NORAD Flight 77 Is Hijacked and Washington-Bound; 9/11 Commission Claims This Never Happens

Shortly after 9/11, NORAD reported that the FAA notified them at this time that Flight 77 "may" have been hijacked and that it appears headed toward Washington. [NORAD, 9/18/01; ASSOCIATED PRESS, 8/19/02 (B); CNN, 9/17/01; WASHINGTON POST, 9/12/01; GUARDIAN, 10/17/01] Apparently, flight controllers at Dulles International Airport discover a plane heading at high speed toward Washington; an alert is sounded within moments that the plane appears to be headed toward the White House. [WASHINGTON POST, 11/3/01] In 2003, the FAA supported this account, but claimed that they had informally notified NORAD earlier. "NORAD logs indicate that the FAA made formal notification about American Flight 77 at 9:24 A.M., but information about the flight was conveyed continuously during the phone bridges before the formal notification." [FAA, 5/22/03] Yet in 2004, the 9/11 Commission claims that both NORAD and the FAA are wrong. The

9/11 Commission explains that the notification NEADS received at 9:24 A.M. was the incorrect information that Flight 11 had not hit the WTC and was headed south for Washington, D.C. Thus, according to the 9/11 Commission, NORAD is never notified by the FAA about the hijacking of Flight 77, but accidentally learns about it at 9:34 A.M. [9/11 COMMISSION REPORT, 6/17/04]

9:24 A.M.*: Langley Fighters Are Ordered to Scramble; but One Pilot Claims the Order Is Only a Battle Stations Alert

The BBC later reports that at this time, Robert Marr, head of NEADS, gives the scramble order to the F-16 fighters based in Langley, Virginia: "North East sectors back on. We ought to be getting the weapons crews back in. Get the scramble order rolling. Scramble." [BBC, 9/1/02] The 9/11 Commission concurs that the scramble order is given now. [9/11 COMMISSION REPORT, 6/17/04] NORAD also has agreed. [NORAD, 9/18/01] However, many media reports have placed it later. [CNN, 9/17/01; WASHINGTON POST, 9/12/01; CNN, 9/17/01; WASHINGTON POST, 9/15/01] A pilot codenamed Honey gives a slightly different account. He claims that at this time a battle stations alert sounds and two other pilots are given the order to climb into their F-16s and await further instructions. Then, Honey, the supervising pilot, talks to the two other pilots. Then, "five or ten minutes later," a person from NORAD calls and Honey speaks to him at the nearby administrative office. He is told that all three of them are ordered to scramble. Honey goes to his living quarters, grabs his flight gear, puts it on, runs to his plane, and takes off. [AMONG THE HEROES, BY JERE LONGMAN, 8/02, P. 64-65] Honey appears to be the codename for Captain Craig Borgstrom, because in another account, Borgstrom is given an alert and then talks to the two other pilots. [ASSOCIATED PRESS, 8/19/02 (C)] A different pilot account has the battle stations warning three minutes earlier, while the 9/11 Commission claims that it happens fifteen minutes earlier. Pilot Major Dean Eckmann recalls, "They go 'active air scramble, vector zero one zero one, max speed.' And then I push us over to the tower frequency and get our departure clearance and they launch us out right away. . . . We can carry M9-Heat Seekers, Side Winders for the M7-Sparrow, plus we have an internal 20mm Vulcan Cannon, and we were pretty much armed with all that. We had a pretty quick response time. I believe it was four to five minutes we were airborne from that point." The BBC reports, "Even while last minute pre-launch checks are being made, the controllers learn that a third plane—American Airlines flight 77 out of Washington—may have been hijacked." Just before the fighters take off, the BBC says, "The pilots get a signal over the plane's transponder—a code that indicates an emergency wartime situation." [BBC, 9/1/02]

9:25 A.M.: FAA Command Center Finally Tells FAA Headquarters About Flight 77

According to the 9/11 Commission, the FAA Command Center advises FAA headquarters that American 77 is lost in Indianapolis flight control's airspace, that Indianapolis has no primary radar track, and is looking for the aircraft. [9/11 COMMISSION REPORT, 6/17/04] The Command Center had learned this 16 minutes earlier at 9:09 A.M. American Airlines headquarters was notified of the same information before 9:00 A.M.

After 9:25 A.M.*: Flight 77 Passenger Call
Reaches Justice Department and Beyond

Theodore (Ted) Olson, the Justice Department's Solicitor General, calls the Justice Department's control center to relate his wife Barbara's call from Flight 77. Accounts vary whether the Justice Department already knows of the hijack or not. [WASHINGTON POST, 9/12/01 (B); CHANNEL 4 NEWS, 9/13/01; NEW YORK TIMES, 9/15/01 (C)] Olson merely says, "They just absorbed the information. And they promised to send someone down right away." He assumes they then "pass the information on to the appropriate people." [FOX NEWS, 9/14/01]

9:26 A.M.*: Rookie FAA Manager Bans All Take Offs Nationwide,
Including Most Military Flights? Mineta Asserts He Issues Order Minutes Later

Time magazine later reports that Jane Garvey, head of the FAA, "almost certainly after getting an okay from the White House, initiate[s] a national ground stop, which forbids takeoffs and requires planes in the air to get down as soon as is reasonable. The order, which has never been implemented since flying was invented in 1903, applie[s] to virtually every single kind of machine that can takeoff—civilian, military, or law enforcement." Military and law enforcement flights are allowed to resume at 10:31 A.M. A limited number of military flights—the FAA will not reveal details—are allowed to fly during this ban. [TIME, 9/14/01] Garvey later calls it "a national ground stop . . . that prevented any aircraft from taking off." [HOUSE COMMITTEE, 9/21/01] Transportation Secretary Norman Mineta later says he was the one to give the order: "As soon as I was aware of the nature and scale of the attack, I called from the White House to order the air traffic system to land all aircraft, immediately and without exception." [STATE DEPARTMENT, 9/20/01] According to Secretary of Transport Norman Mineta "At approximately 9:45 . . . I gave the FAA the final order for all civil aircraft to land at the nearest airport as soon as possible." [9/11 COMMISSION, 5/23/04] At the time, 4,452 planes are flying in the continental U.S. A later account states that Ben Sliney, the FAA's National Operations Manager, makes the decision without consulting his superiors, like Jane Garvey, first. It would be remarkable if Sliney was the one to make the decision, because 9/11 is Sliney's first day on the job as National Operations Manager, "the chess master of the air traffic system." [USA TODAY, 8/13/02] When he accepted the job a couple of months earlier, he had asked, "What is the limit of my authority?" The man who had promoted him replied, "Unlimited." [USA TODAY, 8/13/02 (B)] Yet another account, by Linda Schuessler, manager of tactical operations at the FAA Command Center where Sliney was located, says, ". . . it was done collaboratively . . . All these decisions were corporate decisions. It wasn't one person who said, 'Yes, this has got to get done.'" [AVIATION WEEK AND SPACE TECHNOLOGY, 12/17/01] About 500 planes land in the next 20 minutes, and then much more urgent orders to land are issued at 9:45 A.M. [USA TODAY, 8/13/02; TIME, 9/14/01; USA TODAY, 8/13/02; HOUSE COMMITTEE, 9/21/01; AVIATION WEEK AND SPACE TECHNOLOGY, 6/3/02; NEWSDAY, 9/23/01; ASSOCIATED PRESS, 8/19/02 (B); NEWSDAY, 9/10/02]

Jane Garvey

9:27 A.M.*: Cheney Given Updates on Unidentified
Flight 77 Heading Toward Washington

Vice President Cheney and National Security Adviser Rice, in their bunker below the White House, are told by an aide that an airplane is headed toward Washington from 50 miles away. The plane is Flight 77. Deputy FAA deputy Monty Belger says, "Well we're watching this target on the radar, but the transponder's been turned off. So we have no identification." They are given further notices when the plane is 30 miles away, then ten miles away, until it disappears from radar (time unknown, but the plane is said to be traveling about 500 mph and was 30 miles away at 9:30 A.M., so 50 miles would be about three minutes before that). [ABC NEWS, 9/11/02] Transportation Secretary Norman Mineta gives virtually the same account before the 9/11 Commission. [9/11 COMMISSION, 5/23/03] However, the 9/11 Commission later claims the plane heading toward Washington is only discovered at 9:32 A.M. [9/11 COMMISSION REPORT, 6/17/04]

9:27 A.M.*: Flight 93 Passenger Tom Burnett
Calls Wife, Mentions Bomb, Knife, and Gun

Tom Burnett calls his wife, Deena, using a cell phone and says, "I'm on United Flight 93 from Newark to San Francisco. The plane has been hijacked. We are in the air. They've already knifed a guy. There is a bomb on board. Call the FBI." Deena connects to emergency 911. [AMONG THE HEROES, BY

Tom Burnett

JERE LONGMAN, 8/02, P. 107, ABC NEWS, 9/12/01; MSNBC, 7/30/02; PITTSBURGH POST-GAZETTE, 10/28/01 (B); TORONTO SUN, 9/16/01] Deena wonders if the call might have been before the cockpit was taken over, because he spoke quickly and quietly as if he was being watched. He also had a headset like phone operators use, so he could have made the call unnoticed. Note that original versions of this conversation appear to have been censored. The most recent account has the phone call ending with, "We are in the air. The plane has been hijacked. They already knifed a guy. One of them has a gun. They're saying there is a bomb onboard. Please call the authorities." [AMONG THE HEROES, BY JERE LONGMAN, 8/02, P. 107] The major difference from earlier accounts, is the mention of a gun. The call wasn't recorded, but Deena's call to 911 immediately afterwards was, and on that call she states, "They just knifed a passenger and there are guns on the plane." [AMONG THE HEROES, BY JERE LONGMAN, 8/02, P. 108] Deena Burnett later says of her husband: "He told me one of the hijackers had a gun. He wouldn't have made it up. Tom grew up around guns. He was an avid hunter and we have guns in our home. If he said there was a gun on board, there was." [LONDON TIMES, 8/11/02 (B)] This is the first of over 30 additional phone calls by passengers inside the plane. [MSNBC, 7/30/02] Passengers are told what happened at the WTC in least five of the phone calls. Five calls show an intent to revolt against the hijackers. [SAN FRANCISCO CHRONICLE, 7/23/04]

9:28 A.M.: CNN Reports U.S. Officials Think Attacks Caused by Terrorists

CNN quotes the Associate Press as reporting that a U.S. official believes the attacks are believed to have been carried out by terrorists. [OTTAWA CITIZEN, 9/11/01]

9:28 A.M.: Myers Updates Clarke
Videoconference on Fighter Response

Counterterrorism "tsar" Richard Clarke, directing a video conference with top officials, asks Joint Chiefs of Staff Vice Chairman Richard Myers, "I assume NORAD has scrambled fighters and AWACS. How many? Where?" Myers replies, "Not a pretty picture, Dick. We are in the middle of Vigilant Warrior, a NORAD exercise, but . . . Otis has launched two birds toward New York. Langley is trying to get two up now [toward Washington]. The AWACS are at Tinker and not on alert." Vigilant Warrior may be a mistaken reference to the on-going war game Vigilant Guardian. Otis Air National Guard Base is in Massachusetts, 188 miles east of New York City; Langley is in Virginia, 129 miles south of Washington. Tinker Air Force Base is in Oklahoma. Clarke asks, "Okay, how long to CAP over D.C.?" CAP means combat air patrol. Myers replies, "Fast as we can. Fifteen minutes?" Note that according to Clarke, Myers is surrounded by generals and colonels as he says this (which contradicts Myers' own accounts of where he is and what he's doing). [AGAINST ALL ENEMIES, BY RICHARD CLARKE, 3/04, P. 5] The first fighters don't reach Washington until 30 minutes or more later.

9:28 A.M. (or Before*): Erratic Flight 93
Movements Noticed by Cleveland Flight Controller

Cleveland flight controller Stacey Taylor has been warned to watch transcontinental flights heading west for anything suspicious. She later recalls, "I hear one of the controllers behind me go, "Oh, my God, oh my God," and he starts yelling for the supervisor. He goes, "What is this plane doing? What is this plane doing?" I wasn't that busy at the time, and I pulled it up on my screen and he was climbing and descending and climbing and descending, but very gradually. He'd go up 300 feet, he'd go down 300 feet. And it turned out to be United 93." (Note the time of this incident is not specified, but presumably it is prior to when Cleveland controllers note Flight 93 descends 700 feet at 9:29 A.M.) [MSNBC, 9/11/02 (B)]

9:28 A.M.*: Cleveland Flight Control Hears
Sounds of Struggle as Flight 93 Is Hijacked

Flight 93 acknowledges a transmission from a Cleveland flight controller. This is the last normal contact with the plane. [9/11 COMMISSION REPORT, 6/17/04] According to the 9/11 Commission, less than a minute later, the controller, and pilots of aircraft in the vicinity, hear "a radio transmission of unintelligible sounds of possible screaming or a struggle from an unknown origin . . ." [NEWSWEEK, 11/25/01; 9/11 COMMISSION REPORT, 6/17/04;

Jason Dahl

GUARDIAN, 10/17/01] Someone, presumably pilot Jason Dahl, is overheard by controllers as he shouts "Mayday!" [NEW YORK TIMES, 7/22/04 (B)] Seconds later, the controller responds: "Somebody call Cleveland?" Then there are more sounds of screaming and someone yelling, "Get out of here, get out of here." [MSNBC, 7/30/02; 9/11 COMMISSION REPORT, 6/17/04; OBSERVER, 12/2/01; TORONTO SUN, 9/16/01; NEWSWEEK, 9/22/01] Then the voices of the hijackers can be heard talking in Arabic. The words are later translated to show they are talking to each other, say-

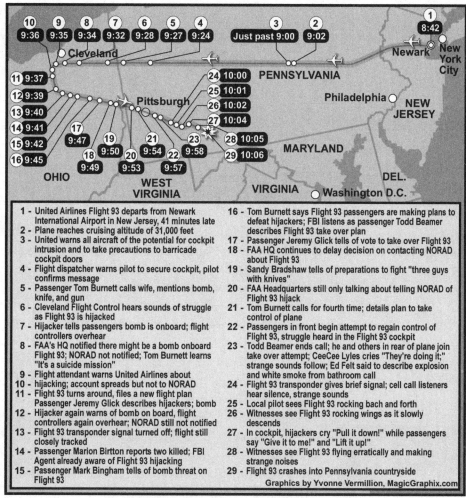

1 - United Airlines Flight 93 departs from Newark International Airport in New Jersey, 41 minutes late
2 - Plane reaches cruising altitude of 31,000 feet
3 - United warns all aircraft of the potential for cockpit intrusion and to take precautions to barricade cockpit doors
4 - Flight dispatcher warns pilot to secure cockpit, pilot confirms message
5 - Passenger Tom Burnett calls wife, mentions bomb, knife, and gun
6 - Cleveland Flight Control hears sounds of struggle as Flight 93 is hijacked
7 - Hijacker tells passengers bomb is onboard; flight controllers overhear
8 - FAA's HQ notified there might be a bomb onboard Flight 93; NORAD not notified; Tom Burnett learns "It's a suicide mission"
9 - Flight attendant warns United Airlines about hijacking; account spreads but not to NORAD
10 - hijacking; account spreads but not to NORAD
11 - Flight 93 turns around, files a new flight plan Passenger Jeremy Glick describes hijackers; bomb
12 - Hijacker again warns of bomb on board, flight controllers again overhear; NORAD still not notified
13 - Flight 93 transponder signal turned off; flight still closely tracked
14 - Passenger Marion Birtton reports two killed; FBI Agent already aware of Flight 93 hijacking
15 - Passenger Mark Bingham tells of bomb threat on Flight 93

16 - Tom Burnett says Flight 93 passengers are making plans to defeat hijackers; FBI listens as passenger Todd Beamer describes Flight 93 take over plan
17 - Passenger Jeremy Glick tells of vote to take over Flight 93
18 - FAA HQ continues to delay decision on contacting NORAD about Flight 93
19 - Sandy Bradshaw tells of preparations to fight "three guys with knives"
20 - FAA Headquarters still only talking about telling NORAD of Flight 93 hijack
21 - Tom Burnett calls for fourth time; details plan to take control of plane
22 - Passengers in front begin attempt to regain control of Flight 93, struggle heard in the Flight 93 cockpit
23 - Todd Beamer ends call; he and others in rear of plane join take over attempt; CeeCee Lyles cries "They're doing it;" strange sounds follow; Ed Felt said to describe explosion and white smoke from bathroom call
24 - Flight 93 transponder gives brief signal; cell call listeners hear silence, strange sounds
25 - Local pilot sees Flight 93 rocking bach and forth
26 - Witnesses see Flight 93 rocking wings as it slowly descends
27 - In cockpit, hijackers cry "Pull it down!" while passengers say "Give it to me!" and "Lift it up!"
28 - Witnesses see Flight 93 flying erratically and making strange noises
29 - Flight 93 crashes into Pennsylvania countryside

Graphics by Yvonne Vermillion, MagicGraphix.com

Key events of Flight 93 (times are based on a *Pittsburgh Post-Gazette* map and otherwise interpolated)

ing, "Everything is fine." [NEWSWEEK, 11/25/01] Later passenger phone calls describe two dead or injured bodies just outside the cockpit; presumably these are the two pilots. [NEW YORK TIMES, 7/22/04 (B)]

9:29 A.M.: President Bush Makes a Scheduled Speech; Proclaims "Terrorist Attack on Our Country"

Still inside Booker Elementary School, President Bush gives a brief speech in front of about 200 students, plus many teachers and reporters. [DAILY MAIL, 9/8/02] He says, "Today we've had a national tragedy. Two airplanes have crashed into the World Trade Center in an apparent terrorist attack on our country." [FEDERAL NEWS SERVICE, 9/11/01] The talk occurs at exactly the time and place stated in his publicly announced advance schedule—making Bush a possible terrorist target. [MSNBC, 9/22/01; *WASHINGTON POST*, 9/12/01; CNN, 9/12/01; *NEW YORK TIMES*, 9/12/01] This is the last most Americans will see of Bush until the evening.

President Bush speaks in front of students and reporters at Booker Elementary School

9:29 A.M.*: Confirmation of Strange Sounds Coming from Flight 93; NORAD Not Notified

Shortly after hearing strange noises from the cockpit of Flight 93, Cleveland flight controllers notice the plane has descended about 700 feet. They try to contact the plane several times, but get no answer. At 9:30 A.M., a controller asks other nearby flights on his frequency if they've heard screaming; several say that they have. [9/11 COMMISSION REPORT, 6/17/04] However, despite these disturbing sounds and lack of contact with the plane, Cleveland doesn't notify anyone else about it.

9:29 A.M.*: Pentagon Command Center Begins High Level Conference Call

Captain Charles Leidig is in command of the National Military Command Center (NMCC), "the military's worldwide nerve center." [CNN, 9/4/02] Telephone links are established with the National Military Command Center (NMCC) located inside the Pentagon (but on the opposite side from where the Pentagon explosion will happen), Canada's equivalent command center, Strategic Command, theater com-

The National Military Command Center

manders, and federal emergency-response agencies. An Air Threat Conference Call is initiated and it lasts for eight hours. At one time or another, President Bush, Vice President Cheney, Defense Secretary Rumsfeld, key military officers, leaders of the FAA and NORAD, the White House, and Air Force One are heard on the open line. [*AVIATION WEEK AND SPACE TECHNOLOGY*, 6/3/02; 9/11 COMMISSION REPORT, 6/17/04] NORAD command director Captain Michael Jellinek claims this happens "immediately" after the second WTC hit. [*AVIATION WEEK AND SPACE TECHNOLOGY*, 6/3/02] However, the 9/11 Commission concludes it starts nearly 30 minutes later, at approximately 9:29 A.M. [9/11 COMMISSION REPORT, 6/17/04] Brigadier General Montague Winfield, who later takes over for Leidig, says, "All of the governmental agencies that were involved in any activity going on in the United States at that point, were in that conference." [ABC NEWS, 9/11/02] The call continues right through the Pentagon explosion; the impact is not felt within the NMCC. [CNN, 9/4/02] However, despite being in the Pentagon, Defense Secretary Rumsfeld doesn't enter the NMCC or participate in the call until 10:30 A.M.

9:30 A.M.: United Flights Are Instructed to Land Immediately; American Follows Suit

United Airlines begins landing all of its flights inside the U.S. (Note: All planes nationwide were already ordered down at 9:26 A.M. and told to land in a reasonable amount of time. Now they're told to land immediately.) American Airlines begins landing all of their flights five minutes later. [WALL STREET JOURNAL, 10/15/01]

9:30 A.M.: Langley Fighters Take Off Toward Washington; They Could Reach City in Six Minutes but Take Half an Hour

The three F-16s at Langley, Virginia get airborne. [NORAD, 9/18/01; ABC NEWS, 9/11/02; WASHINGTON POST, 9/12/01; 9/11 COMMISSION REPORT, 6/17/04] The pilots' names are Major Brad Derrig, Captain Craig Borgstrom, and Major Dean Eckmann, all from the North Dakota Air National Guard's 119th Fighter Wing stationed at Langley. [ASSOCIATED PRESS, 8/19/02 (C); ABC NEWS, 9/11/02] If the assumed NORAD departure time is correct, the F-16s would have to travel slightly over 700 mph to reach Washington before Flight 77 does. The maximum speed of an F-16 is 1,500 mph. [ASSO-CIATED PRESS, 6/16/00] Even traveling at 1,300 mph, these planes could have reached Washington in six minutes—well before any claim of when Flight 77 crashed. Yet it is claimed they are accidentally directed over the Atlantic Ocean instead, and they will only reach Washington about 30 minutes later. NORAD commander Major General

Major Dean Eckmann

Larry Arnold admits in 2003 testimony that had the fighters been going at full speed, "it is physically possible that they could have gotten over Washington" before Flight 77. But asked if the fighters would have had shoot down authorization had they reached the hijacked plane, Arnold says no, claiming that even by this time in the morning it is only "through hindsight that we are certain that this was a coordinated attack on the United States." [9/11 COMMISSION, 5/23/03]

9:30 A.M.*: Clarke Asks Cheney's Bunker for Air Force One Fighter Escort and Shoot Down Authorization; Neither Happen for Some Time

As President Bush begins a speech in Florida, counterterrorism "tsar" Richard Clarke orders all U.S. embassies overseas closed and orders all military bases to an alert level named combat Threatcon. Over the next few minutes, Clarke discusses with aides where Bush should go from Sarasota, Florida. He telephones PEOC, the command bunker containing Vice President Cheney and National Security Adviser Rice, and says, "Somebody has to tell the President he can't come right back here [to Washington]. Cheney, Condi, somebody, Secret Service concurs. We do not want them saying where they are going when they take off. Second, when they take off, they should have fighter escort. Three, we need to authorize the Air Force to shoot down any aircraft—including a hijacked passenger flight—that looks like it is threatening to attack and cause large-scale death on the ground. Got it?" [AGAINST ALL ENEMIES, BY RICHARD CLARKE, 3/04, PP. 5–7] However, when Bush departs on Air Force One about an hour later, there are no fighter escorts, and none appear for an hour or so. In addition, if Clarke requests authorization for a shootdown order at this time, it is apparently ignored; neither President Bush nor Vice President Cheney give shootdown authorization for at least another 30 minutes.

9:30 A.M.*: Dulles Flight Controllers Track Flight 77; Timing Disputed

Radar tracks Flight 77 as it closes within 30 miles of Washington. [CBS NEWS, 9/21/01] Todd Lewis, flight controller at Washington's Dulles Airport, later recalls, ". . . my colleagues saw a target moving quite fast from the northwest to the southeast. So she—we all started watching that target, and she notified the supervisor. However, nobody knew that was a commercial flight at the time. Nobody knew that was American 77. . . . I thought it was a military flight." [MSNBC, 9/11/02 (B)] Another account is similar, saying that just before 9:30 A.M., a Dulles Airport controller sees an aircraft without a transponder traveling almost 500 mph headed toward Washington. [USA TODAY, 8/13/02] In yet another account, Danielle O'Brien, the Dulles flight controller said to be the first to spot the blip, claims she doesn't spot it until it is around 12 to 14 miles from Washington. [ABC NEWS, 10/24/01; ABC NEWS, 10/24/01 (B)] There are also accounts that Vice President Cheney is told around 9:27 A.M. that radar is tracking Flight 77, 50 miles away from Washington. The 9/11 Commission says the plane isn't discovered until 9:32 A.M.

9:30 A.M.*: FAA Emergency Operations Center Is Finally Operational

The FAA's Emergency Operations Center gets up and running, five minutes after the FAA issues an order grounding all civilian, military, and law enforcement aircraft. [TIME, 9/14/01] This center's role in the crisis response remains unclear.

9:30 A.M.*: Who Warns Who of Flight 77's Impending Approach to D.C.?

Chris Stephenson, head flight controller at Washington's Reagan National Airport tower, says that he is called by the Secret Service around this time. He is told an unidentified aircraft is speeding toward Washington. Stephenson looks at the radarscope and sees Flight 77 about five miles to the west. He looks out the tower window and sees the plane turning to the right and descending. He follows it until it disappears behind a building in nearby Crystal City, Virginia. [USA TODAY, 8/12/02] However, according to another account, just before 9:30 A.M., a controller in the same tower has an unidentified plane on radar, "heading toward Washington and without a transponder signal to identify it. It's flying fast, she says: almost 500 mph. And it's heading straight for the heart of the city. Could it be American Flight 77? The FAA warns the Secret Service." [USA TODAY, 8/13/02] In short, it is unclear whether the Secret Service warns the FAA, or vice versa.

9:30 A.M.*: Delta Flight Mistakenly Suspected by Cleveland Flight Control

Flight controllers mistakenly suspect that Delta Flight 1989, flying west over Pennsylvania, has been hijacked. The controllers briefly suspect the sound of hijackers' voices in Flight 93 is coming from this plane, only a few miles away. *USA Today* reports the flight "joins a growing list of suspicious jets. Some of their flight numbers will be scrawled on a white dry-erase board throughout the morning" at FAA headquarters. Miscommunications lead to further suspicion of Flight 1989 even after the source of the hijackers' message is confirmed to come from Flight 93. At some point, the Cleveland Airport flight control tower is evacuated for fear Flight 1989 will crash into it. Flight 1989 lands in Cleveland at 10:10 A.M. Eventually, about 11 flights will be labeled suspicious, with four of them actually hijacked. [USA TODAY, 8/13/02 (B); MSNBC, 9/18/02 (B)] The 9/11 Commission later has another explanation as to why Flight 1989 is suspected. They claim that at 9:41 A.M., Boston flight control identifies Flight 1989 as a possible hijacking strictly because it is a transcontinental 767 that had departed from Logan Airport. Although NEADS never loses track of the flight, it launches fighters from Ohio and Michigan to intercept it soon after 10:00 A.M. [9/11 COMMISSION REPORT, 6/17/04]

9:30–9:37 A.M.*: Langley Fighters Fly East to Ocean Instead of North to Washington; Explanations Differ

The three Langley fighters are airborne, but just where they go and how fast are in dispute. There are varying accounts that the fighters are ordered to Washington, New York, Baltimore, or no destination at all. The 9/11 Commission reports that, in fact, the pilots don't understand there is an emergency and head east. They give three reasons. "First, unlike a normal scramble order, this order did not include a distance to the target, or the target's location. Second, a 'generic' flight plan incorrectly led the Langley fighters to believe they were ordered to fly due east (090) for 60 miles. The purpose of the generic flight plan was to quickly get the aircraft airborne and out of local airspace. Third, the lead pilot and local FAA controller incorrectly assumed the flight plan instruction to go '090 for 60' was newer guidance that superseded the original scramble order." [9/11 COMMISSION REPORT, 6/17/04] However, the *Wall Street Journal* gives a different explanation, surprisingly from 9/11 Commission testimony. "Once they got in the air, the Langley fighters observed peacetime noise restrictions requiring that they fly more slowly than supersonic speed and takeoff over water, pointed away from Washington, according to testimony before the [9/11 Commis-

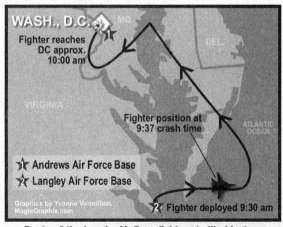

Route of the LangleyAir Base fighters to Washington

sion]." The fighters that departed to New York City over 30 minutes earlier at 8:52 A.M. traveled faster than supersonic because they realized they were in a national emergency. [WALL STREET JOURNAL, 3/22/04] In 2003 testimony, NORAD Commander Major General Larry Arnold explains that the fighters head over the ocean because NORAD is "looking outward" and has to have clearance to fly over land. [9/11 COMMISSION, 5/23/03] Yet, in contrast to these accounts, the BBC reports that just before takeoff at 9:24 A.M., the pilots are specifically told that Flight 77 may have been hijacked, and they get a cockpit signal indicating they are in an emergency wartime situation. All the above accounts concur that, for whatever reason, the fighters go too far east. They don't reach Washington until roughly around 10:00 A.M.

Between 9:30–10:00 A.M.*: Cockpit Voice Recording Begins

Apparently the only cockpit voice recording recovered undamaged from any of the 9/11 crashes is from Flight 93. It recorded on a 30-minute reel, which means that the tape is continually overwritten and only the final 30-minutes of any flight is recorded, though in practice sometimes the tape is slightly longer. Flight 93's recording lasts 31 minutes and begins at this time. [CNN, 4/19/02; *AMONG THE HEROES*, BY JERE LONGMAN, 8/02, P. 206–207, *HARTFORD COURANT*, 4/19/04] According to one account, it begins seconds before the plane is hijacked. [*WASHINGTON POST*, 11/17/01] However, the version of the tape later played for the victims' relatives begins "too late to pick up the sounds of the hijackers' initial takeover." [NBC, 4/18/02]

After 9:30 A.M.*: Secret Service Finally Rushes Bush Out of School

Kevin Down, a Sarasota police officer, recalls that immediately after President Bush's speech concludes, "The Secret Service agent [runs] out from the school and [says] we're under terrorist attack we have to go now." [BBC, 9/1/02] The motorcade departs a few minutes later.

After 9:31 A.M.*: Flight 93 Attendant Is Stabbed

A few minutes after 9:31 A.M., a hijacker on board Flight 93 can be heard on the cockpit voice recorder ordering a woman to sit down. A woman, presumably a flight attendant, implores, "Don't, don't." She pleads, "Please, I don't want to die." Patrick Welsh, the husband of flight attendant Debbie Welsh, is later told that a flight attendant was stabbed early in the takeover, and it is strongly implied it was his wife. She was a first-class attendant, and he says, "knowing Debbie," she would have resisted. [*AMONG THE HEROES*, BY JERE LONGMAN, 8/02, P. 207]

9:32 A.M.: Stock Exchange Closes

The New York Stock Exchange closes. It is a short distance from the WTC. [MSNBC, 9/22/01]

9:32 A.M.: Cheney Is Notified That Flight 77 Is Headed To Washington

Vice President Cheney pointing a finger inside the PEOC bunker. Footage of the World Trade Center plays on the televisions in the background (exact time is unknown).

According to the 9/11 Commission, the Dulles Airport terminal control facility in Washington has been looking for unidentified primary radar blips since 9:21 A.M. and now finds one. Several Dulles flight controllers "observed a primary radar target tracking eastbound at a high rate of speed" and notify Reagan Airport. FAA personnel at both Reagan and Dulles airports notify the Secret Service. The identity or aircraft type is unknown. [9/11 COMMISSION REPORT, 6/17/04] However, other accounts place the discovery of this plane by Dulles around 9:24 A.M. or 9:30 A.M., and Vice President Cheney is told radar is tracking Flight 77 at 9:27 A.M.

9:32 A.M.*: Flight 93 Hijacker Tells Passengers
Bomb Is Onboard; Flight Controllers Overhear

A hijacker says over the radio to Flight 93's passengers: "Ladies and gentlemen, here is the captain, please sit down. Keep remaining sitting. We have a bomb aboard." Apparently, Cleveland flight controllers can understand about a minute of screams, before a voice again says something about a "bomb on board." A hijacker says in broken English that they are returning to the airport. [NEWSWEEK, 9/22/01; PITTSBURGH POST-GAZETTE, 10/28/01; MSNBC, 9/3/02] According to the 9/11 Commission's account, the hijacker's voice says, "Keep remaining sitting. We have a bomb on board." The controller understands, but chooses to respond, "Calling Cleveland [flight control], you're unreadable. Say again, slowly." Apparently there's no answer. The controller notifies his supervisor, who soon passes the notice to FAA headquarters. [9/11 COMMISSION REPORT, 6/17/04]

9:33 A.M.: Planes Warned Away from Washington

The BBC reports that pilot Major Dean Eckmann gets a message as he's flying from Langley, Virginia. "They said—all airplanes, if you come within (I believe it was) 30 miles of Washington, D.C., you will be shot down." [BBC, 9/1/02] It's not clear who "they" are and what authority they have. However, fighters are not actually given shootdown orders until later, if at all.

9:33–9:37 A.M.*: Eyewitness Reports Indicate
There Was No Loss of Control on Flight 77

Radar data shows Flight 77 crossing the Capitol Beltway and headed toward the Pentagon. However, the plane, flying more than 400 mph, is too high when it nears the Pentagon at 9:35 A.M., crossing the Pentagon at about 7,000 feet up. [CBS NEWS, 9/21/01; BOSTON GLOBE, 11/23/01] The plane then makes a difficult high-speed descending turn. It makes a "downward spiral, turning almost a complete circle and dropping the last 7,000 feet in two-and-a-half minutes. The steep turn is so smooth, the sources say, it's clear there [is] no fight for control going on." [CBS NEWS, 9/21/01] It gets very near the White House during this turn. "Sources say the hijacked jet . . . [flies] several miles south of the restricted airspace around the White House." [CBS NEWS, 9/21/01] The *Daily Telegraph* later writes, "If the airliner had approached much nearer to the White House it might have been shot down by the Secret Service, who are believed to have a battery of ground-to-air Stinger missiles ready to defend the president's home. The Pentagon is not similarly defended." [DAILY TELEGRAPH, 9/16/01] White House spokesman Ari Fleischer suggests the plane goes even closer to the White House, saying, "That is not the radar data that we have seen. The plane was headed toward the White House." [CBS NEWS, 9/21/01] If Flight 77 passed within a few miles of the White House, why couldn't it have been shot down by the weapons on the White House?

9:34 A.M.: FAA's Headquarters Notified There Might Be a Bomb Onboard Flight 93; NORAD Not Notified

According to the 9/11 Commission, word of Flight 93's hijacking reaches FAA headquarters. By this time, headquarters has established an open line of communication with the FAA Command Center at Herndon, Virginia. It had instructed the center to poll all flight control centers about suspect aircraft. So, at this time, the Command Center passes on Cleveland's message: "United 93 may have a bomb on board." The Command Center continually updates FAA headquarters on Flight 93 until it crashes. [9/11 COMMISSION REPORT, 6/17/04]

9:34 A.M.: FAA Mentions in Passing to NORAD That Flight 77 Is Missing

According to the 9/11 Commission, NEADS contacts Washington flight control to ask about Flight 11. A manager there happens to mention, "We're looking—we also lost American 77." The commission claims, "No one at FAA Command Center or headquarters ever asked for military assistance with American 77." [9/11 COMMISSION REPORT, 6/17/04] Yet, 38 minutes earlier, flight controllers determined Flight 77 was off course, out of radio contact, and had no transponder signal. They'd warned American Airlines headquarters within minutes. By some accounts, this is the first time NORAD is told about Flight 77, but other accounts have them warned around 9:25 A.M.

9:34 A.M.: Flight 93 Passenger Burnett Calls Again, Learns "It's a Suicide Mission"

Tom Burnett calls his wife Deena a second time. He says, "They're in the cockpit." He has checked the pulse of the man who was knifed (later identified as Mark Rothenberg, sitting next to him in seat 5B) and determined he is dead. She tells him about the hits on the WTC. He responds, "Oh my God, it's a suicide mission." As they continue to talk, he tells her the plane has turned back. By this time, Deena is in constant communication with the FBI and others, and a police officer is at her house. [AMONG THE HEROES, BY JERE LONGMAN, 8/02, P. 110]

9:34 A.M.*: Bush Leaves Booker Elementary School for Sarasota Airport; Possible Threat En Route

President Bush's motorcade leaves Booker Elementary School and heads toward Sarasota-Bradenton International Airport. [WASHINGTON TIMES, 10/8/02; DAILY TELEGRAPH, 12/16/01; WALL STREET JOURNAL, 3/22/04] A few days after 9/11, Sarasota's main newspaper reports, "Sarasota barely skirted its own disaster. As it turns out, terrorists targeted the president and Air Force One on Tuesday, maybe even while they were on the ground in Sarasota and certainly not long after. The Secret Service learned of the threat just minutes after Bush left Booker Elementary." [SARASOTA HERALD-TRIBUNE, 9/16/01] Kevin Down, a Sarasota police officer at the scene, recalls, "I thought they were actually anticipating a terrorist attack on the President while we were en route." [BBC, 8/30/02] ABC News reporter Ann Compton, who is part of the motorcade, recalls, "It was a mad-dash motorcade out to the airport." [BBC, 9/1/02] A year later, Chief of Staff Andrew Card says, "As we were heading to Air Force One, we did hear about the Pentagon attack, and we also learned, what turned out to be a mistake, but we learned that the Air Force One package could in fact be a target." [MSNBC, 9/9/02] Real threat or not, this only increases the strangeness that Bush was not immediately evacuated as some of his security recommended at 9:03 A.M.

9:35 A.M.: Flight 93 Attendant Warns United Airlines About Hijacking; Account Spreads but Not to NORAD

The San Francisco United Airlines maintenance center receives a call from an unnamed flight attendant on Flight 93 saying that the flight has been hijacked. The information is quickly passed on. [9/11 COMMISSION, 1/27/04] Within ten minutes, "everyone" in the United Airlines crisis center "now [knows] that a flight attendant on board had called the mechanics desk to report that one hijacker had a bomb strapped on and another was holding a knife on the crew." [WALL STREET JOURNAL, 10/15/01]

9:35 A.M.*: Silent Flight 93 Climbs and Drops; NORAD Still Not Notified

When Flight 93 is over Youngstown, Ohio, Stacey Taylor and other Cleveland flight controllers see it rapidly climb 6,000 feet above its assigned altitude of 35,000 feet and then rapidly descend. The plane drops so quickly toward Cleveland that the flight controllers worry they might be the target. Other accounts say the climb occurs around 9:35 A.M. Controllers continue to try to contact the plane but still get no response. [GUARDIAN, 10/17/01; USA TODAY, 8/13/02; 9/11 COMMISSION REPORT, 6/17/04]

9:35 A.M.*: Treasury Department Evacuates; Pentagon and Other Washington Department Do Not

The Treasury Department is evacuated a few minutes before Flight 77 crashes. [9/11 COMMISSION, 1/26/04] Yet, CNN notes that "after the Federal Aviation Administration (FAA) warned the military's air defense command that a hijacked airliner appeared to be headed toward Washington, the federal government failed to make any move to evacuate the White House, Capitol, State Department, or the Pentagon." [CNN, 9/16/01] A Pentagon representative says, "The Pentagon was simply not aware that this aircraft was coming our way." Even Defense Secretary Rumsfeld and his top aides in the Pentagon remain unaware of any danger up to the moment of impact. [NEWSDAY, 9/23/01] Senators and congresspeople are in the Capitol building, which is not evacuated until 9:48 A.M. Only Vice President Cheney, National Security Adviser Rice, and possibly a few others are evacuated to safety a few minutes after 9:03 A.M. Yet, since at least the Flight 11 crash, "military officials in a command center [the National Military Command Center] on the east side of the [Pentagon] [are] urgently talking to law enforcement and air traffic control officials about what to do." [NEW YORK TIMES, 9/15/01] The White House is evacuated at 9:45 A.M.

9.36 A.M.: Military Cargo Plane
Asked to Identify Flight 77

Reagan Airport flight control instructs a military C-130 (Golfer 06) that has just departed Andrews Air Force Base to intercept Flight 77 and identify it. [*GUARDIAN*, 10/17/01; *NEW YORK TIMES*, 10/16/01 (D)] Remarkably, this C-130 is the same C-130 that is 17 miles from Flight 93 when it later crashes into the Pennsylvania countryside. [*MINNEAPOLIS STAR-TRIBUNE*, 9/11/02; PITTSBURGH CHANNEL. 9/15/01] The pilot, Lieutenant Colonel Steve O'Brien, claims he took off around 9:30 A.M., planning to return to Minnesota after dropping supplies off in the Caribbean. He later describes his close encounter: "When air traffic control asked me if we had him [Flight 77] in sight, I told him that was an understatement—by then, he had pretty much filled our windscreen. Then he made a pretty aggressive turn so he was moving right in front of us, a

A C-130 cargo airplane

mile and a half, two miles away. I said we had him in sight, then the controller asked me what kind of plane it was. That caught us up, because normally they have all that information. The controller didn't seem to know anything." O'Brien reports that the plane is either a 757 or 767 and its silver fuselage means it is probably an American Airlines plane. "They told us to turn and follow that aircraft—in 20-plus years of flying, I've never been asked to do something like that." [*MINNEAPOLIS STAR-TRIBUNE*, 9/11/02] The 9/11 Commission reports that it is a C-130H and the pilot specifically identifies the hijacked plane as a 757. Seconds after impact, he reports, "Looks like that aircraft crashed into the Pentagon, sir." [9/11 COMMISSION REPORT, 6/17/04]

9:36 A.M.*: Flight 93 Turns Around,
Files a New Flight Plan

Flight 93 files a new flight plan with a final destination of Washington, reverses course and heads toward Washington. [PITTSBURGH POST-GAZETTE, 10/28/01; *GUARDIAN*, 10/17/01; MSNBC, 9/3/02; *AMONG THE HEROES*, BY JERE LONGMAN, 8/02, P. 219] Radar shows the plane turning 180 degrees. [CNN, 9/13/01 (B)] The new flight plan schedules the plane to arrive in Washington at 10:28 A.M. [*AMONG THE HEROES*, BY JERE LONGMAN, 8/02, P. 78]

9:36 A.M.*: Cleveland Flight Control Wants NORAD Notified;
FAA Command Center Says People Are Working on It

According to the 9/11 Commission, at about this time Cleveland flight control specifically asks the FAA Command Center whether someone has requested the military to launch fighters toward Flight 93. Cleveland offers to contact a nearby military base. The Command Center replies that FAA personnel well above them in the chain of command have to make that decision and are working on the issue. [9/11 COMMISSION REPORT, 6/17/04] Cleveland overheard a hijacker say there was a "bomb on board" at 9:32 A.M. and passed the message to FAA higher ups.

Before 9:37 A.M.*: "Sheer Coincidence" Brings
Emergency Rescue and Secret Service Near to Pentagon

In response to an emergency 911 telephone call, the Arlington County Emergency Communications Center dispatches several units to deal with an apartment fire in Rosslyn, Virginia—within the vicinity of the Pentagon. Because this fire is in a high-rise building, nine different fire and medical service units are dispatched. However, the first engine crew to arrive radios to the other units that the fire has gone out. Consequently, by "sheer coincidence," at the time when the Pentagon is hit, there are a significant number of available fire and medical service units already on the road nearby. [ARLINGTON COUNTY AFTER-ACTION REPORT, 7/02; *FIRE ENGINEERING*, 11/02] Additionally, Secret Service personnel are concentrated around the heliport a short distance from where Flight 77 will hit: "President Bush was scheduled to fly from Florida that afternoon, and his helicopter, Marine One, would carry him to the Pentagon. That meant Secret Service everywhere and their cars blocking the driveway." [SCRIPPS HOWARD NEWS SERVICE, 8/1/02]

Before 9:37 A.M.*: Flight 77 Turns, Then Disappears from Radar

Washington flight controllers are watching Flight 77's radar blip. Just before radar contact is lost, FAA headquarters is told, "The aircraft is circling. It's turning away from the White House." [*USA TODAY*, 8/13/02] Then the blip disappears. Its last known position is six miles from the Pentagon and four miles from the White House. The plane is said to be traveling 500 mph, or a mile every seven seconds. [CBS NEWS, 9/21/01; NEWHOUSE NEWS SERVICE, 1/25/02; ABC NEWS, 9/11/02; *USA TODAY*, 8/13/02]

Before 9:37 A.M.*: Rumsfeld Said to Make
Eerie Predictions, but Witness Who Gives Account Is Long Gone

Representative Christopher Cox later claims he is still meeting with Defense Secretary Rumsfeld. They are still discussing missile defense, apparently completely oblivious of the approaching Flight 77. Watching television coverage from New York City, Rumsfeld says to Cox, "Believe me, this isn't over yet. There's going to be another attack, and it could be us." According to the *Daily Telegraph*, Flight 77 hits the building "moments later." [*DAILY TELEGRAPH*, 12/16/01] In another telling, Cox claims that Rumsfeld says, "If we remain vulnerable to missile attack, a terrorist group or rogue state that demonstrates the capacity to strike the U.S. or its allies from long range could have the power to hold our entire country hostage to nuclear or other blackmail. And let me tell you, I've been around the block a few times. There will be another event." Rumsfeld repeats that sentence for emphasis. According to Cox, "Within minutes of that utterance, Rumsfeld's words proved tragically prophetic." Cox also claims, "I escaped just minutes before the building was hit." [REP. COX STATEMENT, 9/11/01] However, Rumsfeld claims that this meeting with Cox ended before the second WTC crash, which occurred at 9:03 A.M. Cox himself said that after being told of the WTC, "[Rumsfeld] sped off, as did I." Cox says he immediately headed to his car, making it impossible for him to still be in the Pentagon "just minutes before" it is hit. [ASSOCIATED

PRESS, 9/11/01] Another account puts Rumsfeld's "I've been around the block a few times. There will be another event" comment two minutes before the first WTC crash at 8:46 A.M., when Rumsfeld reportedly makes other predictive comments. [ASSOCIATED PRESS, 9/16/01 (C)]

9:37 A.M.: Fireman Dodges Flight 77; Immediately Notifies Superior About Crashed Jumbo Jet

Fireman Alan Wallace is busy with a safety crew at the Pentagon's heliport pad. As Wallace is walking in front of the Pentagon, he looks up and sees Flight 77 coming straight at him. It is about 25 feet off the ground, with no landing wheels visible, a few hundred yards away, and closing fast. He runs about 30 feet and dives under a nearby van. [WASHINGTON POST, 9/21/01] The plane is traveling at about 460 mph, and flying so low that it clips the tops of streetlights. [CBS NEWS, 9/21/01] Using the radio in the van, he calls his fire chief at nearby Fort Myer and says, "We have had a commercial carrier crash into the west side of the Pentagon at the heliport, Washington Boulevard side. The crew is OK. The airplane was a 757 Boeing or a 320 Airbus." [SCRIPPS HOWARD NEWS SERVICE SERVICE, 8/1/02]

9:37 A.M.: Flight 77 Crashes into Reinforced Section of the Pentagon

Flight 77 crashes into the Pentagon. Approximately 125 people on the ground are later determined killed or missing. [NORAD, 9/18/01; CNN, 9/17/01; *GUARDIAN*, 10/17/01; *USA TODAY*, 8/13/02; ABC NEWS, 9/11/02; CBS NEWS, 9/11/02

Explosion at the Pentagon after the crash of Flight 77

(B); ASSOCIATED PRESS, 8/19/02 (B); MSNBC, 9/3/02] Flight 77 strikes the only side of the Pentagon that had recently been renovated—it was "within days of being totally [renovated]." [DEFENSE DEPARTMENT, 9/15/01] "It was the only area of the Pentagon with a sprinkler system, and it had been reconstructed with a web of steel columns and bars to withstand bomb blasts. The area struck by the plane also had blast-resistant windows—two inches thick and 2,500 pounds each—that stayed intact during the crash and fire. While perhaps, 4,500 people normally would have been working in the hardest-hit areas, because of the renovation work only about 800 were there. . . ." More than 25,000 people work at the Pentagon. [LOS ANGELES TIMES, 9/16/01 (C)]

9:37 A.M.: Witnesses See Military Cargo Plane Near Flight 77; Pilot Implies He's Far Away

A C-130 transport plane that has been sent to follow Flight 77 is trailing only a short distance behind the plane as it crashes. This curious C-130, originally bound for Minnesota, is the same C-130 that will be 17 miles from Flight 93 when it later crashes into the Pennsylvania countryside. [MINNEAPOLIS STAR-TRIBUNE, 9/11/02; PITTSBURGH CHANNEL, 9/15/01] A number of people see this plane fly remarkably close to Flight 77:

- Kelly Knowles says that seconds after seeing Flight 77 pass, she sees a "second plane that seemed to be chasing the first [pass] over at a slightly different angle." [DAILY PRESS, 9/15/01]

- Keith Wheelhouse says the second plane was a C-130; two other witnesses aren't certain. [DAILY PRESS, 9/15/01] Wheelhouse "believes it flew directly above the American Airlines jet, as if to prevent two planes from appearing on radar, while at the same time guiding the jet toward the Pentagon." As Flight 77 descends toward the Pentagon, the second plane veers off west. [DAILY PRESS, 9/14/01]

- *USA Today* reporter Vin Narayanan, who saw the Pentagon explosion, says, "I hopped out of my car after the jet exploded, nearly oblivious to a second jet hovering in the skies." [USA TODAY, 9/17/01]

- *USA Today* Editor Joel Sucherman sees a second plane but gives few details. [EWEEK, 9/13/01] Brian Kennedy, press secretary for a congressman, and others, also see a second plane. [SACRAMENTO BEE, 9/15/01]

- An unnamed worker at Arlington National Cemetery "said a mysterious second plane was circling the area when the first one attacked the Pentagon." [PITTSBURGH POST-GAZETTE, 12/20/01]

- John O'Keefe is driving a car when he sees the Pentagon crash. "The first thing I did was pull over onto the shoulder, and when I got out of the car I saw another plane flying over my head. . . . Then the plane—it looked like a C-130 cargo plane—started turning away from the Pentagon, it did a complete turnaround." [NEW YORK LAW JOURNAL, 9/12/01]

- The pilot of the C-130, Lieutenant Colonel Steve O'Brien, is later interviewed, but his account differs from the on-the-ground eyewitnesses. He claims that just before the explosion, "With all of the East Coast haze, I had a hard time picking him out," implying he is not nearby. He also says that just after the explosion, "I could see the outline of the Pentagon," again implying he is not nearby. He then asks "the controller whether [I] should set up a low orbit around the building," but he is told "to get out of the area as quickly as possible." "I took the plane once through the plume of smoke and thought if this was a terrorist attack, it probably wasn't a good idea to be flying through that plume." [MINNEAPOLIS STAR-TRIBUNE, 9/11/02]

9:37 A.M.: Langley Fighters Still Short of Washington; Where and Why Is Not Clear

Accounts differ as to how far from Washington the F-16 fighters scrambled from Langley are when Flight 77 crashes. The Langley, Virginia, base is 129 miles from Washington. NORAD originally claimed that, at the time of the crash, the fighters are 105 miles away, despite having taken off seven minutes earlier. [NORAD, 9/18/01] The 9/11 Commission claims that at 9:36 A.M., NEADS discovers that Flight 77 is only a few miles from the White House and is dismayed to find the fighters have headed east over the ocean. They are ordered to Washington immediately, but are still about 150 miles away. This is farther away than the base from which they took off. [9/11 COMMISSION REPORT, 6/17/04] The F-16 pilot codenamed Honey (who is apparently Captain Craig Borgstrom) offers a different explanation. As previously mentioned, he says they are flying toward New York, when they see a black column of smoke coming from Washington, about 30 or 40 miles to the west. He is then asked over the radio by NEADS if he can confirm the Pentagon is burning. He confirms it. The F-16s are then ordered to set up a defensive perimeter above Washington. [AMONG THE HEROES, BY JERE LONGMAN, 8/02, P. 76; NEW YORK OBSERVER, 2/11/04] The maximum speed of an F-16 is 1500 mph. [ASSOCIATED PRESS, 6/16/00] Had the fighters traveled straight to Washington at 1300 mph, they would have reached Washington at least one minute before Flight 77.

9:37 A.M.: Rumsfeld Either Being Briefed by CIA or with Clarke Video Conference When Pentagon Is Hit

There are conflicting accounts of what Defense Secretary Rumsfeld does in the 35 minutes between the second WTC crash and the Pentagon crash. In his 9/11 Commission testimony, he covers the time with the phrase "shortly thereafter:" "I was in my office with a CIA briefer and I was told that a second plane had hit the other tower. Shortly thereafter, at 9:38 A.M., the Pentagon shook with an explosion of then unknown origin." [9/11 COMMISSION, 3/23/04] In the book *Bush at War,* Bob Woodward writes, "Aware of the attacks on the World Trade Center, Rumsfeld had been proceeding with his daily intelligence briefing in his office" when the Pentagon gets hit. [BUSH AT WAR, BY BOB WOODWARD, 11/02, P. 22.] However, according to counterterrorism "tsar" Richard Clarke, Rumsfeld joins a video conference at 9:10 A.M., shortly after the second WTC hit, and stays with the conference, possibly from his office. After being told the Pentagon has been hit, Clarke says, "I can still see Rumsfeld on the screen, so the whole building didn't get hit". The military response to the 9/11 crisis is being coordinated in the NMCC, apparently located only around 200 feet away, directly below Rumsfeld's office. [DEFENSE DEPARTMENT, 9/15/01 (B); REUTERS, 9/11/01]

9:37 A.M.: United Flights Are Told to Bar Cockpit Entry

Captain Jim Hosking, piloting United Flight 890 from Japan to Los Angeles, is sent a warning message to his cockpit printer. It reads, "There has been a terrorist attack against United Airlines and American Airlines aircraft. We are advised there may be additional hijackings in progress. Shut down all access to the flight deck. Unable to elaborate further." He tells his first officer, "Get out the crash axe." Other pilots are receiving similar messages around this time. [USA TODAY, 8/13/02]

9:37 A.M.*: Flight 93 Passenger Jeremy Glick Describes Hijackers, Bomb

Jeremy Glick calls his wife, Lyz, from Flight 93. He describes the hijackers as Middle Eastern- and Iranian-looking. According to Glick, three of them put on red headbands, stood up, yelled, and ran into the cockpit.

He had been sitting in the front of the coach section, but he was then sent to the back with most of the passengers. Glick says the hijackers claimed to have a bomb, which looked like a box with something red around it. Family members immediately call emergency 911 on another line. New York State Police are patched in midway through the call. Glick finds out about the WTC towers. Two others onboard also learn about the WTC at about this time. Glick's phone remains connected until the very end of the flight. [AMONG THE HEROES, BY

Jeremy Glick

JERE LONGMAN, 8/02, P. 143; MSNBC, 7/30/02; PITTSBURGH POST-GAZETTE, 10/28/01 (B); TORONTO SUN, 9/16/01]

After 9:37 A.M.*: FBI Confiscates Film of Pentagon Crash

An employee at a gas station servicing military personnel later says the gas station's security cameras located across the street from the Pentagon should have recorded the moment of impact. However, he says, "I've never seen what the pictures looked like. The FBI was here within minutes and took the film." [RICHMOND TIMES-DISPATCH, 12/11/01] A security camera atop a hotel close to the Pentagon also records the impact. Hotel employees watch the film several times before the FBI confiscates the video. [GERTZ FILE, 9/21/01] This film footage has never been released.

After 9:37 A.M.*: Rumsfeld Reportedly Rushes to Help
Pentagon Crash Victims, but Accounts Are Contradictory and Problematic

By all accounts, Defense Secretary Rumsfeld is in his Pentagon office when Flight 77 crashes, though accounts differ as to what he's doing there. Rumsfeld later relates what he does next: "I was sitting here and the building was struck, and you could feel the impact of it very clearly, and I don't know what made me do anything I did, to be honest with you. I just do it instinctive. I looked out the window, saw nothing here, and then went down the hall until the smoke was too bad, then to a stairwell down and went outside and saw what had happened. Asked a person who'd seen it, and he told me that a plane had flown into it. I had been aware of a plane going into the World Trade Center, and I saw people on the grass, and we just, we tried to put them in stretchers and then move them out across the grass towards the road and lifted them over a jersey wall so the people on that side could stick them into the ambulances. I was out there for awhile, and then people started gathering, and we were able to get other people to do that, to hold IVs for people. There were people lying on the grass with clothes blown off and burns all over them. Then at some moment I decided I should be in here figuring out what to do, because your brain begins to connect things, and there were enough people there to worry about that. I came back in here, came into this office. There was smoke in here by then." [DEFENSE DEPARTMENT, 10/12/01] Versions of this story appear elsewhere. [MINNEAPOLIS STAR-TRIBUNE, 9/12/01; CNN, 12/5/01; ABC NEWS, 9/11/02; DEFENSE DEPARTMENT, 5/9/03] Rumsfeld says the crash site is "around the corner" from his fourth floor office [ABC NEWS, 9/11/02], but, in fact, the crash site is on the opposite site of the huge Pentagon. [REUTERS, 9/11/01] Rumsfeld says he reaches the crash site "moments after" the crash, which would be an impressive feat given the over 2000 feet distance. [9/11 COMMISSION, 3/23/04] One report even has Rumsfeld pull budget analyst Paul

Gonzalez to safety from the burning wreckage. [*DAILY TELEGRAPH*, 9/16/01 (B)] However, Gonzalez later offers his own detailed recollections of pulling other people to safety, which fail to involve Rumsfeld in any way. [*WASHINGTON POST*, 3/11/02] Deputy Defense Secretary Torie Clarke, in the Pentagon at the time, says Rumsfeld is "one of the first people" outside [DEFENSE DEPARTMENT, 9/15/01 (C)], and remains outside for "about half an hour." [DEFENSE DEPARTMENT, 9/15/01 (B)] A Pentagon spokesperson has Rumsfeld helping for "15 min-

utes or so . . ." [REUTERS, 9/11/01] In another account, he loads the wounded onto stretchers for 15 minutes. [SCRIPPS HOWARD NEWS SERVICE, 9/11/01] Rumsfeld reportedly helps at the crash site until a security agent urges him to leave. [*WASHINGTON POST*, 1/27/02] However, in his 2004 testimony to the 9/11 Com-mission, he no longer mentions helping the wounded, merely saying, "I went outside to determine what had happened. I was not there long because I was back in the Pentagon with a crisis action team shortly before or after 10:00 A.M." [9/11 COMMISSION, 3/23/04] There are no photographs or eyewitness accounts of Rumsfeld outside the Pentagon that morning, except for one photograph of him walking down a sidewalk with some aides. In counterterrorism "tsar" Richard Clarke's account, Rumsfeld never leaves a video conference for very long, except to move from one secure teleconferencing studio to another elsewhere in the Pentagon. [*AGAINST ALL ENEMIES*, BY RICHARD CLARKE, 3/04, PP. 8–9]

Defense Secretary Rumsfeld walking outside the Pentagon on the morning of 9/11

After 9:37 A.M.*: Cheney Tells Bush to Stay Away from Washington

Having learned that the Pentagon had been hit, Vice President Cheney telephones President Bush, who is on his way to the airport, and tells him that the White House has been "targeted." Bush says he wants to return to Washington, but Cheney advises him not to "until we could find out what the hell was going on." According to *Newsweek*, this call takes place in a tunnel on the way to the PEOC underground bunker. Cheney reaches the bunker "shortly before 10:00 A.M." [*NEWSWEEK*, 12/31/01] The 9/11 Commission's account largely follows *Newsweek*'s. He reaches the tunnel around the time of the Pentagon crash and lingers by a television and secure telephone as he talks to Bush. The commission has Cheney enter the bunker just before 10:00, but they note, "There is conflicting evidence as to when the Vice President arrived in the shelter conference room." [9/11 COMMISSION REPORT, 6/17/04] Indeed, in other accounts, including those of Richard Clarke and Transportation Secretary Norman Mineta, Cheney reaches the bunker before the Flight 77 crash at 9:37 A.M. [*AGAINST ALL ENEMIES*, BY RICHARD CLARKE, 3/04, PP. 3–4; ABC NEWS, 9/11/02; 9/11 COMMISSION, 5/23/03] Regardless of Cheney's location, as Cheney and Bush talk on the phone, Bush once again refrains from making any decisions or orders about the crisis. [9/11 COMMISSION REPORT, 6/17/04]

After 9:37 A.M.*: Andrews Pilots Aware of Crisis but Still on Ground

After the Pentagon is hit, fighters at nearby Andrews Air Force Base are still preparing to launch. At some unknown point, flight squad commander Lieutenant Colonel Marc Sasseville assembles three F-16 pilots and gives them a curt briefing. He recalls saying, "I have no idea what's going on, but we're flying. Here's our frequency. We'll split up the area as we have to. Just defend as required. We'll talk about the rest in the air." All four of them dress up and get ready. One officer at Andrews recalls, "After the Pentagon was hit, we were told there were more [airliners] coming. Not 'might be'– they were coming." Meanwhile, a "flood" of calls from the Secret Service and local FAA flight control centers pour into Andrews, as the fighter response is coordinated. [AVIATION WEEK AND SPACE TECHNOLOGY, 9/9/02] However, the loading of missiles onto the fighters is very time consuming, and when these fighters finally take off nearly an hour later, they will launch without the missiles installed.

Between 9:37–9:45 A.M.*: Clarke Orders Combat Air Patrols
over All Major Cities; Order Apparently Not Passed On

At an indeterminate time after Flight 77 hits the Pentagon, counterterrorism "tsar" Richard Clarke is given a note by the head of the Secret Service. The note reads, "Radar shows aircraft headed this way. I'm going to empty out the [White House]." The Secret Service knows this because they have equipment that can see what the FAA's radar is seeing around Washington. However, the note is too late: Flight 77 has already crashed. At almost the same time, another aide says to Clarke, "A plane just hit the Pentagon." He replies, "I can still see Rumsfeld on the screen, so the whole building didn't get hit. No emotion in here. We are going to stay focused." He orders an aide, "Find out where the fighter planes are. I want Combat Air Patrol over every major city in this country. Now!" [AGAINST ALL ENEMIES, BY RICHARD CLARKE, 3/04, PP. 7–8; AUSTRALIAN, 3/27/04] NORAD does give this nationwide order around 9:49 A.M., but bases had been calling into NORAD and asking in vain for permission to send up fighters since the second WTC crash. [AVIATION WEEK AND SPACE TECHNOLOGY, 6/3/02; 9/11 COMMISSION REPORT, 6/17/04] Other cities generally remain unprotected until after 11:00 A.M. [TOLEDO BLADE, 12/9/01] The Secret Service order to evacuate the White House takes place at 9:45 A.M.

Between 9:37–9:58 A.M.*: Seven Planes Unaccounted For

New York City Mayor Rudolph Giuliani is told by his chief of staff that the White House knows of seven planes that are unaccounted for. He is told that the Pentagon has been hit, but also hears erroneous reports that the Sears Tower and other buildings have been hit. [9/11 COMMISSION, 5/19/04]

9:39 A.M.: Rumsfeld Is Wanted at NMCC
Teleconference but Cannot Be Reached

Captain Charles Leidig, a low ranking officer temporarily in charge of the NMCC, is handling the NMCC's crisis teleconference. He mentions reports of a crash into the opposite side of the Pentagon, and requests that Defense Secretary Rumsfeld be added to the conference. [9/11 COMMISSION REPORT, 6/17/04] As one magazine has noted, "On September 11, the normal scramble-approval procedure was for an FAA official to contact the [NMCC] and request Pentagon air support. Someone in the NMCC would call NORAD's command center and ask about availability of aircraft, then seek approval from the Defense Secretary—Donald H. Rumsfeld—to launch fighters." [AVIATION WEEK AND SPACE TECHNOLOGY, 6/3/02] Rather than join the NMCC conference, Rumsfeld has already gone out of the Pentagon to see the crash site, and remains out of contact for some time. It is unknown if Rumsfeld had a cell phone or pager, and if so, why he cannot be reached.

9:39 A.M.: Media Reports Pentagon Explosion

Two seconds after 9:39 A.M., reporter Jim Miklaszewski states on NBC News, "Moments ago, I felt an explosion here at the Pentagon." [TELEVISION ARCHIVE, WDCN 9:30] However, no media outlets record video footage of the Pentagon crash, and the cause of the crash remains unknown for some minutes afterward.

9:39 A.M.*: Flight 93 Hijacker Again Warns of Bomb on Board, Flight Controllers Again Overhear; NORAD Still Not Notified

The Flight 93 hijackers (probably inadvertently) transmit over the radio: "Hi, this is the captain. We'd like you all to remain seated. There is a bomb on board. And we are going to turn back to the airport. And they had our demands, so please remain quiet." [BOSTON GLOBE, 11/23/01; MSNBC, 9/3/02; AMONG THE HEROES, BY JERE LONGMAN, 8/02, P. 209; 9/11 COMMISSION REPORT, 6/17/04] The controller responds, "United 93, understand you have a bomb on board. Go ahead," but there is no response. There was a very similar "bomb on board" warning from the same flight at 9:32 A.M. The 9/11 Commission indicates that these are separate incidents. [9/11 COMMISSION REPORT, 6/17/04] Cleveland flight control apparently continues to wait for FAA superiors to notify NORAD. Earlier in the morning, Boston flight control directly contacted NORAD and local air force bases when they determined Flight 11 was hijacked.

9:40 A.M.*: Flight 93 Transponder Signal Turned Off; Flight Still Closely Tracked

The transponder signal from Flight 93 ceases. [MSNBC, 9/3/02; MSNBC, 9/11/02 (B); CNN, 9/17/01; 9/11 COMMISSION REPORT, 6/17/04] However, the plane can be—and is—tracked using primary radar by Cleveland flight controllers and at United headquarters. Altitude can no longer be determined, except by visual sightings from other aircraft. The plane's speed begins to vary wildly, fluctuating between 600 and 400 mph before eventually settling around 400 mph. [AMONG THE HEROES, BY JERE LONGMAN, 8/02, P. 77, 214; 9/11 COMMISSION REPORT, 6/17/04]

9:40 A.M.*: FAA Command Center Identifies Ten Possible Hijacked Planes

Newark, New Jersey, flight controller Bob Varcadapane is talking on the phone with the FAA Command Center. He is told that the Command Center is still suspicious of at least ten planes for one reason or another, all possible hijackings. [MSNBC, 9/11/02 (B)]

9:41 A.M.: Flight 93 Passenger Birtton Reports Two Killed

Flight 93 passenger Marion Birtton calls a friend. She tells him two people have been killed and the plane has been turned around. [PITTSBURGH POST-GAZETTE, 10/28/01]

9:41 A.M.*: FBI Agent Already Aware of Flight 93 Hijacking

Newark, New Jersey, flight controller Greg Callahan is talking on the phone to an FBI agent. The agent says about Flight 93: "We suspect that this aircraft has now been taken over by hostile forces." The agent describes the sharp turn it has made over eastern Ohio and that it is now heading back over southwestern Pennsylvania. Callahan says he could tell the plane is on a course for Washington. [MSNBC, 9/11/02 (B)] The FBI has been in contact with Deena Burnett and informed of what her husband, Flight 93 passenger Tom Burnett, has been saying since at least 9:34 A.M. [AMONG THE HEROES, BY JERE LONGMAN, 8/02, P. 110] It is unclear where in the chain of command details of these Flight 93 calls reach, and the 9/11 Commission has not clarified the issue of what the FBI knew and when.

9:42 A.M.: Passenger Mark Bingham Tells of Bomb Threat on Flight 93

From Flight 93, Mark Bingham calls his mother and says, "I'm on a flight from Newark to San Francisco and there are three guys who have taken over the plane and they say they have a bomb." [PITTSBURGH POST-GAZETTE, 10/28/01 (B)] In an alternate version, he says, "I'm in the air, I'm calling you on the Airfone. I'm calling you from the plane. We've been taken over. There are three men that say they have a bomb." [TORONTO SUN, 9/16/01; BOSTON GLOBE, 11/23/01]

Mark Bingham

9:43 A.M.: Bush Learns of Attack on Pentagon as Motorcade Reaches Sarasota Airport

President Bush's motorcade arrives at Sarasota's airport and pulls up close to Air Force One. As the motorcade nears the airport, he learns a plane has hit the Pentagon. Bush immediately boards the plane. [WASHINGTON TIMES, 10/8/02; DAILY TELEGRAPH, 12/16/01] Congressman Dan Miller and others hurry up the rear steps of the plane while Bush enters through the exposed front stairs. Bush pauses in the doorway to wave to photographers. The *St. Petersburg Times* notes this raises "further questions about security [on 9/11]." [ST. PETERSBURG TIMES, 7/4/04] Security then does an extra-thorough search of all the baggage of the other passengers, delaying takeoff until 9:55 A.M. [ST. PETERSBURG TIMES, 9/8/02 (B)]

9:44 A.M.: NMCC Conference Thinks Flight 1989, Not Flight 93, Is Fourth Hijack

NORAD briefs the NMCC teleconference on the possible hijacking of Delta Flight 1989. Four minutes later, a representative from the White House bunker containing Vice President Cheney asks if there are any indications of other hijacked planes. Captain Charles Leidig, temporarily in charge of the NMCC, mentions the Delta flight and comments, "that would be the fourth possible hijack." Flight 1989 is in the same general Ohio region as Flight 93, but NORAD doesn't scramble fighters toward either plane at this time. [9/11 COMMISSION REPORT, 6/17/04]

9:45 A.M.: United Headquarters Learns Flight 77 Has Crashed into the Pentagon

United Airlines headquarters receives a report that an aircraft has crashed into the Pentagon. They learn it is Flight 77. [9/11 COMMISSION, 1/27/04]

9:45 A.M.: Tom Burnett Says Flight 93 Passengers
Are Making Plans to Defeat Hijackers

Tom Burnett calls his wife, Deena, for the third time. She tells him about the crash at the Pentagon. Tom speaks about the bomb he'd mentioned earlier, saying, "I don't think they have one. I think they're just telling us that." He says the hijackers are talking about crashing the plane into the ground. "We have to do something." He says that "a group of us" are making a plan. [AMONG THE HEROES, BY JERE LONGMAN, 8/02, P. 111] This indicates there would have been at least 19 minutes advance notice that a passenger takeover was likely, if the contents of these phone calls are being passed on to the right authorities. Note that by Burnett's second call at 9:34 A.M., the FBI was already listening in. [TORONTO SUN, 9/16/01]

9:45 A.M.: FBI Listens as Passenger
Todd Beamer Describes Flight 93 Take Over Plan

After having some trouble getting authorization to use an Airfone to call his family, passenger Todd Beamer is able to speak to Verizon phone representative Lisa Jefferson, with the FBI listening in. He talks for about 15 minutes. Beamer says he has been herded to the back of the plane along with nine other passengers and five flight attendants. A hijacker, who says he has a bomb strapped to his body, is guarding them. Twenty-seven passengers are being guarded by a hijacker in first class, which is separated from the rest of the aircraft by a curtain. One hijacker has gone into the cockpit. One passenger is dead (that leaves one passenger unaccounted for—presumably the man who made a call from the bathroom, thought to be Edward Felt). The two pilots are apparently dead. [PITTSBURGH POST-GAZETTE, 9/16/01; NEWSWEEK, 9/22/01; PITTSBURGH POST-GAZETTE, 10/28/01; PITTSBURGH POST-GAZETTE, 10/28/01 (B); BOSTON GLOBE, 11/23/01] A conflicting version states that 27 passengers were in the back, and that Beamer saw four hijackers instead of just three. [BOSTON GLOBE, 11/23/01] It is not clear if Tom Burnett's first class section group is in contact with Todd Beamer's coach section group or if there are two independent plans to take over the plane.

9:45 A.M.*: White House Finally Evacuated

The White House begins a general evacuation. This comes about 30 minutes after the probable time Vice President Cheney has been evacuated from the White House. [NEW YORK TIMES, 9/12/01; MSNBC, 9/22/01; WASHINGTON POST, 1/27/02; DAILY TELEGRAPH, 12/16/01; ASSOCIATED PRESS, 8/19/02 (B)] Initially the evacuation is orderly, but soon the Secret Service agents are yelling that everyone should run. [ABC NEWS, 9/11/02]

9:45 A.M.*: Senior FAA Manager, on His First Day
on the Job, Orders All Planes Out of the Sky Nationwide

Ben Sliney, FAA's National Operations Manager, orders the entire nationwide air traffic system shut down. All flights at U.S. airports are stopped. Around 3,950 flights are still in the air. Sliney makes the decision without consulting FAA head Jane Garvey, Transportation Secretary Norman Mineta, or other bosses, but they quickly approve his actions. It's Sliney's first day on the job. [USA TODAY, 8/13/02; USA TODAY, 8/13/02 (B); MSNBC,

9/22/01; CNN, 9/12/01; *NEW YORK TIMES*, 9/12/01; ASSOCIATED PRESS, 8/12/02; ASSOCIATED PRESS, 8/19/02 (B); *NEWSDAY*, 9/10/02; *USA TODAY*, 8/13/02; *WASHINGTON POST*, 9/12/01] Seventy-five percent of the planes land within one hour of the order. [*USA TODAY*, 8/12/02 (C)] The *Washington Post* has reported that Mineta told Monty Belger at the FAA: "Monty, bring all the planes down," even adding, "[Expletive] pilot discretion." [*WASHINGTON POST*, 1/27/02] However, it is later reported by a different *Post* reporter that Mineta did not even know of the order until 15 minutes later. This reporter "says FAA officials had begged him to maintain the fiction." [*SLATE*, 4/2/02]

Ben Sliney

9:45 A.M.*: Bush Aides Debate Where to Fly Air Force Once

According to the 9/11 Commission, Chief of Staff Andrew Card, the lead Secret Service agent, the President's military aide, and Air Force One pilot Colonel Mark Tillman, confer on a possible destination for Air Force One around this time. According to witnesses, some support President Bush's desire to return to Washington, but the others advise against it. The issue is still not decided when Air Force One takes off around 9:55 A.M. [9/11 COMMISSION REPORT, 6/17/04]

Between 9:45–9:55 A.M.*: Clarke Initiates Continuity of Government Plans; Hears Shoot Down Talk from Cheney Bunker

At some point after the White House is evacuated, counterterrorism "tsar" Richard Clarke institutes Continuity of Government plans. Important government personnel, especially those in line to succeed the President, are evacuated to alternate command centers. Additionally, Clarke gets a phone call from the PEOC command center where Vice President Cheney and National Security Adviser Rice are positioned. An aide tells Clarke, "Air Force One is getting ready to take off with some press still on board. [President Bush will] divert to an air base. Fighter escort is authorized. And . . . tell the Pentagon they have authority from the President to shoot down hostile aircraft, repeat, they have authority to shoot down hostile aircraft." However, acting Joint Chiefs of Staff Chairman Richard Myers wants the rules of engagement clarified before the shootdown order is passed on, so Clarke orders that pilots be given guidelines before receiving shootdown authorization. [*AGAINST ALL ENEMIES*, BY RICHARD CLARKE, 3/04, PP. 8–9] Clarke's account that Cheney is giving shootdown authorization well before 10:00 A.M. matches Transportation Secretary Norman Mineta's account of seeing Cheney giving what he interprets as a shootdown order before the Pentagon crash. [9/11 COMMISSION, 5/23/03] However, the 9/11 Commission later asserts that Cheney doesn't make the shootdown decision until about 10:00 A.M. [9/11 COMMISSION REPORT, 6/17/04]

9:46 A.M.: NMCC Teleconference Still Looking to Include Rumsfeld and Myers

Defense Secretary Rumsfeld's office, and acting Joint Chiefs of Staff Chairman Myers' office, report to the NMCC teleconference that they are still trying to track down Rumsfeld and Myers, respectively, and bring them into the conference. [9/11 COMMISSION REPORT, 6/17/04] Rumsfeld is apparently outside the Pentagon looking at the Flight 77 crash site, though counterterrorism "tsar" Richard Clarke suggests

Rumsfeld is elsewhere in the Pentagon for much of the time. Myers' whereabouts in the period after the Pentagon crash have not been fully explained. Rumsfeld and Myers do not enter the NMCC until about 10:30 A.M.

9:46 A.M.*: Flight 93 Hijackers Bring the Pilot Back In

According to the later recovered Flight 93 cockpit voice recording, around this time one hijacker in the cockpit says to another, "Let the guys in now." A vague instruction is given to bring the pilot back in. It's not clear if this is a reference to an original pilot or a hijacker pilot. Investigators aren't sure if the original pilots were quickly killed or allowed to live. [*AMONG THE HEROES*, BY JERE LONGMAN, 8/02, P. 208]

9:47 A.M.: Internal Collapse at WTC South Tower Reported

A man who is on the 105th floor of the South Tower calls emergency 9-1-1 to report that floors below his location, "in the 90-something floor," have collapsed. The 9-1-1 operator types a record of this call into the Special Police Radio Inquiry Network (SPRINT) data link, which will be passed on to the New York fire department's Emergency Medical Service (EMS). It isn't known when the call is made exactly, but the EMS Dispatch computer apparently receives the call record at this time. However, because it is classified as a "supplement message," it is not yet read by anyone. The police dispatcher dealing with the area around the WTC also receives the call record, but misinterprets it as meaning that the floor the person is on has collapsed. EMS dispatchers are dealing with an enormous volume of calls as well as performing many other tasks under extreme pressure during the crisis, so a report later concludes that the EMS operators didn't have the time to review the information before the collapse of the South Tower at 9:59, and the fire chiefs never received the information. [MCKINSEY REPORT, AUGUST 19, 2002]

9:47 A.M.*: Passenger Jeremy Glick Tells of Vote to Take Over Flight 93

On Flight 93, Jeremy Glick is still on the phone with his wife, Lyz. He tells her that the passengers are taking a vote if they should try to take over the plane or not. [*PITTSBURGH POST-GAZETTE*, 10/28/01; *PITTSBURGH POST-GAZETTE*, 10/28/01 (B)] He later says that all the men on the plane have voted to attack the hijackers. [*TORONTO SUN*, 9/16/01] When asked about weapons, he says they don't have guns, just knives. This appears to contradict an earlier mention of guns. His wife gets the impression from him that the hijacker standing nearby, claiming to hold the bomb, would be easy to overwhelm. [*AMONG THE HEROES*, BY JERE LONGMAN, 8/02, P. 153–154]

9:48 A.M.: Capitol Building Finally Evacuates

The Capitol building in Washington begins evacuation. Congress is in session, but apparently the chambers are not filled with congresspeople. [*GUARDIAN*, 7/22/04; ASSOCIATED PRESS, 8/19/02 (B)] Senator Tom Daschle, Majority Leader of the Senate, later states, "Some capitol policemen broke into the room and said, 'We're under attack. I've got to take you out right away.'" Speaker of the House Dennis Hastert, third in line of succession to the presidency behind Vice President Cheney, is in the Capitol building with other congresspeople. Only after this time are Hastert and others in the line of succession moved to secure locations. Some time after this, Hastert and other leaders are flown by helicopter to secret bunkers. [ABC NEWS, 9/11/02]

9:49 A.M.: Pittsburgh Flight Control Tower Evacuates

The FAA orders the Pittsburgh control tower evacuated. Shortly before the order, Cleveland flight controllers called Pittsburgh flight control to say that a plane is heading toward Pittsburgh and the pilot refuses to communicate. The plane is Flight 93. [PITTSBURGH POST-GAZETTE, 9/23/01 (B)]

9:49 A.M.: FAA Headquarters Continues to Delay Decision on Contacting NORAD About Flight 93

According to the 9/11 Commission, the FAA Command Center has just twice warned FAA headquarters that United 93 is now "29 minutes out of Washington, D.C." Someone at headquarters says to someone at the Command Center, "They're pulling Jeff [last name unknown] away to go talk about United 93." Command Center replies, "Uh, do we want to think about, uh, scrambling aircraft [NORAD fighters]?" FAA headquarters replies, "Uh, God, I don't know." Command Center says, "Uh, that's a decision somebody's gonna have to make probably, in the next ten minutes." FAA headquarters answers, "Uh, ya know, everybody just left the room." [9/11 COMMISSION REPORT, 6/17/04] This is 13 minutes since Cleveland flight control had asked the Command Center in vain to contact NORAD about Flight 93.

9:49 A.M.: Fighters Ordered to Scramble Nationwide

In the words of the 9/11 Commission, the commander of NORAD (General Ralph Eberhart) directs "all air sovereignty aircraft to battle stations fully armed." [9/11 COMMISSION REPORT, 6/17/04] Apparently, this means all fighters with air defense missions are to be armed and ready to scramble. This may be connected to counterterrorism "tsar" Richard Clarke's claim that after the Pentagon is hit, he orders an aide, "Find out where the fighter planes are. I want Combat Air Patrol over every major city in this country. Now!" [AGAINST ALL ENEMIES, BY RICHARD CLARKE, 3/04, PP. 7–8; AUSTRALIAN, 3/27/04] Another account says calls to bases to scramble don't begin until about 10:01 A.M. [TOLEDO BLADE, 12/9/01] It has not been explained why this order wasn't given much earlier. Calls from Air Force bases across the country offering to help had started "pouring into NORAD" shortly after 9:03 A.M., when televised reports made an emergency situation clear. [AVIATION WEEK AND SPACE TECHNOLOGY, 6/3/02] With a couple of exceptions, other fighters do not actually start taking off until about 11:00 A.M.

General Ralph Eberhart

9:50 A.M.: Sandy Bradshaw Tells
of Preparations to Fight "Three Guys with Knives"

**Sandy
Bradshaw**

Sandy Bradshaw calls her husband from Flight 93. She says, "Have you heard what's going on? My flight has been hijacked. My flight has been hijacked with three guys with knives." [BOSTON GLOBE, 11/23/01] She tells him that some passengers are in the rear galley filling pitchers with hot water to use against the hijackers. [PITTSBURGH POST-GAZETTE, 10/28/01; PITTSBURGH POST-GAZETTE, 10/28/01 (B)]

Between 9:50–10:40 A.M.*: Numerous False
Reports of Terrorist Acts in Washington

There are numerous false reports of additional terror attacks. Before 10:00 A.M., some hear reports on television of a fire at the State Department. At 10:20 A.M., and apparently again at 10:33 A.M., it is publicly reported this was caused by a car bomb. [OTTAWA CITIZEN, 9/11/01; BROADCASTING AND CABLE, 8/26/02; DAILY TELEGRAPH, 12/16/01] At 10:23 A.M., the Associated Press reports, "A car bomb explodes outside the State Department, senior law enforcement officials say." [BROADCASTING AND CABLE, 8/26/02] Counterterrorism "tsar" Richard Clarke hears these reports at this time and asks Deputy Secretary of State Richard Armitage in the State Department to see if the building he's in has been hit. Armitage goes outside the building, finds out there's no bomb, and calls his colleagues to inform them that the reports are false. Reports of a fire on the Capitol Mall are also floated and quickly found to be false. [ABC NEWS, 9/15/02 (B); AGAINST ALL ENEMIES, BY RICHARD CLARKE, 3/04, PP. 8-9] There are numerous other false reports over the next hour, including explosions at the Capitol building and *USA Today* headquarters. [BROADCASTING AND CABLE, 8/26/02] For instance, CNN reports an explosion at Capitol Hill at 10:12 A.M. CNN then announces this is untrue 12 minutes later. [OTTAWA CITIZEN, 9/11/01]

9:52 A.M.*: Lynne Cheney Joins Husband in White House Bunker;
Vice President Repeatedly Hangs up Clarke Telephone

According to the 9/11 Commission, Lynne Cheney joins her husband, Vice President Cheney, in the PEOC (Presidential Emergency Operations Center) bunker below the White House. [9/11 COMMISSION REPORT, 6/17/04] She had been at a downtown office around 9:00 A.M. when she was escorted by the Secret Service to the White House. [NEWSWEEK, 12/31/01] Counterterrorism "tsar" Richard Clarke describes the

Lynne Cheney

people in the PEOC as "decidedly more political" than those in his bunker below the other wing of the White House. In addition to Cheney and his wife, most of the day the PEOC contains National Security Adviser Rice, political adviser Mary Matalin, Cheney's Chief of Staff I. Lewis "Scooter" Libby, Deputy White House Chief of Staff Josh Bolten, and White House Communications Director Karen Hughes. Clarke is told later in the day by someone else in the PEOC, "I can't hear the crisis conference [led by Clarke] because Mrs. Cheney keeps turning down the volume on you so she can hear CNN . . . and the vice president keeps hanging up the open line to you." Clarke notes that the "right-wing ideologue" Lynne Cheney frequently offers her advice and opinions during the crisis. [AGAINST ALL ENEMIES, BY RICHARD CLARKE, 3/04, P. 18]

9:53 A.M.: NSA Intercepts al-Qaeda Phone Call Predicting Fourth Attack

The National Security Agency (NSA) reportedly intercepts a phone call from one of bin Laden's operatives in Afghanistan to a phone number in the Republic of Georgia. The caller says he has "heard good news" and that another target is still to come (presumably, the target Flight 93 is intended to hit). [CBS NEWS, 9/4/02] Since the 9/11 crisis began, NSA translators have been told to focus on Middle Eastern intercepts and translate them as they are received instead of oldest first, as is the usual practice. This call is translated in the next hour or two, and Defense Secretary Rumsfeld hears about it just after noon. [CBS NEWS, 9/4/02; A PRETEXT FOR WAR, BY JAMES BAMFORD, 6/04, P. 54]

9:53 A.M.: Hijackers Fear Passenger Retaliation

According to Flight 93's cockpit voice recording, the hijackers grow concerned that the passengers might retaliate. One urges that the plane's fire axe be held up to the cockpit door's peephole to scare the passengers. [AMONG THE HEROES, BY JERE LONGMAN, 8/02, P. 209–210]

9:53 A.M.: FAA Headquarters Still Only Talking About Telling NORAD of Flight 93 Hijack

According to the 9/11 Commission, FAA headquarters informs the FAA Command Center that the Deputy Director for Air Traffic Services is talking to Deputy Administrator Monty Belger about scrambling aircraft after Flight 93. Headquarters is informed that the flight is 20 miles northwest of Johnstown, Pennsylvania. [9/11 COMMISSION REPORT, 6/17/04] Incredibly, FAA headquarters has known since 9:34 A.M. about hijackers talking about a bomb on board the flight, and more evidence has since been passed on confirming a hijacking in progress. Still, reportedly, no one tells NORAD anything about the plane.

9:54 A.M.: Tom Burnett Calls for Fourth Time; Details Plan to Take Control of Plane

Tom Burnett calls his wife, Deena, for the fourth and last time. In early reports of this call, he says, "I know we're all going to die. There's three of us who are going to do something about it." [TORONTO SUN, 9/16/01; BOSTON GLOBE, 11/23/01] However, in a later, more complete, account, he sounds much more upbeat. "It's up to us. I think we can do it." "Don't worry, we're going to do something." He specifically mentions they plan to regain control of the airplane over a rural area. [AMONG THE HEROES, BY JERE LONGMAN, 8/02, P. 118]

9:55–10:15 A.M.*: Langley Fighters Finally Reach Washington; Accounts of Timing Are Contradictory

The three F-16s scrambled after Flight 77 from Langley, Virginia at 9:30 A.M. finally reach Washington and the burning Pentagon. The 129 mile distance could theoretically be covered by the fighters in six minutes, but they've taken a wide detour over the ocean. The exact time they arrive is very unclear. NORAD originally claimed they arrive as soon as 9:49 A.M., but the 9/11 Commission implies they don't arrive until shortly after 10:00 A.M., though no exact time is specified. [CNN, 9/17/01; NORAD, 9/18/01; *NEW YORK TIMES*, 9/15/01; CBS NEWS, 9/14/01; 9/11 COMMISSION REPORT, 6/17/04] Press accounts of when the first fighters reach Washington are highly contradictory. Early news accounts of fighters arriving from Andrews Air Force Base "within minutes," "a few moments," or "just moments" after the Pentagon crash appear to have been accounts of these Langley fighters, since they apparently arrive before Andrews fighters do. [*DAILY TELEGRAPH*, 9/16/01; *DENVER POST*, 9/11/01; ABC NEWS, 9/11/02] Yet other newspaper accounts inaccurately deny fighters from Andrews were deployed [*USA TODAY*, 9/16/01], and some deny Andrews even had fighters at all. [*USA TODAY*, 9/16/01 (B)] Defense officials initially claimed, "There were no military planes in the skies over Washington until 15 to 20 minutes after the Pentagon was hit"—in other words, 9:53 A.M. to 9:58 A.M. [*SEATTLE POST-INTELLIGENCER*, 9/14/01] ABC News reports that by 10:00 A.M., "Dozens of fighters are buzzing in the sky" over Washington. [ABC NEWS, 9/11/02] Whereas the *New York Times* reports, "In the White House Situation Room and at the Pentagon, the response seemed agonizingly slow. One military official recalls hearing words to the effect of, 'Where are the planes?' " The Pentagon insists it had air cover over its own building by 10 A.M., 15 minutes after the building was hit. However, witnesses, including a reporter for the *New York Times* who was headed toward the building, did not see any until closer to 11." [*NEW YORK TIMES*, 9/16/01 (B)] It is likely, though not completely certain, that fighters would have reached Washington before Flight 93 did, had the plane not crashed.

After 9:55 A.M.*: Langley Fighters Receive Vague Order to Protect White House

The Langley F-16s headed Washington are told that all planes in the U.S. have been ordered to land (that command was given at 9:45 A.M.). According to the *New York Times*, at some point after this, someone from the Secret Service gets on the radio and tells the pilots, "I want you to protect the White House at all costs." [*NEW YORK TIMES*, 10/16/01] F-16 pilot Honey (who is apparently Captain Craig Borgstrom) gives a similar, though less dramatic, account. At some point after the F-16s had set up a defensive perimeter over Washington, the lead pilot (again, Borgstrom) receives a garbled message about Flight 93 that isn't heard by the other two pilots. "The message seemed to convey that the White House was an important asset to protect." Honey says he is later told the message is, "Something like, 'Be aware of where it is, and it could be a target.'" Another pilot, codenamed Lou, says Honey tells him, "I think the Secret Service told me this." [*AMONG THE HEROES*, BY JERE LONGMAN, 8/02, P. 76] Both Lou and Honey state they are never given clear and direct orders to shoot down any plane that day. [*AMONG THE HEROES*, BY JERE LONGMAN, 8/02, P. 222]

9:56 A.M.*: Air Force One Gets Airborne Without Fighter Escort

President Bush departs from the Sarasota, Florida airport on Air Force One. [*NEW YORK TIMES*, 9/16/01 (B); *DAILY MAIL*, 9/8/02; *WASHINGTON POST*, 1/27/02; ASSOCIATED PRESS, 9/12/01; ABC NEWS, 9/11/02; 9/11 COMMISSION REPORT, 6/17/04; *WALL STREET JOURNAL*, 3/22/04; CBS NEWS, 9/11/02 (B); *DAILY TELEGRAPH*, 12/16/01] Amazingly, his plane takes off without any fighters protecting it. "The object seemed to be simply to get the President airborne and out of the way," says an administration official. [DAILY TELEGRAPH, 12/16/01] There are still 3,520 planes in the air over the U.S. [USA TODAY, 8/13/02 (B)] About half of the planes in the Florida region where Bush's plane is are still airborne. [ST. PETERSBURG TIMES, 9/7/02]

Air Force One

Apparently, fighters don't meet up with Air Force One until about an hour later. Counterterrorism "tsar" Richard Clarke claims to have heard around 9:50 A.M. from the bunker containing Vice President Cheney that fighter escort had been authorized. [AGAINST ALL ENEMIES, BY RICHARD CLARKE, 3/04, PP. 8–9]

After 9:56 A.M.*: Bush and Cheney Confer on Actions to Be Taken

After flying off in Air Force One, President Bush talks on the phone to Vice President Cheney. Cheney recommends that Bush authorize the military to shoot down any plane under control of the hijackers. "I said, 'You bet,'" Bush later recalls. "We had a little discussion, but not much." [NEWSDAY, 9/23/01; USA TODAY, 9/16/01; WASHINGTON POST, 1/27/02; CBS NEWS, 9/11/02] The 9/11 Commission claims that Cheney tells Bush three planes are still missing and one has hit the Pentagon. [9/11 COMMISSION REPORT, 6/17/04] Bush later says that he doesn't make any major decisions about how to respond to the 9/11 attacks until after Air Force One takes off, [WALL STREET JOURNAL, 3/22/04] which fits with this account of Bush approving shootdown authorization shortly after take off.

9:56–10:40 A.M.*: Air Force One Takes Off, Then Flies
in Circles While Bush and Cheney Argue

Air Force One takes off and quickly gains altitude. One passenger later says, "It was like a rocket. For a good ten minutes, the plane was going almost straight up." [CBS NEWS, 9/11/02 (B)] Once the plane reaches cruising altitude, it flies in circles. Journalists on board sense this because the television reception for a local station generally remains good. "Apparently Bush, Cheney, and the Secret Service argue over the safety of Bush coming back to Washington." [SALON, 9/12/01 (B); DAILY TELEGRAPH, 12/16/01] For much of the day Bush is plagued by connectivity problems in trying to call Cheney and others. He is forced to use an ordinary cell phone instead of his secure phone. [9/11 COMMISSION REPORT, 6/17/04]

9:57 A.M.: Passengers Begin Attempt to Regain Control of Flight 93

One of the hijackers in the cockpit asks if anything is going on, apparently meaning outside the cockpit. "Fighting," the other says. [AMONG THE HEROES, BY JERE LONGMAN, 8/02, P. 210] An analysis of the cockpit flight recording suggests that the passenger struggle actually starts in the front of the plane (where Mark Bingham and Tom Burnett are sitting) about a minute before a struggle in the back of the plane (where Todd Beamer is sitting). [OBSERVER, 12/2/01] Officials later theorize that the Flight 93 passengers reach the cockpit using a food cart as a battering ram and a shield. They claim digital enhancement of the cockpit voice recorder reveals the sound of plates and glassware crashing around 9:57 A.M. [NEWSWEEK, 11/25/01]

9:57 A.M. and After*: Passengers and Hijackers Struggle in the Flight 93 Cockpit

"In the cockpit! In the cockpit!" is heard. The hijackers are reportedly heard telling each other to hold the door. In English, someone outside shouts, "Let's get them." The hijackers are also praying "Allah o akbar" (God is great). One of the hijackers suggests shutting off the oxygen supply to the cabin (which apparently would not have had any effect since the plane was already below 10,000 feet). A hijacker says, "Should we finish?" Another one says, "Not yet." The sounds of the passengers get clearer, and in unaccented English "Give it to me!" is heard. "I'm injured," someone says in English. Then something like "roll it up" and "lift it up" is heard. Passengers' relatives believe this sequence proves that the passengers did take control of the plane. [MSNBC, 7/30/02; *DAILY TELEGRAPH*, 8/6/02; *NEWSWEEK*, 11/25/01; *OBSERVER*, 12/2/01; *AMONG THE HEROES*, BY JERE LONGMAN, 8/02, P. 270–271]

9:58 A.M.: Todd Beamer Ends Call; He and Others in Rear of Plane Join Takeover Attempt

Todd Beamer

Todd Beamer ends his long phone call with a Verizon phone company representative saying that they plan "to jump" the hijacker in the back of the plane who has the bomb. In the background, the phone operator already could hear an "awful commotion" of people shouting, and women screaming, "Oh my God," and "God help us." He lets go of the phone but leaves it connected. His famous last words are said to nearby passengers: "Are you ready guys? Let's roll" (alternate version: "You ready? Okay. Let's roll"). [*AMONG THE HEROES*, BY JERE LONGMAN, 8/02, P. 204; *NEWSWEEK*, 9/22/01; *PITTSBURGH POST-GAZETTE*, 10/28/01 (B)] Sounds of fighting in the back of the plane where Beamer is can be heard about a minute after such sounds in the front. [*OBSERVER*, 12/2/01]

9:58 A.M.: CeeCee Lyles Cries "They're Doing It"; Strange Sounds Follow

CeeCee Lyles

CeeCee Lyles says to her husband, "Aah, it feels like the plane's going down." Her husband Lorne says, "What's that?" She replies, "I think they're going to do it. They're forcing their way into the cockpit" (an alternate version says, "They're getting ready to force their way into the cockpit"). A little later she screams, then says, "They're doing it! They're doing it! They're doing it!" Her husband hears more screaming in the background, then he hears a "whooshing sound, a sound like wind," then more screaming, and then the call breaks off. [*AMONG THE HEROES*, BY JERE LONGMAN, 8/02, P. 180; *PITTSBURGH POST-GAZETTE*, 10/28/01; *PITTSBURGH POST-GAZETTE*, 10/28/01 (B)]

9:58 A.M.: Flight 93 Passengers Run to First Class

Sandy Bradshaw tells her husband, "Everyone's running to first class. I've got to go. Bye." She had been speaking with him since 9:50 A.M. [*PITTSBURGH POST-GAZETTE*, 10/28/01 (B); *BOSTON GLOBE*, 11/23/01]

9:58 A.M.: Ed Felt Said to Describe Explosion and White Smoke from Bathroom Call

A man dials emergency 9-1-1 from a bathroom on the plane, crying, "We're being hijacked, we're being hijacked!" [TORONTO SUN, 9/16/01] He then reports, "he heard some sort of explosion and saw white smoke coming from the plane and we lost contact with him." [ABC NEWS, 9/11/01 (B); ABC NEWS, 9/11/01 (C); ASSOCIATED PRESS, 9/12/01 (B)] One minute after the call begins, the line goes dead. [PITTSBURGH CHANNEL, 12/6/01] Investigators believe this was Edward Felt, the only passenger not accounted for on phone calls. He was sitting in first class, so he probably was in the bathroom near the front of the plane. At one point, he appears to have peeked out the bathroom door during the call. [AMONG THE HEROES, BY JERE LONGMAN, 8/02, PP. 193–194, 196] The mentions of smoke and explosions on the recording of his call are now denied. [AMONG THE HEROES, BY JERE LONGMAN, 8/02, P. 264] The person who took Felt's call is not allowed to speak to the media. [MIRROR, 9/13/02]

9:58 A.M.: Fighters to New York City Possibly Scrambled 56 Minutes Late, According to Giuliani and Early Reports

According to New York City Mayor Rudolph Giuliani's 9/11 Commission testimony in 2004, about one minute before the first WTC tower falls, he is able to reach the White House by phone. Speaking to Chris Henick, deputy political director to President Bush, Giuliani learns the Pentagon has been hit and he asks about fighter cover over New York City. Henick replies, "The jets were dispatched 12 minutes ago and they should be there very shortly, and they should be able to defend you against further attack." [9/11 COMMISSION, 5/19/04] If this is true, it means fighters scramble from the Otis base around 9:46 A.M., not at 8:52 A.M., as most other accounts have claimed. While Giuliani's account may seem wildly off, it is consistent with reports shortly after 9/11. In the first few days, acting Joint Chiefs of Staff Chairman Richard Myers, and a NORAD spokesman, Marine Corps Major Mike Snyder, claimed no fighters were scrambled anywhere until after the Pentagon was hit. [MYERS' CONFIRMATION TESTIMONY, 9/13/01; BOSTON GLOBE 9/15/01] This story only changed on the evening of September 14, 2001, when CBS reported, "contrary to early reports, U.S. Air Force jets did get into the air on Tuesday while the attacks were under way." [CBS NEWS, 9/14/01]

Before 9:59 A.M.*: Giuliani Apparently Told WTC Towers Will Collapse When Fire Chiefs Think Otherwise

Between 9:25 A.M. and 9:45 A.M., one senior New York fire chief recommends to the Fire Department Chief of Department that there might be a WTC collapse in a few hours, and, therefore, fire units probably shouldn't ascend much above the sixtieth floor (presumably this assumes the collapse would be gradual so those on lower floors would still have time to evacuate). This advice is not followed or not passed on. Apparently, no other senior fire chiefs mention or foresee the possibility of the WTC towers falling. [9/11 COMMISSION REPORT, 5/19/04] However, New York City Mayor Rudoph Giuliani recounts, "I went down to the scene and we set up headquarters at 75 Barclay Street, which was right there, with the police commissioner, the fire commissioner, the head of emergency management, and we were operating out of there when we were told that the World Trade Center was going to collapse. And it did collapse before we could actually get out of the building, so we were trapped in the building for ten, 15 minutes, and finally found an exit and got out, walked north, and took a lot of people with us." [ABC NEWS, 9/11/01 (D)] As can be seen by another account of similar events, this happens before the first WTC tower falls, not the second. [9/11 COMMISSION, 5/19/04] It is not clear who tells Giuliani to evacuate when no fire chiefs were considering the possibility of an imminent collapse.

9:59 A.M.*: Fighter over New York City
Never Receives Formal Shootdown Order

An F-16 flies over New York City on September 12, 2001. Smoke is still rising from the World Trade Center.

According to Major Daniel Nash, pilot of one of the two fighters first scrambled on 9/11 at 8:52 A.M., their fighters over New York City are never given a shootdown order by the military that day. He recalls that around the time of the collapse of the South Tower, "The New York controller did come over the radio and say if we have another hijacked aircraft we're going to have to shoot it down." [BBC, 9/1/02] However, he says this is an off-the-cuff personal statement, not connected to the chain of command. [CAPE COD TIMES, 8/21/02]

9:59 A.M.: White House Finally Requests Continuity of Government Plans,
Air Force One Escort, and Fighters for Washington

The 9/11 Commission reports, "An Air Force Lieutenant Colonel working in the White House Military Office [joins] the [NMCC] conference and state[s] that he had just talked to Deputy National Security Adviser Steve Hadley. The White House request[s]: (1) the implementation of Continuity of Government measures, (2) fighter escorts for Air Force One, and (3) the establishment of a fighter combat air patrol over

The South Tower of the World Trade Center collapses

Washington, D.C." [9/11 COMMISSION REPORT, 6/17/04] Counterterrorism "tsar" Richard Clarke gave the Continuity of Government orders a few minutes before from inside the White House. This is consistent with Bush's claim that he doesn't make any major decisions about the 9/11 attacks until shortly before 10:00 A.M.

9:59 A.M.: South Tower of WTC Collapses

The South Tower of the World Trade Center collapses. It was hit by Flight 175 at 9:03 A.M., 57 minutes earlier. [WASHINGTON POST, 9/12/01; MSNBC, 9/22/01; ASSOCIATED PRESS, 8/19/02 (B); ABC NEWS, 9/11/02; NEW YORK TIMES, 9/12/01 (B); USA TODAY, 12/20/01]

9:59 A.M.*: Clarke Told Some Hijackers Have al-Qaeda Connections

Counterterrorism "tsar" Richard Clarke is told in private by Dale Watson, counterterrorism chief at the FBI, "We got the passenger manifests from the airlines. We recognize some names, Dick. They're al-Qaeda." Clarke replies, "How the fuck did they get on board then?" He is told, "Hey, don't shoot the messenger, friend. CIA forgot to tell us about them." As they are talking about this, they see the first WTC tower collapse on television. [AGAINST ALL ENEMIES, BY RICHARD CLARKE, 3/04, PP. 13–14] Some hijacker names, including Mohamed Atta's, were identified on a reservations computer over an hour earlier.

Between 9:59–10:28 A.M.*: Firefighters Don't Hear Any Message to Evacuate North Tower

At some point between the collapse of the two WTC towers, it is claimed that fire chiefs order the firefighters to come down. It has not been reported exactly who issued this order or when. Witnesses claim that scores of firefighters, unaware of the danger, were resting on lower floors in the minutes before the second tower collapsed. "Some firefighters who managed to get out said they had no idea the other building had already fallen, and said that they thought that few of those who perished knew." At least 121 firefighters in the remaining tower die. The Fire Department blames a faulty radio repeater. However, the Port Authority claims later transcripts of radio communications show the repeaters worked. [NEW YORK TIMES, 11/9/02 (B)]

A person falling from the World Trade Center. Over fifty people jumped or fell from the North Tower, none from the South Tower.

After 9:59 A.M.: WTC Building 7 Evacuated; Exact Timing Unclear

According to a soldier on temporary duty there, WTC Building 7 is evacuated before the second tower was hit, i.e. before 9:03 A.M. [D.C. MILITARY, 10/18/01] However, a firefighter who arrived there after the second tower was hit was told that the building was being evacuated due to reports of a third plane [JEMS AND FIRERESCUE SUPPLEMENT, 3/02], indicating that two planes had already crashed.

After 9:59 A.M.*: WTC Building 7 Appears Damaged

WTC Building 7 appears to have suffered significant damage at some point after the WTC Towers had collapsed, according to firefighters at the scene. Firefighter Butch Brandies tells other firefighters that nobody is to go into Building 7 because of creaking and noises coming out of there. [FIREHOUSE MAGAZINE, 8/02] According to Deputy Chief Peter Hayden, there is a bulge in the southwest corner of the building between floors 10 and 13. [FIREHOUSE MAGAZINE, 4/02] Battalion Chief John Norman later recalls, "At the edge of the south face you could see that it was very heavily damaged." [FIREHOUSE MAGAZINE, 5/02] Deputy Chief Nick Visconti also later recalls recounts, "A big chunk of the lower

Damage to World Trade Center Building 7

floors had been taken out on the Vesey Street side." Captain Chris Boyle recalls, "On the south side of 7 there had to be a hole 20 stories tall in the building, with fire on several floors." [FIREHOUSE MAGAZINE, 8/02] The building will collapse hours later.

After 9:59 A.M.*: Clarke Orders Securing of Buildings, Harbors, and Borders

Some time after the first WTC tower collapse, counterterrorism "tsar" Richard Clarke orders all landmark buildings and all federal buildings in the U.S. evacuated. He also orders all harbors and borders closed. [AGAINST ALL ENEMIES, BY RICHARD CLARKE, 3/04, PP. 14–15] The Sears Tower in Chicago begins evacuation around 10:02 A.M. Other prominent buildings are slower to evacuate. [OTTAWA CITIZEN, 9/11/01]

10:00 A.M.: Flight 93 Transponder Gives Brief Signal

The transponder for Flight 93 briefly turns back on. The plane is at 7,000 feet. The transponder stays on until about 10:03 A.M. It is unclear why the transponder signal briefly returns. [MSNBC, 9/11/02 (B); GUARDIAN, 10/17/01]

10:00 A.M.*: Elizabeth Wainio Says "They're Rushing the Cockpit"

Flight 93 passenger Elizabeth Wainio says to her stepmother, "Mom, they're rushing the cockpit. I've got to go. Bye," then hangs up. This may have been a delayed reaction to events, since her stepmother says that in their ten-minute call, Elizabeth was in a trancelike state, appeared to have resigned herself to death, was breathing in a strange manner, and even said she felt she was leaving her body. The timing for this call is also approximate, and variously reported as taking place just before or just after 10:00 A.M. [MSNBC, 7/30/02; PITTSBURGH POST-GAZETTE, 10/28/01 (B)]

10:00 A.M.: Hijackers Respond to Passenger Revolt

According to the 9/11 Commission, the hijacker pilot, presumably Ziad Jarrah, has been rolling the plane sharply to the left and right in an attempt to prevent passengers from reaching the cockpit. At this time, he stabilizes the plane and asks another hijacker, "Is that it? Shall we finish it off?" Another voice answers, "No. Not yet. When they all come, we finish it off." The pilot starts pitching the nose of the airplane up and down. A few seconds later a passenger's voice can be heard saying, "In the cockpit. If we don't we'll die!" Another voice says, "Roll it!" which some speculate could be a reference to pushing a foot cart into the cockpit door. By 10:01, the pilot stops the pitching and says, "Allah o akbar! Allah o akbar!" (meaning "God is great"), then asks, "Is that it? I mean, shall we put it down?" Another hijacker responds, "Yes, put it in it, and pull it down." [NEW YORK TIMES, 7/22/04 (B); SAN FRANCISCO CHRONICLE, 7/23/04]

Between 10:00–10:06 A.M.*: Flight 93 Cell Call Listeners Hear Silence, Strange Sounds

During this time, there apparently are no calls from Flight 93. Several cell phones that are left on record only silence. For instance, although Todd Beamer does not hang up, nothing more is heard after he puts down the phone, suggesting things are quiet in the back of the plane. [*AMONG THE HEROES*, BY JERE LONGMAN, 8/02, P. 218] The only exception is Richard Makely, who listens to Jeremy Glick's open phone line after Glick goes to attack the hijackers. A reporter summarizes Makely explaining that, "The silence last[s] two minutes, then there [is] screaming. More silence, followed by more screams. Finally, there [is] a mechanical sound, followed by nothing." [*SAN FRANCISCO CHRONICLE*, 9/17/01] The second silence lasts between 60 and 90 seconds. [*AMONG THE HEROES*, BY JERE LONGMAN, 8/02, P. 219] Near the end of the cockpit voice recording, loud wind sounds can be heard. [CNN, 4/19/02; *AMONG THE HEROES*, BY JERE LONGMAN, 8/02, P. 270–271] "Sources claim the last thing heard on the cockpit voice recorder is the sound of wind—suggesting the plane had been holed." [*MIRROR*, 9/13/02] There was at least one passenger, Don Greene, who was a professional pilot. Another passenger, Andrew Garcia, was a former flight controller. [*NEWSWEEK*, 9/22/01; *PITTSBURGH POST-GAZETTE*, 10/28/01 (B); *DAILY TELEGRAPH*, 8/6/02]

Between 10:00–10:15 A.M.*: Bush and Cheney Said to Confer on Shootdown Orders, 9/11 Commission Doubts Their Account

According to a 9/11 Commission staff report, Vice President Cheney is told that a combat air patrol has been established over Washington. Cheney then calls President Bush to discuss the rules of engagement for the pilots. Bush authorizes the shoot down of hijacked aircraft at this time. [9/11 COMMISSION REPORT, 6/17/04] According to a *Washington Post* article, which places the call after 9:55 A.M., "Cheney recommended that Bush authorize the military to shoot down any such civilian airliners—as momentous a decision as the president was asked to make in those first hours." Bush then talks to Defense Secretary Rumsfeld to clarify the procedure, and Rumsfeld passes word down the chain of command. [*WASHINGTON POST*, 1/27/02] Cheney and Bush recall having this phone call, and National Security Adviser Rice recalls overhearing it. However, as the commission notes, "Among the sources that reflect other important events that morning there is no documentary evidence for this call, although the relevant sources are incomplete. Others nearby who were taking notes, such as the Vice President's Chief of Staff, [I. Lewis "Scooter"] Libby, who sat next to him, and [Lynne] Cheney, did not note a call between the President and Vice President immediately after the Vice President entered the conference room." The commission also apparently concludes that no evidence exists to support the claim that Bush and Rumsfeld talked about such procedures at this time. [9/11 COMMISSION REPORT, 6/17/04] Commission Chairman Thomas Kean says, "The phone logs don't exist, because they evidently got so fouled up in communications that the phone logs have nothing. So that's the evidence we have." Commission Vice Chairman Lee Hamilton says of the shootdown order, "Well, I'm not sure it was carried out." [9/11 COMMISSION, 6/17/04 (C); *NEW YORK DAILY NEWS*, 6/18/04] *Newsweek* reports that it "has learned that some on the commission staff were, in fact, highly skeptical of the vice president's account and

Vice President Cheney talks to President Bush from his bunker (exact time is unknown)

made their views clearer in an earlier draft of their staff report. According to one knowledgeable source, some staffers "flat out didn't believe the call ever took place." According to a 9/11 Commission staffer, the report "was watered down" after vigorous lobbying from the White House. [NEWSWEEK, 6/20/04] An account by Canadian Captain Mike Jellinek (who was overseeing NORAD's Colorado headquarters, where he claims to hear Bush give a shootdown order), as well as the order to empty the skies of aircraft, appears to be discredited. [TORONTO STAR, 12/9/01]

Between 10:00–10:30 A.M.*: Rumsfeld Returns to the
Pentagon and Speaks to Bush; Rumsfeld's Whereabouts Murky

Rumsfeld returns from the Pentagon crash site "by shortly before or after 10:00 A.M." Then he has "one or more calls in my office, one of which was with the President," according to his testimony before the 9/11 Commission. [9/11 COMMISSION, 6/17/04 (B)] The commission later concludes that Rumsfeld's call with President Bush has little impact: "No one can recall any content beyond a general request to alert forces." The possibility of shooting down hijacked planes is not mentioned. [9/11 COMMISSION REPORT, 6/17/04] Then Rumsfeld goes to the Executive Support Center before finally entering the NMCC at 10:30 A.M. Acting Joint Chiefs of Staff Chairman Richard Myers repeats all these details. [9/11 COMMISSION, 6/17/04 (B)] The Executive Support Center has secure video facilities [WASHINGTON TIMES, 2/23/04], so it is possible Rumsfeld joins or rejoins the video conference that counterterrorism "tsar" Richard Clarke claims Rumsfeld is a part of much of the morning.

10:01 A.M.: Local Pilot Sees Flight 93 Rocking Back and Forth

Bill Wright is piloting a small plane when a flight controller asks him to look around outside his window, according to his later claims. He sees Flight 93 three miles away—close enough that Wright can see the United Airlines colors. Flight control asks him the plane's altitude, and then commands him to get away from the plane and land immediately. Wright sees the plane rock back and forth three or four times before he flies from the area. [PITTSBURGH CHANNEL, 9/19/01] According to the 9/11 Commission, the FAA Command Center tells FAA headquarters that a nearby plane had seen Flight 93 "waving his wings." The commission says, "The aircraft had witnessed the radical gyrations in, what we believe was the hijackers' effort to defeat the passenger assault." [9/11 COMMISSION REPORT, 6/17/04] This presumably is a reference to Wright.

10:01 A.M.: Toledo Fighters Ordered Scrambled
Toward Flight 1989 Instead of Flight 93

The FAA orders F-16 fighters to scramble from Toledo, Ohio. Although the base has no fighters on standby alert status, it manages to put fighters in the air 16 minutes later, a "phenomenal" response time—but still ten minutes after the last hijacked plane has crashed. [TOLEDO BLADE, 12/9/01] The 9/11 Commission concludes these fighters, along with fighters from Michigan, are scrambled to go after Delta

Flight 1989. (Delta Flight 1989 was not out of contact with air traffic controllers, and was not hijacked.) Meanwhile, according to the 9/11 Commission, no fighters are ever scrambled to intercept Flight 93. [9/11 COMMISSION REPORT, 6/17/04]

10:02 A.M.: Secret Service Warns Cheney Hijackers Are Headed Toward Washington

Vice President Cheney and other leaders now in the White House bunker begin receiving reports from the Secret Service of a presumably hijacked aircraft heading toward Washington. The Secret Service is getting this information about Flight 93 through links to the FAA. However, they are looking at a projected path, not an actual radar return, so they do not realize that the plane crashes minutes later. [9/11 COMMISSION REPORT, 6/17/04]

10:02 A.M.: Cockpit Voice Recording Ends Early?

The cockpit voice recording of Flight 93 was recorded on a 30-minute reel, which means that the tape is continually overwritten and only the final 30-minutes of any flight would be recorded. The government later permits relatives to hear this tape. Apparently, the version of the tape played to the family members begins at 9:31 A.M. and runs for 31 minutes, ending one minute before, according to the government, the plane crashes. [CNN, 4/19/02; *AMONG THE HEROES*, BY JERE LONGMAN, 8/02, P. 206-207] The *New York Observer* comments, "Some of the relatives are keen to find out why, at the peak of this struggle, the tape suddenly stops recording voices and all that is heard in the last 60 seconds or so is engine noise. Had the tape been tampered with?" [NEW YORK OBSERVER, 6/17/04]

10:02 A.M.: 9/11 Commission Details the Moments Before Flight 93 Crash

According to the 9/11 Commission, a Flight 93 hijacker says, "Pull, it down! Pull it down!" The airplane rolls onto its back as one of the hijackers shouts, "Allah o akbar! Allah o akbar!" The commission comments, "The hijackers remained at the controls but must have judged that the passengers were only seconds from overcoming them." Presumably the plane crashes seconds later. [SAN FRANCISCO CHRONICLE, 7/23/04] However, there are questions as to whether the voice recording actually ends at this time. Furthermore, there is a near complete disconnect between these quotes and the quotes given in previous accounts of what the cockpit recording revealed. For instance, in other accounts, passenger voices saying, "Give it to me!," "I'm injured," and "Roll it up" or "Lift it up" is heard just before the recording ends. [MSNBC, 7/30/02; DAILY TELEGRAPH, 8/6/02; NEWSWEEK, 11/25/01; OBSERVER, 12/2/01; AMONG THE HEROES, BY JERE LONGMAN, 8/02, PP. 270-271]

10:03 A.M.: NMCC Learns of Flight 93 Hijacking, NORAD Still Not Told

According to the 9/11 Commission, the NMCC learns about the Flight 93 hijacking at this time. Since the FAA has not yet been patched in to the NMCC's conference call, the news comes from the White House. The White House learned about it from the Secret Service, and the Secret Service learned about it from the FAA. NORAD apparently is still unaware. Four minutes later, a NORAD representative on the conference call states, "NORAD has no indication of a hijack heading to Washington, D.C., at this time." [9/11 COMMISSION REPORT, 6/17/04]

10:03–10:10 A.M.*: Flight 93 Crashes;
Seven-Minute Discrepancy on Exact Timing of Crash

Exactly when Flight 93 crashes remains unclear. According to NORAD, Flight 93 crashes at 10:03 A.M. [NORAD, 9/18/01] The 9/11 Commission gives an exact time of 11 seconds after 10:03 A.M. They claim this "time is supported by evidence from the staff's radar analysis, the flight data recorder, NTSB [National Transportation Safety Board] analysis, and infrared satellite data." They do note that "[t]he precise crash time has been the subject of some dispute." [9/11 COMMISSION REPORT, 6/17/04] However, a seismic study authorized by the U.S. Army to determine when the plane crashed concluded that the crash happened at 10:06:05 A.M. [SEISMIC STUDY, 2002; SAN FRANCISCO CHRONICLE, 12/9/02] The discrepancy is so puzzling that the *Philadelphia Daily News* publishes an article on the issue, titled "Three-Minute Discrepancy in Tape." It notes that leading seismologists agree on the 10:06 A.M. time, give or take a couple of seconds. [PHILADEL-PHIA DAILY NEWS, 9/16/02] The *New York Observer* notes that, in addition to the seismology study, "The FAA gives a crash time of 10:07 A.M. In addition, the *New York Times*, drawing on flight controllers in more than one FAA facility, put the time at 10:10 A.M. Up to a seven-minute discrepancy? In terms of an air disaster, seven minutes is close to an eternity. The way our nation has historically treated any airline tragedy is to pair up recordings from the cockpit and air-traffic control and parse the timeline down to the hundredths of a second. However, as [former Inspector General of the Transportation Department] Mary Schiavo points out, "We don't have an NTSB (National Transportation Safety Board) investigation here, and they ordinarily dissect the timeline to the thousandth of a second." [NEW YORK OBSERVER, 2/11/04] (Note that this work uses 10:06 A.M. as the most likely time of the crash, detailed below).

Before 10:06 A.M.*: Witnesses See Flight 93 Rocking Wings as It Slowly Descends

In the tiny town of Boswell, about ten miles north and slightly to the west of Flight 93's crash site, Rodney Peterson and Brandon Leventry notice a passenger jet lumbering through the sky at about 2,000 feet. They realize such a big plane flying so low in that area is odd. They see the plane dip its wings sharply to the left then to the right. The wings level off and the plane keeps flying south, continuing to descend slowly. Five minutes later, they hear news that the plane has crashed. Other witnesses also later describe the plane flying east-southeast, low, and wobbly. [AMONG THE HEROES, BY JERE LONGMAN, 8/02, PP. 205–206; NEW YORK TIMES, 9/14/01] "Officials initially say that it looks like the plane was headed south when it hit the ground." [CLEVELAND NEWSCHANNEL 5, 9/11/01]

Before 10:06 A.M.*: Fighters Trailing Flight 93?

Shortly after 9/11, NORAD claims that there is a fighter 100 miles away from Flight 93 when it crashes. However, no details, such as who the pilot is, or which base or direction the fighter is coming from, are ever given by NORAD. [NORAD, 9/18/01] CBS television reports that two F-16 fighters were tailing the flight and within 60 miles of the plane when it crashed. [CBS NEWS, 9/14/02; *INDEPENDENT*, 8/13/02] Shortly after 9/11, an unnamed New England flight controller ignores a ban on controllers speaking to the media; he reportedly claims "that an F-16 fighter closely pursued Flight 93 . . . the F-16 made 360-degree turns to remain close to the commercial jet." He adds that the fighter pilot "must've seen the whole thing." He reportedly learned this from speaking to controllers nearer to the crash. [ASSOCIATED PRESS, 9/13/01; *NASHUA TELEGRAPH*, 9/13/01] However, a Cleveland flight controller named Stacey Taylor later claims to have not seen any fighters on radar around the crash. [MSNBC, 9/11/02 (B)] Major General Paul Weaver, director of the Air National Guard, had previously claimed that no military planes were sent after Flight 93. [*SEATTLE TIMES*, 9/16/01] A different explanation by ABC News says, "The closest fighters are two F-16 pilots on a training mission from Selfridge Air National Guard Base" near Detroit, Michigan. These are ordered after Flight 93, even though they reportedly aren't armed with any weapons. It is claimed they are supposed to crash into Flight 93 if they cannot persuade it to land. [ABC NEWS, 8/30/02; ABC NEWS, 9/11/02] However, these fighters apparently are not diverted from Michigan until after Flight 93 crashes at 10:06 A.M.

Before 10:06 A.M.*: Witnesses See Flight 93 Flying Erratically and Making Strange Noises

Numerous eyewitnesses see and hear Flight 93 just before its crash:

- Terry Butler, at Stoystown: He sees the plane come out of the clouds, low to the ground. "It was moving like you wouldn't believe. Next thing I knew it makes a heck of a sharp, right-hand turn." It banks to the right and appears to be trying to climb to clear one of the ridges, but it continues to turn to the right and then veers behind a ridge. About a second later it crashes. [*ST. PETERSBURG TIMES*, 9/12/01]

- Ernie Stuhl, the mayor of Shanksville: "I know of two people—I will not mention names—that heard a missile. They both live very close, within a couple of hundred yards . . . This one fellow's served in Vietnam and he says he's heard them, and he heard one that day." He adds that based on what he has learned, F-16s were "very, very close." [*PHILADELPHIA DAILY NEWS*, 11/15/01]

Accounts of the plane making strange noises:

- Laura Temyer of Hooversville: "I didn't see the plane but I heard the plane's engine. Then I heard a loud thump that echoed off the hills and then I heard the plane's engine. I heard two more loud thumps and didn't hear the plane's engine anymore after that." (She insists that people she knows in state law enforcement have privately told her the plane was shot down, and that decompression sucked objects from the aircraft, explaining why there was a wide debris field.) [*PHILADELPHIA DAILY NEWS*, 11/15/01]

- Charles Sturtz, a half-mile from the crash site: The plane is heading southeast and has its engines running. No smoke can be seen. "It was really roaring, you know. Like it was trying to go someplace, I guess." [WPXI CHANNEL 11, 9/13/01]

- Michael Merringer, two miles from the crash site: "I heard the engine gun two different times and then I heard a loud bang . . ." [ASSOCIATED PRESS, 9/12/01 (B)]

- Tim Lensbouer, 300 yards away: "I heard it for ten or 15 seconds and it sounded like it was going full bore." [PITTSBURGH POST-GAZETTE, 9/12/01 (B)]

Accounts of the plane flying upside down:

- Rob Kimmel, several miles from the crash site: He sees it fly overhead, banking hard to the right. It is 200 feet or less off the ground as it crests a hill to the southeast. "I saw the top of the plane, not the bottom." [AMONG THE HEROES, BY JERE LONGMAN, 8/02, P. 210–211]

- Eric Peterson of Lambertsville: He sees a plane flying overhead unusually low. The plane seemed to be turning end-over-end as it dropped out of sight behind a tree line. [PITTSBURGH POST-GAZETTE, 9/12/01]

- Bob Blair of Stoystown: He sees the plane spiraling and flying upside down, not much higher than the treetops, before crashing. [DAILY AMERICAN, 9/12/01]

Accounts of a sudden plunge and more strange sounds:

- An unnamed witness says he hears two loud bangs before watching the plane take a downward turn of nearly 90 degrees. [CLEVELAND NEWSCHANNEL 5, 9/11/01]

- Tom Fritz, about a quarter-mile from the crash site: He hears a sound that "wasn't quite right" and looks up in the sky. "It dropped all of a sudden, like a stone," going "so fast that you couldn't even make out what color it was." [ST. PETERSBURG TIMES, 9/12/01]

- Terry Butler, a few miles north of Lambertsville: "It dropped out of the clouds." The plane rose slightly, trying to gain altitude, then "it just went flip to the right and then straight down." [PITTSBURGH POST-GAZETTE, 9/12/01]

- Lee Purbaugh, 300 yards away: "There was an incredibly loud rumbling sound and there it was, right there, right above my head—maybe 50 feet up. . . . I saw it rock from side to side then, suddenly, it dipped and dived, nose first, with a huge explosion, into the ground. I knew immediately that no one could possibly have survived." [INDEPENDENT, 8/13/02]

Upside down and a sudden plunge:

- Linda Shepley: She hears a loud bang and sees the plane bank to the side. [ABC NEWS, 9/11/01] She sees the plane wobbling right and left, at a low altitude of roughly 2,500 feet, when suddenly the right wing dips straight down, and the plane plunges into the earth. She says she has an unobstructed view of Flight 93's final two minutes. [PHILADELPHIA DAILY NEWS, 11/15/01]

- Kelly Leverknight in Stony Creek Township of Shanksville: "There was no smoke, it just went straight down. I saw the belly of the plane." It sounds like it is flying low, and it's heading east. [DAILY AMERICAN, 9/12/01; ST. PETERSBURG TIMES, 9/12/01]

- Tim Thornsberg, working in a nearby strip mine: "It came in low over the trees and started wobbling. Then it just rolled over and was flying upside down for a few seconds . . . and then it kind of stalled and did a nose dive over the trees." [WPXI CHANNEL 11, 9/13/01]

Some claim that these witness accounts support the idea that Flight 93 is hit by a missile. [PHILADELPHIA DAILY NEWS, 11/15/01] While this theory certainly can be disputed, it is worth noting that some passenger planes hit by missiles continued to fly erratically for several minutes before crashing. For instance, a Korean Airline 747 was hit by two Russian missiles in 1983, yet continued to fly for two more minutes. [KAL COCKPIT VOICE RECORDER TRANSCRIPT]

Before 10:06 A.M.*: Flight 93 Breaks Up Prior to Crash?

Flight 93 apparently starts to break up before it crashes, because debris is found very far away from the crash site. [PHILADELPHIA DAILY NEWS, 11/15/01] The plane is generally obliterated upon landing, except for one half-ton piece of engine found some distance away. Some reports indicate that the engine piece was found over a mile away. [INDEPENDENT, 8/13/02] The FBI reportedly acknowledges that this piece was found "a considerable distance" from the crash site. [PHILADELPHIA DAILY NEWS, 11/16/01] Later, the FBI will cordon off a three-mile wide area around the crash, as well as another area six to eight miles from the initial crash site. [CNN, 9/13/01] One story calls what happened to this engine "intriguing, because the heat-seeking, air-to-air Sidewinder missiles aboard an F-16 would likely target one of the Boeing 757's two large engines." [PHILADELPHIA DAILY NEWS, 11/15/01] Smaller debris fields are also found two, three, and eight miles away from the main crash site. [INDEPENDENT, 8/13/02; MIRROR, 8/13/02] Eight miles away, local media quote residents speaking of a second plane in the area and burning debris falling from the sky. [REUTERS, 9/13/01 (C)] Residents outside Shanksville reported "discovering clothing, books, papers, and what appeared to be human remains. Some residents said they collected bags-full of items to be turned over to investigators. Others reported what appeared to be crash debris floating in Indian Lake, nearly six miles from the immediate crash scene. Workers at Indian Lake Marina said that they saw a cloud of confetti-like debris descend on the lake and nearby farms minutes after hearing the explosion. . . ." [PITTSBURGH POST-GAZETTE, 9/13/01] Moments after the crash, Carol Delasko initially thinks someone had blown up a boat on Indian Lake: "It just looked like confetti raining down all over the air above the lake." [PITTSBURGH TRIBUNE-REVIEW, 9/14/01] Investigators say that far-off wreckage "probably was spread by the cloud created when the plane crashed and dispersed by a ten mph southeasterly wind." [DELAWARE NEWS JOURNAL, 9/16/01] However, much of the wreckage is found sooner than that wind could have carried it, and not always southeast.

10:06 A.M.: Flight 93 Crashes into Pennsylvania Countryside

Flight 93 crashes into an empty field just north of the Somerset County Airport, about 80 miles southeast of Pittsburgh, 124 miles or 15 minutes from Washington, D.C. [NORAD, 9/18/01; GUARDIAN, 10/17/01; PITTS-

The Flight 93 crash site later in the morning of 9/11

BURGH POST-GAZETTE, 10/28/01; MSNBC, 9/3/02; USA TODAY, 8/13/02; ASSOCIATED PRESS, 8/19/02 (B); CNN, 9/12/01] A U.S. Army authorized seismic study times the crash at five seconds after 10:06 A.M. [SEISMIC STUDY, 2002; SAN FRANCISCO CHRONICLE, 12/9/02] As mentioned previously, the timing of this crash is disputed and it may well occur at 10:03 A.M., 10:07 A.M., or 10:10 A.M.

Before and After 10:06 A.M.*: Witnesses See
Low-Flying, Small White Jet at Flight 93 Crash

A second plane, described "as a small, white jet with rear engines and no discernible markings," is seen by at least six witnesses flying low and in erratic patterns, not much above treetop level, over the crash site within minutes of the United flight crashing. [INDEPENDENT, 8/13/02]

- Lee Purbaugh: "I didn't get a good look but it was white and it circled the area about twice and then it flew off over the horizon." [MIRROR, 9/13/02]

- Susan Mcelwain: Less than a minute before the Flight 93 crash rocked the countryside, she sees a small white jet with rear engines and no discernible markings swoop low over her minivan near an intersection and disappear over a hilltop, nearly clipping the tops of trees lining the ridge. [BERGEN RECORD, 9/14/01] She later adds, "There's no way I imagined this plane—it was so low it was virtually on top of me. It was white with no markings but it was definitely military, it just had that look. It had two rear engines, a big fin on the back like a spoiler on the back of a car and with two upright fins at the side. I haven't found one like it on the Internet. It definitely wasn't one of those executive jets. The FBI came and talked to me and said there was no plane around. . . . But I saw it and it was there before the crash and it was 40 feet above my head. They did not want my story—nobody here did." [MIRROR, 9/13/02]

- Dennis Decker and/or Rick Chaney, say: "As soon as we looked up [after hearing the Flight 93 crash], we saw a midsized jet flying low and fast. It appeared to make a loop or part of a circle, and then it turned fast and headed out." Decker and Chaney described the plane as a Learjet type, with engines mounted near the tail and painted white with no identifying markings. "It was a jet plane, and it had to be flying real close when that 757 went down. If I was the FBI, I'd find out who was driving that plane." [BERGEN RECORD, 9/14/01]

- Jim Brandt sees a small plane with no markings stay about one or two minutes over the crash site before leaving. [PITTSBURGH CHANNEL, 9/12/01]

- Tom Spinelli: "I saw the white plane. It was flying around all over the place like it was looking for something. I saw it before and after the crash." [MIRROR, 9/13/02]

The FBI later says this was a Fairchild Falcon 20 business jet, directed after the crash to fly from 37,000 feet to 5,000 feet and obtain the coordinates for the crash site to help rescuers. [PITTSBURGH POST-GAZETTE, 9/16/01; PITTSBURGH CHANNEL, 9/15/01] The FBI also says there was a C-130 military cargo aircraft flying at 24,000 feet about 17 miles away, but that plane wasn't armed and had no role in the crash. [PITTSBURGH CHANNEL, 9/15/01; PITTSBURGH POST-GAZETTE, 9/16/01] Note that this is the same C-130 that flies very close to Flight 77 right as that planes crashes into the Pentagon.

Looking from above straight down onto the Flight 93 crash site

After 10:06 A.M.*: Fighter Said to Fly Past Flight 93 Crash Site

"Up above, a fighter jet streak[s] by," just after Flight 93 crashes, according to ABC News. [ABC NEWS, 9/15/02] It isn't clear what evidence this ABC News claim is based on. There are other accounts of a fighter or fighters in the area before the crash, mentioned previously.

After 10:06 A.M.*: Michigan Fighters Diverted Toward Flight 1989

At some point after Flight 93 crashes, NORAD diverts "unarmed Michigan Air National Guard fighter jets that happened to be flying a training mission in northern Michigan since the time of the first attack." [ASSOCIATED PRESS, 8/30/02] The 9/11 Commission concludes these fighters and fighters from Ohio are scrambled for Delta Flight 1989, a flight that was never hijacked or even out of contact. Meanwhile, reportedly, no fighters are scrambled after Flight 93 at all, which has already crashed. [9/11 COMMISSION REPORT, 6/17/04]

After 10:06 A.M.*: Clarke Updated on Fighter Situation, Told Flight 93 Still Headed Toward Washington

Counterterrorism "tsar" Richard Clarke is told by an aide, "Secret Service reports a hostile aircraft ten minutes out." Two minutes later, he is given an update: "Hostile aircraft eight minutes out." In actual fact, when Flight 93 crashes at 10:06 A.M., it's still about 15 minutes away from Washington. Clarke is also told that there are 3,900 aircraft still in the air over the Continental U.S. (which is roughly accurate); four of those aircraft are believed to be piloted by terrorists (which is inaccurate by this time). Joint Chiefs of Staff Vice Chairman Richard Myers then reports, "We have three F-16s from Langley over the Pentagon. Andrews is launching fighters from the D.C. Air National Guard. We have fighters aloft from the Michigan Air National Guard, moving east toward a potential hostile over Pennsylvania. Six fighters from Tyndall and Ellington are en route to rendezvous with Air Force One over Florida. They will escort it to Barksdale." [AGAINST ALL ENEMIES, BY RICHARD CLARKE, 3/04, PP. 8–9, NORAD, 9/18/01] However, fighters do not meet up with Air Force One until about an hour later. Franklin Miller, a senior national security official who worked alongside Clarke on 9/11, and another official there, later fail to recall hearing any aide warning that a plane could be only minutes away. [NEW YORK TIMES, 3/30/04 (B)] The time of this incident is not given, but the Michigan fighters are not diverted until after 10:06 A.M. If this takes place after 10:06 A.M., it would parallel similar warnings about Flight 93 after it has already crashed provided to Vice President Cheney elsewhere in the White House.

After 10:06 A.M.*: Clarke Told of Flight 93 Crash

Counterterrorism "tsar" Richard Clarke is told by an aide, "United 93 is down, crashed outside of Pittsburgh. It's odd. Appears not to have hit anything much on the ground." The timing of this event is unclear. [AGAINST ALL ENEMIES, BY RICHARD CLARKE, 3/04, PP. 14–15]

After 10:06 A.M.*: Bush, Told of Flight 93 Crash, Wonders If It Was Shot Down

President Bush is told that Flight 93 crashed a few minutes after it happened, but the exact timing of this notice is unclear. Because of Vice President Cheney's earlier order, he asks, "Did we shoot it down or did it crash?" Several hours later, he is assured that it crashed. [WASHINGTON POST, 1/27/02]

After 10:06 A.M.*: Al-Qaeda Agents Heard Saying "We've Hit the Targets"

According to *Newsweek*, "shortly after the suicide attacks," U.S. intelligence picks up communications among bin Laden associates relaying the message: "We've hit the targets." [NEWSWEEK, 9/13/01]

10:07 A.M.: Cleveland Flight Control Updates NEADS

According to the 9/11 Commission, NEADS finally receives a call from Cleveland flight control about Flight 93. Cleveland passes on the plane's last known latitude and longitude. NEADS is unable to locate it on radar because it has already crashed. By the commission's account, this is NORAD's first notification about the Flight 93 hijacking, even though Cleveland realized Flight 93 was hijacked at 9:32 A.M., 35 minutes earlier, and notified FAA headquarters at 9:34 A.M., 33 minutes earlier. [9/11 COM-MISSION REPORT, 6/17/04]

10:08 A.M.: Guards Surround the White House
Armed agents deploy around the White House. [CNN, 9/12/01]

10:08 A.M.: Military Cargo Pilot Asked to Verify Flight 93 Crash
Cleveland flight controller Stacey Taylor has asked a nearby C-130 pilot to look at Flight 93's last position and see if they can find anything. Remarkably, this C-130 pilot is the same pilot who was asked by flight control to observe Flight 77 as it crashed in Washington earlier. He tells Taylor that he saw smoke from the crash shortly after the hijacked plane went down. [9/11 COMMISSION REPORT, 6/17/04; MSNBC, 9/11/02 (B); GUARDIAN, 10/17/01]

10:08 A.M.: FAA Informed That Flight 93 Has Crashed,
Confirms Crash Nine Minutes Later
According to the 9/11 Commission, the FAA Command Center reports to FAA headquarters at this time that Flight 93 has crashed in the Pennsylvania countryside. "It hit the ground. That's what they're speculating, that's speculation only." The Command Center confirms that Flight 93 crashed at 10:17 A.M. [9/11 COMMISSION REPORT, 6/17/04; MSNBC, 9/11/02 (B); GUARDIAN, 10/17/01]

10:10 A.M.: Langley Fighters Told They Cannot Shoot Down Hijacked Planes
According to the 9/11 Commission, the NEADS Mission Crew Commander is sorting out the orders given to the Langley fighter pilots. The Commander does not know that Flight 93 had been heading toward Washington nor that it had crashed. He explicitly instructs the Langley fighters that they cannot shoot down aircraft—they have "negative clearance to shoot" aircraft over Washington. Authorization to shoot down hijacked civilian aircraft only reaches NEADS at 10:31 A.M. Even then, the authorization is not passed on to the pilots. [9/11 COMMISSION REPORT, 6/17/04]

10:10 A.M.*: Military Put on High Alert
All U.S. military forces are ordered to Defcon Three (or Defcon Delta), "The highest alert for the nuclear arsenal in 30 years." [ABC NEWS, 9/11/02; CNN, 9/4/02; AGAINST ALL ENEMIES, BY RICHARD CLARKE, 3/04, P. 15; DAILY TELE-GRAPH, 12/16/01] Rumsfeld claims that he makes the recommendation, but it is hard to see how he can do this, at least at this time. He later asserts that he discusses the issue with acting Joint Chiefs of Staff Chairman Richard Myers in the NMCC first. However, they do not arrive at the PEOC until about 10:30 A.M. [9/11 COMMISSION, 3/23/04] At 10:15 A.M., the massive blast doors to U.S. Strategic Command, headquarters for NORAD in Cheyenne Mountain, Colorado, are closed for the first time in response to

the high alert. [BBC, 9/1/02; *AVIATION WEEK AND SPACE TECHNOLOGY*, 6/3/02] In another account, acting Joint Chiefs of Staff Chairman Richard Myers gives the Defcon order by himself. President Bush later contradicts both accounts, asserting that he gives the order. [*WALL STREET JOURNAL*, 3/22/04]

Between 10:10–10:15 A.M.*: Cheney, Told That Flight 93 Is Still Heading to Washington, Orders It Shot Down

The Secret Service, viewing projected path information about Flight 93, rather than actual radar returns, does not realize that Flight 93 has already crashed. Based on this erroneous information, a military aide tells Vice President Cheney and others in the White House bunker, that the plane is 80 miles away from Washington. Cheney is asked for authority to engage the plane, and he quickly provides authorization. The aide returns a few minutes later and says the plane is 60 miles out. Cheney again gives authorization to engage. A few minutes later and presumably after the flight has crashed or been shot down, White House Deputy Chief of Staff Joshua Bolten suggests Cheney contact President Bush to confirm the engage order. Bolten later tells the 9/11 Commission that he had not heard any prior discussion on the topic with Bush, and wanted to make sure Bush knew. Apparently, Cheney calls Bush and obtains confirmation. [9/11 COMMISSION REPORT, 6/17/04] However, there is controversy over whether Bush approved a shootdown before this incident or whether Cheney gave himself the authority to make the decision on the spot. As *Newsweek* notes, it is moot point in one sense, since the decision was made on false data and there is no plane to shoot down. [*NEWSWEEK*, 6/20/04]

10:13 A.M.: Washington Buildings Evacuate

More prominent buildings in Washington begin evacuation. The United Nations building evacuates first; many federal buildings follow later. [CNN, 9/12/01; *NEW YORK TIMES*, 9/12/01] Counterterrorism "tsar" Richard Clarke apparently began arranging these evacuations a short time before this. [*AGAINST ALL ENEMIES*, BY RICHARD CLARKE, 3/04, PP. 14–15]

10:13–10:23 A.M.*: Projected Flight 93 Arrival into Washington; Could It Have Been Shot Down?

The 9/11 Commission later concludes that if Flight 93 had not crashed, it would probably have reached Washington around this time. The commission notes that there are only three fighters over Washington at this time, all from Langley, Virginia. However, the pilots of these fighters were never briefed about why they were scrambled. As the lead pilot explained, "I reverted to the Russian threat . . . I'm thinking cruise missile threat from the sea. You know, you look down and see the Pentagon burning and I thought the bastards snuck one by us. . . . You couldn't see any airplanes, and no one told us anything." The pilots knew their mission was to identify and divert aircraft flying within a certain radius of Washington, but did not know that the threat came from hijacked planes. In addition, the commission notes that NEADS did not know where Flight 93 was when it crashed, and wonders if they would have determined its location and passed it on the pilots before the plane reached Washington. They conclude, "NORAD officials have maintained that they would have intercepted and shot down United 93. We are not so sure." [9/11 COMMISSION REPORT, 6/17/04]

10:14 A.M.: Cheney Gives 'Engage' Order to NMCC to Relay to Fighters

According to the 9/11 Commission, beginning at this time, the White House repeatedly conveys to the NMCC that Vice President Cheney confirmed fighters were cleared to engage the inbound aircraft if they could verify that the aircraft was hijacked. However, the authorization fails to reach the pilots. [9/11 COMMISSION REPORT, 6/17/04]

10:15 A.M.: Pentagon Section Collapses

The front section of the Pentagon that had been hit by Flight 77 collapses. [CNN, 9/12/01; *NEW YORK TIMES*, 9/12/01; TELEVISION ARCHIVE, WDCC 10:00] A few minutes prior to its collapse, firefighters saw warning signs and sounded a general evacuation tone. No firefighters were injured. [*NFPA JOURNAL*, 11/1/01]

10:15 A.M.: NEADS Learns Flight 93 Is Down

According to the 9/11 Commission, NEADS calls Washington flight control at this time. Asked about Flight 93, flight control responds, "He's down." It is clarified that the plane crashed "somewhere up northeast of Camp David." [9/11 COMMISSION REPORT, 6/17/04]

10:17 A.M.*: FAA Out of the Loop; Finally Joins NMCC Teleconference

The National Military Command Center (NMCC) has been conducting an interagency teleconference, to coordinate the nation's response to the hijackings since 9:29 A.M. Yet the 9/11 Commission reports that the FAA is unable to join the call until this time, apparently due to technical difficulties. NORAD asked three times before the last hijacked plane crashed for the FAA to provide a hijacking update to the teleconference. None were given, since no FAA representative was there. When an FAA representative finally joins in, that person has no proper experience, no access to decision makers, and no information known to senior FAA officials at the time. Furthermore, the highest-level Defense Department officials rely on this conference and do not talk directly with senior FAA officials. As a result, the leaders of NORAD and the FAA are effectively out of contact with each other during the entire crisis. [9/11 COMMISSION REPORT, 6/17/04]

10:20 A.M.: United Headquarters Learns Flight 93 Has Crashed

United Airlines headquarters receives confirmation that Flight 93 has crashed from the airport manager in Johnstown, Pennsylvania. [9/11 COMMISSION, 1/27/04] Cleveland flight control had confirmation of the crash at 10:08 A.M.

10:24 A.M.*: All International Flights into U.S. Ordered Diverted by FAA

Jane Garvey, head of the FAA, orders the diversion of all international flights to the U.S. Most flights are diverted to Canada. [TIME, 9/14/01; MSNBC, 9/22/01; CNN. 9/12/01; NEW YORK TIMES, 9/12/01]

10:28 A.M.: WTC North Tower Collapses

The World Trade Center's North Tower collapses. It was hit by Flight 11 at 8:46, 102 minutes earlier. [MSNBC, 9/22/01; CNN, 9/12/01; NEW YORK TIMES, 9/12/01; ASSOCIATED PRESS, 8/19/02 (B); SEISMIC STUDY, 2002: The death toll could have been much worse—an estimated 15,000 people made it out of the WTC to safety after 8:46 A.M. [ST. PETERSBURG TIMES, 9/8/02]

The North Tower collapses in a matter of seconds

Before 10:30 A.M.*: Myers Finally Enters NMCC; Prior Whereabouts Disputed

Acting Joint Chiefs of Staff Chairman Richard Myers enters the NMCC, though exactly when this happens remains unclear. According to his own statements, he was on Capitol Hill, in the offices of Senator Max Cleland (D) from the time just prior to the first WTC attack until around the time the Pentagon was hit. However, counterterrorism "tsar" Richard Clarke claims Myers takes part in a video conference for much of the morning. Defense Secretary Rumsfeld, who enters the NMCC around 10:30 A.M., claims that as he entered, Myers "had just returned from Capitol Hill." [DEFENSE DEPARTMENT, 3/23/04] In Myers' testimony before the 9/11 Commission, he fails to mention where he was or what he was doing from the time of the Pentagon crash until about 10:30 A.M., except to say, "I went back to my duty station. And we—what we started doing at that time was to say, 'OK, we've had these attacks. Obviously they're hostile acts. Not sure at that point who perpetrated them.'" [9/11 COMMISSION, 6/17/04 (B)] These discrepancies in Myers' whereabouts remain unresolved.

10:30 A.M.*: Missing Rumsfeld Finally Enters NMCC

Defense Secretary Donald Rumsfeld, missing for at least 30 minutes, finally enters the NMCC, where the military's response to the 9/11 attacks is being coordinated. [9/11 COMMISSION REPORT, 6/17/04; CNN, 9/4/02] Rumsfeld later claims that he only started to gain a situational awareness of what was happening after arriving at the NMCC. [9/11 COMMISSION REPORT, 6/17/04] Rumsfeld was in his office only 200 feet away from the NMCC until the Pentagon crash at 9:37 A.M. His activities during this period are unclear. He went outside to the Flight 77 crash site and then stayed somewhere else in the Pentagon until his arrival at the NMCC. Brigadier General Montague Winfield later says, "For 30 minutes we couldn't find him. And just as we began to worry, he walked into the door of the [NMCC]." [ABC NEWS, 9/11/02] Winfield himself apparently only shows up at the NMCC around 10:30 A.M. as well.

10:30 A.M.*: Medevac Helicopter Provides
Scare for Bunkered Cheney, Others

Vice President Cheney and others in the White House bunker are given a report of another airplane heading toward Washington. Cheney's Chief of Staff, I. Lewis "Scooter" Libby, later states, "We learn that a plane is five miles out and has dropped below 500 feet and can't be found; it's missing." Believing they only have a minute or two before the plane crashes into Washington, Cheney orders fighters to engage the plane, saying, "Take it out." However, reports that this is another hijacking are mistaken. It is learned later that day that a Medevac helicopter five miles away was mistaken for a hijacked plane. [*NEWSWEEK*, 12/31/01; 9/11 COMMISSION REPORT, 6/17/04]

10:31 A.M.: Military and Law Enforcement Flights Resume

The FAA allows "military and law enforcement flights to resume (and some flights that the FAA can't reveal that were already airborne)." All civilian, military, and law enforcement flights were ordered at 9:26 A.M. to land as soon as reasonably possible. [*TIME*, 9/14/01] Civilian flights remain banned until September 13. Note that the C-130 cargo plane that witnessed the Flight 77 crash and which came upon the Flight 93 crash site right after it had crashed was apparently not subject to the grounding order issued about an hour earlier.

10:31 A.M.: NEADS Does Not Pass Along NORAD Shoot Down Order

According to the 9/11 Commission, NORAD Commander Major General Larry Arnold instructs his staff to broadcast the following message over a NORAD chat log: "10:31 Vice President [Cheney] has cleared us to intercept tracks of interest and shoot them down if they do not respond, per CONR CC [General Arnold]." NEADS first learns of the shootdown order from this message. However, NEADS does not pass the order to the fighter pilots in New York City and Washington. NEADS leaders later say they do not pass it on because they are unsure how the pilots should proceed with this guidance. [9/11 COMMISSION REPORT, 6/17/04] The pilots flying over New York City claim they are never given a formal shoot-down order that day.

10:32 A.M.: Air Force One Threatened? Some Doubt Entire Story

Vice President Cheney reportedly calls President Bush and tells him of a threat to Air Force One and that it will take 40–90 minutes to get a protective fighter escort in place. Many doubt the existence of this threat. For instance, Representative Martin Meehan (D) says, "I don't buy the notion Air Force One was a target. That's just PR, that's just spin." [*WASHINGTON TIMES*, 10/8/02] A later account calls the threat "completely untrue," and says Cheney probably made the story up. A well-informed, anonymous Washington official says, "It did two things for [Cheney]. It reinforced his argument that the President should stay out of town, and it gave George W. an excellent reason for doing so." [*DAILY TELEGRAPH*, 12/16/01]

10:35 A.M.*: Bush Heads for Louisiana on Air Force One

Air Force One turns toward a new destination of Barksdale Air Force Base, near Shreveport, Louisiana, in response to a decision that Bush should not go directly to Washington. [CBS NEWS, 9/11/02 (B); *WASHINGTON POST*, 1/27/02]

Before 10:36 A.M.*: Andrews Fighters Ordered to Shoot
Down Threatening Planes Over Washington

A Secret Service agent again contacts Andrews Air Force Base and commands, "Get in the air now!" It's not clear if this is treated as an official scramble order, or how quickly fighters respond to it. According to fighter pilot Lieutenant Colonel Marc Sasseville, almost simultaneously, a call from someone else in the White House declares the Washington area "a free-fire zone. That meant we were given authority to use force, if the situation required it, in defense of the nation's capital, its property and people." [*AVIATION WEEK AND SPACE TECHNOLOGY*, 9/9/02] Apparently, this second call is made to General David Wherley, flight commander of the Air National Guard at Andrews. He had contacted the Secret Service after hearing reports that it wanted fighters airborne. One Secret Service agent, using two telephones at once, relays instructions to Wherley from another Secret Service agent in the White House who has been given the instructions from Vice President Cheney. Wherley's fighters are to protect the White House and shoot down any planes that threaten Washington. Wherley gives Lieutenant Colonel Marc Sasseville, lead pilot, the authority to decide whether to execute a shootdown, President Bush and Vice President Cheney later claim they were not aware that any fighters had scrambled from Andrews at the request of the Secret Service. [9/11 COMMISSION REPORT, 6/17/04] Sasseville and Lucky will take off at 10:42 A.M.

10:38 A.M.*: Fighters Training in
North Carolina Relaunch from Andrews

The 9/11 Commission claims that the first fighters from Andrews Air Force Base scramble at this time and are flying patrol over Washington by 10:45 A.M. [9/11 COMMISSION REPORT, 6/17/04] The three F-16s flying on a training mission in North Carolina, 200 miles away have finally been recalled to their home base at Andrews. As soon as lead pilot Major Billy Hutchison lands and checks in via radio, he is told to take off again immediately. His fighter apparently has no weapons whatsoever. The two other fighters only have training rounds for their guns, and very little fuel. "Hutchison was probably airborne shortly after the alert F-16s from Langley arrive over Washington, although [the] pilots admit their timeline-recall 'is fuzzy.' The officer who sent Hutchison off "told him to 'do exactly what ATC asks you to do.' Primarily, he was to go ID [identify] that unknown [aircraft] that everybody was so excited about [Flight 93]. He blasted off and flew a standard departure route, which took him over the Pentagon." Flight 93 crashed half an hour before this; it is unclear how the Andrews base could still not know it crashed by this time. The pilots later say that, had all else failed, they would have rammed into Flight 93, had they reached it in time. [*AVIATION WEEK AND SPACE TECHNOLOGY*, 9/9/02]

10:39 A.M.: Cheney Brings Rumsfeld Up to Date, But Errs on Pilot Knowledge About Shootdown Order

Vice President Cheney tries to bring Defense Secretary Rumsfeld up to date over the NMCC's conference call, as Rumsfeld has just arrived there minutes before. Cheney explains that he has given authorization for hijacked planes to be shot down and that this has been told to the fighter pilots. Rumsfeld asks, "So we've got a couple of aircraft up there that have those instructions at the present time?" Cheney replies, "That is correct. And it's my understanding they've already taken a couple of aircraft out." Then Rumsfeld says, "We can't confirm that. We're told that one aircraft is down but we do not have a pilot report that they did it." Cheney is incorrect that this command has reached the pilots. [9/11 COMMISSION REPORT, 6/17/04]

10:42 A.M.*: Andrews Fighters Finally Take Off, but Without Missiles

Two F-16s take off from Andrews Air Force Base lightly armed with nothing more than "hot" guns and non-explosive training rounds. Lead pilot Lieutenant Colonel Marc Sasseville flies one; the other pilot is only known by the codename Lucky. [AVIATION WEEK AND SPACE TECHNOLOGY, 9/9/02] These fighters had been at another base that morning, waiting to be armed with AIM-9 missiles, a process that takes about an hour. [AVIATION WEEK AND SPACE TECHNOLOGY, 6/3/02] Since they took off without the missiles, presumably they could have taken off unarmed much earlier. (The first call for them to scramble came not long after 9:00 A.M.). Two more F-16s, armed with AIM-9 missiles, take off twenty-seven minutes later, at 11:09 A.M. These are apparently piloted by Major Dan Caine and Captain Brandon Rasmussen. [AVIATION WEEK AND SPACE TECHNOLOGY, 9/9/02; 9/11 COMMISSION, 6/17/04 (B)] F-16s from Richmond, Virginia and Atlantic City, New Jersey arrive over Washington a short time later. [AVIATION WEEK AND SPACE TECHNOLOGY, 9/9/02] The Andrews fighters are apparently the only fighters in the U.S. scrambled before 11:00 with official shootdown authorization, but the first Andrews fighters into the air have no missiles. It is unclear if the Andrews fighters relaunching a few minutes earlier had shootdown orders, but they had no weapons either. It appears the Andrews fighters launching at 11:09 A.M. are the first fighters in the U.S. with both shootdown orders and missiles to use.

10:42 A.M.*: Status of Three Planes Unknown; False Rumors Persist of More Terrorist Activity

Around this time (roughly), the FAA tells the White House that it still cannot account for three planes in addition to the four that have crashed. It takes the FAA another hour and a half to account for these three aircraft. [TIME, 9/14/01] Vice President Cheney later says, "That's what we started working off of, that list of six, and we could account for two of them in New York. The third one we didn't know what had happened to. It turned out it had hit the Pentagon, but the first reports on the Pentagon attack suggested a helicopter and then later a private jet." [LOS ANGELES TIMES, 9/17/01] Amongst false rumors during the day are reports of a bomb aboard a United Airlines jet that just landed in Rockford, Illinois. "Another plane disappears from radar and might have crashed in Kentucky. The reports are so serious that [FAA head Jane] Garvey notifies the White House that there has been another crash. Only later does she learn the reports are erroneous." [USA TODAY, 8/13/02 (B)]

10:53 A.M.: Election Postponed

New York's primary elections, already in progress, are postponed. [CNN, 9/12/01]

10:55 A.M.*: Air Force One Takes Evasive Action from False Alarm

Colonel Mark Tillman, pilot of Air Force One, is told there is a threat to President Bush's plane. Tillman has an armed guard placed at his cockpit door while the Secret Service double-checks the identity of everyone on board. Air traffic controllers warn that a suspect airliner is dead ahead, according to Tillman: "Coming out of Sarasota there was one call that said there was an airliner off our nose that they did not have contact with." Tillman takes evasive action, pulling his plane high above normal traffic. [CBS NEWS, 9/11/02 (B)] Reporters on board notice the rise in elevation. [DALLAS MORNING NEWS, 8/28/02; SALON, 9/12/01 (B)] The report is apparently a false alarm.

Between 10:55–11:41 A.M.*: Fighter Escort Finally Reaches Air Force One? Reports Conflict

No fighters escort President Bush's Air Force One until around this time, but accounts conflict. At 10:32 A.M., Vice President Cheney said it would take until about 11:10 A.M. to 12:00 A.M. to get a fighter escort to Air Force One. [WASHINGTON POST, 1/27/02] However, according to one account, around 10:00 A.M., Air Force One "is joined by an escort of F-16 fighters from a base near Jacksonville, Florida." [DAILY TELEGRAPH, 12/16/01] Another report states, "At 10:41 [A.M.] . . . Air Force One headed toward Jacksonville to meet jets scrambled to give the presidential jet its own air cover." [NEW YORK TIMES, 9/16/01 (B)]

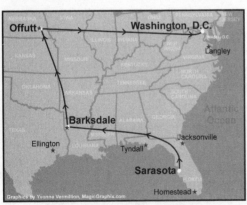

The travels of President Bush on Air Force One. Relevant air force bases are marked.

But apparently, when Air Force One takes evasive action around 10:55 A.M., there is still no fighter escort. NORAD commander Major General Larry Arnold later says, "We scrambled available airplanes from Tyndall [near Tallahassee and not near Jacksonville, Florida] and then from Ellington in Houston, Texas," but he does not say when this occurs. [CODE ONE MAGAZINE, 1/02] In yet another account, two F-16s eventually arrive, piloted by Shane Brotherton and Randy Roberts, from Ellington, not from any Florida base. [CBS NEWS, 9/11/02] The *St. Petersburg Times*, after interviewing people on Air Force One, estimate the first fighters, from Texas, arrive between

11:00 and 11:20. [ST. PETERSBURG TIMES] By 11:30 A.M., there are six fighters protecting Air Force One. [SARA-SOTA MAGAZINE, 9/19/01] The BBC, however, reports that the Ellington, Texas fighters are scrambled at 11:30 A.M., and quotes ABC reporter Ann Compton, inside Air Force One, saying fighters appear out the windows at 11:41 A.M. [BBC, 9/1/02] Given that two of the seven bases said to have fighters on alert on 9/11 are in Florida (Homestead Air Station, 185 miles from Sarasota; and Tyndall Air Station, 235 miles from Sarasota), why a fighter escort does not reach Air Force One earlier remains unclear. Philip Melanson, author of a book on the Secret Service, comments, "I can't imagine by what glitch the protection was not provided to Air Force One as soon as it took off. I would have thought there'd be something in place whereby one phone call from the head of the security detail would get the fighters in the air immediately." [ST. PETERSBURG TIMES, 7/4/04]

President Bush (center, bending) and others look out the windows of Air Force One as their fighter escort arrives

11:00 A.M.*: Skyscrapers, Tourist Attractions Closed

More skyscrapers and tourist attractions are evacuated, including Walt Disney World, Philadelphia's Liberty Bell and Independence Hall, Seattle's Space Needle, and the Gateway Arch in St. Louis. [TIMES UNION, 9/11/01]

11:00 A.M.*: All Flights over U.S. Soil Complying with Controllers

The FAA Command Center is told that all the flights over the United States are accounted for and pilots are complying with controllers. There are 923 planes still in the air over the U.S. Every commercial flight in U.S. airspace—about a quarter of the planes still in the air—is within 40 miles of its destination. Others are still over the oceans, and many are heading toward Canada. [USA TODAY, 8/13/02 (B)]

11:00 A.M.*: Customs Claims to Determine the Names of All 19 Hijackers

Robert Bonner, the head of Customs and Border Protection, later testifies, "We ran passenger manifests through the system used by Customs—two were hits on our watch list of August 2001." (This is presumably a reference to hijackers Khalid Almihdhar and Nawaf Alhazmi, watch-listed on August 23, 2001.) "And by looking at the Arab names and their seat locations, ticket purchases and other passenger information, it didn't take a lot to do a rudimentary link analysis. Customs officers were able to ID 19 probable hijackers within 45 minutes. I saw the sheet by 11 A.M. And that analysis did indeed correctly identify the terrorists." [NEW YORK OBSERVER, 2/11/04] However, Bonner appears to be at least somewhat incorrect: for two days after the attacks the FBI believes there are only 18 hijackers, and the original list contains some erroneous Arab-sounding names on the flight manifests, such as Adnan Bukhari and Ameer Bukhari. [CNN, 9/13/01 (D)] Some hijacker names, including Mohamed Atta's, were identified on a reservations computer around 8:30 A.M., and Richard Clarke was told some of the names were al-Qaeda around 10:00 A.M.

11:08 A.M.: Korean Air Flight Mistakenly Believed Hijacked

A message sent from Korean Air Flight 85 is misinterpreted to indicate a possible hijacking. At 1:24 P.M., the pilots accidentally issue a hijacking alert as the plane nears Alaska on its way to Anchorage. Two fighters tail the plane, and notify it that it will be shot down unless it avoids populated areas. Strategic sites are evacuated across Alaska. The plane eventually lands safely in Whitehorse, Canada, at 2:54 P.M. [USA TODAY, 8/12/02 (B)]

11:30 A.M.: Television Speculation al-Qaeda Is Responsible

General Wesley Clark, former Supreme Commander of NATO, says on television, "This is clearly a coordinated effort. It hasn't been announced that it's over. . . . Only one group has this kind of ability and that is Osama bin Laden's." [OTTAWA CITIZEN, 9/11/01]

11:30 A.M.*: Congressmen Meet with Bush

Two congressmen, Dan Miller and Adam Putnam, are on Air Force One. They've been receiving periodic updates on the crisis from President Bush's adviser Karl Rove. At this time, they're summoned forward to meet with the president. Bush points out the fighter escort, F-16s from a base in Texas, has now arrived. He says that a threat had been received from someone who knew the plane's code name. As mentioned above, there are doubts that any such threat ever occurred. [ST. PETERSBURG TIMES, 7/4/04]

11:45 A.M.: Air Force One Lands at Louisiana Air Force Base

Air Force One lands at Barksdale Air Force Base near Shreveport, Louisiana. "The official reason for landing at Barksdale was that President Bush felt it necessary to make a further statement, but it isn't unreasonable to assume that—as there was no agreement as to what the President's movements should be—it was felt he might as well be on the ground as in the air."

[SALON, 9/12/01 (B); NEW YORK TIMES, 9/16/01 (B); DAILY TELEGRAPH, 12/16/01; CBS NEWS, 9/11/02]

12:00 Noon*: Bush Provided Tight Security in Louisiana

President Bush arrives at the Barksdale Air Force Base headquarters in a Humvee escorted by armed outriders. Reporters and others are not allowed to say where they are. Bush remains in this location for approximately one hour, recording a brief message and talking on the phone. [DAILY TELEGRAPH, 12/16/01]

President Bush arrives at Barksdale Air Force Base in Louisiana

12:00 Noon*: Senator Hatch Repeats Intelligence Community's Conclusion That Osama Bin Laden Is Responsible

Senator Orrin Hatch (R), a member of both the Senate Intelligence and Judiciary Committees, says he has just been "briefed by the highest levels of the FBI and of the intelligence community." He says,

"They've come to the conclusion that this looks like the signature of Osama bin Laden, and that he may be the one behind this." [SALON, 9/12/01 (B)]

12:05 P.M.: Rumsfeld Finds Evidence of al-Qaeda Role Not Good Enough

CIA Director Tenet tells Defense Secretary Rumsfeld about an intercepted phone call from earlier in the day at 9:53 A.M. An al-Qaeda operative talked of a fourth target just before Flight 93 crashed. Rumsfeld wrote notes to himself at the time. According to CBS, "Rumsfeld felt it was 'vague,' that it 'might not mean something,' and that there was 'no good basis for hanging hat.' In other words, the evidence was not clear-cut enough to justify military action against bin Laden." [CBS NEWS, 9/4/02] More evidence suggesting an al-Qaeda link comes several hours later.

12:15 P.M.: Border Crossings Security Tightened

It is announced that U.S. borders with Canada and Mexico are on the highest state of alert, but no decision has been made about closing borders. [CNN, 9/12/01]

12:16 P.M.: U.S. Airspace Cleared of All Civilian Aircraft

U.S. airspace is clear of aircraft except for military and emergency flights. Only a few transoceanic flights are still landing in Canada. [USA TODAY, 8/12/02 (C)] At 12:30 P.M., the FAA reports about 50 (non-civilian) planes still flying in U.S. airspace, but none are reporting problems. [CNN, 9/12/01; NEW YORK TIMES, 9/12/01]

12:36 P.M.: Bush Records Second Speech; Aired About 30 Minutes Later

President Bush records a short speech that is played by the networks at 1:04 P.M. [SALON, 9/12/01 (B); WASHINGTON TIMES, 10/8/02] In a speech at the Louisiana base, President Bush announces that security measures are being taken and says: "Make no mistake, the United States will hunt down and punish those responsible for these cowardly acts." [MSNBC, 9/22/01; CNN, 9/12/01; NEW YORK TIMES, 9/12/01] He also states, "Freedom itself was attacked this morning by a faceless coward. And freedom will be defended." [ABC NEWS, 9/11/02]

12:58 P.M.*: Bush Argues with Cheney, Others About Where He Should Go Next

President Bush spends most of his time at Barksdale Air Force Base arguing on the phone with Vice President Cheney and others over where he should go next. "A few minutes before 1 P.M.," he agrees to fly to Nebraska. As earlier, there are rumors of a "credible terrorist threat" to Air Force One that are said to prevent his return to Washington. [DAILY TELEGRAPH, 12/16/01]

1:02 P.M.: Rumsfeld Calls for War

Defense Secretary Rumsfeld later claims that he says to President Bush on the phone, "This is not a criminal action. This is war." [WASHINGTON TIMES, 2/23/04]

1:02 P.M.*: Giuliani Orders Evacuation of Southern Manhattan

New York Mayor Rudolph Giuliani orders an evacuation of Manhattan south of Canal Street. [MSNBC, 9/22/01; ASSOCIATED PRESS, 8/19/02 (B)]

1:04 P.M.: Military on High Alert Worldwide

President Bush announces that the U.S. military has been put on high alert worldwide. [CNN, 9/12/01; ASSOCIATED PRESS, 8/19/02 (B)] Apparently, this occurs in a televised speech that was actually recorded half an hour earlier.

1:27 P.M.: State of Emergency in Washington

A state of emergency is declared in Washington. [CNN, 9/12/01; NEW YORK TIMES, 9/12/01]

1:30 P.M.*: Air Force One Leaves Louisiana; Flies to Nebraska

President Bush leaves Louisiana on Air Force One, and flies to Nebraska's Offutt Air Force Base, where the U.S. Strategic Command is located. [DAILY TELEGRAPH, 12/16/01; SALON, 9/12/01 (B); MSNBC, 9/22/01; CNN, 9/12/01] He travels with Chief of Staff Andrew Card, senior adviser Karl Rove, communications staffers Dan Bartlett, Ari Fleischer, Gordon Johndroe, and a small group of reporters. [SALON, 9/12/01 (B)]

1:44 P.M.: U.S. Military Deployed at Sea, in Skies

The Pentagon announces that aircraft carriers and guided missile destroyers have been dispatched toward New York and Washington. Around the country, more fighters, airborne radar (AWACs), and refueling planes are scrambling. NORAD is on its highest alert. [MSNBC, 9/22/01; CNN, 9/12/01]

2:00 P.M.*: Fighter Pilot Told Flight 93 Was Shot Down

F-15 fighter pilot Major Daniel Nash returns to base around this time, after chasing Flight 175 and patrolling the skies over New York City. He says that when he gets out of the plane, "he [is] told that a military F-16 had shot down a fourth airliner in Pennsylvania ..." [CAPE COD TIMES, 8/21/02; AVIATION WEEK AND SPACE TECHNOLOGY, 6/3/02]

2:40 P.M.*: Rumsfeld Wants to Blame Iraq

Defense Secretary Rumsfeld is provided information from the CIA indicating that three of the hijackers were suspected al-Qaeda operatives. Notes composed by aides who were with Rumsfeld in the National Military Command Center on 9/11 are leaked nearly a year later. According to the notes, information shows, "One guy is [an] associate of [USS] *Cole* bomber." (This is a probable reference to Khalid Almihdhar or Nawaf Alhazmi.) Rumsfeld has also been given information indicating an al-Qaeda operative had advanced details of the 9/11 attack. According to the aide's notes, Rumsfeld wants the "best info fast. Judge whether good enough hit S.H. [Saddam Hussein] at same time. Not only UBL. [Osama bin Laden] Go massive. Sweep it all up. Things related and not." [CBS NEWS, 9/4/02; *A PRETEXT FOR WAR*, BY JAMES BAMFORD, 6/04, P. 285]

2:50 P.M.*: Bush Arrives in Nebraska; Enters Strategic Command Center

Air Force One lands at Offutt Air Force Base near Omaha, Nebraska. President Bush stays on the plane for about ten minutes before entering the United States Strategic Command bunker at 3:06 P.M. [SALON, 9/12/01] Bush is taken into a bunker far underground designed to withstand a nuclear blast. There, he uses an advanced strategic command and communications center to teleconference with other top leaders. [*DAILY TELEGRAPH*, 12/16/01; *WASHINGTON TIMES* 10/8/02; *DAILY MAIL*, 9/8/02; SALON, 9/12/01; ASSOCIATED PRESS, 8/19/02 (B)]

3:00 P.M.*: Bush Meets with Top Officials via Video Conference Call

President Bush begins a video conference call from a bunker beneath Offutt Air Force Base. He and Chief of Staff Andrew Card visually communicate directly with Vice President Cheney, National Security Adviser Rice, Defense Secretary Rumsfeld, Deputy Secretary of State Richard Armitage, CIA Director Tenet, Transportation Secretary Norman Mineta, counterterrorism "tsar" Richard Clarke, and others. [*DAILY TELEGRAPH*, 12/16/01; *WASHINGTON TIMES*, 10/8/02; ABC NEWS, 9/11/02] According to Clarke, Bush begins the meeting by saying, "I'm coming back to the White House as soon as the plane is fueled. No discussion." Clarke leads a quick review of what has already occurred, and issues that need to be quickly addressed. CIA Director Tenet states that al-Qaeda is clearly behind the 9/11 attacks. Defense Secretary Rumsfeld states that about 120 fighters are now above U.S. cities. [*AGAINST ALL ENEMIES*, BY RICHARD CLARKE, 3/04, PP. 21–22] The meeting ends at 4:15 P.M. [*DAILY TELE-GRAPH*, 12/16/01; *WASHINGTON TIMES*, 10/8/02]

President Bush takes part in a video teleconference at Offutt Air Force Base. Chief of Staff Andy Card sits on his left, and Admiral Richard Mies sits on his right.

3:55 P.M.: Bush Said to Be at Undisclosed Location

White House adviser Karen Hughes briefly speaks to the media and says President Bush is at an undisclosed location, taking part in a video conference. This is possibly the only in-person media appearance by any Bush administration official since the attacks and until a news conference by Defense Secretary Rumsfeld at 6:40 P.M. [CNN, 9/12/01]

4:00 P.M.: CNN Blames bin Laden for Attacks

CNN reports U.S. officials say there are "good indications" that bin Laden is involved in the attacks, based on "new and specific" information developed since the attacks. [CNN, 9/12/01]

4:00 P.M.*: Bush Determined to Return to Washington; Rove Later Misinforms Public About Threat to Bush

President Bush has just told his advisers that he is returning to Washington as soon as the plane is fueled—"No discussion" [AGAINST ALL ENEMIES, BY RICHARD CLARKE, 3/04, PP. 21–22] Yet, Bush adviser Karl Rove later says that at this time President Bush is hesitant to return to Washington because, "They've accounted for all four [hijacked] planes, but they've got another, I think, three or four or five planes still outstanding." [NEW YORKER, 9/25/01] However, the FAA points out there are no such reports and that Bush had been quickly informed when domestic U.S. skies were completely cleared at 12:16 P.M. [WALL STREET JOURNAL, 3/22/04]

Several small fires burn inside World Trade Center Building 7

4:10 P.M.: WTC Building 7 Burning

World Trade Center Building 7 is reported to be on fire. [CNN, 9/12/01]

After 4:15 P.M.*: Leaders Determine to Crush Taliban

After President Bush leaves his video conference, other top leaders continue to discuss what steps to take. Counterterrorism "tsar" Richard Clarke asks what to do about al-Qaeda, assuming they are behind the attacks. Deputy Secretary of State Richard Armitage states, "Look, we told the Taliban in no uncertain terms that if this happened, it's their ass. No difference between the Taliban and al-Qaeda now. They both go down." Regarding Pakistan, the Taliban's patrons, Armitage says, "Tell them to get out of the way. We have to eliminate the sanctuary." [AGAINST ALL ENEMIES, BY RICHARD CLARKE, 3/04, PP. 22–23]

4:30 P.M.*: WTC Building 7 Area Is Evacuated

The area around WTC Building 7 is evacuated at this time. [KANSAS CITY STAR, 3/28/04] New York fire department chief officers, who have surveyed the building, have determined it is in danger of collapsing. Several senior firefighters have described this decision-making process. According to fire chief Daniel Nigro, "The biggest decision we had to make was to clear the area and create a collapse zone around the severely damaged [WTC Building 7]. A number of fire officers and companies assessed the damage to the building. The appraisals indicated that the building's integrity was in serious doubt." [FIRE ENGINEERING, 9/02]

4:33 P.M.*: Air Force One Leaves Nebraska; Heads Toward Washington

President Bush leaves Offutt Air Force Base in Nebraska for Washington. [MSNBC, 9/22/01; CNN, 9/12/01; DAILY TELEGRAPH, 12/16/01; WASHINGTON TIMES, 10/8/02]

5:20 P.M.*: WTC Building 7 Collapses; Cause Remains Unclear

Building 7 of the WTC complex, a 47-story tower, collapses. No one is killed. [MSNBC, 9/22/01; CNN, 9/12/01; WASHINGTON POST, 9/12/01; ASSOCIATED PRESS, 8/19/02 (B)] Many questions will arise over the cause of this collapse in the coming weeks and months. Building 7, which was not hit by an airplane, is the first modern, steel-reinforced high-rise to collapse because of fire. [*CHICAGO TRIBUNE*, 11/29/01; STANFORD REPORT, 12/3/01; *NEW YORK TIMES*, 3/2/02] Some later suggest that that diesel fuel stored in several tanks on the premises may have contributed to the building's collapse. The building contained a 6,000-gallon tank between its first and second floors and another four tanks, holding as much as 36,000 gallons, below ground level. There were also three smaller tanks on higher floors. [*CHICAGO TRIBUNE*, 11/29/01; *NEW YORK TIMES*, 2/3/02; *NEW YORK OBSERVER*, 3/25/02; FEMA STUDY, 5/1/02] However, the cause of the collapse is uncertain. A 2002 government report concludes: "The specifics of the fires in WTC 7 and how they caused the building to collapse remain unknown at this time. Although the total diesel fuel on the premises contained massive potential energy, the best hypothesis has only a low probability of occurrence." [FEMA STUDY, 5/1/02] Some reports indicate that the building may have been deliberately destroyed. Shortly after the collapse, CBS news anchor Dan Rather comments that the collapse is "reminiscent of . . . when a building was deliberately destroyed by well-placed dynamite to knock it down." [CBS NEWS, 9/11/01] In a PBS documentary broadcast in 2002, the World Trade Center's lease-holder Larry Silverstein talks about a phone call from the Fire Department commander he had on 9/11. Silverstein recalls saying to the commander about the building: "You know, we've had such terrible loss of life, maybe the smartest thing to do is pull it. And they made that decision to pull and then we watched the building collapse." [PBS, 9/10/02] It is unclear what Silverstein meant by the phrase "decision to pull."

World Trade Center Building 7 has collapsed.

6:54 P.M.*: Bush Returns to White House

President Bush arrives at the White House, after exiting Air Force One at 6:42 P.M. and flying across Washington in a helicopter. [ABC NEWS, 9/11/02; SALON, 9/12/01 (B); *WASHINGTON TIMES*, 10/8/02; CNN, 9/12/01; *DAILY TELEGRAPH*, 12/16/01; ASSOCIATED PRESS, 8/19/02 (B)]

7:00 P.M.*: Powell Returns from Peru

Secretary of State Powell returns to Washington from Lima, Peru. He is finally able to speak to President Bush for the first time since the 9/11 attacks began when they both arrive at the White House at about the same time. Powell later says of his flight, "And the worst part of it, is that because of the communications problems that existed during that day, I couldn't talk to anybody in Washington." [ABC NEWS, 9/11/02] The *Daily Telegraph* later theorizes, "Why so long? In the weeks before September 11, Washington was full of rumors that Powell was out of favor and had been quietly relegated to the sidelines . . ." [*DAILY TELEGRAPH*, 12/16/01]

President Bush addresses the nation from the White House

8:30 P.M.: Bush Gives Third Speech to Nation, Declares "Bush Doctrine"

President Bush addresses the nation on live television. [CNN, 9/12/01] In what will later be called the Bush Doctrine, he states, "We will make no distinction between the terrorists who committed these acts and those who harbor them." [WASHINGTON POST, 1/27/02]

9:00 P.M.*: Bush Meets with Advisers, Declares War Without Barriers

President Bush meets with his full National Security Council in the PEOC beneath the White House for about 30 minutes. He then meets with a smaller group of key advisers. Bush and his advisers have

President Bush meets with the National Security Council in a bunker below the White House. In the far row from left to right, are Attorney General Ashcroft, President Bush, Chief of Staff Andrew Card, CIA Director Tenet, and counterterrorism "tsar" Clarke. In the near row, Secretary of State Colin Powell can be seen waving his hand, and National Security Advisor Rice sits to his right.

already decided bin Laden is behind the attacks. CIA Director Tenet says that al-Qaeda and the Taliban in Afghanistan are essentially one and the same. Bush says, "Tell the Taliban we're finished with them." [WASHINGTON POST, 1/27/02] According to counterterrorism "tsar" Richard Clarke, Bush states, "We will make no distinction between the terrorists who committed these acts and those who harbor them." He goes on to say, "I want you all to understand that we are at war and we will stay at war until this is done. Nothing else matters. Everything is available for the pursuit of this war. Any barriers in your way, they're gone. Any money you need, you have it. This is our only agenda." When Rumsfeld points out that international law only allows force to prevent future attacks and not for retribution, Bush yells, "No. I don't care what the international lawyers say, we are going to kick some ass." [AGAINST ALL ENEMIES, BY RICHARD CLARKE, 3/04, PP. 23–24]

10:49 P.M.: Ashcroft Claims to Already Understand Hijacking Procedure

It is reported that Attorney General Ashcroft has told members of Congress that there were three to five hijackers on each plane armed only with knives. [CNN, 9/12/01]

11:30 P.M.*: Bush Sees New "Pearl Harbor"

Before going to sleep, President Bush writes in his diary, "The Pearl Harbor of the 21st century took place today. . . . We think it's Osama bin Laden." [WASHINGTON POST, 1/27/02]

PART V

THE POST-9/11 WORLD

18.

Post-9/11 Afghanistan

A complete accounting of all that has happened in Afghanistan since 9/11, including a detailed accounting of the events of the Afghan war begun in October 2001, is beyond the scope of this book.

This chapter mostly focuses on the mystery of how Osama bin Laden, after apparently being cornered several times, managed to escape during the Afghan war. Could it be that the U.S. did not actually try its hardest to catch him? A U.S. official stated that "casting our objectives too narrowly" risked "a premature collapse of the international effort [to overthrow the Taliban] if by some lucky chance Mr. bin Laden was captured." [MIRROR, 7/8/02] And, in fact, in early November 2001, the U.S. combat commander in Afghanistan, General Tommy Franks, said that apprehending bin Laden was not one of the missions of the war. [USA TODAY, 11/8/01]

There are many curious failures to catch bin Laden and other top al-Qaeda and Taliban leaders in late 2001. Perhaps the strangest is a reported U.S.-condoned airlift of thousands of Pakistanis surrounded by Northern Alliance forces, the allies of the United States. Large numbers of Taliban and al-Qaeda fighters escaped in this airlift, possibly including members of bin Laden's immediate family.

The hunt for bin Laden greatly cooled down in early 2002. Special Forces troops with Middle Eastern expertise were pulled out of Afghanistan to prepare for the Iraq war. Their replacements were troops with expertise in Spanish cultures. [USA TODAY, 3/28/04] Since then, there are reports of top Taliban leaders living openly, and in luxury, in Afghanistan and Pakistan.

Was the attempt to capture bin Laden merely beset by bad luck and failure? Or could it be that the U.S. government sought a quick "victory" in Afghanistan, even if it failed to capture the man most responsible for the 9/11 attacks? Could it be that the U.S. government's "war" in Afghanistan merely provided political cover while it planned the war against its real target—Iraq?

One mystery related to the failure to defeat al-Qaeda and the Taliban is how warlords and illegal drug production continue to flourish in post-Taliban Afghanistan. The country remains dominated by opium warlords, with the country's nominal leader actually directly controlling little more than the capital of Kabul, if that. The Taliban are still a force in many parts of Afghanistan, and the country continues to provide a safe haven for al-Qaeda, despite the continued presence of 10,000 or so U.S. troops there. [NEW YORKER, 4/5/04]

Illegal drug production has boomed since the U.S. invasion of Afghanistan. The latest record harvest is 30 percent higher than the year before. [INDEPENDENT, 4/18/04] Experts note a connection between such drug production and terrorism. For example, Avaz Yuldashov of the Tajikistan Drug Control Agency comments, "There's absolutely no threat to the labs inside Afghanistan. . . . Some of them even work outside, in the open air. . . . Drug trafficking from Afghanistan is the main source of support for international terrorism now. That's quite clear." [PHILADELPHIA INQUIRER, 5/10/04] Why hasn't the production of illegal drugs, which largely end up for sale on the streets of the U.S. and Europe, been vigorously combated by the U.S. and allied forces stationed there?

Today (in mid-2004), it is unclear whether bin Laden is still alive. Many reports suggest that he is somewhere near the Pakistan-Afghanistan border. Al-Qaeda's numbers are growing, rather than diminishing. As the months since 9/11 have passed, both bin Laden and his spokesmen have repeatedly promised that al-Qaeda is actively preparing for the next attacks. In 2002, a London-based Saudi newspaper, *al-Majallah* reported that bin Laden had obtained 48 nuclear "suitcases" (nuclear weapons that are transportable by truck) from Russia. Some Pakistani nuclear engineers have reportedly admitted assisting al-Qaeda officials in the construction of nuclear weapons. In October 2003, a recording of a voice said to be bin Laden's promised, "We will continue the martyrdom operations inside and outside the United States . . . " [OSAMA'S REVENGE, BY PAUL WILLIAMS, 6/04, PP. 41, 91, 104-05, 133-36]

In short, it is quite possible that 9/11, however horrific, was only a prelude to the attack bin Laden and al-Qaeda is currently plotting. If so, the failure to eliminate bin Laden and the rest of the al-Qaeda leadership when they were trapped in Afghanistan in late 2001 will be proven a disaster.

September 21, 2001: U.S. Denies Plans for Afghanistan Regime Change

A secret report to NATO allies says the U.S. privately wants to hear allied views on "post-Taliban Afghanistan after the liberation of the country." However, the U.S. is publicly claiming it has no intentions to overthrow the Taliban. [GUARDIAN, 9/21/01] For instance, four days later, Press Secretary Ari Fleischer denies that military actions there are "designed to replace one regime with another." [STATE DEPARTMENT, 12/26/01]

Late September–Early October 2001: Bin Laden Reportedly Agrees to Face International Tribunal; U.S. Not Interested?

Leaders of Pakistan's two Islamic parties are negotiating bin Laden's extradition to Pakistan to stand trial for the 9/11 attacks during this period, according to a later *Mirror* article. Under the plan, bin Laden will be held under house arrest in Peshawar and will face an international tribunal, which will decide whether to try him or hand him over to the U.S. According to reports in Pakistan (and the *Daily Telegraph*), this plan has been approved by both bin Laden and Taliban leader Mullah Omar. [MIRROR, 7/8/02] Based on the first priority in the U.S.'s new "war on terror" proclaimed by President Bush, the U.S. presumably would welcome this plan. For example, Bush had just announced, "I want justice. And there's an old poster out West, I recall, that says, 'Wanted: Dead or Alive.'" [ABC NEWS, 9/17/01] Yet, Bush's ally in the war on terror, Pakistani President Musharraf, rejects the plan (stating that his reason for doing so was because he "could not guarantee bin Laden's safety"). Based on a U.S. official's later statements, it appears that the U.S. did not want the deal: "Casting our objectives too narrowly" risked "a premature collapse of the international effort [to overthrow the Taliban] if by some lucky chance Mr. bin Laden was captured." [MIRROR, 7/8/02]

October 7, 2001: U.S. Begins Bombing in Afghanistan

The U.S. begins bombing Afghanistan in the first strike of its "war on terror." [MSNBC, 11/01] Most documentary evidence suggests the U.S. was not planning this bombing before 9/11. However, former Pakistani Foreign Secretary Niaz Naik has claimed that in July 2001 senior U.S. officials told him that a military action to overthrow the Taliban in Afghanistan would, as the BBC put it, "take place before the snows started falling in Afghanistan, by the middle of October at the latest." [BBC, 9/18/01]

October 15, 2001: Russian Newspaper Calls
Afghanistan War U.S. Political Power Move

According to the *Moscow Times,* the Russian government sees the upcoming U.S. conquest of Afghanistan as an attempt by the U.S. to replace Russia as the dominant political force in Central Asia, with the control of oil as a prominent motive: "While the bombardment of Afghanistan outwardly appears to hinge on issues of fundamentalism and American retribution, below the surface, lurks the prize of the energy-rich Caspian basin into which oil majors have invested billions of dollars. Ultimately, this war will set the boundaries of U.S. and Russian influence in Central Asia—and determine the future of oil and gas resources of the Caspian Sea." [*MOSCOW TIMES,* 10/15/01] The U.S. later appears to gain military influence over Kazakhstan, the Central Asian country with the most resource wealth, and closest to the Russian heartland (see chapter 16).

October 19, 2001: U.S. Ground Attacks Begin in Afghanistan

U.S. Special Forces begin ground attacks in Afghanistan. [MSNBC, 11/01] However, during the Afghanistan war, U.S. ground soldiers are mainly employed as observers, liaisons, and spotters for air power to assist the Northern Alliance—not as direct combatants. [*CHRISTIAN SCIENCE MONITOR,* 3/4/02 (B)]

October 25, 2001: Afghani Resistance Leader Killed

Abdul Haq, a leader of the Afghani resistance to the Taliban, is killed. According to some reports, he "seemed the ideal candidate to lead an opposition alliance into Afghanistan to oust the ruling Taliban." [*OBSERVER,* 10/28/01] Four days earlier, he had secretly entered Afghanistan with a small force to try to raise rebellion, but was spotted by Taliban forces and surrounded. He calls former National Security Adviser Robert McFarlane (who had supported him in the past) who then calls the CIA and asks for immediate assistance to rescue Haq. A battle lasting up to twelve hours ensues. (The CIA had previously rejected Haq's requests for weapons to fight the Taliban, and so his force is grossly underarmed.) [*SYDNEY MORNING HERALD,* 10/29/01] The CIA refuses to send in a helicopter to rescue him, alleging that the terrain is too rough, even though Haq's group is next to a hilltop once used as a helicopter landing point. [*OBSERVER,* 10/28/01; *LOS ANGELES TIMES,* 10/28/01 (B)] An unmanned surveillance aircraft eventually attacks some of the Taliban forces fighting Haq, but not until five hours after Haq has been captured. The Taliban executes him. [*WALL STREET JOURNAL,* 11/2/01] Vincent Cannistraro, a former CIA director of counterterrorism, and others suggest that Haq's position was betrayed to the Taliban by the ISI. Haq was already an enemy of the ISI, who may have killed his family. [*VILLAGE VOICE,* 10/26/01; *USA TODAY,* 10/31/01; *TORONTO STAR,* 11/5/01; KNIGHT RIDDER, 11/3/01]

Early November 2001: Al-Qaeda Convoy Flees Kabul

Many locals in Afghanistan reportedly witness a remarkable escape of al-Qaeda forces from Kabul around this time. One local businessman says, "We don't understand how they weren't all killed the night before because they came in a convoy of at least 1,000 cars and trucks. It was a very dark night, but it must have been easy for the American pilots to see the headlights. The main road was jammed from eight in the evening until three in the morning." This convoy was thought to have contained al-Qaeda's top officials. [LONDON TIMES, 7/22/02]

Early November 2001: Al-Qaeda Fighters, bin Laden Said to Move into Jalalabad Without Hindrance

Since late October, U.S. intelligence reports began noting that al-Qaeda fighters and leaders were moving into and around the Afghanistan city of Jalalabad. By early November, bin Laden is said to be there. Knight-Ridder Newspapers reports that "American intelligence analysts concluded that bin Laden and his retreating fighters were preparing to flee across the border. However, the U.S. Central Command, which was running the war, made no move to block their escape. 'It was obvious from at least early November that this area was to be the base for an exodus into Pakistan,' said one intelligence official, who spoke only on condition of anonymity. 'All of this was known, and frankly we were amazed that nothing was done to prepare for it.'" The vast majority of leaders and fighters are eventually able to escape into Pakistan. [KNIGHT RIDDER, 10/20/02]

November 9, 2001: The Taliban Loses Control of Northern Afghanistan

The Taliban abandon the strategic northern Afghan city of Mazar-i-Sharif, allowing the Northern Alliance to take control. [ASSOCIATED PRESS, 8/19/02] The Taliban abandons the rest of Northern Afghanistan in the next few days, except the city of Kunduz, where most of the Taliban flee. Kunduz falls on November 25, but not before most of the thousands of fighters there are airlifted out. [NEW YORKER, 1/21/02]

November 13, 2001: Kabul Falls to Northern Alliance; Rest of Country Soon Follows

Kabul, Afghanistan's capital, falls to the Northern Alliance. The Taliban will abandon the rest of the country over the next few weeks. [BBC, 11/13/01] As *New Yorker* reports, "The initial American aim in Afghanistan had been not to eliminate the Taliban's presence there entirely but to undermine the regime and al-Qaeda while leaving intact so-called moderate Taliban elements that would play a role in a new postwar government. This would insure that Pakistan would not end up with a regime on its border dominated by the Northern Alliance." The surprisingly quick fall of Kabul ruins this plan. [NEW YORKER, 1/21/02]

November 14, 2001: Al-Qaeda Convoy Flees to Tora Bora; U.S. Fails to Attack

The Northern Alliance captures the Afghan city of Jalalabad. [SYDNEY MORNING HERALD, 11/14/01] That night, a convoy of 1,000 or more al-Qaeda and Taliban fighters escapes from Jalalabad and reaches the fortress of Tora Bora after hours of driving and then walking. Bin Laden is believed to be with them, riding in one of "several hundred cars" in the convoy. The U.S. bombs the nearby Jalalabad airport, but apparently does not attack the convoy. [KNIGHT RIDDER, 10/20/02; CHRISTIAN SCIENCE MONITOR, 3/4/02 (B)]

November 14–November 25, 2001:
U.S. Secretly Authorizes Airlift of Pakistani and Taliban Fighters

At the request of the Pakistani government, the U.S. secretly allows rescue flights into the besieged Taliban stronghold of Kunduz, in Northern Afghanistan, to save Pakistanis fighting for the Taliban (and against U.S. forces) and bring them back to Pakistan. Pakistan's President "Musharraf won American support for the airlift by warning that the humiliation of losing hundreds—and perhaps thousands—of Pakistani Army men and intelligence operatives would jeopardize his political survival." [NEW YORKER, 1/21/02] Dozens of senior Pakistani military officers, including two generals, are flown out. [NOW WITH BILL MOYERS, 2/21/03] In addition, it is reported that the Pakistani government assists 50 trucks filled with foreign fighters to escape the town. [NEW YORK TIMES, 11/24/01] Many news articles at the time suggest an airlift is occurring. [See, e.g., INDEPENDENT, 11/16/01; NEW YORK TIMES, 11/24/01; BBC, 11/26/01; INDEPENDENT, 11/26/01; GUARDIAN, 11/27/01; MSNBC, 11/29/01] Significant media coverage fails to develop, however. The U.S. and Pakistani governments deny the existence of the airlift. [STATE DEPARTMENT, 11/16/01; NEW YORKER, 1/21/02] On December 2, when asked

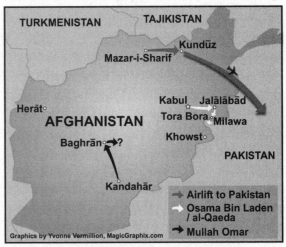

The main routes al-Qaeda and the Taliban escape
U.S. and Northern Alliance forces

to assure that the U.S. did not allow such an airlift, Rumsfeld says, "Oh, you can be certain of that. We have not seen a single—to my knowledge, we have not seen a single airplane or helicopter go into Afghanistan in recent days or weeks and extract people and take them out of Afghanistan to any country, let alone Pakistan." [MSNBC, 12/2/01] Reporter Seymour Hersh believes that Rumsfeld must have given approval for the airlift. [NOW WITH BILL MOYERS, 2/21/03] However, the *New Yorker* magazine reports, "What was supposed to be a limited evacuation apparently slipped out of control and, as an unintended consequence, an unknown number of Taliban and al-Qaeda fighters managed to join in the exodus." A CIA analyst says, "Many of the people they spirited away were in the Taliban leadership" who Pakistan wanted for future political negotiations. U.S. intelligence was "supposed to have access to them, but it didn't happen," he says. According to Indian intelligence, airlifts grow particularly intense in the last three days before the city falls on November 25. Of the 8,000 remaining al-Qaeda, Pakistani, and Taliban, about 5,000 are airlifted out and 3,000 surrender. [NEW YORKER, 1/21/02] Hersh later claims that "maybe even some of bin Laden's immediate family were flown out on the those evacuations." [NOW WITH BILL MOYERS, 2/21/03]

November 15, 2001: Al-Qaeda Leader Reported Dead in Bombing Raid

Al-Qaeda leader Mohammed Atef is believed to have been killed in a U.S. bombing raid on Afghanistan. Atef is considered al-Qaeda's military commander, and one of its top leaders. [STATE DEPARTMENT, 11/16/01; ABC NEWS, 11/17/01]

**Al-Qaeda leader
Mohammed Atef**

Mid-November 2001: Afghan Politician Says
U.S. Policy Prevented bin Laden Capture

Ismail Khan's troops and other Northern Alliance fighters are reportedly ready to take back Pashtun areas from Taliban control at this time. Khan, governor of Herat province and one of Afghanistan's most successful militia leaders, later maintains that "we could have captured all the Taliban and the al-Qaeda groups. We could have arrested Osama bin Laden with all of his supporters." [USA TODAY, 1/2/02] However, according to Khan, his forces hold back at the request of the U.S., who allegedly do not want the non-Pashtun Northern Alliance to conquer Pashtun areas. British newspapers at the time report bin Laden is surrounded in a 30-mile area, but the conquest of Kandahar takes weeks without the Northern Alliance and bin Laden slips away (other accounts put him at Tora Bora). [CNN, 11/18/01 (B)]

November 16, 2001: Al-Qaeda, Taliban
Leaders Reportedly Escape Afghanistan

According to *Newsweek*, approximately 600 al-Qaeda and Taliban fighters, including many senior leaders, escape Afghanistan on this day. There are two main routes out of the Tora Bora cave complex to Pakistan. The U.S. bombed only one route, so the 600 escaped without being attacked using the other route. Hundreds continue to use the escape route for weeks, generally unbothered by U.S. bombing or Pakistani border guards. U.S. officials later privately admit they lost an excellent opportunity to close a trap. [NEWSWEEK, 8/11/02 (B)] On the same day, the media reports that the U.S. is studying routes bin Laden might use to escape Tora Bora [LOS ANGELES TIMES, 11/16/01], but the one escape route is not closed, and apparently bin Laden and others escape into Pakistan using this route several weeks later. High-ranking British officers will later privately complain, "American commanders had vetoed a proposal to guard the high-altitude trails, arguing that the risks of a firefight, in deep snow, gusting winds, and low-slung clouds, were too high." [NEW YORK TIMES, 9/30/02 (B)]

November 21, 2001: Opium Boom in Afghanistan

The *Independent* runs a story with the title: "Opium Farmers Rejoice at the Defeat of the Taliban." Massive opium planting is underway all across Afghanistan. [INDEPENDENT, 11/21/01] Four days later, the *Observer* runs a story headlined, "Victorious Warlords Set to Open the Opium Floodgates." It states that farmers are being encouraged by warlords allied with the U.S. to plant "as much opium as possible." [OBSERVER, 11/25/01]

November 25, 2001: U.S. Troops Arrive in Kandahar amid Talk of a Secret Deal

U.S. troops land near the Taliban stronghold of Kandahar, Afghanistan. [ASSOCIATED PRESS, 8/19/02] Apparently, as the noose tightens around Kandahar, new Afghanistan head Hamid Karzai makes a deal with the Taliban, giving them a general amnesty in return for surrender of the city. Taliban's leader Mullah Omar is allowed to escape "with dignity" as part of the deal. However, the U.S. says it will not abide by the deal and Karzai then says he will not let Omar go free after all. Taliban forces begin surrendering on December 7. [SYDNEY MORNING HERALD, 12/8/01] Omar escapes.

November 25, 2001 (B): Bin Laden Reportedly Gives Last Public Speech to Followers

It is believed bin Laden makes a speech before a crowd of about 1,000 followers in the village of Milawa, Afghanistan. This village is on the route from Tora Bora to the Pakistani border, about eight to ten hours by walking. In his last known public appearance, bin Laden encourages his followers to leave Afghanistan, so they could regroup and fight again. [KNIGHT RIDDER, 10/20/02] It is believed he leaves the country a few days later. [DAILY TELEGRAPH, 2/23/02]

November 28, 2001: Bin Laden Reportedly Escapes Tora Bora by Helicopter

A U.S. Special Forces soldier stationed in Fayetteville, North Carolina later (anonymously) claims that the U.S. has bin Laden pinned in a certain Tora Bora cave on this day, but fails to act. Special Forces soldiers allegedly sit by waiting for orders and watch two helicopters fly into the area where bin Laden is believed to be, load up passengers, and fly toward Pakistan. No other soldiers have come forward to corroborate the story, but bin Laden is widely believed to have been in the Tora Bora area at the time. [FAYETTEVILLE OBSERVER, 8/2/02] However, other reports indicate that bin Laden may have left the Tora Bora region by this time. *Newsweek* separately reports that many locals "claim that mysterious black helicopters swept in, flying low over the mountains at night, and scooped up al-Qaeda's top leaders." [NEWSWEEK, 8/11/02 (B)] Perhaps coincidentally, on the same day this story is reported, months after the fact, the media also will report a recent spate of strange deaths at the same military base in Fayetteville. Five soldiers and their wives all die since June 2002 in apparent murder-suicides. At least three were Special Forces soldiers recently returned from Afghanistan. [INDEPENDENT, 8/2/02]

Early December 2001: Battle for Tora Bora Is Called "Charade"

The *Daily Telegraph* later reports on the battle for Tora Bora around this time: "In retrospect, and with the benefit of dozens of accounts from the participants, the battle for Tora Bora looks more like a grand charade." Eyewitnesses express shock that the U.S. pinned in Taliban and al-Qaeda forces, thought to contain many high leaders, on three sides only, leaving the route to Pakistan open. An intelligence chief in Afghanistan's new government says, "The border with Pakistan was the key, but no one paid any attention to it. In addition, there were plenty of landing areas for helicopters had the Americans acted decisively. Al-Qaeda escaped right out from under their feet." [DAILY TELEGRAPH, 2/23/02] It is believed that up to 2,000 were in the area when the battle began. The vast majority successfully flees, and only 21 al-Qaeda fighters are finally captured. [*CHRISTIAN SCIENCE MONITOR*, 3/4/02 (B)] The U.S. relies on local forces "whose loyalty and enthusiasm were suspect from the start" to do most of the fighting. [KNIGHT RIDDER, 10/20/02] Some of the local commanders drafted to help the U.S. had ties to bin Laden going back to the 1980s. [*NEW YORK TIMES*, 9/30/02 (B)] These forces actually help al-Qaeda escape. An Afghan intelligence officer says he is astounded that Pentagon planners did not consider the most obvious exit routes and put down light U.S. infantry to block them. It is later widely believed that bin Laden escapes along one of these routes on November 30 or December 1, walking out with about four loyal followers. [*CHRISTIAN SCIENCE MONITOR*, 3/4/02; *CHRISTIAN SCIENCE MONITOR*, 3/4/02 (B)] Al-Qaeda's number two leader, Ayman al-Zawahiri, also escapes the area. [KNIGHT RIDDER, 10/20/02]

December 17, 2001: Northern Alliance Declares Victory at Tora Bora; Afghan War Considered Over

Northern Alliance forces declare that the battle of Tora Bora, with a ground assault begun on December 5, has been won. The Afghan war is widely considered finished. However, in retrospect, many consider the battle a failure because most of the enemy escapes, and the Taliban will later regroup. [*CHRISTIAN SCIENCE MONITOR*, 3/4/02 (B)]

December 22, 2001: Hamid Karzai Assumes Power in Afghanistan

See chapter 16.

December 24, 2001: Taliban Free, "Living in Luxury"

The *Guardian* reports that many in Afghanistan intelligence say former top Taliban officials are living openly in villas in Afghanistan and Pakistan. At least four top leaders who had been caught have been simply released. One intelligence source claims to know the exact location of many, and says they could be rounded up within hours. A former Taliban minister now working with the Northern Alliance also claims, "Some are living in luxury in fine houses, they are not hiding in holes. They could be in jail by tonight if the political will existed." The U.S. claims it is working hard to find and catch these leaders. [*GUARDIAN*, 12/24/01]

Taliban leader Mullah Mohammed Omar

January 6, 2002: Mullah Omar Escapes Capture by U.S. Military

The U.S. allegedly locates former Taliban leader Mullah Omar and 1,500 of his soldiers in the remote village of Baghran, Afghanistan. After a six-day siege, and surrounded by U.S. helicopters and troops, Omar and four bodyguards supposedly escape the dragnet in a daring chase on motorcycles over dirt roads. His soldiers are set free in return for giving up their weapons, in a deal brokered by local leaders. Yet it remains unclear if Omar was ever in the village in the first place. [*OBSERVER*, 1/6/02]

February 2002: U.S. Deploys Resources from Afghanistan to Iraq

Senator Bob Graham (D) later claims that this month he is told by a senior military commander, "Senator, we have stopped fighting the war on terror in Afghanistan. We are moving military and intelligence personnel and resources out of Afghanistan to get ready for a future war in Iraq." [COUNCIL ON FOREIGN RELATIONS, 3/26/04] *New Yorker* magazine also reports on a redeployment of resources to Iraq at this time. [NEW YORKER, 10/20/03]

February 25, 2002: Captured Taliban Leader Ignored by CIA

Time magazine reports that the second highest Taliban official in U.S. custody, Mullah Haji Abdul Samat Khaksar, has been waiting for months for the CIA to talk to him. Two weeks after *Time* informed U.S. officials that Khaksar wanted to talk, he still has not been properly interviewed. He says he has useful information, and may be able to help locate former Taliban leader Mullah Omar. *Time* notes that "he claims to have information about al-Qaeda links to the ISI." [TIME, 2/25/02] "The little that Khaksar has divulged to an American general and his intelligence aide—is tantalizing. . . . He says that the ISI agents are still mixed up with the Taliban and al-Qaeda," and that the three groups have formed a new group to get the U.S. out of Afghanistan. He also says that "the ISI recently assassinated an Afghan in the Paktika province who knew the full extent of ISI's collaboration with al-Qaeda." [TIME, 2/19/02]

April 1, 2002: Afghan Opium Crackdown Fails

"American officials have quietly abandoned their hopes to reduce Afghanistan's opium production substantially this year and are now bracing for a harvest large enough to inundate the world's heroin and opium markets with cheap drugs." They want to see the new Afghan government make at least a token effort to destroy some opium, but it appears that the new government is not doing even that. Afghan leader Hamid Karzai had announced a total ban on opium cultivation, processing, and trafficking, but it appears to be a total sham. The new harvest is so large that it could be "enough opium to stockpile for two or two and a half more years." [NEW YORK TIMES, 4/1/02] Starting this month, Karzai's government offers farmers $500 for every acre of poppies they destroy, but farmers can earn as much as $6,400 per acre for the crop. The program is eventually cancelled when it runs out of money to pay farmers. [ASSOCIATED PRESS, 3/27/03]

April 17, 2002: Failure to Capture bin Laden in Afghan War Is "Gravest Error"

The *Washington Post* reports that, "The Bush administration has concluded that Osama bin Laden was present during the battle for Tora Bora late last year and that failure to commit U.S. ground troops to hunt him was its gravest error in the war against al-Qaeda," allowing bin Laden to escape. The newspaper claims that while the administration has failed to acknowledge the mistake publicly, "inside the government there is little controversy on the subject." [WASHINGTON POST, 4/17/02] The next day, Defense Secretary Rumsfeld denies this, and states he did not know at the time of the assault, "nor do I know today of any evidence that he was in Tora Bora at the time or that he left Tora Bora at the time or even where he is today." [USA TODAY, 4/18/02]

June 20, 2002: Afghan Council Appears
Manipulated in Selecting Warlords

The long-awaited *loya jirga*, or grand council, is concluded in Afghanistan. This council was supposed to be a traditional method for the Afghan people to select their leaders, but most experts conclude that the council is clearly rigged. [BBC, 8/1/02] Half of the delegates walk out in protest. [CNN, 6/18/02] One delegate states, "This is worse than our worst expectations. The warlords have been promoted and the professionals kicked out. Who calls this democracy?" Delegates complain, "This is interference by foreign countries," obviously meaning the U.S. The *New York Times* publishes an article ("The Warlords Win in Kabul") pointing out that the "very forces responsible for countless brutalities" in past governments are back in power. [NEW YORK TIMES, 6/21/02]

July 6, 2002: Afghan Vice President Assassinated

Afghan Vice President Hajji Abdul Qadir is assassinated by Afghan warlords. Some believe that Qadir was assassinated by opium warlords upset by Qadir's efforts to reduce the rampant opium farming and processing that has taken place since the U.S. occupation. Qadir had been overseeing a Western-backed eradication program, and had recently complained that the money meant to be given to reward farmers for not planting opium was in fact not reaching the farmers. Additionally, Qadir "had long been suspected of enriching himself through involvement in the opium trade." [NEW YORK TIMES, 7/8/02 (B), CHICAGO TRIBUNE, 7/8/02]

Late July 2002: Taliban General Reportedly
Captured, but Released After Questioning

U.S. Special Forces apprehend Mullah Akhter Mohammed Osmani, a top general and one of the six most-wanted Taliban, in Kandahar. He is flown to a detention center north of Kabul for interrogation, but is released a few weeks later and escapes to Pakistan. Contradicting the statements of many soldiers in Kandahar, the Defense Intelligence Agency says it "has no knowledge that Mullah Akhter Mohammed Osmani was ever in U.S. custody in Afghanistan. Given Osmani's high profile and [U.S.] interest in detaining him, misidentification by experienced personnel is unlikely." [WASHINGTON TIMES, 12/18/02]

August 11, 2002: Afghanis Directly Producing and
Exporting Heroin "in Broad Daylight"

In the past, Afghanistan had mostly exported raw opium, but now many new refineries are converting the opium into heroin. The British government has spent £20 million to eradicate opium, but the program is marred by corruption and largely seen as a failure. The new heroin factories are said to be "working in broad daylight." There has been a rash of bombings and assassinations in Afghanistan as various factions fight over drug profits. Reporters for a British newspaper are able to determine the precise location of some of these factories, but the U.S. led forces in Afghanistan are doing nothing to stop them. [OBSERVER, 8/11/02]

August 15, 2002: U.S. General Believes Troops Will Remain in Afghanistan for Long Time

General Tommy Franks, commander of U.S. troops in Central Asia, says, "It does not surprise me that someone would say, 'Oh gosh, the military is going to be in Afghanistan for a long, long time.' Sure we will be." He likens the situation to South Korea, where the U.S. has stationed troops for over 50 years. A few days earlier, Joint Chiefs of Staff Chairman Richard Myers said the war on terrorism "could last years and years." [CBS NEWS, 8/16/02]

October 8, 2002: War in Afghanistan Still Not Over

Many in the U.S. have the impression that the war in Afghanistan is over, and U.S. allied forces conquered the country. However, the U.S. ambassador says, "The war is certainly not over. Military operations are continuing, especially in the eastern part of the country and they will continue until we win." Most of the country is controlled by warlords who are now being supplied with weapons with weapons and money by the U.S. government. [DAILY TELEGRAPH, 10/8/02]

December 9, 2002: Special Forces in Afghanistan Back Away from Risky Operations

U.S. commanders have rejected as too risky many special operations missions to attack Taliban and al-Qaeda fighters in Afghanistan. After Army Green Beret A-Teams received good intelligence on the whereabouts of former Taliban leader Mullah Mohammed Omar, commanders turned down the missions as too dangerous. Soldiers traced the timidity to an incident in June 2002 called Operation Full Throttle, which resulted in the death of 34 civilians. [WASHINGTON TIMES, 12/9/02]

March 14, 2003: Afghanistan Becomes Number One Heroin Producer

The Afghani government warns that unless the international community hands over the aid it promised, Afghanistan will slip back into its role as the world's premier heroin producer. The country's foreign minister warns Afghanistan could become a "narco-mafia state." [BBC, 3/17/03] A United Nations study later in the month notes that Afghanistan is once again the world's number one heroin producer, producing 3,750 tons in 2002. Farmers are growing more opium poppies than ever throughout the country, including areas previously free of the crop. [ASSOCIATED PRESS, 3/27/03]

Spring 2004: Conditions In Afghanistan Deteriorate

It is reported that conditions in Afghanistan have deteriorated significantly in nearly every respect. According to Lakhdar Brahima, UN special envoy to Afghanistan, the situation "is reminiscent to what was witnessed after the establishment of the mujahedeen government in 1992." Abdul Rasul Sayyaf, a member of the Wahhabi sect of Islam who opposed the presence of U.S. troops in Saudi Arabia, along with several other warlords accused of atrocities in the mid-1990s have returned to power and are effectively ruling the country. Several hold key positions within the government. They "continue to maintain their own private armies and . . . are reaping vast amounts of money from Afghanistan's illegal opium trade. . . ." The U.S., while claming to support Afghan President Karzai, is relying on these warlords to "help" hunt down Taliban and al-Qaeda factions, although the success rate is abysmal, and much of the intelligence provided by the warlords is faulty. The Taliban has begun to regroup, and now essentially controls much of the southern and eastern regions of the country. [FOREIGN AFFAIRS, MAY/JUNE 2004]

19.

Post-9/11 Investigations

Investigations of 9/11 began on that day; many still continue. In addition to ongoing media investigations, there have been at least three major investigations undertaken by the government.

PENTTBOM, the FBI's investigation, which began on the day of the attacks and continues. PENTTBOM stands for Pentagon, Twin Towers, and bomb.

The 9/11 Congressional Inquiry, which had a limited mandate to look at U.S. intelligence community failures only, and very limited resources. This inquiry was conducted in 2002 and released its final report in 2003.

The 9/11 Commission, known formally as the National Commission on Terrorist Attacks upon the United States. (It is also referred to as the Independent Commission.) This commission was established in early 2003, and completed its investigation in July 2004. Its final report was being released as this book went to press, and unfortunately the results of that could not be fully integrated (though it is discussed in the Afterword). However, the lion's share of their conclusions have already come out in a series of staff statements that are incorporated here.

Each of these investigations has had major flaws, caused by political considerations and biases. For example, the FBI was responsible for many of the most egregious failures before 9/11, so the PENTTBOM investigation is certain to be beset by intrinsic conflicts of interest. For instance, the Bureau's consistent handling of Saudi suspects Osama Basnan and Omar al-Bayoumi indicates the tendency to draw politically expedient conclusions, rather than investigate troubling questions that could impact our relationship with "friendly" nations. Both Basnan and al-Bayoumi were investigated prior to 9/11, but the FBI apparently did not consider them suspects until a large media outcry in late 2002 forced a reexamination. When the public outcry died down, so too, apparently, did the investigation. The investigation was apparently re-instated in mid-2003 when the 9/11 Congressional Inquiry again focused attention on the two men. The FBI finally concluded Basnan and al-Bayoumi had nothing to do with the 9/11 plot, but there are many who had important contacts with the two in San Diego who apparently have never been interviewed by the FBI. [SAN DIEGO MAGAZINE, 9/03]

The 9/11 Congressional Inquiry was technically conducted by 37 members of the House and Senate Intelligence committees, but in reality its co-chairmen, Bob Graham (D) and Porter Goss (R), exercised "near total control over the panel." [ST. PETERSBURG TIMES, 9/29/02] These co-chairs were not impartial observers of the events of 9/11. Both men had travelled to Pakistan for talks on bin Laden shortly before 9/11. They both were meeting with Mahmood Ahmed, the head of Pakistan's ISI, when the 9/11 attacks occurred. Some media reports make the claim that Mahmood ordered $100,000 to head hijacker Mohamed Atta. [AUSTRALIAN, 10/10/01; AGENCE FRANCE-PRESSE, 10/10/01; WALL STREET JOURNAL, 10/10/01] Yet, despite Goss and Graham's direct expertise with Pakistan, the country of Pakistan is hardly mentioned at all in the hundreds of pages of the Inquiry's final report.

The 9/11 Commission has suffered the same conflict-of-interest and political bias problems as the other two investigations. All of the 9/11 Commission's top members are political figures with partisan agendas and conflicts of interest.

Despite all of these investigations, a remarkable amount of information remains classified. The usual excuse is that releasing information would interfere with the trial of Zacarias Moussaoui, the only person on trial in the U.S. for an alleged involvement in the 9/11 attacks. Evidence being withheld includes, but is by no means limited to:

- The flight manifests of all four hijacked flights (there are conflicting accounts)

- The boarding cards of the 19 alleged hijackers

- The seat numbers of all the 19 alleged hijackers (some accounts conflict)

- Video footage of the hijackers passing the security checks before entering the hijacked planes (the 9/11 Commission has released the video of the Flight 77 hijackers only)

- DNA identification of recovered hijacker remains

- The cockpit voice recordings of the Flight 77 and 93 black boxes (Flight 93's recording has been played in private to some victims' relatives)

- The radar recordings from 9/11

- The complete recordings of the conversations between the flight towers and the hijackers

- Known but confiscated video footage of the crash into the Pentagon

Why has so much information been withheld? What possible "national security" secrets could be at stake if some of this material were released? Is the Moussaoui trial no more than a convenient excuse to withhold vital information from the public? If so, what will happen when the Moussaoui trial is completed? Most important, why has the Bush administration consistently opposed *any* investigation into the events surrounding 9/11, and obfuscated and delayed in producing information requested by the various investigations?

September 11, 2001: Suspected Terrorists Found with "Calling Cards"

An unnamed, young, Middle Eastern man flying from Amsterdam, Netherlands, to Detroit, Michigan is arrested after his plane is diverted to Toronto, Canada. He is apparently found to be carrying a flight jacket, Palestinian Authority travel documents, and a picture of himself in a flight crew uniform in front of a fake backdrop of the WTC. [*TORONTO STAR*, 9/15/01 (B), *TORONTO SUN*. 9/15/01; *OTTAWA CITIZEN*, 9/17/01 (B)] Apparently, the man, who identifies himself as an aircraft maintenance engineer in Gaza, Palestine, was supposed to have arrived in the U.S. a few days before but was delayed for unknown reasons. [CBS NEWS. 9/14/01 (B)] A second man was arrested a few days earlier while trying to enter Canada carrying a similar photo. He also possessed maps and directions to the WTC. Both men are soon handed to the U.S. [*TORONTO STAR*, 9/15/01 (B)] A similar picture of suspected Egyptian al-Qaeda terrorist Mohammed Zeki Majoub, arrested in Canada in June 2000, in front of a fake WTC backdrop was found in the luggage of one of the U.S. hijackers. [ASSOCIATED PRESS, 3/1/01; *TORONTO SUN*, 9/15/01] Canadian officials "believe the photos could be calling cards used by the terrorists to identify those involved in plotting the attacks." [*TORONTO SUN*, 9/15/01]

September 11, 2001 (B): Weapons Found on Additional Planes: Inside Jobs?

Later in the day, weapons are found planted on board three other U.S. airplanes. A U.S. official says of the hijackings: "These look like inside jobs." *Time* magazine reports, "Sources tell *Time* that U.S. officials are investigating whether the hijackers had accomplices deep inside the airports' 'secure' areas." [*TIME*, 9/22/01] Penetrating airport security does not appear to have been that difficult: Argenbright, the company in charge of security at all the airports used by the 9/11 hijackers, had virtually no security check on any of their employees, and even hired criminals and illegal immigrants. Security appears to have particularly abysmal at Boston's Logan Airport, even after 9/11. [CNN, 10/12/01; *BOSTON GLOBE*, 10/1/01]

September 11, 2001 (C): Germans Learn of 30 People Traveling for 9/11 Plot

A few hours after the attacks, German intelligence intercepts a phone conversation between followers of bin Laden that leads the FBI to search frantically for two more teams of suicide hijackers, according to U.S. and German officials. The Germans overhear the terrorists refer to "the 30 people traveling for the operation." The FBI scours flight manifests and any other clues for more conspirators still at large. [*NEW*

YORK TIMES, 9/29/01] Two days later, authorities claim to have identified teams of as many as 50 infiltrators who supported or carried out the strikes. About 40 are accounted for as dead or in custody; ten are missing. They also believe a total of 27 suspected terrorists received some form of pilot training. This corresponds with many analyses that the attacks required a large support network. [*LOS ANGELES TIMES*, 9/13/01] Yet there is no evidence that any accomplices in the U.S. shortly before 9/11 have since been arrested or charged.

September 11–13, 2001: Hijackers Leave a Clear Trail of Evidence

Investigators find a remarkable number of possessions left behind by the hijackers:

- Two of Mohamed Atta's bags are found on 9/11. They contain a handheld electronic flight computer, a simulator procedures manual for Boeing 757 and 767 aircraft, two videotapes relating to "air tours" of the Boeing 757 and 747 aircraft, a slide-rule flight calculator, a copy of the Koran, Atta's passport, his will, his international driver's license, a religious cassette tape, airline uniforms, a letter of recommendation, "education related documentation" and a note to other hijackers on how to mentally prepare for the hijacking. [ASSOCIATED PRESS, 10/5/01; *SYDNEY MORNING HERALD*, 9/15/01; *BOSTON GLOBE*, 9/18/01; *INDEPENDENT*, 9/29/01]

- Marwan Alshehhi's rental car is discovered at Boston's Logan Airport containing an Arabic language flight manual, a pass giving access to restricted areas at the airport, documents containing a name on the passenger list of one of the flights, and the names of other suspects. The name of the flight school where Atta and Alshehhi studied, Huffman Aviation, is also found in the car. [*LOS ANGELES TIMES*, 9/13/01]

- A car registered to Nawaf Alhazmi is found at Washington's Dulles Airport on September 12. Inside is a copy of Atta's letter to the other hijackers, a cashier's check made out to a flight school in Phoenix, four drawings of the cockpit of a 757 jet, a box cutter-type knife, maps of Washington and New York, and a page with notes and phone numbers. [*ARIZONA DAILY STAR*, 9/28/01; COX NEWS SERVICE, 10/21/01; *DIE ZEIT*, 10/1/02]

- A rental car is found in an airport parking lot in Portland, Maine. Investigators are able to collect fingerprints and hair samples for DNA analysis. [*PORTLAND PRESS HERALD*, 10/14/01]

- A Boston hotel room contains airplane and train schedules. [*SYDNEY MORNING HERALD*, 9/15/01]

- FBI agents carry out numerous garbage bags of evidence from a Florida apartment where Saeed Alghamdi lived. [CNN, 9/17/01 (B)]

- Two days before 9/11, a hotel owner in Deerfield Beach, Florida, finds a box cutter left in a hotel room used by Marwan Alshehhi and two unidentified men. The owner checks the nearby trash and finds a duffel bag containing Boeing 757 manuals, three illustrated martial arts books, an 8-inch stack of East Coast flight maps, a three-ring binder full of handwritten notes, an English-German dictionary, an airplane fuel tester, and a protractor. The FBI seizes all the items when they are notified on September 12 (except the binder of notes, which the owner apparently threw away). [MIAMI HERALD, 9/16/01; ASSOCIATED PRESS, 9/16/01 (B)]

- In an apartment rented by Ziad Jarrah and Ahmed Alhaznawi, the FBI finds a notebook, videotape, and photocopies of their passports. [MIAMI HERALD, 9/15/01]

- In a bar the night before 9/11, after making predictions of a terrorist attack on America the next day, terrorists leave a business card and a copy of the Koran at the bar. The FBI also recovers the credit card receipts from when they paid for their drinks and lap dances. [ASSOCIATED PRESS, 9/14/01]

- A September 13 security sweep of Boston airport's parking garage uncovers items left behind by the hijackers: a box cutter, a pamphlet written in Arabic and a credit card. [WASHINGTON POST, 9/16/01]

- A few hours after the attacks, suicide notes that some of the hijackers wrote to their parents are found in New York. Credit card receipts showing that some of the hijackers paid for flight training in the U.S. are also found. [LOS ANGELES TIMES, 9/13/01]

- A FedEx bill is found in a trash can at the Comfort Inn in Portland, Maine, where Atta stayed the night before 9/11. The bill leads to Dubai, United Arab Emirates, allowing investigators to determine much of the funding for 9/11. [NEWSWEEK, 11/11/01; LONDON TIMES, 12/1/01]

The hijackers past whereabouts can even be tracked by their pizza purchases. An expert points out: "Most people pay cash for pizza. These [hijackers] paid with a credit card. That was an odd thing." [SAN DIEGO UNION-TRIBUNE, 9/3/02] "In the end, they left a curiously obvious trail—from martial arts manuals, maps, a Koran, Internet and credit card fingerprints. Maybe they were sloppy, maybe they did not care, maybe it was a gesture of contempt of a culture they considered weak and corrupt." [MIAMI HERALD, 9/22/01] Note the *New Yorker's* quote of a former high-level intelligence official: "Whatever trail was left was left deliberately—for the FBI to chase." [NEW YORKER, 10/1/01]

September 12, 2001: Threat to Air Force One? Stories Conflict

Press Secretary Ari Fleischer explains that Bush went to Nebraska because "[t]here was real and credible information that the White House and Air Force One were targets." The next day, William Safire of the *New York Times* writes, and Bush's Political Strategist Karl Rove confirms, that the Secret Service believed "'Air Force One may be next,' and there was an 'inside' threat which 'may have broken the secret codes [i.e., showing a knowledge of Presidential procedures].'" [NEW YORK TIMES, 9/13/01] By September 27,

Fleischer begins to backpedal on the claim that there were specific threats against Air Force One and/or the President, and news stories flatly contradict it. [WASHINGTON POST, 9/27/01] A well-informed, anonymous Washington official says, "It did two things for [Cheney]. It reinforced his argument that the President should stay out of town, and it gave George W. an excellent reason for doing so." [DAILY TELEGRAPH, 12/16/01] By 2004, a Bush spokesperson says there was no threat, but Cheney continues to maintain that there may have been. Cheney also claims the Secret Service passed him word of the threat, but two Secret Service agents working that day deny their agency played any role in receiving or passing on such a threat. The threat was allegedly based on the use of the word "Angel," the code word for Air Force One, but Secret Service agents later note that the code word was not an official secret, but a radio shorthand designation that had been made public well before 2001. [WALL STREET JOURNAL, 3/22/04]

September 12, 2001 (B): "Inside Help" for Terrorists at Airports?

Billie Vincent, a former FAA security director, suggests the hijackers had inside help at the airports. "These people had to have the means to take control of the aircrafts. And that means they had to have weapons in order for those pilots to relinquish control. Think about it, they planned this thing out to the last detail for months. They are not going to take any risks at the front end. They knew they were going to be successful before they started . . . It's the only thing that really makes sense to me." [MIAMI HERALD, 9/12/01 (B)]

September 12, 2001 (C): Hijacker's Passport Found Near WTC

The passport of hijacker Satam Al Suqami is found a few blocks from the WTC. [ABC NEWS, 9/12/01 (C). ASSOCIATED PRESS, 9/16/01; ABC NEWS, 9/16/01] The *Guardian* says, "the idea that Mohamed Atta's passport had escaped from that inferno unsinged [tests] the credulity of the staunchest supporter of the FBI's crackdown on terrorism." [GUARDIAN, 3/19/02] (Note that, as in this *Guardian* account, the passport is frequently mistakenly referred to as Atta's passport.)

September 13, 2001: White House Announces bin Laden–9/11 Connection

The White House announces that there is "overwhelming evidence" that bin Laden is behind the attacks. [MSNBC, 9/13/01]

September 13, 2001 (B): Wide Flight 93 Debris Field
Spurs Rumors Flight Was Shot Down

Investigators say they have found debris from the Flight 93 crash far from the main crash site. A second debris field centers around Indian Lake about three miles from the crash scene, where eyewitnesses report seeing falling debris only moments after the crash. More debris is found in New Baltimore, some eight miles away. Later in the day, the investigators say all that debris likely was blown there. [CNN, 9/13/01; PITTSBURGH TRIBUNE-REVIEW, 9/13/01] Another debris field is found six miles away, and human remains are found miles away. After all of this is discovered, the FBI still "stresses" that "no evidence had surfaced" to support the idea that the plane was shot down. [PITTSBURGH POST-GAZETTE, 9/13/01] A half-ton piece of one of the engines is found 2,000 yards away from the main crash site. This was the single heaviest piece recovered from the crash. [PHILADELPHIA DAILY NEWS, 12/28/01; INDEPENDENT, 8/13/02] Days later, the FBI says the wide debris field was probably the result of the explosion on impact. The *Independent* nevertheless later cites the wide debris field as one of many reasons why widespread rumors remain that the plane was shot down. [INDEPENDENT, 9/20/01]

September 13, 2001 (C): 18 Hijackers Named, Hanjour Follows One Day Later

The FBI says there were 18 hijackers, and releases their names. [CNN, 9/13/01 (C)] The next day, it is revealed there is one more hijacker—Hani Hanjour. [CNN, 9/14/01; ASSOCIATED PRESS, 9/14/01 (B)] A few days later, it is reported that Hanjour's "name was not on the American Airlines manifest for [Flight 77] because he may not have had a ticket." [WASHINGTON POST, 9/16/01 (B)]

September 13, 2001: "Series of Circumstances" Said to Make Hijacked Passenger Cell Phone Calls Possible

It is reported that the many phone calls made by passengers from the hijacked flights are normally technically impossible to make. A major cell phone carrier spokeswoman claims, "Those were a series of circumstances that made those calls go through, which would not be repeated under normal circumstances." Supposedly, the calls worked because they were made when the planes were close to the ground and they were kept short. [WIRED, 9/13/01] However, many of the cell phone calls were made from high cruising altitudes and lasted ten minutes or more. The *New York Times* later reports, "According to industry experts, it is possible to use cell phones with varying success during the ascent and descent of commercial airline flights, although the difficulty of maintaining a signal appears to increase as planes gain altitude. Some older phones, which have stronger transmitters and operate on analog networks, can be used at a maximum altitude of ten miles, while phones on newer digital systems can work at altitudes of five to six miles. A typical airline cruising altitude would be 35,000 feet, or about 6.6 miles." [SLATE, 9/14/01] A spokesperson for the AT&T phone company notes that cell phone networks are not designed for calls from high altitudes. She suggests that "it was almost a fluke that the calls reached their destinations." [WIRELESS REVIEW, 11/1/01]

September 13–14, 2001: Flight 93 and Flight 77 Black Boxes Found

All the "black boxes" for Flights 93 and 77 are found. However, Flight 93's two boxes are deemed severely damaged, and it is not known if the data can be recovered. [REUTERS, 9/13/01 (B), BBC, 9/15/01] In December, the FBI reveals they know the contents, but only release select quotes. [CNN, 12/21/01] Flight 93's recording is eventually played in private to victims' relatives, and also to the 9/11 Commission. FBI

Director Mueller will later say that the boxes for Flight 77 provided altitude, speed, headings, and other information, but the voice recorder contained "nothing useful." [CBS NEWS, 2/23/02]

September 14, 2001: Officials Deny Flight 93 Shot Down

Officials deny that Flight 93 was shot down, but propose the theory that the hijackers had a bomb on board and blew up the plane. [PITTSBURGH TRIBUNE-REVIEW, 9/14/01] Later in the month, it is reported that the "FBI has determined from the on site investigation that no explosive was involved." [ASSOCIATED PRESS, 9/25/01]

September 14, 2001 (B): Conflicting Accounts About Planes Near Flight 93's Crash

Officials admit that two planes were near Flight 93 when it crashed, which matches numerous eyewitness accounts. For example, local man Dennis Decker says that immediately after hearing an explosion, "We looked up, we saw a midsized jet flying low and fast. It appeared to make a loop or part of a circle, and then it turned fast and headed out. If you were here to see it, you'd have no doubt. It was a jet plane, and it had to be flying real close when that 757 went down . . . If I was the FBI, I'd find out who was driving that plane." [BERGEN RECORD, 9/14/01] Later the same day, the military says it can "neither confirm nor deny" the nearby planes. [PITTSBURGH TRIBUNE-REVIEW, 9/14/01] Two days later, they claim there were two planes near, but that they were a military cargo plane and business jet, and neither had anything to do with the crash. [PITTSBURGH POST-GAZETTE, 9/16/01] Supposedly, the business jet was requested to fly low over the crash site, 25 minutes after all aircraft in the U.S. had been ordered to land. However, the story appears physically impossible since the FBI says this jet was at 37,000 feet and asked to descend to 5,000 feet. [PITTSBURGH CHANNEL, 9/15/01] That would have taken many minutes for that kind of plane, and witnesses report seeing the plane flying very low even before the crash. [BERGEN RECORD, 9/14/01] Another explanation of a farmer's plane 45 minutes later is put forth, but that also does not fit the time at all. [PITTSBURGH CHANNEL, 9/15/01] Deputy Defense Secretary Paul Wolfowitz states: "We responded awfully quickly, I might say, on Tuesday [9/11], and, in fact, we were already tracking in on that plane that crashed in Pennsylvania. I think it was the heroism of the passengers on board that brought it down. But the Air Force was in a position to do so if we had had to." [DEFENSE DEPARTMENT, 9/14/01] The next day, the Director of the Air National Guard denies that any plane was scrambled after Flight 93. [SEATTLE TIMES, 9/16/01] That in turn contradicts what Vice President Cheney will say later. [WASHINGTON POST, 1/27/02]

September 14, 2001 (C): Contrary Accounts of Flight 93's Speed Raises Questions

It is initially reported that Flight 93 is traveling fairly slowly when it crashed. "It slammed into the ground at a speed law enforcement authorities said might have approached 300 mph" [NEW YORK TIMES, 9/14/01] "Flight 93 slammed into the earth nose-first at over 200 mph, according to estimates by the National Transportation Safety Board and other experts." [DELAWARE NEWS JOURNAL, 9/16/01] However, by 2002 it is being reported that the plane crashed going nearly 600 mph. [AMONG THE HEROES, BY JERE LONGMAN, 8/02, P. 212]

"It could have even broken the sound barrier for a while," says Hank Krakowski, director of flight operations control at United's system control center on September 11. [NEW YORK TIMES, 3/27/02] The design limits of the plane are 287 mph when flying below 10,000 feet. [AMONG THE HEROES, BY JERE LONGMAN, 8/02, P. 208]

September 14, 2001 (D): Gruesome WTC Remains Found—but No Black Boxes

Some gruesome remains are discovered in the WTC ruins. Investigators find a pair of severed hands bound together with plastic handcuffs on a nearby building. They are believed to have belonged to a flight attendant. [NEWSDAY, 9/15/01] There are reports of whole rows of seats with passengers in them being found, as well as much of the cockpit of one of the planes, complete with the body of one of the hijackers, and the body of another stewardess, whose hands were tied with wire. [ANANOVA, 9/13/01; NEW YORK TIMES, 9/15/01 (D)] Yet, contradicting the claim that a hijacker's body was found, only in February 2003 are the remains of two hijackers identified. While all of these bodies and plane parts are supposedly found, not one of the four black boxes for these two airplanes is ever found. A National Transportation Safety Board spokesperson says, "It's extremely rare that we don't get the recorders back. I can't recall another domestic case in which we did not recover the recorders." [CBS NEWS, 2/23/02] The black boxes are considered "nearly indestructible," are placed in the safest parts of the aircraft, and are designed to survive impacts much greater than the WTC impact. They can withstand heat of up to 2,000 degrees Fahrenheit for one hour, and can withstand an impact of an incredible 3,400 G's. [ABC NEWS, 9/17/01]

September 15–17, 2001: Did Some Hijackers Get U.S. Military Training?

A series of articles suggest that at least seven of the 9/11 hijackers trained in U.S. military bases. [NEW YORK TIMES, 9/15/01 (E); NEWSWEEK, 9/15/01] Ahmed Alnami, Ahmed Alghamdi, and Saeed Alghamdi even listed the Naval Air Station in Pensacola, Florida as their permanent address on their driver's licenses. [PENSACOLA NEWS JOURNAL, 9/17/01] Hamza Alghamdi was also connected to the Pensacola base. [WASHINGTON POST, 9/16/01] A defense official confirms that Saeed Alghamdi is a former Saudi fighter pilot who attended the Defense Language Institute in Monterey, California. [LOS ANGELES TIMES, 9/15/01; GANNETT NEWS SERVICE, 9/17/01] Abdulaziz Alomari attended Brooks Air Force Base Aerospace Medical School in San Antonio, Texas. [GANNETT NEWS SERVICE, 9/17/01] A defense official confirms Mohamed Atta is a former Saudi fighter pilot who graduated from the U.S. International Officers School at Maxwell Air Force Base, Alabama. [LOS ANGELES TIMES 9/15/01; WASHINGTON POST 9/16/01] The media stops looking into the hijackers' possible U.S. military connections after the Air Force makes a not-very-definitive statement, saying that while the names are similar, "we are probably not talking about the same people." [WASHINGTON POST, 9/16/01]

September 16, 2001: Bin Laden Denies Involvement in 9/11 Attacks

Confirming earlier reports [REUTERS, 9/13/01 (D)], bin Laden denies any involvement in the 9/11 attacks. In a statement to Al Jazeera, he states, "I would like to assure the world that I did not plan the recent attacks, which seems to have been planned by people for personal reasons." [CNN, 9/17/01 (C)] The U.S.

claims that he confesses his role in a video message two months later, but the contents of that video are highly disputed.

September 16, 2001 (B): Usual Investigative Procedures Not Followed in Examining Flight 93 Wreckage

A report suggests the crash site of Flight 93 is being searched and recorded in 60 square-foot grids. [DELAWARE NEWS JOURNAL, 9/16/01] This approach is preferred by the two forensic scientists in charge of the crash, who say that doing so can help determine who was where when the plane crashed, and possibly how it crashed. However, almost a year later it comes out that this approach is not followed: "The FBI overruled them, instead dividing the site into five large sectors. It would be too time-consuming to mark tight grids, and would serve no real investigative purpose, the bureau decided. There was no mystery to solve about the crash. Everybody knew what happened to the plane." [AMONG THE HEROES, JERE LONGMAN, 8/02, PP. 262–63] While the military may suggest there is no mystery, some articles have suggested the plane was shot down. (For example, [PHILADELPHIA DAILY NEWS, 11/15/01; INDEPENDENT, 8/13/02].) In addition, at the time of this decision, investigators were still considering the possibility a bomb might have destroyed the plane.

September 16–23, 2001: People with Hijacker Names and Identifying Details Are Still Alive

Reports appear in many newspapers suggesting that some of the people the U.S. says were 9/11 hijackers are actually still alive:

- Hamza Alghamdi: No media outlet has claimed that Hamza Alghamdi is still alive, but his family says the FBI photo "has no resemblance to him at all." [WASHINGTON POST, 9/25/01]

- Saeed Alghamdi is alive and flying airplanes in Tunisia. [LOS ANGELES TIMES, 9/21/01; DAILY TELEGRAPH, 9/23/01; BBC, 9/23/01] He says he studied flight training in a Florida flight schools for parts of the years, 1998, 1999, 2000, and 2001. [ARAB NEWS, 9/18/01] The *Daily Telegraph* notes, "The FBI had published [Saeed Alghamdi's] personal details but with a photograph of somebody else, presumably a hijacker who had 'stolen' his identity. CNN, however, showed a picture of the real Mr. Alghamdi." [DAILY TELEGRAPH, 9/23/01] If this account is true, as of mid-2004 the FBI is still using the wrong photograph of Alghamdi.

- Salem Alhazmi is alive and working at a petrochemical plant in Yanbou, Saudi Arabia. [LOS ANGELES TIMES, 9/21/01; DAILY TELEGRAPH, 9/23/01] He says his passport was stolen by a pickpocket in Cairo three years ago and that pictures and details such as date of birth are of him. [GUARDIAN, 9/21/01 (C), WASHINGTON POST, 9/20/01; SAUDI GAZETTE, 9/29/02]

- Ahmed Alnami is alive and working as an administrative supervisor with Saudi Arabian Airlines, in Riyadh, Saudi Arabia. [LOS ANGELES TIMES, 9/21/01] He had never lost his passport and

found it "very worrying" that his identity appeared to have been stolen. [DAILY TELEGRAPH, 9/23/01]
However, there is another "Ahmed Alnami" who is ten years younger, and appears to be dead,
according to his father. [ABC NEWS, 3/15/02] Ahmed Alnami's family says his FBI picture is correct.
[WASHINGTON POST, 9/25/01]

- Abdulaziz Alomari is alive and working as a pilot for Saudi Arabian Airlines. [NEW YORK TIMES,
  9/16/01; INDEPENDENT, 9/17/01; BBC, 9/23/01] He claims that his passport was stolen in 1995 while he
  was living in Denver, Colorado. [LOS ANGELES TIMES, 9/21/01] "They gave my name and my date of
  birth, but I am not a suicide bomber. I am here. I am alive." [DAILY TELEGRAPH, 9/23/01; LONDON TIMES,
  9/20/01]

- Marwan Alshehhi may be alive in Morocco. [SAUDI GAZETTE, 9/18/01; KHALEEJ TIMES, 9/20/01] Family and
  neighbors do not believe he took part in the attacks. [REUTERS, 9/18/01]

- Mohand Alshehri: The Saudi government has claimed that Mohand Alshehri is alive and that
  he was not in the U.S. on 9/11, but no more details are known. [ASSOCIATED PRESS, 9/29/01 (B)]

- The brothers Waleed M. Alshehri and Wail Alshehri are alive. A Saudi spokesman said, "This
  is a respectable family. I know his sons, and they're both alive." The father is a diplomat who
  has been stationed in the U.S. and Bombay, India. [LOS ANGELES TIMES, 9/21/01; ARAB NEWS, 9/19/01]
  There is a second pair of Saudi brothers named Wail and Waleed M. who may have been the
  real hijackers. Their father says they have been missing since December 2000. [ABC NEWS, 3/15/02;
  ARAB NEWS, 9/17/01] The still-living Waleed M. Alshehri is a pilot with Saudi Airlines, studying in
  Morocco. [LOS ANGELES TIMES, 9/21/01; ASSOCIATED PRESS, 9/22/01] He acknowledges that he attended
  flight training school at Dayton Beach in the United States. [BBC, 9/23/01; DAILY TRUST, 9/24/01] He
  was interviewed by U.S. officials in Morocco, and cleared of all charges against him (though
  apparently the FBI photos are still of him). [EMBRY RIDDLE AERONAUTICAL UNIVERSITY PRESS RELEASE, 9/21/01]
  The still living Wail Alshehri is also apparently a pilot. [LOS ANGELES TIMES, 9/21/01] He claims that
  he saw his picture on CNN and recognized it from when he studied flying in Florida. But he
  also says that he has no brother named Wail. [AS-SHARQ AL-AUSAT, 9/22/01]

- Mohamed Atta's father says he spoke to his son on the phone on September 12, 2001. [NEW
  YORK TIMES, 9/19/01; CHICAGO TRIBUNE, 9/20/01]

- Khalid Almihdhar: On September 19, the Federal Deposit Insurance Corp. distributes a
  "special alert" to its member banks asking for information about the attackers. The list
  includes "Al-Midhar, Khalid. Alive." The Justice Department later calls this a "typo." [ASSOCI-
  ATED PRESS, 9/20/01; COX NEWS SERVICE, 10/21/01] The BBC says, "There are suggestions that another
  suspect, Khalid Almihdhar, may also be alive." [BBC, 9/23/01] The *Guardian* says Almihdhar is
  believed to be alive, but investigators are looking into three possibilities. Either his name was

stolen for a hijacker alias, or he allowed his name to be used so that U.S. officials would think he died, or he died in the crash. [GUARDIAN, 9/21/01 (B)]

• Majed Moqed was last seen by a friend in Saudi Arabia in 2000. This friend claims the FBI picture does not look like Moqed. [ARAB NEWS, 9/22/01]

The Saudi government insists that five of the Saudis mentioned are still alive. [NEW YORK TIMES, 9/21/01] On September 20, FBI Director Mueller says, "We have several others that are still in question. The investigation is ongoing, and I am not certain as to several of the others." [NEWSDAY, 9/21/01] On September 27, after all of these revelations mentioned above are revealed in the media, FBI Director Mueller states, "We are fairly certain of a number of them." [SOUTH FLORIDA SUN-SENTINEL, 9/28/01] On September 20, the *London Times* reported, "Five of the hijackers were using stolen identities, and investigators are studying the possibility that the entire suicide squad consisted of impostors." [LONDON TIMES, 9/20/01] The mainstream media briefly doubted some of the hijackers' identities. For instance, a story in the *Observer* on September 23 put the names of hijackers like Saeed Alghamdi in quotation marks. [OBSERVER, 9/23/01] However, the story will die down after the initial reports, and it is hardly noticed when Mueller states on November 2, 2001, "We at this point definitely know the 19 hijackers who were responsible," and claims that the FBI is sticking with the names and photos released in late September. [ASSOCIATED PRESS, 11/03/02]

September 17, 2001: Knife Found at Flight 93 Crash Scene

A confidential FBI bulletin states a "badly damaged" commercially manufactured cigarette lighter with a concealed knife blade has been recovered at the Flight 93 crash scene. The knife was about two and three-fourths inches long, with a knife blade of about two and a half inches. [LOS ANGELES TIMES, 9/18/01]

September 17, 2001 (B): White House Meeting Leads to Cover-up?

In a later 9/11 Commission hearing, Commissioner Bob Kerrey says that NORAD gives a briefing at the White House on the day. He adds, "[A]nd it feels like something happened in that briefing that produced almost a necessity to deliver a story that's different than what actually happened on that day." [9/11 COMMISSION, 6/17/04] The next day, NORAD releases a timeline of 9/11 events detailing fighter response times. The 9/11 Commission later strongly disputes many details from NORAD's timeline. For instance, the timeline claims that NORAD is notified about the hijacking of Flight 93 at 9:16 A.M., but the commission concludes that when the plane crashes after 10:00 A.M., NORAD still has not been notified. [NORAD, 9/18/01; 9/11 COMMISSION REPORT, 6/17/04]

September 19, 2001: Unverified Reports
of Additional Flights to Be Hijacked

The FBI claims on this day that there were six hijacking teams on the morning of 9/11. [NEW YORK TIMES, 9/19/01 (B), GUARDIAN, 10/13/01] A different report claims investigators are privately saying eight. [INDEPENDENT,

9/25/01] However, the reports below suggest there may have been as many as eight aborted flights, leading to a potential total of 12 hijackings:

- Knives of the same type used in the successful hijackings were found taped to the backs of fold-down trays on a Continental Airlines flight from Newark. [GUARDIAN, 9/19/01]

- The FBI is investigating American Airlines Flight 43, which was scheduled to leave Boston about 8:10 A.M. bound for Los Angeles but was canceled minutes before takeoff due to a mechanical problem. [BBC, 9/18/01 (C), CHICAGO TRIBUNE, 9/18/01; GUARDIAN, 9/19/01] Another version claims the flight left from Newark and made it as far as Cincinnati before being grounded in the nationwide air ban. [NEW YORK TIMES, 9/19/01 (B)]

- Knives and box cutters were found on two separate canceled Delta Airlines planes later that day, one leaving Atlanta for Brussels and the other leaving from Boston. [TIME, 9/22/01; INDEPENDENT, 9/25/01]

- On September 14, two knives were found on an Air Canada flight that would have flown to New York on 9/11 if not for the air ban. [CNN, 10/15/01]

- Two men arrested on 9/11 may have lost their nerve on American Airlines Flight 1729 from Newark to San Antonio via Dallas that was scheduled to depart at 8:50, and was later forced to land in St. Louis. Alternately, they may have been planning an attack for September 15, 2001. [NEW YORK TIMES, 9/19/01 (B)]

- There may have been an attempt to hijack United Airlines Flight 23 flying from Boston to Los Angeles around 9:00 A.M. Shortly after 9:00 A.M., United Airlines flight dispatcher Ed Ballinger sent out a warning about the first WTC crash to the flights he was handling. Because of this warning, the crew of Flight 23 told the passengers it had a mechanical problem and immediately returned to the gate. Ballinger was later told by authorities that six men initially wouldn't get off the plane. When the men finally disembarked, they disappeared into the crowd and never returned. Later, authorities checked their luggage and found copies of the Koran and al-Qaeda instruction sheets. [CHICAGO DAILY HERALD, 04/14/04] In mid-2002, a NORAD deputy commander says "we don't know for sure" if Flight 23 was to have been hijacked. [GLOBE AND MAIL, 6/13/02]

- Knives were found stashed in the seats on a plane due to leave Boston that was delayed due to technical problems and then canceled. [GUARDIAN, 10/13/01]

- A box cutter knife was found under a seat cushion on American Airlines Flight 160, a 767 that would have flown from San Diego to New York on the morning of 9/11 but for the air ban. [CHICAGO TRIBUNE, 9/23/01]

The FBI is said to be seeking a number of passengers who failed to board the same, rescheduled flights when the grounding order on commercial planes in the U.S. was lifted. [BBC, 9/18/01 (C)] The *Independent* points out suspicions have been fueled "that staff at U.S. airports may have played an active role in the conspiracy and helped the hijackers to circumvent airport security." They also note, "It is possible that at least some of the flights that have come under scrutiny were used as decoys, or as fallback targets." [INDEPENDENT, 9/25/01]

September 19, 2001 (B): Atta's Father Claims Son Was Framed

Mohamed Atta's father holds a press conference in Cairo and makes a number of surprising claims. He believes that the Mossad, Israel's spy agency, did the 9/11 attacks, and stole his son's identity. He claims that Atta was a mama's boy prone to airsickness, a dedicated architecture student who rarely mentioned politics, and a victim of an intricate framing. He says that Atta spoke to him on the phone on September 12 about "normal things," one day after he was supposed to be dead. Atta called his family about once a month, yet never told them he was in the U.S., continuing to say he was studying in Germany. Atta's family never saw him after 1999, and Atta canceled a trip to visit them in late 2000. His father even shows a picture of his son, claiming he looks similar but not the same as the terrorist Atta. [NEWSWEEK, 9/24/01 (B), NEW YORK TIMES, 9/19/01; CHICAGO TRIBUNE, 9/20/01] He also says that the man pictured in published photos from an airport surveillance camera had a heavier build than his son. [CAIRO TIMES, 9/20/01] A year later, he still believes his son is alive. [GUARDIAN, 9/2/02]

September 21, 2001: Algerian Pilot Mistakenly Arrested

Lotfi Raissi, an Algerian pilot living in Britain, is arrested and accused of helping to train four of the hijackers. An FBI source says, "We believe he is by far the biggest find we have had so far. He is of crucial importance to us." [LAS VEGAS REVIEW JOURNAL, 9/29/01] However, in April 2002 a judge dismisses all charges against him. U.S. officials originally said, "They had video of him with Hani Hanjour, who allegedly piloted the plane that crashed into the Pentagon; records of phone conversations between the two men; evidence that they had flown a training plane together; and evidence that Raissi had met several of the hijackers in Las Vegas. It turned out, the British court found, that the video showed Raissi with his cousin, not Mr. Hanjour, that Raissi had mistakenly filled in his air training logbook and had never flown with Hanjour, and that Raissi and the hijackers were not in Las Vegas at the same time. The U.S. authorities never presented any phone records showing conversations between Raissi and Hanjour. It appears that in this case the U.S. authorities handed over all the information they had . . . " [CHRISTIAN SCIENCE MONITOR, 3/27/02] Raissi later says he will sue the British and American governments unless he is given a "widely publicized apology" for his months in prison and the assumption of "guilty until proven innocent." [REUTERS, 8/14/02] In September 2003, he does sue both governments for $20 million. He also wins a undisclosed sum from the British tabloid *Mail on Sunday* for printing false charges against him. [GUARDIAN, 9/16/03; ARIZONA REPUBLIC, 10/14/03; BBC, 10/7/03]

September 21, 2001 (B): Report Suggests There
Are Confiscated Videos of Pentagon Crash

A report suggests, "Federal investigators may have video footage of the deadly terrorist attack on the Pentagon. A security camera atop a hotel close to the Pentagon may have captured dramatic footage of the hijacked Boeing 757 airliner as it slammed into the western wall of the Pentagon. Hotel employees sat watching the film in shock and horror several times before the FBI confiscated the video as part of its investigation. It may be the only available video of the attack. The Pentagon has told broadcast news reporters that its security cameras did not capture the crash. The attack occurred close to the Pentagon's heliport, an area that normally would be under 24-hour security surveillance, including video monitoring." [GERTZ FILE, 9/21/01] In a later report, an employee at a gas station across the street from the Pentagon that services only military personnel says the gas station's security cameras should have recorded the moment of impact. However, he says, "I've never seen what the pictures looked like. The FBI was here within minutes and took the film." [RICHMOND TIMES-DISPATCH, 12/11/01] A later release of five tiny and grainy images of the crash from a Pentagon security camera shows the government's claim that no security cameras captured the crash was untrue.

September 24, 2001: Terrorists Reportedly
Stole Pilot Uniforms, Sat in Cockpits

Fox News claims that up to 12 other Middle Eastern men dressed in pilot uniforms were on other flights scheduled to take off on the morning of 9/11. Hijackings on all these flights were foiled when an unexpected ban on new flights prevented them from taking off. An FBI source says they had been invited into the cockpits under the impression that they were guest pilots from other airlines. It is standard practice to give guest pilots the spare seat in the cockpit known as the jump seat. [FOX NEWS, 9/24/01] Flight 93's cockpit voice recording has apparently shown that "one of the four hijackers had been invited into the cockpit area before the flight took off." Many pilot uniforms had gone missing prior to 9/11. It is claimed that Mohamed Atta was given a guided tour of Boston's Logan Airport the week before 9/11 when he turned up in a pilot uniform saying he was with Saudi Airlines. [HERALD SUN, 9/25/01]

September 25, 2001: Several 9/11 Passengers Have
Possible Connections to Pilotless Aircraft Program

As details of the passengers on the four hijacked flights emerge, some are shown to have curious connections to the defense company Raytheon, and possibly its Global Hawk pilotless aircraft program. Stanley Hall (Flight 77) was director of program management for Raytheon Electronics Warfare. One Raytheon colleague calls him "our dean of electronic warfare." [ASSOCIATED PRESS, 9/25/01] Peter Gay (Flight 11) was Raytheon's vice president of operations for Electronic Systems and had been on special assignment to a company office in El Segundo, California. [ASSOCIATED PRESS, 9/25/01] Raytheon's El Segundo's Electronic Systems division is one of two divisions making the Global Hawk. [ISR JOURNAL, 3/02] Kenneth

Waldie (Flight 11) was a senior quality control engineer for Raytheon's electronic systems. David Koval-cin (Flight 11) was a senior mechanical engineer for Raytheon's electronic systems. [CNN, 9/01] Herbert Homer (Flight 175) was a corporate executive working with the Department of Defense. [CNN, 9/01 (B), NORTHEASTERN UNIVERSITY VOICE, 12/11/01] A surprising number of passengers, especially on Flight 77, have military connections. For instance, William E. Caswell was a Navy scientist whose work was so classified that his family knew very little about what he did each day. Says his mother, "You just learn not to ask questions." [CHICAGO TRIBUNE, 9/16/01]

September 28, 2001: Text of Atta Note Is Made Public, Authenticity Is Disputed

The text of a handwritten, five-page document found in Atta's luggage is made public. [OBSERVER, 9/30/01 (C)] The next day, the *Independent* strongly questions if the note is genuine. It points out the "note suggests an almost Christian view of what the hijackers might have felt" and is filled with "weird" comments that Muslims would never say, such as "the time of fun and waste is gone." If the note "is genuine, then the [hijackers] believed in a very exclusive version of Islam—or were surprisingly unfamiliar with their religion." [INDEPENDENT, 9/29/01] Another copy of the document was discovered in a vehicle parked by a Flight 77 hijacker at Washington's Dulles airport. A third copy of essentially the same document was found in the wreckage of Flight 93. Therefore, the letter neatly ties most of the hijackers together. [CBS NEWS, 9/28/01] The *Guardian* says, "The finds are certainly very fortunate, though some might think them a little too fortunate." [GUARDIAN, 10/1/01] Interestingly, an FBI affidavit of the contents of Atta's baggage written on September 14, 2001, and released on October 4 fails to mention the how-to letter.

September 29, 2001: No Video Cameras in Boston's Logan Airport; Footage from Other Airports Remains Classified

It is reported that Boston's Logan Airport has no cameras in its terminals, gate areas, or concourses. It is possibly the only major airport in the U.S. not to have such cameras. The two other airports used by the hijackers to launch the 9/11 attacks had security cameras, but none of the footage has been released. [BOSTON HERALD, 9/29/01] It was previously reported that FBI agents had "examined footage from dozens of cameras at the three airports where the terrorists boarded the aircraft." [LOS ANGELES TIMES, 9/13/01]

October 2001: FBI Recovers Hijacker E-Mails

Reports this month indicate that many hijacker e-mails have been recovered. *USA Today* reports many unencrypted e-mails coordinating the 9/11 plans written by the hijackers in Internet cafes have been recovered by investigators. [USA TODAY, 10/1/01] FBI sources say, "[H]undreds of e-mails linked to the hijackers in English, Arabic and Urdu" have been recovered, with some messages including "operational details" of the attack. [WASHINGTON POST, 10/4/01] "A senior FBI official says investigators have obtained hundreds of e-mails in English and Arabic, reflecting discussions of the planned September 11 hijackings."

[WALL STREET JOURNAL, 10/16/01] However, in April 2002, FBI Director Mueller says no documentation of the 9/11 plot has been found. By September 2002, the *Chicago Tribune* reports, "Of the hundreds, maybe thousands, of e-mails sent and received by the hijackers from public Internet terminals, none is known to have been recovered." [CHICAGO TRIBUNE, 9/5/02] The texts of some e-mails sent by Atta from Germany are published a few months later. [CHICAGO TRIBUNE, 2/25/03]

October 1, 2001: Taliban Possibly Trained Pilots in Afghanistan

It is reported that "a worldwide hunt is under way for 14 young Muslims said to have been trained in secret to fly Boeing airliners at an air base in Afghanistan. A senior pilot for the Afghan state-owned airline Ariana has told how he and four colleagues were forced by the Taliban regime to train the men who are now thought to be hiding in Europe and the United States. The fourteen men, seven of whom are said to speak fluent English, are described as 'dedicated Muslim fanatics' who spoke of being involved in a holy war. They are thought to have left Afghanistan a year ago. All had close links with the Taliban and some had fought for the regime." [EVENING STANDARD, 10/1/01]

October 2, 2001: Remote Controlled Passenger Airplane Flew Before 9/11, Despite Claims to the Contrary

It is reported that the U.S. company Raytheon landed a 727 six times in a military base in New Mexico without any pilots on board. This was done to test equipment making future hijackings more difficult, by allowing ground control to take over the flying of a hijacked plane. [ASSOCIATED PRESS, 10/2/01 (C); DER SPIEGEL, 10/28/01] Several Raytheon employees with possible ties to this remote control technology program appear to have been on the hijacked 9/11 flights. Earlier in the year, a specially designed Global Hawk plane flew from the U.S. to Australia without pilot or passengers. [ITN, 4/24/01] However, most media reports after 9/11 suggest such technology is currently impossible. For instance, the *Observer* quotes an expert who says that "the technology is pretty much there" but still untried. [OBSERVER, 9/16/01] An aviation-security expert at *Jane's Defence Weekly* says this type of technology belongs "in the realms of science fiction." [FINANCIAL TIMES, 9/18/01 (B), ECONOMIST, 9/20/01] Even President Bush appears to deny the technology currently exists. He gives a speech after 9/11 in which he mentions that the government would give grants to research "new technology, probably far in the future, allowing air traffic controllers to land distressed planes by remote control." [NEW YORK TIMES, 9/28/01]

October 4, 2001: Blair Presents Case for al-Qaeda 9/11 Involvement

British Prime Minister Tony Blair publicly presents a paper containing evidence that al-Qaeda is responsible for the 9/11 attacks. [LOS ANGELES TIMES, 10/5/01; LOS ANGELES TIMES, 10/4/01] Secretary of State Powell and other U.S. officials had promised on September 23 that the U.S. would present a paper containing such evidence. [LOS ANGELES TIMES, 9/24/01] However, the U.S. paper is never released. Apparently, the British paper is meant to serve as a substitute. [NEW YORKER, 5/27/02] In the speech, Blair claims, "One of bin Laden's closest lieutenants has said clearly that he helped with the planning of the September 11 attacks and admitted the involvement of the al-Qaeda organization" and that "there is other intelligence, we cannot disclose, of an even more direct nature indicating guilt" of al-Qaeda in the attacks. [CNN, 10/4/01; TIME, 10/5/01] There has been no confirmation or details since of these claims. Even though most of the evidence in the British paper comes from the U.S., pre-attack warnings, such as the August 6, 2001 memo to Bush titled "Bin Laden Determined to Strike in U.S.," are not included. In fact, Blair's paper states, "incorrectly, that no such information had been available before the attacks: 'After 11 September we learned that, not long before, bin Laden had indicated he was about to launch a major attack on America.'" [NEW YORKER, 5/27/02]

October 10, 2001: Famous Arab Commentator Says al-Qaeda Could Not Have Conducted 9/11 Attacks

Mohammed Heikal, longtime Egyptian journalist, former government spokesman, and the "Arab world's foremost political commentator," expresses disbelief that bin Laden and al-Qaeda could have conducted the 9/11 attack without the U.S. knowing. "Bin Laden has been under surveillance for years: every telephone call was monitored and al-Qaeda has been penetrated by American intelligence, Pakistani intelligence, Saudi intelligence, Egyptian intelligence. They could not have kept secret an operation that required such a degree of organization and sophistication." [GUARDIAN, 10/10/01]

October 10, 2001 (B): Baggage Handling Company Cleared of Wrongdoing

It is reported that Globe Aviation Services Corp., in charge of the baggage handlers for Flight 11 and all other American Airlines flights at Boston's Logan Airport, have been cleared of any wrongdoing. Globe Aviation supervisors claim that none of the employees working that day was in the U.S. illegally. Supposedly, no weapons were detected, but a baggage handler for Globe Aviation and an American Airlines has told the FBI that one of the hijackers—believed to be either Wail or Waleed Alshehri—was carrying one wooden crutch under his arm when he boarded Flight 11. Crutches are apparently routinely scanned through X-ray machines. [BOSTON GLOBE, 10/10/01 (B)]

October 11, 2001: FBI Claims 11 Terrorists Unaware They Were on Suicide Mission

According to an FBI report, "FBI investigators have officially concluded that 11 of the 19 terrorists who hijacked the aircraft on September 11 did not know they were on a suicide mission." "Unlike the eight 'lead' attackers, who were all trained pilots, they did not leave messages for friends and family indicating they knew their lives were over," and they did not have copies of Atta's final prayer note. Personal items found suggest the men thought they were taking part in a conventional hijacking and were preparing for the possibility of prison. [OBSERVER, 10/14/01] This is later contradicted by video filmed in Afghanistan in March 2001 showing several of the 13 non-lead hijackers proclaiming their willingness to die on an upcoming suicide mission.

October 11, 2001 (B): Ashcroft Takes Over All Terrorist Prosecutions

The Ashcroft-led Justice Department assumes control of all terrorist prosecutions from the U.S. Attorneys office in New York, which has had a highly successful record of accomplishment in prosecuting terrorist cases connected to bin Laden. [NEW YORK TIMES, 10/11/01]

October 16, 2001: Some Flight Control
Transcripts Released, but Sections Are Missing

The government releases flight control transcripts of three of the four hijacked planes [NEW YORK TIMES, 10/16/01 (B), NEW YORK TIMES, 10/16/01 (C), NEW YORK TIMES, 10/16/01 (D)]. Strangely, Flight 93 is left out. Yet even the three released transcripts are incomplete (for instance, Flight 77's ends at least 20 minutes before it crashes), and certain events that are part of the official story do not show up on these transcripts.

October 27, 2001: Officials Furious over NSC Lack of Cooperation

Furious government intelligence officials accuse the NSA of destroying data pertinent to the 9/11 investigation. They claim that possible leads are not being followed because of the NSA's lack of cooperation. [BOSTON GLOBE, 10/27/01]

November 15, 2001: Newspaper Questions Whether Flight 93 was Shot Down

For the first time, a major newspaper publishes an article strongly suggesting Flight 93 was shot down. The *Philadelphia Daily News* quotes numerous eyewitnesses who believe the plane was shot down. The FBI has reported a half-ton piece of an engine was found "a considerable distance" from the main crash site. "That information is intriguing to shoot-down theory proponents, since the heat-seeking, air-to-air Sidewinder missiles aboard an F-16 would likely target one of the Boeing 757's two large engines." The article concludes, "No one has fully explained why the plane went down, or what exactly happened during an eight-minute gap from the time all cell phone calls from the plane stopped and the time it crashed." [PHILADELPHIA DAILY NEWS, 11/15/01]

November 21, 2001: Flight 77 Remains Identified,
Hijackers' Identities Not Confirmed

The remains of all but one of the people on board Flight 77, including the hijackers, are identified. However, the identities of the hijackers have still not been confirmed through their remains [WASHINGTON POST, 11/21/01; MERCURY, 1/11/02], and the FBI never provides DNA profiles of the hijackers to medical examiners for identification. Strangely, the official position is that there was a giant fireball on impact that not only destroyed the airplane, but actually vaporized the metal. A rescue worker states: "The only way you could tell that an aircraft was inside was that we saw pieces of the nose gear. The devastation was horrific." [NFPA JOURNAL, 11/1/01] As of mid-2004, there still have been no reports that the hijackers' remains have been identified by their DNA, except possibly for two unnamed hijackers.

November 23, 2001: Report Suggests Hijackers Snuck into Cockpits

The *Boston Globe* reports information strongly suggesting that at least one hijacker was inside the cockpits on every flight before the 9/11 hijackings began. An airplane captain theorizes how they took control: "The most likely scenarios are something that was swift, where the pilots couldn't have changed their transponder code and called the controllers. You think four times in one morning one of those crews would have done that. That means they had to be upon them before they could react." On practice flights before 9/11, the hijackers repeatedly obtained access to cockpits by various methods. Perhaps the most important method was jumpseating, which allows certified airline pilots to use a spare seat in the cockpit when none is available in the passenger cabin. Airlines reciprocate to help pilots get home or to the city of their originating flight. Officials say they do not believe any of the hijackers were jumpseating on 9/11 despite media reports to the contrary. However, since 9/11 the FAA has banned the practice unless a pilot works for the airline in whose cockpit that person wants to ride. [*BOSTON GLOBE*, 11/23/01] The 9/11 Commission later concludes that the hijackers didn't use jumpseating because they couldn't find any paperwork relating to jumpseat requests.

December 13, 2001: U.S. Releases bin Laden Video; Authenticity Questioned

The U.S. releases a video of bin Laden that seems to confirm his role in the 9/11 attack. [*GUARDIAN*, 12/13/01] However, a number of strange facts about this video soon emerge. For example, all previous videos had been made with the consent of bin Laden, and usually released to the Arabic television channel Al Jazeera. This video was supposedly recorded without his knowledge, found in a house in Afghanistan, and then passed to the CIA by an unknown person or group. Experts point out that it would be possible to fake such a video. So many people doubt the video's authenticity that Bush soon makes a statement, saying it was "preposterous for anybody to think this tape was doctored. Those who contend it's a farce or a fake are hoping for the best about an evil man." Some observers point out that bin Laden is wearing a ring on his right hand. In previous films, he had worn no jewelry apart from a watch. [*GUARDIAN*, 12/15/01] The German television show "Monitor" conducts an independent translation that questions the translation given by the U.S. military. According to Professor Gernot Rotter, scholar of Islamic and Arabic Studies at the University of Hamburg, "This tape is of such poor quality that many passages are unintelligible. And those that are intelligible have often been taken out of context, so that you can't use that as evidence. The American translators who listened to the tape and transcribed it obviously added things that they wanted to hear in many places." [*MONITOR*, 12/20/01] There are reports that bin Laden had from four to ten look-alike doubles at the time. [AGENCE FRANCE-PRESSE, 10/7/01; *LONDON TIMES*, 11/19/01]

December 21, 2001: FBI Won't Release Flight 93 Black Box Information

The FBI reveals that it knows what is on the Flight 93 black boxes, but refuses to release the transcript or audio recording. Families of the victims have requested to hear the cockpit voice recording, but the FBI says, "[W]e do not believe that the horror captured on the cockpit voice recording will console them in any way." [CNN, 12/21/01] Accuracy in Media immediately submits a Freedom of Information Act request to have the transcript released, but the FBI turns it down because a release "could reasonably be expected to interfere with enforcement proceedings." The *Philadelphia Daily News* asks, "What enforcement proceedings?" and suggests the FBI may be covering up a shoot down of the plane. [PHILADELPHIA DAILY NEWS, 12/28/01] The recordings are later played, but only in private to victims' relatives and the 9/11 Commission.

December 25, 2001: Experts: WTC Collapse Investigation "Inadequate"

The *New York Times* reports that "some of the nation's leading structural engineers and fire-safety experts" believe the investigation into the collapse of the WTC is "inadequate," and "are calling for a new, independent and better-financed inquiry that could produce the kinds of conclusions vital for skyscrapers and future buildings nationwide." Experts critical of the investigation include "some of those people who are actually conducting it." They point out that the current team of 20 or so investigators has no subpoena power, inadequate financial support, and little staff support. Additionally, it has been prevented from interviewing witnesses and frequently prevented from examining the disaster site, and has even been unable to obtain basic information like detailed blueprints of the buildings that collapsed. The decision to recycle the steel columns, beams, and trusses from the WTC rapidly in the days immediately after 9/11 means definitive answers may never be known. [NEW YORK TIMES, 12/25/01] Incredibly, some of the steel is reforged into commemorative medallions selling for $30 apiece. [ASSOCIATED PRESS, 1/30/02]

January 2002 (F): 9/11 Flight Control Recording Completely Destroyed

Shortly before noon on 9/11, about sixteen people at the New York Air Route Traffic Control Center recorded their version of the response to the 9/11 attack. At least six are air traffic controllers who dealt with two of the hijacked airliners. But officials at the center never tell higher-ups about the tape. Around this time, a quality-assurance manager, whose name has not been released, crushes the cassette recording in his hand, shreds the tape, and drops the pieces into different trashcans. This manager later asserts that keeping the tape would have been a violation of union rules and accident procedures. When he destroyed the tape, he had already received an e-mail from the FAA instructing officials to safeguard all records that specifically stated, "If a question arises whether or not you should retain data, RETAIN IT." Most, but not all, of the air traffic controllers involved make written statements about three weeks after 9/11, but it isn't clear how these might differ with what was on the tape. The unidentified manager is later said to be disciplined for this incident, though it isn't clear how. [WASHINGTON POST, 5/6/04]

January 4, 2002: Firefighter Magazine Scolds WTC Investigation

A firefighter trade magazine with ties to the New York Fire Department calls the investigation into the collapse of the WTC a "half-baked farce." The article points out that the probe has not looked at all aspects of the disaster and has had limited access to documents and other evidence. "The destruction and removal of evidence must stop immediately." It concludes that a growing number of fire protection engineers have theorized that "the structural damage from the planes and the explosive ignition of jet fuel in themselves were not enough to bring down the towers." [NEW YORK DAILY NEWS, 1/4/02; FIRE ENGINEERING, 1/02]

January 13, 2002: Former German Minister Believes CIA Is Responsible for 9/11

Andreas von Bülow, former German Minister for Research and Technology and a long-time member of German parliament, suggests in an interview that the CIA could have been behind the 9/11 attacks. He states: "Whoever wants to understand the CIA's methods, has to deal with its main task of covert operations: Below the level of war, and outside international law, foreign states are to be influenced by inciting insurrections or terrorist attacks, usually combined with drugs and weapons trade, and money laundering. . . . Since, however, it must not under any circumstances come out that there is an intelligence agency behind it, all traces are erased, with tremendous deployment of resources. I have the impression that this kind of intelligence agency spends 90 percent of its time this way: creating false leads. So that if anyone suspects the collaboration of the agencies, he is accused of paranoia. The truth often comes out only years later." [DER TAGESSPIEGEL, 1/13/02] In an example of covering tracks, Ephraim Halevy, head of Israel's Mossad from 1998 till 2002, claims, "Not one big success of the Mossad has ever been made public." [CBS NEWS, 2/5/03]

January 24, 2002: Cheney and Bush Pressure Senator to Avoid 9/11 Inquiry

Senate Majority Leader Tom Daschle (D) later claims that on this day, Vice President Cheney calls him and urges that no 9/11 inquiry be made. Bush repeats the request on January 28, and Daschle is repeatedly pressured thereafter. *Newsweek* summarizes one of these conversations: "Bush administration officials might say they're too busy running the war on terrorism to show up. Press the issue . . . and you risk being accused of interfering with the mission." [NEWSWEEK, 2/4/02] Cheney later disagrees: "Tom's wrong. He has, in this case, let's say a misinterpretation." [REUTERS, 5/27/02]

Early March 2002: French Author Claims Plane Crash into Pentagon Was Staged

The book *l'Effroyable Imposture (The Horrifying Fraud)* is published in France. The book denies that a passenger airliner crashed into the Pentagon on 9/11. It is written by Thierry Meyssan, "president of the Voltaire Network, a respected independent think tank whose left-leaning research projects have until now been considered models of reasonableness and objectivity." [GUARDIAN, 4/1/02] The book is widely denounced by the media (See, for example, [AGENCE FRANCE-PRESSE, 3/21/02; LONDON TIMES, 5/19/02; NATIONAL POST, 8/31/02; BALTIMORE SUN, 9/12/02]). One reporter heavily criticizes the book even while admitting never to have read it. [LA WEEKLY, 7/19/02] In France, however, the book sets a publishing record for first-month sales. [TIME (EUROPE VERSION), 5/20/02] One of Meyssan's theories is that people within the U.S. government wanted to hit the Pentagon for its propaganda effect, but did not want to create a lot of damage or kill important people like Defense Secretary Rumsfeld. They note that the crash hit the one section under construction, thus greatly reducing the loss of life. [AGENCE FRANCE-PRESSE, 3/21/02; *LONDON TIMES*, 5/19, 02] Furthermore, the wall at point of impact was the first and only one to be reinforced and have blast-resistant windows installed as part of an upgrade plan. [*NFPA JOURNAL*, 11/1/01]

March 2, 2002: Diesel Tank May Have Destroyed Building and Secret Files on 9/11

A *New York Times* article theorizes that a diesel fuel tank was responsible for the collapse of Building 7 near the WTC. It collapsed on 9/11 even though it was farther away than many other buildings that remained standing. It was the first time a steel-reinforced high-rise in the U.S. had ever collapsed in a fire. The fuel tank had been installed in 1999 as part of a new "command center" for Mayor Rudolph Giuliani. [*NEW YORK TIMES*, 3/2/02; *DOW JONES NEWS*, 9/10/02] Curiously, given all the Wall Street scandals later in the year, Building 7 housed the SEC files related to numerous Wall Street investigations, as well as other federal investigative files. All the files for approximately 3,000 to 4,000 SEC cases were destroyed. Some were backed up in other places, but many were not, especially those classified as confidential. [*NATIONAL LAW JOURNAL*, 9/17/01] Lost files include documents that could show the relationship between Citigroup and the WorldCom bankruptcy. [*THE STREET*, 8/9/02] The Equal Employment Opportunity Commission estimates over 10,000 cases will be affected. [*NEW YORK LAW JOURNAL*, 9/14/01] The Secret Service also lost investigative files. Says one agent: "All the evidence that we stored at 7 World Trade, in all our cases, went down with the building." [TECH TV, 7/23/02] It is also eventually revealed that there was a secret CIA office in Building 7. [CNN, 11/4/01] A few days later, the head of the WTC collapse investigation says he "would possibly consider examining" the collapse of Building 7, but by this time all the rubble has already been removed and destroyed. [HOUSE OF REPRESENTATIVES TESTIMONY, 3/6/02]

March 7, 2002: Plane Crashing into Pentagon Is Shown in Photos

A series of photos surface purporting to show a plane crashing into the Pentagon on 9/11. It is not clear who released the photos, but the Pentagon asserts that they are authentic, and were taken by a Pentagon security camera. The release of these pictures comes within days of the publication of the book *l'Effroyable Imposture* that disputes the claim that Flight 77 hit the Pentagon. "Officials could not immediately explain why the date typed near the bottom of each photograph is September 12 and the time is written as 5:37 P.M." [FOX NEWS, 3/8/02]

March 27, 2002: Cockpit Recordings Raise Doubts About Flight 93 Events

New York Times reporter Jere Longman writes an article based on recent leaks to him about Flight 93's cockpit flight recording. (Later, relatives of the victims are given a single chance to listen to the recording). He claims that earlier reports of a 911 call from a bathroom reporting smoke and an explosion are incorrect. He names the passenger-caller as Edward Felt and notes that the dispatcher who took the call and Felt's wife both deny the smoke and explosion story. There were messages from both passengers and hijackers on the plane speaking of a bomb. [PITTSBURGH POST-GAZETTE, 10/28/01 (B)] Longman also claims that one passenger, Tom Burnett, told his wife there were guns on the plane. [NEW YORK TIMES, 3/27/02] Previously, it had been widely reported that Tom Burnett told his wife he did not see any guns. [MSNBC, 9/14/01] Note that the passengers appeared doubtful that the terrorists had either real guns or bombs, but there is a March 2002 report of a gun being used on Flight 11.

April 18, 2002: Private Showing of Flight 93 Recordings Fails to Quell Confusions

The FBI allows relatives of passengers on Flight 93 to listen to and see a written transcript of the cockpit recordings. Seventy people do so. But the FBI says the relatives are not allowed to make recordings, because the tape might be used in the trial of Zacarias Moussaoui. [GUARDIAN, 4/19/02] The *San Francisco Chronicle* responds: "Is there even a dollop of logic in that explanation? It's like saying we can't watch video of the planes crashing into the World Trade Center because that video might be used in a trial." [SAN FRANCISCO CHRONICLE, 6/3/02] *New York Times* reporter Jere Longman writes the book *Among the Heroes* based on his access to the recordings and interviews with officials and relatives. New details of their struggle on board emerge, but the government still has not officially stated if the passengers took over the plane or not. [DAILY TELEGRAPH, 8/6/02; MSNBC, 7/30/02]

April 19, 2002: FBI Claims Hijacker Computer Use Offered No Evidence

FBI Director Mueller states: "In our investigation, we have not uncovered a single piece of paper either here in the United States or in the treasure trove of information that has turned up in Afghanistan and elsewhere that mentioned any aspect of the September 11 plot." He also claims that the attackers used "extraordinary secrecy" and "investigators have found no computers, laptops, hard drives or other storage media that may have been used by the hijackers, who hid their communications by using hundreds of pay phones and cell phones, coupled with hard-to-trace prepaid calling cards." [FBI SPEECH TRANSCRIPT, 4/19/02; LOS ANGELES TIMES, 4/22/02] However, before 9/11 CIA Director Tenet told the Senate that al-Qaeda is "embracing the opportunities offered by recent leaps in information technology" [CIA, 03/21/00]; the FBI broke the al-Qaeda computer encryption before February 2001 [UPI, 2/13/01]; witnesses report seeing the hijackers use computers for e-mail at public libraries in Florida and Maine [SOUTH FLORIDA SUN-SENTINEL, 9/16/01 (B), BOSTON HERALD, 10/5/01]; in October 2001 there were many reports that hundreds of e-mails discussing the 9/11 plot had been found; Moussaoui's laptop was found to contain important information, etc. . . .

May 1, 2002: Head of Congressional Probe Resigns

L. Britt Snider, ex-CIA official and the head of the joint congressional investigation into 9/11, resigns. Apparently there were many conflicts between Snider and his own staff, as well as with Congress. It is later revealed the final straw occurred when Snider tried to hire a CIA employee who had failed an agency polygraph test as an inquiry staffer. The hearings were expected to start in late May, but the resignation is one reason why the first public hearings are delayed until September. [LOS ANGELES TIMES, 5/2/02; LOS ANGELES TIMES, 10/19/02] Snider is replaced by Eleanor Hill. She is widely credited for turning around an inquiry "hampered by infighting, politics, leaks and duelling agendas." [MIAMI HERALD, 7/14/02; WASHINGTON POST, 9/25/02 (B)]

May 1, 2002 (B): Investigation into Cause of Building Collapse on 9/11 Is Inconclusive

FEMA releases its report of the WTC collapses. It concludes, "[W]ith the information and time available, the sequence of events leading to the collapse of each tower could not be definitively determined." On Building 7: "The specifics of the fires in WTC 7 and how they caused the building to collapse remain unknown at this time." [FEMA STUDY, 5/1/02]

May 21, 2002: Fraudulent Consular Staff Admits to Providing Hijackers with Visas

Abdulla Noman, a former employee of the U.S. consulate in Jeddah, Saudi Arabia, where 15 of the 19 9/11 hijackers got their visas, says that he took money and gifts to provide fraudulent visas to foreigners. He pleads guilty and is convicted. About 50 to 100 visas were improperly issued by Noman from September 1996 until November 2001, when he was arrested. However, a former visa officer in Jeddah, Michael Springmann, has claimed in the past that the Jeddah office was notorious for purposefully giving visas to terrorists to train in the U.S. [ASSOCIATED PRESS, 5/21/02]

May 22, 2002: Illegal Status of Terrorists Points to U.S. Immigration Failure

A study indicates that at least half of the 48 Muslim radicals linked to terrorist plots in the U.S. since 1993 manipulated or violated immigration laws to enter this country and then stay here. Even when the terrorists did little to hide violations of visa requirements or other laws, INS officials failed to enforce the laws or to deport the offenders. The terrorists used a variety of methods. At the time they committed their crimes, 12 of the 48 were illegal immigrants. At least five others had lived in the U.S. illegally, and four others had committed significant immigration violations. Others were here legally but should have been rejected for visas because they fit U.S. immigration profiles of people who are likely to overstay their visas. [USA TODAY, 5/22/02] Experts later strongly suggest that the visa applications for all 15 of the Saudi Arabian 9/11 hijackers should have been rejected due to numerous irregularities.

May 23, 2002: Bush Opposes Special Inquiry into Terrorist Warnings

President Bush says he is opposed to establishing a special, independent commission to probe how the government dealt with terror warnings before 9/11. [CBS NEWS, 5/23/02] He later changes his stance in the face of overwhelming support for the idea, and then sabotages an agreement that Congress had reached to establish the commission.

May 27, 2002: Organization of 9/11 Attacks Is Still Not Clear

The *New Yorker* reports that a senior FBI official acknowledges there has been "no breakthrough" in establishing how the 9/11 suicide teams were organized and how they operated. Additionally, none of the thousands of pages of documents and computer hard drives captured in Afghanistan has enabled investigators to broaden their understanding of how the attack occurred, or even to bring an indictment of a conspirator. [NEW YORKER, 5/27/02]

June 22, 2002: 9/11 Inquiry Member Appears Biased in Defending FBI

Internal FBI documents show that Thomas Kelley, in charge of matters relating to the FBI in the joint congressional intelligence 9/11 inquiry, blocked an inquiry into the FBI's role in Waco. For instance, an internal FBI memo from December 2000 states that Kelley "continued to thwart and obstruct" the Waco investigation to the point that a special counsel was forced to send a team to search FBI headquarters for documents Kelley refused to turn over. [WASHINGTON POST, 6/22/02]

July 23, 2002: New York Declares
Records of Firefighters' Actions Secret

The New York City government decides that the audio and written records of the Fire Department's actions on 9/11 should never be released to the general public. The *New York Times* has been trying to get copies of the materials, which include firsthand accounts given to Fire Department officials by scores of firefighters and chiefs. The city claims the firefighters were told their accounts would be kept confidential, but senior fire officials say they were never told that their remarks would be kept confidential. [NEW YORK TIMES, 7/23/02]

August 2, 2002: FBI Questions Members of
Congressional Committees About 9/11 Leaks

The *Washington Post* reveals that FBI agents have questioned nearly all 37 members of the Senate and House intelligence committees about 9/11-related information leaks. They have asked them to submit to lie detector tests but most have refused. Congress people express "grave concern" for this historically unprecedented move. A law professor states, "Now the FBI can open dossiers on every member and staffer and develop full information on them. It creates a great chilling effect on those who would be critical of the FBI." [WASHINGTON POST, 8/2/02] Senator John McCain (R) suggests that "the constitutional separation of powers is being violated in spirit if not in the letter. 'What you have here is an organization compiling dossiers on people who are investigating the same organization. The administration bitterly complains about some leaks out of a committee, but meanwhile leaks abound about secret war plans for fighting a war against Saddam Hussein. What's that about? There's a bit of a contradiction here, if not a double standard.'" [WASHINGTON POST, 8/3/02] Later the search for the source of the leak intensifies to unprecedented levels as the FBI asks 17 senators to turn over phone records, appointment calendars and schedules that would reveal their possible contact with reporters. [WASHINGTON POST, 8/24/02] Most, if not all, turn over the records, even as some complain that the request breaches the separation of powers between the executive and legislative branches. One senator says the FBI is "trying to put a damper on our activities and I think they will be successful." [ASSOCIATED PRESS, 8/29/02] In January 2004, it is reported that the probe is now focusing on Republican Senator Richard Shelby. There has been no further word or indictments since. [WASHINGTON POST, 1/22/04]

August 3, 2002: U.S. Pilots Believe 9/11 Conspirators Used Utmost Professional Skill

A Portuguese newspaper reports on an independent inquiry into 9/11 by a group of military and civilian U.S. pilots that challenges the official version of events. The group's press statement says, "The so-called terrorist attack was in fact a superbly executed military operation carried out against the [U.S.], requiring the utmost professional military skill in command, communications, and control. It was flawless in timing, in the choice of selected aircraft to be used as guided missiles and in the coordinated delivery of those missiles to their preselected targets." A member of the inquiry team, a U.S. Air Force officer who flew over 100 sorties during the Vietnam War, says: "Those birds (airliners) either had a crack fighter pilot in the left seat, or they were being maneuvered by remote control." [PORTUGAL NEWS, 8/3/02; PORTUGAL NEWS, 8/8/02]

August 4, 2002: Firefighters Saw Only Limited Fire in South Tower

A "lost tape" of radio messages from firefighters inside the WTC on 9/11 is made public. Supposedly, "city fire officials simply delayed listening" to this tape until after the official report on the fire department's response to the attacks was published, and they still refuse to allow any officials to discuss the contents. The tape reveals that two firefighters were able to reach the crash site on the 78th floor of the South Tower. While there, "Chief Palmer could see only two pockets of fire, and called for a pair of engine companies to fight them." [NEW YORK TIMES, 8/4/02; GUARDIAN, 8/5/02]

August 12, 2002: FAA Releases No New Information About 9/11

A group of FAA flight controllers hold a press conference to talk about the 9/11 events for the first time. However, virtually no new information is disclosed. As the *Boston Globe* put it, "questions about detailed communications from the hijacked planes was avoided, with FAA officials saying that information remains under investigation." [BOSTON GLOBE, 8/13/02]

August 13, 2002: Electronic Warfare Methods May Have Brought Flight 93 Down

The *Independent* carries a story entitled, "Unanswered Questions: The Mystery of Flight 93," a rare critique of the official version of events around that plane's crash. Most of the information is a summation of what was reported before. However, there is one interesting new theory. Theorizing why witnesses did not see smoke from the faltering plane, the article points to the 1996 research of Harvard academic Elaine Scarry, "showing that the Air Force and the Pentagon have conducted extensive research on 'electronic warfare applications' with the possible capacity to intentionally disrupt the mechanisms of an aeroplane in such a way as to provoke, for example, an uncontrollable dive. Scarry also reports that U.S. Customs aircraft are already equipped with such weaponry; as are some C-130 Air Force transport planes. The FBI has stated that, apart from the enigmatic Falcon business jet, there was a C-130 military cargo plane within 25 miles of the passenger jet when it crashed. According to the Scarry findings, in 1995 the Air Force installed 'electronic suites' in at least 28 of its C-130s—capable, among other things, of emitting lethal jamming signals." [INDEPENDENT, 8/13/02]

September 5, 2002: Senator Decries Lack of Government Cooperation in 9/11 Congressional Inquiry

Richard Shelby of Alabama, the ranking Republican on the Senate Intelligence Committee, expresses doubts that the committee's 9/11 Congressional Inquiry will be able to accomplish anything, and he supports an independent investigation. "Time is not on our side," he says, since the investigation has a built-in deadline at the end of 2002. "You know, we were told that there would be cooperation in this investigation, and I question that. I think that most of the information that our staff has been able to get that is real meaningful has had to be extracted piece by piece." He adds that there is explosive information that has not been publicly released. "I think there are some more bombs out there . . . I know that." [NEW YORK TIMES, 9/10/02 (B)]

September 11, 2002: One Year After 9/11, Details of Plot Are Still Very Mysterious

On the first anniversary of the 9/11 attacks, the *New York Times* writes, "One year later, the public knows less about the circumstances of 2,801 deaths at the foot of Manhattan in broad daylight than people in 1912 knew within weeks about the *Titanic,* which sank in the middle of an ocean in the dead of night." John F. Timoney, the former police commissioner of Philadelphia, says: "You can hardly point to a cataclysmic event in our history, whether it was the sinking of the *Titanic,* the Pearl Harbor attack, [or] the Kennedy assassination, when a blue-ribbon panel did not set out to establish the facts and, where appropriate, suggest reforms. That has not happened here." The *Times* specifically points to a failure by New York City Mayor Bloomberg to conduct a real investigation into the WTC attack response. Bloomberg stated in August 2002, "Every single major event is different from all others. The training of how you would respond to the last incident is not really important." [NEW YORK TIMES, 9/11/02 (B)] The *Chicago Tribune* made similar comments a week earlier, pointing out that despite the "largest investigation in history," "Americans know little more today about the September 11 conspiracy, or the conspirators, than they did within a few weeks of the attacks." [CHICAGO TRIBUNE, 9/5/02]

September 18, 2002: First 9/11 Inquiry Hearing Amidst Protests
About Lack of Government Cooperation

The 9/11 Congressional Inquiry holds its first public hearing. The inquiry was formed in February 2002 but suffered months of delays. The day's testimonies focuses on intelligence warnings that should have led the government to believe airplanes could be used as bombs. [9/11 CONGRESSIONAL INQUIRY, 9/18/02] However, the *Washington Post* reports, "lawmakers from both parties . . . [protest] the Bush administration's lack of cooperation in the congressional inquiry into September 11 intelligence failures and [threaten] to renew efforts to establish an independent commission." Eleanor Hill, the joint committee's staff director, testifies that "According to [CIA Director Tenet], the president's knowledge of intelligence information relevant to this inquiry remains classified even when the substance of that intelligence information has been declassified." She adds that "the American public has a compelling interest in this information and that public disclosure would not harm national security." [WASHINGTON POST, 9/19/02] Furthermore, the committee believes that "a particular al-Qaeda leader may have been instrumental in the attacks" and U.S. intelligence has known about this person since 1995. Tenet "has declined to declassify the information we developed [about this person] on the grounds that it could compromise intelligence sources and methods and that this consideration supersedes the American public's interest in this particular area." [9/11 CONGRESSIONAL INQUIRY, 9/18/02] A few days later, The *New York Times* reveals this leader to be Khalid Shaikh Mohammed, the mastermind of the 9/11 attacks. [NEW YORK TIMES, 9/22/02] An FBI spokesman says the FBI had offered "full cooperation" to the committee. A CIA official denies that the report is damning: "The committee acknowledges the hard work done by intelligence community, the successes it achieved . . . " [MSNBC, 9/18/02]

September 18, 2002 (B): 9/11 Victims' Relatives
Raise Questions About Agencies' Conduct

Two relatives of 9/11 victims testify before the Congressional 9/11 inquiry. Kristen Breitweiser, whose husband Ronald died at the WTC, asks how the FBI was so quickly able to assemble information on the hijackers. She cites a *New York Times* article stating that agents descended on flight schools within hours of the attacks. "How did the FBI know where to go a few hours after the attacks?" she asks. "Were any of the hijackers already under surveillance?" [MSNBC, 9/18/02] She adds, "Our intelligence agencies suffered an utter collapse in their duties and responsibilities leading up to and on September 11th. But their negligence does not stand alone. Agencies like the Port Authority, the City of NY, the FAA, the INS, the Secret Service, NORAD, the Air Force, and the airlines also failed our nation that morning." [9/11 CONGRESSIONAL INQUIRY, 9/18/02] Stephen Push states, "If the intelligence community had been doing its job, my wife, Lisa Raines, would be alive today." He cites the government's failure to place Khalid Almihdhar and Nawaf Alhazmi on a terrorist watch list until long after they were photographed meeting with alleged al-Qaeda operatives in Malaysia. [MSNBC, 9/18/02]

September 20, 2002: Bush Changes Course, Backs 9/11 Commission

In the wake of damaging Congressional 9/11 inquiry revelations, President Bush reverses course and backs efforts by many lawmakers to form an independent commission to conduct a broader investigation than the current Congressional inquiry. *Newsweek* reports that Bush had virtually no choice. "There was a freight train coming down the tracks," says one White House official. [NEWSWEEK, 9/22/02] But as one of the 9/11 victim's relatives says, "It's carefully crafted to make it look like a general endorsement but it actually says that the commission would look at everything except the intelligence failures." [CBS NEWS, 9/20/02] Rather than look into such failures, Bush wants the commission to focus on areas like border security, visa issues, and the "role of Congress" in overseeing intelligence agencies. The White House also refuses to turn over documents showing what Bush knew before 9/11. [NEWSWEEK, 9/22/02]

October 5, 2002: FBI Refuses to Allow FBI Informant to Testify Before 9/11 Inquiry

The *New York Times* reports that the FBI is refusing to allow Abdussattar Shaikh, the FBI informant who lived with hijackers Nawaf Alhazmi and Khalid Almihdhar in the second half of 2000, to testify before the 9/11 Congressional Inquiry. The FBI claims the informer would have nothing interesting to say. The Justice Department also wants to learn more about the informant. [NEW YORK TIMES, 10/5/02] The FBI also tries to prevent Shaikh's handler Steven Butler from testifying, but Butler does end up testifying before a secret session on October 9, 2002. Shaikh does not testify at all. [WASHINGTON POST, 10/11/02 (B)] Butler's testimony uncovers many curious facts about Shaikh. [NEW YORK TIMES, 11/23/02; U.S. NEWS AND WORLD REPORT, 11/29/02; CONGRESSIONAL INQUIRY, 7/24/03; SAN DIEGO UNION-TRIBUNE, 7/25/03]

October 10, 2002: Bush Backtracks on Support for Independent 9/11 Investigation

A tentative congressional deal to create an independent commission to investigate the 9/11 terrorist attacks falls apart hours after the White House objected to the plan (it appears Vice President Cheney called Republican leaders and told them to renege on the agreement [NEW YORK TIMES, 11/2/02]). Bush had pledged to support such a commission a few weeks earlier, but doubters who questioned his sincerity appear to have been proven correct. Hours after top Republican leaders announced at a press conference that an agreement had been reached, House Republican leaders said they wouldn't bring the legislation to the full House for a vote unless the commission proposal was changed. There are worries that if the White House can delay the legislation for a few more days until Congress adjourns, it could stop the creation of a commission for months, if not permanently. [NEW YORK TIMES, 10/11/02] Another deal is made a few weeks later and the commission goes forward.

October 17, 2002: NSA Denies Having Indications of 9/11 Planning

NSA Director Michael Hayden testifies before a Congressional inquiry that the "NSA had no [indications] that al-Qaeda was specifically targeting New York and Washington . . . or even that it was planning an attack on U.S. soil." Before 9/11, the "NSA had no knowledge . . . that any of the attackers were in the United States." Supposedly, a post-9/11 NSA review found no intercepts of calls involving any of the 19 hijackers. [REUTERS, 10/17/02; *USA TODAY*, 10/18/02; NSA DIRECTOR TESTIMONY, 10/17/02] Yet, in the summer of 2001, the NSA intercepted communications between Khalid Shaikh Mohammed, the mastermind of the 9/11 attacks, and hijacker Mohamed Atta, when he was in charge of operations in the U.S. [*INDE-PENDENT*, 6/6/02; *INDEPENDENT*, 9/15/02] What was said between the two has not been revealed. The NSA also intercepted multiple phone calls from Abu Zubaida, bin Laden's chief of operations, to the U.S. in the days before 9/11. But who was called or what was said has not been revealed. [ABC NEWS, 2/18/02]

October 21, 2002: 13 Hijackers Were Never Interviewed by U.S. Consular Officials

The General Accounting Office, the nonpartisan investigative arm of Congress, releases a report asserting that at least 13 of the 19 9/11 hijackers were never interviewed by U.S. consular officials before being granted visas to enter the U.S. This contradicts previous assurances from the State Department that 12 of the hijackers had been interviewed. It also found that, for 15 hijackers whose applications could be found, none had filled in the documents properly. Records for four other hijackers (the four non-Saudis, including Ziad Jarrah and Mohamed Atta) could not be checked because they were accidentally destroyed. [*WASHINGTON POST*, 10/22/02] The State Department maintains that visa procedures were properly followed. In December 2002, Senators Jon Kyl (R) and Pat Roberts (R) state in a report that "if State Department personnel had merely followed the law and not granted non-immigrant visas to 15 of the 19 hijackers in Saudi Arabia . . . 9/11 would not have happened." [ASSOCIATED PRESS, 12/19/02]

October 23, 2002: Handling of Hijackers' Visa Applications Denounced

Visa applications for the 15 Saudi Arabian hijackers are made public, and six separate experts agree: "All of them should have been denied entry [into the U.S.]." Joel Mowbray, who first breaks the story for the conservative *National Review*, says he is shocked by what he saw: "I really was expecting al-Qaeda to have trained their operatives well, to beat the system. They didn't have to beat the system, the system was rigged in their favor from the get-go." A former U.S. consular officer says the visas show a pattern of criminal negligence. Some examples: "Abdulaziz Alomari claimed to be a student but didn't name a school; claimed to be married but didn't name a spouse; under nationality and gender, he didn't list anything." "Khalid Almihdhar . . . simply listed 'Hotel' as his U.S. destination—no name, no city, no state but no problem getting a visa." Only one actually gave a U.S. destination, and one stated his destination as "no." Only Hani Hanjour had a slight delay in acquiring his visa. His first application was flagged because he wrote he wanted to visit for three years when the legal limit is two. When he returned two weeks later, he simply changed the form to read "one year" and was accepted. The experts agree that even allowing for chance, incompetence, and human error, the odds were that only a few should have been approved. [*NEW YORK POST*, 10/9/02; ABC NEWS, 10/23/02]

November 5, 2002: Study on 9/11 Pentagon Damage Kept Secret

The *New York Times* reports that the official Pentagon study assessing the structural effect of the 9/11 attack on the Pentagon was completed in July 2002 but has not been released, and may never be released. The report "was specifically intended to consider Pentagon security in the light of new terrorist threats . . . Some, confused over what could be considered sensitive in the report, have expressed outrage that the lessons it may hold for other buildings could be squandered." Engineers outside the investigation say the implications are considerable, since the design of the Pentagon is much more similar to other major buildings elsewhere than the design of the WTC. If the report were released, it is likely building codes would be changed and many lives saved in the long term. [NEW YORK TIMES, 11/5/02]

November 11, 2002: Box Cutters and
Pepper Spray Were Banned by Airlines on 9/11

It is revealed that while the government did not ban box cutters, the airlines' own rules did. It had been widely reported the hijackers used box cutters because they were legal. It now appears pepper spray was also banned, and like box cutters, should have been confiscated. There is evidence the hijackers used pepper spray as well. It has been reported that nine of the hijackers were given special security screenings on 9/11, and six of those had their bags checked for weapons. [ASSOCIATED PRESS, 11/11/02]

November 15, 2002: Congress Starts New 9/11 Investigation

Congress approves legislation creating an independent commission—the National Commission on Terrorist Attacks Upon the United States—to "examine and report on the facts and causes relating to the September 11th terrorist attacks" and "make a full and complete accounting of the circumstances surrounding the attacks." President Bush signs it into law November 27, 2002. [STATE DEPARTMENT, 11/28/02] Bush originally opposed an independent commission, but he changes his mind over the summer after political pressure. The Democrats concede several important aspects of the commission (such as subpoena approval) after the White House threatens to create a commission by executive order, over which it would have more control. Bush will appoint the commission chairman and he sets a strict time frame (18 months) for the investigation. [CNN, 11/15/02] The commission will only have a $3 million budget. Senator Jon Corzine (D) and others wonder how the commission can accomplish much with such a small budget. [ASSOCIATED PRESS, 1/20/03] The budget is later increased.

November 27, 2002: Kissinger Named Chairman of New 9/11 Commission

President Bush names Henry Kissinger as Chairman of the 9/11 Commission. Congressional Democrats appoint George Mitchell, former Senate majority leader and peace envoy to Northern Ireland and the Middle East, as vice chairman. Their replacements and the other eight members of the commission are chosen by mid-December. Kissinger served as Secretary of State and National Security Adviser for Presidents Nixon and Ford. [NEW YORK TIMES, 11/29/02] Kissinger's ability to remain independent is met with skepticism. [PITTSBURGH POST-GAZETTE, 12/3/02; WASHINGTON POST, 12/17/02; CHICAGO SUN-TIMES, 12/13/02; CNN, 11/30/02; SYDNEY MORNING HERALD, 11/29/02]. He has a very controversial past. For instance, "Documents recently released by the CIA, strengthen previously-held suspicions that Kissinger was actively involved in the establishment of Operation Condor, a covert plan involving six Latin American countries including Chile, to assassinate thousands of political opponents." He is also famous for an "obsession with secrecy." [BBC, 4/26/02] It is even difficult for Kissinger to travel outside the U.S. Investigative judges in Spain, France, Chile, and Argentina seek to question him in several legal actions related to his possible involvement in war crimes, particularly in Latin America, Vietnam, Cambodia, Bangladesh, Chile, and East Timor. [BBC, 4/18/02; VILLAGE VOICE, 8/15–21/01; CHICAGO TRIBUNE, 12/1/02] The *New York Times* suggests, "Indeed, it is tempting to wonder if the choice of Mr. Kissinger is not a clever maneuver by the White House to contain an investigation it long opposed." [NEW YORK TIMES, 11/29/02] The *Chicago Tribune* notes that "the president who appointed him originally opposed this whole undertaking." Kissinger is "known more for keeping secrets from the American people than for telling the truth" and asking him "to deliver a critique that may ruin friends and associates is asking a great deal." [CHICAGO TRIBUNE, 12/5/02] Both Kissinger and Mitchell resign a short time later rather than reveal the clients they work with.

December 11, 2002: Mitchell Resigns from New 9/11 Commission

George Mitchell resigns as Vice Chairman of the recently-created 9/11 investigative commission. Lee Hamilton, an Indiana congressman for more than 30 years and chairman of the committee which investigated the Iran-Contra affair, is named as his replacement. [CNN, 12/11/02] Mitchell cites time constraints as his reason for stepping down, but he also does not want to sever ties with his lawyer-lobbying firm, Piper Rudnick, or reveal his list of clients. Recent clients include two Mideast governments—Yemen and the United Arab Emirates. [NEWSWEEK, 12/15/02] Committee Chairman Henry Kissinger resigns two days later.

December 11, 2002 (B): 9/11 Congressional Inquiry Blames Bush and Tenet

The 9/11 Congressional Inquiry concludes its seven month investigation of the performance of government agencies before the 9/11 attacks. A report hundreds of pages long has been written, but only nine pages of findings and 15 pages of recommendations are released at this time, and those have blacked out sections. [LOS ANGELES TIMES, 12/12/02] After months of wrangling over what has to be classified, the final report is released in July 2003. In the findings released at this time, the inquiry accuses the Bush administration of refusing to declassify information about possible Saudi Arabian financial links to U.S.-based terrorists, criticizes the FBI for not adapting into a domestic intelligence bureau after the attacks and says the CIA lacked an effective system for holding its officials accountable for their actions. Asked if 9/11 could have been prevented, Senator Bob Graham (D), the committee chairman, gives "a conditional yes." Graham says the Bush administration has given Americans an "incomplete and distorted picture" of the foreign assistance the hijackers may have received. [ABC NEWS, 12/10/02] Graham further says,

"There are many more findings to be disclosed" that Americans would find "more than interesting," and he and others express frustration that information that should be released is being kept classified by the Bush administration. [ST. PETERSBURG TIMES, 12/12/02] Many of these findings remain classified after the Inquiry's final report is released. Senator Richard Shelby (R), the vice chairman, singles out six people as having "failed in significant ways to ensure that this country was as prepared as it could have been": CIA Director Tenet; Tenet's predecessor, John Deutch; former FBI Director Louis Freeh; NSA Director Michael Hayden; Hayden's predecessor, Lieutenant General Kenneth Minihan; and former Deputy Director Barbara McNamara. [WASHINGTON POST, 12/11/02 (B); 9/11 CONGRESSIONALL INQUIRY, 12/11/02] Shelby says that Tenet should resign. "There have been more failures on his watch as far as massive intelligence failures than any CIA director in history. Yet he's still there. It's inexplicable to me." [REUTERS, 12/10/02; PBS NEWSHOUR, 12/11/02] But the *Los Angeles Times* criticizes their plan of action: "A list of 19 recommendations consists largely of recycled proposals and tepid calls for further study of thorny issues members themselves could not resolve." [LOS ANGELES TIMES, 12/12/02]

December 11, 2002 (C): Senator Claims
Foreign Governments Were Involved in 9/11

In discussing the report of the Senate Select Committee on Intelligence on 9/11, Senator Bob Graham (D), the committee chairman, says he is "surprised at the evidence that there were foreign governments involved in facilitating the activities of at least some of the [9/11] terrorists in the United States. . . . To me that is an extremely significant issue and most of that information is classified, I think overly classified. I believe the American people should know the extent of the challenge that we face in terms of foreign government involvement. I think there is very compelling evidence that at least some of the terrorists were assisted not just in financing—although that was part of it—by a sovereign foreign government and that we have been derelict in our duty to track that down. . . . It will become public at some point when it's turned over to the archives, but that's 20 or 30 years from now. "[PBS NEWSHOUR, 12/11/02] In March 2003; *Newsweek* says its sources indicate Graham is speaking about Saudi Arabia, and that leads pointing in this direction have been pursued. Graham also says that the report contains far more miscues than have been publicly revealed. "There's been a cover-up of this," he says. [NEWSWEEK, 3/1/03 (B)]

December 13, 2002: Kissinger Resigns from New 9/11 Commission

Henry Kissinger resigns as head of the new 9/11 Commission. [ASSOCIATED PRESS, 12/13/02; ASSOCIATED PRESS, 12/13/02] Two days earlier, the Bush administration argued that Kissinger was not required to disclose his private business clients. [NEW YORK TIMES, 12/12/02] However, the Congressional Research Service insists that he does, and Kissinger resigns rather than reveal his clients. [MSNBC, 12/13/02; SEATTLE TIMES, 12/14/02] It is reported that Kissinger is (or has been) a consultant for Unocal, the oil corporation, and was involved in plans to build pipelines through Afghanistan (see chapter 5). [WASHINGTON POST, 10/5/98; SALON, 12/3/02]

Kissinger claims he did no current work for any oil companies or Mideast clients, but several corporations with heavy investments in Saudi Arabia, such as ABB Group, a Swiss-Swedish engineering firm, and Boeing Corp., pay him consulting fees of at least $250,000 a year. A Boeing spokesman said its "long-standing" relationship with Kissinger involved advice on deals in East Asia, not Saudi Arabia. Boeing sold $7.2 billion worth of aircraft to Saudi Arabia in 1995. [NEWSWEEK, 12/15/02] In a surprising break from usual procedures regarding high-profile presidential appointments, White House lawyers never vetted Kissinger for conflicts of interest. [NEWSWEEK, 12/15/02] The *Washington Post* says that after the resignations of Kissinger and Mitchell, the commission "has lost time" and "is in disarray, which is no small trick given that it has yet to meet." [WASHINGTON POST, 12/14/02]

December 16, 2002: Ex-Governor Kean replaces
Kissinger as Chairman of New 9/11 Commission

President Bush names former New Jersey governor Thomas Kean as the Chairman of the 9/11 Commission after his original choice, Henry Kissinger, resigned. [WASHINGTON POST, 12/17/02] In an appearance on NBC, Kean promises an aggressive investigation. "It's really a remarkably broad mandate, so I don't think we'll have any problem looking under every rock. I've got no problems in going as far as we have to in finding out the facts." [ASSOCIATED PRESS, 12/17/02] However, Kean plans to remain President of Drew University and devote only one day a week to the commission. He also claims he would have no conflicts of interest, stating: "I have no clients except the university." [WASHINGTON POST, 12/17/02] However, he has a

Thomas Kean

history of such conflicts of interest. Multinational Monitor has previously stated: "Perhaps no individual more clearly illustrates the dangers of university presidents maintaining corporate ties than Thomas Kean," citing the fact that he is on the Board of Directors of Aramark (which received a large contract with his university after he became president), Bell Atlantic, United Health Care, Beneficial Corporation, Fiduciary Trust Company International, and others. [MULTINATIONAL MONITOR, 11/97]

December 16, 2002 (B): Members of 9/11
Commission Have Potential Conflicts of Interest

The ten members of the new 9/11 Commission are appointed by this date, and are: Republicans Thomas Kean (Chairman), Slade Gorton, James Thompson, Fred Fielding, and John Lehman, and Democrats Lee Hamilton (Vice Chairman), Max Cleland, Tim Roemer, Richard Ben-Veniste, and Jamie Gorelick. [NEW YORK TIMES, 12/17/02; WASHINGTON POST, 12/15/02; ASSOCIATED PRESS, 12/16/02; CHICAGO TRIBUNE, 12/12/02] Senators Richard Shelby (R) and John McCain (R) had a say in the choice of one of the Republican positions. They and many 9/11 victims' relatives wanted former Senator Warren Rudman (R), who cowrote an acclaimed report about terrorism before 9/11. But Senate Republican leader Trent Lott blocks Rudman's appointment and chooses John Lehman instead. [ST. PETERSBURG TIMES, 12/12/02; ASSOCIATED

**Richard
Ben-Veniste**

PRESS, 12/13/02; REUTERS, 12/16/02] It slowly emerges over the next several months that at least six of the ten commissioners have ties to the airline industry. [CBS NEWS, 3/5/03] Henry Kissinger and his replacement Thomas Kean both caused controversy when they were named. In addition, the other nine members of the commission are later shown to all have potential conflicts of interest.

Republican commissioners:

Fred Fielding also works for a law firm lobbying for Spirit Airlines and United Airlines. [ASSOCIATED PRESS, 2/14/03; CBS NEWS, 3/5/03]

Slade Gorton has close ties to Boeing, which built all the planes destroyed on 9/11, and his law firm represents several major airlines, including Delta Airlines. [ASSOCIATED PRESS, 12/12/02; CBS NEWS, 3/5/03]

John Lehman, former secretary of the Navy, has large investments in Ball Corp., which has many U.S. military contracts. [ASSOCIATED PRESS, 3/27/03 (B)]

James Thompson, former Illinois governor, is the head of a law firm that lobbies for American Airlines, and he has previously represented United Airlines. [ASSOCIATED PRESS, 1/31/03; CBS NEWS, 3/5/03]

Democratic commissioners:

Richard Ben-Veniste represents Boeing and United Airlines. [CBS NEWS, 3/5/03] Ben-Veniste also has other curious connections, according to a 2001 book on CIA ties to drug running written by Daniel Hopsicker, which has an entire chapter called "Who is Richard Ben-Veniste?" Lawyer Ben-Veniste, Hopsicker says, "has made a career of defending political crooks, specializing in cases that involve drugs and politics." Ben-Veniste has been referred to in print as a "Mob lawyer," and was a long-time lawyer for Barry Seal, one of the most famous drug dealers in U.S. history who also is alleged to have had CIA connections. [BARRY AND THE BOYS, DANIEL HOPSICKER, 9/01, PP. 325–30]

Max Cleland, former U.S. senator, has received $300,000 from the airline industry. [CBS NEWS, 3/5/03]

James Gorelick is a director of United Technologies, one of the Pentagon's biggest defense contractors and a supplier of engines to airline manufacturers. [ASSOCIATED PRESS, 3/27/03 (B)]

Lee Hamilton sits on many advisory boards, including those to the CIA, the president's Homeland Security Advisory Council, and the U.S. Army. [ASSOCIATED PRESS, 3/27/03 (B)]

Tim Roemer represents Boeing and Lockheed Martin. [CBS NEWS, 3/5/03]

January–July 2003: Bush Administration Delays Release of 9/11 Congressional Inquiry Report

The 9/11 Congressional Inquiry is originally expected to release its complete and final report in January 2003, but the panel spends seven months negotiating with the Bush administration about what material could be made public, and the final report is not released until July 2003. [WASHINGTON POST, 7/27/03] The administration originally wanted two thirds of the report to remain classified. [ASSOCIATED PRESS, 5/31/03] Former Senator Max Cleland, (D), member of the 9/11 Commission, later claims, "The administration sold the connection (between Iraq and al-Qaeda) to scare the pants off the American people and justify the war. There's no connection, and that's been confirmed by some of bin Laden's terrorist followers . . . What you've seen here is the manipulation of intelligence for political ends. The reason this report was delayed for so long—deliberately opposed at first, then slow-walked after it was created—is that the administration wanted to get the war in Iraq in and over . . . before (it) came out. Had this report come out in January [2003] like it should have done, we would have known these things before the war in Iraq, which would not have suited the administration." [UPI, 7/25/03]

January 22, 2003: House of 9/11 Suspect Finally Searched by FBI

The FBI conducts a very public search of a Miami, Florida, house belonging to Mohammed Almasri and his Saudi family. Having lived in Miami since July 2000, on September 9, 2001 they said they were returning to Saudi Arabia, hurriedly put their luggage in a van, and sped away, according to neighbors. A son named Turki Almasri was enrolled at Huffman Aviation in Venice, Florida, where hijackers Atta and Marwan Alshehhi also studied. [WASHINGTON POST, 1/23/03; PALM BEACH POST, 1/23/03] Neighbors repeatedly called the FBI after 9/11 to report their suspicions, but the FBI only began to search the house in October 2002. The house had remained abandoned, but not sold, since they left just before 9/11. [WASHINGTON POST, 1/23/03; PALM BEACH POST, 1/22/03; PALM BEACH POST, 1/23/03; SOUTH FLORIDA SUN-SENTINEL, 1/22/03] The FBI returned for more thorough searches in January 2003, with some agents dressed in white biohazard suits. [WASHINGTON POST, 1/23/03] U.S. Representative Robert Wexler (D), later says, "This scenario is screaming out one question: Where was the FBI for 15 months?" The FBI determines there is no terrorism connection, and apologizes to the family. [UPI, 1/24/03] An editorial notes the "ineptitude" of the FBI in not reaching family members over the telephone, as reporters were easily able to do. [PALM BEACH POST, 2/1/03]

January 27, 2003: 9/11 Commission Starts Off with Little Funding

The 9/11 Commission, officially titled the National Commission on Terrorist Attacks Upon the United States, holds its first meeting in Washington. The commission has $3 million and only a year and a half to explore the causes of the attacks. By comparison, a 1996 federal commission to study legalized gambling was given two years and $5 million. [ASSOCIATED PRESS, 1/27/03] Two months later the Bush administration grudgingly increases the funding to $12 million total. Philip Zelikow, currently the director of the Miller Center of Public Affairs at the University of Virginia and formerly in the National Security Council during the first Bush administration, is also appointed executive director of the commission. [ASSOCIATED PRESS, 1/27/03] Zelikow cowrote a book with National Security Adviser Rice. [9/11 COMMISSION, 3/03] A few days later, Vice Chairman Lee Hamilton says, "The focus of the commission will be on the future. We want to make recommendations that will make the American people more secure. . . . We're not interested in trying to assess blame, we do not consider that part of the commission's responsibility." [UPI, 2/6/03]

Philip Zelikow

Late February 2003: DNA Identifies Passenger
Remains, but Hijacker DNA Is Not Tested

Medical examiners match human remains to the DNA of two of the hijackers that flew on Flights 11 and/or 175 into the WTC. The names of the two hijackers are not released. The FBI gave the examiners DNA profiles of all ten hijackers on those flights a few weeks earlier. Genetic profiles of five hijackers from Flight 77 and the four from Flight 93 that did not match any of the passengers' profiles have been given to the FBI, but the FBI has not given any DNA profiles with which to match them. [CNN. 2/27/03]

March 26, 2003: Bush Turns Down
Increased Budget for 9/11 Commission

Time reports that the 9/11 Commission has requested an additional $11 million to add to the $3 million for the commission, and the Bush administration has turned down the request. The request will not be added to a supplemental spending bill. A Republican member of the commission says the decision will make it "look like they have something to hide." Another commissioner notes that the recent commission on the Columbia shuttle crash will have a $50 million budget. Stephen Push, a leader of the 9/11 victims' families, says the decision "suggests to me that they see this as a convenient way for allowing the commission to fail. They've never wanted the commission and I feel the White House has always been looking for a way to kill it without having their finger on the murder weapon." The administration has suggested it may grant the money later, but any delay will further slow down the commission's work. Already, commission members are complaining that scant progress has been made in the four months since the commission started, and they are operating under a deadline. [TIME, 3/26/03] Three days later, it is reported that the Bush administration has agreed to extra funding, but only $9 million, not $11 million. The commission agrees to the reduced amount. [WASHINGTON POST, 3/29/03] The *New York Times* criticizes such penny-pinching, saying, "Reasonable people might wonder if the White House, having failed in its initial attempt to have Henry Kissinger steer the investigation, may be resorting to budgetary starvation as a tactic to hobble any politically fearless inquiry." [NEW YORK TIMES, 3/31/03]

March 27, 2003: Security Clearance of
9/11 Commission Members Stalled

It is reported that "most members" of the 9/11 Commission still have not received security clearances. [WASHINGTON POST, 3/27/03] For instance, Slade Gorton, picked in December 2002, is a former senator with a long background in intelligence issues. Fellow commissioner Lee Hamilton says, "It's kind of astounding that someone like Senator Gorton can't get immediate clearance. It's a matter we are concerned about." The commission is said to be at a "standstill" because of the security clearance issue, and cannot even read the classified findings of the previous 9/11 Congressional inquiry. [SEATTLE TIMES, 3/12/03]

March 28, 2003: Independence of 9/11 Commission Called Into Question

An article highlights conflicts of interest amongst the commissioners on the 9/11 Commission. It had been previously reported that many of the commissioners had ties to the airline industry, but a number have other ties. "At least three of the ten commissioners serve as directors of international financial or consulting firms, five work for law firms that represent airlines and three have ties to the U.S. military or defense contractors, according to personal financial disclosures they were required to submit." Bryan Doyle, project manager for the watchdog group Aviation Integrity Project says, "It is simply a failure on the part of the people making the selections to consider the talented pool of non-conflicted individuals." Commission chairman Thomas Kean says that members are expected to steer clear of discussions that might present even the appearance of a conflict. [ASSOCIATED PRESS, 3/28/03]

March 31, 2003: U.S. Government Draws Harsh Criticism at First 9/11 Commission Hearing

The 9/11 Commission has its first public hearing. The *Miami Herald* reports, "Several survivors of the attack and victims' relatives testified that a number of agencies, from federal to local, are ducking responsibility for a series of breakdowns before and during September 11." [MIAMI HERALD, 3/31/03] The *New York Times* suggests that the 9/11 Commission would never have been formed if it were not for the pressure of the 9/11 victims' relatives. [NEW YORK TIMES, 4/1/03] Some of the relatives strongly disagreed with statements from some commissioners that they would not place blame. For instance, Stephen Push states, "I think this commission should point fingers. . . . Some of those people [who failed us] are still in responsible positions in government. Perhaps they shouldn't be." [UPI, 3/31/03] The most critical testimony comes from 9/11 relative Mindy Kleinberg, but her testimony is only briefly reported on by a few newspapers. [UPI, 3/31/03; NEWSDAY, 4/1/03; NEW YORK TIMES, 4/1/03; NEW YORK POST, 4/1/03; NEW JERSEY STAR-LEDGER, 4/1/03] In her testimony, Kleinberg says, "It has been said that the intelligence agencies have to be right 100 percent of the time and the terrorists only have to get lucky once. This explanation for the devastating attacks of September 11th, simple on its face, is wrong in its value. Because the 9/11 terrorists were not just lucky once: They were lucky over and over again." She points out the insider trading based on 9/11 foreknowledge, the failure of fighters to catch the hijacked planes in time, hijackers getting visas in violation of standard procedures, and other events, and asks how the hijackers could have been lucky so many times. [9/11 COMMISSION, 3/31/03]

April 24, 2003: 9/11 Committee Member Barred from Viewing Intelligence Material

9/11 Commissioner Tim Roemer tries to review the transcripts of the 9/11 Congressional Inquiry. However, he learns that he has no permission to see them, even though he served on the Inquiry and had read the material before. [WASHINGTON POST, 4/26/03] Roemer says the arrangement is outrageous: "No entity, individual, or organization should sift through or filter our access to material." [ASSOCIATED PRESS, 4/30/03]

July 8, 2003: 9/11 Commission
Denounces Lack of Cooperation

A status report released by the 9/11 Commission shows that various government agencies are not cooperating fully with the investigation. Neither the CIA nor the Justice Department have provided all requested documents. Lack of cooperation on the part of the Department of Defense "[is] becoming particularly serious," and the commission has received no responses whatsoever to requests related to national air defenses. The FBI, State Department, and Transportation Department receive generally positive reviews. [ASSOCIATED PRESS, 7/9/03] Commissioner Tim Roemer complains, "We're not getting the kind of cooperation that we should be. We need a steady stream of information coming to us . . . Instead, we're getting a trickle." [GUARDIAN, 7/10/03] Chairman Thomas Kean is also troubled by the Bush administration's insistence on having a Justice Department official present during interviews with federal officials. [ASSOCIATED PRESS, 7/9/03] The 9/11 Commission is eventually forced to subpoena documents from the Defense Department and FAA.

July 24, 2003: 9/11 Congressional Inquiry
Says Almost Every Government Agency Failed

The 9/11 Congressional Inquiry's final report comes out. [9/11 CONGRESSIONAL INQUIRY, 7/24/03; 9/11 CONGRESSIONAL INQUIRY, 7/24/03 (B)] Officially, the report was written by the 37 members of the House and Senate intelligence committees, but in practice, co-chairmen Bob Graham (D) and Porter Goss (R) exercised "near

**Senators Bob Graham
and Porter Goss**

total control over the panel, forbidding the inquiry's staff to speak to other lawmakers." [ST. PETERSBURG TIMES, 9/29/02] Both Republican and Democrats in the panel complained how the two co-chairmen withheld information and controlled the process. [PALM BEACH POST, 9/21/02] The report was finished in December 2002 and some findings were released then, but the next seven months were spent in negotiation with the Bush administration over what material had to remain censored. The Inquiry had a very limited mandate, focusing just on the handling of intelligence before 9/11. It also completely ignores or censors out all mentions of intelligence from foreign governments. Thomas Kean, the chairman of 9/11 Commission says the Inquiry's mandate covered only "one-seventh or one-eighth" of what his newer investigation will hopefully cover. [WASHINGTON POST, 7/27/03] The report blames virtually every government agency for failures:

- *Newsweek*'s main conclusion is: "The investigation turned up no damning single piece of evidence that would have led agents directly to the impending attacks. Still, the report makes it chillingly clear that law-enforcement and intelligence agencies might very well have uncovered the plot had it not been for blown signals, sheer bungling—and a general failure to understand the nature of the threat." [NEWSWEEK, 7/28/03]

- According to the *New York Times*, the report also concludes, "the FBI and CIA had known for years that al-Qaeda sought to strike inside the United States, but focused their attention on the possibility of attacks overseas." [NEW YORK TIMES, 7/26/03]

- CIA Director Tenet was "either unwilling or unable to marshal the full range of Intelligence Community resources necessary to combat the growing threat." [WASHINGTON POST, 7/25/03]

- U.S. military leaders were "reluctant to use . . . assets to conduct offensive counterterrorism efforts in Afghanistan" or to "support or participate in CIA operations directed against al-Qaeda." [WASHINGTON POST, 7/25/03]

- "There was no coordinated . . . strategy to track terrorist funding and close down their financial support networks" and the Treasury Department even showed "reluctance" to do so. [WASHINGTON POST, 7/25/03]

- According to the *Washington Post*, the NSA took "an overly cautious approach to collecting intelligence in the United States and offered 'insufficient collaboration' with the FBI's efforts." [WASHINGTON POST, 7/25/03]

Many sections remain censored, especially an entire chapter detailing possible Saudi support for 9/11. The Bush administration insisted on censoring even information that was already in the public domain. [NEWSWEEK, 5/25/03 (B)] The Inquiry attempted to determine "to what extent the President received threat-specific warnings" but received very little information. The was a focus on learning what was in Bush's briefing on August 6, 2001 but the White House refused to release this information, citing "executive privilege." [WASHINGTON POST, 7/25/03 (B), NEWSDAY, 8/7/03]

July 24, 2003 (B): 9/11 Congressional Inquiry Suggests Hijackers Received Considerable Assistance Inside U.S.

The 9/11 Congressional Inquiry's final report concludes that at least six hijackers received "substantial assistance" from associates in the U.S., though its "not known to what extent any of these contacts in the United States were aware of the plot." These hijackers came into contact with at least 14 people who were investigated by the FBI before 9/11, and four of these investigations were active while the hijackers were present. But in June 2002, FBI Director Mueller testified: "While here, the hijackers effectively operated without suspicion, triggering nothing that would have alerted law enforcement and doing nothing that exposed them to domestic coverage. As far as we know, they contacted no known terrorist sympathizers in the United States." CIA Director Tenet made similar comments at the same time, and another FBI official stated, "[T]here were no contacts with anybody we were looking at inside the United States." These comments are untrue, because one FBI document from November 2001 uncovered by the Inquiry concludes that the six lead hijackers "maintained a web of contacts both in the

United States and abroad. These associates, ranging in degrees of closeness, include friends and associ-ates from universities and flight schools, former roommates, people they knew through mosques and religious activities, and employment contacts. Other contacts provided legal, logistical, or financial assistance, facilitated U.S. entry and flight school enrollment, or were known from [al-Qaeda]-related activities or training." [9/11 CONGRESSIONAL INQUIRY, 7/24/03] The declassified sections of the 9/11 Congressional Inquiry's final report show the hijackers have contact with:

- Mamoun Darkazanli, investigated several times starting in 1991; the CIA makes repeated efforts to turn him into an informer

- Mohammed Haydar Zammar, investigated by Germany since at least 1997, the Germans periodically inform the CIA what they learn

- Osama Basnan; U.S. intelligence is informed of his terror connections several times in early 1990s but fails to investigate

- Omar al-Bayoumi, investigated in San Diego from 1998–1999

- Anwar Al Aulaqi, investigated in San Diego from 1999–2000

- Osama "Sam" Mustafa, owner of a San Diego gas station, and investigated beginning in 1991

- Ed Salamah, manager of the same gas station, and an uncooperative witness in 2000

- An unnamed friend of Hani Hanjour, whom the FBI tries to investigate in 2001

- An unnamed associate of Marwan Alshehhi, investigated beginning in 1999

Hijackers Nawaf Alhazmi and Khalid Almihdhar, who had contact with Basnan, al-Bayoumi, Aulaqi, Mustafa and Salamah, "maintained a number of other contacts in the local Islamic community during their time in San Diego, some of whom were also known to the FBI through counterterrorist inquiries and investigations," but details of these individuals and possible others are still classified. [9/11 CONGRES-SIONAL INQUIRY, 7/24/03] None of the above figures have been arrested or even publicly charged of any terror-ist crime, although Zammar is in prison in Syria.

July 28, 2003: Bush Opposes Release of Full 9/11 Congressional Inquiry Report

In the wake of the release of the 9/11 Congressional Inquiry's final report, pressure builds to release most of the still-censored sections of the report, but on this day Bush says he is against the idea. [ASSOCI-ATED PRESS, 7/29/03 (B), NEW YORK TIMES, 7/29/03] Through an obscure rule the Senate could force the release of the material with a majority vote [USA TODAY, /5/29/03], but apparently the number of votes in favor of this idea falls just short. MSNBC reports that "the decision to keep the passage secret . . . created widespread sus-picion among lawmakers that the administration was trying to shield itself and its Saudi allies from embarrassment. . . . Three of the four leaders of the joint congressional investigation into the attacks have said they believed that much of the material on foreign financing was safe to publish but that the administration insisted on keeping it secret." [MSNBC, 7/28/03] Senator Richard Shelby (R), one of the main authors of the report, states that "90, 95 percent of it would not compromise, in my judgment, anything in national security." Bush ignores a reporter's question on Shelby's assessment. [ASSOCIATED PRESS, 7/29/03 (B)] Even the Saudi government claims to be in favor of releasing the censored material so it can better respond to criticism. [MSNBC, 7/28/03] All the censored material remains censored; however, some details of the most controversial censored sections are leaked to the media.

October—November 2003: 9/11 Commission Subpoenas FAA and Pentagon for Missing Documents

The 9/11 Commission unanimously agrees to subpoena the FAA after it refuses to produce records relating to FAA notification to U.S. air defenses concerning the hijacked planes on 9/11. The panel states, "This disturbing development at one agency has led the commission to reexamine its general policy of relying on document requests rather than subpoenas." [ASSOCIATED PRESS, 10/15/03] The commission also votes to subpoena the Pentagon for documents related to NORAD's fighter response on 9/11. The commission says it is "especially dismayed" by incomplete document production on the part of NORAD. The commission explains, "In several cases we were assured that all requested records had been produced, but we then discovered, through investigation, that these assurances were mistaken." [ASSOCIATED PRESS, 11/7/03]

November 12, 2003: 9/11 Commission and White House Agree to Terms of Access

Senators of both parties have been accusing the White House of stonewalling the 9/11 Commission by blocking its demands for documents despite threats of a subpoena. [ASSOCIATED PRESS, 10/27/03] On this day, the White House and the 9/11 Commission strike a deal. The main issue is access to the Presidential Daily Briefings given to President Bush. Under the deal, only some of the ten commissioners will be allowed to examine classified intelligence documents, and their notes will be subject to White House review. Some 9/11 victims' relatives complain that the agreement gives the White House too much power. The Family Steering Committee complains, "All ten commissioners should have full, unfettered, and unrestricted access to all evidence." It urges the public release of "the full, official, and final written agreement." [ASSOCIATED PRESS, 11/13/03] Commissioner Max Cleland is unsatisfied with the deal and resigns a short time later.

December 9, 2003: Bob Kerry Replaces Max Cleland on 9/11 Commission

Bob Kerrey, the former Nebraska senator who also served as the ranking Democrat on the Senate Intelligence Committee, is appointed to the 9/11 Commission, replacing Max Cleland, who leaves the commission to accept a position on the board of the Export-Import Bank. [WASHINGTON POST, 12/10/03] Just before resigning, Cleland called the Bush administration's attempts to stonewall and "slow walk" the commission a "national scandal." He criticized the commission for cutting a deal with the White House that compromised their access to information, and said, "I'm not going to be part of looking at information only partially. I'm not going to be part of just coming to quick conclusions. I'm not going to be part of political pressure to do this or not do that. I'm not going to be part of that. This is serious." [SALON, 11/21/03]

February 2004: Bush Administration Fails to Act on 9/11 Inquiry Recommendations

The 9/11 Congressional Inquiry, which ended in late 2002, made 19 urgent recommendations to make the nation safer against the terrorist attack attempts. However, more then one year later, the White House has only implemented two of the recommendations. Furthermore, investigative leads have not been pursued. Senator Bob Graham (D), complains, "It is incomprehensible why this administration has refused to aggressively pursue the leads that our inquiry developed." He is also upset that the White House classified large portions of the final report. [NEW YORK OBSERVER, 2/11/04]

February–April 2004: Bush Administration Withholds Clinton Documents from 9/11 Commission

The Bush administration withholds thousands of documents from the Clinton administration that had already been cleared by Clinton's general counsel Bruce Lindsey for release to the 9/11 Commission. [NEW YORK TIMES, 4/2/04] In April, after a public outcry, the Bush administration grants access to most of the documents. [WASHINGTON POST, 4/3/04; FOX NEWS, 4/4/04] However, they continue to withhold approximately 57 documents. According to the commission, the documents being withheld by the Bush White House include references to al-Qaeda, bin Laden, and other issues relevant to the panel's work. [WASHINGTON POST, 4/8/04]

February 9, 2004: Full 9/11 Commission Allowed To View PDB Summaries

The 9/11 Commission gets greater access to classified intelligence briefings under a new agreement with the White House. The 10-member panel had been barred from reviewing notes concerning the presidential daily briefings taken by three of its own commissioners and the commission's director in December 2003. The new agreement allows all commission members the opportunity to read White House-edited versions of the summaries. The White House had faced criticisms for allowing only some commissioners to see the notes. Still, only three commissioners are allowed to see the original, unclassified documents. [ASSOCIATED PRESS, 2/10/04]

February 11, 2004: Hijackers Said to Use Short Knives, Not Box Cutters

It is reported the 9/11 Commission now believes that the hijackers used short knives instead of box cutters. The *New York Observer* comments, "Remember the airlines' first reports, that the whole job was pulled off with box cutters? In fact, investigators for the commission found that box cutters were reported on only one plane [Flight 77]. In any case, box cutters were considered straight razors and were always illegal. Thus the airlines switched their story and produced a snap-open knife of less than four inches at the hearing. This weapon falls conveniently within the aviation-security guidelines pre-9/11. [NEW YORK OBSERVER, 2/11/04] It was publicly revealed that box cutters were illegal on 9/11 in late 2002. [ASSO-CIATED PRESS, 11/11/02]

March 21, 2004: Victims' Relatives Demand That
9/11 Commission Executive Director Resign

The 9-11 Family Steering Committee and 9-11 Citizens Watch demand the resignation of Philip Zelikow, executive director of the 9/11 Commission. The demand comes shortly after former counterterrorism "tsar" Richard Clarke told the *New York Times* that Zelikow was present when he gave briefings on the threat posed by al-Qaeda to National Security Adviser Rice from December 2000 to January 2001. The Family Steering Committee, a group of 9/11 victims' relatives, writes, "It is clear that [Zelikow] should never have been permitted to be a member of the commission, since it is the mandate of the commission to identify the source of failures. It is now apparent why there has been so little effort to assign individual culpability. We now can see that trail would lead directly to the staff director himself." Zelikow has been interviewed by his own commission because of his role during the transition period. But a spokesman for the commission claims that having Zelikow recluse himself from certain topics is enough to avoid any conflicts of interest. [NEW YORK TIMES, 3/20/04; UPI 3/23/04]

March 24, 2004: Counterterrorism "Tsar"
Clarke Gives High-Profile Testimony

Just a few days after releasing a new book, former counterterrorism "tsar" Richard Clarke testifies before the 9/11 Commission. His opening statement consists of little more than an apology to the relatives of the 9/11 victims. He says, "Your government failed you, those entrusted with protecting you failed you, and I failed you. For that failure, I would ask . . . for your understanding and forgiveness." Under questioning, he praises the Clinton administration, saying, "My impression was that fighting terrorism, in general, and fighting al-Qaeda, in particular, were an extraordinarily high priority in the Clinton administration—certainly no higher priority." But he's very critical of the Bush administration, stating, "By invading Iraq . . . the president of the United States has greatly undermined the war on terrorism." He says that under Bush before 9/11, terrorism was "an important issue, but not an urgent issue . . . . [CIA Director] George Tenet and I tried very hard to create a sense of urgency by seeing to it that intelligence reports on the al-Qaeda threat were frequently given to the president and other high-level officials. But although I continue to say it was an urgent problem, I don't think it was ever treated that way." He points out that he made proposals to fight al-Qaeda in late January 2001. While the gist of them were implemented after 9/11, he complains, "I didn't really understand why they couldn't have been done in February [2001]." He says that with a more robust intelligence and covert action program, "we might have been able to nip [the plot] in the bud." [WASHINGTON POST 3/24/04; NEW YORK TIMES, 3/24/04; 9/11 COM-MISSION 3/24/04] It soon emerges that President Bush's top lawyer places a telephone call to at least one of the Republican members of the commission just before Clarke's testimony. Critics call that an unethical interference in the hearings. [WASHINGTON POST, 4/1/04 (B)] Democratic commissioner Bob Kerrey complains, "To call commissioners and coach them on what they ought to say is a terrible mistake." [NEW YORK DAILY NEWS, 4/2/04]

April 8, 2004: Rice Testifies Before the 9/11 Commission

National Security Adviser Rice testifies before the 9/11 Commission under oath and with the threat of perjury. The Bush administration originally opposed her appearance, but relented after great public demand. [INDEPENDENT, 04/03/04] In her statement she repeats her claim that "almost all of the reports [before 9/11] focused on al-Qaeda activities outside the United States. . . . The information that was specific enough to be actionable referred to terrorists operation overseas." Moreover, she stresses that the "kind of analysis about the use of airplanes as weapons actually was never briefed to us." But she concedes, "In fact there were some reports done in '98 and '99. I think I was—I was certainly not aware of them . . ." [9/11 COMMISSION, 4/8/04]

During heated questioning several subjects are discussed:

- Why didn't counterterrorism "tsar" Richard Clarke brief President Bush on al-Qaeda before September 11? Clarke says he had wished to do so, but Rice states, "Clarke never asked me to brief the President on counterterrorism." [9/11 COMMISSION, 4/8/04]

- What was the content of the briefing President Bush received on August 6, 2001? While Rice repeatedly underlines that it was "a historical memo . . . not threat reporting," Commissioners Richard Ben-Veniste and Tim Roemer ask her why then it cannot be declassified. [9/11 COMMISSION, 4/8/04] Two days later the White House finally publishes it, and it is shown to contain more than just historical information.

- Did Rice tell Bush of the existence of al-Qaeda cells in the U.S. before August 6, 2001? Rice says that she does not remember whether she "discussed it with the President." [9/11 COMMISSION, 4/8/04]

- Were warnings properly passed on? Rice points out, "The FBI issued at least three nation-wide warnings to federal, state, and law enforcement agencies, and specifically stated that although the vast majority of the information indicated overseas targets, attacks against the homeland could not be ruled out. The FBI tasked all 56 of its U.S. field offices to increase surveillance of known suspected terrorists and to reach out to known informants who might have information on terrorist activities." But Commissioner Jamie Gorelick remarks, "We have no record of that. The Washington field office international terrorism people say they never heard about the threat, they never heard about the warnings." [9/11 COMMISSION, 4/8/04]

Rice does not apologize to the families of the victims, as Clarke did weeks earlier. The Associated Press comments, "The blizzard of words in Condoleezza Rice's testimony Thursday did not resolve central points about what the government knew, should have known, did and should have done before the September 11 terrorist attacks." [ASSOCIATED PRESS, 4/8/04 (C)] The *Washington Post* calls "her testimony an ambitious feat of jujitsu: On one hand, she made a case that 'for more than 20 years, the terrorist threat gathered, and America's response across several administrations of both parties was insufficient.' At the same time, she argued that there was nothing in particular the Bush administration itself could have done differently that would have prevented the attacks of September 11, 2001—that there was no absence of vigor in the White House's response to al-Qaeda during its first 233 days in office. The first thesis is undeniably true; the second both contradictory and implausible." [WASHINGTON POST, 4/9/04]

April 29, 2004: Bush and Cheney Privately Meet with 9/11 Commission; Decline to Provide Testimony Under Oath

President Bush and Vice President Cheney appear for three hours of private questioning before the 9/11 Commission. (Former President Clinton and former Vice President Al Gore met privately and separately with the commission earlier in the month. [WASHINGTON POST, 430/04; NEW YORK TIMES, 4/30/04]) The commission permits Bush and Cheney to appear together, in private, and not under oath. The testimony is not recorded. Commissioners can take notes, but the notes are censored by the White House. [NEWSWEEK, 4/2/04; KNIGHT RIDDER, 3/31/04; NEW YORK TIMES, 4/3/04] The commission drew most of their questions from a list submitted to the White House before the interview, but few details about the questions or the answers given are available. [WASHINGTON POST, 04/29/04] Two commissioners, Lee Hamilton and Bob Kerrey, leave the session early for other engagements. They claim they had not expected the interview to last more than the previously agreed upon two-hour length. [NEW YORK TIMES, 5/1/04]

May 11, 2004: Administration Gives Top Prisoner Access to Some, Denies Custody to Others

In a secret agreement with the White House, the 9/11 Commission obtains the right to question at least two top al-Qaeda leaders in U.S. custody. The two men are believed to be Khalid Shaikh Mohammed and Ramzi bin al-Shibh, two accused masterminds of the 2001 attacks. [BALTIMORE SUN, 5/12/04] The results of the commission's questioning of these suspects are published in a 9/11 Staff Statement released in June 2004. [9/11 COMMISSION REPORT, 6/16/04 (B)] However, in an ironic twist, during a 9/11-related lawsuit hearing held in June, U.S. authorities refuse to acknowledge whether or not they have Khalid Shaikh Mohammed in custody. [ASSOCIATED PRESS, 4/23/04; ASSOCIATED PRESS, 6/15/04] Insurance companies representing 9/11 victims had requested that the U.S. Justice Department to serve a summons against Khalid Shaikh

Mohammed, but a judge rules that the U.S. government does not have to disclose whether it is holding alleged terrorists in custody. [ASSOCIATED PRESS, 4/23/04; ASSOCIATED PRESS, 6/15/04]

June 4, 2004: Victims' Families Listen to 9/11 Phone Recordings

When the recording of flight attendant Betty Ong is played in public before the 9/11 Commission in January 2004, family members demand that the FBI honor the family member's rights under the Victims Assistance Act to hear any and all phone calls made from the hijacked airplanes. So, on this date, about 130 victim's relatives gather in Princeton, New Jersey, and hear previously unavailable calls. But Justice Department only plays what it decided are "relevant" calls. However, attendees are ordered not to disclose what they hear lest it compromise the prosecution of Zacarias Moussaoui. [CNN, 5/28/04; ASSOCIATED PRESS, 6/5/04; NEW YORK OBSERVER, 6/17/04] Some family members nonetheless later discuss what they have heard. Witnesses describe one recording of two American Airlines managers who are told details of flight attendant Amy Sweeney's call from Flight 11 shortly after the first hijacking has begun. Rather than report news of a possible hijacking to other government agencies so they can learn what to do in case there is a crisis, the managers say things like, "Don't spread this around. Keep it close," and "Keep it quiet." [NEW YORK OBSERVER, 6/17/04]

June 14, 2004: FBI 9/11 Investigation Continues

The *Washington Post* reports that the FBI's 9/11 investigation still continues, though at a reduced level. Originally, the investigation, named PENTTBOM, was staffed by about 70 full time FBI agents and analysts. The team now has only about ten members. Some observers complain the FBI has not done enough. Mary Galligan, who headed the investigation until early 2004, emphasizes how much is still unknown about the plot. She says, "There is still information coming in, and we still have so many unanswered questions." [WASHINGTON POST, 6/14/04]

June 16, 2004: 9/11 Commission Gives
Account of Prisoner Interrogations

The 9/11 Commission releases a new report on how the 9/11 plot developed. Most of their information appears to come from interrogations of prisoners Khalid Shaikh Mohammed, the 9/11 mastermind, and Ramzi bin al-Shibh, a key member of the al-Qaeda Hamburg cell. In this account, the idea for the attacks appears to have originated with Mohammed. In mid-1996, he met bin Laden and al-Qaeda leader Mohammed Atef in Afghanistan. He presented several ideas for attacking the U.S., including a version of the 9/11 plot using ten planes (presumably an update of Operation Bojinka's second phase plot—see chapter 3). Bin Laden does not commit himself. In 1999, bin Laden approves a scaled-back version of the idea, and provides four operatives to carry it out: Nawaf Alhazmi, Khalid Almihdhar, Khallad bin Attash, and Abu Bara al Taizi. Attash and al Taizi drop out when they fail to get U.S. visas. Alhazmi and Almihdhar prove to be incompetent pilots, but the recruitment of Mohamed Atta and the others in the Hamburg al-Qaeda cell solves that problem. Bin Laden wants the attacks to take place between May and July 2001, but the attacks are ultimately delayed until September. [9/11 COMMISSION, 6/16/04 (B)] However, information such as these accounts resulting from prisoner interrogations is seriously doubted by some experts, because it appears they only began cooperating after being coerced or tortured. For instance, it is said that Mohammed was "waterboarded," a technique in which his head is pushed under water until he nearly drowns. Information gained under such duress often is unreliable. Additionally, there is a serious risk that the prisoners might try to intentionally deceive. [NEW YORK TIMES, 6/17/04] The commission itself expresses worry that Mohammed could be trying to exaggerate the role of bin Laden in the plot to boost bin Laden's reputation in the Muslim world. [9/11 COMMISSION REPORT, 6/16/04 (B)] Most of what these prisoners have said is uncorroborated from other sources. [NEW YORK TIMES, 6/17/04 (E)]

20.

Other Post-9/11 Events

This chapter includes information deemed relevant to the aftermath of 9/11, but which does not neatly fit into the other chapters. As with other issues, it is impossible to include all potentially relevant information due to space considerations. Volumes could be written on the post-9/11 terror threats and responses alone. Most of the entries here connect to 9/11 in some manner. Themes include media coverage of 9/11 and its aftermath—including the development of, and challenges to, the "official" story.

Shortly after the 9/11 attacks, mainstream media editorials throughout the country rang with cries for justice, retribution, and—above all—unity with our fellow Americans and behind our country's leaders. President Bush, who promised swift retribution against those who caused the attack and a safer nation, was hailed as a hero, and even his most strident political foes rallied behind him in a national show of unity.

Unity was paramount and dissent unpatriotic. Congress swiftly granted Bush the authority to use "all necessary military force" against the perpetrators of 9/11, their supporters, and their protectors. War in Afghanistan soon followed. A new cabinet-level Office of Homeland Security was created. The Patriot Act, granting nearly unprecedented powers to law enforcement officials, sailed through Congress with virtually no opposition or debate. Media outlets, asked to refrain from coverage of certain issues, silently complied. Those few who dared to question Bush administration requests were criticized as unpatriotic, and those who dared to ask questions about how 9/11 happened were derided as "conspiracy nuts."

As the weeks and months passed, however, the questions grew. Although the Justice Department had detained hundreds of terrorist "suspects," none of the suspects would prove to be culpable in the 9/11 attacks. Government employees who revealed the existence of pre-9/11 warnings ignored by their superiors were subjected to internal investigations while their superiors were promoted, and, in one case, even awarded with a presidential citation and cash bonus. As noted in chapter 18, despite the declaration of victory in Afghanistan, Taliban, and al-Qaeda forces are reconstituting there. In 2003, 84 U.S. targets were attacked by terrorists. [STATE DEPARTMENT, 6/22/04]

Today (in mid-2004), bin Laden remains at large (if in fact he is still alive). Nearly 1,000 U.S. soldiers and many more Iraqis have been killed in the continuing chaos and violence there. Rumors of another terrorist attack on U.S. soil are growing, and bin Laden's representatives have announced that their plans for the next, even bigger attack, are 90 percent complete. [OSAMA'S REVENGE, BY PAUL WILLIAMS, 6/04, PP. 91, 123–24, 146]

Some serious questions about how well protected the U.S. is are beyond the scope of this chapter. However, in brief, some worrying facts are that:

- Airports are said to be unacceptably vulnerable to terrorism. [ASSOCIATED PRESS, 6/8/04];

- Terrorist watch lists remain unconsolidated. [UPI, 4/30/03]

- Basic background checks on air security personnel remain undone. [TIME, 7/8/03]

- The Treasury Department has assigned five times as many agents to investigate Cuban embargo violations as it has to track al-Qaeda's finances. [ASSOCIATED PRESS, 4/30/04]

- The White House has spurned a request for 80 more investigators to track and disrupt the global financial networks of terrorist groups. [NEW YORK TIMES, 4/4/04 (F)]

- Terrorist cases have been fizzling out in U.S. courts. [LOS ANGELES TIMES, 12/9/03]

- Experts have concluded that the Iraq War has diverted resources from the war on terrorism and made the U.S. less secure. [MSNBC, 7/29/03; SALON, 7/31/03]

- There has been a huge increase in government spending to train and respond to terrorist attack, but *Time* Magazine reports that the geographical spread of "funding appears to be almost inversely proportional to risk." [TIME, 3/21/04]

- Several high-profile studies have concluded that despite its frequent "bear any burden" rhetoric, the Bush administration has grossly underfunded domestic security. [NEW REPUBLIC, 3/3/03; NEW YORK TIMES, 7/25/03]

- Community-based "first responders" lack basic equipment, including protective clothing and radios. [NEW REPUBLIC, 3/3/03; NEW YORK TIMES, 7/25/03]

- Spending on computer upgrades, airport security, more customs agents, port security, border controls, chemical plant security, bioweapon vaccinations, and much more, is far below needed levels and often below promised levels. [NEW REPUBLIC, 3/3/03]

Are we safer today than we were on September 10, 2001? How can reforms be implemented to ensure that the failures culminating in that horrific day will not happen again, when we still do not know exactly what our internal failures *were?* Has the implementation of PNAC's global domination strategy (see chapter 16), which requires a permanent Persian Gulf presence, made us stronger, or left us more vulnerable?

September 11, 2001: Television News Footage
of Gleeful Palestinians Shown Out of Context

Television news coverage on 9/11 repeatedly shows images of Palestinians rejoicing over the 9/11 attack. According to Mark Crispin Miller, a Professor of Media Studies at New York University who investigated the issue, the footage was filmed during the funeral of nine people killed the day before by Israeli authorities. He said, "to show it without explaining the background, and to show it over and over again is to make propaganda for the war machine and is irresponsible." [AGENCE FRANCE-PRESSE, 9/18/01; *AUSTRALIAN*, 9/27/01]

September 11, 2001 (B): All-Republican "Shadow Government" Formed

It is later revealed that only hours after the 9/11 attacks, a U.S. "shadow government" is formed. Initially deployed "on the fly," executive directives on government continuity in the face of a crisis dating back to the Reagan administration are put into effect. Approximately 100 midlevel officials are moved to underground bunkers and stay there 24 hours a day. Officials rotate in and out on a 90-day cycle. When its existence is revealed, some controversy arises because the shadow government includes no Democrats. In fact, top congressional Democrats are unaware of it until journalists break the story months later. [*WASHINGTON POST*, 3/1/02; CBS NEWS, 3/2/02]

September 12, 2001: Planned Terrorism Exercise
May Have Sped Up Response to 9/11 Attack

Before 9/11, New York City is scheduled to have a biological terrorism exercise on this day called Tripod II. As Mayor Rudolph Giuliani later testifies, "hundreds of people . . . from FEMA, from the federal government, from the state, from the State Emergency Management Office" come to New York to take part in the exercise. Presumably many have already arrived when the 9/11 attacks occur. Giuliani notes that the equipment for the exercise is already there, so when his emergency bunker (in WTC Building 7) is destroyed in the attacks, he moves his response center to the planned site of the Tripod exercise. [9/11 COMMISSION, 5/19/04; *NEW YORK MAGAZINE*, 10/15/01]

September 14, 2001: Account of Fighter
Response Times Changes Significantly

CBS News announces that "contrary to early reports, U.S. Air Force jets did get into the air on Tuesday while the attacks were under way." According to this new account, the first fighters get airborne toward New York City at 8:52 A.M. [CBS NEWS, 9/14/01] A day earlier before this announcement, acting Chairman of the Joint Chiefs of Staff Richard Myers in congressional testimony stated that the first fighters got airborne only after the Pentagon was hit at 9:37 A.M. [MYERS' CONFIRMATION TESTIMONY, 9/13/01] NORAD spokesman Marine Corps Major Mike Snyder also claimed no fighters launched anywhere until after the Pentagon was hit. [BOSTON GLOBE, 9/15/01] Four days later, the official NORAD timeline is changed to include this new account. [NORAD, 9/18/01] New York City Mayor Rudolph Giuliani later testifies before the 9/11 Commission that he found out from the White House at about 9:58 A.M. that the first fighters were not launched toward New York City until twelve minutes earlier—9:46 A.M. [9/11 COMMISSION, 5/19/04] This would correspond to Myers' and Snyder's accounts that no fighters are scrambled until after the Pentagon is hit. But the 9/11 Commission later agrees with this CBS report and by their account the first fighters launch around 8:52. [9/11 COMMISSION REPORT, 6/17/04]

September 14, 2001 (B): Lack of Debate
About Poor Fighter Response on 9/11

The *Miami Herald* reports, "Forty-five minutes. That's how long American Airlines Flight 77 meandered through the air headed for the White House, its flight plan abandoned, its radar beacon silent . . . Who was watching in those 45 minutes? 'That's a question that more and more people are going to ask,' said one controller in Miami. 'What the hell went on here? Was anyone doing anything about it? Just as a national defense thing, how are they able to fly around and no one go after them?'" [MIAMI HERALD, 9/14/01] In the year after this article and a similar one in the *Village Voice* [VILLAGE VOICE, 9/13/01], there will be only one other U.S. article questioning slow fighter response times, and that article notes the strange lack of articles on the topic. [SLATE, 1/16/02] The fighter response issue finally makes news in 9/11 Commission hearings in 2004.

September 14, 2001 (C): Congress to Bush: Use All Necessary Military Force

Congress authorizes President Bush to use all necessary military force against the perpetrators of the 9/11 attacks, their sponsors, and those who protected them. [STATE DEPARTMENT, 12/26/01] In March 2003, President Bush informs Congress that Iraq is being attacked for its support of 9/11, despite the lack of any evidence for such a connection (see chapter 15).

September 16, 2001: EPA Misleads Public About Health Risks at WTC Site

The Environmental Protection Agency (EPA) and Occupational Safety and Health administration (OSHA) release a joint statement asserting that the air in downtown New York City is safe to breathe.

"New samples confirm previous reports that ambient air quality meets OSHA standards and consequently is not a cause for public concern," the agencies claim. [EPA, 9/16/01] However, the government's statements are based on ambient air quality tests using outdated technologies. [ST. LOUIS POST-DISPATCH, 1/14/04] Furthermore, it is later learned that the press release was heavily edited under pressure from the White House's Council on Environmental Quality (CEQ). Critical passages in the original draft were either deleted or modified to downplay public health risks posed by contaminants that were released into the air during the collapse of the World Trade Center. [NEWSDAY, 8/26/03; EPA, 8/21/2003] In late October, the *New York Daily News* obtains internal EPA documents containing information that had been withheld from the public. One document says that "dioxins, PCBs, benzene, lead and chromium are among the toxic substances detected . . . sometimes at levels far exceeding federal levels." [NEW YORK DAILY NEWS, 10/21/01] Later, in October, it is reported that thousands of rescue workers and residents are experiencing respiratory problems that experts attribute to the toxic smoke flume and ultra fine dust. [CNN, 10/29/01; NEWSDAY, 10/30/01; BBC, 10/30/01; *NEW YORK POST*, 10/29/01]

September 17, 2001: Stock Exchange Reopens; Economic Costs of Attack Are High

The New York Stock Exchange, closed since the 9/11 attacks, reopens. The economy slowly returns to normal. The attacks caused more than $20 billion in property damage to buildings in New York City and Washington. The work stoppage and other loss of economic output costs about another $47 billion, making the attacks the costliest man-made disaster in U.S. history. [ABC NEWS, 9/10/02]

September 20, 2001: Tom Ridge Named Homeland Security Secretary

Bush announces the new cabinet-level Office of Homeland Security, to be led by Pennsylvania Governor Tom Ridge. [ASSOCIATED PRESS, 8/19/02] In November, Ridge becomes secretary of a new Homeland Security Department.

Tom Ridge

September 20, 2001 (B): FBI Translator Sees Pattern of Deliberate Failure

Sibel Edmonds is hired as a Middle Eastern languages translator for the FBI. As she later tells CBS's *60 Minutes*, she immediately encounters a pattern of deliberate failure in her translation department. Her boss says, "Let the documents pile up so we can show it and say that we need more translators and expand the department." She claims that if she was not slowing down enough, her supervisor would delete her work. Meanwhile, FBI agents working on the 9/11 investigation would call and ask for urgently needed translations. Senator Charles Grassley (R) says of her charges, "She's credible and the reason I feel she's very credible is because people within the FBI have corroborated a lot of her story." He points out

Sibel Edmonds

that the speed of such translation might make the difference between a terrorist bombing succeeding or failing. [CBS NEWS, 10/25/02; *NEW YORK POST*, 10/26/02] In January 2002, FBI officials tell government auditors that translator shortages have resulted in "the accumulation of thousands of hours of audio tapes and pages" of material that had not been translated. [*WASHINGTON POST*, 6/19/02] Edmonds files a whistleblower lawsuit against the FBI for these and other charges in March 2002. However, the case is later dismissed because all evidence related to proving the charges is classified. [CNN, 07/07/04]

September 21, 2001: Congress Approves Aid Package for Airline Industry, 9/11 Victims

Congress approves a $15 billion federal aid package for the battered U.S. airline industry, and sets up a government fund to compensate 9/11 victims' relatives. [*LOS ANGELES TIMES*, 9/22/01] However, relatives are only allowed to sue terrorists, and if they sue anyone else, they are not entitled to any compensation money. The law also limits the airlines' liability to the limits of their insurance coverage—around $1.5 billion per plane. [*LOS ANGELES TIMES*, 1/17/02] Nevertheless, some later sue entities that make them ineligible for the fund, such as the Port Authority, owner of the WTC.

September 23, 2001–Present: 9/11 Skeptics Derided as "Conspiracy Nuts"

The first of many mainstream articles ridiculing 9/11 "conspiracy theories" is published. [*INDEPENDENT*, 9/23/01] Early articles of this type generally deride Middle Eastern views blaming Israel. [ASSOCIATED PRESS, 10/3/01 (C), *WASHINGTON POST*, 10/13/01; *DALLAS MORNING NEWS*, 11/19/01] Later articles mostly deride Western theories blaming President Bush, and criticize the Internet and Congresswoman Cynthia McKinney for spreading these ideas. [*CHICAGO SUN-TIMES*, 2/8/02; ABC NEWS, 4/17/02; *ORLANDO SENTINEL*, 5/18/02; *TORONTO SUN*, 5/19/02] The title of one article, "Conspiracy Nuts Feed On Calamity," expresses the general tone of these articles. [*ATLANTA JOURNAL AND CONSTITUTION*, 5/22/02] An *Ottawa Citizen* article mockingly includes a Do-It-Yourself Conspiracy Theory section, where you can fill in the blanks for your own personal 9/11 theory. The article calls 9/11 conspiracy theories "delirious," "dangerous," and "viruses," while admitting, "[I]t's true that some of the events surrounding the September 11 attacks are hard to explain." [*OTTAWA CITIZEN*, 9/1/02] Another article attempts discredit theories that oil was a motive for the U.S. to attack Afghanistan by interspersing them with theories that space aliens were behind the 9/11 attacks. [*DAILY TELEGRAPH*, 9/5/02]

October 2001–September 2002: Vital Army Translators Dismissed for Homosexuality

Nine Army linguists, including six trained to speak Arabic, are dismissed from the military's Defense Language Institute in Monterey, California, because they are gay. At the same time, the military claims it is facing a critical shortage of translators and interpreters for the war on terrorism. [ASSOCIATED PRESS, 11/14/02] The *Miami Herald* comments: "The message is unmistakable: We find gay people more frightening than Osama bin Laden, whose stated goal is our destruction." [*MIAMI HERALD*, 11/22/02]

October 2001: Anthrax Letters Kill Five, Heighten Terrorist Fears

A total of four letters containing anthrax are mailed to NBC, the *New York Post*, Democratic Senator Daschle, and Democratic Senator Leahy. The letters sent to the senators both contain the words "Death to America, Death to Israel, Allah is Great." Twenty-three people are infected and five people die. Panic sweeps the nation. On October 16, the Senate office buildings are shut down, followed by the House of Representatives, after 28 congressional staffers test positive for exposure to anthrax. A number of hoax letters containing harmless powder turn up. [SOUTH FLORIDA SUN-SENTINEL, 12/01] Initially it is suspected that either al-Qaeda or Iraq are behind the anthrax letters. [OBSERVER, 10/14/01; BBC, 10/16/01; LONDON TIMES, 10/27/01] However, further investigation leads the U.S. government to conclude that, "everything seems to lean toward a domestic source . . . Nothing seems to fit with an overseas terrorist type operation." [WASHINGTON POST, 10/27/01; ST. PETERSBERG TIMES, 11/10/01] In August 2002, the FBI names Steven Hatfill, a bioweapons researcher who worked for the U.S. government, as a "person of interest" in the case. [ASSOCI-

Anthrax letter sent to Senator Tom Daschle

ATED PRESS, 8/1/02; LONDON TIMES, 8/2/02] Though he undergoes intense scrutiny by the FBI, he is never charged with any crime. As of mid-2004, no one has been charged in relation to the anthrax letter attacks.

October 10, 2001: U.S. Television Networks "Doing Too Much of the Government's Bidding"

The U.S. government asks the major U.S. television networks to refrain from showing unedited video messages taped by Osama bin Laden. They agree. A *Newsweek* article is critical of the decision, pointing out "all but one [of these networks] are controlled by major conglomerates that have important pending business with the government." The article openly questions if the media is "doing too much of the government's bidding" in reporting on 9/11. Says one expert, "I'm not saying that everything is a horrible paranoid fantasy, but my sense is there's an implicit quid pro quo here. The industry seems to be saying to the administration, 'we're patriotic, we're supporting the war, we lost all of this advertising, now free us from [business] constraints.'" [NEWSWEEK, 10/13/01]

October 20, 2001: Few Detained in U.S. Have Real Terror Ties

The *New York Times* reports that, although 830 people have been arrested in the 9/11 terrorism investigation (a number that eventually reaches between 1,200 and 2,000), there is no evidence that anyone now in custody was a conspirator in the 9/11 attacks. Furthermore, "none of the nearly 100 people still being sought by the [FBI] is seen as a major suspect." Of all the people arrested, only four—Zacarias Moussaoui, Ayub Ali Khan, Mohammed Azmath, and Nabil al-Marabh—are likely connected to al-Qaeda. [NEW YORK TIMES, 10/21/01 (C)] Three of those are later cleared of ties to al-Qaeda (though the clearance of al-Marabh is a mystery—see chapter 11). Al-Marabh is eventually convicted of the minor charge of entering the United States illegally. [CANADIAN BROADCASTING CORP., 8/27/02; WASHINGTON POST, 6/12/02] On September

12, 2002, after a year in solitary confinement (the first four months of which he was unable to contact a lawyer), Mohammed Azmath pleads guilty to one count of credit card fraud. He is released with time served. Ayub Ali Khan, whose real name is apparently Syed Gul Mohammed Shah, is given a longer sentence for credit card fraud, but is released and deported by the end of 2002. [VILLAGE VOICE, 9/25/02; NEW YORK TIMES, 12/31/02] By December 2002, only six of the original 1,000-plus detainees are known to still be in custody, and none charged with any terrorist acts. [WASHINGTON POST, 12/12/02]

November 10, 2001: Bush Dismisses 9/11 Conspiracy Theories

In a speech to the United Nations General Assembly, President Bush states, "We must speak the truth about terror. Let us never tolerate outrageous conspiracy theories concerning the attacks of September the 11th; malicious lies that attempt to shift the blame away from the terrorists, themselves, away from the guilty." [WHITE HOUSE, 11/1/01]

Early December 2001: Al-Qaeda "Puppet Master" Disappears in Britain

Al-Qaeda leader Abu Qatada disappears, despite being under surveillance in Britain. He has been "described by some justice officials as the spiritual leader and possible puppet master of al-Qaeda's European networks." [TIME, 7/7/02] Qatada had already been sentenced to death in absentia in Jordan, and is wanted at the time by the U.S., Spain, France, and Algeria as well. [GUARDIAN, 2/14/02] In October 2001, the media had strongly suggested that Qatada would soon be arrested for his known roles in al-Qaeda plots, but no such arrest occurred. [LONDON TIMES, 10/21/01] In November, while Qatada was still living openly in Britain, a Spanish judge expressed disbelief that Qatada hadn't been arrested already, as he has previously been connected to a Spanish al-Qaeda cell that may have met with Mohamed Atta in July 2001. [OBSERVER, 11/25/01] *Time* magazine will later claim that just before new anti-terrorism laws go into effect in Britain, Abu Qatada and his family are secretly moved to a safe house by the British government, where he is lodged, fed, and clothed by the government. "The deal is that Abu Qatada is deprived of contact with extremists in London and Europe but can't be arrested or expelled because no one officially knows where he is," says a source, whose claims were corroborated by French authorities. The British reportedly do this to avoid a "hot potato" trial. [TIME, 7/7/02] A British official rejects these assertions: "We wouldn't give an awful lot of credence [to the story]." [GUARDIAN, 7/8/02] Some French officials tell the press that Qatada was allowed to disappear because he is actually a British intelligence agent. [OBSERVER, 2/24/02 (B)] Qatada is later arrested on October 23, 2002, in London but there still has not been a trial, or signs of a coming trial, as of mid-2004. [LONDON TIMES, 10/25/02]

December 22, 2001: Shoe Bomber Arrested, Claims Allegiance to bin Laden

British citizen Richard Reid is arrested for allegedly trying to blow up a Miami-bound jet using explosives hidden in his shoe. [ASSOCIATED PRESS, 8/19/02] He later pleads guilty to all charges, and declares himself a follower of bin Laden. [CBS NEWS, 10/4/02] He may have ties to Pakistan. [*WASHINGTON POST*, 3/31/02] It is later believed that Reid and others in the shoe bomb plot reported directly to 9/11 mastermind Khalid Shaikh Mohammed. [CNN 1/30/03] It has been suggested that Mohammed has ties to the ISI, and that Reid is a follower of Ali Gilani, a religious leader believed to be working with the ISI (see chapter 12).

January 4, 2002: U.S. Doctors Information About Terrorists

It is reported that the State Department said Mohamed Atta "wanted to learn to fly, but didn't need to take off and land" when this information clearly refers to Zacarias Moussaoui (although that story isn't exactly true for him either). The Defense Department even releases a photo purporting to be bin Laden in Western cloth-

Defense Department pamphlet with doctored photo of bin Laden

ing, with his hair cut short and beard shaved off. An expert says "Frankly, this is sloppy," and the article calls these propaganda efforts "worthy of the tabloids." [ASSOCIATED PRESS, 1/4/02]

January 29, 2002: Bush Sees an "Axis of Evil"

President Bush's State of the Union speech describes an "axis of evil" consisting of Iraq, Iran, and North Korea. Adviser Richard Perle cautioned against these same three countries a month before 9/11. Bin Laden is not mentioned in the speech. [CNN, 1/29/02] The speech is followed by a new public focus on Iraq and a downplaying of bin Laden.

February 10, 2002: Driver's License Examiner Dies in Suspicious Circumstances

Katherine Smith is killed one day before her scheduled appearance in court on charges she helped five Muslim terrorists get illegal drivers licenses. According to witnesses, she veered into a utility pole when a fire erupted in her car. She was burned beyond recognition. The FBI later determines that gasoline was poured on her clothing before she died in the fire and find that arson was the cause of death. [OAK RIDGER, 2/14/02] A suicide note was found, but prosecutors say they are looking for murder suspects. One of the five Muslims, Sakhera Hammad, was found with a pass in his wallet giving access to the restricted areas of the WTC, dated September 5, 2001. Hammad claims he was a plumber and worked on the WTC's sprinkler system that day, but the company with exclusive rights to all WTC plumbing work has never heard of him. Smith was being investigated by the FBI; the five later plead guilty to charges of fraud. [ASSOCIATED PRESS, 2/13/02; REUTERS, 2/15/02; GO MEMPHIS, 2/12/02; MEMPHIS COMMERCIAL APPEAL, 2/21/02] One month later, the coroner who examines her body is targeted by a bomb, which is defused. Then in June the coroner is attacked, bound with barbed wire, and left with a bomb tied to his body, but he survives. [MEM-PHIS COMMERCIAL APPEAL, 3/14/02]

February 20, 2002: Pentagon Office Designed
for Telling Lies Revealed; Declared Closed

The Pentagon announces the existence of the new Office of Strategic Influence (OSI), which "was qui-etly set up after September 11." The role of this office is to plant false stories in the foreign press, phony e-mails from disguised addresses, and other covert activities to manipulate public opinion. The new office proves so controversial that it is declared closed six days later. [CNN, 2/20/02; CNN, 2/26/02] It is later reported that the "temporary" Office of Global Communications will be made permanent (it is unknown when this office began its work). This office seems to serve the same function as the earlier OSI, minus the covert manipulation. [WASHINGTON POST, 7/20/02] Defense Secretary Rumsfeld later states that after the OSI was closed, "I went down that next day and said fine, if you want to savage this thing fine I'll give you the corpse. There's the name. You can have the name, but I'm gonna keep doing every sin-gle thing that needs to be done and I have." [DEFENSE DEPARTMENT, 11/18/02]

February 28, 2002: Majority of Muslims Believe bin Laden Not Responsible for 9/11

A Gallup poll conducted in Muslim nations shows 18 percent believe that Arabs were responsible for 9/11 and 61 percent do not. 86 percent in Pakistan say Arabs were not responsible. [GUARDIAN, 2/28/02] Even Pakistani President Musharraf has said bin Laden was not the mastermind, though he says that he probably supported it. [REUTERS, 8/4/02]

March 22, 2002: FBI Translator Charges That Evidence of
Turkish Spies in Pentagon and State Department Was Suppressed

Translator Sibel Edmonds later claims that she is fired by the FBI on this day after repeatedly raising suspicions about a co-worker named Jan (or Can) Dickerson. When Dickerson was hired in November 2001, she had connections to a Turkish intelligence officer and had worked with a Turkish organization, both of which were being investigated by the FBI's own counter-intelligence unit. Edmonds claims that Dickerson insisted that she alone translate documents relating to the investigation of this organization and official. When Edmonds reviewed Dickerson's translations, she found information that the Turkish officer had spies inside the State Department and Pentagon was not being translated. Dickerson then tried to recruit Edmonds as a spy, and when Edmonds refused, Dickerson threatened to have her killed. After Edmonds's boss and others in the FBI failed to respond to her complaints, she wrote to the Justice Department's inspector general's office in March: "Investigations are being compromised. Incorrect or misleading translations are being sent to agents in the field. Translations are being blocked and circum-vented." Edmonds is then fired and she sues the FBI. The FBI eventually concludes Dickerson had left out significant information from her translations. A second FBI whistleblower, John Cole, also claims to know of security lapses in the screening and hiring of FBI translators. [WASHINGTON POST, 6/19/02; COX NEWS SERVICE, 8/14/02; CBS NEWS, 7/13/03] The supervisor who told Edmonds not to make these accusations and also encouraged her to go slow in her translations is later promoted. [CBS NEWS, 10/25/02]

April, June, or August 2002: Al Jazeera Reporter Claims
to Conduct Interview with 9/11 Masterminds

It is originally reported that Al Jazeera reporter Yosri Fouda interviews 9/11 mastermind Khalid Shaikh Mohammed and 9/11 associate Ramzi bin al-Shibh at a secret location in Karachi, Pakistan in either June [LONDON TIMES, 9/8/02] or August. [GUARDIAN, 9/9/02] Details and audio footage of the interview come out between September 8 and 12, 2002. The video footage of the interview al-Qaeda promised to hand over is never given to Al Jazeera. [ASSOCIATED PRESS, 9/8/02] Both figures claim the 9/11

Ramzi bin al-Shibh

attacks were originally going to target nuclear reactors, but "decided against it for fear it would go out of control." Interviewer Fouda is struck that Mohammed and bin al-Shibh remember only the hijackers' code names, and have trouble remembering their real names. [AUSTRALIAN, 9/9/02] Mohammed, who calls himself the head of al-Qaeda's military committee and refers to bin al-Shibh as the coordinator of the "Holy Tuesday" operation, reportedly acknowledges "[a]nd, yes, we did it." [MASTERMINDS OF TERROR, BY YOSRI FOUDA AND NICK FIELDING, 5/03 PP. 38] These interviews "are the first full admission by senior figures from bin Laden's network that they carried out the September 11 attacks." [LONDON TIMES, 9/8/02] Some, however, call Fouda's claims into doubt. For example, the *Financial Times* states: "Analysts cited the crude editing of [Fouda's interview] tapes and the timing of the broadcasts as reasons to be suspicious about their authenticity. Dia Rashwan, an expert on Islamist movements at the Al-Ahram Centre for Strategic Studies in Cairo, said: 'I have very serious doubts [about the authenticity of this tape]. It could have been a script written by the FBI.'" [FINANCIAL TIMES, 9/11/02] Mohammed is later variously reported to be arrested in June 2002, killed or arrested in September 2002, and then arrested in March 2003. After this last arrest report, for the first time Fouda claims this interview took place in April, placing it safely before the first reports of Mohammed's capture. [GUARDIAN, 3/4/03; CANADA AM, 3/6/03] Bin al-Shibh also gets captured several days after Fouda's interview is broadcast, and some reports say he is captured because this interview allows his voice to be identified. [CBS NEWS, 10/9/02; OBSERVER, 9/15/02] As a result, Fouda has been accused of betraying al-Qaeda, and now fears for his life. [INDEPENDENT, 9/17/02] As the *Washington Post* states, "Now al Jazeera is also subject to rumors of a conspiracy." [WASHINGTON POST, 9/15/02] Yet after being so reviled by al-Qaeda supporters, Fouda is later given a cassette said to be a bin Laden speech. [MSNBC, 11/18/02] U.S. officials believe the voice on that cassette is "almost certainly" bin Laden, but one of the world's leading voice-recognition institutes said it is 95 percent certain the tape is a forgery. [BBC, 11/18/02; BBC, 11/29/02]

April 11, 2002: Congresswoman Suspects Bush Knew of 9/11 in Advance

Congresswoman Cynthia McKinney (D) calls for a thorough investigation into whether President Bush and other government officials may have been warned of the 9/11 attacks but did nothing to prevent them, the first national-level politician to do so. She states "News reports from *Der Spiegel* to the *London Observer*, from the *Los Angeles Times* to MSNBC to CNN, indicate that many different warnings were received by the administration. . . . I am not aware of any evidence showing that President Bush or members of his administration have personally profited from the attacks of 9/11. . . . On the other hand, what is undeniable is that corporations close to the administration have directly benefited from the increased defense spending arising from the aftermath of September 11. The Carlyle Group, Dyn-Corp, and Halliburton certainly stand out as companies close to this administration." [ATLANTA JOURNAL AND CONSTITUTION, 4/12/02] McKinney's comments are criticized and ridiculed by other politicians and the media. For instance, Congressman Mark Foley (R) states, "She has said some outrageous things but this has gone too far . . . Maybe there should be an investigation as she suggests—but one focused on her." Senator Zell Miller (D) says her comments were dangerous and irresponsible. [WASHINGTON POST, 4/12/02] An editorial in her home state calls her the "most prominent nut" promoting 9/11 "conspiracy theories." [ATLANTA JOURNAL AND CONSTITUTION, 4/15/02] One columnist says she is possibly "a delusional paranoiac" or "a socialist rabble-rouser who despises her own country." [ORLANDO SENTINEL, 4/21/02] White House Press Secretary Ari Fleischer said McKinney "must be running for the hall of fame of the Grassy Knoll Society." [WASHINGTON POST, 4/12/02] One month after McKinney's comments, the Bush administration comes under fire after reports reveal it had been warned five weeks before 9/11 about possible al-Qaeda plane hijackings, and McKinney claims vindication. [MCKINNEY WEBSITE, 5/16/02]

May 17, 2002: Fear of Being Unpatriotic Affects Media Coverage After 9/11

CBS anchorman Dan Rather tells the BBC that he and other journalists haven't been properly investigating since 9/11. He says, "There was a time in South Africa that people would put flaming tires around people's necks if they dissented. And in some ways the fear is that you will be necklaced here, you will have a flaming tire of lack of patriotism put around your neck. Now it is that fear that keeps journalists from asking the toughest of the tough questions." [GUARDIAN, 5/17/02]

May 20–24, 2002: Government Terrorist Warnings Believed Political

The Bush administration issues a remarkable series of terror warnings that many believe are politically motivated. Vice President Cheney warns it is "not a matter of if, but when" al-Qaeda will next attack the U.S. [CNN, 5/20/02] Homeland Security Director Tom Ridge says the same thing. Defense Secretary Rumsfeld says terrorists will "inevitably" obtain weapons of mass destruction. FBI Director Mueller says more suicide bombings are "inevitable." [WASHINGTON POST, 5/22/02] Authorities also issue separate warnings that al-Qaeda terrorists might target apartment buildings nationwide, banks, rail and transit systems, the Statue of Liberty, and the Brooklyn Bridge. USA Today titles an article, "Some Question Motives Behind Series of Alerts." [USA TODAY, 5/24/02] David Martin, CBS's national security correspondent, says, "Right now they're putting out all these warnings to change the subject from what was known prior to September 11 to what is known now." It had been revealed the week before that Bush received a briefing in August 2001 entitled, "Bin Laden Determined to Strike in U.S." [WASHINGTON POST, 5/27/02] Remarkably, even Press Secretary Ari Fleischer says the alerts were issued "as a result of all the controversy that took place last week." [*VILLAGE VOICE*, 5/23/02; *WASHINGTON TIMES*, 5/22/02] *Time* notes, "Though uncorroborated and vague, the terror alerts were a political godsend for an administration trying to fend off a bruising bipartisan inquiry into its handling of the terrorist chatter last summer. After the wave of warnings, the Democratic clamor for an investigation into the government's mistakes subsided." [*TIME*, 5/27/02]

May 30, 2002: Agent Claims FBI Obstructed
Efforts to Stop Terrorist Money Flows

FBI Agent Robert Wright announces he is suing the FBI over a publishing ban. He has written a book but the FBI will not allow him to show it to anyone. He delivers a tearful press conference at the National Press Club describing his lawsuit against the FBI for deliberately curtailing investigations that might have prevented the 9/11 attacks. Unfortunately he has been ordered to not reveal specifics publicly. [FOX NEWS, 5/30/02] Wright claims the FBI shut down his 1998 criminal probe into alleged terrorist-training camps in Chicago and Kansas City. He uses words like "prevented," "thwarted," "obstructed," "threatened," "intimidated," and "retaliation" to describe the actions of his superiors in blocking his attempts to shut off money flows to al-Qaeda, Hamas and other terrorist groups. He also alleges that for years the U.S. was training Hamas terrorists to make car bombs to use against Israel, one of the U.S.'s closest allies. [LA WEEKLY, 8/2/02 LA WEEKLY, 8/2/02 (B)] Wright's book still has not been released as of mid-2004.

June 4, 2002: Officer with Possible Unique 9/11
Knowledge Is Reprimanded for Criticizing Bush

Air Force Lieutenant Colonel Steve Butler is suspended from his post at the Defense Language Institute in Monterey, California, and is told he could face a court martial for writing a letter to a local newspaper calling President Bush a "joke" and accusing him of allowing the 9/11 attacks to happen. The military prohibits public criticism of superiors. [BBC, 6/5/02; *MONTEREY COUNTY HERALD*, 6/5/02] What is not reported is that he may have had unique knowledge about 9/11: A hijacker named Saeed Alghamdi trained at the Defense Language Institute and Butler was Vice Chancellor for Student Affairs there (note that this is not the same person as the Steven Butler who later testifies before the 9/11 Congressional inquiry). [GANNETT NEWS SERVICE, 9/17/01] Later in the month the Air Force announces "the matter is resolved" and Butler will not face a court-martial, but it is unknown if he faced a lesser punishment. [KNIGHT RIDDER, 6/14/02]

August 9, 2002: FBI Ended Investigation into Training Camps for Terrorists

FBI agent Robert Wright is a whistleblower who claims the FBI shut down a criminal investigation into the operation of terror-training camps in Chicago and Kansas City, years before the 9/11 attacks. It is reported that Wright submitted a complaint to the Inspector General's Office of the Justice Department, which probes agency wrongdoing and mistakes. Amazingly, he was turned away, and told to take his cause to Congress. The Inspector General's Office claims it does "not have the resources to conduct an investigation of this anticipated size and scope." Yet they've conducted similar investigations in the past, including a full-blown investigation into the FBI's alleged mishandling of evidence in the probe of Timothy McVeigh, the convicted Oklahoma City bomber. [LA WEEKLY, 8/9/02]

August 15, 2002: CNN General Manager Concedes Media Censored Itself

Rena Golden, the executive vice-president and general manager of CNN International, claims that the press has censored itself over 9/11 and the Afghanistan war. "Anyone who claims the U.S. media didn't censor itself is kidding you. It was not a matter of government pressure but a reluctance to criticize anything in a war that was obviously supported by the vast majority of the people. And this isn't just a CNN issue—every journalist who was in any way involved in 9/11 is partly responsible." [PRESS GAZETTE, 8/15/02] These comments echo criticisms by Dan Rather in May 2002.

August 28, 2002: UN Calls Attempt to Thwart al-Qaeda Money Flow a Failure

The *Washington Post* reports, "A global campaign to block al-Qaeda's access to money has stalled, enabling the terrorist network to obtain a fresh infusion of tens of millions of dollars and putting it in a position to finance future attacks, according to a draft UN report." In the months immediately following 9/11, more than $112 million in assets was frozen. Since then, only $10 million more has been frozen, and most of the original money has been unfrozen due to lack of evidence. Private donations to the group, estimated at $16 million a year, are believed to "continue, largely unabated." The U.S. and other governments are not sharing information about suspected terrorists, and known terrorists are not being put on suspected terrorist lists. [WASHINGTON POST, 8/29/02]

September 8–11, 2002: Interview with al-Qaeda Leaders Is Broadcast

Details of an Al Jazeera interview with al-Qaeda leaders Khalid Shaikh Mohammed and Ramzi bin al-Shibh are widely publicized. [LONDON TIMES, 9/8/02; AUSTRALIAN, 9/9/02; GUARDIAN, 9/9/02] But there are numerous doubts about this interview. The possibility has been raised that the broadcast of Ramzi bin al-Shibh's voice in the interview helps in his capture a few days later. [CBS NEWS, 10/9/02; OBSERVER, 9/15/02] Al Jazeera also broadcasts footage of hijacker Abdulaziz Alomari speaking against the U.S. filmed in Afghanistan in early 2001.

September 10, 2002: Port Authority Sued by Relatives and Insurance Companies

Right before a one-year deadline, the Port Authority, the government body that owns the WTC complex, is sued by five insurance companies, one utility and 700 relatives of the WTC victims. The insurance companies and utility are suing because of safety violations connected to the installation of diesel fuel tanks in 1999 that many blame for the collapse of WTC Building 7. [DOW JONES NEWS, 9/10/02] The relatives' lawsuit is much more encompassing, and even blames the Port Authority for the Flight 93 hijacking (the Port Authority owns Newark airport, where the flight originated). The relatives' lawsuit is likely to lie dormant for at least six months as evidence is collected. Relatives are also considering suing the airlines, security companies and other entities. [NEWSWEEK, 9/13/02]

September 11, 2002: Story of Bush's 9/11
Conduct Changes for 9/11 Anniversary

On the first anniversary of the 9/11 attacks, the story of what Bush did on that day is significantly rewritten. In actual fact, when Chief of Staff Andrew Card told Bush about the second plane crash into the WTC, Bush continued to sit in a Florida elementary school classroom and hear a story about a pet goat for at least seven more minutes, as video footage later broadcast in the 2004 movie *Fahrenheit 9/11* shows. But one year later, Card claims that after he told Bush about the second WTC crash, "it was only a matter of seconds" before Bush "excused himself very politely to the teacher and to the students, and he left the Florida classroom." [SAN FRANCISCO CHRONICLE, 9/11/02] In a different account, Card says, "Not that many seconds later the president excused himself from the classroom." [MSNBC, 9/9/02] An interview with the classroom teacher claims that Bush left the class even before the second WTC crash: "The president bolted right out of here and told me: 'Take over.'" When the second WTC crash occurred, she claims her students are watching television in a nearby media room. [NEW YORK POST, 9/12/02]

September 12, 2002: Major Paper First to Give Room for 9/11 Skeptics

For the first time, a mainstream U.S. newspaper looks at the people who believe there was government complicity or criminal incompetence in 9/11 and does not immediately dismiss them. The *San Francisco Examiner* quotes a number of 9/11 skeptics and lets them speak for themselves. "While different theorists focus on different aspects of the attacks, what they seem to have in common is they would like an independent investigation into 9/11." [SAN FRANCISCO EXAMINER, 9/12/02]

October 2002: State Department Restarts Propaganda Activities

The State Department's propaganda office, closed in 1996, is reopened. Called the Counter-Disinformation/Misinformation Team, this office supposedly only aims its propaganda overseas to counter propaganda from other countries. [ASSOCIATED PRESS, 3/10/03]

October 6, 2002: Christian Fundamentalists
Believed to Influence Bush Foreign Policy

60 Minutes airs a program on the religious support for Bush's expansionist Middle Eastern policies. [CBS NEWS, 10/6/02] A *Guardian* editorial from around the same time suggests that "Christian millenarians" who are "driven by visions of messiahs and Armageddon" have formed an alliance with "secular, neoconservative Jewish intellectuals, such as Richard Perle and Paul Wolfowitz" and are strongly influencing Bush's foreign policy. [GUARDIAN, 9/17/02] A later *Washington Post* article also sees the support of evangelical Christians and right-wing Jewish groups as instrumental in defining U.S. Middle East policy. [WASHINGTON POST, 2/9/03]

October 12, 2002: Bali Bombing Kills Over 200

A car bomb detonates in front of a discotheque at Kuta Beach, on the Indonesian resort island of Bali, starting a fire that rages through a dozen buildings, killing 202 people. No group claims responsibility, but Jemmah Islamiyah, a radical Islamic organization in Indonesia, is suspected. [NEW YORK TIMES, 10/13/02; NEW YORK TIMES, 10/14/02; BBC, 2/19/03]

October 17, 2002: None Punished at Agencies for 9/11 Failures

The directors of the U.S.'s three most famous intelligence agencies, the CIA, FBI and NSA, testify before a Congressional inquiry on 9/11. [9/11 CONGRESSIONAL INQUIRY, 10/17/02; NSA DIRECTOR CONGRESSIONAL TESTIMONY, 10/17/02] All three say no individual at their agencies has been punished or fired for any of missteps connected to 9/11. This does not satisfy several on the inquiry, including Senator Carl Levin (D), who says "People have to be held accountable." [WASHINGTON POST, 10/18/02]

October 27, 2002: Author Gore Vidal Says Bush Used 9/11 as Pretext

The *Observer* reports, "America's most controversial writer Gore Vidal has launched the most scathing attack to date on George W. Bush's Presidency, calling for an investigation into the events of 9/11 to discover whether the Bush administration deliberately chose not to act on warnings of al-Qaeda's plans. Vidal's highly controversial 7,000 word polemic titled 'The Enemy Within' . . . argues that what he calls a 'Bush junta' used the terrorist attacks as a pretext to enact a preexisting agenda to invade Afghanistan and crack down on civil liberties at home. Vidal states, 'Apparently, "conspiracy stuff" is now shorthand for unspeakable truth.'"

November 17, 2002: Toronto Paper Says Bush Did Nothing to Stop 9/11

A *Toronto Star* editorial entitled "Pursue the Truth About September 11" strongly criticizes the government and media regarding 9/11: "Getting the truth about 9/11 has seemed impossible. The evasions, the obfuscations, the contradictions and, let's not put too fine a point on it, the lies have been overwhelming. . . . The questions are endless. But most are not being asked—still—by most of the media most of the time. . . . There are many people, and more by the minute, persuaded that, if the Bushies didn't cause 9/11, they did nothing to stop it." [TORONTO STAR, 11/17/02]

November 24, 2002: Rumsfeld Creates New Propaganda Agencies

The *Los Angeles Times* reports that Defense Secretary Rumsfeld is creating new agencies to make "information warfare" a central element of any U.S. war. For instance, Rumsfeld created a new position of deputy undersecretary for "special plans"—a euphemism for deception operations. "Increasingly, the administration's new policy—along with the steps senior commanders are taking to implement it—

blurs or even erases the boundaries between factual information and news, on the one hand, and public relations, propaganda and psychological warfare, on the other. And, while the policy ostensibly targets foreign enemies, its most likely victim will be the American electorate." [LOS ANGELES TIMES, 11/24/02 (B)]

November 25, 2002: Bush Creates Department of Homeland Security

President Bush signs legislation creating the Department of Homeland Security. Homeland Security Director Tom Ridge is promoted to Secretary of Homeland Security. The Department will consolidate nearly 170,000 workers from 22 agencies, including the Coast Guard, the Secret Service, the federal security guards in airports, and the Customs Service. [NEW YORK TIMES, 11/26/02 (C), LOS ANGELES TIMES, 11/26/02] However, the FBI and CIA, the two most prominent anti-terrorism agencies, will not be part of Homeland Security. [NEW YORK TIMES, 11/20/02] The department wants to be active by March 1, 2003, but "it's going to take years to integrate all these different entities into an efficient and effective organization." [NEW YORK TIMES, 11/20/02; LOS ANGELES TIMES, 11/26/02] Some 9/11 victims' relatives are angry over sections inserted into the legislation at the last minute. Airport screening companies will be protected from lawsuits filed by family members of 9/11 victims. Kristen Breitweiser, whose husband died in the WTC, says, "We were down there lobbying last week and trying to make the case that this will hurt us, but they did it anyway. It's just a slap in the face to the victims." [NEW YORK TIMES, 11/26/02 (B)]

November 28, 2002: Terrorist Attacks in Kenya

Three suicide bombers detonate their explosives outside a resort hotel in Mombasa, Kenya. Terrorists also fire shoulder-launched missiles unsuccessfully at a passenger jet. [NEW YORK TIMES, 11/30/02] The death toll reaches 16. [CNN, 12/1/02] Al-Qaeda purportedly claims responsibility a few days later. [CNN, 12/2/02]

December 12, 2002: Most Detainees After 9/11 Released

The vast majority of the more than 900 people the federal government acknowledges detaining after the 9/11 attacks have been deported, released or convicted of minor crimes unrelated to terrorism, according to government documents. An undisclosed number—most likely in the dozens—are or were held as material witnesses. The Justice Department reports that only six of the 765 people arrested for immigration violations are still held by the INS. An additional 134 people were charged with criminal offenses, with 99 found guilty through pleas or trials. [CHICAGO SUN-TIMES, 12/12/02; WASHINGTON POST, 12/12/02] *Newsweek* reports that of the "more than 800 people" rounded up since 9/11, "only 10 have been linked in any way to the hijackings" and "probably will turn out to be innocent." [NEWSWEEK, 10/29/02]

January 10, 2003: Government Employees Responsible for 9/11 Failures Are Promoted

FBI Director Mueller personally awards Marion (Spike) Bowman with a presidential citation and cash bonus of approximately 25 percent of his salary. [SALON, 3/3/03 (B)] Bowman, head of the FBI's National

Security Law Unit and the person who refused to seek a special warrant for a search of Zacarias Moussaoui's belongings before the 9/11 attacks is among nine recipients of bureau awards for "exceptional performance." The award comes shortly after a 9/11 Congressional inquiry report saying Bowman's unit gave Minneapolis FBI agents "inexcusably confused and inaccurate information" that was "patently false." [MINNEAPOLIS STAR-TRIBUNE, 12/22/02] Bowman's unit also blocked an urgent request by FBI agents to begin searching for Khalid Almihdhar after his name was put on a watch list. In early 2000, the FBI acknowledged serious blunders in surveillance Bowman's unit conducted during sensitive terrorism and espionage investigations, including agents who illegally videotaped suspects, intercepted e-mails without court permission, and recorded the wrong phone conversations. [ASSOCIATED PRESS, 1/10/03] As Senator Charles Grassely (R) and others have pointed out, not only has no one in government been fired or punished for 9/11, but several others have been promoted:

- Pasquale D'Amuro, the FBI's counterterrorism chief in New York City before 9/11, is promoted to the bureau's top counterterrorism post. [TIME, 12/30/02]

- FBI Supervisory special agent Michael Maltbie, who removed information from the Minnesota FBI's application to get the search warrant for Moussaoui, is promoted to field supervisor. [SALON, 3/3/03 (B)]

- David Frasca, head of the FBI's Radical Fundamentalist Unit, is "still at headquarters," Grassley notes. [SALON, 3/3/03 (B)] Frasca received the Phoenix memo warning al-Qaeda terrorists could use flight schools inside the U.S., and then a few weeks later he received the request for Moussaoui's search warrant. "The Phoenix memo was buried; the Moussaoui warrant request was denied." [TIME, 5/27/02] Even after 9/11 Frasca continued to "threw up roadblocks" in the Moussaoui case. [NEW YORK TIMES, 5/27/02]

- President Bush later names Barbara Bodine the director of Central Iraq shortly after the U.S. conquest of Iraq. Many in government are upset about the appointment because of her blocking of the USS *Cole* investigation, which some say could have uncovered the 9/11 plot. She failed to admit she was wrong or apologize. [WASHINGTON TIMES, 4/10/03] However, she is fired after about a month, apparently for doing a poor job.

- An FBI official who tolerates penetration of the translation department by Turkish spies and encourages slow translations just after 9/11 is promoted. [CBS NEWS, 10/25/02]

The CIA has promoted two unnamed top leaders of its unit responsible for tracking al-Qaeda in 2000, when the agency mistakenly failed to put the two suspected terrorists on the watch list. "The leaders were promoted even though some people in the intelligence community and in Congress say the counterterrorism unit they ran bore some responsibility for waiting until August 2001 to put the suspect pair on the interagency watch list." CIA Director Tenet has failed to fulfill a promise given to Congress in late 2002 that he would name the CIA officials responsible for 9/11 failures. [NEW YORK TIMES, 5/15/03]

January 13, 2003: British Paper Criticizes U.S. Media for Insufficiently Informing Public

The *Guardian* reports on the state of journalism in the U.S.: "The worldwide turmoil caused by President Bush's policies goes not exactly unreported, but entirely de-emphasized. *Guardian* writers are inundated by e-mails from Americans asking plaintively why their own papers never print what is in these columns . . . If there is a Watergate scandal lurking in [the Bush] administration, it is unlikely to be [*Washington Post* journalist Bob] Woodward or his colleagues who will tell us about it. If it emerges, it will probably come out on the web. That is a devastating indictment of the state of American newspapers." [GUARDIAN, 1/13/03]

January 15, 2003: U.S. Has Gone Mad, Says Novelist

Famous spy novelist John le Carré, in an essay entitled, "The United States of America Has Gone Mad," says "The reaction to 9/11 is beyond anything Osama bin Laden could have hoped for in his nastiest dreams. As in McCarthy times, the freedoms that have made America the envy of the world are being systematically eroded." He also comments, "How Bush and his junta succeeded in deflecting America's anger from bin Laden to Saddam Hussein is one of the great public relations conjuring tricks of history." [LONDON TIMES, 1/15/03]

September 6, 2003: British Cabinet Minister Hints U.S. Government Knew of 9/11 in Advance

British government minister Michael Meacher publishes an essay entitled, "The War on Terrorism is Bogus." Meacher is a long time British Member of Parliament, and served as Environmental Minister for six years until three months before releasing this essay. The *Guardian*, which publishes the essay, states that Meacher claims "the war on terrorism is a smoke screen and that the U.S. knew in advance about the September 11 attack on New York but, for strategic reasons, chose not to act on the warnings. He says the U.S. goal is 'world hegemony, built around securing by force command over the oil supplies' and that this Pax Americana 'provides a much better explanation of what actually happened before, during and after 9/11 than the global war on terrorism thesis.' Mr. Meacher adds that the U.S. has made 'no serious attempt' to catch the al-Qaeda leader, Osama bin Laden." [GUARDIAN, 9/6/03] Meacher provides no personal anecdotes based on his years in Tony Blair's cabinet, but he cites numerous mainstream media accounts to support his thesis. He emphasizes the Project for a New American Century 2000 report

Michael Meacher

as a "blueprint" for a mythical "global war on terrorism," "propagated to pave the way for a wholly differ-ent agenda—the U.S. goal of world hegemony, built around securing by force command over the oil sup-plies" in Afghanistan and Iraq (see chapter 16). [*GUARDIAN*, 9/6/03 (B)] Meacher's stand causes a controversial debate in Britain, but the story is almost completely ignored by the mainstream U.S. media.

September 12, 2003: Bush Administration Is Sued for Having Foreknowledge of 9/11 Attacks

9/11 victim's relative Ellen Mariani sues the U.S. government for what she claims is their foreknowledge of 9/11. "I'm 100 percent sure that they knew," she says. In doing so, she is ineligible for government compensation from what she calls the "shut-up and go-away fund." She believes she would have

received around $500,000. According to a statement by her lawyer, her lawsuit against President Bush, Vice President Cheney, the CIA, Defense Department, and other administration members "is based upon prior knowledge of 9/11; knowingly failing to act, prevent or warn of 9/11; and the ongoing obstruction of justice by covering up the truth of 9/11; all in violation of the laws of the United States." As the *Toronto Star* points out, this interesting story has been "buried" by the mainstream media, at least initially. Coverage has been limited mostly to Philadelphia where the case was filed and

Ellen Mariani

New Hampshire where Mariani lives. [ASSOCIATED PRESS, 12/24/03; *PHILADELPHIA INQUIRER*, 9/23/03; *PHILADELPHIA INQUIRER*, 12/3/03; AL JAZEERA, 12/9/03; *TORONTO STAR*, 11/30/03; *VILLAGE VOICE*, 12/3/03]

December 22, 2003: Most 9/11 Victims' Families Accept Settlement, Some Sue

The deadline arrives for 9/11 victims' relatives to apply for government compensation. [*TORONTO STAR*, 12/23/03] By receiving an award from the fund, families give up their right to sue the airlines, airports, security companies, or other U.S. organizations that may be faulted for negligence and inadequate secu-rity measures. [CBS NEWS, 3/7/02; *USA TODAY*, 7/13/03; *WASHINGTON POST*, 9/10/03] Relatives may still sue "knowing par-ticipants in the hijacking conspiracy" without losing compensation. [*USA TODAY*, 7/13/03] Ninety-seven per cent of the 2,973 eligible families apply to the fund; compensation averages about $2.1 million per fam-ily. However, 70 families decide to forego the fund, and instead sue various government agencies and private companies for alleged negligence. [*NEW YORK TIMES*, 6/16/04; *GUARDIAN* 6/16/04] Widow Beverly Eckert explains her decision: "I am suing because unlike other investigative avenues . . . my lawsuit requires all testimony be given under oath and fully uses powers to compel evidence. The victims' fund was not cre-ated in a spirit of compassion . . . . Lawmakers capped the liability of the airlines at the behest of lob-byists who descended on Washington while the September 11 fires still smoldered." [*USA TODAY*, 12/19/03]

March 1, 2004: *New Pearl Harbor* Book Is Released

A 9/11 book, *The New Pearl Harbor: Disturbing Questions about the Bush Administration and 9/11*, writ-ten by theology professor David Ray Griffin, is released. The *Daily Mail* calls it "explosive." Well-known historian Howard Zinn calls the book: "the most persuasive argument I have seen for further investigation of the Bush administration's relationship to that historic and troubling event." The book suggests there is evidence that the Bush administration may have arranged the 9/11 attacks or deliberately allowed them to happen. It questions why no military fighter jets were sent up to intercept the hijacked planes after the terrorists first struck. It also explores the question of whether the Pentagon was really hit by Flight 77, and suggests that explosives could have assisted the collapse of the World Trade Center. [DAILY MAIL, 6/5/04; DEMOCRACY NOW, 5/26/04] The book sells well, but is virtually ignored by the mainstream U.S. news media. Those who do report on the book generally deride it. For example, *Publishers Weekly* states, "Even many Bush opponents will find these charges ridiculous, though conspiracy theorists may be haunted by the suspicion that we know less than we think we do about that fateful day." [PUBLISHERS WEEKLY, 3/22/04]

March 11, 2004: Al-Qaeda Bombings in Madrid

A series of train bombings in Madrid, Spain, kills approximately 200 people. Basque separatists are initially blamed, but evidence later points to people loosely associated with al-Qaeda. Former counterterrorism "tsar" Richard Clarke says later in the month, "If we catch [bin Laden] this summer, which I expect, it's two years too late. Because during those two years when forces were diverted to Iraq . . . al-Qaeda has metamorphosized into a hydra-headed organization with cells that are operating autonomously like the cells that operated in Madrid recently." [USA TODAY, 3/28/04]

March 21, 2004: Counterterrorism "Tsar" Clarke Goes Public with Complaints Against Bush Response to Terrorism

Richard Clarke, counterterrorism "tsar" from 1998 until October 2001; ignites a public debate by accusing Bush of doing a poor job fighting al-Qaeda before 9/11. In a prominent *60 Minutes* interview, he says, "I find it outrageous that the President is running for re-election on the grounds that he's done such great things about terrorism. He ignored it. He ignored terrorism for months, when maybe we could have done something to stop 9/11 . . . . I think he's done a terrible job on the war against terrorism." He adds, "We had a terrorist organization that was going after us! Al-Qaeda. That should have been the first item on the agenda. And it was pushed back and back and back for months." He complains that he was Bush's chief adviser on terrorism, yet he never got to brief Bush on the subject until after 9/11. [CBS NEWS, 3/20/04; CBS NEWS, 3/21/04; GUARDIAN, 3/23/04; SALON, 3/24/04] The next day, his book Against All Enemies is released and becomes a best seller. [WASHINGTON POST 3/22/04] He testifies before the 9/11 Commission a few days later.

Late March 2004: Clarke Attacked by Republicans

Republicans attack Richard Clarke in the wake of his new book and 9/11 Commission testimony, while Democrats defend him. [NEW YORK TIMES, 3/25/04] Senator John McCain (R) calls the attacks "the most vigorous offensive I've ever seen from the administration on any issue." [WASHINGTON POST, 3/28/04] Republicans on the 9/11 Commission criticize him while Democrats praise him. The White House violates a long-standing confidentiality policy by authorizing Fox News to air remarks favorable to Bush that Clarke had made anonymously at an administration briefing in 2002. National Security Adviser Rice says to the media, "There are two very different stories here. These stories can't be reconciled." However, in what the *Washington Post* calls a "masterful bit of showmanship," Clarke replies that he emphasized the positives in 2002 because he was asked to, but did not lie. [FOX NEWS, 3/24/04; WASHINGTON POST, 3/25/04; WASHINGTON POST, 3/26/04 (B)] Republican Senate leader Frist asks "If [Clarke] lied under oath to the United States Congress" in closed testimony in 2002. [WASHINGTON POST, 3/27/04 (B)] However, a review of declassified citations from Clarke's 2002 testimony provides no evidence of contradiction, and White House officials familiar with the testimony agree that any differences are matters of emphasis, not fact. [WASHINGTON POST, 4/4/04 (B)] Republican leaders threaten to release his 2002 testimony, and Clarke claims he welcomes the release. The testimony remains classified. [ASSOCIATED PRESS, 3/26/04; ASSOCIATED PRESS, 3/28/04] Clarke also calls on Rice to release all e-mail communications between the two of them before 9/11; this is not released either. [GUARDIAN, 3/29/04] Vice President Cheney calls Clarke "out of the loop" on terrorism. A *Slate* editorial calls Cheney's comment "laughably absurd. Clarke wasn't just in the loop, he was the loop" [SLATE, 3/23/04] Even Clarke's later political opponent Rice says Clarke was very much involved. [NEW YORK TIMES, 3/25/04 (D)] Clarke responds by pointing out that he voted Republican in 2000 and he pledges under oath not to seek a post if Senator John Kerry wins the 2004 Presidential election. [WASHINGTON POST, 3/24/04] According to Reuters, a number of political experts conclude, "The White House may have mishandled accusations leveled by their former counterterrorism adviser Richard Clarke by attacking his credibility, keeping the controversy firmly in the headlines into a second week." [REUTERS 3/29/04]

May 2004: Previously Public Information About FBI Whistleblower Is Now Classified

The Justice Department retroactively classifies information it gave to Congress in 2002 regarding FBI translator Sibel Edmonds. Senator Charles Grassley (R) says, "What the FBI is up to here is ludicrous. To classify something that's already been out in the public domain, what do you accomplish? . . . This is about as close to a gag order as you can get." The *New York Times* reports that some of the information discussed is "so potentially damaging if released publicly" that it has to be classified. Topics like what languages Edmonds translated, what types of cases she handled, and where she worked is now classified, even though much of this has been widely reported on shows like CBS's *60 Minutes.* [NEW YORK TIMES, 5/20/04] In late 2002, the Justice Department invoked the rarely used "state secrets privilege" to limit what she could say. [SALON, 3/26/04]

June 3, 2004: CIA Director George Tenet Resigns

Citing personal reasons, CIA Director Tenet announces he will be stepping down in the next month. President Bush praises Tenet's service, but there is widespread agreement that significant intelligence failures occurred during Tenet's tenure, most strikingly 9/11 itself. Sources also suggest that Tenet, originally a Clinton appointee, has been made a convenient scapegoat for Bush administration intelligence failures in Iraq and elsewhere. [CNN, 6/3/04; *INDEPENDENT*, 6/4/04]

June 25, 2004: Michael Moore's *Fahrenheit 9/11* Movie Highlights 9/11 Issues

Fahrenheit 9/11, a film by well-known filmmaker and author Michael Moore, is released in the U.S. Amongst other things, this film reveals connections between the Bush family and prominent Saudis including the bin Laden family. [*NEW YORK TIMES*, 5/6/04; *NEW YORK TIMES*, 5/17/04; *TORONTO STAR*, 6/13/04] It reviews evidence the White House helped members of Osama bin Laden's family and other Saudis fly out of the U.S. in the days soon after 9/11. [*NEW YORK TIMES*, 5/17/04; *NEW YORK TIMES*, 6/18/04; *TORONTO STAR*, 6/13/04; *LOS ANGELES TIMES*, 6/23/04; *NEWSWEEK*, 6/30/04] It introduces to the mainstream damning footage of President Bush continuing with a photo-op for seven minutes after being told of the second plane hitting the WTC on 9/11. [*NEW YORK TIMES*, 6/18/04; *WASHINGTON POST*, 6/19/04; *NEWSWEEK*, 6/20/04; *LOS ANGELES TIMES*, 6/23/04] Disney refused to let its Miramax division distribute the movie in the United States, supposedly because the film was thought too partisan. [*NEW YORK TIMES*, 5/6/04; *GUARDIAN*, 6/2/04; *LOS ANGELES TIMES*, 6/11/04; *AGENCE FRANCE-PRESSE*, 6/23/04] The film won the top award at the prestigious Cannes film festival—the first documentary to do so in nearly 50 years. [*BBC*, 5/24/04; *GUARDIAN*, 5/24/04; *AGENCE FRANCE-PRESSE*, 6/23/04] It is generally very well received, with most U.S. newspapers rating it favorably. [*AGENCE FRANCE-PRESSE*, 6/23/04; *EDITOR AND PUBLISHER*, 6/27/04] The film is an instant hit and is seen by tens of millions. [*ASSOCIATED PRESS*, 6/27/04; *BBC*, 6/28/04; *CNN*, 6/28/04; *CBS NEWS*, 6/28/04] There are some criticisms that it distorts certain facts, such as exaggerating the possible significance of Bush and bin Laden family connections, and gripes about a $1.4 billion number representing the money flowing from Saudi companies to the Bush family. However, the *New York Times* claims that the public record corroborates the film's main assertions. [*NEW YORK TIMES*, 5/17/04; *NEW YORK TIMES*, 6/18/04; *NEWSWEEK*, 6/30/04]

Fahrenheit 9/11

July 6, 2004: FBI Translator Whistleblower Lawsuit Dismissed

Former FBI translator Sibel Edmonds has her lawsuit against the Justice Department and FBI dismissed. Edmonds originally sued the FBI in March 2002 for being fired from her post shortly after revealing shortcomings in the translating group of the FBI. In October 2002, Attorney General Ashcroft asked a judge to throw out Edmonds's lawsuit against the Justice Department. He said he was applying the state secrets privilege in order "to protect the foreign policy and national security interests of the United States" and preserve diplomatic relations with (unspecified) nations. [*ASSOCIATED PRESS*, 10/18/02 (B)] The judge in the case, appointed by George W. Bush, agrees with the government's position, stating " . . . the plaintiff's case must be dismissed, albeit with great consternation, in the interests of national security." He says he cannot explain his decision in any further detail because the explanation itself would expose classified information. During his deliberation on the case, Judge Walton met privately on two occasions with the government's defense lawyers, but neither Edmonds nor her lawyer were allowed to attend these discussions. [CNN, 07/07/04; ASSOCIATED PRESS, 07/06/04] A government report looking into Edmonds's allegations of wrongdoing in the FBI's translation department is released a few days later, but the FBI classifies the entire document. Not even Edmonds is allowed to see the contents. [WASH-INGTON POST, 7/9/04]

Afterword

As chance would have it, this book went to press the same week the 9/11 Commission's Final Report was released to widespread praise by relieved government officials, and most mainstream media, as a rare, successful example of bipartisan cooperation. (Note that while new information contained in the Final Report could not be incorporated here, it will be incorporated into the online version of this timeline, available at www.complete911timeline.org.)

The 9/11 Commission's fundamental finding—i.e., the reason that the U.S. government did not prevent the attack—was as follows: "We believe the 9/11 attacks revealed four kinds of failures: in imagination, policy, capabilities, and management." [9/11 COMMISSION FINAL REPORT, 7/24/04]

The *Washington Post* reported, "Though openly dreaded for months by many Republicans and quietly feared by the White House, the report was much gentler on the Bush administration than they feared. Rather than focus criticism on the Bush administration, the commission spread the blame broadly and evenly across two administrations, the FBI, and Congress." [WASHINGTON POST, 7/23/04] More to the point, as former counterterrorism "tsar" Richard Clarke astutely noted in a *New York Times* editorial, "Honorable Commission, Toothless Report," because the commission wanted a unanimous report from a bipartisan group, "it softened the edges and left it to the public to draw many conclusions." [NEW YORK TIMES, 7/25/04]

The *Washington Post* commented, "In many respects, the panel's work has been closer to the fact-finding, conspiracy-debunking Warren Commission of the mid-1960s, which investigated the assassination of President John F. Kennedy, than to the reform-oriented Church Commission, which exposed assassination plots and CIA abuses during the mid-1970s." [WASHINGTON POST, 7/18/04 (B)] Yet, a closer review reveals that the Final Report fails to provide the benefits of either approach.

First, while purporting to "provide the most complete account we can of the events of September 11, what happened and why" [9/11 COMMISSION FINAL REPORT, 7/24/04], the commission declined to address *most* of the unanswered questions about "what happened and why." It failed to provide evidence (or, in many cases, even summary conclusions) to debunk the wide array of theories that continue to develop in the absence of clear answers. Sadly for the families of those who lost their lives on 9/11, the fundamental question for which they pushed so hard for an independent investigation remains unanswered in the name of "bipartisanship." *We still do not know what happened—or why.*

Second, while asserting a four-pronged failure as the *reason* that the attacks were not prevented, the commission declined to identify a single person who committed or contributed to these failures. *No one has been held accountable.*

Lastly, though the commission chairman and vice chairman had both previously declared that the attacks were preventable [CBS NEWS, 7/21/04], the commission failed to address this fundamental question: *Were the attacks preventable?*

What happened?

Incredibly, a significant number of facts reported in the mainstream media over the past few years (and included in this book) are simply ignored by the commission. Virtually every strange anomaly mentioned in this book that wasn't already widely known, from U.S. generals being warned not to fly the night before 9/11 to reports of hijackers training at U.S. military bases, were neither confirmed nor debunked. These "dots" simply didn't fit the desired narrative. The Final Report bears no mention of the myriad foreign intelligence warnings received prior to 9/11. The multiple instances indicating individual foreknowledge across the U.S. in the days preceding the attacks are not mentioned. The highly unusual trading on United Airlines and American Airlines stocks was brushed aside in a single footnote. Nabil al-Marabh wasn't mentioned. There was no mention whatsoever addressing the mysterious Israeli "art student spy ring," or any other related Israeli issue.

While one might think that the information was not included because the commission did not have access to the necessary information, we know that is not the case. We know, based on media reports, that some information *was* given to the commission—but was not included in their report. For instance, United Airlines flight dispatcher Ed Ballinger told a reporter what he was going to tell the commission a few days before he went to testify. He explained how officials told him the hijacking of Flight 23 was narrowly averted that morning. Korans and al-Qaeda instructions were found in the baggage of some passengers who fled when the flight failed to take off and returned to its gate. [CHICAGO DAILY HERALD, 4/14/04] The commission quoted Ballinger about ten times on other topics, but failed to mention Flight 23 at all. [9/11 COMMISSION REPORT, 7/24/04]

When the questions were directly addressed, few answers were given. For instance, the report seems to imply that the hijackers received help from associates inside the U.S., but no specifics were given. The funding for the hijackers and the operation also remain unknown.

More troubling is how the commission dealt with Pakistan and Iran. From this, we not only can see problems with the commission and its report, but also deep-seated problems in the way the media and politicians report and respond to information.

Commission Chairman Thomas Kean said one month before the release of the report, "We believe. . . . that there were a lot more active contacts, frankly, [between al-Qaeda and] Iran and with Pakistan than there were with Iraq." [TIME, 7/16/04]

Based on this comment, one might think the commission would have further explored al-Qaeda ties with both Iran and Pakistan. But in fact, most of the attention has been directed at Iran. The Commission's report mentioned that around ten of the hijackers passed through Iran in late 2000 and early 2001. At least some Iranian officials turned a blind eye to the passage of al-Qaeda agents, but there was no evidence that the Iranian government had any foreknowledge or involvement in the 9/11 plot. [TIME, 7/16/04; REUTERS, 7/18/04]

In the wake of these findings, President Bush stated of Iran, "As to direct connections with September 11, we're digging into the facts to determine if there was one." This put Bush at odds with his own CIA, which saw no Iran-9/11 ties. [LOS ANGELES TIMES, 7/20/04] Bush has considered Iran part of his "axis of evil," and there has been talk of the U.S. attacking or overthrowing the Iranian government. [REUTERS, 7/18/04] Provocative articles soon appeared, such as a *Daily Telegraph* article headlined, "Now America Accuses Iran of Complicity in World Trade Center Attack." [DAILY TELEGRAPH, 7/18/04]

However, much more significant ties between al-Qaeda and Pakistan were downplayed—or ignored—both by the commission and by most media outlets:

- UPI reported that the 9/11 Commission was given a document from a high-level, anonymous source claiming that the Pakistani "ISI was fully involved in devising and helping the entire [9/11 plot]." The document blamed General Hamid Gul, a former ISI Director, as being a central participant in the plot. It noted that Gul is a self-avowed "admirer" of bin Laden. An anonymous, ranking CIA official said the CIA considered Gul to be "the most dangerous man" in Pakistan. A senior Pakistani political leader said, "I have reason to believe Hamid Gul was Osama bin Laden's master planner." The document further suggested that Pakistan's appearance of fighting al-Qaeda is merely an elaborate charade, and top military and intelligence officials in Pakistan still closely sympathize with bin Laden's ideology. [UPI, 7/22/04]

- As the 9/11 Commission report was being released, Michael Meacher, a British Member of Parliament, and a Cabinet Minister in Tony Blair's government until 2003, wrote in the *Guardian*, "Significantly, [Saeed] Sheikh is . . . the man who, on the instructions of General Mahmood Ahmed, the then head of Pakistan's Inter-Services Intelligence (ISI), wired $100,000 before the 9/11 attacks to Mohamed Atta, the lead hijacker. It is extraordinary that neither Ahmed nor Sheikh have been charged and brought to trial on this count. Why not?" [GUARDIAN, 7/22/04]

- Daniel Ellsberg, the "Pentagon Papers" whistleblower the during the Nixon presidency, stated, "It seems to me quite plausible that Pakistan was quite involved in [9/11] . . . To say Pakistan is, to me, to say CIA because . . . it's hard to say that the ISI knew something that the CIA had no knowledge of." [GUARDIAN, 7/22/04]

- A month before the 9/11 Commission's final report was released, the *Los Angeles Times* revealed that some on the commission saw a very significant terrorist role for Pakistan. For instance, one senior commission staff member stated that the Pakistanis were involved with the Taliban and al-Qaeda "up to their eyeballs." The commission's findings were said to be only the "tip of the iceberg" on what was discovered about these connections. [LOS ANGELES TIMES, 6/20/04]

Yet, in fact, the 9/11 Commission's Final Report only mentions the ISI by name three times. The only significant mention is a brief comment that the ISI was the Taliban's "primary patron." (The report also notes that details of the 9/11 plot were widely known by the Taliban leadership, but fails to ask: if the Taliban knew, and the ISI was the patron of the Taliban, did the ISI know?) ISI Director Mahmood Ahmed is mentioned only twice, both in the context of post-9/11 diplomacy. Saeed Sheikh is not mentioned at all. [9/11 COMMISSION FINAL REPORT, 7/24/04]

Indeed, far from criticizing Pakistan, the commission praised the country for its support in the war on terrorism, and suggested that the U.S. should greatly increase its foreign aid there. [ASSOCIATED PRESS, 7/24/04]

It's not hard to see that emphasizing possible connections between Iran and 9/11 could help further the stated U.S. foreign policy goal of achieving regime change in Iran. But Pakistan is a close U.S. ally. Investigating possible ties between Pakistan and 9/11 could open a Pandora's Box of troubling questions.

In short, the commission didn't address the most troublesome facts of 9/11, and the facts they did highlight appear to only further existing political objectives.

Who is responsible?

An anonymous, former high-ranking CIA officer noted, "You can't have a disaster the size of 9/11 without someone having been asleep at the switch. . . . I don't see how we can come away from [the attacks] without a single person having been disciplined." [BBC, 7/28/04]

Shortly after the 9/11 Commission report was released, a letter by FBI Director Robert Mueller regarding FBI whistleblower Sibel Edmonds was leaked to the media. As mentioned in the book, Edmonds has made some very serious allegations about the FBI, including important missed 9/11 warnings and the existence of a foreign spy ring inside U.S. government agencies. The letter revealed that a highly classified Justice Department report on Edmonds concluded that her allegations "were at least a contributing factor in why the FBI terminated her services." The report also criticized the FBI's failure to adequately pursue her allegations of espionage. An anonymous official stated that the report concluded that some of her allegations were shown to be true, others could not be corroborated because of a lack of evidence, and none of her accusations were disproved. [NEW YORK TIMES, 7/29/04]

Edmonds has said, "My translations of the 9/11 intercepts included [terrorist] money laundering, detailed and date-specific information . . . if they were to do real investigations, we would see several significant high-level criminal prosecutions in [the US] . . . and believe me, they will do everything to cover this up." [GUARDIAN, 7/22/04]

Yet the 9/11 Commission praised Mueller even as it severely faulted the FBI for failures before 9/11. Commission Chairman Thomas Kean said, "We think he's doing exactly the right thing." [ASSOCI-ATED PRESS, 7/29/04] When Mueller publicly testified before the 9/11 Commission in April 2004, nearly every commissioner prefaced his or her questions with praise for Mueller's work. The commission had a total of one hour to question him, but actually the hearing ended early because the commissioners ran out of questions to ask. [NEW YORK TIMES, 4/14/04 (D)] Despite claiming interest in Edmonds' allegations, the Commissioners did not publicly question Mueller about them, and failed to mention her allegations in the Commission's final report (her name is mentioned twice in the report, but only in the names of footnoted documents). [CONNECTICUT POST, 4/15/04; NEW YORK TIMES, 4/14/04 (D); 9/11 COMMISSION FINAL REPORT, 7/24/04]

Edmonds remains gagged (as do other FBI whistleblowers such as Robert Wright). All facts about her job at the FBI, even which languages she translated, have been declared "state secrets." *Slate* notes, "For linguists and other analysts looking at what happened to Sibel Edmonds, the system of rewards and penalties is all too clear. The lesson they draw: Keep your head down; just do your job; if you see others doing their job badly, even if to the detriment of national security, don't get involved." [SLATE, 7/29/04]

The different treatment of suggestions of Iranian vs. Pakistani involvement in 9/11, and Sibel Edmonds' troubles with the FBI, are but two examples of what's wrong with the 9/11 Commission's, the government's, and media's coverage of these issues.

Conclusion

If the 9/11 Commission's suggestions becomes widely accepted, there may be some bureaucratic structural changes made, but as Richard Clarke points out, these changes are unlikely in and of themselves to stop another 9/11 from happening. As the Pakistan and Sibel Edmonds examples above illustrate, there are deeper problems at work than just bureaucratic communication inefficiencies, inadequate funding, and the like. For instance, in the case of Pakistan, fundamental contradictions in the U.S. relationship with that country are not being addressed. In Edmonds' case, the FBI and commission have buried serious complaints that might lead to reform.

Sadly, the hoped for answers from the 9/11 Commission have not been provided. In addition to the many conflicts of interest of the commissioners themselves, as mentioned previously in this book, a look at the background of the commission's staff show that a majority come from the CIA, FBI, Pentagon, White House, FAA, etc. . . . the very government entities that the commission was supposed to investigate.

Senator Bob Graham saw still-classified evidence pointing to foreign government involvement in 9/11, but he complained, "It will become public at some point when it's turned over to the archives, but that's 20 or 30 years from now." [PBS *NEWSHOUR*, 12/11/02] Author James Bamford first brought to public attention Operation Northwoods—a proposal by the Joint Chiefs of Staff of the U.S. military to stage a series of fake terror attacks against the U.S. populace to drum up support for a war against Cuba in the 1960s. Bamford commented, "Here we are, 40 years afterward, and it's only now coming out. You just wonder what is going to be exposed 40 years from now." [*INSIGHT*, 7/30/01]

If the "toothless" commission's "softened" version of events becomes accepted, we may never know the full truth at all, or if we do, it might take 20, 30, 40, or more years to come out.

But there is another possibility. We need to find out the full truth behind 9/11 now, while it still can affect history. Those responsible for perpetrating the 9/11 attacks and allowing them to happen need to brought to justice and held accountable. We need to chase after that truth, where ever such a path may lead. If we don't, we will never find the right solutions because we misunderstand the problems.

We, the general public, are not helpless. Even the 9/11 Commission would never have happened had the public and especially the 9/11 victim's relatives demanded it. We deserve better than that investigation, and we need to demand more.

Bibliography

As noted in the Introduction, this book is based on the Complete 9/11 Timeline found at the Cooperative Research website (www.cooperativeresearch.org). The bibliography for this book is available at www.complete911timeline.org/bibliography.

Alternatively, readers can access the online version of the timeline set forth in this book at www.complete911timeline.org. Each entry in the on-line timeline contains dynamic hyperlinks, allowing readers to view the original articles, government documents, videos, or other documentation in its entirety. The online timeline is a dynamic "document," which changes as new information is uncovered. Therefore, some articles cited in this book may be replaced with more current articles as the timeline changes and grows.

If you are interested in future developments on all the topics mentioned in this book, please use the website to learn more.

Acknowledgments

This book was not written in the usual sense, by a single author, but rather is the combined work of literally thousands of people. Since I began putting the contents of my research on the Internet, I have received thousands of e-mails containing criticism, encouragement, corrections, and/or suggestions. All these comments led to improvements, and I'm sorry I can not mention every single person by name.

As mentioned in the Introduction, I grew interested in this topic after seeing numerous intriguing websites about 9/11. Rather than list some and slight others by not including them, I suggest anyone interested in this topic to go to my website, www.complete911timeline.org, a part of www.cooperativeresearch.org, and click on the Links page. There you can see the many websites that inspired the research behind this book. I highly encourage everyone to go to those websites to learn more. This book simply would not have been possible had it not been for the Internet and all the research found there done by others. There is a growing number of ordinary people who are dissatisfied with the lack of a definitive investigation and definitive answers to many 9/11 mysteries. The closest thing to a real 9/11 investigation exists on the Internet, and thanks to search engines and accessible databases like Lexis Nexus, it is possible for anyone, anywhere, to contribute.

This book would never have come to be had it not been for the website, and the website would never have existed without a tremendous amount of help from many people. The www.cooperativeresearch.org site was developed to encourage people to play an active role in scrutinizing the activities of all individuals, groups, and institutions that wield significant political and economic power. It is run by Derek Mitchell, and he has spent thousands of hours helping my timeline research reach a larger audience. Hanns Brown, Michael Bevin, Mike Thompson, and many more have helped with the technical aspects. Thanks also to other websites who have helped my research reach a larger audience, including those run by Alastair Thompson at www.scoop.co.nz, and Fred Burks, who has made excellent summaries at www.wanttoknow.info.

There are many who sent me information or pointed me to important resources. Allan Duncan sent literally thousands of articles. Peter Jung did the same. Andreas Westphalen has been my "German connection," keeping me abreast of 9/11 developments in Germany and helping translate important articles into English. Nikos Levis and Nico Haupt also helped translate articles and provided information, as did many others. Thanks to this group, there is little written in the media about 9/11 that escapes my attention.

Some volunteers actually wrote or cowrote individual timeline entries. Allan Duncan wrote many entries while I was off traveling (as well as co-writing an essay and providing other additional help). Andreas Westphalen, Matthew Everett, Teresa La Loggia, and many others also wrote some of the exact words that appear in this book. Others helped with proofreading. Vicki Smith, Melissa Ennen, Melissa Kavonic, and others spent many hours on this thankless task.

The fact that this book came to be published as quickly as it did was a near miracle. I had expected the process of turning my research into a book to take about a year, but events fell into place that enabled the book to be published much sooner, provided it could be completed in two months. Peter Lance deserves all the credit for creating that opportunity.

I put out a call for help to reach this deadline, and dozens of people responded with substantial help. The publication date would never have been met had it not been due to a surge of help from this group, including (but not limited to): Coen Ackers, Ed S. Crouse, Deborah McCarty, Mari Eliza, Matthew Everett, Feargus, Tim Gale, Jeremy R. Hammond, Laura Lindsey, Al Matthews, Milton MacPhail, Pat Smith, Chuck Vermillion, Brett Willmot, and Rita Wiltsie.

A few deserve special thanks. Thanks to Sandra McConville for her idea of creating titles for each entry. Thanks to Matt Carmody and Greg Bolton, for particularly timely help on thankless proofreading tasks. Thanks to Debra Sokolowski, for her truly non-stop, sleepless efforts. The excellent maps in this book are due to the fast work of Yvonne Vermillion.

That call for help also brought Teresa La Loggia. This book would be but a shadow of itself had it not been for her efforts. I only half-jokingly consider her efforts to the equal to that of ten other people. She repeatedly took time off of work and went nights without sleep to help with the book. The index and much more would never have been included had it not been for her unexpected but greatly appreciated help.

Thanks to my great friend Bill Shadow. He was unlucky to be on hand when much last-minute help was needed, but time and time again he came through quickly.

Finally, thanks to Judith Regan, Cal Morgan, Cassie Jones, Kris Tobiassen, and the others at Regan-Books for daring to take this book on and see it to completion.

There is a reason why this book is credited to "Paul Thompson and the Center for Cooperative Research." I consider all the people mentioned above to be my co-authors. Thanks again to everyone for all the hard work. Together, we are forging new paths to uncover the truth through collaborative Internet-based research.

Photography Credits

Index